Coastal and Estuarine Studies

Coastal and Estuarine Studies

formerly Lecture Notes on Coastal and Estuarine Studies

38

R.T. Cheng (Ed.)

Residual Currents and Long-term Transport

Springer-Verlag

New York Berlin Heidelberg London

Paris Tokyo Hong Kong Barcelona

Editor

R.T. Cheng
U.S. Geological Survey
Water Resources Division
Menlo Park
CA 94025, USA

ISBN 0-387-97376-1 Springer-Verlag New York Berlin Heidelberg
ISBN 3-540-97376-1 Springer-Verlag Berlin Heidelberg New York

Library of Congress Cataloging-in-Publication Data
Residual currents and long-term transport / R.T. Cheng (ed.) – 1st ed. p. cm. – (Coastal and estuarine studies ; 38.) Includes bibliographical references and index.
ISBN 0-387-97376-1 (acid-free paper)
1. Ocean currents. 2. Marine sediments. 3. Estuarine oceanography. I. Cheng, Ralph T.
II. Series. GC239.R47 1990 551.47'01–dc20 90-41794

Printed in Germany

Printing and binding: Druckhaus Beltz, Hemsbach/Bergstr.
2837/3140-543210 – Printed on acid-free paper

ACKNOWLEDGEMENT

Chapters in this volume of Coastal and Estuarine Studies are compiled from papers presented at the 1988 International Conference on Physics of Shallow Estuaries and Bays. The Conference was held at Asilomar Conference Center, Pacific Grove, California, November 29 through December 2, 1988. The theme of the Conference was **Residual Currents and Long-term Transport** in estuaries and bays, and thus the title of this volume. The Conference was sponsored by the U. S. Geological Survey, Department of the Interior. Additional financial support was received from the U. S. National Science Foundation. The material published in this volume is based upon activities supported by the U. S. Geological Survey and by the National Science Foundation (under Agreement No. OCE-8900251). Any opinion, findings and conclusions or recommendations expressed in this publication are those of the authors and do not necessarily reflect the views of the U. S. Geological Survey nor of the U. S. National Science Foundation. Monterey Bay Aquarium, California was open to the participants in the evening of December 1, where the Conference banquet was held. The hospitality of Monterey Bay Aquarium is gratefully acknowledged.

I am most grateful to Prof. J. (Co) van de Kreeke, without whose continuing support, these Conferences on Physics of Shallow Estuaries and Bays could not have come to pass, and without his encouragement I would not have assumed the chores for organizing the 1988 "Physics" Conference. I am also grateful to the Organizing Committee whose members are :

Mr. W.H. McAnally, Jr., U.S. Army Corps of Engineers, USA;
Prof. Jilan Su, National Oceanic Administration, The People's Republic of China;
Prof. J. van de Kreeke, University of Miami, USA;
Prof. J.T.F. Zimmerman, Netherlands Institute for Sea Research, the Netherlands.

The Organizing Committee provided me constant support and continuing advice including the selection of papers before the Conference, organizing the Conference program, and the recommendations of reviewers after the Conference. Each paper included in this volume has been subjected to at least two technical reviews. I would like to thank all reviewers whose constructive suggestions have greatly improved the quality of many manuscripts that appear in this volume.

My special appreciation goes to Ms. Carolyn L. McLeod, my able secretary, who not only designed the beautiful and fitting Conference logo, but also handled a large volume of pre- and post-Conference correspondence. She is responsible for typing all the equations in the manuscripts, and for reformatting the manuscripts sent from the authors to the present final form. Without her help, the publication of this volume would be much delayed.

In organizing such an International Conference, there were a great deal of details that required careful planning and execution. I was most fortunate to have been aided by Messrs. Jon R. Burau, Jeffrey W. Gartner, and Lawrence H. Smith of the U. S. Geological Survey, and by Mr. Thomas Wakeman of the U. S. Army Corps of Engineers. These gentlemen have attended to all the details of the Conference and have provided the Conference a friendly atmosphere ideally suited for the "work hard and play hard" philosophy. I am very grateful for their contributions to this Conference.

Ralph T. Cheng
Water Resources Division
U. S. Geological Survey
Menlo Park, California 94025

CONTENTS

Part I

Introduction

Introduction

Ralph T. Cheng

Water Resources Division
U. S. Geological Survey
Menlo Park, California 94025

Estuaries, bays, and contiguous coastal seas of the world are the most valuable and yet most vulnerable marine ecosystems. For centuries, society has placed enormous value on coastal areas for living, working, and recreation. In nearly all regions of the world, the largest population is distributed along the coastlines. The marine ecosystems provide food, shelter, and spawning grounds for fisheries, and refuge and sanctuary for wildlife. Dramatic increases in the population living in coastal regions have changed the pattern of land use and the movement of sediments. Obviously, these changes have not come without a price. Accompanying the coastal population increase is competition for the use of estuarine and coastal resources: the diversion of fresh water for irrigation and the discharge of waste water into these systems. The changing patterns of water use have resulted in deterioration of water quality, and in irreversible impacts on many marine ecosystems. All of these alterations directly affect the economic development and the quality of life in adjacent regions.

In order to optimize the management of and to protect these important resources, we must deepen our understanding of the estuarine and coastal ecosystem processes, expand our knowledge to better assess the effects of human activities on these natural resources, develop our ability to evaluate the changing trend of these processes, and we must be ready to take corrective action in anticipation of adverse consequences. To reach these goals, the major areas of concern include: 1) alterations of freshwater inflows to and circulation in the systems, 2) loading and dispersion of toxics, nutrients, sediments, and contaminants, 3) habitat loss and degradation, and 4) decline in living resources in these coastal ecosystems. A comprehensive strategy and research agenda should aim at an inter-disciplinary and/or multi-disciplinary approach to address the complex issuses of marine ecosystems. Of all aspects in the inter- or multi-disciplinary research program, understanding the physical processes such as circulation, mixing, and transport of salt, nutrient, and sediment in coastal embayments and shallow seas is foremost fundamental. A series of biennial International Conferences on the Physics of Shallow Estuaries and Bays was organized so that research scientists could meet to discuss their latest research findings. Because of the implied importance of these studies to ecosystem processes whose time scales are typically much longer than a few tidal periods, we placed the emphasis of the 1988 "Physics" conference on the exchange of knowledge and research results in the areas of residual currents and long-term transport processes in estuaries and bays.

In the keynote address, David Aubrey stressed the importance of inter-disciplinary research. Although based on circumstances in the United States, limited funding situations facing research communities around the world are believed to be universal. Becasue resources are limited, research scientists should oblige themselves to the delicate balance between "doing the right things" and "doing the things right". Here, "doing the right things" means asking the right questions, and conducting the right research in order to get the right answers. "Doing the things right" refers to using the proper methods and techniques in the course of investigation. In an ideal research situation, both of these criteria should be taken as givens. In practice, the required task cannot often be carried out as desired, and resource limitations necessitate some compromises in the methods of solution.

Coastal and Estuarine Studies, Vol. 38
R. T. Cheng (Ed.)
Residual Currents and Long-term Transport

The complexity of conducting research in residual currents and long-term transport is deeply rooted in the fact that the residual variables are the results of nonlinear interactions of tidal variables, and are not directly measurable. Values of residual variables, which are commonly an order of magnitude smaller than the tidal variables, are derived from either observations or calculations of tidal variables. Yet, the magnitude and spatial and temporal variations in residual variables exert most significant impact on the interactive physical, bio-chemical, and sedimentary processes. In some cases, these processes are characterized by spatial and temporal scales an order or orders of magnitude greater than that of the tidal variables. The micro-scale turbulence determines the macro-scale mixing and dispersion characteristics, and the macro-scale processes affect long-term residual properties. Solutes, pollutants, contaminants, sediments, and nutrients are transported predominantly by tidal currents, along with mixing and dispersion with adjacent water parcels. Research on micro-scale turbulent mixing is fundamental to further understanding the macroscopic processes, detailed treatment of this subject is beyond the scope of this volume. Macro-scale transport processes, which take place within a few tidal cycles, are reasonably well understood. It is the accumulated, long-term results of these processes which are of great importance because of their direct impact on the relationships within estuarine and coastal ecosystem processes.

Following a Lagrangian point of view, large tidal excursions often mask the importance of the smaller residual currents. When the transport of solutes in the system is convection-dominated by tidal circulation, the Lagrangian residual currents have been shown to be responsible for the transport of dissolved substances in periods of weeks or months. Another area of great interest is the study of the transport of suspended sediments which involves the examination of fine-scale turbulent motions near the sediment-water interface, the advection and dispersion of suspended particulates, and the erosion and deposition of sediment over a tidal time scale up to a geological time scale.

An agenda for research in shallow water physical oceanography should include the establishment of a list of "right questions"; the answers to these questions would enhance understanding of the physical processes in the system, and would aid protection and management of our valuable estuarine and coastal marine resources. The central subject of this volume, i.e., the residual currents and long-term transport processes in estuaries and coastal systems, represents one of the many important research topics in our research agenda. These processes appear to have direct control over freshwater inflows, contaminant loadings, dispersion and transport of sediments and nutrients, and causes of declining living resources. Judging from the collection of papers compiled in this volume, there is a clear indication that the authors have "done the right things" by raising the right questions. Although many accomplishments have been reported and much progress made in recent years, the difficulty in conducting research of residual currents and long-term transport processes dictates that these research subjects to remain on our research agenda for some years to come.

Part II of this volume is a collection of papers that provide a theoretical formulation of residual currents and mass transport. Notably, the Lagrangian viewpoint for this type of analysis has gained continued recognition. The analyses are still limited to weakly nonlinear dynamics, but they have been extended to include three-dimensional flows, stratification, and ebb-flood asymmetry. Complementary to theoretical analyses are the increased use of numerical hydrodynamic models as research tools for furthering our understanding of residual currents. Adoption of this approach is the direct result of the revolutionary advances achieved in computing hardware and the subsequent changes introduced in the design of modeling software. Three-dimensional models are often used to generate a realistic representation of flow field, with sufficient detail, in all three-dimensions of an embayment. Obviously, this trend will continue, and three-dimensional, numerical hydrodynamic models are expected to become more practical and useful tools in future research. For example, a large number of particles can be tracked in a numerical model to determine properties of residual currents and dispersion characteristics. Another example is the use of numerical models to assess the impact of long-term sea-level rise in shallow estuaries.

Salinity variations have a profound impact on numerous important marine ecosystem parameters. Salinity data are often used as the basis in regression and correlation analyses with other water quality parameters. Because salt is a conservative substance, understanding the mixing and transport of salt is essential and fundamental in any attempt to deal with the transport of non-conservative solutes. Part III of this volume is a collection of papers whose focus is on salt mixing and transport. Effects on mixing characteristics due to shear, ebb-flood asymmetry, and stratification are discussed. The formation of a coastal boundary layer, and the interactions and exchanges of coastal water with water parcels within coastal shallow embayments are important mechanisms that eventually control the long-term trend and variations of salinity in the system. Although detailed studies of turbulent mixing in these waters are needed, research results in this area are evidently missing in this collection of papers. Detailed field measurements of turbulence and turbulent mixing characteristics should rank high in future research agenda. The deficiency in detailed field data is mitigated in part by results generated with large scale computer models in studies of mixing characteristics. A large number of particles are released and followed, in a Lagrangian sense, to simulate the mixing and dispersion processes. It appears that numerical modeling has assumed a new role in the research of mixing processes in coastal waters.

In many estuaries and bays, the changes in the patterns of sediment transportation are due to changes introduced in the patterns of land use in the coastal regions. To greater or lesser extent, researchers around the world have recognized the potential importance that the movements of sediments might have on the ecosystem. Part IV of this volume includes reviews, laboratory experiments, and recent findings on the erosion and deposition of fine sediments, and on the transport of sediments and fluid muds. These topics are intimately related to both the micro- and macro-scale properties of fluid flows. Laboratory experiments are extremely useful for providing further insight into the basic mechanisms that determine sediment transport processes, and for providing evidence to develop conceptual models of these processes. Specifically, fluid flows that take place within a boundary layer have a profound effect on the control of sediment erosion and deposition. Without the understanding of these basic mechanisms, computation and modeling of sediment transport cannot be achieved. Complementary to laboratory experiments, field observations are equally important in order to test the validity of these conceptual models. Some examples of such observations are given for the North Sea, for Hangzhou Bay, China, and for Seto Inland Sea, Japan. Again, a computer model has been used as an alternative research tool for an estimation of the re-distribution of fine sediments, although this approach must be backed by extensive field observations and field data.

As mentioned previously, one of the difficulties in research is that direct field measurements of residual variables are not yet possible. Part V of this volume is a collection of papers presenting observational results and evidence of residual currents that are generated by density forcing, wind forcing, and topographical forcing. In most instances, the properties of residual currents and the associated transport processes have significant implications for long-term variability in marine ecosystem characteristics. Nearly all results of residual variables are derived from field observations of tidal variables by applying a suitable filter. Aside from the OSCR HF radar measurements, descriptions of the use of new current measuring techniques are generally lacking in this volume. This clearly suggests the need for further refinement of current measuring devices and current measurement techniques. Obviously, the density-difference-driven residual currents play a major role in many estuaries and bays. Important distinctions between the Eulerian residual currents and Lagrangian residual currents are again made clear, and their implications in long-term mass transport can be quite significant. Estuaries are as unique as individuals. In this part of the volume, several examples are drawn from the dynamic responses of a small Australian estuary, and from the responses to wind and river inflows in a few small estuaries in North America. Other examples include the interactions of estuarine and continental shelf waters, the effect of salinity intrusion, and the changes introduced in residual circulation patterns due to drought. The authors have demonstrated the unique "estuariality" for a variety of embayments.

Finally, as stated in the beginning, one of the important motivations for the study of residual currents and long-term transport processes is to enhance the understanding of the interactive physical and bio-chemical processes in ecosystems. The transition to multi-disciplinary studies is usually quite difficult and challenging. In Part VI, two papers exemplify and discuss the important role that physical processes might play in marine ecosystems.

Studies of residual currents and long-term transport are by no means complete; much additional research is needed. The subjects pursued by the authors in these proceedings represent a partial research agenda that reflect the current thinking of the authors in "doing the right things". This Editor leaves it to the readers to to decide whether the authors have carried their research to the level of "doing the things right".

Interdisciplinary Estuarine Research: A Challenge for the Future

David G. Aubrey

Woods Hole Oceanographic Institution
Woods Hole, MA 02543

ABSTRACT

Growing public concern worldwide about our coastal bays and estuaries recently has focused attention on these ecosystems, but without a defensible, coherent scientific plan responsive research efforts are less efficient and less cost-effective than otherwise might be possible. The scientific community has both an opportunity and an obligation to respond to this renewed public interest. If the scientific response is neither immediately forthcoming nor cohesive, the governmental and public focus on these highly stressed environments will run its course without provision of adequate scientific input. Many physical studies of estuaries are local, focusing on small-scale issues rather than on fundamental physics or global truths. Emphasis on understanding of physics of basic processes has been neglected in favor of more descriptive, locally-tuned models. The estuarine scientific community has a long history of independence and autonomy in research projects, lacking a coherent thrust (despite the preparation of many planning documents).

Needed in the future is an improved balance between site-specific, local studies and process-oriented studies of basic physics that provide more general knowledge independent of site. Also required are institutional mechanisms for deriving a coherent, workable, and international estuarine research plan cognizant of both site-specific and general needs, and for nurturing such plans through the increasingly competitive processes of funding. The individual researcher must accept increased responsibility for furthering the goals of the estuarine community, and in assuring the highest quality of scientific merit in estuarine programs. Without such scientific credibility, estuarine research cannot compete successfully with other oceanographic programs for increasingly tight funding.

I. Introduction

The impact of human activities on the quality of coastal waters and ecosystems has been widely discussed and amply documented (see, for example, NRC, 1971; Ketchum, 1972; NERC, 1983; National Ocean Pollution Program Office, 1988). Public outcry to improve the estuarine environment has been expressed frequently in the press, in local scientific workshops (e.g., Beach, 1988), and in governmental actions (e.g., Committee on Merchant Marine and Fisheries, 1989). The Long Island hypoxia event of the summer of 1988 had well-publicized dramatic economic consequences on tourism, fisheries, public perception, and the resource itself. The spill of toxic pesticides into the Rhine River in 1986 aroused concern over river/estuarine damage and the international flavor of such spills (e.g., Capel et al., 1988). Degradation of harbors in Hong Kong has raised consciousness that we cannot continue to abuse our coastal waters (Morton, 1988). Major national programs have been created by legislative action in response to public outcry to address problems of pollution in coastal waters (such as the U.S. Coastal and Estuarine Assessment Program).

The scientific response to these public and governmental concerns has been indecisive. Partly for historical reasons and partly for structural reasons, the estuarine research community has been unable to respond as powerfully and effectively as it might. In this article thoughts are presented as to why we must act coherently and promptly, why the estuarine community finds itself in the position it is now in, what the community can do to rectify this situation, when we should act, where we should proceed, and who might take the lead.

II. The Need for Community Action

The final years of the 1980's and the decade of the 1990's are becoming known as the time of global science programs. Issues of a global scope, such as ozone depletion and greenhouse intensification, have focused global scientific thought on planetary-scale problems. A plethora of planning documents, budget documentation, and position papers has inundated the scientific community, targeting these global issues. National and international programs are proliferating, designed to address some of these global issues. Dominant amongst these U.S. global programs in the marine environment are the World Ocean Circulation Experiment (WOCE), the Global Ocean Flux Study (GOFS), and TOPEX/POSEIDON, with analogous international components (see, for instance, GOSP Working Group, 1987). International groups such as the International Council for Scientific Unions (ICSU) have initiated their own global focus, such as the International Geosphere/Biosphere Program (IGBP). Other international players include the World Meteorological Organization (WMO: world climate research program), United Nations Educational, Scientific, and Cultural Organization (UNESCO: Intergovernmental Oceanographic Commission), United Nations Environmental Programme (UNEP), and the Intergovernmental Panel on Climate Change. Priorities have been set for the U.S. Global Change Research Program, with science priorities identified for seven categories, within each of which specific research areas are targeted. Land/ocean water fluxes are amongst the highest priority listed. Virtually every major U.S. funding group is represented in the global science plan.

In response to the global focus, many disciplinary groups have developed research strategies and plans for the next decade. Perhaps closest to the interests of estuarine scientists and engineers is the coastal physical oceanography group, which has prepared a series of documents showing how their scientific interests meld with the global programs (e.g., COPO, 1988, 1990). To the detriment of estuarine researchers, the COPO experimental boundary is defined to exclude estuarine research, concentrating instead on open shelves. The estuarine community is left with little representation in the global physical oceanography programs for the next decade. Estuarine engineers similarly have had little impact on the definition and conduct of these major global programs. Geologists have had perhaps a little more input, in their contribution to the Continental Margin Integrated Sediment Studies (CMISST), where estuaries and riverine processes are at least mentioned (e.g., Nittrouer _et al_., 1988).

This paper is not a call for estuarine scientists and engineers to begin speaking global-eze. However, it is important for shallow water physicists to understand the funding climate, and the competition to be faced when formulating broad-scope research plans for the next decade. For a disciplinary group that lacks a history of research cohesion, the road to an integrated program is certain to include some bumps.

On the other hand, arguments for the need for improved prediction and understanding of estuaries are compelling and well-stated. We all agree that we don't know enough about these resource areas, that mankind is damaging those resources, and that the resources impacted are valuable (in some variously defined sense). That the estuaries and embayments of the globe are important for navigation, fisheries, ecological value, recreation, and other culturally and economically valuable pursuits, is accepted. The U.S. emphasis on estuaries in joint Environmental Protection Agency (EPA)/NOAA ventures is built from these strong arguments for activity. Thus, estuarine scientists and engineers have a strong, widely-accepted basis from which to start.

However, as Schubel (1984) and others have pointed out, most governmental effort has been spent on developing strategies to stop pollution and to enhance aesthetic values and living resources. These efforts have not worked out well, in part because of the paucity of basic research funded to support these large government programs.

Without the strong foundation of basic research (new knowledge), management of these systems is bound to be inadequate. What basic research is funded commonly lacks a well-defined peer review process.

While peer review has its share of weak points, lack of peer review is certain to fail to assure that the best scientific effort is always provided. These large, expensive inter-agency and international efforts must include a reasonable balance of basic research following a clear, well-defined scientific agenda. Such an agenda is either lacking or not being followed, in many instances. Without the overall agenda to guide individual projects, the projects may be neither cost effective, scientifically defensible, nor have goals that are attainable. Officer et al. (1981) showed in their study of basic research in estuarine processes, that some 77 percent of the refereed publications came from the academic community. However, this same academic community received only some 37 percent of the research dollars devoted to estuaries. Clearly the academic contribution is not matched by the allocation of available dollars. What Congress and perhaps other governments see is a huge price tag, rather than the details of how those monetary resources are allocated to address long-term, basic scientific and engineering questions.

The environmental issues are well-defined. The need and value are accepted. The public is concerned. Why do we still lack the resources to address these pressing scientific and engineering questions? Has the estuarine community missed an opportunity? Where were we when the funding and research decisions were made? Why do questions addressing physics of estuaries have such a weak showing in the Global Research Plan?

III. Why Are Estuarine Scientists and Engineers in this Research Predicament?

The estuarine research community has not organized as have other oceanographic disciplines. For instance, compared to deep sea researchers, who are bound by well-defined funding sources and limited access to deep water, the estuarine community has no coherent agenda.

Why this is so is partly a function of the evolution of the estuarine research community. Several reasons stand out:

- Estuarine researchers receive funding from many different agencies, including states, municipalities, and industry, as well as federal government. Hence, the research requirements are bound to be more site-specific, and no national or international agenda need be followed.

- Little federal money is available for basic research in estuaries by academic scientists. In the U.S., estuaries are felt to be the mandate of the more "applied" agencies such as NOAA and EPA, rather than NSF or even the Office of Naval Research (ONR). This lack of leadership for estuarine basic research has resulted in lack of an accepted research agenda.

- Because of the local overtones of funding for estuarine research, the problems addressed tend to be parochial rather than global. Each estuarine researcher is trying to address site-specific problems, with resources that are inadequate to address or solve the global issues.

- A proliferation of estuarine researchers has resulted from this local orientation, and from the ease of doing field work. To work in an estuary, one dons hip-boots or rows a small boat. These simple tools are within the budgets of most academic groups. The result historically has been a non-cohesive mix of individual researchers studying different systems with often little impetus to communicate with one another.

• Estuarine questions are inherently interdisciplinary, yet few budgets will allow major programs funding the requisite level of diversity of researchers. To protect one's own budget, there is a tendency to maintain the funding sources for one's sub-discipline, sometimes at the expense of the broader field.

• As discussed above, the monies that are allocated towards estuaries are not spent on basic research to a significant extent. Hence, the researcher is forced to perform more "applied" research instead of more basic research into estuarine physics according to a national or international agenda.

• As opposed to more prosaic research fields, estuarine researchers are constantly queried about management questions about one's chosen estuary that are immediate in a social context. Consequently, the researcher can be committed too heavily to management issues at the expense of basic research, and interaction with other sub-disciplines or focus on basic research issues are put on the back burner.

• Estuarine physicists are often viewed as a necessary evil in interdisciplinary studies of shallow embayments, since life sciences and other sub-disciplines commonly must know the pathways for water and sediment. However, this view is not reciprocal, because many physicists don't realize that the pressing estuarine issues are biological or chemical in nature, not physical. The public would take less interest in estuarine research if problems were purely physical (navigation, for instance) and not aesthetic, life-threatening, or resource-threatening. Much concern is focused on the input, fate, and effects of contaminants in the coastal ocean--for which physics is a required, but peripheral, discipline.

• The complex and rich spectra of space and time scales associated with estuarine processes complicate their prediction. The answers are elusive and hard to define in a simple document to sell to funding interests. The result is lack of a cohesive plan that the community as a whole can use to foster increased funding for this important area.

• Lack of a single organizational spokesperson or group has reduced impact of the community as a whole. Once again, the diversity of researchers has detracted from the community's ability to impact the science strategy and funding decisions in the nations, since the diversity inhibits consensus.

IV. What Should the Estuarine Community Do?

Although the picture painted above may appear bleak, there is a clear path to follow. NASULGC's Marine Division's Estuarine Committee (1986) and many other review groups have provided some guidance for the community, in the form of specific recommendations for action.

For the physics community in particular, some specific recommendations stand out:

• Establish a consensus amongst physicists studying shallow bays and estuaries regarding a research agenda for the next decade on basic estuarine processes. A coherent plan is needed, one that can be communicated effectively to science managers to increase the competitiveness of this community.

• Recognize that estuarine research is inherently interdisciplinary, and create the research agenda with the driving needs in mind. Formulate programs that are truly interdisciplinary, and not just window dressing. To be truly interdisciplinary, the questions must be well-posed by scientists of all disciplines, and a strategy developed to lead to models that incorporate each discipline at its suitable and compatible space and time scales. Theory, sampling, and modeling requirements must be compatible in scale and in complexity.

• Encourage peer review of estuarine research funding, rather than the looser funding relationships that exist now, at least in some countries. This lack of peer review is especially pervasive in the United States, perhaps less so in other countries.

• Develop instrumentation to enter the next decade. In general, estuarine instrumentation is inadequate to make the interdisciplinary measurements required to address the fundamental science questions on adequate space and time scales. Instrumentation is commonly old, lacks modern technological improvements, and is too expensive. New developments must be made, and proven technology from other disciplines need to be applied more readily to estuaries (telemetry, microprocessing, miniaturization, etc.). The Marine Technology Society (1987) discussed some of the pressing issues in estuarine instrumentation.

• Although field measurement programs are expensive, it is true that observations lag models in many areas, particularly physics. This condition exists for many reasons, partly cost of field experimentation, lack of adequate instrumentation, and over-selling of the rapidly advancing art of numerical modeling at the expense of field verification. This imbalance must be corrected, and proper attention given to design of field experiments to address specific hypotheses when the rare opportunity for field experimentation does present itself.

• Mechanisms need to be identified to develop a community consensus on an international estuarine research agenda. We have plenty of documents to begin from; now we need to prioritize as a community and make a case for basic research in estuaries as an international marine science priority. These arguments must be made to research agencies as well as to our peers in other disciplines, with whom we will be competing for the same research dollars. However, estuarine researchers have a distinct advantage: the world knows of the importance of these environments. We need only make a stronger case for basic research in this area.

Estuarine researchers must develop a planning document that is developed on community consensus. This planning document must be presented in national and international forums, and be discussed at prestigious scientific meetings, not just local chapter meetings. These documents must pass muster under peer review by other communities having conflicting needs for the same research dollars. The agenda must be persuasive (important science that can be accomplished), timely, and soundly based scientifically. Without these qualities the agenda stands no chance. The case must be made also to funding agencies, including those "applied" agencies that currently address estuarine issues but with too little attention to basic academic science and engineering.

These recommendations will get the estuarine physics community into the mainstream of national and international scientific consensus-building. Once the case is laid out for the peer groups and programmatic agencies, representing a community consensus and not just a parochial view, positive steps will have been taken towards achieving scientific parity with other well-organized oceanographic interest groups.

V. Where Does This Research Take Place?

In meetings where estuarine research is discussed, too frequently discussions degrade to the logistical issues of WHERE to do the research. We need to deflect the emphasis away from WHERE, to WHAT research needs to be conducted. We must identify the critical research agenda, and then decide WHERE. Possibly the WHERE will be answered by funding opportunities, but the WHAT must come first before we can take advantage of such funding opportunities to their fullest extent. The research agenda must be process-oriented. By focusing first on WHERE, the community tends to lose the process orientation in favor of more parochial issues.

VI. When?

The estuarine community, especially perhaps the physics community, needs to move now to become part of the science of the 1990's. The research agenda already has been set for the next decade, largely without the input of the estuarine community as a whole. We are fortunate that a few niches have been included in the Global

Science Plan that can be filled by estuarine physicists. We need to act now to define a set of priorities to fill those research niches, priorities that are backed by solid consensus within the estuarine community. This research agenda is overdue, and should be completed early in the decade, before funding decisions are made to accompany the science plan, prolonging the drought in research funds that has plagued the estuarine community for the past decade or longer.

VII. Who Will Lead?

The process by which the international estuarine physical research agenda is to be set is not so clear. However, it is clear that both the research community and the individual have their own responsibilities to meet if the estuarine research crisis is to be relieved. Various national groups have provided documents to clarify the research agendas (e.g., NRC, 1977; NERC, 1983; NOAA, 1988, 1989), however, wide acceptance by the international community is lacking.

It appears that an umbrella organization is needed. Possibilities include groups such as the American Geophysical Union, American Society of Limnologists and Oceanographers, the Oceanography Society, the American Society of Civil Engineers, the Estuarine Research Federation, and so on. Most listed above are from the U.S. and may not be able to develop an international consensus. However, many international groups lack the scientific credibility to serve as focal point, although certainly ICSU, IOC, and SCOR hold some promise if politics can be avoided. The selection of an appropriate umbrella for developing the consensus is a critical juncture; a wrong choice could derail the effort. The selected group must possess and maintain scientific credibility, be international in scope, have sufficient resources to pursue the funding successfully against competing oceanographic disciplines that have a longer history of organization and cooperation, and should not have a particular viewpoint it wishes to impose on the community.

While a scientific umbrella organization is critical, there is an individual challenge as well. Both perspectives (individual and community) are essential to the success of our setting of resource priorities and funding levels. Each individual has some specific tasks:

- Lay aside provinciality of one's own research for the sake of the broader scientific issues. Estuarine research will always represent a balance between process-oriented studies and site-specific studies. The effort described above will not remove the site-specific opportunities from the researcher, but rather broaden the scope of available funding bases. When formulating the research agenda, each researcher must be careful to lay aside biases and individual research priorities for the sake of the field as a whole. This recommendation does not imply that one's biases and personal research agenda cannot be pursued, however they might have to be pursued under existing guises rather than new programs.

- Individual scientists and scientific groups must provide guidance to NSF, ONR, and other research agencies to convince them of the timeliness and priorities of estuarine science. In general, these funding agencies respond to community consensus, and strong backing of that consensus. It is incumbent on each of us to communicate that consensus, and pave the way for expanded emphasis on estuarine research.

- We must convince colleagues and agencies that basic estuarine research must be a research priority. Once we convince them of the scientific importance and quality of such research, the funding will follow.

- We must continue to educate "applied" agencies and our federal and local governments of the need for basic estuarine research. Since so much money is allocated to management of these resources, the implication is often that there is adequate money for research. Such is not the case. Management of these resources with inadequate scientific understanding represents mismanagement that may cost mankind dearly in the future.

- We must continue to serve on policy-making and oversight boards for national and international programs. In the United States, accepting positions to National

Academy of Science committees (standing, ad hoc, or study boards) is one way to assure that estuarine research remains a priority. Other review panels serve the same purpose; while time-consuming and often inconvenient, participation provides not only a community service but also the opportunity to represent the estuarine viewpoint.

The challenge is laid out to the estuarine research community, particularly the physicists. It is up to us to act as a community to develop by consensus-building a scientifically defensible, competitive, doable research agenda for the next decade. We need to integrate this agenda into existing programs, particularly the Global Change Research program, and increase the priority of this research in direct competition with other oceanographic interest groups. The choice is clear, now the community must act.

Acknowledgments

I thank Dr. Ralph Cheng and other convenors of the Third Symposium on Physics of Shallow Bays and Estuaries for their invitation to present this paper as a Keynote Address in Asilomar, California, in December, 1988. I thank Bruce Tripp for his input on the early notes for my presentation. This research and background behind this talk were funded by the Coastal Research Center of the Woods Hole Oceanographic Institution and by the Department of Commerce, Office of Sea Grant, under a Grant to the Woods Hole Oceanographic Institution under grant number NA-80-AA-D-00077. Although the symposium was international in scope, the author has emphasized research agendas and programs from the United States, where his primary experience lies. Such an emphasis does not imply that the U.S. model is either preferable or unique.

This paper appears as Woods Hole Oceanographic Institution Contribution Number 7329.

VIII. References

Beach, V., 1988: South Carolina Estuaries: under siege? South Carolina Sea Grant Consortium, Charleston, SC, 84 pp.

Capel, P. D., W. Giger, P. Reichert, and O. Wanner, 1988: Accidental input of pesticides into the Rhine River. Environmental Science and Technology, v. 22, p. 992-997.

COPO, 1988: Coastal Physical Oceanography: towards a national plan. NSF-sponsored meeting, Gulf Park, MS, 118 pp.

COPO, 1990: Shelf-Deep Ocean Exchange: a coastal physical oceanography workshop. Norfolk, VA, July 17-18, 1989, 22 pp.

Committee on Earth Sciences, 1990: Our changing planet: a U.S. strategy for Global Change Research. Federal Coordinating Council for Science, Engineering, and Technology, Executive Office of the President, OSTP, 38 pp.

Committee on Merchant Marine and Fisheries, 1989: Coastal waters in Jeopardy: reversing the decline and protecting America's coastal resources. Serial No. 100-E, U.S. Govt. Printing Office, Washington, D.C., 48 pp.

Duke, T. W. (Ed.), 1982: Impact of man on the coastal environment. U.S. Environmental Protection Agency, Office of Research and Development, EPA-600/8-82-021, 114 pp.

Farrington, J. W., A. C. Davis, B. W. Tripp, D. K. Phelps, and W. B. Galloway, 1987: "Mussel Watch" - measurements of chemical pollutants in bivalves as one indicator of coastal environmental quality. In T. P. Boyle (ed.), New Approaches to Monitoring Aquatic Ecosystems, ASTM STP 940, Amer. Soc. Testing and Materials, Philadelphia, p. 125-139.

GOSP Interagency Working Group, 1987: The U.S. Global Ocean Science Program: a strategy for understanding the role of the oceans in global change. JOI, Inc., Washington, D.C., 38 pp.

Ketchum, B. H. (ed.), 1972: The Water's Edge--Critical Problems of the Coastal Zone. MIT Press, Cambridge, MA, 393 pp.

Lauff, G. H. (ed.), 1967: Estuaries: Publication no 83, AAAS, Washington, D.C., 757 pp.

Marine Division's Estuarine Committee, 1986: On the importance of estuarine research. National Association of State Universities and Land Grant Colleges, Washington, D.C.

Marine Technology Society, 1987: Technology limitations in estuarine research: a conference. New England Section of the Marine Technology Society, held at the University of Rhode Island Narragansett Bay Campus, May, 1987, informal proceedings.

Morton, B., 1988: Hong Kong's first marine ecological disaster. Marine Pollution Bulletin, v. 19, p. 299-300.

National Environment Research Council, 1983: Research on estuarine processes. Report of a multidisciplinary workshop held at the University of East Anglia, 14-17 September, 1982, 31 pp.

National Ocean Pollution Program Office, 1988: National Marine Pollution Program, Federal plan for pollution research, development, and monitoring. U.S. Dept. of Commerce, NOAA, U.S. Govt. Printing Office, Washington, D.C., 205 pp., app.

National Oceanic and Atmospheric Administration, 1989: NOAA's coastal ocean program: an integrated systems approach to problems confronting our Nation's coastal waters. U.S. Department of Commerce, Washington, D.C., 20230, 17 pp.

National Research Council, 1971: Marine environmental quality--suggested research for understanding man's effect on the oceans. National Academy of Sciences, Washington, D.C., 107 pp.

National Research Council, 1977: Estuaries, Geophysics, and the Environment. National Academy of Sciences, Washington D.C., 127 pp.

Nittrouer, C. A. et al., 1988: Continental Margin Integrated Sedimentation Studies (CMISST). EOS, v. 69, p. 58-68.

Officer, C. B., L. E. Cronin, R. B. Biggs, and J. H. Ryther, 1981: A perspective on estuarine and coastal research funding. Environmental Science and Technology, v. 15, p. 1282-1285.

Schubel, J. R., 1984: On improving estuarine management. Testimony to House Subcommittee on Oceanography, 26 Sept.

Schubel, J. R. and D. J. Hirschberg, 1978: Estuarine graveyards, climatic change, and the importance of the estuarine environment. Estuarine Interactions, p. 285-303.

Smith, R. A., R. B. Alexander, and M. G. Wolman, 1987: Water-quality trends in the nation's rivers. Science, v. 235, p. 1607-1615.

Part II

Residual Currents

The Dynamics of Long-Term Mass Transport in Estuaries

John M. Hamrick

Virginia Institute of Marine Science
The College of William and Mary
Gloucester Point, VA 23062

ABSTRACT

The dynamics of long-term mass transport in estuaries having weakly nonlinear long wave dynamics are investigated. Low pass filtered long-term mass transport equations are derived for conditions of weak and strong vertical stratification. The dynamics of the residual mass transport velocity, or lowest order approximation to the Lagrangian residual velocity, are investigated by perturbation analyses of the hydrodynamics equations. For weak vertical stratification conditions, analytical results showing the influence of topography, the earth's rotation, and channel curvature on the distribution of the residual mass transport velocity field are presented. The calculation of the residual mass transport velocity field from field current meter measurements and numerical hydrodynamic model output is also discussed.

I. Introduction

It is now recognized that the management of estuarine resources necessitates the ability to predict and understand the transport and transformation of dissolved and suspended biogeochemical constituents over time periods on the order of several years. From the modeling point of view, the disparity between short time scale transport processes, associated with tides, for example, and the desired long-term evaluation of water quality, makes the integration of mass transport equations at the short time steps necessary to resolve short time scale transport processes unattractive. Likewise, from the point of view of achieving an understanding of the pathways of long-term mass transport in estuaries, a formulation and solution of the mass transport problem, which filters out short term advective and diffusive transport variations, while effectively retaining the correlations or interactions of short term variations is desirable. Both of these considerations have led to the formulation of time-averaged or filtered mass transport equations.

Early attempts at formulating time-averaged or filtered long-term mass transport equations were primarily ad hoc in approach, and resulted in physically unjustifiable temporal subscale transport parameterization. Typically in these equations, the time-averaged concentration field is advectively transported by some type of time-averaged velocity field, with the nonzero averaged subscale advective transport completely ignored or unjustifiably parameterized as diffusive or dispersive transport depending upon the spatial dimensions of the equations. A recent example of this approach, starting from the three-dimensional advection-diffusion equation in conservative form using a stretched vertical coordinate, used the Eulerian residual transport velocity field obtained from simple tidal cycle time averaging as the long-term advective transport velocity, (Hydroqual, 1987). Calibration to field data was relied upon to determine the magnitude of diffusion coefficients necessary to represent temporal subscale transport processes. An alternate approach to the formulation of time-averaged or filtered long-term mass transport equations has been the utilization of perturbation analyses of the three-dimensional advection-diffusion equation for conditions under which the instantaneous advective transport field is dominated by a velocity field resulting

Coastal and Estuarine Studies, Vol. 38
R. T. Cheng (Ed.)
Residual Currents and Long-term Transport

from weakly nonlinear long wave motions, associated with tidal and/or atmospheric forcings. Using this approach, Feng et al. (1986) derived a two-dimensional in the horizontal, steady, time averaged mass transport equation from the unsteady depth averaged advection-dispersion equation. Depending upon the scaling of the horizontal shear dispersion terms, two equations were obtained. In an equation for strong shear dispersion transport, the advective transport field is the lowest order barotropic Lagrangian residual or residual mass transport velocity presented by Longuet-Higgins (1969). In an equation for weak shear dispersion transport, an additional next order correction, termed the Lagrangian drift velocity, is made to the advective transport field. Subsequently, Hamrick (1986a) derived an unsteady form of the equation of Feng et al., appropriate for strong shear dispersion from the three-dimensional advective-diffusion equation, and provided a rigorous definition of vertical stratification conditions for which the equation is appropriate.

A three-dimensional, unsteady, expected value, long-term mass transport equation, appropriate for conditions of strong vertical stratification, was derived using the perturbation approach, by Hamrick (1987), with a similar steady time averaged equation presented by Feng (1987). In both of these equations, the advective transport field is the three-dimensional lowest order Lagrangian residual or residual mass transport velocity, and is consistent with the three-dimensional Eulerian-Lagrangian transformation presented by Longuet-Higgins (1969). The parameterization of vertical diffusive transport in Hamrick's (1987) and Feng's (1987) equations use respectively the expected value and time averaged values of the instantaneous vertical turbulent diffusion coefficient, while Hamrick's equation also includes horizontal turbulent diffusion parameterized in a similar fashion.

Along a complementary line of investigation, beginning with Longuet-Higgins (1969), numerous researchers have recognized that the Lagrangian residual velocity field, and in particular its lowest order approximation termed the residual mass transport velocity field, is the appropriate representation of long-term advective mass transport in estuaries and coastal seas having weakly nonlinear long wave dynamics. The ability to predict and understand the long-term transport pathways in estuaries now becomes strongly dependent upon knowledge of the dynamics of the residual mass transport velocity which is composed of the Eulerian residual velocity and the Stokes drift velocity. Until the last decade, dynamic theories of residual circulation in estuaries have been dominated by the two-dimensional vertical plane density driven Eulerian residual circulation theory formalized by Hansen and Rattray (1965) and most recently extended by Hamrick (1979) and Oey (1984), and to a lesser extent the two-dimensional horizontal plane density driven Eulerian residual circulation theory proposed by Fischer (1972) and extended by Imberger (1976), Hamrick (1979) and Smith (1980).

Over the last decade, the scope of study of residual circulation dynamics has expanded to consider Eulerian residual circulation induced by nonlinear interaction of long wave velocity and surface elevation fields, as well as the Stokes drift velocity field associated with the long wave motion. For two-dimensional motion in the vertical plane, Ianniello (1977, 1979) presented analytical solutions for the Eulerian residual velocity field, the Stokes drift velocity and the residual mass transport velocity for homogeneous density and a single constituent tidal forcing. Subsequently, Najarian et al. (1984) reported the results of a numerical study of the two dimensional geometry considered by Ianniello. Two-dimensional in the horizontal plane, tide induced Eulerian and Lagrangian residual circulation in estuaries has primarily been studied with the aid of numerical hydrodynamic models, with recent studies including Cheng and Casulli (1982), Oey et al. (1985), Feng et al. (1986), and Smith and Cheng (1987). In contrast, an extensive body of analytical results for two-dimensional in the horizontal plane Eulerian and Lagrangian residual circulation, associated with topographic influences on the nonlinear interaction of long wave velocity and surface elevation fields in coastal seas, exist with recent reviews given by Zimmerman (1981) and Robinson (1983).

The purpose of the present paper is to present some results on the dynamics of long-term mass transport in estuaries. The paper is organized in the following manner. The hydrodynamic and mass transport equations in a scaled form consistent with weakly nonlinear long wave dynamics and suitable for perturbation analysis are presented. This is followed by a discussion of the relationship between temporal filtering and the multiple time scale perturbation expansion to be used in the long-term mass transport analysis. Some linearization considerations that are useful in subsequent analytical treatments and in understanding the scaling of the vertical turbulent viscosity and diffusion coefficients are also presented. Filtered long-term mass transport equations for weak and strong vertical stratification conditions are next presented as are the two-dimensional barotropic and the three-dimensional residual mass transport velocity fields. The paper concludes with an analysis of the dynamics of the residual mass transport velocity field, including the presentation of some analytical results for the two-dimensional barotropic field and a discussion of the general features of the three dimensional field. Methodologies for numerically determining the three-dimensional mass transport velocity and analyzing field current meter data are also presented.

II. Hydrodynamic and Mass Transport Equations

The scaled hydrodynamic and mass transport equations in a horizontal curvilinear orthogonal and vertically stretched coordinate system are:

$$\partial_t u_\alpha + F\left(\frac{u_\beta}{m_\beta}\partial_\beta u_\alpha + \frac{w}{h}\partial_z u_\alpha\right) - F_R f e_{\alpha\beta} u_\beta - \frac{F}{m_\alpha m_\beta}(u_\beta \partial_\alpha m_\beta - u_\alpha \partial_\beta m_\alpha) u_\beta$$

$$- \frac{1}{h^2}\partial_z(N_v \partial_z u_\alpha) + \frac{1}{Fm_\alpha}\partial_\alpha\left[g(h-h_o) + g\beta \frac{F^2}{F_D^2} h \int_z^1 s dz\right]$$

$$- g\beta \frac{F^2}{F_D^2}\frac{s}{m_\alpha}\partial_\alpha (h_o - zh) + 0(h_s\lambda_s) = 0 \tag{1}$$

$$\partial_t h + \frac{F}{m}\partial_\beta\left(\frac{m}{m_\beta} h u_\beta\right) + \partial_z w = 0 \tag{2}$$

$$\partial_t s + F\left(\frac{u_\beta}{m_\beta}\partial_\beta s + \frac{w}{h}\partial_z s\right) - \frac{F^\epsilon}{h^2}\partial_z(K_v \partial_z s) + 0(h_s\lambda_s) = 0 \tag{3}$$

with m_β being the square root of the diagonal components of the metric tensor, m the square root of the metric tensor determinant, u_α the components of the horizontal velocity vector, w the physical vertical velocity in the stretched vertical coordinate, $0 \le z \le 1$, h the total water column depth, h_o the still water column depth, s the salinity, β the volumetric expansion coefficient, and N_v and K_v vertical turbulent diffusion coefficients for momentum and mass. The scaling parameters are: F the long wave Froude number or ratio of free surface displacement, $h-h_o$, scale to still water depth scale, F_R the rotational Froude number, and F_D the densimetric long wave Froude number. It is noted that Eqs.(1-3) assume dimensionless form by setting f, g and β to unity, while dimensional forms are obtained by setting the three scaling parameters to unity.

Horizontal velocity has been scaled by FC, with C being the shallow water wave speed, $(gh)^{1/2}$. Time and horizontal distance have been scaled by the long wave motion frequency, ω_s, and wave number λ_s, with $C = \omega_s/\lambda_s$. The vertical turbulent viscosity, N_v, has been scaled by $\omega_s h_s^2$, thus including frictional effects at lowest order. The vertical turbulent diffusivity, K_v, has been scaled by $F^\epsilon \omega_s h_s^2$. The exponent, ϵ, is used to adjust the scaling to accommodate different vertical stratification conditions with a value of zero for weak vertical stratification and of two for strong vertical stratification. The rational for this variable scaling will be discussed further in a following section. The omitted terms of order $h_s \lambda_s$ in Eqs.(1) and (3) are the horizontal diffusive momentum and mass fluxes, assuming that the scale of the horizontal diffusion coefficients does not exceed $\omega_s h_s/\lambda_s$, a conservatively large value.

III. Scaling, Filtering and Linearization Considerations

III.1 Multiple time scaling and filtering

For weakly nonlinear long wave dynamics, the long wave Froude number, F, is much less than unity, and may be taken as the small parameter in a regular perturbation analysis of Eqs.(1-3). It is expedient to introduce a multiple time scale expansion of the time derivative,

$$\partial_t = \partial_{t_o} + F^2 \partial_{t_2} \tag{4}$$

which serves to isolate short-term variations in the fast time variable, t_o, from long-term variations in the slow time variable, t_2. Correspondingly, the time scale associated with long-term variations is $F^{-2}\omega_s^{-1}$, corresponding to weeks or months for long-term variability as opposed to hours or days for short term variability.

Previous derivations of long-term mass transport equations have utilized simple time averaging over the period of the long wave motion for single frequency motion (Feng et al. 1986), time averaging over the synodic period for deterministic multiple period long wave motion (Hamrick, 1987), and ensemble averaging for a continuous spectrum of mixed deterministic and random long wave motion, (Hamrick, 1987). A more general approach for the analysis of continuous time variation and discrete time variation, characteristic of both field data and numerical model output, is to employ a filtering operation rather than a time or ensemble average operation.

The multiple time scale expansion of the time derivative, Eq.(4), applied to the salinity, for example, is equivalent to the time domain convolution operation

$$\Delta^* s = (\Delta_o + F^2 \Delta_2)^* s \tag{5}$$

or the frequency domain multiplication operation

$$\hat{\Delta}\,\hat{s} = (\hat{\Delta}_o + F^2 \hat{\Delta}_2)\hat{s} \tag{6}$$

where, $*$, denotes convolution and, $\hat{}$, denotes the Fourier transform. The Fourier transforms of the derivative operators may be defined as

$$\hat{\Delta} = \quad i\omega;\ 0 \le |\omega| \le \infty \tag{7a}$$

$$\hat{\Delta}_o = \begin{cases} 0;\ 0 \le |\omega| \le \infty \\ i\omega;\ \omega_p < |\omega| \le \infty \end{cases} \tag{7b}$$

$$\hat{\Delta}_2 = \begin{cases} i\omega;\ 0 \le |\omega| \le \omega_p \\ 0;\ \omega_p < |\omega| \le \infty \end{cases} \tag{7c}$$

with ω_p defining the intermediate frequency scale satisfying, $F^2\omega_s < \omega_p < \omega_s$. A low pass filter operator is now defined in the time domain by

$$\langle s \rangle = \phi^* s \tag{8}$$

with the Fourier transformation of the filtering operation defined theoretically by the ideal frequency domain filter response function

$$\hat{\phi} = \begin{cases} 1;\ 0 \le |\omega| \le \omega_p \\ 0;\ \omega_p \le |\omega| \le \infty \end{cases} \tag{9}$$

where ω_p is now identified as the upper limit of the filter pass band. When the filter operation is applied to the time derivative, the result is

$$\langle \partial_t s \rangle = F^2 \partial_{t_2} \langle s \rangle \tag{10}$$

representing the slow or long-term variation of low pass filtered salinity. For application to digital processing of discrete data, the ideal filter, Eq.(9) is replaced with a filter tapered about the frequency ω_p.

The choice of ω_p, the upper limit of the filter is related to the dynamics of the situation of interest. For weakly nonlinear wave motion, the lowest order motion at order F^0 should not pass through the filter. The next order motions at order F^1 represent nonlinear corrections including higher harmonics and higher frequency compounds and steady and lower frequency compounds. It is these steady and lower frequency compound components and the slowly varying density and atmospheric forced motions, also presumed of order F^1, that are the residual circulation and must pass through the low pass filter, thus determining ω_p.

III.2 Vertical diffusivity scaling and linearization

In this section linearizations of the vertical diffusive fluxes of momentum and mass are presented, which are useful in subsequent analytical solution procedures and provide insight into the scaling of the momentum and mass diffusion coefficients. Beginning with the vertical momentum flux, the vertical boundary conditions are

$$\frac{N_v}{h} \partial_z u_\alpha = \tau_{b\alpha} = c_b |u_\beta| u_\alpha; \; z = z_b \tag{11a}$$

$$\frac{N_v}{h} \partial_z u_\alpha = \tau_{s\alpha}; \quad z = 1 \tag{11b}$$

$$c_b^{1/2} = \frac{\kappa}{\ln\left(\dfrac{z_b + z_o}{z_o}\right)} , \tag{12}$$

where $\tau_{b\alpha}$ and $\tau_{s\alpha}$ are the bottom and free surface kinematic stresses respectively and c_b is the bottom friction coefficient, which depends on z_b, the height above the bottom boundary at which the boundary condition is applied, z_o, the roughness height, and the von Karman constant, κ. It is noted that as z_b approaches zero, the friction coefficient c_b approaches infinity and the bottom boundary condition becomes the no slip condition. For stratified turbulent estuarine flow a simple, but realistic, relationship for the vertical turbulent viscosity or momentum diffusivity is

$$\frac{N_v}{h} = S_N \left(\frac{\Lambda}{h}\right)^2 |\partial_z u_\beta| \quad , \tag{13}$$

where S_N is a Richardson number dependent stability function such as that given by Mellor and Yamada (1982), and Λ is the mixing length.

The bottom boundary condition, Eq.(11) and the form of N_v given by Eq.(13) are strongly nonlinear, in both temporal and spatial dependence. For analytical solution of Eq.(1), a constant momentum diffusivity is defined by

$$\frac{N_v}{h} = q_N \tag{14}$$

where q_N has dimensions of velocity. This allows a linearized bottom boundary condition

$$q_N \partial_z u_\alpha = \tau_{b\alpha} = \gamma q_N u_\alpha, \qquad \text{at } z = 0 \tag{15}$$

$$\frac{1}{\gamma} = (z_m + z_o) \ln\left(\frac{z_m + z_o}{z_o}\right), \tag{16}$$

where γ is a slip coefficient chosen such that the constant diffusivity velocity profile matches the logarithmic profile at a height z_m. Requiring that the depth integrated, low pass filtered, turbulent energy production by vertical shear be equivalent between the nonlinear and linearized formulations gives

$$q_N = \frac{< c_b \, |u_\beta(z_b)|^3 + \int_{z_b}^{1} S_N \left(\frac{\Lambda}{h}\right)^2 |\partial_z u_\beta|^3 \, dz >}{< \gamma \, |u_\beta(0)|^2 + \int_{o}^{1} |\partial_z u_\beta|^2 \, dz >} \tag{17}$$

allowing the magnitude of q_N to be determined such that bulk energy conservation is insured.

The preceeding results can be used to support the scaling of the vertical turbulent diffusivity, N_v. In essence the scaling of N_v is chosen consistent with the scaling of the vertical momentum flux whose scaling is in turn set by the bottom boundary stress. This is immediately evident from Eqs.(14) and (15), which indicate that N_v scales as

$$N_{vs} = q_{Ns}h_s = \frac{c_b}{\gamma} u_s = \frac{c_b}{\gamma} FC \tag{18}$$

in the limit of no vertical shear. At tidal forcing frequencies and for representative matching and roughness heights the scale given by Eq.(18) is consistent with the scale $\omega_s h_s^2$.

For the vertical salinity flux, the boundary conditions are

$$\frac{K_v}{h} \partial_z s = 0 \ ; \ z = 0,\ 1 \tag{19}$$

where the vertical turbulent mass diffusivity, K_v, is given by

$$\frac{K_v}{h} = S_K \left(\frac{\Lambda}{h}\right)^2 |\partial_z u_\beta| \tag{20}$$

with S_K being a Richardson number dependent stability function. For analytical solution of Eq.(3), a constant mass diffusivity is defined by

$$\frac{K_v}{h} = q_K \tag{21}$$

where q_K has units of velocity. Requiring that the depth integrated, low pass filtered, turbulent energy production by the vertical buoyancy or salinity flux be equivalent between the nonlinear and linearized formulations gives

$$q_K = \frac{< \int_0^1 S_k \left(\frac{\Lambda}{h}\right)^2 |\partial_z u_\beta| (-\partial_z s) dz >}{<s(1) - s(0)>}, \tag{22}$$

which allows the magnitude of q_K to be determined such that bulk energy conservation is insured.

The results represented by Eq.(22) can now be used to justify the variable scale, $F^\epsilon \omega_s h_s^2$, introduced for the vertical turbulent mass diffusivity, K_v. For weak vertical stratification ϵ was chosen as unity with K_v being scaled by $\omega_s h_s^2$, the same scale used for the vertical turbulent viscosity. The limiting case of weak vertical stratification is vertical homogeniety for which Eq.(22) fails to define q_N. Thus, for weak vertical stratification the scale for K_v can be chosen consistent with the scale for N_v based on similarity of mass and momentum diffusion in homogeneous flow. The limiting case of strong stratification is a two layer system with sufficient interfacial stability as to eliminate diffusive mass flux across the interface, with (22) giving q_K equal to zero, however, a finite but very small scale for K_v is necessary in the strong stratification case. To meet this condition, ϵ is chosen as two, giving a scale of $F^2 \omega_s h_s^2$, which serves to include vertical diffusive transport in the long-term mass transport equation for strong stratification.

IV. Filtered Long-term Mass Transport Equations

IV.1 Weak vertical stratification equation

For weak vertical stratification conditions, Eq.(3) in conservative form incorporating Eq.(4), is

$$(\partial_{t_o} + F^2\partial_{t_2})(hs) + \frac{F}{m} \partial_\beta\left(\frac{m}{m_\beta} hu_\beta s\right)$$

$$+ F\partial_z(ws) - \frac{1}{h} \partial_z(K_v \partial_z s) + O(h_s \lambda_s) = 0 \tag{23}$$

with the boundary conditions

$$\frac{K_v}{h}\,\partial_z s = 0 \;;\; z = 0,\ 1. \tag{24}$$

The order F^o approximation to Eq.(23) is

$$\partial_{t_o} s_o - h_o^{-2}\,\partial_z(K_v \partial_z s_o) = 0 \tag{25}$$

with the zero flux boundary conditions given by Eq.(24). For arbitrary temporal and spatial distribution of K_v and an arbitrary initial condition, the solution for s_o will exponentially decay on a time scale of h_o^2/K_v, asymptotically approaching its vertically averaged, t_o time domain filtered value, $<\bar{s}_o>$, where, $\bar{\ }$, demotes the depth average over $0 \le z \le 1$. Without loss of generality $<\bar{s}_o>$ may be replaced by $<\bar{s}>$, the slowly or long-term varying filtered, depth averaged salinity for which a mass transport equation is sought.

The asymptotic order F approximation to Eq.(23) is, in nonconservative form,

$$\partial_{t_o} s_1 + \frac{u_{\beta o}}{m_\beta}\partial_\beta <\bar{s}> - \frac{1}{h_o^2}\partial_z(K_v \partial_z s_1) = 0 \tag{26}$$

subject to the no flux boundary condition, Eq.(24). Integrating Eq.(26) over the depth gives:

$$\partial_{t_o} \bar{s}_1 + \frac{\bar{u}_{\beta o}}{m_\beta}\partial_\beta <\bar{s}> = 0\ , \tag{27}$$

which may be integrated to give

$$\bar{s}_1 = \bar{s}_1\ (t_o = 0) - \int_o^{t_o} \bar{u}_{\beta o} dt_o \frac{1}{m_\beta}\ \partial_\beta <\bar{s}> = \frac{-\bar{\xi}_{\beta o}}{m_\beta}\ \partial_\beta <\bar{s}>\ , \tag{28}$$

where $\bar{\xi}_\beta$ is the barotropic displacement. For $\bar{s}_1$, given by Eq.(28) to be bounded in the t_o time domain, $\bar{u}_{\beta o}$, the order F^o barotropic or external mode horizontal velocity cannot pass the low pass filter, $<\bar{u}_{\beta o}> = 0$, which also results in $<\bar{s}_1> = 0$, consistent with the assumption $<\bar{s}_o> = <\bar{s}>$. This allows $\bar{u}_{\beta o}$ to be identified as the higher frequency barotropic or external velocity field associated with tidal and strong higher frequency atmospheric forcings. The salinity field, $\bar{s}_1$, simply represents the advection of the long-term varying salinity field, $<\bar{s}>$, by the short-term varying velocity field $\bar{u}_{\beta o}$.

Subtracting Eq.(27) from Eq.(26) gives

$$\partial_{t_o} s_1' + \frac{u_{\beta o}'}{m_\beta}\ \partial_\beta <\bar{s}> - \frac{1}{h_o^2}\partial_z(K_v \partial_z s_1') = 0 \tag{29}$$

the equation governing the order F vertical salinity variation or stratification. It is noted the vertical shear or internal mode velocity field, $u_{\beta o}'$, may include a long-term or slowly varying portion resulting in a slowly varying vertical stratification governed by

$$\frac{<u_{\beta o}'>}{m_\beta}\ \partial_\beta <\bar{s}> - \frac{1}{h_o^2}\partial_z <K_v \partial_z s_1'> = 0. \tag{30}$$

Hamrick (1986b), presented analytic solutions of Eqs.(29) and (30), for constant K_v.

The depth integrated, low pass filtered, order F^2 approximation to Eq.(23), in conservative form, is,

$$h_o \partial_{t_2} <\bar{s}> + \frac{1}{m}\partial_\alpha\left[\frac{m}{m_\alpha}\ h_o(<\bar{u}_{\alpha_1}><\bar{s}> + <h_1 \bar{u}_{\alpha o}><\bar{s}> + <\bar{u}_{\alpha o}\bar{s}_1> + <\overline{u_{\alpha o}' s_1'}>)\right] = 0. \tag{31}$$

The third divergence term in Eq.(31) may be rewritten using the solution for s_1, Eq.(28), to give

$$\partial_\alpha(\frac{m}{m_\alpha}h_o\langle\bar{u}_{\alpha o}\bar{s}_1\rangle) = \partial_\alpha[\frac{m}{m_\alpha m_\beta}e_{\alpha\beta}\partial_\beta(h_o\bar{A}_z)] \tag{32}$$

$$\bar{A}_z = \langle\bar{u}_{1o}\int_0^{t_o}\bar{u}_{2o}dt_o\rangle \tag{33}$$

where $h_o A_z$ is the vertical component of a vector potential. The fourth divergence term in Eq.(31) represents shear dispersion and may be written as

$$\overline{\langle u'_{\alpha o}s'_1\rangle} = -\frac{\langle D_{\alpha\beta}\rangle}{m_\beta}\partial_\beta\langle\bar{s}\rangle \tag{34}$$

with $\langle D_{\alpha\beta}\rangle$ being the low pass filtered generalization of the ensemble averaged shear dispersion coefficient tensor given by Hamrick (1986b).

It is useful to express Eq.(31) in expanded form using conventional notation, the results being

$$\begin{aligned} & m_x m_y h_o\partial_{t_2}\langle s\rangle + \partial_x(m_y h_o\bar{u}_L\langle\bar{s}\rangle) + \partial_y(m_y h_o\bar{v}_L\langle\bar{s}\rangle) \\ & - \partial_x(\frac{m_y}{m_x}h_o\langle D_{xx}\rangle\partial_x\langle\bar{s}\rangle + h_o\langle D_{xy}\rangle\partial_y\langle\bar{s}\rangle) \\ & - \partial_y(h_o\langle D_{yx}\rangle\partial_x\langle s\rangle + \frac{m_x}{m_y}h_o\langle D_{yy}\rangle\partial_y\langle\bar{s}\rangle) = 0, \end{aligned} \tag{35}$$

the filtered long-term mass transport equation for weak vertical stratification. The advective transport field, (u_L, v_L) is given by

$$\bar{u}_L = \langle\bar{u}_1\rangle + \langle\frac{h_1\bar{u}_o}{h_o}\rangle + \frac{1}{h_o m_y}\partial_y(h_o\bar{A}_z) \tag{36}$$

$$\bar{v}_L = \langle\bar{v}_1\rangle + \langle\frac{h_1\bar{v}_o}{h_o}\rangle - \frac{1}{h_o m_x}\partial_x(h_o\bar{A}_z) \tag{37}$$

$$\bar{A}_z = \langle\bar{u}_o\int_o^{t_o}\bar{v}_o dt_o\rangle, \tag{38}$$

which is identified as the lowest order approximation to the barotropic Lagrangian residual velocity (Longuet-Higgins, 1969), and may be appropriately termed the barotropic residual mass transport velocity. The residual mass transport velocity, $(\bar{u}_L,\bar{v}_L)$ is the sum of the Eulerian residual velocity, $(\langle\bar{u}_1\rangle,\langle\bar{v}_1\rangle)$, and the Stokes drift velocity. The Stokes drift velocity may be further divided into the wave transport velocity, $(\langle h_1\bar{u}_o\rangle/h_o, \langle h_1\bar{v}_o\rangle/h_o)$, since h_1 is the long wave surface displacement, and the vector potential transport velocity, the last terms in Eqs.(38) and (39).

The Eulerian residual velocity and the wave transport velocity may be combined to form the barotropic Eulerian residual transport velocity which satisfies the depth averaged, low pass filtered, order F approximation of the continuity equation, Eq.(2). In addition, the residual mass transport velocity satisfies the zero divergence condition

$$\partial_x(m_y h_o\bar{u}_L) + \partial_y(m_x h_o\bar{v}_L) = 0. \tag{39}$$

The weak stratification, long-term mass transport model is completed by noting that Eq.(30) may be used to determine the slowly or long-term varying vertical stratification, $\langle s_1\rangle$, after Eq.(35) has been used to determine the depth averaged, filtered salinity, $\langle\bar{s}\rangle$. Analytic solutions of Eq.(30) may be obtained using a constant vertical diffusion coefficient defined by Eqs.(21) and (22).

IV.2 Strong vertical stratification equation

For strong vertical stratification conditions, Eq.(3) in conservative form incorporating Eq.(4), is

$$(\partial_{t_o} + F^2\partial_{t_2})(hs) + \frac{F}{m}\partial_\beta(\frac{m}{m_\beta} hu_\beta s)$$
$$+ F\partial_z(ws) - \frac{F^2}{h}\partial_z(K_v\partial_z s) + 0(h_s\lambda_s) = 0 \quad (40)$$

with the boundary conditions given by Eq.(24). The order F^o approximation to Eq.(40) has the solution, $s_o = \langle s_o \rangle$, and without loss of generality, $s_o = \langle s \rangle$, the low pass filtered salinity. The order F, nonconservative approximation to Eq.(40) is

$$\partial_{t_o} s_1 + \frac{u_{\beta o}}{m_\beta}\partial_\beta\langle s\rangle + \frac{w_o}{h_o}\partial_z\langle s\rangle = 0, \quad (41)$$

which may be integrated to give

$$s_1 = s_1(t_o{=}0) - \int_0^{t_o} u_{\beta o} dt_o \frac{1}{m_\beta}\partial_\beta\langle s\rangle - \int_0^{t_o} w_o dt_o \frac{1}{h_o}\partial_z\langle s\rangle$$

$$s_1 = s_1(t_o{=}0) - \xi_{\beta o}\frac{1}{m_\beta}\partial_\beta\langle s\rangle - \zeta_o\frac{1}{h_o}\partial_z\langle s\rangle \quad (42)$$

with $\xi_{\beta o}$ and ζ_o being the horizontal and vertical displacements. For s_1 to be bounded in the t_o time domain, the order F^o velocity field cannot pass the low pass filter. The order F^o velocity field may be identified as the higher frequency three dimensional velocity field associated with tidal and strong higher frequency atmospheric forcings.

The low pass filtered, order F^2 approximation to Eq.(40), in conservative form is

$$h_o\partial_{t_2}\langle s\rangle + \frac{1}{m}\partial_\alpha[\frac{m}{m_\alpha}h_o(\langle u_1\rangle\langle s\rangle + \langle\frac{h_1 u_o}{h_o}\rangle\langle s\rangle + \langle u_o s_1\rangle)]$$
$$+ \partial_z(\langle w_1\rangle\langle s\rangle + \langle w_o s_1\rangle) - \frac{1}{h_o}\partial_z(\langle K_v\rangle\partial_z\langle s\rangle) = 0. \quad (43)$$

The third horizontal divergence term, and the second vertical divergence term may be rewritten, using the solution for s_1 given by Eq.(42), to give

$$\partial_\alpha(\frac{m}{m_\alpha}h_o\langle u_o s_1\rangle) = \partial_\alpha[\frac{m}{m_\alpha m_\beta}e_{\alpha\beta}\partial_\beta(h_o B_z) - \frac{m}{m_\alpha}e_{\alpha\beta}\partial_z B_\beta] \quad (44)$$

$$\partial_z\langle w_o s_1\rangle = \partial_z[\frac{1}{m}e_{\alpha\beta}\partial_\alpha(m_\beta B_\beta)] \quad (45)$$

$$B_1 = \langle u_{20}\int_0^{t_o} w_o dt_o\rangle, \quad B_2 = \langle w_o\int_0^{t_o} u_{10} dt_o\rangle, \quad B_3 = \langle u_{10}\int_0^{t_o} u_{20} dt_o\rangle \quad (46)$$

where B_β and B_z are the horizontal and vertical components, respectively, of a vector potential.

Expressing Eq.(43) in expanded form gives

$$h_o\partial_{t_2}\langle s\rangle + \frac{1}{m_x m_y}\partial_x(m_y h_o u_L\langle s\rangle) + \frac{1}{m_x m_y}\partial_y(m_x h_o v_L\langle s\rangle)$$
$$+ \partial_z(w_L\langle s\rangle) - \frac{1}{h_o}\partial_z(\langle K_v\rangle\partial_z\langle s\rangle) = 0 \quad , \quad (47)$$

the filtered long-term mass transport equation for strong vertical stratification. The advective transport field is given by

$$u_L = \langle u_1\rangle + \langle\frac{h_1 u_o}{h_o}\rangle + \frac{1}{h_o m_y}\partial_y(h_o B_z) - \frac{1}{h_o}\partial_z B_y \quad (48a)$$

$$v_L = \langle v_1\rangle + \langle\frac{h_1 v_o}{h_o}\rangle + \frac{1}{h_o}\partial_z B_x - \frac{1}{h_o m_y}\partial_x(h_o B_z) \quad (48b)$$

$$w_L = \langle w_1\rangle + 0 + \frac{1}{m_x m_y}\partial_x(m_y B_y) - \frac{1}{m_x m_y}\partial_y(m_x B_x) \quad (48c)$$

$$B_x = \langle v_o \int_o^{t_o} w_o dt_o \rangle \tag{49a}$$

$$B_y = \langle w_o \int_o^{t_o} u_o dt_o \rangle \tag{49b}$$

$$B_z = \langle u_o \int_o^{t_o} v_o dt_o \rangle , \tag{49c}$$

which is identified as the lowest order approximation to the three-dimensional Lagrangian residual velocity, and may be appropriately termed the three-dimensional residual mass transport velocity. The residual mass transport velocity, (u_L, v_L, w_L) is the sum of the Eulerian residual velocity, $(\langle u_1 \rangle, \langle v_1 \rangle, \langle w_1 \rangle)$, and the three-dimensional Stokes drift velocity. The Stokes drift velocity may be further divided into the horizontal wave transport velocity, $(\langle h_1 u_o \rangle / h_o, \langle h_1 v_o \rangle / h_o, 0)$ and the vector potential transport velocity. The vector potential transport velocity is composed of the terms in Eq.(48), which involve (B_x, B_y, B_z), the components of the vector potential B and is equivalent to curl B since h_o^z is equivalent to m_z.

The Eulerian residual velocity and the horizontal wave transport velocity may be combined to form the Eulerian residual transport velocity which satisfies the low pass filtered, order F approximation of the continuity equation, Eq.(2). In addition, the residual mass transport velocity satisfies the zero divergence condition

$$\partial_x (m_y h_o u_L) + \partial_y (m_x h_o v_L) + \partial_z w_L = 0 , \tag{50}$$

which allows the long-term mass transport equation, Eq.(47), to alternately be written in nonconservative form.

The parameterization of the vertical diffusive salinity transport in the long-term mass transport equation, Eq.(47), using the low-pass filtered vertical diffusivity requires some elaboration. If the lowest order and most significant contribution to the velocity field is a single dominant frequency tidal forcing, Eq.(20), indicates that the temporal structure of K_v will involve steady and second harmonic of the dominant frequency components. For this simple forcing, the lowest order salinity will be steady in time, while Eq.(42), indicates that the next order salinity will be oscillatory at the dominant frequency. Thus, it is readily shown that for these temporal structures of the vertical diffusivity and salinity, the low pass filter of the vertical salinity flux gives the parameterization in the filtered long-term mass transport equation, Eq.(47).

V. Dynamics of the Residual Mass Transport Velocity Field

A definitive analysis of the dynamics of the residual mass transport velocity field, for either weak or strong vertical stratification conditions, for a prototype estuary would require numerical integration of the hydrodynamic and mass transport equations and subsequent manipulation and filtering of the output. An alternative approach is to rigorously analyze the hydrodynamic equations using the perturbation techniques employed in deriving the two filtered long-term mass transport equations in the preceeding section, with an objective of gaining insight into the significant dynamic influences.

V.1 Weak vertical stratification analysis

The major simplification to the hydrodynamic equations, Eqs.(1) and (2), for the condition of weak vertical stratification results from the vertical uniformity and short-term temporal independence of the order F^o salinity. For weak vertical stratification conditions, the order F^o approximations to the momentum and continuity equations, Eqs.(1) and (2), are:

$$\partial_{to} u_{\alpha o} + F_R f e_{\alpha\beta} u_{\beta o} - \frac{1}{h_o^2} \partial_z (N_v \partial_z u_{\alpha o}) + \frac{g}{m_\alpha} \partial_\alpha h_1 + g\beta \frac{F}{F_D^2} h_o \frac{(1-z)}{m_\alpha} \partial_\alpha \langle s \rangle = 0 \tag{51}$$

$$\partial_{t_0} h_1 + \frac{1}{m} \partial_\alpha (\frac{m}{m_\alpha} h_o u_{\alpha o}) + \partial_z w_o = 0 \tag{52}$$

where for the present, the order of F/F_D^2 is unspecified. Eqs.(51) and (52) may be integrated over the depth to give equations governing the external or barotropic mode. The presence of the slowly or long-term varying salinity gradient in the momentum equations is in conflict with the requirement that the order F^o barotropic or external velocity field not pass the low pass filter. Since the order of F/F_D^2 is in general intermediate between F^o and F^1, it is reasonable to rescale the barotropic portion of the horizontal salinity gradient forcing and the portion of the bottom boundary stress associated with it to order F. The resulting external equations are:

$$\partial_{to} \bar{u}_{\alpha o} + F_R f e_{\alpha\beta} \bar{u}_{\beta o} + g \frac{1}{m_\alpha} \partial_\alpha h_1 + \frac{\tau}{h_o} (\tau_{b\alpha o} - \tau_{s\alpha o}) = 0 \tag{52}$$

$$\partial_{t_o} h_1 + \frac{1}{m} \partial_\alpha (\frac{m}{m_\alpha} h_o \bar{u}_{\alpha o}) = 0 \ . \tag{54}$$

Before continuing with the analysis of the external equations, it is necessary to consider the internal or shear and baroclinic mode equations and the specification of the bottom boundary stress.

The internal or shear and baroclinic mode equations are obtained by subtracting the depth integrals of Eqs.(51) and (52) from the original equations, the results being

$$\partial_{to} u'_{\alpha o} + F_R f e_{\alpha\beta} u'_{\beta o} - \frac{1}{h_o^2} \partial_z (N_v \partial_z u'_{\alpha o})$$
$$+ \frac{1}{h_o} (\tau_{s\alpha o} - \tau_{b\alpha o}) + g\beta \frac{F}{F_D{}^2} \frac{1}{2} \frac{1}{h_o} \frac{(1-2z)}{m\alpha} \partial_\alpha <\bar{s}> = 0 \tag{55}$$

$$\frac{1}{m} \partial_\alpha (\frac{m}{m_\alpha} h_o u'_{\alpha o}) + \partial_z w_o = 0. \tag{56}$$

The baroclinic portion of the horizontal salinity gradient forcing has been retained at this order such that the strong vertical shear associated with it might be incorporated into the shear dispersion transport in the long-term mass transport equation. Hamrick (1986b) has presented analytical solutions to Eq.(55) for a constant vertical turbulent viscosity, N_v, and with the horizontal velocity represented as

$$u'_{\alpha o} + <u'_{\alpha o}> + \int_{-\infty}^{\infty} u'_{\alpha o} e^{i\omega t_o} d\omega. \tag{57}$$

Using N_v given by Eqs.(14) and (17), the solutions for no surface stress are:

$$<u'_{\alpha o}> = - \delta_{\alpha\beta} (\Gamma_3 - r_3 \Gamma_1 - r_4 \Gamma_2)_{\sigma=0} \, g \frac{\beta h_o^2}{q_N} \frac{F}{F_D{}^2} \frac{1}{m_\beta} \partial_\beta <s>$$
$$- e_{\alpha\beta} (\Gamma_4 + r_4 \Gamma_1 - r_3 \Gamma_2)_{\sigma=0} \, g \frac{\beta h_o^2}{q_N} \frac{F}{F_D^2} \frac{1}{m_\beta} \partial_\beta <\bar{s}> \tag{58}$$

$$U'_{\alpha o} = [-\delta_{\alpha\beta} (r_1 \Gamma_1 + r_2 \Gamma_2) - e_{\alpha\beta} (r_1 \Gamma_2 - r_2 \Gamma_1)] \bar{U}_{\beta o} \tag{59}$$

where the Γ's are functions of z given by

$$\Gamma_1 = 2 \sum_{n=1}^{\infty} \frac{(n^2\pi^2 + i\sigma) \cos(n\pi z)}{(n^2\pi^2 + i\sigma)^2 + \epsilon^2} \tag{60a}$$

$$\Gamma_2 = 2\epsilon \sum_{n=1}^{\infty} \frac{\cos(n\pi z)}{(n^2\pi^2 + i\sigma)^2 + \epsilon^2} \tag{60b}$$

$$\Gamma_3 = 2\sum_{n=1}^{\infty} \frac{[1-\cos(n\pi)]\cos(n\pi z)}{n^4\pi^4 + \epsilon^2} \tag{60c}$$

$$\Gamma_4 = 2\epsilon\sum_{n=1}^{\infty} \frac{[1-\cos(n\pi)]\cos(n\pi z)}{n^2\pi^2\,(n^4\pi^4 + \epsilon^2)} \tag{60d}$$

$$\sigma = \frac{\omega h_o}{q_N} = \frac{\omega h_o^2}{N_v} \tag{61}$$

$$\epsilon = F_R \frac{f h_o}{q_N} = F_R \frac{h_o^2}{N_v}\ . \tag{62}$$

The r coefficients are functions of σ, ϵ and the bottom slip parameter, γ. They are determined by requiring that $\tau_{b\alpha}$ satisfy Eq.(15), which in turn defines the Fourier transform of the bottom boundary stress as

$$T_{b\alpha o} = (\delta_{\alpha\beta} r_1 - e_{\alpha\beta} r_2) q_N \bar{U}_{\beta o} \tag{63}$$

where $\bar{U}_{\beta o}$ is the Fourier transform of the external or barotropic mode horizontal velocity.

The external or depth mean order F^o hydrodynamic problem may now be analyzed by Fourier transforming Eqs.(53) and (54) and using Eq.(63), to give

$$[\delta_{\alpha\beta}(i\omega + r_1 \frac{q_N}{h_o}) - e_{\alpha\beta}(F_R f + r_2 \frac{q_N}{h_o})]\ \bar{U}_{\beta o} - \frac{g}{m_\alpha}\partial_\alpha H_1 = 0 \tag{64}$$

$$i\omega H_1 + \frac{1}{m}\partial_\alpha(\frac{m}{m_\alpha} h_o \bar{U}_{\alpha o}) = 0 \tag{65}$$

for the case of no surface stress. These equations may be combined into a single elliptic equation for the Fourier transform of the free surface displacement, H_1. The elliptic equation may be solved numerically for realistic bottom topography and shoreline geometry, however, analytical solutions of the two-dimensional problem with variable bottom topography are difficult, if not impossible to obtain. To gain some general insight into the role of topography, channel curvature and the earth's rotation in determining the external barotropic flow dynamics, and ultimately the residual mass transport velocity dynamics, Eqs.(64) and (65) can be solved by perturbation techniques for slight variations in topography, slight channel curvature and a narrow channel width resulting in a weak geostrophic effect.

Rescaling the lateral horizontal coordinate, $x_2 = y$, and the lateral velocity, $u_{20} = v_o$, by a small parameter δ such that the channel width is of order δ/λ_s, and representing the topography and channel curvature by

$$m_x = 1 - \delta_c \frac{b}{R}\frac{y}{b} \tag{66}$$

$$m_y = 1 \tag{67}$$

$$h_o = h_{oo} + \delta_h h_{o1} \tag{68}$$

where y is zero along the channel centerline whose radius of curvature R varies with x along the channel. The spatially averaged still water depth is h_{oo} while h_{o1} represents a slight lateral, y, variation having zero mean when averaged across the constant channel width, b. The solutions of Eqs.(64) and (65) to lowest order in the perturbation parameters, in dimensional form are:

$$\bar{U}_o = \bar{U}_{oo} + (\frac{\nu}{i\omega+\nu})\frac{h_{o1}}{h_{oo}}\bar{U}_{oo} + \frac{b}{R}\bar{U}_{oo}\frac{y}{b}$$
$$- (\frac{i\omega}{i\omega+\nu})\frac{fb}{(gh_{oo})^{1/2}}\frac{H_{10}}{h_{oo}}(gh_{oo})^{1/2}\frac{y}{b} \tag{69}$$

$$\bar{V}_o = \left(\frac{i\omega+2\nu}{i\omega+\nu}\right) \frac{\omega b}{(gh_{oo})^{1/2}} \frac{H_{10}}{h_{oo}} (gh_{oo})^{1/2} \frac{1}{b} \int_{-b/2}^{y} \frac{h_{01}}{h_{oo}} dy$$

$$+ \frac{i\omega b}{(gh_{oo})^{1/2}} \frac{b}{R} \frac{H_{10}}{h_{oo}} (gh_{oo})^{1/2} \left[\left(\frac{y}{b}\right)^2 - \frac{1}{4}\right] \tag{70}$$

$$H_{11} = H_o - \frac{fb}{(gh_{oo})^{1/2}} \frac{\bar{U}_{oo}}{(gh_{oo})^{1/2}} h_{oo} \frac{y}{b} \tag{71}$$

$$\nu = r_1 \frac{q_N}{h_{oo}} \tag{72}$$

where $\bar{U}_{oo}$ and H_{10} are the solutions of

$$(i\omega+\nu)\, \bar{U}_{oo} + g\partial_x H_{10} = 0 \tag{73}$$

$$i\omega H_{10} + \partial_x (h_{oo} \bar{U}_{oo}) = 0 \tag{74}$$

the equations for a constant depth, straight, nonrotating channel. Eqs.(69-71) capture the essential dynamic influences of lateral topography, channel curvature and the earth's rotation. As indicated by the solutions, the longitudinal velocity distribution across the channel is influenced by all three effects, while the lateral velocity results only from topography and curvature, and the lateral variation of the free surface displacement is associated only with the geostrophic effect.

To continue with the analysis, the order F hydrodynamic equations must be solved for the barotropic Eulerian residual horizontal velocity, $\langle u_{\alpha 1}\rangle$. For shallow depth and strong turbulence as represented by a large value of $q_N = N_v/h$, the parameters σ and ϵ defined by Eqs.(61) and (62) are small, indicating that the magnitude of the internal modes are small relative to the external, thus allowing the vertical integrals of quadratic products of the shear modes to be at lowest approximation neglected. Since the barotropic portion of the order F/F_D^2 horizontal salinity gradient forcing and its associated contribution to the bottom boundary stress have been rescaled to this order, it is reasonable to neglect higher order F^2/F_D^2 horizontal salinity gradient forcings. The resulting set of filtered equations for the order F barotropic Eulerian residual velocity are:

$$- F_R f e_{\alpha\beta} \langle \bar{u}_{\beta 1}\rangle + \left\langle \frac{\bar{u}_{\beta o}}{m_\beta} \partial_\beta \bar{u}_{\alpha o}\right\rangle - \frac{1}{m_\alpha m_\beta} \langle (\bar{u}_{\beta o} \partial_\alpha m_\beta + \bar{u}_{\alpha o} \partial_\beta m_\alpha) \bar{u}_{\beta o}\rangle$$

$$- (\delta_{\alpha\beta} r_1 - e_{\alpha\beta} r_2) \frac{1}{h_o^2} q_N \langle h_1 \bar{u}_{\beta o}\rangle$$

$$+ (\delta_{\alpha\beta} r_1 - e_{\alpha\beta} r_2) \frac{1}{h_o} q_N \langle \bar{u}_{\beta 1}\rangle + \frac{g}{m_\alpha} \partial_\alpha \langle h_2\rangle$$

$$+ (\delta_{\alpha\beta}(1-2r_3) + e_{\alpha\beta} r_2) g\beta \frac{1}{F_D^2} \frac{\bar{h}_o}{2} \frac{1}{m_\beta} \partial_\beta \langle s\rangle = 0, \tag{75}$$

$$\frac{1}{m} \partial_\alpha \left[\frac{m}{m_\alpha} h_o \left(\langle \bar{U}_{\alpha 1}\rangle + \frac{\langle h_1 u_{\alpha o}\rangle}{h_o}\right)\right] = 0 \; . \tag{76}$$

The most promising solution strategy for Eqs.(75) and (76) is to eliminate $\langle h_2\rangle$ between the components of Eq.(75) forming a vorticity equation, and introduce the Eulerian residual transport stream function defined by

$$h_o \langle \bar{u}_{\alpha 1}\rangle + \langle h_1 \bar{u}_{\alpha o}\rangle = \frac{e_{\alpha\beta}}{m_\beta} \partial_\beta \langle \psi_1\rangle \tag{77}$$

The solution of the resulting equation for, $\langle \psi \rangle$, readily allows c [illegible]f the Eulerian residual transport velocity.

The equation for the Eulerian residual transport stream function was solved by perturbation methods for the narrow channel rescaling and the slight channel curvature and topography conditions specified by Eqs.(66-68). The longitudinal salinity gradient was also assumed independent of lateral position. Making use of the solutions of the order F^o baratropic problem, Eqs.(69-71), the longitudinal, x, component of the Eulerian residual transport velocity is in dimensional form

$$\bar{u}_{ET} = \frac{-Q_f}{gh_{oo}}(1 + \frac{h_{01}}{h_{oo}} + \frac{y}{R}) - g\frac{bh_{00}^2}{2\nu}(1-2r_3)\partial_x<s> (\frac{h_{01}}{h_{oo}})$$

$$+ \frac{1}{2} \mathrm{Re}[\frac{i\omega}{(i\omega+\nu)} \frac{H_{10}\bar{U}_{oo}^*}{h_{oo}} - \frac{(5i\omega+2\nu)}{(i\omega+\nu)} \frac{H_{10}^*\bar{U}_{oo}}{h_{oo}}]\frac{h_{01}}{h_{oo}}$$

$$- \frac{fb}{2} \frac{|U_{oo}|^2}{gh_{oo}} + \frac{3\omega^2}{(\omega^2+\nu^2)} |\frac{H_{10}}{h_{oo}}|^2 \frac{y}{b} \quad . \tag{78}$$

The above results show the important influences of topography, channel curvature and the earth's rotation on the Eulerian residual transport velocity. The fresh water river discharge and longitudinal salinity gradient driven portions are stronger in the seaward and landward directions, respectively in deeper regions of the lateral transect. Curvature serves to intensify the seaward river discharge toward the inner bank of a curved section. The geostrophic influence on the tidal rectification induced portion results in landward transport to the right side facing landward. The topographic influence of the tidal rectified portion is not immediately apparent.

The barotropic residual mass transport velocity, u_L, defined by Eqs.(36-38), may now be obtained by adding the vector potential transport velocity. The longitudinal vector potential transport velocity, evaluated using Eqs.(69) and (70) is

$$\frac{1}{h_o}\partial_y(h_o\bar{A}_z) = \frac{b}{R} \mathrm{Re}(\frac{\bar{U}_{oo}{}^*H_{10}}{h_{oo}})\frac{y}{b} + \frac{1}{2}\mathrm{Re}[\frac{(i\omega+2\nu)}{(i\omega+\nu)} \frac{\bar{U}_{00}^*H_{10}}{h_{oo}}] \frac{h_{01}}{h_{oo}} \quad . \tag{79}$$

Combining Eqs.(78) and (79) gives the longitudinal barotropic residual mass transport velocity,

$$\bar{u}_L = \frac{-Q_f}{bh_{oo}}(1 + \frac{h_{01}}{h_{oo}} + \frac{b}{R} \frac{y}{b}) - g\frac{\beta}{2\nu} h_{oo}^2(1-2r_3)\partial_x<s> (\frac{h_{01}}{h_{oo}})$$

$$- \frac{3}{2}\mathrm{Re}(\frac{i\omega}{i\omega+\nu} \frac{H_{10}^*\bar{U}_{oo}}{h_{oo}})\frac{h_{01}}{h_{oo}} + \frac{b}{R}(\frac{\bar{U}_{oo}{}^*H_{10}}{h_{oo}})\frac{y}{b}$$

$$- \frac{fb}{2} \frac{|U_{oo}|^2}{gh_{oo}} + \frac{3\omega^2}{\omega^2+\nu^2} |\frac{H_{10}}{h_{oo}}|^2) \frac{y}{b} \tag{80}$$

in dimensional form. Eq.(79) indicates that the residual vector potential transport velocity is influenced by topography and curvature as would be expected from the results for the lateral order F^o barotropic velocity, Eq.(70). Combining the residual vector potential transport velocity with the Eulerian residual transport velocity to form the residual mass transport velocity results in modifying the topographic influenced portion of tidal rectification component and adding a curvature influenced portion. For the topographic influenced portion in the limit of small friction, the real operator produces a negative quantity for a simple landward propagating wave, thus the transport velocity is seaward in shallower regions. The real operator in the curvature influenced portion also produced a negative quantity resulting in seaward transport toward the inner bank of a curving channel.

V.2 Strong vertical stratification analysis

For the case of strong vertical stratification, the constraint that the order F^o three-dimensional velocity not pass the low pass filter requires that the salinity gradient forcing terms in Eq.(1) not enter until order F, which is readily accomplished by specifying F_D to be of order unity. The order F^o approximations to Eqs.(1) and (2) are identical to Eq.(51), with the salinity gradient term absent and Eq.(52). The internal shear mode solutions to the horizontal momentum equations are given by Eq.(59), while for slight topography and curvature, and a narrow channel width the external or barotropic solutions, Eqs.(69-71), are also applicable. Thus, the condition of strong vertical stratification does not modify the order F^o tidal driven long wave motion. For the functions, Γ, given by Eq.(60), very slowing varying with horizontal position, the Fourier transform of the vertical velocity is

$$W_o = \int_0^z [-i\omega(r_1\Gamma_1 + r_2\Gamma_2)H_1 + (r_1\Gamma_2 - r_2\Gamma_1)h_o\bar{\Omega}_{zo} + (r_1\Gamma_2 - r_2\Gamma_1)(\frac{\bar{V}_o}{m_x}\partial_x h_o - \frac{\bar{U}_o}{m_y}\partial_y h_o)]dz \tag{81}$$

where $\bar{\Omega}_{zo}$ is the Fourier transform of the order F^o barotropic vorticity. Since the internal shear modes of the horizontal velocity exhibit cos(nπz) modal structure, the corresponding modes of the vertical velocity exhibit sin(nπz) structure, which satisfies the boundary conditions on w_o in the stretched vertical coordinate exactly.

The analysis for strong vertical stratification could continue by presenting the order F^1 filtered approximations to Eqs.(1) and (2), which govern the three-dimensional Eulerian residual velocity field. However, such a lengthy exercise is beyond the scope of this work, and instead only the solution strategy will be briefly discussed, before moving to a qualitative discussion of the three-dimensional residual mass transport velocity field. Since the order F^o internal solutions, Eqs.(59) and (81) are in a modal or spectral form with cosine and sine basis functions, the order F, filtered approximations to Eqs.(1) and (2) can be readily expanded in the same basis functions, for the three dimensional Eulerian residual transport velocity. The salinity would be expanded in the cosine functions as would the filtered long-term mass transport equation, Eq.(47). Semi-analytical solutions may then be possible for low order expansions while numerical solutions would be necessary for higher order expansions. For situations where, F_D, the densimetric long wave Froude number is less than unity, the salinity gradient driving force will likely be dominant in determining the vertical structure, while geostrophic, topographic and curvature influences will make significant contributions to the horizontal structure.

The general features of the three-dimensional vector potential transport velocity field may be briefly discussed at this point. The longitudinal component, from Eq.(48), is

$$u_{vp} = \frac{1}{h_o m_y}\partial_y(h_o B_z) - \frac{1}{h_o}\partial_z B_y \tag{82}$$

with B_z and B_y given by Eq.(49). Expressing the Fourier transforms of the order F^o three-dimensional velocity field using cosine and sine basis modal forms allows B_z and B_y to be written as

$$\begin{aligned} B_z = &< \int_{-\infty}^{\infty} \bar{U}_0 e^{i\omega t_o} d\omega \int_{\infty}^{\infty} \frac{\bar{V}_0}{i\omega} e^{i\omega t_o} d\omega > \\ &+ \sum_{n=1}^{\infty} < \int_{-\infty}^{\infty} \bar{U}_0 e^{i\omega t_o} d\omega \int_{-\infty}^{\infty} \frac{\bar{V}_0^{(n)}}{i\omega} e^{i\omega t_o} d\omega \\ &+ \int_{-\infty}^{\infty} U_0^{(n)} e^{i\omega t_o} d\omega \int_{\infty}^{\infty} \frac{V_o}{i\omega} e^{i\omega t_o} d\omega > \cos(n\pi z) \\ &+ \sum_{m=1}^{\infty}\sum_{n=1}^{\infty} < \int_{-\infty}^{\infty} U_0^{(m)} e^{i\omega t_o} d\omega \int_{\infty}^{\infty} \frac{V_0^{(n)}}{i\omega} e^{i\omega t_o} d\omega > \cos(m\pi z)\cos(n\pi z) \end{aligned} \tag{83}$$

$$\begin{aligned} B_y = &\sum_{n=1}^{\infty} < \int_{-\infty}^{\infty} W_0^{(n)} e^{i\omega t_o} d\omega \int_{\infty}^{\infty} \frac{\bar{U}_0}{i\omega} e^{i\omega t_o} d\omega > \sin(n\pi z) \\ &+ \sum_{m=1}^{\infty}\sum_{n=1}^{\infty} < \int_{-\infty}^{\infty} W_0^{(m)} e^{i\omega t_o} d\omega \int_{\infty}^{\infty} \frac{U_0^{(n)}}{i\omega} e^{i\omega t_o} d\omega > \sin(n\pi z)\cos(n\pi z). \end{aligned} \tag{84}$$

It is readily seen that B_z has a barotropic component equivalent to A_z given by Eq.(38). The behavior of this barotropic component will likely dominate B_z and the single series contribution will dominate the double series contribution. Thus a low order approximation of the portion of u_{vp} associated with the lateral gradient of B_z will be strongly influenced by topography and dominated by a barotropic component followed next by a two layer shear mode.

The vertical structure of B_y will be dominated by the lowest mode of the single series which when differentiated with respect to z in forming its contribution to u_{vp} in Eq.(82) will result in a two layer structure. Using W_o given by Eq.(81), this two layer structure will likely have landward flow near the surface in opposition to the classical salinity gradient driven flow.

The four temporal representations of B_z and B_y in Eqs.(83) and (84) and a similar expression for B_x suggest how the filtering operations necessary to obtain these quantities might be carried out. For current meter records or numerical hydrodynamic model output, the three-dimensional velocity time series are fast Fourier transformed and filtered in the frequency domain, which is equivalent to setting, for example U_o, equal to zero over the band, $|\omega| \leq \omega_p$. The time integration is also performed in the frequency domain by dividing the transformed quantities by $i\omega$. The multiplication of the two transform representations is best done after inverting each filtered and time integrated series back to the time domain. The final series is again fast Fourier transformed, filtered in the frequency domain and inverted to give slowly or long-term varying series for the vector potential components. The same procedure can be applied to determine the Eulerian residual and wave transport velocities. For values of quantities at discrete spatial points, local linear basis functions are the natural spatial representations, with spatial gradients determined by finite difference differentiation.

VI. Summary and Conclusion

An approach for analyzing the dynamics of long-term mass transport in estuaries has been presented. The primary results are a pair of filtered long-term mass transport equations for conditions of weak and strong vertical stratification and weakly nonlinear long wave dynamics. The advective transport field in these equations has been shown to be the two and three-dimensional lowest order approximations to the Lagrangian residual velocity field, termed the residual mass transport velocity field. Analytical solutions were presented for the internal shear and baroclinic mode and external barotropic mode order F^o velocity fields. These results were used to obtain an analytical solution for the weak stratification case, two-dimensional barotropic residual mass transport velocity showing the influences of topography, channel curvature and the earth's rotation. The dynamics of the three-dimensional residual mass transport velocity were discussed in a qualitative manner. A filtering procedure for analyzing current meter measurements and numerical hydrodynamic model output to determine the residual mass transport velocity was also presented.

Acknowledgement

Contribution number 1558 from the Virginia Institute of Marine Science, The College of William and Mary, Gloucester Point, VA.

VII. References

Cheng, R. T., and V. Casulli, 1982: On Lagrangian residual currents with applications in South San Francisco Bay, Ca. Water Resour. Res., **18**, 1652-1662.

Feng, S., R. T. Cheng, and P. Xi, 1986: On tide induced Lagrangian residual current and residual transport. Part 2. Water Resour. Res., **22**, 1635-1646.

Feng S., 1987: A three-dimensional weakly nonlinear model of tide-induced Lagrangian residual current and mass transport, with an application to the Bohai Sea, in

J.C. Nihoul and P. M. Jamart, (eds). Three-Dimensional Models of Marine and Estuarine Dynamics. Elsevier, Amsterdam.

Fischer, H. B., 1972: Mass transport mechanisms in partially stratified estuaries. J. Fluid Mech., **53**, 671-687.

Hamrick, J. M., 1979: Salinity intrusion and gravitational circulation in partially stratified estuaries. PhD dissertation. University of California, Berkeley.

Hamrick, J. M., 1986a: Subtidal circulation and transport in estuaries. In Advancements in Aerodynamics, Fluid Mechanics and Hydraulics. American Society of Civil Engineers, New York.

Hamrick. J. M., 1986b: Long-term dispersion in unsteady skewed free surface flow. Estuarine. Coastal and Shelf Sci., **23**, 807-845.

Hamrick, J. M., 1987: Time averaged estuarine mass transport equations, in Proceeding of the National Conference on hydraulic Engineering, American Society if Civil Engineers, New York.

Hansen, D. V. and M. Rattray, Jr., 1965: Gravitational circulation in straits and estuaries. J. of Marine Res., **23**, 104-122.

Hydroqual, Inc., 1987: A steady state coupled hydrodynamic, water quality model of the eutrophication and anoxia process in Chesapeake Bay.

Ianniello, J. P., 1977: Tidally induced residual currents in estuaries of constant breadth and depth. J. Mar. Res., **35**, 755-786.

Ianniello, J. P., 1979: Tidally induced residual currents in estuaries of variable breadth and depth. J. Phys. Oceanogr., **9**, 962-974.

Imberger, J., 1976: Dynamics of a longitudinally stratified estuary, in Proceeding 15th Int. Conf. on Coastal Engr.

Longuet-Higgins, M. S., 1969: On the transport of mass by time-varying ocean currents. Deep-Sea Res., **16**, 431-447.

Mellor, G. L. and T. Yamada, 1982: Development of a turbulence closure model for geophysical fluid problems. Rev.Geophys., **20**, 851-875.

Najarian, T. O., D. -P. Wang, and P. -S. Huang, 1984: Lagrangian transport model for estuaries. J. Waterways, Port, Coastal and Ocean Engr., **110**, 321-333.

Oey, L. -Y., 1984: On steady salinity distribution and circulation in partially mixed and well mixed estuaries. J. Phys.Oceanogr., **14**, 629-645.

Oey, L. -Y., G. L. Mellor, and R. I. Hires, 1985: Tidal modeling of the Hudson-Raritan estuary. Estuarine, Coastal and Shelf Sci., **20**, 511-527.

Robinson, I. S., 1983: Tidally induced residual flows, in B. Johns (ed), Physical Oceanography of Coastal and Shelf Seas, Elsevier, Amsterdam.

Smith, L. H. and R. T. Cheng, 1987: Tidal and tidally averaged circulation characteristics of Suisun Bay, CA. Water Resour. Res., **23**, 143-155.

Smith, R., 1980: Buoyancy effects upon longitudinal dispersion in wide well-mixed estuaries. Philos. Trans, R. Soc. London, **A 296**, 467-496.

Zimmerman, J. T. F., 1981: Dynamics, diffusion and geomorphological significance of tidal residual eddies. Nature, **290**, 549-555.

On the Lagrangian Residual Velocity and the Mass-Transport in a Multi-Frequency Oscillatory System

Shizuo Feng

Institute of Physical Oceanography
Ocean University of Qingdao
China

ABSTRACT

Using a three-dimensional weakly nonlinear baroclinic shallow water model, the Lagrangian residual velocity associated with a multi-frequency tidal system has been analyzed. The first-order Lagrangian residual velocity, the mass-transport velocity, has been shown to be the sum of the mass-transport velocities derived from the respective constituents of astronomical tides, and both the wind-driven (barotropic) and the density-driven (baroclinic) components. The second-order perturbation Lagrangian residual velocity, i.e., the Lagrangian drift velocity, has been shown to involve a series of nonlinear interactions between the products of the respective constitutuents of tides, and reflects the periodicities of all the constituents of astronomical tides contained in the multi-frequency tidal system through the initial phases. As an example, the Lagrangian drift velocity induced by an M_2-S_2 tidal system is analyzed in detail for a more thorough understanding of the mechanism of nonlinear interactions of the second-order dynamics. A coupled set of nonlinear field equations for the mass-transport velocity and the zeroth-order apparent concentration has been derived and used to describe and understand the shallow water residual circulation along with the intertidal transport processes coupled by the wind stress over the sea surface, the heat flux across the water surface, the horizontal gradient of water density, and the tidal body force resulting from the nonlinear interaction among the multi-frequency astronomical tidal variables. An application of the model to the summer circulation in the Bohai Sea, China gives some heuristic results. The tide-induced component of the residual circulation in the Bohai Sea is more appropriately associated with an M_2-K_1 tidal system than an M_2-tidal system alone.

1. Introduction

In recent years, studies of environmental hydrodynamics of tidal estuaries and coastal seas have focused on residual circulation and long-term transport processes. Intertidal transports of temperature, salinity, solutes, pollution, sediment, nutrients, fish eggs, and other tracers are fundamental to the interactive physical, chemical, and biological processes in an ecological system. In shallow, coastal seas, such as the Bohai Sea, China, the dominant observable motion is tidal currents and the apparent dominating transport mechanism appears to be convection due to tidal currents. Over a seasonal time scale, however, the overall ecological balance and long-term transport processes are mainly dependent on the residual circulation, rather than tidal currents. It is becoming increasingly clear that the residual circulation should be described by means of the net displacements of water parcels, or the Lagrangian mean velocity, but not by an Eulerian mean velocity. The Lagrangian mean velocity has led to the concept of a Lagrangian residual velocity, and the research on the Lagrangian residual velocity and the coupled intertidal transport processes is a relatively recent and difficult topic. Numerous

Coastal and Estuarine Studies, Vol. 38
R. T. Cheng (Ed.)
Residual Currents and Long-term Transport

investigators, including Longuet-Higgins (1969), Moore (1970), Tee (1976), Zimmerman (1979), Ianniello (1979), Cheng and Casulli (1982), Backhaus (1985) and Hamrick (1987), have made important contributions to this topic. A weakly nonlinear theory of Lagrangian residual currents and intertidal transports has been investigated by the author and his collaborators (Feng et al., 1986 a, b, c; Cheng et al., 1986; Feng, 1986, 1987, 1988 a, b; Zheng, 1988; Wei, 1988; Sun et al., 1989; and Cheng et al., 1989).

To the author's knowledge, most studies of the Lagrangian residual flow and long-term transports refer to a single-frequency tidal system, such as an M_2-nonlinear system in coastal seas and tidal estuaries because the most important tidal harmonics is, normally, the lunar semi-diurnal constituent, M_2. Other principal astronomical tidal harmonics are given in Table 1 (Hansen, 1962). The tidal motions in coastal seas or tidal estuaries are genearlly caused by the ocean tides, and dynamically characterized by the shallow water constituents and residual motions resulting from the nonlinear couplings of the astronomical tides or, more generally, of the lower-order constituents (Feng, 1977).

In this paper, a weakly nonlinear baroclinic shallow water system is analyzed using a perturbation method. The Lagrangian residual velocity induced by a multi-frequency tidal system is derived and examined, up to the second order, and compared with that induced by a single-frequency tidal system. A generalized set of field equations for the first-order Lagrangian residual velocity, and the zeroth-order apparent concentration is proposed and used to describe a time-averaged water circulation coupled with the intertidal transport processes in coastal seas or tidal estuaries. The system is driven by the wind stress over the sea surface, the heat flux across the sea surface, the horizontal gradient of water density, and the tidal body force due to the nonlinear interactions of the constituents of the multi-frequency astronomical tidal system. This set of equations is solved numerically to model the summer circulation in the Bohai Sea, China, using a two-frequency representation. The calculated circulation pattern is compared with that induced by an M_2 system, and verified by field data. For a more thorough understanding of the second-order dynamics, the Lagrangian drift velocity induced by an M_2-S_2 tidal system is discussed in detail.

Table 1. The principal terms of astronomical tide

Name	Frequency (°/hour)	Relative Coefficient
M_2	28.98	0.9085
S_2	30.00	0.4227
K_1	15.04	0.5305
O_1	13.94	0.3771

II. A Baroclinic Shallow Water System

The governing equations describing a baroclinic shallow water system are the continuity equation

$$\nabla \cdot \vec{U} = 0 \quad , \tag{1}$$

the x-momentum equation,

$$\frac{\partial u}{\partial t} + \vec{U} \cdot \nabla u - fv = -g\frac{\partial \zeta}{\partial x} + \frac{\partial}{\partial z}\left(\nu \frac{\partial u}{\partial z}\right) - g\beta \frac{\partial}{\partial x} \int_z^\zeta c dz' \quad , \tag{2}$$

the y-momentum equation,

$$\frac{\partial v}{\partial t} + \vec{U}\cdot\nabla v + fu = -g\frac{\partial \zeta}{\partial y} + \frac{\partial}{\partial z}(\nu\frac{\partial v}{\partial z}) - g\beta\frac{\partial}{\partial y}\int_z^\zeta c dz' , \tag{3}$$

the conservation equation

$$\frac{\partial c}{\partial t} + \vec{U}\cdot\nabla c = \frac{\partial}{\partial z}(K\frac{\partial c}{\partial z}) , \tag{4}$$

and the associated boundary conditions at $z = \zeta$

$$w = \frac{\partial \zeta}{\partial t} + u\frac{\partial \zeta}{\partial x} + v\frac{\partial \zeta}{\partial y} , \tag{5}$$

$$(\tau_x, \tau_y) = \nu\frac{\partial(u,v)}{\partial z} , \tag{6}$$

$$K\frac{\partial c}{\partial z} = \Gamma , \tag{7}$$

and at $z = -h$,

$$\vec{U} = 0 , \tag{8}$$

$$\partial c/\partial z = 0 , \tag{9}$$

and the expression of the "apparent concentration"

$$c = (S-S_*) + \sum_{j=1}^{m}(\beta_j/\beta)(C_j-C_{j*}) - (\alpha/\beta)(\theta - \theta_*) , \tag{10}$$

where t is time;
- (x, y, z) are the Cartesian coordinates on an f-plane;
- $\nabla = \vec{i}\partial/\partial x + \vec{j}\partial/\partial y + \vec{k}\partial/\partial z$ is a vector operator;
- $(\vec{i}, \vec{j}, \vec{k})$ are unit vectors;
- $\vec{U} = \vec{i}u + \vec{j}v + \vec{k}w$ is the velocity vector;
- ζ and h are the water surface elevation and the water depth measured from mean sea level, respectively;
- f is the Coriolis parameter;
- ρ is the water density;
- ν and K are the eddy viscosity and the eddy diffusivity, respectively;
- g is the gravitational acceleration;
- (τ_x, τ_y) are the wind stress components in the (x, y) directions at the water surface divided by the water density.
- $\Gamma = -(\alpha/\beta)\, Q/\rho C_p$, where α is the coefficient of thermal expansion;
  - β is the fractional increase in density per unit increase in salinity;
  - Q is the heat flux across the sea surface;
  - C_p is the specific heat;

c is the "apparent concentration", which is formally similar to the "apparent temperature" defined by Fofonoff (1962). C_j is the concentration of j-tracer, β_j is the corresponding constant; the notation "*" indicates the corresponding reference constant. A generalized linearized equation of state containing salinity, S, and temperature, θ, and C_j (j = 1, 2, ...m), $(\rho - \rho_*) = \beta c$ is used, and θ, S and C_j satisfy respectively the same comservation equation as given in Eq.(4) (Nihoul, 1975).

In order to close this basic set of equations, we suppose that the eddy coefficients, such as ν and K, are the function of spatial and temporal coordinates only. The other nonlinear terms of this set of equations are in the order of magnitude characterized by a nondimensional parameter $\kappa = \zeta_c/h_c = U_c/\sqrt{gh_c}$, where the subscript c indicates the characteristic value of the corresponding quantity.

Let us assume that (i) $O(\kappa) < 1$; (ii) $O[\frac{\partial}{\partial z}(k\frac{\partial c}{\partial z})] = \kappa^2$; and

$$\text{(iii)}\ O(g\beta\frac{\partial}{\partial x}\int_z^{\zeta} c dz',\ g\beta\frac{\partial}{\partial y}\int_z^{\zeta} c dz') = O(\tau_x,\ \tau_y) = \kappa\ ;$$

(Feng, 1987; Feng et al., 1986c; Feng et al., 1984). These assumptions imply that (i) the dynamic system is weakly nonlinear, and a perturbation technique with the small parameter κ can be used to solve the jth-order solution of this system ($j = 0, 1, \ldots$) (Feng, 1977); (ii) the transport mechanisim is dominated by convection, and thus the zeroth-order apparent concentration, C_0, is independent of time (Feng, 1987); (iii) the astronomical tides, such as M_2, S_2, K_1, O_1, are independent of the effects due to baroclinic forcing and due to the wind stress forcing (Feng et al., 1986c; Feng, 1988a).

The streakline of a marked water parcel can be described by a displacement vector $\vec{\xi} = \vec{X} - \vec{X}_0$, or

$$\vec{\xi} = \int_{t_0}^{t} \vec{U}(\vec{X}_0 + \vec{\xi},\ t')dt' \tag{11}$$

where $(\vec{X}_0,\ t_0)$ indicates that the marked water parcel is released from the spatial point $\vec{X}_0$ at instant t_0.

In the studies of long-term transport processes, one of the most important quantities is the net displacement of the marked water parcel released from $\vec{X}_0$ at t_0 and moving over several tidal periods T,

$$\vec{\xi}_r = \int_{t_0}^{t_0+T} \vec{U}(\vec{X}_0 + \vec{\xi},\ t')dt',$$

or equivalently, the Lagrangian mean velocity

$$\vec{U}_{LR} = (1/T) \int_{t_0}^{t_0+T} \vec{U}(\vec{X}_0 + \vec{\xi},\ t)dt \quad . \tag{12}$$

Introducing the tidal excursion ξ_c as the scale of $\vec{\xi}$ and L as the horizontal scale of the flow field, and supposing $(\xi_c/L) = \kappa$, the Lagrangian residual velocity can be defined by the Lagrangian mean velocity, $\vec{U}_{LR}$ as given in Eq.(12) (Zimmerman, 1979; Cheng and Casulli, 1982; Feng, 1987).

III. First-order Dynamics

III.1 Mass-transport velocity

Taking an appropriate averaging time for removing, at least approximately, all dominant tidal oscillations in an m-frequency tidal system, and substituting

$\vec{U}(\vec{X}_0+\vec{\xi}, t)$ and Eq.(11) into Eq.(12) yields the first-order Lagrangian residual velocity, or the mass-transport velocity, $\vec{U}_L$ as

$$\vec{U}_L = \vec{U}_E + \vec{U}_S = \sum_{k=0}^{m} {}^k\vec{U}_L \tag{13}$$

where $(\vec{U}_E, \vec{U}_S) = \sum_{k=0}^{m} ({}^k\vec{U}_E, {}^k\vec{U}_S)$ and ${}^k\vec{U}_L = {}^k\vec{U}_E + {}^k\vec{U}_S$;

${}^k\vec{U}_E$ and ${}^k\vec{U}_S$ are the Eulerian residual velocity and the Stokes' drift velocity, respectively. K, K= 1, 2, ...m, indicates the corresponding quantities induced by the k-constituent of astronomical tide; except ${}^0\vec{U}_E$ contains a wind-driven barotropic component and a density-driven baroclinic component. It is noted that ${}^0\vec{U}_S = 0$ because the Stokes' drift velocity is independent of the wind stress and the baroclinic effect (Feng, 1988a; Feng et al., 1986c).

Eq.(13) reveals that the mass-transport velocity $\vec{U}_L$ in an m-frequency baroclinic tidal system can be written as the sum of all k-constituent induced mass-transport velocities and including both the wind-driven barotropic and the density-driven baroclinic components.

III. 2. A set of field equations governing $\vec{U}_L$ and C_0

By tidally time-averaging the Eqs. (1)-(9), and noting

$$<\psi_0\ \phi_0> = < \sum_{k=1}^{m} {}^k\psi_0 \sum_{k=1}^{m} {}^k\phi_0> = \sum_{k=1}^{m} <{}^k\psi_0\, {}^k\phi_0>,$$

where ψ_0 and ϕ_0 represent the zeroth-order variables, and < > is a tidally time-averaging operator, the set of field equations governing $\vec{U}_L$ and C_0 is derived as follows:

$$\nabla\cdot\vec{U}_L = 0 \tag{14}$$

$$-fv_L = -g\frac{\partial<\zeta_1>}{\partial x} + \frac{\partial}{\partial z}(\nu\frac{\partial u_L}{\partial z}) - g\beta\frac{\partial}{\partial y}\int_z^0 C_0 dz' + \Pi_1 \tag{15}$$

$$fu_L = -g\frac{\partial<\zeta_1>}{\partial y} + \frac{\partial}{\partial z}(\nu\frac{\partial v_L}{\partial z}) - g\beta\frac{\partial}{\partial y}\int_z^0 C_0 dz' + \Pi_2 \tag{16}$$

$$\vec{U}_L\cdot\nabla C_0 = \frac{\partial}{\partial z}(<K>\frac{\partial C_0}{\partial z}) \tag{17}$$

at z = 0:

$$w_L = 0 \quad ; \tag{18}$$

$$\nu\frac{\partial(u_L, v_L)}{\partial z} = (<\tau_x>, <\tau_y>) \quad ; \tag{19}$$

$$K(\partial C_0/\partial z) = <\Gamma> \quad ; \text{ and} \tag{20}$$

at z = -h:

$$\vec{U}_L = 0 \qquad ; \qquad (21)$$

$$\partial C_0/\partial z = 0 \qquad ; \qquad (22)$$

where $(\Pi_1, \Pi_2) = \sum_{k=1}^{m} ({}^k\Pi_1, {}^k\Pi_2)$ (23)

$${}^k\Pi_1 = -g\frac{\partial}{\partial x}<\frac{1}{2}{}^k\vec{\xi}_0\cdot\nabla{}^k\zeta_0> + \frac{\partial}{\partial z}<(\frac{5}{2}\frac{\partial{}^k\xi_0}{\partial x} + \frac{\partial{}^k\eta_0}{\partial y})\,\nu\,\frac{\partial{}^k u_0}{\partial z}$$

$$+ (\frac{\partial{}^k\xi_0}{\partial y} + \frac{1}{2}\frac{\partial{}^k\eta_0}{\partial x})\,\nu\frac{\partial{}^k v_0}{\partial z}> + \frac{\partial}{\partial z}<\nabla\nu\cdot{}^k\vec{\xi}_0\frac{\partial{}^k u_0}{\partial z}>$$

$$- <\frac{\partial{}^k\vec{\xi}_0}{\partial z}\cdot\nabla(\nu\frac{\partial{}^k u_0}{\partial z})> - \frac{1}{2}<\frac{\partial^{2\,k}\xi_0}{\partial z\partial x}(\nu\frac{\partial{}^k u_0}{\partial z}) + \frac{\partial^{2\,k}\eta_0}{\partial z\partial x}(\nu\frac{\partial{}^k v_0}{\partial z})$$

$${}^k\xi_0\frac{\partial^2}{\partial z\partial x}(\nu\,\frac{\partial{}^k u_0}{\partial z}) + {}^k\eta_0\frac{\partial^2}{\partial z\partial x}(\nu\,\frac{\partial{}^k v_0}{\partial z})>$$

$${}^k\Pi_2 = -g\frac{\partial}{\partial y}<\frac{1}{2}{}^k\vec{\xi}_0\cdot\nabla{}^k\zeta_0> + \frac{\partial}{\partial z}<(\frac{5}{2}\frac{\partial{}^k\eta_0}{\partial y} + \frac{\partial{}^k\xi_0}{\partial x})\,\nu\,\frac{\partial{}^k v_0}{\partial z}$$

$$+ (\frac{\partial{}^k\eta_0}{\partial x} + \frac{1}{2}\frac{\partial{}^k\xi_0}{\partial y})\,\nu\frac{\partial{}^k u_0}{\partial z}> + \frac{\partial}{\partial z}<\nabla\nu\cdot{}^k\vec{\xi}_0\frac{\partial{}^k v_0}{\partial z}>$$

$$- <\frac{\partial{}^k\vec{\xi}_0}{\partial z}\cdot\nabla(\nu\frac{\partial{}^k v_0}{\partial z})> - \frac{1}{2}<\frac{\partial^{2\,k}\xi_0}{\partial z\partial y}(\nu\frac{\partial{}^k u_0}{\partial z}) + \frac{\partial^{2\,k}\eta_0}{\partial z\partial y}(\nu\frac{\partial{}^k v_0}{\partial z})$$

$$+ {}^k\xi_0\frac{\partial^2}{\partial z\partial y}(\nu\,\frac{\partial{}^k u_0}{\partial z}) + {}^k\eta_0\frac{\partial^2}{\partial z\partial y}(\nu\,\frac{\partial{}^k v_0}{\partial z})>$$

where $\vec{U}_L = \vec{i}u_L + \vec{j}v_L + \vec{k}w_L$, and ζ_0 and ζ_1 are respectively the zeroth- and first-order solutions of ζ.

This set of field equations, [Eqs.(14)-(23)], is an extension of the corresponding set of field equations proposed in the previous papers (Feng, 1987; Feng _et al_., 1986c; Feng, 1988). It can be used to describe the (residual) circulation and long-term transport processes in coastal seas and tidal estuaries, and the nonlinear coupling between them. If $({}^k\Pi_1, {}^k\Pi_2) = C_j = 0$, the set of equations derived here will be reduced to the classical set of equations for the wind-driven and thermohaline circulation in coastal seas (Ramming and Kowalik, 1980). In addition to the wind stress, and the baroclinic forcing, (Π_1, Π_2) is a tide-induced body force, termed "tidal body force" (Feng, 1987), and represents the nonlinear couplings between the Lagrangian displacements of astronomical tides and tidal currents or tidal elevation. Using the "velocity-splitting method", the three-dimensional velocity field in a m-frequency tidal system can easily be solved for the barotrophic flows, or for the diagnosis mode of the baroclinic flows (Song, 1987).

III.3. An application to the Bohai Sea, China

To test the dynamic model formulation, the summer residual circulation and associated transport processes of water mass in the Bohai Sea, China are examined using numerical solutions. The Bohai Sea is a semi-enclosed coastal shallow sea, which is located at the north-eastern corner of China and connected to the Huanghai Sea by the Bohai Strait (Fig. 1). The summer residual circulation in the Bohai Sea is driven mainly by the warm water branch of the Huanghai Sea and the nonlinear M_2-tide, which enter the Bohai Sea from the Huanghai Sea through the Bohai Strait (Feng et al. 1986c). The depth-averaged field of the mass-transport velocity calculated by Feng et al., (1986c) and Zheng (1988) has revealed some interesting features and explained some observations. Here a supplement to these is that a clockwise gyre in the Bohai Bay might explain the existence of calcite and particularly muscovite in the Bohai Bay since the calcite and muscovite are the characteristic minerals of the Huanghe River [Figs. 2, 3, and 4, copied from Zheng (1988)]. It should be noted the Zheng's numerical modeling was based on the set of field equations governing the M_2-induced mass-transport velocity (Feng, 1987), i.e., the set of field equations Eqs.(14)-(23) with wind stress and baroclinic terms set to zero, and $\Gamma = 0$ and $(\Pi_1, \Pi_2) = ({}^{M_2}\Pi_1, {}^{M_2}\Pi_2)$, where the notation "$M_2$" inidcates that the tidal body force is generated by M_2 alone.

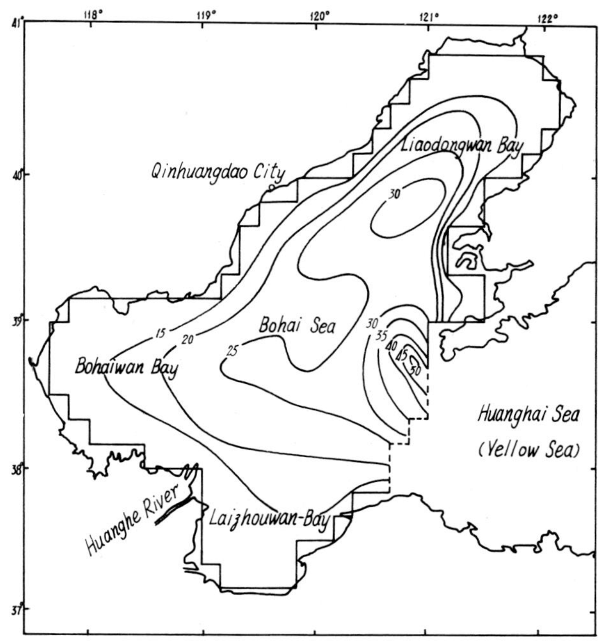

FIGURE 1. Depth distribution of the Bohai Sea, China

Besides M_2, as a representative of the diurnal constituent, K_1 is also important for the Bohai Sea, particularly in some regions, such as the region near Qinhuangdao city where there exists an amphidromic point of M_2-tide (Feng et al., 1986c). In order to model the summer residual circulation in the Bohai Sea, an M_2- "K_1" system has been used where "K_1" is an approximation to K_1 constituent having the same amplitude as K_1 and the frequency to be half that of M_2 frequency (Wei, 1988). Thus, the tidal body force, (Π_1, Π_2), expressed by Eq.(23) is greatly simplified. The depth-averaged field of the mass-transport velocity has been obtained by depth-averaging the three-dimensional solution of the mass-transport velocity, and is depicted in Fig. 5. The three-dimensional mass-transport velocity was calculated based on Eqs.(14)-(16), (18), (19) and (21), and omitting the baroclinic force and the wind stress in Eqs. (15), (16) and (19), and by using the "velocity-splitting method" (for further details, see Zheng, 1988, and Wei, 1988). In the region near Qinhuangdao City, where there is an amphidromic point of M_2-tide, a principal difference between the two calculated residual circulation patterns appears (Cf. Fig. 5 and Fig. 2). The longshore currents are opposite in direction, which is just what we expected. In fact, a north-eastward Eulerian residual flow along the coast has been pointed out by Prof. Shi (1983), Fig. 6, through field observations. In order to make a comparison between computed and observed results, the three-dimensional Eulerian residual velocity has been calculated by means of the formula

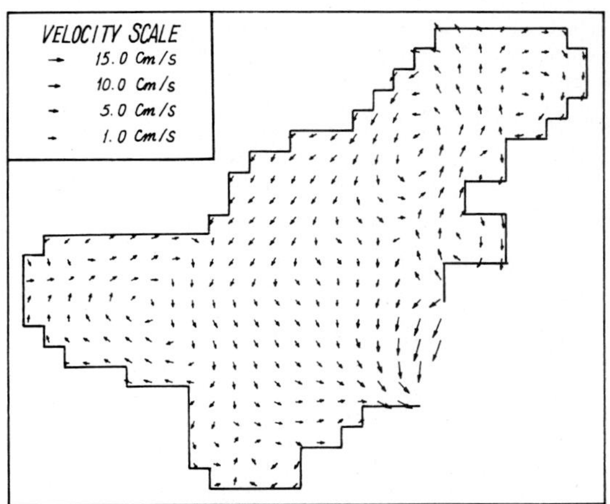

FIGURE 2. M_2-induced depth-averaged mass-transport velocity in the Bohai Sea, associated with the part of the Huanghai warm water coming from the Huanghai Sea through the Bohai Strait.

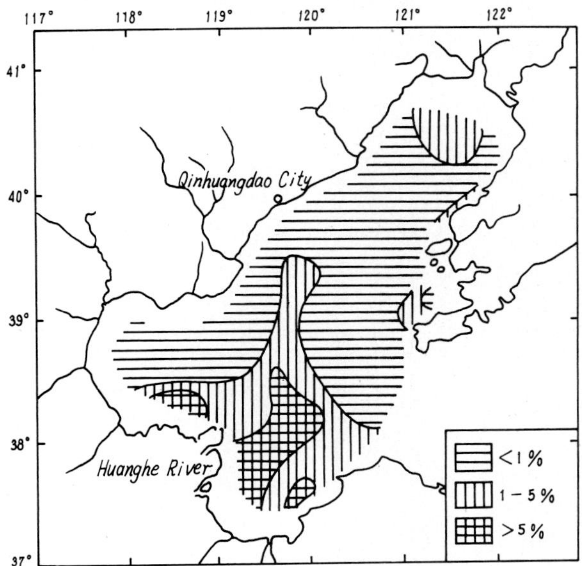

FIGURE 3. Distribution of muscovite (white mica) in the Bohai Sea. The muscovite is one of the characteristic minerals of the Huanghe River.

$\vec{U}_E = \vec{U}_L - \vec{U}_S$, and then the depth-averaged Eulerian residual velocity is derived by depth-averaging $\vec{U}_E$ as shown in Fig. 6, and they are in good qualitative agreement. Thus, while "K_1" is an approximate constituent, it is heuristic and interesting to note that the tide-induced component of the (summer) residual circulation in the Bohai Sea should more appropriately be generated by an M_2-K_1 system than by an M_2-system alone.

IV. The Lagrangian Drift Velocity

In a single-frequency tidal system, the Lagrangian drift velocity, $\vec{U}_{LD}$, a second-order perturbation term of the Lagrangian residual velocity has been derived by Cheng et al. (1986) and Feng et al. (1986a) for a two-dimensional system, by Feng (1986, 1987) for a three-dimensional barotropic system, and by Feng et al. (1986c) for a three-dimensional baroclinic system. In this paper, the Lagrangian drift velocity induced by a multi-frequency tidal system will be examined. To avoid confusion with the main issues, the effects due to the sea surface wind stress and baroclinic forcing are not included.

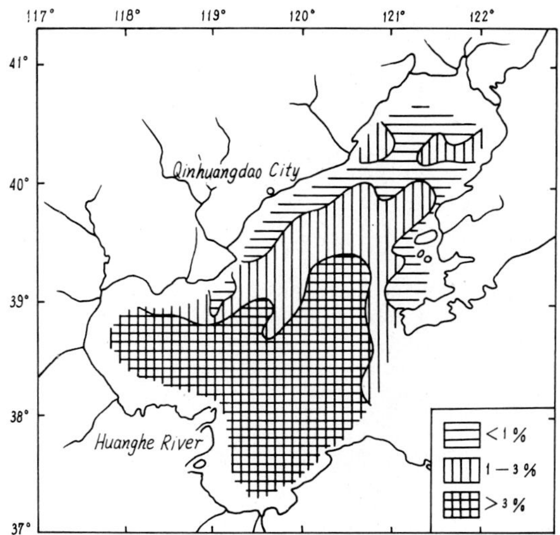

FIGURE 4. Distribution of calcite in the Bohai Sea. The calcite is one of the characterisitc minerals of the Huanghe River.

IV.1. An M_2-S_2 nonlinear system

For simplicity, the derivation of the Lagrangian drift velocity induced by an M_2-S_2 nonlinear system is summarized as follows

$$\vec{U}_{LD} = -\vec{\xi}_0(t_0)\cdot\nabla\vec{U}_L + \vec{U}_L\cdot\nabla\vec{\xi}_0(t_0) \tag{24}$$

where

$$\vec{U}_L = {}^M\vec{U}_L + {}^S\vec{U}_L \tag{25}$$

$$\vec{\xi}_0(t_0) = {}^M\vec{\xi}_0(t_0) + {}^S\vec{\xi}_0(t_0)$$

$${}^k\vec{\xi}_0(t_0) = -\frac{1}{\sigma_k}{}^k\vec{U}_0'' \cos(\sigma_k t_0) + \frac{1}{\sigma_k}{}^k\vec{U}_0'\sin(\sigma_k t_0)$$

where $(^{k}\vec{U}_0', {}^{k}\vec{U}_0'')$ are the harmonic coefficients of the velocity of the k-constituent of astronomical tide, which are inlayed in the expression for the zeroth-order tidal

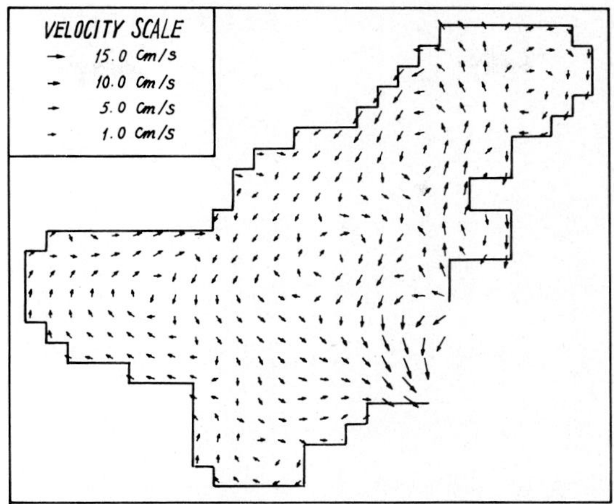

FIGURE 5. (M_2, "K_1")-induced depth-averaged mass-transport velocity in the Bohai Sea, associated with the part of the Huanghai warm water coming from the Huanghai Sea through the Bohai Strait.

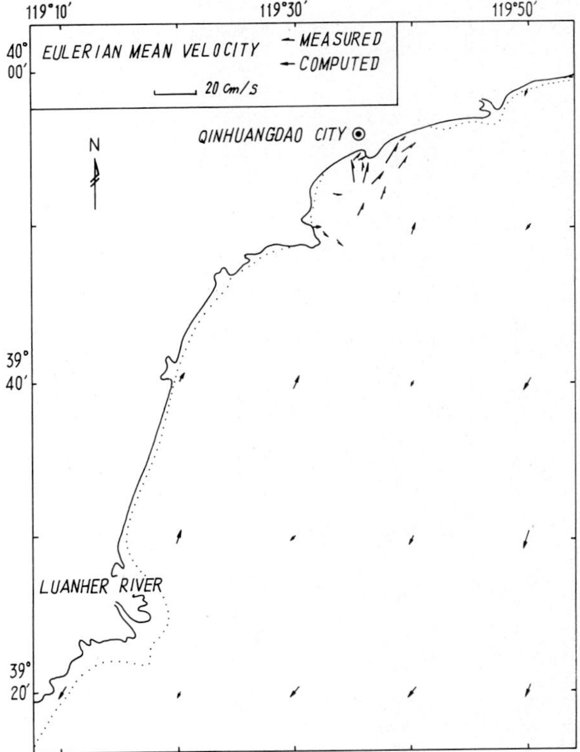

FIGURE 6. A comparison between the computed (M_2, "K_1")-induced depth-averaged Eulerian mean velocity and field observations.

velocity-- ${}^{k}\vec{U}_0 = {}^{k}\vec{U}_0' \cos(\sigma_k t) + {}^{k}\vec{U}_0'' \sin(\sigma_k t)$; σ is the frequency; and K = M or K = S refer to M_2 or S_2 quantities, respectively.

Substituting Eq.(25) into Eq.(24), we have

$$\vec{U}_{LD} = {}^{M}\vec{U}_{LD} + {}^{S}\vec{U}_{LD} + {}^{I}\vec{U}_{LD} \tag{26}$$

where

$${}^{k}\vec{U}_{LD} = -{}^{k}\vec{\xi}_0(t_0)\cdot\nabla{}^{k}\vec{U}_L + {}^{k}\vec{U}_L\cdot\nabla{}^{k}\vec{\xi}_0(t_0) \quad (k = M, \text{ or } k = S) \quad ,$$

$${}^{I}\vec{U}_{LD} = {}^{MS}\vec{U}_{LD} + {}^{SM}\vec{U}_{LD} \quad ,$$

$${}^{MS}\vec{U}_{LD} = -{}^{M}\vec{\xi}_0(t_0)\cdot\nabla{}^{S}\vec{U}_L + {}^{S}\vec{U}_L\cdot\nabla{}^{M}\vec{\xi}_0(t_0)$$

$${}^{SM}\vec{U}_{LD} = -{}^{S}\vec{\xi}_0(t_0)\cdot\nabla{}^{M}\vec{U}_L + {}^{M}\vec{U}_L\cdot\nabla{}^{S}\vec{\xi}_0(t_0)$$

Eq.(26) shows that the Lagrangian drift velocity induced by an M_2-S_2 system, $\vec{U}_{LD}$, contains not only the sum of the M_2- and S_2- induced Lagrangian drift velocities, ${}^{M}\vec{U}_{LD} + {}^{S}\vec{U}_{LD}$, but also ${}^{I}\vec{U}_{LD}$--a product of nonlinear interaction between M_2 and S_2 terms. Thus, while the mass-transport velocity--the first-order Lagrangian residual velocity expressed in Eq. (13), or Eq. (25), has the additive property, the Lagrangian drift velocity, $\vec{U}_{LD}$, is a second-order perturbution of the Lagrangian residual velocity. it cannot be expressed as the sum of the Lagrangian drift velocities due to M_2 and S_2.

Substituting the expression for ${}^{k}\vec{\xi}_0(t_0)$ into Eq.(24), the Lagrangian drift velocity can be also written as

$$\vec{U}_{LD} = {}^{M\sigma}\vec{U}_{LD} + {}^{S\sigma}\vec{U}_{LD} \tag{27}$$

where

$${}^{k\sigma}\vec{U}_{LD} = {}^{k\sigma}\vec{U}_{LD}' \cos(\sigma_k t_0) + {}^{k\sigma}\vec{U}_{LD}'' \sin(\sigma_k t_0) \quad , \quad (k = M, \text{ or } k = S)$$

$${}^{k\sigma}\vec{U}_{LD}' = {}^{k\sigma}\vec{U}_{LD}' + {}^{I}\vec{U}_{LD}', \qquad {}^{k\sigma}\vec{U}_{LD}'' = {}^{k\sigma}\vec{U}_{LD}'' + {}^{I}\vec{U}_{LD}'',$$

$${}^{k}\vec{U}_{LD}' = (1/\sigma_k)[\,{}^{k}\vec{U}_0''\cdot\nabla{}^{k}\vec{U}_L - {}^{k}\vec{U}_L\cdot\nabla{}^{k}\vec{U}_0'']$$

$${}^{k}\vec{U}_{LD}'' = (1/\sigma_k)[-{}^{k}\vec{U}_0'\cdot\nabla{}^{k}\vec{U}_L + {}^{k}\vec{U}_L\cdot\nabla{}^{k}\vec{U}_0']$$

$${}^{I}\vec{U}_{LD}' = (1/\sigma_k)[\,{}^{k}\vec{U}_0''\cdot\nabla{}^{q}\vec{U}_L - {}^{q}\vec{U}_L\cdot\nabla{}^{k}\vec{U}_0'']$$

$${}^{I}\vec{U}_{LD}'' = (1/\sigma_k)[-{}^{k}\vec{U}_0'\cdot\nabla{}^{q}\vec{U}_L + {}^{q}\vec{U}_L\cdot\nabla{}^{k}\vec{U}_0']$$

in which q = S if k = M, or q = M if k = S.

Eq.(27) shows the dependence of the Lagrangian drift velocity on the phases of both M_2 and S_2 when the water parcel is labelled and released. As the initial phases vary from 0 to 2π, the horizontal projections of ${}^{M\sigma}\vec{U}_{LD}$ and ${}^{S\sigma}\vec{U}_{LD}$ trace out a respective ellipse on a hodograph plane. A comparison between Eq. (27) and Eq. (26)

shows the difference between ${}^{k\sigma}\vec{U}_{LD}$ and ${}^{k}\vec{U}_{LD}$ (k = M, S) due to the nonlinear interaction.

IV.2. Generalized theory

Substituting $\vec{U}(\vec{X}_0 + \vec{\xi}, t)$ and Eq.(11) into Eq.(12), the Lagrangian drift velocity, or simply the Lagrangian drift, $\vec{U}_{LD}$, induced by an m-frequency tidal system can be derived as

$$\vec{U}_{LD} = -\vec{\xi}_0(t_0)\cdot\nabla\vec{U}_L + \vec{U}_L\cdot\nabla\vec{\xi}_0(t_0) + \vec{U}_L' \tag{28}$$

where $\vec{U}_L' = \vec{U}_E' + \vec{U}_S'$, and $\vec{U}_L'$ is independent of t_0 and is the additional term when compared to Eq.(24). $\vec{U}_E' = \langle\vec{U}_2(\vec{X}_0, t)\rangle$, a second-order velocity, is the time average of $\vec{U}_2$, which contains in general not only the second-order constituents but also the residual motions. Thus there is normally an additional Eulerian Residual velocity, $\vec{U}_E' = \langle\vec{U}_2(\vec{X}_0, t)\rangle$, which is due to the nonlinear couplings between the zeroth- and first-order constituents, and itself is of second order. $\vec{U}_S$ is composed of three parts, which are respectively extracted from

$$\langle\vec{\xi}_0(t)\cdot(\nabla\vec{U}_1)_0\rangle, \quad \langle\vec{\xi}_1(t)\cdot(\nabla\vec{U}_0)_0\rangle, \text{ and}$$

$$\frac{1}{2}\langle\xi_0^2\left(\frac{\partial^2\vec{u}_0}{\partial x^2}\right)_0 + \eta_0^2\left(\frac{\partial^2\vec{u}_0}{\partial y^2}\right)_0 + \lambda_0^2\left(\frac{\partial^2\vec{u}_0}{\partial z^2}\right)_0 + 2\xi_0\eta_0\left(\frac{\partial^2\vec{u}_0}{\partial x\partial y}\right)_0 + 2\xi_0\lambda_0\left(\frac{\partial^2\vec{u}_0}{\partial x\partial z}\right)_0 + 2\eta_0\lambda_0\left(\frac{\partial^2\vec{u}_0}{\partial y\partial z}\right)_0\rangle$$

where $\vec{\xi}_0 = \vec{i}\xi_0 + \vec{j}\eta_0 + \vec{k}\lambda_0$, and the notation $(\ )_0$ indicates the term is evaluated at $\vec{X}_0$.

An expanded form of Eq.(28) can be obtained as

$$\vec{U}_{LD} = \sum_{k=1}^{m} {}^{k\sigma}\vec{U}_{LD} + \vec{U}_K' + \vec{U}_S' \tag{29}$$

where

$${}^{k\sigma}\vec{U}_{LD} = {}^{k\sigma}\vec{U}_{LD}\cos(\sigma_k t_0) + {}^{k\sigma}\vec{U}_{LD}\sin(\sigma_k t_0)$$

$${}^{k\sigma}\vec{U}_{LD} = (1/\sigma_k)[{}^{k}\vec{U}_0\cdot\nabla\vec{U}_L - \vec{U}_L\cdot\nabla{}^{k}\vec{U}_0'']$$

$${}^{k\sigma}\vec{U}_{LD} = (1/\sigma_k)[-{}^{k}\vec{U}_0\cdot\nabla\vec{U}_L - \vec{U}_L\cdot\nabla{}^{k}\vec{U}_0']$$

Eq.(29) shows the dependence of the Lagrangian drift velocity induced by an m-frequency tidal system on all phases of k-constituent (k = 1, 2, ...m) when the water parcel is released, and as these phases vary from 0 to 2π, the horizontal projections of ${}^{k\sigma}\vec{U}_{LD}$ (k = 1, 2, ...m) trace out a respective ellipse on a hodograph plane. It is noted that the Lagrangian drift velocity does not have the additive property, because it involves the nonlinear interactions among the k-contituent (k = 1, 2, ...m).

IV.3. An example

It has been recognized that the Stokes' drift velocity induced by an M_2-tidal current velocity field without shear and with a constant phase is identically zero, and thus the Lagrangian drift velocity is also zero if the Eulerian residual velocity is supposed to be zero (Feng, et al., 1986a). This result seems to suggest that the phase variation of tidal current plays an influential role in the mechanism which generates Lagrangian residual flow. This statement is only valid for a single-frequency oscillatory system.

Consider a coupled M_2-S_2 induced unidirectional ideal flow without shear for which $v = 0$ and $\partial u/\partial y = 0$, and the phase of the M_2-tidal current is further required to be constant, or ${}^{M}u_0''/{}^{M}u_0' = \tan(\alpha_M)$, where α_M is the constant phase angle. A direct substitution yields

$$ {}^{M}u_L = {}^{M}u_S = 0 $$

and

$$ {}^{M\sigma}U_{LD} = \frac{\sec\alpha_M}{\sigma_M}\left[{}^{M}u_0'\frac{\partial\,{}^{S}u_L}{\partial x} - {}^{S}u_L\frac{\partial\,{}^{M}u_0'}{\partial x}\right]\cos(\sigma_M t_0 - \beta_M) $$

where $\tan\beta_M = -\cot\alpha_M$, and to avoid confusing the main issues, ${}^{M}U_E$ has been supposed to be zero.

It becomes immediately obvious that although the tidal current velocity shear is zero and the phase of M_2-tidal current is constant, ${}^{M\sigma}U_{LD}$ is an ellipse on a hodograph plane, or more precisely, a straight line, or a degenerate ellipse. This Lagrangian residual ellipse and M_2-tidal current ellipse are 90° out of phase.

V. Conclusions

The Lagrangian residual velocity and associated intertidal transport processes in a three-dimensional, weakly nonlinear, multi-frequency oscillatory, baroclinic shallow water system have been examined and analyzed using a perturbation method. The first-order dynamics show that the mass-transport velocity is composed of a tide-induced component, a wind-driven barotropic component and a density-driven baroclinic component, and the tide-induced mass-transport velocity is the sum of the mass-transport velocities by the respective constituents of astronomical tide contained in the multi-frequency tidal system. A set of generalized field equations governing the mass-transport velocity and the zeroth-order apparent concentration is presented and used to describe the long-term mean circulation and transport processes in coastal seas and tidal estuaries. The present theory has been applied to the Bohai Sea, China with the aid of numerical solutions of the model equations. A comparison between the model and the field observations implies that the tide-induced residual circulation in the Bohai Sea resembles more closely to an M_2-K_1 tidal system than to an M_2-tidal system. The secon-order dynamics show that the Lagrangian drift velocity induced by a multi-frequency tidal system involves nonlinear coupling terms between the products of the respective constitutents of tides. The dependence on the periodicities of all the astronomical tidal constituents contained in the multi-frequency oscillatory system show up in the Lagrangian residual ellipses on a hodograph plane.

VI. References

Backhaus, J. O., 1985: A Three-dimensional model for the simulation of shelf sea dynamics. Dt. Hydrogr. Z., **38**, 185-187.

Cheng, R. T., and V. Casulli, 1982: On Lagrangian residual currents with applications in South San Francisco Bay, California. Water Resour. Res., **18**. 1652-1662.

Cheng, R. T., S. Feng, and P. Xi, 1986: On Lagrangian residual ellipse, In: Intern. Conf. on Physics of Shallow Estuaries and Bays, J. van de Kreeke (Editor), Lecture Notes on Coastal and Estuarine Studies, **16**, 102-113, Springer-Verlag.

Cheng, R. T., S. Feng, and P. Xi, 1989: On inter-tidal transport equation, In Estuarine Circulation, B. J. Neilson, J. Brubaker, and A. Kuo (Editors), Humana Press, Inc., 133-156.

Feng, S., 1977: A three-dimensional nonlinear model of tides. Scientia Sinica, Vol. 20, 4, 436-446.

Feng, S., 1986: A three-dimensional weakly nonlinear dynamics on tide-induced Lagrangian residual current and mass-transport. Chinese J. of Oceanology and Limnology, Vol. 4, 2, 139-158.

Feng, S., 1987: A three-dimensional weakly nonlinear model of tide-induced Lagrangian residual current and mass-transport, with an application to the Bohai Sea, In: Three-dimensional Models of Marine and Estuarine Dynamics, J. C. J. Nihoul and B. M. Jamart (Editors), Elsevier Oceanography Series, 45, 471-488, Elsevier.

Feng, S., 1988a: Fundamentals of hydrodynamics of shallow water circulation and transport, Proc. of Conf. on Engineering Mechanics and Termophysics, 199-208, Qinghua Univeristy Press (in Chinese).

Feng, S., 1988b: The focus of marine environment hydrodynamic research, J. of Shandong College of Oceanology, Vol. 18, 2 (2), 1-7 (in Chinese).

Feng, S., R. T. Cheng, and P. Xi, 1986a: On tide-induced Lagrangian residual current and residual transport, Part I, Residual current, Water Resour. Res., Vol. 22, 12, 1623-1634.

Feng, S., R. T. Cheng, and P. Xi, 1986b: On tide-induced Lagrangian residual current and residual transport, Part II, Residual transport with application in South San Francisco Bay, California, Water Resour. Res., Vol. 22, 12, 1635-1646.

Feng, S., R. T. Cheng, W. Sun, P. Xi, and L. Song, 1986c: Lagrangian residual current and Long-term transport process in a weakly nonlinear baroclinic system, Proc. of Intern. Symp. on Physics of Shallow Bays, Estuaries and Continental Shelves, China Ocean Press (in press).

Feng, S., P. Xi, and S. Zhang, 1984: The baroclinic residual circulation in shallow seas. Chinese J. of Oceanology and Limnology, Vol. 2, 1, 49-60.

Fofonoff, N. P., 1962: Dynamics of ocean currents, In: M. N. Hill (Editor), The Sea, Vol. I, Physical Oceanography, Section III, Interscience Publishers, 323-396.

Hamrick, J. M., 1987: Time averaged estuarine mass transport equations, Hydraulic Engineering, In: R. M Ragan (Editor), Proc. of the 1987 National Conf. on Hydraulic Engineering, 624-629, ASCE.

Hansen, W., 1962: Tides, In: M. N. Hill (Editor), The Sea, vol. I, Physical Oceanography. Interscience Publishers, CH. 23, Interscience, 764-801.

Ianniello, J. P.: 1979: Tidally induced residual currents in estuaries of variable breadth and depth. J. Phys. Oceanogr., **9**, 963-974.

Longuet-Higgins, M. S., 1969: On the transport of mass by Time-Varying Ocean currents. Deep Sea Res., **16**, 431-447.

Moore, D., 1970: The mass transport velocity induced by free oscillations at a single frequency. Geophysical Fluid Dynamics, **1**, 237-247.

Nihoul, J. C. J., 1975: Modeling of Marine System, Elsevier, 272 pp.

Ramming, H. G., and Z. Kowalik, 1980: Numerical Modeling of Marine Hydrodynamics, Elsevier Oceanography Series, Vol. 26, Elsevier.

Song, L., 1987: A hydrodynamic velocity-splitting model with depth-varying eddy viscosity in shallow seas. Acta Oceanologica Sinica, vol. 6, Supp. I, 135-151.

Sun, W., P. Xi, and L. Song, 1989: Numerical calculation of the three-dimensional tide-induced Lagrangian residual circulation in the Bohai Sea. J. of Ocean University of Qingdao, vol. 19, 2(1), 27-36.

Tee, K. T., 1976: Tide-induced residual current, a 2-d numerical nonlinear tidal model. J. Mar. Res., **34**, 603-628.

Wei, G., 1988: On Lagrangian residual currents and long-term transports by tides at both frequency, M.S. thesis, Ocean Univ. Press (in Chinese), 54 pp.

Zheng, L., 1988: A numerical study on the three-dimensional hydrodynamic equations for the mass-transport velocity, with an application to the Bohai Sea, M.S. thesis, Ocean Univ. Press (in Chinese), 74 pp.

Zimmerman, J. T. F., 1979: On the Euler-Lagrangian transformation and the Stokes drift in the presence of oscillaotry and residual currents. Deep Sea Res., 26A, 505-520.

Residual Circulation in Shallow Estuaries: Shear, Stratification and Transport Processes

David A. Jay

Geophysics Program
University of Washington
Seattle, WA 98195

ABSTRACT

It is a useful conceptual device to consider the residual properties in shallow, stratified estuaries to be the result of an ebb-flood asymmetry. This asymmetry has two aspects: a barotropic non-linearity associated with finite amplitude of the tide, and an internal non-linearity caused by the baroclinic pressure gradient and the tidal variations in stratification generated thereby. While some stratification may be present in broad shallow estuaries, a substantial buoyancy flux and baroclinic pressure gradient are usually only present if the system is narrow enough to be nearly two-dimensional in the vertical and along-channel directions. The interaction of the stratification and the baroclinic forcing normally causes the velocity profiles in such systems to be more uniform in the vertical on flood than on ebb. A stratified tidal flow differs fundamentally from a neutral one, because the internal asymmetry may be strong in the outer part of the flow, even when the stratification is quite weak. Furthermore, the internal ebb-flood asymmetry is the primary factor determining the form of the residual velocity profile. An analogous statement can be made regarding the salt balance: the salinity distribution in narrow estuaries can only be maintained by shear and stratification in the mean and tidal flows. Landward transport of salt by the vertical integral of any tidal constituent is, in these systems, exactly balanced by an outward transport associated with the Stokes drift compensation flow for that constituent. This distinguishes narrow estuaries from broad shallow systems, in which the salt balance may be maintained in the horizontal by a vertically uniform flow. This occurs when transport in one horizontal direction maintains the salinity gradient in the normal direction. A broad estuary that is also stratified may be much more complex than either narrow estuaries or wide, neutrally stratified estuaries, because both types of transport mechanisms may operate. The flow is then fully three dimensional.

I. Introduction

Theories about residual circulation in shallow estuaries have generally focussed on estuarine circulation as a two-dimensional phenomenon. One important class of theories has concentrated on narrow, channelized flows that are nearly two dimensional in x and z (the along channel and vertical coordinates). Hansen and Rattray (1965) determined velocity and density profiles associated with the steady, density-driven circulation in a channel of uniform depth and width. Tidal non-linearities were neglected and tidal advection of salt treated parametrically. Ianniello (1977, 1979) considered the residual circulation caused by a finite amplitude long wave tide in a neutrally stratified channel flow over arbitrary but small-scale topographic features. Density forcing was neglected. More recently Jay (1987) and Jay and Smith (1989a,b) have extended Ianniello's long wave methods to

Coastal and Estuarine Studies, Vol. 38
R. T. Cheng (Ed.)
Residual Currents and Long-term Transport

consider both density forcing and finite amplitude tidal effects in a uniform channel. Different assumptions about vertical mixing lead to three types of narrow, shallow estuaries: highly stratified, weakly stratified and partially mixed (Jay and Smith, 1988).

Other studies have concentrated on systems where vertical variability is small, so that the flow is two-dimensional in the horizontal (that is, in x and y). Dyke (1980) considered the residual flow arising from the Stokes drift in a wide shallow estuary of simple form, including the possibility of an internal Kelvin wave tide. If the topography is simple, then the Coriolis influence on the barotropic residual circulation is subtle unless the width exceeds the Rossby radius. Important lateral effects can occur in the baroclinic residual flow on a much smaller scale than in the barotropic residual, however, because of the disparity between the internal and barotropic Rossby radii. Zimmerman (1978, 1980a) has shown that the interaction of complex topography with finite amplitude tides can have an entirely new class of effects on residual circulation in wide shallow estuaries. In these horizontally extensive systems, topographic features with a scale of about one tidal excursion play a potent role in generating tidal eddies and residual vorticity.

It is perhaps a fair summary to say that estuarine researchers are struggling to attain a more complete understanding of nearly two-dimensional flows and are beginning to contemplate the complexities of fully three-dimensional systems; i.e. those that are broad, stratified and have complex topography. It is the purpose of this paper to use the conceptual model of the residual circulation as arising out of the ebb-flood asymmetry to draw more clearly several important distinctions between narrow and wide shallow estuaries. These differences pertain to both the mass and salt balances, and have important implications for understanding more complex, three-dimensional estuaries.

In narrow, shallow estuaries, there is both a definite axis along which the tide propagates and a dominant transport orientation defined by the seaward movement of freshwater along this axis. It will be shown that the residual vorticity generation mechanisms so prominent in broad estuaries are necessarily absent in narrow systems (Section II). This result comes from scaling the vertically-integrated continuity equation and is equivalent to saying that the along-channel transport variations in one-dimensional long waves scale with the integral length scale, not with the local scale. Any reduction in along-channel length scale simply reduces the velocity scale proportionally. The Stokes drift must also take a simple form in these systems, despite the presence of complex topography of arbitrary scale, and truncation of the series representation of the Stokes drift at a low order is justified. Landward transport of water and salt by the Stokes drift may still be a prominent feature in these systems, insofar as barotropic non-linearity contributes to the residual shear. Dynamically, this tidal non-linear generation of residual flow is related to an ebb-flood asymmetry in surface elevation, as is evident in scaling tidal non-linearities by $\epsilon = a/H$ (surface tidal amplitude over mean depth).

The narrowness of such estuaries, however, allows the buoyancy flux from riverflow to be substantial. This buoyancy flux and the resulting horizontal density gradient and stratification are often a far more potent generator of ebb-flood asymmetry and thus of residual circulation than the barotropic non-linearity alone (Section III). The effects of ebb-flood asymmetry in shear and stratification can also be examined from the perspective of their effects on salt transport. This is conveniently done by considering the problem of transport of salt through a cross-section of a narrow, two-dimensional (in x and z) estuary from a Lagrangian point of view (Section IV). Generalized Lagrangian Mean (GLM) theory (Andrews and McIntyre, 1978a) allows the Lagrangian transport through this section to be expressed in Eulerian variables (Jay, 1990a). The key GLM theorem (Dunkerton, 1980) allows (for steady waves with ϵ finite but small) the expression of the wave cycle average of tidal advection of tidal salinity gradient variations in terms of the Stokes drift and the mean salinity gradient. The GLM representation of tidal advection shows that in a narrow channel, vertically integrated tidal advection cannot result in landward transport of salt, because this landward transport will always be exactly compensated by the vertical average of the zero-frequency Stokes drift compensation flow. Thus, salt, pollutant or suspended sediment balances in narrow estuaries can only be maintained against outward riverflow by transport modes associated with the shear and stratification in the mean flow, tidal flow and overtides.

The vital role of shear modes in maintaining the salinity distribution differentiates narrow estuaries from broad, horizontally extensive systems in which topographic generation of residual vorticity dominate the residual circulation. The salinity gradient in one direction in these systems can be maintained by salt transport associated with salinity gradients in the normal direction. Stratification and vertical shear may be relatively small. The salt balance in large, complex estuaries such as Puget Sound and Long Island Sound is almost certainly maintained by both shear and horizontal, topographic dispersion mechanisms. The former should dominate over shallow sills and the latter in some peripheral embayments, while both are likely to be important in the larger basins. Such systems are substantially more complex than either of the types of two-dimensional estuaries considered here.

II. Barotropic Processes and Conservation of Mass

It will be shown herein that mass conservation in narrow, channelized tidal flows (those much less than a tidal excursion in width) fundamentally constrains the response of the flow to topography in a way that does not occur in broad, shallow estuaries (those one or more tidal excursions wide). Consider first flow in a narrow, two-dimensional (in x and z) estuary. The vertical uniformity of the velocity profile in a hypothetical, frictionless long wave flows suggests use of mass conservation:

$$\frac{\partial Q_x}{\partial x} + b\,\frac{\partial \zeta}{\partial t} = 0 \tag{1a}$$

where:

Q_x is the along-channel transport in m^3/s;
b is width;
t is time;
z is surface elevation.

to derive the velocity scale U:

$$U = \epsilon c = \epsilon (gH)^{1/2} \quad , \tag{1b}$$

where:

c = long wave celerity;
H = scale depth.

While the horizontal velocity u is not vertically uniform in a long wave with friction, the log profile observed in neutrally stratified long wave flows is still reasonably scaled by U. Even in the presence of stratification, model studies show U is an adequate estimate of the velocity scale (Jay and Smith, 1989b). The above analysis implicitly assumes a channel of constant width and depth, so that the only x length scale is the inverse wave number ℓ. U/ℓ is then clearly the scale for $\partial u/\partial x$.

Most estuaries are short relative to ℓ and estuarine topography is usually at an even smaller scale. What happens if realistic topography is added to the problem? One suspects immediately that the scale for $\partial u/\partial x$ should change, but this is incorrect. ζ remains dominantly at well-defined tidal (and overtide) frequencies and Eq.(1a) still determines the scale for $\partial u/\partial x$. Consider more formally the dominant component of the flow, that at the basic tidal frequency. Take the total flow to be the sum of a component (scaled by U) associated with the integral scale ℓ and a topographic perturbation with topographic velocity scale U_t and length scale L_t. U_t must be reduced relative to U such that Eq.(1a) is satisfied for the topographic perturbation (because it is by the definition of U satisfied for the long wave component). This leaves the scale for $\partial u/\partial x$ unchanged in the presence of topography. Indeed, if this were not true, the many vertically integrated models of tidal flow in channels that appear in the literature would be greatly in error, because they explicitly or implicitly rely on the above scaling. Ianniello (1979,

1981) came to a similar conclusion by a different route when he extended his original uniform-channel model of vertical shear in the tidally generated residual flow (Ianniello, 1977) to include topographic effects.

The discussion in Section IV will require reference to the O(1) tidal frequency and $O(\epsilon)$ residual parts of Eq.(1a). These are, for a steady tidal wave motion:

$$\frac{\partial}{\partial x}\int_{-h}^{\bar{\zeta}} bu'dz + b\frac{\partial \zeta'}{\partial t} = 0 \qquad (2a)$$

$$\frac{\partial}{\partial x}\int_{-h}^{\bar{\zeta}} b\bar{u}dz + b\frac{\partial}{\partial x}\,\overline{u'\zeta'} = 0 \qquad (2b)$$

where a prime indicates a tidal frequency variable; an overbar indicates a tidal cycle average (Eulerian) variable; and h is depth below mean water level. Eqs.(2a,b) are based on use of a Taylor series expansion (truncated after the second term) of the part of the integral between the mean free surface $\bar{\zeta}$ and the actual surface ζ. The next term (the first neglected) in this series scales as $O(\epsilon^2)$ but vanishes for an unstressed free surface, and subsequent terms become small rapidly relative to the term retained.

Interpretation of Eqs.(2a,b) requires definition of the Stokes correction and determination of its vertical integral. The Stokes correction for u is (Andrews and MacIntyre, 1978a) the difference between the Eulerian mean value of u (wave cycle average at a vector point **x** fixed in space, (variables in bold face are vector quantities) and the Lagrangian mean [that following the vector particle trajectory $\chi(\mathbf{x},t)$]. In two-dimensional channel flow for $\epsilon \ll 1$:

$$\bar{U}^S = \bar{u}^L - \bar{u} = \overline{u(\mathbf{x}+\chi)} - \overline{u(\mathbf{x})} = \overline{\chi_{\mathbf{x}}\frac{\partial u'}{\partial x}} + \overline{\chi_{\mathbf{z}}\frac{\partial u'}{\partial z}} + O(\epsilon^2) \qquad (3)$$

Because $\partial u/\partial x$ is bounded in narrow estuaries because Eq.(1a), the higher order terms are $O(\epsilon^2)$ in magnitude relative to the first term and may be neglected. A similar procedure may be used to define the Stokes correction for any variable ϕ. Using ϕ instead of u in Eq.(3), and integrating from seabed to the mean free surface, the integrated Stokes correction for small (but finite) amplitude waves is, in two dimensions:

$$\int_{-h}^{\bar{\zeta}} \bar{\phi}^S dz = \zeta' Q'|_{\bar{s}} + \frac{\partial}{\partial x}\int_{-h}^{\bar{\zeta}} \overline{\chi_x \phi'}dz \qquad (4a)$$

For $\phi = u$, we may write the integrated Stokes drift Q^S as

$$\bar{Q}^S = \overline{\zeta' u'|_{\zeta'}} \qquad (4b)$$

This is Longuet-Higgins' (1969) result for a flow that is two-dimensional in x and z. So long as the flow is two dimensional in this sense, higher order terms may be neglected even in the presence of topography of arbitrary scale, because $\partial u'/\partial x$ is bounded. Thus, the traditional interpretation of the second term in Eq.(2b) as the integrated, along-channel Stokes drift is fully justified.

Along-channel integration of Eq.(2b) from x to the head of the tide (where the wave motion vanishes and the only flow is the riverflow Q^R) allows determination of the total transport in either Eulerian or Lagrangian terms (Ianniello, 1977; Jay and Smith, 1989b):

$$\bar{Q}_x = \int_{-h}^{\zeta} b\bar{u}dz = \bar{Q}^R - \bar{Q}^S \qquad (5a)$$

$$\bar{Q}_x^L = \bar{Q}^R \qquad (5b)$$

The river flow is directed outward, so $\overline{Q}^R$ is negative. The Stokes drift is always in the direction of the wave motion. Even if a reflected tidal wave is present, friction requires it to be smaller than the incident wave. Thus, $\overline{Q}^S$ is positive. The mean Eulerian transport [$\overline{Q}_x$ in Eq.(5a)] is, therefore, directed outward and is the sum of the riverflow and a compensation flow for the Stokes Drift. More detailed examination of the dynamics of narrow estuaries shows that the inward Stokes drift sets up a surface slope that drives a compensating flow exactly sufficient to balance $\overline{Q}^S$, thus the minus sign in Eq.(5a) (Ianniello, 1977). Allen _et al_. (1980) and Jay (1987) among others, have shown that this residual slope varies over the tidal month as the tidal range (and thus the Stokes drift) varies.

The mean Lagrangian transport [$\overline{Q}^L_x$ in Eq.(5b)] is simply the riverflow, and the net transport associated with the wave is apparently zero. The net wave transport is in fact equal to the "pseudomomentum" as defined by Andrews and MacIntyre (1978b). For small amplitude, incompressible flows in which the Coriolis force is unimportant, the pseudomomentum is the Stokes drift (Leibovich, 1980). However, the strong constraint of the channel geometry (not the wave itself) causes the Eulerian, zero-frequency Stokes drift compensation flow discussed in the previous paragraph, which cancels the effect of the Stokes drift.

A fundamentally different situation pertains in broad shallow estuaries. Integrated mass conservation requires

$$\frac{\partial Q_x}{\partial x} + \frac{\partial Q_y}{\partial y} + \frac{\partial \zeta}{\partial t} = 0 \tag{6}$$

where y is the cross-channel axis, and

Q_x = integrated mass transport in the x direction, in m^2/s;
Q_y = integrated mass transport in the y direction, in m^2/s.

Eq.(6) suggests the possibility that $\mathbf{Q} = (Q_x, Q_y)$ may be nearly non-divergent in the horizontal ($\nabla_H \cdot \mathbf{Q} \approx 0$). Thus, $\partial Q_x/\partial x$ may be balanced by $\partial Q_y/\partial y$, and both may be large relative to $\partial\zeta/\partial t$. The frequency of ζ no longer determines the $\partial u/\partial x$ scale.

Zimmerman's (1978, 1980a) arguments concerning the generation of residual vorticity by topography can still, however, be used to constrain $\partial u'/\partial x$ in a useful manner. He has argued that the topographic scale most effective in generating residual vorticity through vortex stretching and bottom friction is the tidal excursion scale $L = \epsilon \ell$. While other shorter length scales also appear in the topography of most tidal systems, let us average the motion over scales less than L and take L to be the minimum scale of interest. Consider the form of Stokes correction under these conditions, still for small amplitude waves ($\epsilon > 0$ but $\ll 1$). The first three terms are, in tensor notation,

$$\overline{Q}^S = \overline{\chi_j \phi'_{,j}} + \frac{1}{2!}\,\overline{\chi_j \chi_k}\;\overline{\phi_{,jk}} + \frac{1}{3!}\,\overline{\chi_j \chi_k \chi_l \phi'_{,jkl}} + \cdots \tag{7}$$

where repeated indices indicate summation; an index j without or before a comma indicates the jth component; an index j after a comma indicates differentiation with respect to x_j. The assumption that $\overline{\phi}$ scales as $\epsilon\phi'$ leads to the conclusion that the second and all other even-numbered terms are only $O(\epsilon^2)$, or $O(\epsilon)$ relative to the first term. The third and all succeeding odd-numbered terms are of the same order as the first term. The series thus converges rather slowly, if at all (Zimmerman, 1980b). This is one fundamental difference between flow processes in narrow and wide estuaries -barotropic non-linearities causing the Stokes drift are stronger in wide estuaries, because Eq.(6) does not constrain $\partial u'/\partial x$ in such systems. Another essential difference is that, unlike Eq.(2b), Eq.(6) cannot be directly integrated to specify $\overline{\mathbf{Q}}^L$ as was done in Eq.(5). That $\overline{\mathbf{Q}}^L$ is strongly constrained in narrow estuaries is the basic cause of the difference between wide and shallow estuaries in salt transport processes discussed in Section IV. It will be shown there that stratification and shear are necessary to the maintenance of the salt balance in narrow estuaries, a constraint which does not apply in wide estuaries. Because tidal processes leading to ebb-flood asymmetry in the stratification and shear are essential to maintenance of the salt balance in narrow estuaries, we next consider the mechanisms leading to this internal or baroclinic ebb-flood asymmetry.

III. Ebb-Flood Asymmetry and the Baroclinic Pressure Gradient

Consideration of the internal or baroclinic part of the ebb-flood asymmetry requires a distinction between highly stratified shallow estuaries, in which an internal, tidal oscillation can occur, and weakly stratified and partially mixed systems, in which such an oscillation is necessarily absent (Fig.1; Jay and Smith, 1988, 1989b). The baroclinic circulation in highly stratified, narrow and shallow estuaries (e.g. a salt wedge advance) is an O(1) phenomenon and the resulting ebb-flood asymmetry is strong throughout the flow. The result is a vigorous two-layer residual flow. Although ebb-flood asymmetry is strong in any system in which riverflow and tidal forcing are balanced such that the system falls near the $F_B = 1$ line in Fig. 1, the systems considered herein are those without an internal tidal oscillation. The baroclinic flow and the ebb-flood asymmetry in partially mixed and weakly stratified narrow shallow estuaries away from the $F_B = 1$ line may be, from the perspective of the flow as a whole, weak. However, as discussed below, the ebb-flood asymmetry in the outer part of the flow may still be strong, and this may be quite important for the dispersion of pollutants concentrated near the surface.

The baroclinic part of the ebb-flood asymmetry in weakly stratified and partially mixed estuaries is associated with the horizontal baroclinic pressure gradient. The following description is based on observations taken under low-flow, weakly stratified conditions in a system, the Columbia River Estuary, that falls

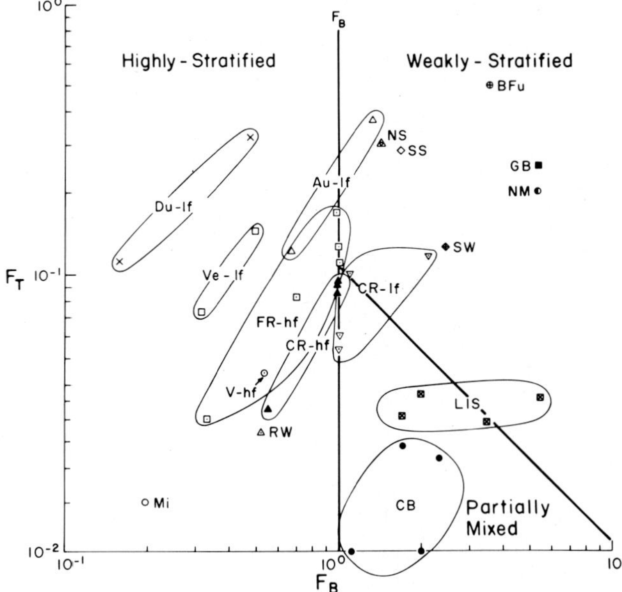

FIGURE 1. Classification of shallow, narrow estuaries on the basis of an external Froude number $F_T = a/h$ and internal Froude number F_B calculated from the density field. Although the dependence of F_B on U and the riverflow U_R varies with the the type of estuary, increasing U increases F_B and increasing U_R decreases F_B in all cases. Estuaries with a strong internal ebb-flood asymmetry (and thus those with strong non-linear generation of residual flow) are those where fluvial and tidal forces are balanced such that is F_B is O(1). The names of the estuaries shown in Fig. 1 are given in Jay and Smith (1988). Published by permission of Springer-Verlag.

near the F_B = 1 line (Jay, 1987). This is a good system in which to analyze the interaction of baroclinic and tidal forcing, because both the riverflow and tides are strong, and the topography is not overly complex. Moreover, regular tidal-monthly transitions from highly to weakly stratified allow observation of both states; ebb-flood asymmetry under highly stratified conditions is discussed in Jay (1990b). Most aspects of ebb-flood asymmetry discussed here have been confirmed in other systems; e.g. the Fraser River Estuary (Geyer, 1988), the Duwamish (Partch and Smith, 1977), the Great Ouse (West, et al., 1984) and the Snohomish (recent observations by the author).

On flood, the baroclinic pressure gradient is in the same direction as the barotropic gradient and enhances it at depth. This tends to reduce the shear in the water column by enhancing landward, flood flow near the bed (Fig. 2). Stratification is weak or absent, because the remaining shear tends to advect denser seawater over lighter water, encouraging mixing. Bottom boundary induced vertical turbulent mixing may therefore penetrate throughout most of the flow. Because shear is relatively weak, the reversal of the tide at the end of flood is nearly simultaneous, top to bottom.

On ebb, the baroclinic pressure gradient opposes the barotropic gradient at depth. The result is substantial shear (Fig. 3). This shear is of the correct sign so that tidal straining of the density field induces stratification by advection of lighter water over denser seawater. The resulting shear and stratification have several important consequences. First, boundary shear stress is reduced on ebb relative to the neutral value, because the shear is no longer concentrated near the bed. This reduces the frictional energy loss at the bed and thus the effective drag coefficient of the flow. It also tends to bring about a landward, near-bed residual flow. In weakly stratified estuaries, the upstream bottom flow is typically weak and may be small relative to the riverflow. In partially mixed systems, the two layer mean exchange is usually much larger than the riverflow. Even very weak stratification (in the tidal-cycle average sense) suffices to sharply reduce the ebb boundary shear stress, and this reduction may also have very important consequences for sediment transport.

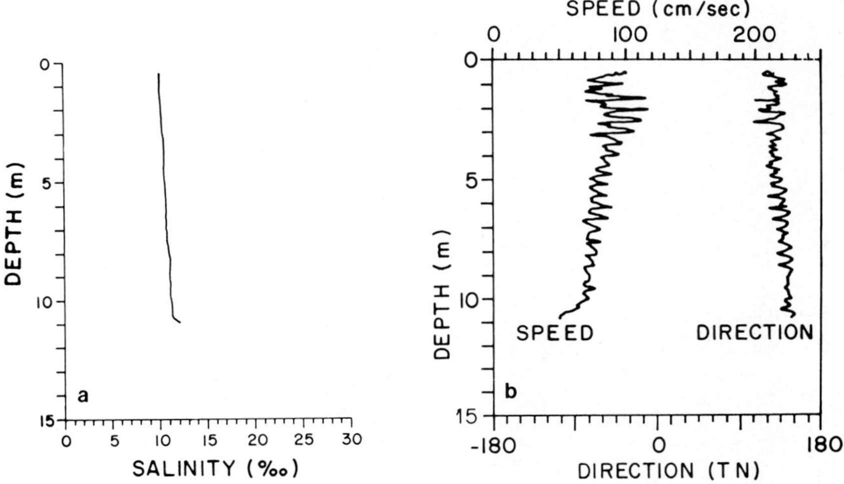

FIGURE 2. Salinity (a) and speed and direction (b) on early flood during a low-flow, spring tide in the Columbia River Estuary, as determined by a vessel-mounted velocity-conductivity-depth profiler system. Much of the noise in the velocity profile is believed to result from boat motion. The velocity profile approaches that of a neutrally stratified log layer, and shear is weak in the outer part of the flow. The tidal straining of the density field by the weak shear that is present is of the correct sign to cause vertical mixing. Published by permission of the American Geophysical Union.

Second, if the gradient Richardson number exceeds the critical value at some point in the water column, the remaining, outer part of the flow is isolated from bottom-boundary-induced turbulence and associated vertical mixing. The detachment of the outer part of the flow may allow very large, top to bottom shears to develop at the end of ebb (Fig. 4), and the surface layer may in extreme cases continue to ebb for two hours or more after the flood has begun at depth. The time of current reversal is then a strong function of depth.

Third, because of the weakness of friction, the force balance for the outer layer is hard to predict after detachment, as it is very susceptible to winds and other perturbing influences (e.g. pressure gradients associated with flow from tidal flats). Thus, the ebb-flood asymmetry may be strong in the outer part of the flow, even if it is weak from the point of view of the flow as a whole. That is, mass and momentum conservation for the integrated flow maybe relatively insensitive to this detachment of a small part of the flow near the surface, so long as the ebb flood asymmetry in bed stress is properly accounted for in the vertically integrated long wave equation. This detachment, however, may have a profound influence on transport of a pollutant or other material concentrated near the surface. Furthermore, the primary effect of weakening the stratification is to thin the detached layer, not to prevent detachment altogether, because (in the absence of wind stress) the shear becomes very small as the free surface is approached. Finally, the current associated with a tidal constituent for which both transport and elevation are exactly harmonic does not necessarily exhibit a simple harmonic structure. Ebb-flood redistribution of the flow in fact causes the currents at any depth to exhibit an enriched overtide spectrum (Jay, 1990b). This may complicate interpretation of current meter records.

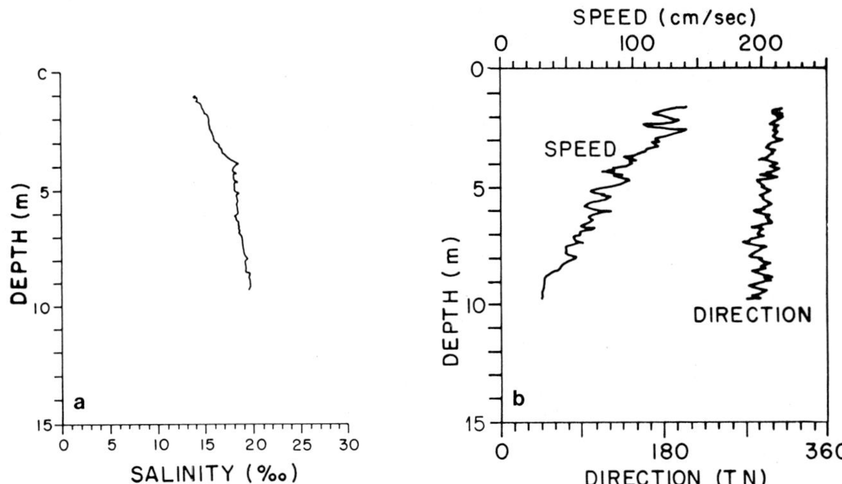

FIGURE 3. Salinity (a) and speed and direction (b) on ebb during a low-flow, spring tide in the Columbia River Estuary, determined as in Figure 2. Stratification and shear are somewhat larger than on flood, and the outer part of the flow (above about 4m) is isolated from the effects of boundary friction. The shear in this part of the flow is considerably enhanced, and is of the correct sign to increase stratification. Published by permission of the American Geophysical Union.

In summary, stratification and the presence of a horizontal baroclinic pressure gradient fundamentally alter estuarine velocity profiles in the many estuaries near the $F_B = 1$ line (Fig. 1) in such a way as to introduce an ebb-flood asymmetry. These differences do not entirely disappear even when the stratification is weak (i.e. in systems far to the right of the $F_B = 1$ line). A stratified estuarine flow is thus fundamentally different than a neutral flow. A strong ebb-flood asymmetry is more likely to be found in a narrow estuary where a substantial buoyancy input occurs, than in a wide system. It is shown in the next section that this difference in velocity and salinity profiles between narrow and wide estuaries is also reflected in the salt balance.

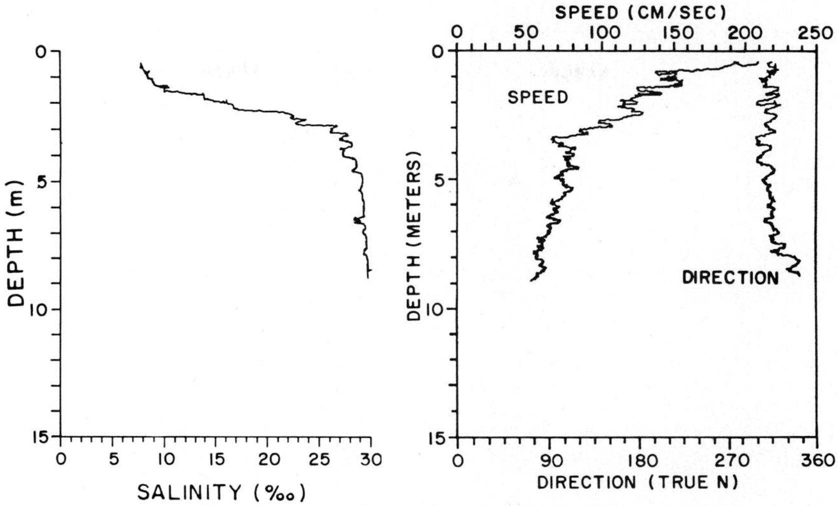

FIGURE 4. Salinity (a) and speed and direction (b) on late ebb during a low-flow, neap tide in the Columbia River Estuary, determined as in Figure 2. The top 3m of the flow is still ebbing strongly, while the flow close to the bed is decelerating rapidly. The time of the following slack water was a strong function of depth.

IV. Estuarine Salt Transport--A Lagrangian Point of View

The essence of the Andrews and MacIntyre (1978a) Generalized Lagrangian Mean (GLM) theory of wave-mean flow interaction is the development of a method to express Lagrangian mean quantities as field variables rather than as functions of initial particle tags. The Lagrangian average is an average over the particle drift, as in Eq.(3). Perhaps the key result from GLM theory for present purposes is the relationship of net tidal transport of a conservative scalar variable C′ (e.g. salt) and stokes advection of $\overline{C}$ for incompressible, steady, small amplitude waves (Dunkerton, 1980):

$$\overline{\mathbf{u}' \cdot \nabla C'} \approx \overline{\mathbf{u}}^{S} \cdot \nabla \overline{C} - \overline{\frac{\partial}{\partial z} T_z}^{S} \tag{8}$$

where T_z is the vertical turbulent exchange of salt. Feng <u>et al</u>. (1986) proved a vertical integral of this result with an Eulerian perturbation analysis and the assumption of weak topographic variations. Their method essentially repeats earlier developments in the atmospheric science literature, as summarized by Middleton and Loder (1989). This result Eq.(8) is not valid for wide estuaries in the presence of small scale topography, because the velocity gradients are sufficiently strong that the Stokes at a point cannot be expressed in terms of the velocity at that point and its derivatives; that is, the series representation of the Stokes drift does not converge, as discussed in Section II. It is valid for narrow, channelized flows that are nearly two-dimensional in x and z. Vertical and horizontal integration of Eq.(8) gives

$$\int_{-h}^{\bar{\zeta}} \overline{u'C'}\,dz \approx \int_{-h}^{\bar{\zeta}} \bar{u}\,\bar{C}dz \tag{9}$$

for narrow estuaries. Vertical integration of a salt balance of the form,

$$\frac{D^L\bar{C}^L}{Dt} = \overline{\frac{\partial}{\partial z}T_z}^L \tag{10}$$

and Eqs.(8) and (9) can be employed (Jay, 1989) to obtain a useful new form for the salt balance in narrow estuaries. The result is independent of the form of T_z, so long as the vertical turbulent exchange vanishes at the bed and free surface. The integrated salt balance takes the form:

$$\begin{aligned}\frac{\partial}{\partial t}\int_x^\infty \int_{-h}^{\bar{\zeta}} \bar{C}\, dz\, dx &= Q^R\langle\bar{C}\rangle + \langle\bar{u}_d\bar{C}_d\rangle H + \overline{\langle u'_d C'_d\rangle} H \\ &= Q^R\langle\bar{C}\rangle + \langle\bar{u}_d\bar{C}_d\rangle H + \langle\bar{u}^S\bar{C}_d\rangle H\end{aligned} \tag{11}$$

where $H = \bar{\zeta} + h$; angle brackets indicate a vertical average; the subscript d indicates a vertical deviation from a vertical average.

Several important conclusions can be drawn from Eq.(11). First, although Eq.(11) is specific to the case of a single tidal constituent, it can easily be generalized to any number of tidal constituents. Each additional constituent, whether of the same species, another basic tidal species or a non-linearly driven overtide, adds only one additional term to Eq.(11). If the second constituent is indicated by a double prime, the additional term is of the form $H\,\overline{\langle u_d'' c_d''\rangle}$. Second, Eq.(11) indicates that the outward salt flux associated with the riverflow can only be balanced by inward fluxes associated with the gravitation circulation (i.e. shear and stratification in the mean flow) and wave fluxes that are associated with time-varying shear and stratification in the tidal flow. While it is not necessary for both of the rightmost terms in Eq.(11) to be directed inward, the gravitational circulation usually transports salt landward and the equivalence of the wave flux to a Stokes drift transport of the mean stratification suggests that both will normally be inward.

Third, vertically integrated modes cannot cause landward transport of salt. The vertical of the Lagrangian mean flow Eq.(5b) is $\bar{Q}^R$, which moves salt only outward. The inward wave flux associated with the vertical integral of any tidal constituent is equivalent to the inward flux by the Stokes drift transport associated with that constituent, and, as such, is exactly balanced by the outward flux driven by the corresponding Stokes drift compensation flow. Thus no truly two-dimensional estuary (in x and z) can then ever be well mixed (i.e. neutrally stratified), because maintenance of the salt balance against the outward riverflow would then be impossible. Proper calculation of the stratification (both mean and tidally varying parts) is essential, therefore, to understanding estuarine circulation.

Fourth, the salt balance is very sensitive to the external forcing. Small changes in tidal range or riverflow may bring about large changes in the shear and stratification, causing, for example the tidal monthly transitions that are a prominent feature in many estuaries (Haas, 1976; Allen _et al_., 1980; Jay and Smith, 1989a,b). Such variations are accompanied by substantial alterations in salinity intrusion length, which may vary directly or even inversely with tidal range depending on the type of estuary. Thus, vertically integrated calculations of the salt balance (i.e. those based only on knowledge of the tidal transport) in a narrow estuary are ill-posed, because they do not account for the various different vertical mixing regimes that can occur as the external forcing is changed.

Finally, the salt balance in wide estuaries is fundamentally different from that in narrow systems. No theorem comparable to Eq.(11) can be formulated for these horizontally extensive bodies, because, as discussed in Section II, the Lagrangian transport cannot be specified in these systems in the manner used for narrow estuaries Eq.(5b). Moreover, the Stokes drift is more complex in wide tidal flows than in narrow estuaries, and horizontal transport by chaotic, eddy dispersion mechanisms may be the dominant mechanism controlling both the salt balance and particle dispersion (Zimmerman, 1986). The horizontal salinity gradient in one direction can thus be maintained by salt transport associated with a gradient in the normal direction. A vertically integrated calculation may adequately represent the salt balance, if the buoyancy flux and thus stratification are so weak that the velocity profile is nearly that of a neutrally stratified flow. If the buoyancy flux is sufficiently strong, however, then the horizontal density gradients will cause asymmetries in flood and ebb velocity profiles that will almost certainly lead to net transports by shear modes. Such a flow cannot readily be represented with a two-dimensional model.

V. Discussion

More than two decades of salt balance calculations in narrow estuaries, beginning with those based on field observations (e.g. Bowden 1963; Hansen 1965) and extending through calculations using field observations and flume results (Winterwerp, 1983) and even numerical model simulations (Oey et al., 1985) have resulted in a large literature on maintenance of the salt balance. These numerous studies have not resulted in any consensus on the nature of the mechanisms maintaining the salt balance for two primary reasons. First, the natural variability of estuarine flows yielded a diversity of results. Second, these studies could not be directly tied to any theoretical model of estuarine circulation, because salt balance calculations relied on salt conservation equations far too complex to be modelled analytically. Use of GLM theory allows the vertically integrated tidal and Stokes transport modes to be subtracted from the integrated salt conservation equation. The resulting salt balance of Eq.(11) is simple enough to be used directly in data analysis and theoretical studies.

The sharp distinction drawn here between narrow and wide estuaries also suggests that the two-parameter, shallow estuary classification system of Fig. 1 is incomplete, because it considers only narrow estuaries. These are classified on the basis of a barotropic Froude number ($F_T = a/h$) and an internal Froude number F_B representing the strength of the non-linearities associated with tidal cycle excursion of the density surfaces. These two parameters together represent the interaction of tidal and density driven circulation in a two-dimensional (x and z) flow. A third parameter, measuring the relative importance of two-dimensional horizontal, salt dispersion processes (in x and y) relative to shear transport in maintaining the salt balance is clearly required.

Furthermore, some parameters of interest are at least weakly three-dimensional ("two and one-half dimensional" in modelers' jargon) in fundamentally two-dimensional flows. This important limitation of all the two-dimensional analyses discussed herein can be explained using wave theory formalism. Use of any two-dimensional model requires, of course, that the distributions of lowest order wave variables (e.g. $\mathbf{u}'$ and C') be nearly two-dimensional, either in x and z or in x and y. Non-linear, second order parameters (e.g. fluxes or flux gradients such as $(\overline{\mathbf{u}' \cdot \nabla C'}) \approx \overline{U}^S \cdot \nabla \overline{C}$) can and often will exhibit some variability in the third dimension without violating in any way the assumptions on which these analyses are based. Thus, for example, cross-channel variability in along-channel salt transport parameters is frequently observed in systems where the basic tidal flow is essentially two-dimensional in x and z. Even linear residual flow variables (e.g. velocity and salinity) can be expected to show more variability in three dimensions than tidal variables, because these residual variables are driven by bilinear correlations between tidal variables. This variability may include a change of sign. This is an extension to three dimensions of Ianniello's (1977) result that a vertically integrated calculation of the Lagrangian tidal residual current in a narrow channel is inadequate, because the sign of the flow reverses with depth. As an example of the inadequacy of two-dimensional (in x and z) flow models of broad estuaries, turning in the bottom (Ekman) boundary layer under the influence of the Coriolis force is a well-known phenomenon in continental shelf sediment transport (Long and Smith, 1976). It is almost certainly important in wide shallow, weakly stratified estuaries, where the entire flow is in the bottom boundary layer.

Finally, many systems resist simplification (even approximately) to two dimensions - what can be said about these more complex flows? The tidal circulation in broad estuaries that are also stratified will often be fully three-dimensional and drive a more complex residual circulation than that found in either type of two-dimensional estuaries discussed above. This occurs because both shear mechanisms and horizontal, topographic dispersion processes will be active and will interact. Taking Puget Sound as an example, shear dispersion mechanisms probably dominate over the entrance sill in Admiralty Inlet, chaotic dispersion processes are likely to be prominent in some of the peripheral bays, and both should occur in the main basin. An example of the interaction of the two mechanisms would be that of an eddy forming initially in the neutrally stratified surface layer in shallow water in the lee of a point, where it would be in direct contact with the seabed. As this eddy is swept into deeper water, it would occupy the surface layer above a pycnocline, wherein shear processes driven by the eddy might substantially enhance mixing. These three-dimensional processes pose extremely challenging problems to both observational and modelling studies.

The preceding paragraphs may appear to be an argument for the exclusive use of three-dimensional numerical methods in the analysis of estuarine residual flows. This is not, however, their intent. Powerful, semi-analytical models have been developed in at least two recent instances for flows that are not fully three-dimensional, but still complex and nonlinear: the Ianniello (1979, 1981) model for barotropic tidal flows over topography and a model of the steady, coupled river meander/sediment transport problem by Nelson (1988). Internal shear mixing and non-local mixing associated with the propagation of internal waves may also be more difficult to represent in a three-dimensional grid than in a simpler, two-layer model. Thus, the modelling technique employed in a given situation should be suited to the processes occurring in the environment to be studied, and correct representation of mixing processes will usually be the controlling factor in choosing a modelling approach. Moreover, a detailed understanding of the various types of weak three-dimensional variability will almost certainly help show the way toward comprehension of fully three-dimensional flows.

VI. Summary and Conclusions

It has been argued herein that narrow estuaries in which the flow is nearly two dimensional in the along-channel (x) and vertical (z) directions are essentially different from those wide systems in which the flow is two dimensional in the horizontal (in x and y). The differences between the two types of systems can be associated with ebb-flood asymmetries in both the barotropic and internal (or baroclinic) circulations. With regard to the barotropic aspects of the ebb-flood asymmetry, the Lagrangian mean transport through a cross section in a narrow estuary is seaward and equal to the riverflow, even in the presence of a finite amplitude tide. This occurs because inward Stokes drift is exactly compensated by an outward Eulerian flow known as the Stokes drift compensation flow. The Stokes drift itself can be specified simply in a narrow estuary, even in the presence small-scale topography, because successive terms in its Taylor series representation rapidly become small. Truncation at a low order is justified. This is related to the fact that $\partial u'/\partial x$ is strictly bounded in these flows. No comparable constraints are available on the Lagrangian transport or the horizontal velocity gradient in wide shallow estuaries. The series representation of the Stokes drift may then converge very slowly, if at all. The non-convergence of the Stokes drift series is symptomatic of the presence of chaotic eddy processes in the flow.

The baroclinic aspect of the ebb-flood asymmetry in narrow estuaries is a product of the net flux of freshwater, which introduces both stratification and a baroclinic pressure gradient. On flood, the baroclinic and barotropic pressure gradients are in the same direction, and the total horizontal pressure gradient increases with depth. The interaction of the pressure gradient with the bottom friction results in a velocity profile with only weak shear. The remaining shear is, however, of the correct sign to limit stratification. On ebb, the barotropic pressure gradient changes sign (relative to the flood), but the baroclinic gradient does not; the magnitude of the total horizontal pressure gradient decreases with depth. The result of the interaction of the pressure gradient with bottom friction is a highly sheared velocity profile. The tidal straining of the density field is of the correct sign to increase stratification. Enhanced ebb stratification often allows the detachment of the outer part of the flow from the frictional influence of the seabed. Thus, even though the internal ebb-flood asymmetry is, from the point of view of the total flow, weak, (so that the representation of the flow as simply harmonic and perfectly reversing is nearly correct) the ebb-flood asymmetry in the outer part of the flow is strong. The harmonic structure of the currents in this part of the flow is more complex (i.e. has more overtides) than that of the surface elevation. This strong ebb-flood asymmetry near the surface may also be very important for dispersion of pollutants that concentrate near the surface. In contrast, the internal ebb-flood asymmetry in broad estuaries is likely to be much weaker than that in narrow estuaries, because the buoyancy flux and, therefore, stratification and horizontal density forcing, are likely to be weaker than in narrow estuaries. Although tidal ellipses may vary considerably in space, ebb-flood asymmetry does not complicate their harmonic structure.

Differences in the salt balance between wide and narrow estuaries reflect both the barotropic and internal aspects of the ebb-flood asymmetry in the two kinds of systems. The Generalized Lagrangian Mean theory of wave-mean flow interaction (Andrews and MacIntyre, 1978a) has been used to show that the salt balance in a narrow (two-dimensional in x and z) estuary can only be maintained by salt transport associated with stratification and shear in the mean and tidal flows. This salt transport is closely associated with the tidal cycle variations in shear and stratification; i.e. with the ebb-flood asymmetry. No such constraint exists in wide, horizontally extensive flows. This difference arises fundamentally because the Lagrangian transport is not, in wide estuaries, equal to the riverflow and cannot be specified without a full dynamical calculation. The salinity gradient in one direction in these systems can be maintained, independent of shear and stratification, by vertically-uniform salt by transport associated with the salinity gradient in the normal direction.

Future numerical and analytical studies will need to consider the three-dimensional variability induced in second-order parameters by nearly two-dimensional tidal flows as a prelude to analysis of fully three-dimensional systems.

Acknowledgements

Dr. Richard Garvine of the University of Delaware and Dr. Wayne R. Geyer of Woods Hole Oceanographic Institute provided verbal discussions of narrow and wide estuaries, respectively. Ruben Passmanter of the University of Utrecht pointed out to me the inadequacy of GLM theory in wide estuaries. This work was supported by NSF grant OCE-8711790.

VII. References

Allen, G. P., J. C. Salomon, P. Bassoulet, Y. Du Penhoat, and C. De Grandpré, 1980: Effects of tides on mixing and suspended sediment transport in macrotidal estuaries. Sed. Geol., **26**, 6990.

Andrews, D. G., and M. E. McIntyre, 1978a: An exact theory of nonlinear waves on a Lagrangian-mean flow, J. Fluid Mech., **89(4)**, 609-646.

Andrews, D. G., and M. E. McIntyre, 1978b: On wave-action and its relatives, J. Fluid Mech., **89(4)**, 647664.

Bowden, K. F., 1963: The mixing processes in a tidal estuary. Intern. J. Air Water Pollution, **7**, 83-95.

Dyke, P. P. G., 1980: On the Stokes drift induced by tidal motions in a wide estuary. Estuar. Coast. Mar. Sci., **11**, 17-25.

Dunkerton, T., 1980: A Lagrangian mean theory of wave, mean-flow interaction with application to nonacceleration and its breakdown. Rev. Geophys. Space Phys., **18(2)**, 387-400.

Feng, S., R. T. Cheng, and P. Xi, 1986: On tide-induced Lagrangian residual current and residual transport 2. residual transport with application in San Francisco Bay, California. Water Resources. Res., **22(12)**, 1635-1646.

Geyer, W. R., 1988: The advance of a salt wedge front: observations and dynamical model: In J. Dronkers and W. van Leussen, Physical Processes in Estuaries, Springer-Verlag, 181-195.

Haas, L. W., 1976: The effect of the spring-neap tidal cycle on the vertical salinity structure of the James, York and Rappahannock Rivers, Virginia, U.S.A., Estuar. Coast. Mar. Sci., **5**, 485-496.

Hansen, D. V., 1965: Salt balance and circulation in partially mixed estuaries, Estuaries, Publication No. 83, American Association for the Advancement of Science, Washington, D.C., pp. 45-51.

Hansen, D. V., and M. Rattray Jr., 1965: Gravitational circulation in straits and estuaries. Mar. Res., **23**, 104-122.

Ianniello, J. P., 1977: Tidally induced residual currents in estuaries of constant breadth and depth. J. Mar. Res., **35**, 755- 774.

Ianniello, J. P., 1979: Tidally induced currents in estuaries of variable breadth and depth. J. Phys. Oceanogr., **9**, 962-974.

Ianniello, J. P., 1981: Comments on tidally induced residual currents in estuaries: dynamics and near-bottom flow characteristics. J. Phys. Oceanogr., **11**, 126-134.

Jay, D. A., 1987: Residual circulation in shallow stratified estuaries, Ph.D. dissertation, University of Washington, Seattle.

Jay, D. A., 1990a, Estuarine salt and sediment balances: a Lagrangian approach. In press, Estuar. Coast. Shelf Sci.

Jay, D. A., 1990b: Anharmonicity and internal ebb-flood asymmetry in shallow estuaries. Advances in Tidal Hydrodynamics (in press).

Jay, D. A., and J. D. Smith, 1988: Residual circulation in and classification of shallow estuaries. J. Dronkers and W. van Leussen, Physical Processes in Estuaries, Springer Verlag, 19-41.

Jay, D. A., and J. D. Smith, 1989a: Residual circulation in shallow estuaries I. Narrow, highly stratified estuaries. J. Geophys Res. (in press).

Jay, D. A., and J. D. Smith, 1989b: Residual circulation in shallow estuaries II. Narrow, weakly stratified and partially mixed estuaries. J. Geophys Res. (in press).

Leibovich, S., 1980: On wave-current interaction theories of Langmuir circulations, J. Fluid Mech., **99(4)**, 715-724.

Long, C. and J. D. Smith, 1976: The effect of turning in the bottom boundary layer on continental shelf sediment transport. Mem. Soc. Roy. Liege 6[e] Ser., **X**, 369-396.

Longuet-Higgins, M. S., 1969: On the transport of mass by time-varying ocean currents, Deep-Sea Res., **16**, 431-447.

Middleton, J. F. and J. W. Loder, 1989: Skew fluxes in polarized wave fields. J. Phys. Oceanogr., **19**, 68-76.

Nelson, J., 1988: Mechanics of flow and sediment transport over nonuniform erodible beds. Ph.D dissertation, University of Washington, Seattle, 227 pp.

Oey, L.-Y., G. L. Mellor, and R. I. Hires, 1985: A three-dimensional simulation of the Hudson-Raritan Estuary. Part III: salt flux analyses. J. of Phys. Oceanogr., **15**, 1711-1720.

Partch, E. and J. D. Smith, 1978: Time-dependent mixing in a salt wedge estuary. Estuar. Coast. Mar. Res., **6**, 3-19.

West, J. R., D.W. Knight and K. Shiono, 1984: A note on flow structure in the Great Ouse Estuary, Estuarine, Coastal and Shelf Sci., **19**, 271-290.

Winterwerp, J. C., 1983: Decomposition of the mass transport in narrow estuaries, Estuar. Coast. Shelf Sci., **16**, 627-638.

Zimmerman, J. F. T., 1978: Topographic generation of residual circulation by oscillatory (tidal) currents. Geophys. Astrophys. Fluid Dyn., **11**, 35-47.

Zimmerman, J. F. T., 1980a: Vorticity transfer by tidal currents over an irregular topography, J. Mar. Res., **38**, 601-630.

Zimmerman, J. F. T., 1980b: On the Euler-Lagrange transformation and the Stokes drift in the presence of oscillatory and residual currents. Deep-Sea Res., **26A**, 505-520.

Zimmerman, J. F. T., 1986: The tidal whirlpool: a review of horizontal dispersion by tidal and residual currents. Netherlands J. Sea Res., **20(2/3)**, 133-154.

Dynamics of Tidally Forced Basin-wide Coastal Eddies

Andre W. Visser[1], Malcolm J. Bowman[1], and William R. Crawford[2]

[1]Marine Sciences Research Center
State University of New York at Stony Brook
Stony Brook, NY 11794

[2]Institute of Ocean Sciences
Department of Fisheries and Oceans
P.O. Box 6000, Sidney, B.C. V8L 4B2

ABSTRACT

The dynamics generating and governing tidally induced meso-scale recirculations in wide coastal sea straits are investigated using the **TIDAL STRESS** formulation. In this formulation we introduce the Stokes' drift into the residual dynamics and present the Lagrangian Tidal Stress estimate in terms of quasi-Eulerian fields. An analysis is presented of these processes active within Dixon Entrance and Hecate Strait off the west coast of Canada using idealized tidal and geomorphological representations. It is shown that the generation of an observed cyclonic basin-wide eddy in eastern Dixon Entrance is consistent with a tidal rectification process active over the strongly sloping escarpment connecting Dixon Entrance and Hecate Strait. It is concluded that a sufficiently strong rectification process active over a limited sector of the eddy circumference may substantially influence the circulation within adjacent bodies of water where perhaps no significant tidal nonlinearities operate.

I. Introduction

This work was prompted by observations of a surface intensified, strong and persistent meso-scale eddy (diameter ≈ 100 km) within eastern Dixon Entrance. Hecate Strait together with Dixon Entrance [cf. Figure (1)] constitute a tidally energetic wide coastal sea strait lying off the west coast of British Columbia, Canada, and south of the Alaskan panhandle. The strait separates the mainland from the Queen Charlotte Islands (principally Graham Island and Moresby Island) and has open connections to the north Pacific both through the western approaches of Dixon Entrance and through Queen Charlotte Sound to the south. The observations of the Dixon Entrance eddy include results from long term current meter moorings and Lagrangian drifters deployments (Crawford and Greisman, 1987). The eddy has also been observed in satellite thermal IR and CZCS images, during periods when climatic conditions are conducive in producing thermally distinct water masses within northern Hecate Strait. Figure (2) shows the sea-surface temperature contours derived from a NOAA-9 AVHRR satellite image taken during July 1986. Solar heating has generated a warm water mass in northern Hecate Strait, some of which is entrained into the Dixon Entrance eddy. This thermally distinct water mass appears to be advected in a cyclonic sense around eastern Dixon Entrance, indicating the dimension of the eddy.

Several modelling experiments, both hydraulic (Bell, 1963) and numerical (Bowman et al., 1989), have suggested that the eddy is at least in part driven by nonlinear tidal effects. Lagrangian residual currents derived from a non-rotating hydraulic model of the region [Figure (3); Bell, 1963] show a basin-wide cyclonic recirculation within eastern Dixon Entrance which is consistent in scale, location

Coastal and Estuarine Studies, Vol. 38
R. T. Cheng (Ed.)
Residual Currents and Long-term Transport

and sense of rotation with the satellite images of the eddy. The dynamical role of tidal rectification is further supported by the persistence of the eddy through periods of different prevailing weather conditions (Crean, 1967). The bathymetry of Dixon Entrance and Hecate Strait [cf. Figure (4)] is essentially bimodal. Dixon Entrance, which is relatively deep (200 to 400 m) and northern Hecate Strait, which is considerably shallower (20 to 50 m), are connected by a steeply sloping escarpment of slope ≈ 0.01. The observational evidence invariably shows that the southern periphery of the Dixon Entrance eddy overlies this escarpment, a region where strong tidal nonlinearities are expected to operate. Numerical tidal modelling (Bowman et al., 1989) supports the hypothesis that the tidal rectification operating over this limited region is responsible for the recirculation within Dixon Entrance. The central question addressed in this paper is; can the tidal rectification over the escarpment, a region representing only ≈ 10% of area of the eddy, drive the basin-wide cyclonic recirculation within Dixon Entrance?

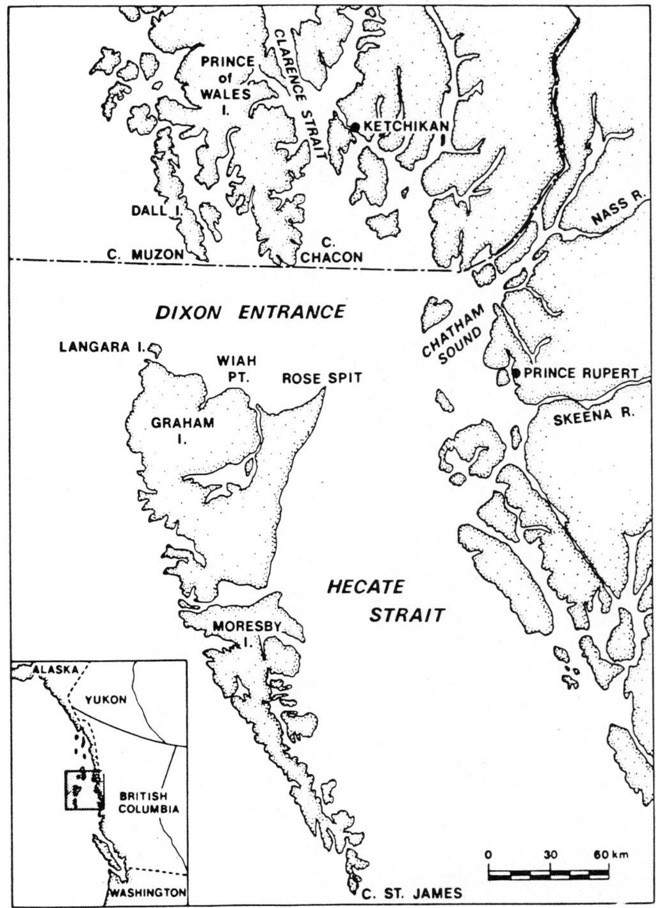

FIGURE 1. Dixon Entrance, Hecate Strait and adjacent waters.

It has long been recognized that the nonlinear interactions inherent in the tidal dynamics of the coastal ocean may transfer energy (plus momentum, vorticity etc.) from tidal oscillations into a mean circulation. Such mechanisms are highly correlated with strong bathymetric features and variations of coastline. The investigation of these processes have generally followed two different approaches viz: nonlinear numerical modelling (e.g., Tee, 1976; Pingree and Maddock, 1977; 1985; Maddock and Pingree, 1978; Bowman et al., 1980), and tidal stress (e.g., Huthnance, 1973; Nihoul and Ronday, 1974; Zimmerman, 1978; Loder, 1980). These approaches differ in that the nonlinear tidal models seek a solution to the

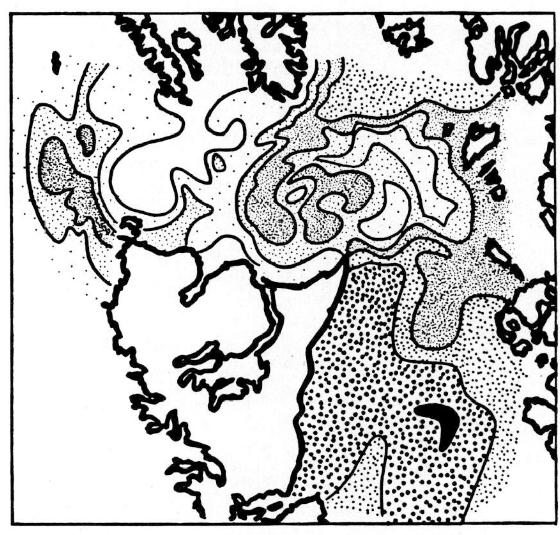

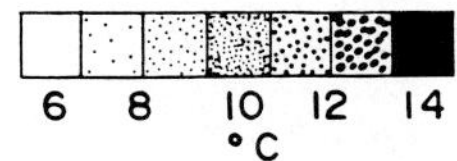

FIGURE 2. Contours of sea-surface temperature derived from a satellite IR image of Dixon Entrance taken on July 26th 1986.

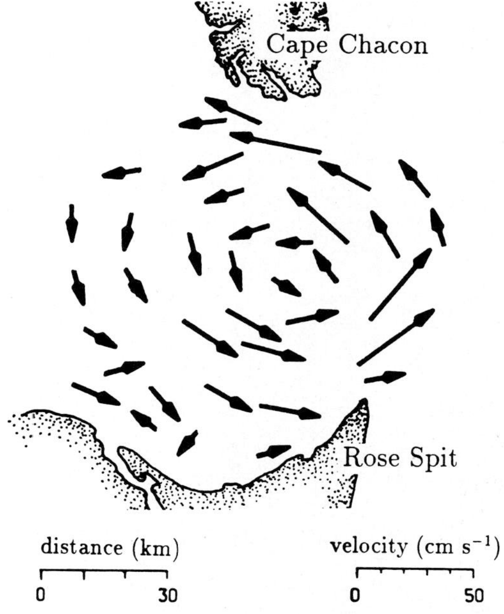

FIGURE 3. The residual surface currents for mean tidal conditions observed in the Rose Point eddy (i.e the Dixon Entrance eddy) of Dixon Entrance from the hydraulic model - Project Hecate (after Bell, 1963). The vectors are derived by connecting the start and end points of drifters scattered over the surface of the model (i.e. the method of Lagrange) for a single M_2 tidal period.

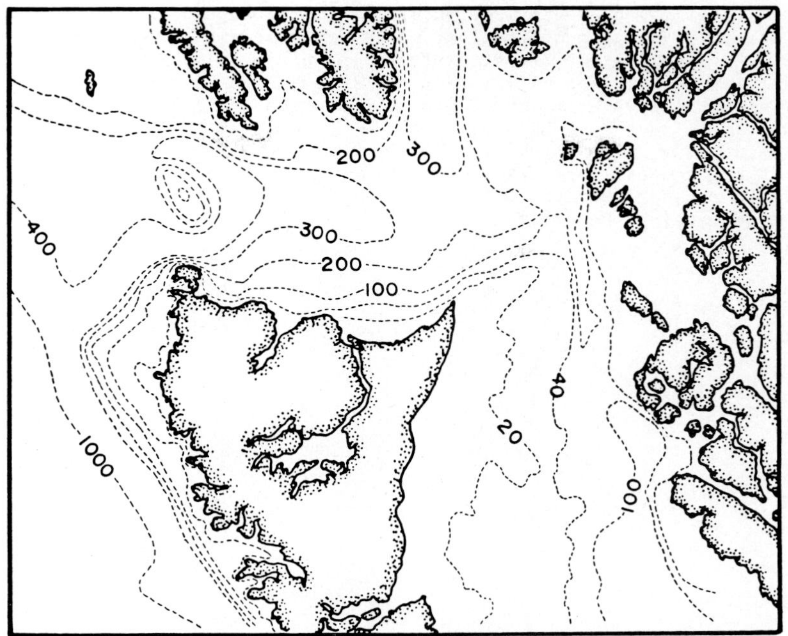

FIGURE 4. The bathymetry of Dixon Entrance and northern Hecate Strait. Depths are shown in meters.

nonlinear time-dependent equations of motion. A time average over a sufficiently long period gives the low frequency solution. The tidal stress approach however seeks a low frequency solution directly from the long-term time-averaged equations of motion. Under appropriate circumstances, this approach separates the nonlinear problem into two coupled linear problems, one at the tidal frequency and one for the residual motion. Thus the causal relationships governing the generation and dynamics of the residual circulation may be readily investigated. The coupling comes about through the tidal average of the nonlinear tidal interactions which act as forcing terms at the residual frequency (hence the terminology **tidal stress**, (Nihoul and Ronday, 1974). We schematically illustrate the tidal stress approach in Figure (5).

Previous investigators have demonstrated that significant differences arises when considering an Eulerian residual as compared to a Lagrangian residual (e.g. Longuet-Higgins, 1969; Maas et al., 1987). The Lagrangian residual is considered more meaningful when estimating a long term measure of a fluid quantity. The Lagrangian residual estimate is generally formulated by considering the Eulerian residual dynamics and a kinematically derived Stokes' estimate (e.g. Loder, 1980). A particular focus of this paper is the formulation for the Lagrangian residual dynamics and how these may be represented in terms of quasi-Eulerian fields. Indeed, the concept of residual dynamics is implied in the tidal stress approach and a Lagrangian tidal stress estimate allows the tidal rectification problem to be viewed as a forced quasigeostrophic phenomenon.

II. Tidal Stress Formulation

We begin with the following nondimensionalized nonlinear equations of motion:

$$\frac{\partial U}{\partial t} + \alpha[\mathbf{U}\cdot\nabla]\mathbf{U} = -\nabla E - \mathbf{z} \times \mathbf{U} - k\mathbf{U} \tag{1}$$

$$\frac{\partial E}{\partial t} + \nabla\cdot[(h + \alpha E)\mathbf{U}] = 0 \tag{2}$$

where $\mathbf{U}(x,y,t)$ is the vertically averaged horizontal velocity, $E(x,y,t)$ is the surface elevation and $h(x,y)$ is the water column depth below the undisturbed water level. We use a linear parameterization of bottom friction where k(x,y) is the time invariant, spatially varying coefficient of friction and is taken to be level. We use a linear parameterization of bottom friction where k(x,y) is the time invariant, spatially varying coefficient of friction and is taken to be proportional to some inverse power of the water column depth (e.g. Huthnance, 1973; Loder, 1980). **z** is the unit vertical vector. Eqs.(1) and (2) are the horizontal momentum and continuity equations respectively. In performing the nondimensionalization, we have used scaling factors appropriate for a long-period, shallow-water wave (i.e. the anticipated tidal wave solution). Dimensional quantities (superscript *) are related to nondimensional quantities as follows:

$$h^* \rightarrow H_o h \quad ; \quad E^* \rightarrow N_o E \quad ; \quad t^* \rightarrow t/f$$

$$(x^*, y^*) \rightarrow R_o(x, y) \quad ; \quad \nabla^* \rightarrow 1/R_o \nabla \quad ; \quad k^* \rightarrow fk$$

$$\mathbf{U}^* \rightarrow \alpha R_o f \mathbf{U} \quad .$$

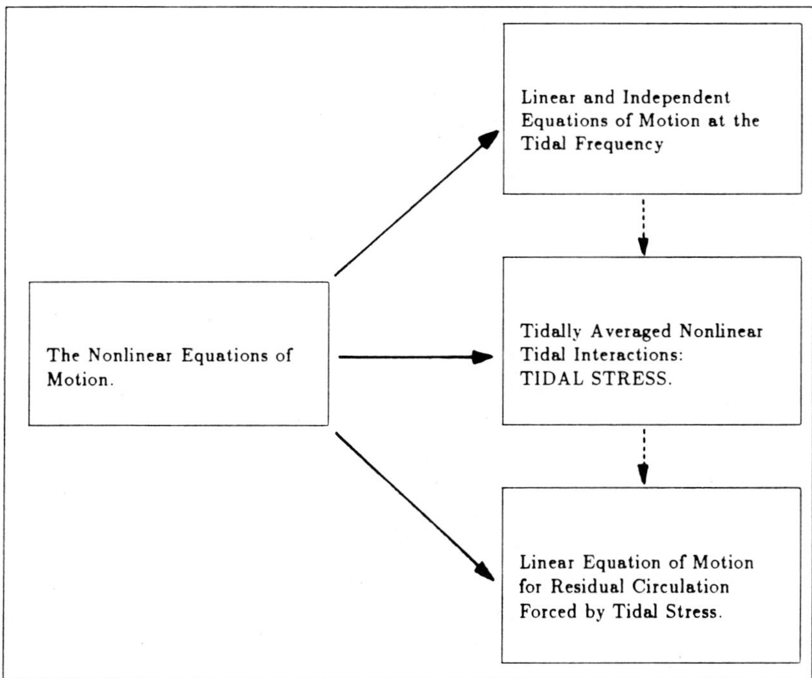

FIGURE 5. The tidal stress decomposition of nonlinear tidal dynamics into coupled linear dynamics for motion at the tidal frequency and residual circulation.

We use the local Coriolis parameter, f, as the frequency scale, and the barotropic Rossby radius, R_o, as the horizontal length scale. In the vertical direction, we have used two scaling parameters, N_o for the surface elevation, and H_o for the depth. The ratio of these parameters i.e., $\alpha = N_o/H_o$, is a small parameter used in the scaling arguments to follow. Note also that $R_o = \sqrt{g H_o}/f$ and the velocity scale becomes $\alpha R_o f$. An important point to note is that having used R_o as the horizontal length scale, nondimensionalized bathymetric gradients may be of order much greater than unity. Let us write the bathymetric length scale as L_B. Then

$$L_B = \left|\frac{h}{\nabla h}\right|_{\text{dimensional}} = R_o \left|\frac{1}{\nabla h}\right|_{\text{nondimensional}}$$

Thus,

$$O[\nabla h] \equiv \beta = R_o/L_B \quad .$$

Furthermore we note that the scale of the tidal excursion may be given by αR_o so that

$$\alpha\, O[\nabla h] = \gamma = \alpha\, R_o/L_B \quad .$$

Thus, while we have written the scaling factor of the nonlinear terms of the equations of motion explicitly as α, there is an additional factor β involved in regions of strong topography. Combining these, we arrive at the commonly used scaling factor for nonlinear tidal interactions, γ (ie. the ratio of the tidal excursion to the bathymetric length scale). For Dixon Entrance and Hecate Strait, typical values are $R_o \approx 300$ km, $H_o \approx 150$ m, and $N_o \approx 1$ m, so that $\alpha \sim 10^{-2}$. For the escarpment separating Dixon Entrance and Hecate Strait $L_B \approx 15$ km so that $\beta \sim 20$ and $\gamma \approx 10^{-1}$.

	Frequency Coefficient n			
	0	1	2	3 ...
fundamental	0^0 γ	1^0 1	2^0 γ	...
$\downarrow \gamma$				
1st nonlinear interaction	$(1^0 - 1^0)$ γ	$(0^0 + 1^0)$ $(2^0 - 1^0)$ γ^2	$(1^0 + 1^0)$ γ	...
$\downarrow \gamma$				
2nd nonlinear interaction	$(0^1 + 0^1)$ $(2^1 - 2^1)$ γ^3	$(0^1 + 1^1)$ $(2^1 - 1^1)$ γ^4	$(0^1 + 2^1)$ γ^3	...
$\downarrow \gamma$				
⋮	⋮	⋮	⋮	⋮ ⋱

FIGURE 6. The cascade of coefficients resulting from quadratic interactions of the idealized tide as determined by the interaction level at particular harmonic frequencies. Each nonlinear interaction generates an inherent factor γ, and the coefficient of the ensuing component is γ x the product of the magnitudes of the interacting components. The notation $(l^i \pm m^j)$ indicates interaction of the l from the i-th level with the $m\sigma$ component from the j-th level leading to a new component at $(l \pm m)\sigma$. Only those interactions which generate the most significant product are listed here.

We now consider an idealized tide composed of the sum of a dominant tidal component $O[1]$ at frequency σ, and a residual (at zero frequency) and over-tide components (at frequencies 2σ, 3σ, ...etc.) each of $O[\gamma]$ or less. The nonlinear self-interactions of the dominant tidal component generate only residual and first over-tide components each of $O[\gamma]$. Furthermore, from the nondimensionalized equations of motion, we see that the scale of the nonlinear interactions are

$O[\gamma]$ multiplied by the scale of the two interacting components. No contribution at the tidal frequency is generated except through a two stage process; the interaction of the residual or first over-tide components, $O[\gamma]$, with the $O[1]$ dominant tide to produce an $O[\gamma^2]$ component at the tidal frequency. Thus, to $O[\gamma^2]$, the tidal dynamics may be considered linear. In a similar manner, nonlinear self interactions at the residual frequency generate residual components $O[\gamma^3]$ (i.e. $O[\gamma^2]$ less than the principal residual component). While important nonlinear interactions $O[\gamma]$ do contribute at the residual frequency, these only involve tide-tide interactions and are independent of any residual motion. Thus the residual motion may also be considered linear to $O[\gamma^2]$. This "cascade" of the significance of the nonlinearly generated components at harmonic frequencies is illustrated in Figure (6).

Thus at the tidal frequency, whose harmonic variables are indicated with the subscript 1, the equations of motion may be written as:

$$\frac{\partial U_1}{\partial t} = -\nabla E_1 - \mathbf{z} \times \mathbf{U}_1 - k\mathbf{U}_1 + O[\gamma^2] \tag{3}$$

$$\frac{\partial E_1}{\partial t} + \mathbf{U}_1 \cdot \nabla h + h\nabla \cdot \mathbf{U}_1 + O[\gamma^2] = 0 \tag{4}$$

We may rewrite the tidal components as

$$\mathbf{U}_1(x,y,t) = \mathrm{Re}\{\mathbf{u}(x,y)e^{i\sigma t}\}$$

$$E_1(x,y,t) = \mathrm{Re}\{\eta(x,y)e^{i\sigma t}\} \tag{5}$$

where $\mathbf{u}(x,y)$ and $\eta(x,y)$ are complex valued variables containing both amplitude and phase information. The tidal dynamics may be rewritten as a tidal wave equation:

$$- i\sigma L^2\eta + hKL\nabla^2\eta + KL(\nabla h \cdot \nabla\eta) + L(\nabla h \times \nabla\eta)$$
$$- h(K^2 - 1)(\nabla k \cdot \nabla\eta) - 2hK(\nabla k \times \nabla\eta) = 0 \quad . \tag{6}$$

For convenience we write $K(x,y) = k + i\sigma$ and $L(x,y) = K^2 + 1$. The tidal velocity field follows from:

$$L\mathbf{u} = -[K\nabla\eta - \mathbf{z} \times \nabla\eta] \quad . \tag{7}$$

Given appropriate boundary conditions, the tidal wave equation may be readily solved by a variety of efficient numerical techniques.

At the residual frequency, whose components are indicated by the subscript ^{0}, we may write:

$$\nabla E_0 + \mathbf{z} \times \mathbf{U}_0 + k\mathbf{U}_0 = -\alpha/\gamma \langle \mathbf{U}_1 \cdot \nabla \mathbf{U}_1 \rangle_0 + O[\gamma^2] \tag{8}$$

$$\mathbf{U}_0 \cdot \nabla h + h\nabla \cdot \mathbf{U}_0 = -\alpha/\gamma \langle \nabla \cdot (E_1\mathbf{U}_1) \rangle_0 + O[\gamma^2] \tag{9}$$

Here, $\langle \ldots \rangle_0$ indicates a long term average and which, for the idealized tide, reduces to an average over the tidal period. The right hand side of Eq.(8), the residual momentum equation, represents one of the tidal stress components. We denote this the **ADVECTIVE TIDAL STRESS**, τ_A, which may be written as:

$$\boldsymbol{\tau}_A = -\alpha/\gamma \langle (\mathbf{U}_1 \cdot \nabla)\mathbf{U}_1 \rangle_0$$
$$= -\alpha/\gamma\ \mathbf{z} \times \langle \mathbf{U}_1 \Omega_1 \rangle_0 - \alpha/(2\gamma)\langle \mathbf{U}_1 \cdot \mathbf{U}_1 \rangle_0 \tag{10}$$

where Ω_1 is the vertical component of the tidal vorticity defined as $\Omega_1 = \mathbf{z} \cdot (\nabla \times \mathbf{U}_1)$. We may now eliminate the pressure gradient term by taking the curl of Eq.(8) to give the residual vorticity equation:

$$\nabla \cdot \mathbf{U}_0 + \mathbf{z} \cdot (\nabla k \times \mathbf{U}_0) + k\Omega_0 = -\alpha/\gamma[\langle \Omega_1 \nabla \cdot \mathbf{U}_1 \rangle_0 + \langle \mathbf{U}_1 \cdot \nabla \Omega_1 \rangle_0] \tag{11}$$

The right hand side of Eq.(11) is the curl of the advective tidal stress. Included in these terms are mechanisms classically associated with tidal rectification; viz the Coriolis and the bottom friction mechanisms (cf. Huthnance, 1973).

Let us now examine the residual equation of continuity. We introduce the residual volume transport, $\mathbf{Q}_0$, which may be shown to be given by:

$$\mathbf{Q}_0 = h\mathbf{U}_0 + \alpha/\gamma \langle E_1\mathbf{U}_1\rangle_0 \tag{12}$$

In this work we use the term RESIDUAL RADIATION TRANSPORT to signify the component $\langle E_1\mathbf{U}_1\rangle_0$ and the associated residual velocity, $\langle E_1\mathbf{U}_1\rangle_0/h$ is denoted herein as the RESIDUAL RADIATION VELOCITY. From the equation of continuity, we may show that $\mathbf{Q}_0$ is nondivergent; thus we may write the residual transport in terms of a residual stream function:

$$\mathbf{Q}_0 = h\mathbf{U}_0 + \alpha/\gamma\langle E_1\mathbf{U}_1\rangle_0 = \mathbf{z} \times \nabla\psi_0 \tag{13}$$

Rewriting the residual vorticity equation in terms of ψ_0 we arrive at:

$$\begin{aligned} kh\nabla^2\psi_0 - k\nabla h\cdot\nabla\psi_0 + h\nabla k\cdot\nabla\psi_0 + \mathbf{z}\cdot(\nabla h \times \nabla\psi_0) \\ = h^2\mathbf{z}\cdot[\nabla \times (\boldsymbol{\tau}_A + \boldsymbol{\tau}_R)] \\ = h^2\mathbf{z}\cdot(k \times \boldsymbol{\tau}_E) \quad . \end{aligned} \tag{14}$$

$\boldsymbol{\tau}_E$ is the EULERIAN TIDAL STRESS and differs from the advective tidal stress, $\boldsymbol{\tau}_A$, by the additional quantity $\boldsymbol{\tau}_R$. We term this the RADIATION TIDAL STRESS. In fact, it may be shown that this tidal stress component is given by:

$$\boldsymbol{\tau}_R = \frac{\alpha}{h\,\gamma}\,\mathbf{z} \times \langle E_1\mathbf{U}_1\rangle_0 + k\langle E_1\mathbf{U}_1\rangle_0 \tag{15}$$

That is, the radiation tidal stress is equivalent to the velocity dependent forces (i.e. the Coriolis and bottom friction forces) acting on the residual radiation velocity component.

To this point only the nonlinearities active in an Eulerian view have been considered. It is recognized that important nonlinear mechanisms may also operate when individual water particles are advected about through space by the tidal currents. When a dynamic balance or a kinematic constraint is formulated, an <u>instantaneous</u> condition governing the fluid motion is expressed, viz in writing the momentum equation for instance, the forces and inertial response that always and every where control the motion of water columns are stated. The time dependence and spatial dependence are generally considered separately, so that at any particular instant the dynamics are considered which relate to field quantities which satisfy the equations of motion at fixed locations in space. That is, the dynamics are investigated in terms of Eulerian field quantities. The time variance follows by either assuming a specific functional form as in the tidal wave dynamics, or by investigating solutions at successive time intervals as with numerical tidal models.

In Lagrangian dynamics, i.e. the dynamics experienced by individual water columns as they move, follow from the fundamental identity that <u>instantaneously</u>, the motion as measured at a fixed location must be identical to the motion of a water column traversing that location at that instant. The real interest concerning the different nature of Eulerian and Lagrangian measures arises when such measures are considered over finite time intervals. Over such an interval, not only does a water column perceive a variation in some quantity due to time variations of the field, it also perceives a variation due to its motion through spatial gradients of the field. In ascribing the tidal average of a perceived quantity as being equal to the tidal average of the associated Eulerian field, we assume that the quantity is well described over all time by the value at the mean tidal position of a water column. However, significant differences may arise when considering the tidal trajectory of a water column about its mean position, and the dynamics it experiences as it moves. For example, the mean velocity of a water column may be significantly affected by its advection around its tidal ellipse through gradients of the tidal velocity field.

Consider the variable $\mathbf{X}_1(\mathbf{X}_0,t)$, defined as the indefinite time integral of the tidal velocity $\mathbf{U}_1$ at the position $\mathbf{X}_0$. $\mathbf{X}_1$ thus describes the lowest order estimate of the tidal trajectory of a particle about its mean tidal position $\mathbf{X}_0$. That is the instantaneous position of the particle may be estimated by:

$$\mathbf{X}(t) \approx \mathbf{X}_0 + \mathbf{X}_1(\mathbf{X}_0,t) \tag{16}$$

If we perform a first order Taylor expansion of the velocity field about $\mathbf{X}_0$, then to first order the velocity experienced by a moving water particle is:

$$\mathbf{U}(t) \approx \mathbf{U}(\mathbf{X}_0,t) + \alpha/\gamma[\mathbf{X}_1(\mathbf{X}_0,t)\cdot\nabla]\mathbf{U}(\mathbf{X}_0,t) \qquad (17)$$

In this work we define the tidal average of this quantity as the Lagrangian residual velocity, $\mathbf{U}_L$.

$$\begin{aligned}\mathbf{U}_L(\mathbf{X}_0) &= \langle\mathbf{U}(t)\rangle_0 \\ &= \langle\mathbf{U}(\mathbf{X}_0,t)\rangle_0 + \alpha/\gamma\langle[\mathbf{X}_1(\mathbf{X}_0,t)\cdot\nabla]\mathbf{U}(\mathbf{X}_0,t)\rangle_0 \\ &= \mathbf{U}_0 + \alpha/\gamma\langle[\mathbf{X}_1\cdot\nabla]\mathbf{U}_1\rangle_0 \end{aligned} \qquad (18)$$

We define the difference between the Lagrangian and Eulerian residual velocities as the Stokes' residual velocity, $\mathbf{U}_S$. That is

$$\begin{aligned}\mathbf{U}_S &= \alpha/\gamma\langle[\mathbf{X}_1\cdot\nabla]\mathbf{U}_1\rangle_0 \\ &= 1/2(\alpha/\gamma)\nabla \times \langle\mathbf{U}_1 \times \mathbf{X}_1\rangle_0 + \alpha/\gamma\langle\mathbf{U}_1\nabla\cdot\mathbf{U}_1\rangle_0\end{aligned} \qquad (19)$$

and a residual transport component associated with this velocity may be written as:

$$h\mathbf{U}_S = 1/2(\alpha/\gamma)\nabla \times \langle h\mathbf{U}_1 \times \mathbf{X}_1\rangle_0 + \alpha/\gamma\langle\mathbf{U}_1 E_1\rangle_0. \qquad (20)$$

Finally we may define the Lagrangian residual transport as:

$$\begin{aligned}\mathbf{Q}_L &= h\mathbf{U}_L = h[\mathbf{U}_0 + \mathbf{U}_S] \\ &= \mathbf{U}_0 + \alpha/\gamma\langle\mathbf{U}_1 E_1\rangle_0 + 1/2(\alpha/\gamma)\nabla \times \langle h\mathbf{U}_1 \times \mathbf{X}_1\rangle_0 \\ &= \mathbf{Q}_0 + 1/2(\alpha/\gamma)\nabla \times \langle h\mathbf{U}_1 \times \mathbf{X}_1\rangle_0\end{aligned} \qquad (21)$$

Note that the radiation component of the residual transport arising from the Stokes' residual may be absorbed into the Eulerian residual transport component. We define the Stokes' residual transport as:

$$\mathbf{Q}_S = 1/2(\alpha/\gamma)\nabla \times \langle h\,\mathbf{U}_1 \times \mathbf{X}_1\rangle_0 \qquad (22)$$

The result expressed in Eq.(21) is essentially the same as given by Longuet-Higgins (1969). It immediately follows from this that $\nabla\cdot\mathbf{Q}_L \equiv 0$. The nondivergence of this residual transport component allows us to write it in terms of a residual stream function, ψ_L. We denote this the Lagrangian stream function. Indeed, the nondivergence of the residual Eulerian transport, $\nabla\cdot\mathbf{Q}_0 = 0$ as discussed in the previous section, allows us to separate the Lagrangian stream function into an Eulerian component, ψ_0, and a Stokes' component, ψ_S. That is :

$$h\mathbf{U}_L = \mathbf{z} \times \nabla\psi_L = \mathbf{z} \times \nabla\{\psi_0 + \psi_S\} \qquad (23)$$

We define the Stokes' residual stream function as:

$$\mathbf{z} \times \nabla\psi_S = 1/2(\alpha/\gamma)\nabla \times \langle h\mathbf{U}_1 \times \mathbf{X}_1\rangle_0 \qquad (24)$$

From this, we may introduce a component of the tidal stress due to the Coriolis force and bottom friction (i.e. the velocity dependent forces), acting on this component. We may denote this the STOKES' TIDAL STRESS and denote it as $\boldsymbol{\tau}_S$. This is analogous to the incorporation of the radiation tidal stress into the Eulerian component. That is, we may write

$$\boldsymbol{\tau}_S = \alpha/(2h\gamma)[\nabla(\mathbf{z}\cdot\langle h\mathbf{U}_1 \times \mathbf{X}_1\rangle_0) + k\nabla \times \langle h\mathbf{U}_1 \times \mathbf{X}_1\rangle_0] + \nabla\phi_S \qquad (25)$$

For completeness we include a potential function, $\nabla\phi_S$, which may for instance, represent a pressure gradient induced by the Stokes' residual transport. The rationale for this definition may be deduced from the form of the radiation tidal stress (Eq.15). Furthermore, by the definition of the Stokes' stream function we may write:

$$[k\mathbf{z} \times \nabla\psi_S - \nabla\psi_S]/h = \boldsymbol{\tau}_S \quad . \tag{26}$$

Taking the curl of this equation, we arrive at:

$$kh\nabla^2\psi_S - k\nabla h\cdot\nabla\psi_S + h\nabla k\cdot\nabla\psi_S + \nabla h \times \nabla\psi_S = h^2\mathbf{z}\cdot[\nabla \times \boldsymbol{\tau}_S] \quad . \tag{27}$$

This is precisely the same form as the Eulerian stream function equation Eq.(14). In fact, taking the sum of these equations, i.e. Eq.(14) + Eq.(27), and noting the additive identity expressed in Eq.(21), it follows immediately that:

$$kh\nabla^2\psi_L - k\nabla h\cdot\nabla\psi_L + h\nabla k\cdot\nabla\psi_L + \nabla h \times \nabla\psi_L = h^2\mathbf{z}\cdot[\nabla \times (\boldsymbol{\tau}_S + \boldsymbol{\tau}_E)] \quad . \tag{28}$$

Thus we have a dynamic expression relating the Lagrangian residual stream function in terms of the curl of tidal stress fields. Indeed we may define a kinematically derived LAGRANGIAN TIDAL STRESS field as:

$$\boldsymbol{\tau}_L = \boldsymbol{\tau}_E + \boldsymbol{\tau}_S \tag{29}$$

Thus the Lagrangian tidal field may be estimated form the principal tidal wave solutions using Eqs.(10), (15) and (25).

III. The Dixon Entrance Model

A highly idealized geomorphological representation of the Dixon Entrance and Hecate Strait region is constructed. This consists of two channel sections of uniform rectangular cross section, one shallow ($h^* = 30$ m) representing Hecate Strait, and one deep ($h^* = 270$ m) representing Dixon Entrance set at right angles to each other. The sections are connected by a steeply sloping escarpment ($L_B = 15$ km) running across the channel from the apex of Graham Island at a slight angle to the channel axes as noted in the inset of Figure (8). We choose an analytic function with continuous first and second derivatives to model the depth profile in this region. The channel width is 60 km except in eastern Dixon Entrance where it is 80 km. Included is a promontory on the northern coast of Dixon Entrance corresponding to Cape Chacon.

We solve the tidal wave equation, Eq.(6), within this model and subject to the following boundary conditions:

1) Along coastal boundaries we use the kinematic constraint that the tidal transport perpendicular to the coastline is zero. That is,

$$K[\mathbf{n}_c\cdot\nabla\eta] + \mathbf{n}_c \times \nabla\eta = 0$$

where $\mathbf{n}_c$ is a unit vector normal to the coast.

2) We choose open boundaries to lie perpendicular to the local channel axis. At these boundaries we impose the following constraint;

$$K[\mathbf{n}_o \times \nabla\eta] - \mathbf{n}_o\cdot\nabla\eta = 0$$

where $\mathbf{n}_o$ is the unit vector perpendicular to the open boundary (i.e. parallel to the local channel axis). In effect, this condition is a cross-channel geostrophic balance and allows only Kelvin waves to propagate into or out of the model. If we choose the open boundaries sufficiently distant from the escarpment and promontories such that any wave modes (notably Poincare waves) generated there decay, then this condition is consistent with a radiation condition for the outgoing wave. Note that for the channel dimensions used, Poincare waves are evanescent and cannot propagate tidal energy over any substantial distance.

In Figure (7) we present the tidal wave solution for the model using tidal amplitude and phase values consistent with Dixon Entrance and Hecate Strait. The solutions compare well with co-range and co-tidal charts of the region (e.g. Godin, 1980).

We use Eqs.(10), (15) and (25) to estimate the tidal stress fields from the principal tidal wave solution. These tidal stress fields are highly correlated to bathymetric gradients and are essentially zero everywhere except over the escarpment separating Dixon Entrance and Hecate Strait. Indeed, it may be shown (Visser, 1989) that under the conditions prescribed herein, the Lagrangian tidal stress is of the form:

$$\boldsymbol{\tau}_L = \frac{\alpha}{2\gamma} \mathrm{Re}\left[\frac{u\, \mathrm{Co}\{v\}}{i\, \sigma}\right] \mathbf{z} \times \nabla k \qquad (30)$$

where u and v are the across and along isobath components of the tidal velocity as expressed in Eq.(7), Co{...} represents complex conjugate, and Re{...} is the real component operator. Thus for the linearized bottom friction parameterization used, the only contributions to the Lagrangian tidal stress appear over the escarpment where $\nabla k \neq 0$. We solve the Lagrangian residual vorticity equation [i.e. Eq.(28)] for the Lagrangian residual stream function forced by the curl of the Lagrangian tidal stress, and subject to the following boundary conditions:

$$\mathbf{n}_c \times \nabla\psi_L = 0$$

$$\mathbf{n}_o \cdot \nabla\psi_L = 0$$

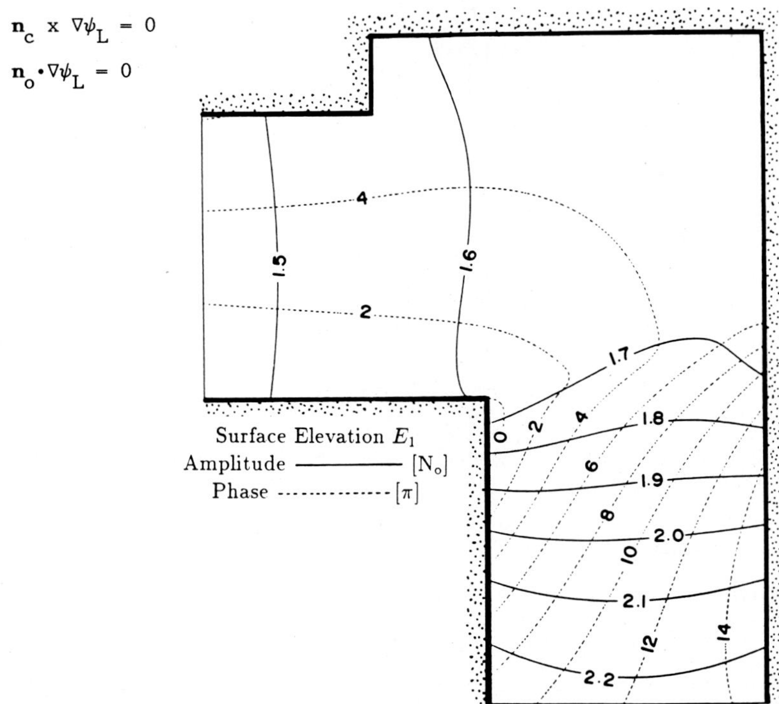

FIGURE 7. The tidal wave solution for the surface elevation, E(x,y), amplitude and phase for the modelled region.

We present the Lagrangian stream function solution and associated contours of isotachs in figure (8). The solution depicts two, closed, basin-wide circulations; an anticyclonic circulation within northern Hecate Strait, and a cyclonic circulation within eastern Dixon Entrance corresponding to the observed Rose Point eddy first described by Crean, 1967. While some evidence exists from numerical and hydraulic models of an anticyclonic eddy in northern Hecate Strait, the low frequency circulation of this region is poorly documented (cf. Crawford et al., 1988). Furthermore, within these shallow waters, we may expect the circulation to be strongly influenced by local variations in bathymetry, such as the 70m-deep trough connecting Dixon Entrance and Hecate Strait, which are not represented in the model. Therefore the idealized geometry used for this region may be somewhat deficient and we hesitate to ascribe any great significance to the solution within this region of the model. Within eastern Dixon Entrance, local bathymetric variations will contribute comparatively less to variations of the water column depth.

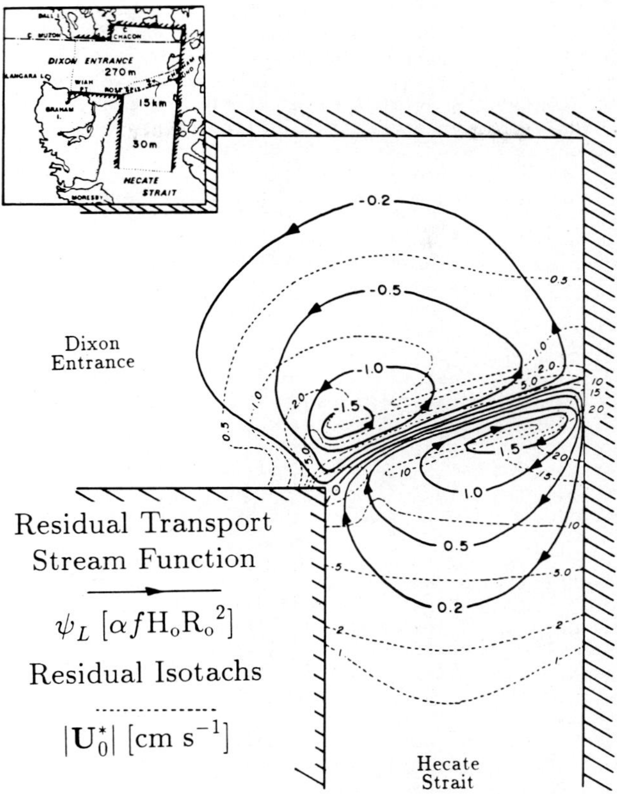

FIGURE 8. The Lagrangian stream function solution and associated dimensional isotachs for the modelled region. The stream function solution, ψ_L, is in dimensions of $\alpha\ f\ H_o\ R_o^2$ and the isotachs are shown in cm/s. The inset indicates the correspondence of the model with the Dixon Entrance region.

The Lagrangian residual stream function solution within this region depicts a basin-wide cyclonic gyre which is consistent in scale, location and sense of rotation with the observed Dixon Entrance eddy. This immediately suggests that the observed eddy is driven, at least in part, by the nonlinear tidal interactions active over the escarpment. The velocities associated with this circulation have a maximum of 120 cm/s over the escarpment, and fall off to 11 cm/s within the bulk of the gyre. While these speeds in mid-gyre are less than those observed in the hydraulic model, they do not compare too unfavorably with the depth-averaged velocities of the current meter results reported by Crawford and Greisman, 1987. In essence we may picture the Dixon Entrance eddy as a flywheel; being spun up along one edge by tidal stress acting over the escarpment. This spin up is balanced by dissipation acting over the entire region of the eddy.

The stream function solution depicted in Figure (8) is mathematically consistent with the form of Eq.(28); essentially a Poisson equation with a localized source along the escarpment. There appears to be an offset of the centers of rotation of the two lobes; the anticyclonic gyre is offset to the west, and the cyclonic gyre is offset to the east. This appears to be a consequence of the alignment of the escarpment with the local channel axis. Similar solutions for the escarpment aligned perpendicular to the channel axis showed a reflected symmetry about the escarpment between the transport stream function in the shallow and deep regions.

IV. Discussion

Perhaps the most important conclusion to be drawn from the results presented here is that a localized rectification process active over a limited region may substantially influence the circulation within adjacent bodies of water. Such relatively large scale, persistent residual circulation may, among other things, significantly effect planktonic biota of coastal regions. For instance, the Dixon Entrance eddy may have significant bearing on the recruitment success of the Dungeness crab (Cancer magister) of the region (cf. Visser, 1989).

In relationship to the residual dynamics, we may conclude that the primitive force balance often used in tidal rectification studies:

$$- k\mathbf{U}_0 = \boldsymbol{\tau}$$

(i.e. the dissipation of residual momentum balances the tidal stress locally, Huthnance, 1973) is perhaps not valid in many cases. A fuller balance involving all the residual dynamics acting over the region of tidal stress generation and adjacent bodies of water is appropriate, particularly in proximity to coastal boundaries. In this context we should note that the formulation outlined in this paper may be considered a lowest order estimation of the residual dynamic balance. A more sophisticated formulation including the advection of residual momentum may be particularly illuminating - bringing the formulation into a quasigeostrophic framework.

Previous investigations of tidally induced jets over strong bathymetric gradients (Loder, 1980; Robinson, 1981; Maas et al., 1987) have indicated that the rectified current only one or two tidal excursions on either side of the topographic feature. Such results, however, apply to one dimensional bathymetric features well removed from coastal boundaries. For the case reported in this paper, the tidal stress pumps water along the escarpment where it encounters a coastal boundary. This results in a difference in sea surface level between one end of the escarpment and the other which provides the forcing function for the circulation outside of the escarpment region. As with the results of previous investigations, the intensity of the rectified current is dependent on the strength of the tidal current and bathymetric gradient, and the ratio of the tidal excursion to the bathymetric length scale. However, for the two dimensional case reported here, the scale of the tidally rectified circulation assumes the length scale of the large scale, two dimensional continuity constraints.

An estimate of the Lagrangian tidal stress may be formulated by defining the Stokes' tidal stress as the velocity-dependent forces (i.e. the Coriolis and bottom fiction forces) acting on the Stokes' residual velocity. For the first order estimate of the Stokes' and Lagrangian residual velocities, the Lagrangian tidal stress may be presented as a time invariant quasi-Eulerian field.

In addition, it appears [particularly in Eq.(30)] that the Lagrangian tidal stress is strongly controlled by variations of bottom friction. We note firstly that the inclusion of bottom friction removes the indeterminacy of the Lagrangian residual circulation as reported by Moore (1970). This may be seen simply as a consequence of a dissipation mechanism existing to balance the input of energy (or momentum, vorticity etc.) from the tidal stress so that a nonzero steady state solution may be found. We also note that the residual stream function solution is critically dependent on the form of the bottom friction parameterization included in the formulation. For instance, if a quadratic bottom friction parameterization is used, then additional tidal stress components arise in regions where the depth is uniform (Heaps, 1978) due to local variations of the tidal current amplitude. These tidal stress contributions are responsible for generating tidally induced headland eddies (Maddock and Pingree, 1978; Robinson, 1981). Finally we note that a vertical structure of the tidal currents is imposed by bottom stress, particularly in shallow water. A vertical structure may also arise through stratification of the water column. This vertical structure translates into a depth dependence of the tidal stress (Maas, 1987; Wright and Loder; 1985, 1988). It appears therefore that the generation of low frequency circulation by tidal rectification is a phenomenon relating not solely to nonlinearities active at tidal scale and frequency, but is also critically dependent on the nonlinearities active at the turbulent scale.

Acknowledgments

The authors wish to thank Dr. W. Emery for use of the Satellite Oceanography Facility at the University of British Columbia (UBC), and Dr. D.G. Wright for useful discussions relating to this work. We also wish to thank Dr. D.-P. Wang for help with the computational algorithm, and G. Carroll and R. Malouf, Jr. for their assistance with the MSRC computer facility. This work was primarily supported by NSF grant OCE86-14050 and the preliminary numerical studies were supported by ONR contract N00014-84-C-0178. This paper is contribution 669 of the Marine Sciences Research Center of the State University of New York at Stony Brook. M. Bowman was partially supported by NCERC grants to L. Mysak while a visitor to UBC, 1985 - 86.

V. References

Bell, W. H., 1963: Surface current studies in the Hecate model. J. Fish. Res. Bd. Can., MS Rept. Ser. (Oceanogr. Limnol.), **159,** 27pp.

Bowman M. J., A. W. Visser and W. R. Crawford, 1989: Vortex motions in Dixon Entrance, British Columbia: evidence from observations and modeling. (in prep)

Bowman M. J., A. C. Kibblewhite and D. E. Ash, 1980: M2 tidal effects in greater Cooks Strait, New Zealand. J. Geophys. Res., **85**, 2728-2742.

Crawford, W. R., W. S. Huggett and M. J. Woodward, 1988: Water transport through Hecate Strait. British Columbia Atmosphere-Ocean, **26**, 301-320.

Crawford, W. R. and P. Greisman, 1987: Investigation of permanent eddies in Dixon Entrance, British Columbia. Continental Shelf Res., **7**, 851-870.

Crean, P. B., 1967: Physical oceanography of Dixon Entrance. J. Fish. Res. Bd. Can., Bull. Ottawa., **156**, 42 pp.

Godin, G., 1980: Cotidal Charts for Canada. Manuscript Report Series, **No. 55,** Marine Science and Information Directorate, Department of Fisheries and Oceans, Ottawa, Canada K1A 0E6.

Heaps, N. S., 1978: Linearized vertically-integrated equations for residual circulation in coastal seas. Dt. Hydrogr. Z., **31**, 147-169.

Hunter, J. R., 1975: A note on quadratic friction in the presence of tides. Est. Coast. Mar. Sci., **3**, 473-475.

Huthnance, J. M., 1973: Tidal current asymmetries over the Norfolk Sandbanks. Est. Coast. Mar. Sci., **1**, 367-387.

Loder, J. W., 1980: Topographic rectification of tidal currents on the sides of Georges Bank. J. Phys. Ocean., **10**, 1399-1416.

Longuet-Higgins, M. S., 1969: On the transport of mass by time-varying ocean currents. Deep-Sea Res., **16**, 431-447.

Maas, L. R. M., 1987: Tide-Topography Interactions in a Stratified Shelf Sea. Ph.D. dissertation. Rijksuniversiteit te Utrecht. 241 pp

Maas, L. R. M., J. T. F. Zimmerman and N. M. Temme, 1987: On the exact shape of the horizontal profile of a topographically rectified tidal flow. Geophys. Astrophys. Fluid Dyn., **38**, 105-129.

Maddock, L. and R. D. Pingree, 1978: Numerical simulations of the Portland tidal eddies. Est. Coast. Mar. Sci., **16**, 353-363.

Moore, D., 1970: The mass transport velocity induced by free oscillations at a single frequency. Geophys. Fluid Dyn., **1**, 237-247.

Nihoul, J. C. J. and F. C. Ronday, 1974: The influence of ''tidal stress'' on the residual circulation. Tellus, **27**, 484-354.

Pingree, R. D. and L. Maddock, 1977: Tidal residuals in the English Channel. J. Mar. Biol. Ass. U.K., **57**, 339-354.

Robinson, I. S., 1981: Tidal vorticity and residual circulation. Deep-Sea Res., **28**, 195-212.

Tee, K. -T., 1976: Tide-induced residual currents, a 2-D nonlinear numerical tidal model. J. Mar. Res., **34**, 603-628.

Visser, A. W., 1989: Tidal Stress and Residual Eddy Dynamics of Wide Coastal Sea Straits. 350 pp Ph.D. dissertation. Marine Sciences Research Center, State University of New York at Stony Brook.

Wright, D. G. and J. W. Loder, 1985: A depth-dependent study of topographic rectification of tidal currents. Geophys. Astrophys. Fluid. Dyn., **31**, 169-220

Wright, D. G. and J. W. Loder, 1988: On the influences of nonlinear bottom friction on the topographic rectification of tidal currents Geophys. Astrophys. Fluid Dyn., **42**, 227-245.

Zimmerman, J. T. F., 1978: Topographic generation of residual circulation by oscillatory tidal currents. Geophys. Astrophys. Fluid Dyn., **11**, 35-47.

Three-Dimensional Modeling of the Tidally Induced Residual Circulation Off Southwest Nova Scotia

Kim-Tai Tee[1] and Denis Lefaivre[2]

[1]Physical and Chemical Sciences
Department of Fisheries and Oceans
Bedford Institute of Oceanography
Dartmouth, Nova Scotia
Canada B2Y 4A2

[2]Institut Maurice Lamontagne
Department of Fisheries and Oceans
P.O. Box 1000
Mont Joli, Quebec
Canada G5H 3Z4

ABSTRACT

A three-dimensional tidal model is used to study the tidally induced horizontal and vertical residual circulation in the Cape Sable area, southwest of Nova Scotia. The computed residual current a few grid spacings away from the open boundary is found to be very insensitive to the specified residual current at the boundary. This is because strong residual current in the Cape Sable area is generated locally and associated closely with local topography.

The model predicts several areas of strong residual upwelling and downwelling. Through the coupling of the upwelling and the strong tidal mixing in the shallow water, this tidally induced vertical transport can be significant for supplying cold, saline and nutrient-rich water from the deep water to the surface. The observed low temperature water near Cape Sable is probably induced by this mechanism. The same mechanism of producing nutrient-rich water may be important in many coastal areas.

I. Introduction

In the last fifteen years, the tidally induced horizontal residual circulations have been studied extensively. Strong residual currents have been predicted and observed in many coastal areas (Tee, 1976, 1977; Pingree and Maddock, 1977; Loder, 1980; Greenberg, 1983). However, only a few simple models have been developed to examine the tidally induced vertical residual circulation. Most of the models are for narrow channels where the cross-channel variation can be neglected (Johns, 1970; Ianniello, 1977, 1979). For the case of tidal current passing around a sharp coastal promontories (such as a rectangular cape), the vertical circulation has been examined using a simple two-level model (Oonishi, 1978). Some simple models have also been developed to examine the cross-sectional circulation on sides of submarine banks (Tee, 1985; Loder and Wright, 1985; Wright and Loder, 1985). These models are simplified by applying the condition of uniformity along isobaths. Here, we examine the tidally induced vertical residual circulation off southwest Nova Scotia (Fig. 1) using a three-dimensional tidal model. One of the objectives of this paper is to show the existence of significant residual upwelling and downwelling in the area. Such tidally induced vertical motion can have important effects on biological and geological activities since nutrient and sediment are transported vertically.

Coastal and Estuarine Studies, Vol. 38
R. T. Cheng (Ed.)
Residual Currents and Long-term Transport

One of the reasons the tidally induced vertical circulation in an offshore area has not been examined accurately is that the study requires a three-dimensional numerical model which is expensive to run and generally requires a large computer. One possible approach is to limit the model to a small area so that accurate resolutions in the horizontal and vertical can be obtained. This approach is applied here. The modelled area near Cape Sable is a portion of the Gulf of Maine and Bay of Fundy system (Fig. 1). A major concern of such limited area model is the open boundary conditions. Another objective of this paper is to examine carefully the effect of open boundary condition on the tidally induced residual circulation computed from a limited area model. In the Cape Sable area, as well as many areas where residual currents are induced by topographic effects, it will be shown that the residual currents are generated locally and not affected significantly by uncertainties associated with the open boundary condition.

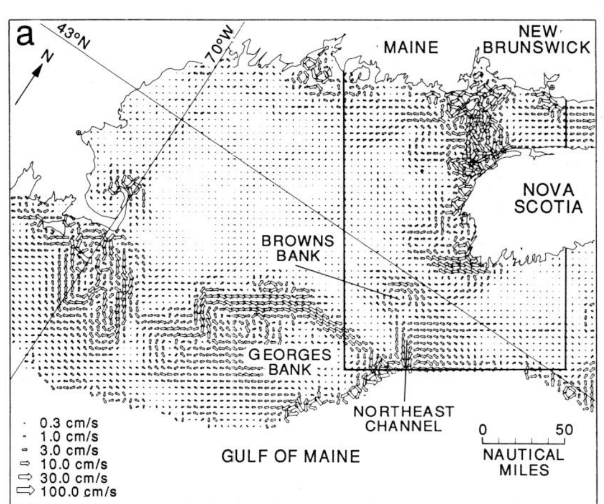

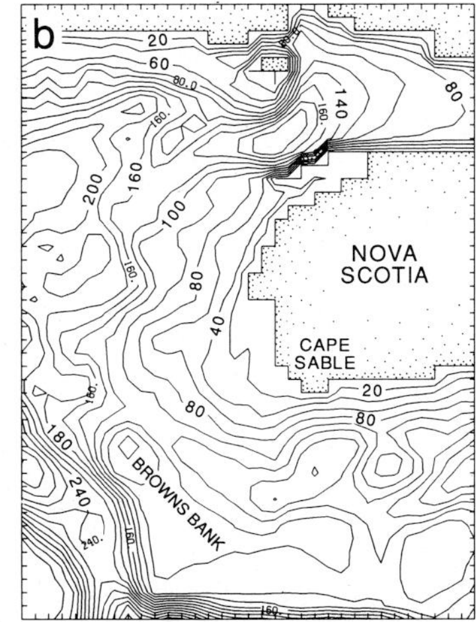

FIGURE 1. (a) The depth-averaged residual current from Greenberg's (1983) two-dimensional tidal model. The limited area domain for the Cape Sable model used in this study is shown by solid lines. (b) Depths of water (m) in the Cape Sable model. The distance between two tick marks in the horizontal and vertical axes is the grid spacing of the model (Δx=Δy=7.047 km).

In section 2, a numerical model is briefly described. The justification of using the limited area model is described in section 3. The existence of significant residual upwelling and downwelling is shown in section 4.

II. Numerical Model

A simple method of calculating the three-dimensional tidal and residual currents has been developed by Tee (1979, 1980, 1987). A comparison of this method with other expensive three-dimensional numerical methods has been described by Tee (1987). In this section, only those aspects of the computation, particularly the open boundary condition, that are related directly to the following solution and discussion are presented.

The Cape Sable numerical model (Fig. 1) has 41 levels in the vertical, 36 grid points on the north-western and south-eastern boundaries, and 47 grid points on the north-eastern and south-western boundaries. A large number of vertical levels is used to resolve a bottom boundary layer. The value of vertical grid spacing increases rapidly from the bottom as n^4 in the boundary layer and as n^2 above that, where n is the number of vertical grid points measured from the bottom. Twenty-one grid points are located in the layer, and the other twenty grid points in the rest of the water column. The sensitivity of residual currents to the number of vertical levels has been examined by Tee (1987). The open boundaries include those along the Gulf of Maine and Bay of Fundy, and along and across Scotia Shelf. The horizontal grid spacing is 7.047 km. The bathymetry of the modeled area is shown in Fig. 1b.

There are two important characteristics of the topography near Cape Sable: one is the convex shape of isobaths around the Cape, and the other is Browns Bank, a shallow submarine bank approximately 50 km long and 30 km wide. The crest of the Bank is ≈50 m below the surface.

II.1 The governing equation

For a Cartesian system with horizontal coordinates x, y and vertical coordinate z, measured vertically upward from the mean sea level, the equations describing the tidal motion of a homogeneous fluid in shallow water are

$$\frac{\partial \mathbf{u}}{\partial t} + \mathbf{u}\cdot\nabla\mathbf{u} + w\frac{\partial \mathbf{u}}{\partial z} + \mathbf{f}x\mathbf{u} = -g\nabla\zeta + \frac{\partial}{\partial z}\mu_w\frac{\partial \mathbf{u}}{\partial z} + \mu_h\nabla^2\mathbf{u} \quad , \tag{1a}$$

$$\frac{\partial \zeta}{\partial t} + \nabla\cdot\int_{-D}^{\zeta}\mathbf{u}dz = 0 \quad , \tag{1b}$$

where $\mathbf{u}$ is the horizontal velocity vector with components u and v in the x and y directions; w is the vertical velocity component; D is the depth of the bottom below mean sea level; ζ is the height of the water surface above mean sea level; f is the Coriolis parameter; $\mu_w = \mu_w(x,y,z)$ is a vertical eddy viscosity coefficient; μ_h is a horizontal eddy viscosity coefficient; g is the acceleration due to gravity; ∇ is the horizontal gradient operator; and ∇^2 is the horizontal Laplacian operator.

By assuming that the advective term is small compared to the other terms in the momentum equation (weakly nonlinear case) we obtain (Tee, 1979) the first order equations for the oscillating current ($\mathbf{u}_1$)

$$\frac{\partial \mathbf{u}_1}{\partial t} + \mathbf{f}x\mathbf{u}_1 = -g\nabla\zeta_1 + \frac{\partial}{\partial z}\mu_w\frac{\partial \mathbf{u}_1}{\partial z} + \mu_h\nabla^2\mathbf{u}_1 \quad , \tag{2a}$$

$$\frac{\partial \zeta_1}{\partial t} + \nabla\cdot D\mathbf{U}_1 = 0 \quad , \tag{2b}$$

and the second-order equations for the residual current ($\mathbf{u}_2$)

$$\overline{\mathbf{u}_1\cdot\nabla\mathbf{u}_1} + \overline{w_1\frac{\partial \mathbf{u}_1}{\partial z}} + \mathbf{f}x\mathbf{u}_2 = -g\nabla\zeta_2 + \frac{\partial}{\partial z}\mu_w\frac{\partial \mathbf{u}_2}{\partial z} + \mu_h\nabla^2\mathbf{u}_2 \quad , \tag{3a}$$

$$\nabla\cdot(D\mathbf{U}_2 + \overline{\zeta_1\mathbf{u}_{1_s}}) = 0 \quad , \tag{3b}$$

where the overbar denotes the time-average over a tidal period, $\mathbf{u}_{1_s}$ is the value of $\mathbf{u}_1$ at the surface (z = 0), and $\mathbf{U}_1$ and $\mathbf{U}_2$ are the depth averaged values of $\mathbf{u}_1$ and $\mathbf{u}_2$.

The depth-averaged momentum equations for the tidal and residual currents can be derived from Eqs.(2a) and (3a) as

$$\frac{\partial \mathbf{U}_1}{\partial t} + \mathbf{f}x\mathbf{U}_1 = -g\nabla\zeta_1 - \frac{\boldsymbol{\tau}_{1b}}{\rho D} + \mu_h\nabla^2\mathbf{U}_1 \quad , \tag{4}$$

$$\overline{\langle\mathbf{u}_1\cdot\nabla\mathbf{u}_1\rangle} + \overline{\langle w_1\frac{\partial \mathbf{u}_1}{\partial z}\rangle} + \mathbf{f}x\mathbf{U}_2 = -g\nabla\zeta_2 - \frac{\boldsymbol{\tau}_{2b}}{\rho D} + \mu_h\nabla^2\mathbf{U}_2 \quad , \tag{5}$$

where $\boldsymbol{\tau}_{1b}$ and $\boldsymbol{\tau}_{2b}$ are respectively the bottom stress for the tidal and residual currents, and angle brackets denote the depth-averaged operator,

$$U_{1,2} = \langle u_{1,2}\rangle \equiv D^{-1}\int_{-D}^{0} u_{1,2}dz \quad . \tag{6}$$

II.2 The vertical eddy viscosity

The vertical eddy viscosity (μ_w) is specified as

$$\mu_w = \begin{cases} \mu_0[1 + R(D + z)]^2 & \text{for } z + D \le D_\mu \\ \mu_m = \mu_0(1 + RD_\mu)^2 & \text{for } z + D \ge D_\mu, \end{cases} \tag{7}$$

where $\mu_0 = 1.4 \times 10^{-6}$ m²/s is the molecular viscosity, D_μ is the thickness of the transition layer between laminar and turbulent flow, and R is a parameter which is determined once μ_m and D_μ are specified. This form of μ_w has been applied by Johns and Dyke (1971, 1972), and Tee (1979, 1980, 1985). There are two adjustable parameters in Eq.(7): (1) the vertical eddy viscosity above the bottom transition layer, μ_m; and (2) the thickness of the transition layer D_μ. The value of μ_m can be estimated from the relationship between vertical eddy viscosity (μ_w) and linear bottom friction coefficient (λ) (Tee, 1980) given by

$$\lambda = -\left\{\frac{1}{D}\int_{-D}^{0}\int_{-D}^{z'}\frac{z}{\mu_w}dzdz'\right\}^{-1}. \tag{8}$$

The bottom friction in the depth-dependent model will approximate that in the depth-independent model if Eq.(8) is satisfied.

The value of λ can be estimated from

$$\lambda(x,y) = \frac{YU_a}{D}, \tag{9}$$

where Y is the drag coefficient taken to be 0.0021, and U_a is the depth-averaged semi-major axis. For given values of λ, (8) is used to compute μ_m. The value of D_μ is 0.15 m, obtained by comparing the vertical variation of the observed and computed tidal current at a station near Cape Sable (Tee et al., 1987).

II.3 Computational procedure

The depth-averaged tidal current is first computed from (4) and (2b) using the approximation $\boldsymbol{\tau}_{1_b} = \rho\lambda\mathbf{U}_1$ where λ is the linear friction coefficient (9). Using the computed $\mathbf{U}_1$ and the given values of μ_w, we can compute the vertical variation of tidal current in the offshore area where the horizontal diffusion term is neglected (Tee, 1979). The computation of depth-averaged tidal current can be improved by updating the bottom stress from the vertical variation of tidal current ($\boldsymbol{\tau}_{1_b} = \rho\mu_w\partial\mathbf{u}_1/\partial z$ at $z = -D$). It was found that the improvement is small (1-2%) if μ_w is chosen according to Eq.(8). Here, the improvement is neglected.

Using the computed $\mathbf{u}_1$, and setting the bottom stress of the residual current ($\boldsymbol{\tau}_{2_b}$) to be $\rho\lambda\mathbf{U}_2$, where $\mathbf{U}_2$ is the depth-averaged residual current, we compute $\mathbf{U}_2$ by transforming the depth-averaged equations for residual current [Eq.(5)] into the vorticity equation, which is then solved numerically using a relaxation method (Tee, 1987). From the known values of $\mathbf{U}_2$, the vertical variation of $\mathbf{u}_2$ in the offshore area can be solved numerically from Eq.(3a) (Tee, 1980). The accuracy of $\mathbf{U}_2$ is then improved by updating the bottom stress ($\boldsymbol{\tau}_{2_b}$) in the depth-averaged momentum equation from $\boldsymbol{\tau}_{2_b} = \rho\mu_w\partial\mathbf{u}_2/\partial z$ at $z = -D$. The computation is iterated until $\boldsymbol{\tau}_{2_b}$ is not significantly improved from the updated $\mathbf{u}_2$. Here, it was found that five iterations were sufficient to obtain an accurate solution (i.e. to within 0.01%).

II.4 The boundary conditions

The boundary conditions at the surface ($z = 0$) are

$$\frac{\partial\mathbf{u}_1}{\partial z} = 0 \quad \text{and} \quad \frac{\partial\mathbf{u}_2}{\partial z} = \frac{\boldsymbol{\tau}_{2_s}}{\rho\mu_w}, \tag{10}$$

where

$$\boldsymbol{\tau}_{2_s} = -\rho\frac{\partial}{\partial z}\mu_w\frac{\partial}{\partial z}\overline{\zeta_1\mathbf{u}_1}\Big|_{z=0}.$$

At the bottom ($z = -D$), and at the coast $\mathbf{u}_1 = \mathbf{u}_2 = 0$. At the open boundary, we can see from Eqs.(2b) and (4) that the depth-averaged tidal current can be computed if the tidal elevation ζ_1 is specified. Note that the horizontal diffusion terms in Eqs.(4) and (5) can be neglected at the open boundary. The vertical variation of tidal current can be computed once the depth-averaged component is known. Thus, there is no need to specify the vertical variation of tidal current at the open boundary. Here, the values of ζ_1 at the open boundary are taken from Greenberg's (1983) large scale model.

Given the values of $\mathbf{u}_1$, the depth-averaged residual currents ($\mathbf{U}_2$) can be computed from Eqs.(3b) and (5) if their values at the open boundary are known. Again, by neglecting a small contribution of the horizontal diffusion, we do not need to specify the vertical variation of residual current. Here, two cases of $\mathbf{U}_2$ at the open boundary are tested: one is taken from Greenberg's two-dimensional nonlinear tidal model, and another is set to be zero. The sensitivity of these two different cases of $\mathbf{U}_2$ at the open boundary on the tidally induced residual current in the interior regions will be examined.

III. Reliability of Residual Currents in the Limited Area Model

III.1 Open boundary problem

Figure 2 shows the depth-averaged residual currents for the case that has $\mathbf{U}_2$ and ζ_1 at the open boundary taken from Greenberg's (1983) large scale model. The basic

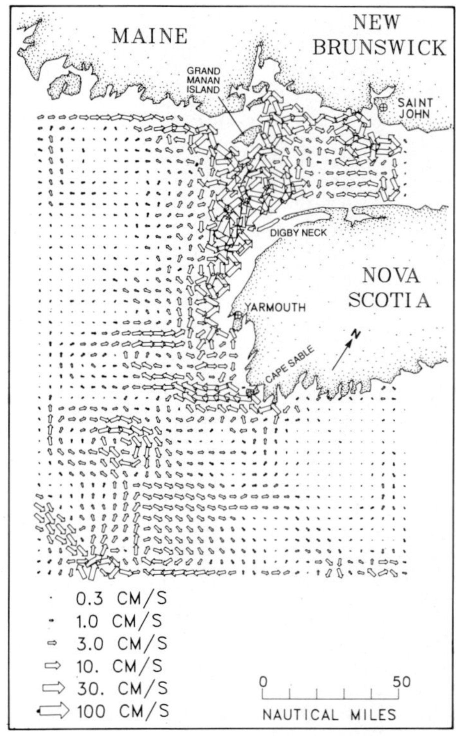

FIGURE 2. Horizontal distribution of the depth-averaged residual current. U_2 at the open boundary is taken from Greenberg's (1983) model.

features in the residual circulation near the Cape Sable area are: (1) the westward jet near the Cape with the return flows at both sides which form an anticyclonic eddy at the northern region and a cyclonic eddy at the southern region, (2) an anticyclonic circulation around Brown's Bank, (3) the eastward flow east of Browns Bank, (4) the eastward flow on the eastern tip of Georges Bank, and (5) the northward flow along the deep Northeast Channel. By comparing Fig. 2 and Fig. 1a which shows the depth-averaged residual current computed from a two-dimensional (x, y) nonlinear large scale tidal model (Greenberg, 1983), we can see that the basic features in the residual current near the Cape are reproduced very well by the limited area model. Note that, by choosing the bottom friction of 3-D model to be approximately equal to that of 2-D model (section II.2), the depth-averaged residual currents for the two models are found to be approximately the same. The agreement in the residual current between the limited area and large scale models strongly indicates that the uncertainties associated with the open boundary condition in the limited area model are small.

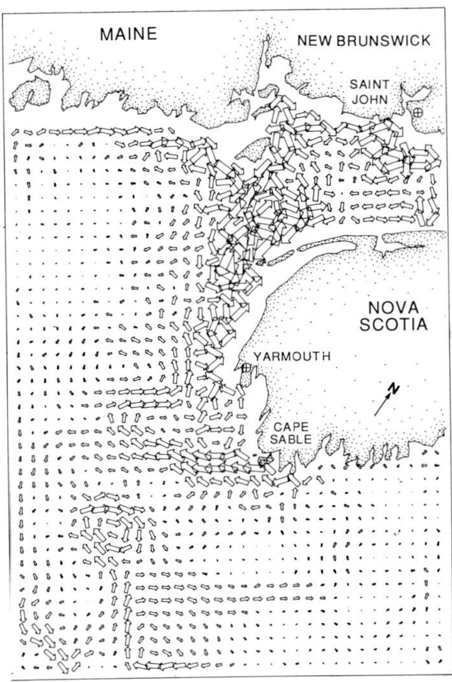

FIGURE 3. Horizontal distribution of the depth-averaged residual current. $\mathbf{U}_2 = 0$ at the open boundary. The velocity and distance scales are the same as those in Fig. 2.

The computed depth-averaged residual current in the 2-D large-scale model is not free of uncertainty. This is because the model involves errors associated with numerical formulation, the depth-independent approximation, and the uncertainty associated with its open boundary conditions. Thus, the values of $\mathbf{U}_2$ and ζ_1 at the open boundary of the limited area model, although taken from a large scale model, may not be completely free of uncertainty. The specification of ζ_1 from Greenberg's model is probably quite accurate because the observed tidal elevations and currents near Cape Sable were reproduced reasonably well (Tee <u>et al</u>., 1988). To examine further the effect of $\mathbf{U}_2$ specified at the open boundary on the residual circulation near Cape Sable, a computation was carried out by keeping ζ_1 at the boundary to be the same (taken from Greenberg's model), but setting $\mathbf{U}_2$ at the open boundary to be zero. Figure 3 shows the depth-averaged residual currents computed using these new boundary conditions. By comparing Fig. 2 and Fig. 3, we can see that the basic features of the residual circulation remain the same. The only significant changes in the currents are those near the open boundary. These changes include the reduction in both the eastward flow at the north eastern corner of Georges Bank and the inflow at the Northeast Channel. The north-westward flow at the west of Browns Bank is reversed to weak south-eastward flow. Despite significant changes occur near the open boundary, we can see that most of the residual currents located three grid spacings away from the boundary, including those strong currents near Browns Bank and Cape Sable remain approximately the same. This result indicates that the residual circulation in the interior region of the Cape Sable model is very insensitive to the specified residual current at the open boundary.

III.2 Problem of horizontal resolution

From Fig. 1 to Fig. 3, we can see that there are large irregular flows at the entrance to the Bay of Fundy between Grand Manan Island and Digby neck, (Fig. 2)

The same sort of irregularities also occur in Isaji and Spaulding's (1984) large-scale depth-averaged model. Because these irregular flows occur in both the large scale and limited-area models, they are not attributed to open boundary conditions. Greenberg has suggested that the irregularities originate from insufficient horizontal resolution in the area. Although only the residual currents in the Cape Sable area are examined here, it is important to examine whether there is any effect on these currents by the strong irregular flow further north.

To reduce the limitation of horizontal resolution, we artificially smooth out the bottom topography and coastal geometry at the entrance to the Bay of Fundy by removing Grand Manan Island and Digby Neck from the model's domain, and allowing the depths to increase gradually from the coast to the maximum depth near the middle of the Bay (Fig. 4a).

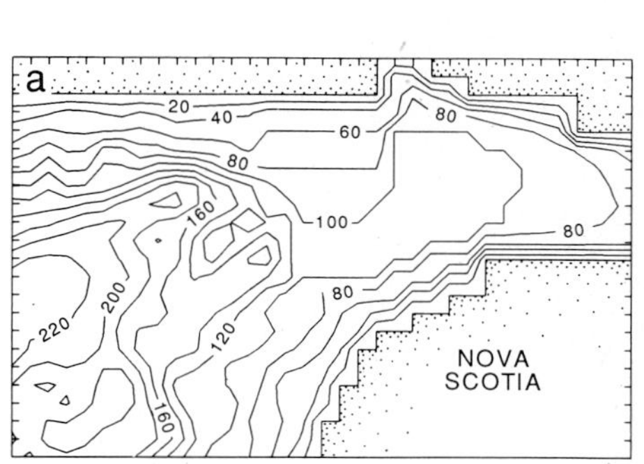

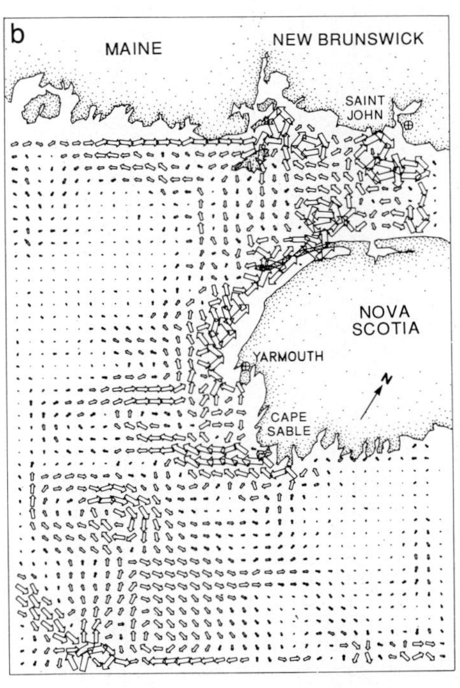

FIGURE 4. The model with smooth bottom topography and coastal geometry near the entrance to the Bay of Fundy. (a) Depth of water (m) for the north-western part of the model which shows the modified bottom topography and coastal geometry in the north-western part of the model. (b) The horizontal distribution of depth-averaged residual current. The velocity and distance scales are the same as those in Fig. 2.

The depth-averaged residual currents computed using the smoothed topography and coastal geometry are shown in Fig. 4b. By comparing Fig. 2 and Fig. 4b, we can see that the strong irregular flow at the entrance in the former computation has now disappeared. Weak south-eastward flow now occurs at the western side of the entrance, and north-westward flow at the eastern side. However, the most significant result from the comparison of the two models (Figs. 2 and 4b) is that the residual currents in the Cape Sable area (about south of Yarmouth, or south of 44° N) are basically unchanged. This result strongly indicates that the residual currents in the Cape Sable area are generated locally, and not affected significantly by distant forcings. This localized effect is examined in more detail in the next subsection.

It has been shown by Tee *et al*. (1987, 1988) that the observed tidal and residual currents near Cape Sable were reproduced reasonably well by the numerical model. Thus, although the effects of horizontal resolution on the tidal and residual currents are large near the entrance to the Bay of Fundy, they are probably small near Cape Sable.

III.3 Localized features and dynamics of circulation

Three numerical experiments have been carried out to show that the circulations around Browns Bank and near Cape Sable are associated with local topographic features. In the first experiment, the depth variation between zero and 100 m around Cape Sable is kepth, and the rest of the area is set at 100 m (Fig. 5a). Figure 5b shows the computed depth-averaged residual current. The anticyclonic circulation around Browns Bank, and other strong residual currents near the southern corner of the model's domain are now filtered out. The major circulations near Cape Sable are the westward jet near the Cape and the weak return flow at the northern and southern regions. By comparing Figs. 5b and 3, we can see that although the smooth return flow in the southern region in this latter computation is different from more complex circulation computed with real topographic variation, the westward

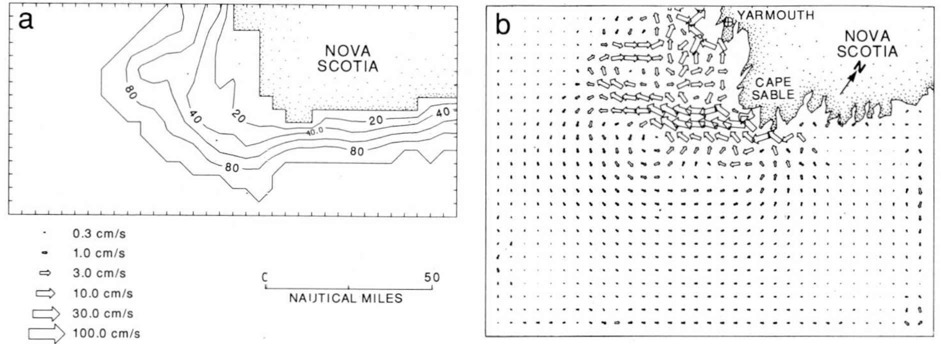

FIGURE 5. The model which keeps the depth variation near Cape Sable, but sets the constant depth of 100 m in the rest of the area. (a) Depth of water (m). (b) The depth-averaged residual current. Only the variables in the vicinity of Cape Sable and Browns Bank (south-eastern part of the model's domain) are shown. The distance between two tick marks in the horizontal and vertical axes is 7.047 km.

jet and the return flow in the northern regions are basically unchanged. This result strongly suggests that the main feature of the circulation near the Cape are associated with the local topography. Part of the dynamics associated with the jet may be similar to those associated with anticyclonic circulation around submarine banks.(Huthnance, 1973; Loder, 1980; Tee, 1985). As fluid columns, whose flow pattern follows closely with the convex shape of Cape Sable, pass over shallow areas near the Cape, they are squeezed and the negative (anticyclonic) vorticity is generated. This generation of negative vorticity on the shallow water occurs whichever direction the tide is flowing, so that the net effect over a tidal cycle is for negative vorticity to be found over the shallow area. This results in the anticyclonic circulation in the shallow area near the Cape, indicated by the westward jet and the clockwise return flow in the shallow area. However, these simple dynamics cannot explain the large portion of the flow that crosses the isobaths from shallow to deep water near the end of the jet, and the subsequent return flow in the southern region. The understanding of these return flows in the southern region requires some further studies.

The second numerical experiment is carried out by setting the depth near the Cape to be a constant of 100 m (Fig. 6a). The depth-averaged residual current is shown in Fig. 6b. We can see that the westward jet is filtered out in this case. However, the remaining features of the residual circulation such as those around Browns and Georges Banks, on the Northeast Channel and the south-eastern part of the model's domain remain more or less unchanged (compare Figs. 6b and 3). This result indicates that the topographic feature near the Cape has little effect on the circulation away from the Cape.

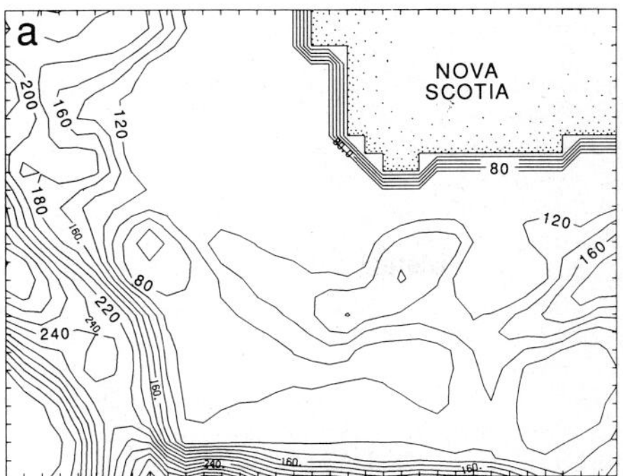

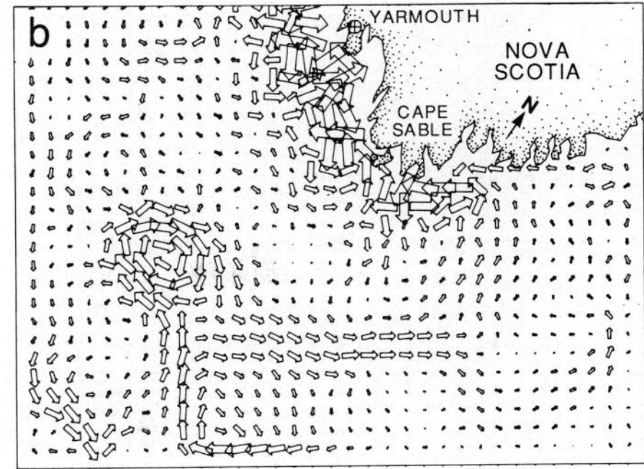

FIGURE 6. The model which sets the constant depth of 100 m near Cape Sable, but keeps the depth variation in the rest of the area. (a) Depth of water (m). (b) The depth-averaged residual current. Only the variables in the vicinity of Cape Sable and Browns Bank are shown. The velocity and distance scales are the same as those in Fig. 5.

In the third experiment, both the depths near Cape Sable and on Georges Bank are set at 100 m (Fig. 7a). The strong eastward jet on the side of Georges Bank, as well as the westward jet near the Cape are now filtered out (Fig. 7b). By comparing Figs. 7b and 6b, we can see that the anticyclonic circulation around Browns Bank, the eastward flow east of the bank, and the inflow along the Northeast Channel remain basically unchanged.

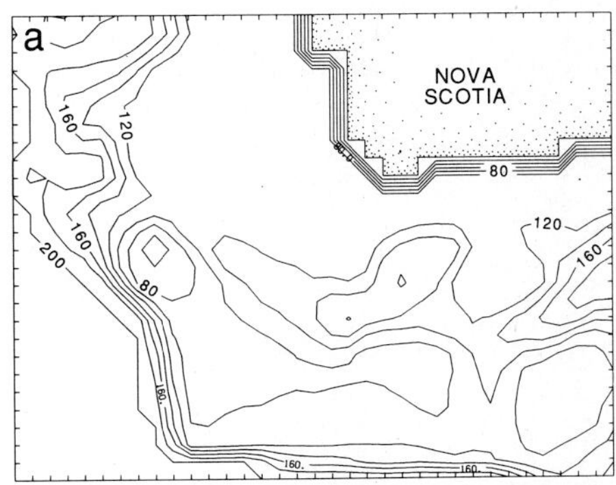

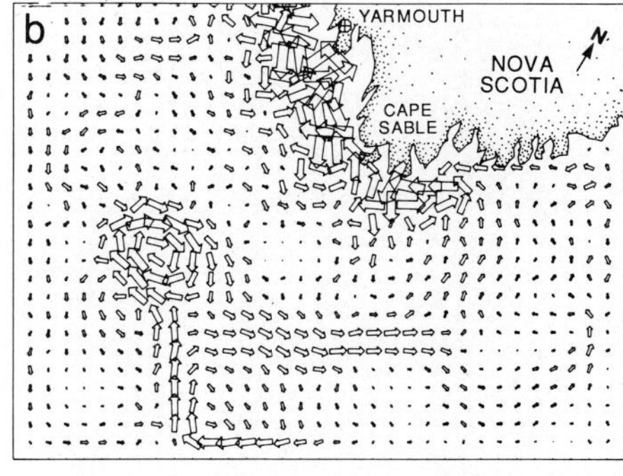

FIGURE 7. The model which sets the constant depth of 100 m near Cape Sable and at Georges Bank. (a) Depth of water (m). (b) The depth-averaged residual current. Only the variables in the vicinity of Cape Sable and Browns Bank are shown. The velocity and distance scales are the same as those in Fig. 5.

The anticyclonic circulation around Browns Bank is induced by topographic rectification of tidal current, the mechanism described earlier for the tidal current passing over shallow water near Cape Sable. The anticyclonic circulation includes eastward flow east of Browns Bank, the inflow in the Northeast Channel, and the return flow at the south-eastern boundary may be associated with another shallow water region, the eastern extension of Browns Bank and located southeast of Cape Sable (Fig. 7b).

In summary, the main residual circulations in the Cape Sable area are found to be associated closely with local topographic features. It is thus not surprising that the circulation in the limited area model is very insensitive to the uncertainty of the open boundary condition.

IV. Residual Upwelling and Downwelling

The continental shelf off southwest Nova Scotia is among the most biologically productive areas on the eastern coast of North America. From the observed shoreward movement of seabed drifter and the low temperature near Cape Sable, Lauzier (1967) suggested that this was an area of upwelling. Garrett and Loucks (1976) suggested that the upwelling might be driven by centrifugal forces associated with the strong tidal currents past the convex coastline. The estimated onshore current near the bottom was 2 cm/s. In a simple dynamical model including the centrifugal forces of the tidal currents, Smith (1983) found that the centrifugal effects produced an upwelling (bottom) current of only 0.5 cm/s. He suggested that the upwelling was induced by along-isobath density variations. However, such density variation in the summer, which is the primary driving force is the diagnostic model for that season, cannot reproduce the observed mean current in the area. In the above two studies, the driving force for the tidally induced residual upwelling includes only the centrifugal term, and not the other terms associated with the complex topography in the area. We shall show in the following that the three-dimensional tidal model which includes realistic topography and variable tidal forcing in the Cape Sable area, produces complex structures of upwelling and downwelling very different from those computed previously. The upwelling near Cape Sable will be shown to be fairly strong.

From the continuity equation of residual current,

$$\nabla\cdot\mathbf{u}_2 + \frac{\partial w_2}{\partial z} = 0 \quad , \tag{11}$$

the vertical velocity at depth z' can be computed from

$$w_2 = w_{20} + \int_{z'}^{0} \nabla\cdot\mathbf{u}_2 dz \tag{12}$$

in which w_{20} is the value of w_2 at $z = 0$. w_{20} can be computed from

$$w_{20} \simeq w_{2_s} - \overline{\frac{\partial w_1}{\partial z}\zeta_1\Big|_{z=0}} \tag{13}$$

where w_{2_s} is the value of w_2 at $z = \zeta_1$. Because

$$w_{2_s} = \overline{\mathbf{u}_{1_s}\cdot\nabla\zeta_1} \quad \text{and} \quad \frac{\partial w_1}{\partial z}\Big|_{z=0} = -\nabla\cdot\mathbf{u}_{1_s} \tag{14}$$

where u_{1_s} is the value of u_1 at surface $(z = 0)$, w_{20} can be computed from

$$w_{20} = \nabla\cdot\overline{\mathbf{u}_{1_s}\zeta_1} \tag{15}$$

From Eqs.(12) and (15), w_2 in the water column can be computed.

Figure 8 shows the spatial variation of w_2 at 30 m below the surface. Upwelling is denoted in the figure by solid curves, and downwelling by dashed curves. Near Cape Sable, upwelling occurs on the eastern side and downwelling on the western side. If upwelling was induced only by centrifugal force, it would exist at all the area around Cape Sable. This result indicates that the models used previously by Garrett and Loucks (1976), and Smith (1983) to predict tidally induced upwelling in the Cape Sable area are too simple.

In Browns Bank, upwelling occurs on north-western and south-eastern sides, and downwelling on north-eastern and south-western sides. The upwelling and downwelling on the western side are significantly higher than those on the eastern side. There are also some significant upwelling and downwelling areas near the eastern tip of

Georges Bank and the Northeast Channel. The reliability of these predicted areas requires more study because they are located very close to the open boundary.

The value of w_2 is small near the surface, and increases generally to a maxiumum value in the middle portion of the water column. Examples of the vertical variation of w_2 at the centre of upwelling and downwelling areas near Cape Sable (the locations are indicated by solid square and triangle in Fig. 8) are shown in Fig. 9.

The maximum values of w_2 are 1.32×10^{-4} m/s at 38 m for the upwelling region, and 0.99×10^{-4} m/s at 33 m for the downwelling region. These vertical transports of water mass are comparable to wind-induced upwelling in a continental shelf. For example, the tidally induced upwelling transport per unit along-shore distance at the eastern side of Cape Sable is ≈ 1 m²/s (average $w_2 \approx 3 - 4 \times 10^{-5}$ m/s, width of upwelling zone $\approx$30 km; Fig. 8). For a north-eastward along-shore wind stress, τ_w, the upwelling transport per unit along-shore distance is $\tau_w/\rho f$, where ρ is density of water and f is the Coriolis parameter. To produce an equivalent amount of upwelling transport induced by tidal rectification ($\approx$1 m²/s), we require a significant value ($\approx$7 m/s) of north-eastward wind speed.

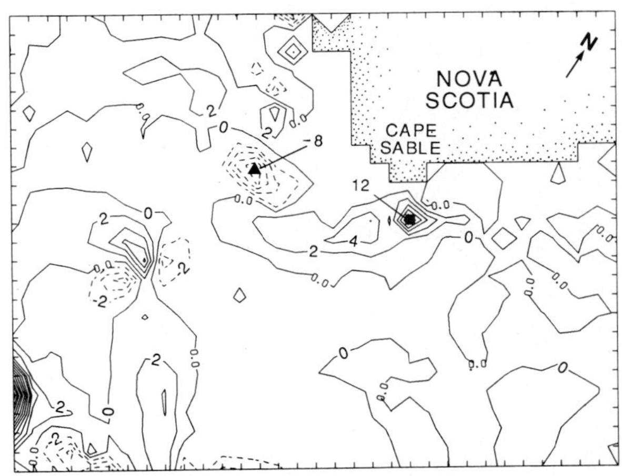

FIGURE 8. The tidally-induced residual vertical current w_2 (10^{-5} m/s) at 30 m below the surface. Upwelling is indicated by solid curves, and downwelling by dashed curves. The contour interval is 2×10^{-5} m/s. The distance between two tick marks in the horizontal and vertical axes is 7.047 km. Only the variables in the vicinity of Cape Sable and Browns Bank are shown.

In a two-dimensional case (on-off shore section), the upwelling (downwelling) is frequently associated with a two-layer circulation which has onshore (offshore) flow in the lower portion of water column and offshore (onshore) flow in the upper portion. This two-layer circulation provides efficient transport of cool, saline and nutrient rich water from lower to surface layers. However, in a three-dimensional case carried out in this study, this two-layer circulation associated with a vertical motion of water particle is not observed. This is because the upward or downward motion of water mass generated from convergence or divergence of the cross-isobath flow are compensated by the divergence or convergence of the along-isobath current. Thus, the water particles in the lower layer of the upwelling region, although they are advected from deep water to shallow water across the isobath (upwelling in the eastern side of Cape Sable, Fig. 8) they may remain in the lower layer and advected by the along-isobath flow away from the upwelling region. The upward transfer of cool and saline water in the three-dimensional case thus may not be as efficient as the two-dimensional case. However, the vertical transfer can be very efficient in the three-dimensional case if the areas where the deep water is advected to are sufficiently shallow and have strong tidal currents. This is because, in such areas, the vertical mixing from the turbulence generated by tidal current at the bottom is very significant. Near Cape Sable, the strong vertical mixing in the shallow water region is indicated by the formation of well-mixed water above 65 m isobath (Garrett et al., 1978; Loder and Greenberg, 1986). These well-mixed regions cover most of the shallow water area around the Cape and on top of Browns Bank. From Figs. 2 and 8, we find that the upwelling is associated with the water movement from deep to shallow areas, and the downwelling from shallow to deep areas. A clear example is the correspondence between upwelling and the north-westward residual flow on the eastern side of Cape Sable. Thus the tidally induced upwelling predicted on the eastern side of Cape Sable, and on the north-western and south-eastern sides of Browns Bank can be quite important. The observed low temperature around Cape Sable in mid-summer (6-7°C) (Bailey et al., 1954; Smith, 1983; Loder and Greenberg, 1986) which is much lower than the temperature induced by heat flux alone in the well-mixed area (≈15°C) is probably induced by the upwelling in the area.

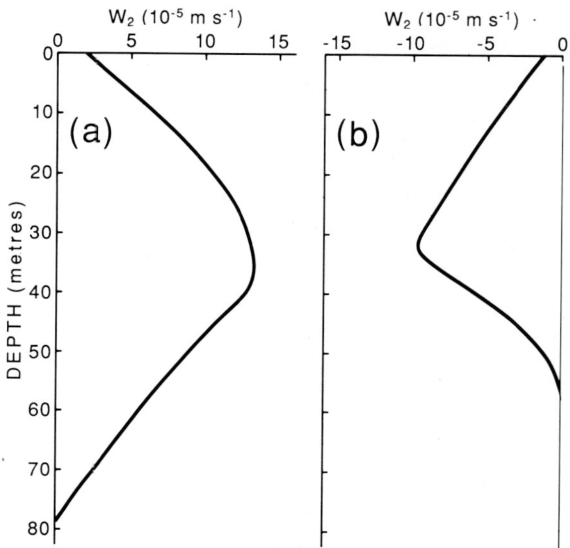

FIGURE 9. The vertical variation of w_2 at the centre of upwelling and downwelling areas near Cape Sable. The locations of the centre are shown by solid square and triangle in Fig. 8.

Because strong tidal currents have been observed in many coastal areas, the mechanism of producing nutrient rich water by coupling residual upwelling and tidal mixing may be also important in these areas. More studies of tidally induced residual upwelling and downwelling should be carried out in the future.

V. Conclusion

A three-dimensional tidal model is used to study the tidally induced horizontal and vertical residual circulation in the Cape Sable area, southwest of Nova Scotia (Fig. 1). The modeled domain covers the eastern part of the Gulf of Maine. Two aspects of the model's result are discussed here. The first is the reliability of the limited area model, and the second is the tidally induced residual upwelling and downwelling.

In the first aspect, the limited-area model is found to be very insensitive to the specified residual current at the open boundary. This is because strong residual currents in the Cape Sable area are generated locally and associated closely with local topographic features.

In the second aspect, the model predicts several areas of strong residual upwelling and downwelling. Significant upwelling areas include the eastern side of Cape Sable, and the north-western and south-eastern sides of Browns Bank (Fig. 8).

The strong downwelling areas includes the western side of Cape Sable, and the north-eastern and south-western sides of Browns Bank. Maximum values of upwelling and downwelling are of the order of 10^{-4} m/s. By coupling the upwelling and strong tidal mixing in shallow water areas, this tidally induced vertical transport of water mass can be very efficient for supplying cold, saline and nutrient rich water from deep water to surface. The observed low temperature near Cape Sable is probably induced by this mechanism. The same mechanism of producing nutrient rich water can be important in many coastal areas.

VI. References

Bailey, W. B., W. Templeman and R. P. Hunt, 1954: The horizontal distribution of temperature and salinity off the Canadian Atlantic Coast, Fish. Res. Bd. Can., MS Rept. Serv. (Biol.) No. 584, 21 pp.

Garrett, C. J. R. and R. H. Loucks, 1976: Upwelling along the Yarmouth shore of Nova Scotia, J. Fish. Res. Bd. Can., **33**, 116-117.

Garrett, C. J. R., J. R. Keeley and D. A. Greenberg, 1978: Tidal mixing versus thermal stratification in the Bay of Fundy and Gulf of Maine, Atmos.-Ocean, **16**, 403-423.

Greenberg, D. A., 1983: Modelling the mean barotropic circulation in the Bay of Fundy and Gulf of Maine. J. Phys. Oceanogr., **13**, 886-904.

Huthnance, J. M., 1973: Tidal current asymmetries over Norfolk sandbanks. Estuarine Coast. Mar. Sci., **1**, 89-99.

Ianniello, J. P., 1977: Tidally-induced residual current in estuaries of constant breadth and depth. J. Mar. Res., **35**, 755-786.

Ianniello, J. P., 1979: Tidally-induced residual current in estuaries of variable breadth and depth. J. Phys. Oceanogr., **9**, 962-974.

Isaji, T., and M. L. Spaulding, 1984: A model of the tidally-induced residual circulation in the Gulf of Maine and Georges Bank. J. Phys. Oceanogr., **14**, 1119-1126.

Johns, B., 1970: On the determination of the tidal structure and residual current system in a narrow channel. Geophys. J. Royal Astron. Soc., **20**, 159-175.

Johns, B., and P. Dyke, 1971: On the determination of the structure of an offshore tidal stream. Geophys J. Royal Astron. Soc., **23**, 287-297.

Johns, and P. Dyke, 1972: The structure of the residual flow in an offshore tidal stream, J. Phys. Oceanogr., **2**, 73-79.

Lauzier, L. M., 1967: Bottom residual drift on the Continental shelf area of the Canadian Atlantic Coast. J. Fish. Res. Bd. Can., **24**, 1845-1858.

Loder, J. W., 1980: Topographic rectification of tidal currents on the sides of Georges Bank. J. Phys. Oceanogr., **10**, 1399-1416.

Loder, J. W., and D.G. Wright, 1985: Tidal rectification and frontal circulation on the sides of Georges Bank. J. Mar. Res., **43**, 581-604.

Loder, J.W., and D.A. Greenberg, 1986: Predicted position of tidal fronts in the Gulf of Maine region. Cont. Shelf Res., **6**, 397-414.

Oonishi, Y., 1978: A numerical study on the tidal residual flow-vertical motion induced by tidal current. J. Oceanogr. Soc. Japan, **34**, 140-159.

Pingree, R. D., and L. Maddock, 1977: Tidal residual in the English Channel. J. Mar. Biol. Assoc. U.K., **57**, 339-354.

Smith, P. C., 1983: The mean and seasonal circulation off southwest Nova Scotia. J. Phys. Oceanogr., **13**, 1034-1054.

Tee, K. T., 1976: Tide-induced residual current, a 2-D non-linear numerical tidal model. J. Mar. Res., **34**, 603-628.

Tee, K. T., 1977: Tide-induced residual current-verification of a numerical model. J. Phys. Oceanogr., **7**, 396-402.

Tee, K. T., 1979: The structure of three-dimensional tide-generating currents. Part I: Oscillating currents. J. Phys. Oceanogr., **9**, 930-944.

Tee, K. T., 1980: The structure of three-dimensional tide-induced current. Part II: Residual current. J. Phys. Oceanogr., **10**, 2035-2057.

Tee, K. T., 1985: Depth-dependent studies of tidally-induced residual currents on the sides of Georges Bank. J. Phys. Oceanogr., **15**, 1818-1846.

Tee, K. T., 1987: Simple models to simulate three-dimensional tidal and residual currents. In: Three-dimensional coastal ocean models, edited by N.S. Heaps, Coastal and Estuarine Science, **4**, 125-147.

Tee, K. T., P. C. Smith, and D. Lefaivre, 1987: Modelling and observations of the residual current off southwest Nova Scotia, Proceedings of the Eighteenth International Liege Colloquium on Ocean Hydrodynamics, Edited by J.C.J. Nihoul and B.M. Jamart, Elsevier Oceanography Series, **45**, 455-470.

Tee, K. T., P. C. Smith, and D. Lefaivre, 1988: Estimation and verification of tidally induced residual currents, J. Phys., Oceanogr., **18**, 1415-1434.

Wright, D. G. and J. W. Loder, 1985: A depth-dependent study of the topographic rectification of tidal current. Geophys. Astrophys. Fluid Dyn., **31**, 169-220.

Residual Currents in the Western Dutch Wadden Sea

H. Ridderinkhof[1] and J.T.T. Zimmerman[1 & 2]

[1]Netherlands Institute for Sea Research
P.O. Box 59, 1790 AB Texel
The Netherlands

[2]Institute of Meteorology and Oceanography
Buys Ballot Laboratory
University Utrecht
The Netherlands

ABSTRACT

It is shown by means of a numerical model of the western Dutch Wadden Sea that the residual current velocity field cocnsists of a throughflow from one tidal basin to another and a field of residual eddies. The first can be explained by the residual momentum balance which gives the driving force of the throughflow in terms of the tidal stress. The latter is more easily understood from the vorticity balance in which bottom frictional generation and damping of vorticity, together with vortex stretching, explain most of the observed features. Near the inlets the rediual eddy pattern has the shape of a highly distorted vorticity quadrupole.

1. Introduction

The western Dutch Wadden Sea is a typical shallow tidal area, characterized by a complex morphology shaped by strong tidal currents. It consists of various basins each of which is drained by an inlet which connects the area with the North Sea (figure 1). The vertical amplitude of the semidiurnal tide ranges from 1.34m for the southernmost inlet (the Marsdiep) to about 1.80m for the northernmost considered here (the Borndiep). Tidal currents easily reach an amplitude in excess of 1m/s in the deeper channels with a strong spatial variability due to the complicated morphology. It is particularly the latter aspect that has been held responsible for the occurrence of complicated patterns of residual currents in the area. These were first discussed by Zimmerman (1976) in an analysis of current measurements in order to explain the strong mixing processes in the area concerned. More recently Ridderinkhof (1988, 1989a, 1989b) in a number of studies has investigated the residual current pattern in the Wadden Sea more closely by means of a fine-grid numerical model in order to incorporate its properties in a transport model for describing water exchange in the area. In fact the the present paper is just an extended abstract of Ridderinkhof's results. For most of the details the reader is referred to the original work which also contains a more extensive list of references.

In a companion paper (Ridderinkhof and Zimmerman, 1989) the Lagrangean characteristics of the tidal currents in the Wadden Sea are discussed, particularly with respect to the mixing processes. Here we exclusively deal with the tidally induced residual Eulerian field. We first give a brief description of the numerical model and its results for the residual circulation pattern. Then we show that the latter can be subdivided in basically two types of flow: a throughflow from one inlet to another driven by that part of the tidal stress that is due to tidal-mean dynamic pressure head between the inlets, and secondly the residual eddy field ocurring all over the area, particularly near the inlets where it is in the form of

Coastal and Estuarine Studies, Vol. 38
R.T. Cheng (Ed.)
Residual Currents and Long-term Transport

a strongly deformed residual vorticity quadrupole.

II. The Numerical Model and Its Residual Current Pattern

The numerical model used is an adapted version of the WAQUA system (Leendertse, 1967; Stelling, 1984) which solves the shallow water equations employing a alternating- direction implicit method on a staggered grid. These equations read

$$\frac{\partial u}{\partial t} + u\frac{\partial u}{\partial x} + v\frac{\partial u}{\partial y} - fv = -g\frac{\partial \zeta}{\partial x} - g\frac{Cu\sqrt{(u^2+v^2)}}{(H+\zeta)^{4/3}} + \nu\nabla^2 u \tag{1}$$

$$\frac{\partial u}{\partial t} + u\frac{\partial u}{\partial x} + v\frac{\partial v}{\partial y} + fu = -g\frac{\partial \zeta}{\partial y} - g\frac{Cv\sqrt{(u^2+v^2)}}{(H+\zeta)^{4/3}} + \nu\nabla^2 v \tag{2}$$

$$\frac{\partial \zeta}{\partial t} + \frac{\partial (H+\zeta)u}{\partial x} + \frac{\partial (H+\zeta)v}{\partial y} = 0 \tag{3}$$

in which u and v are the horizontal velocity components, z the variation of the waterlevel, H the depth, f the Coriolis parameter, g the acceleration of gravity, ν the horizontal eddy viscosity and C the bottom friction coefficient (Manning's factor). The grid size of the model is 500m, sufficiently fine to reproduce most of the complex morphological structures.

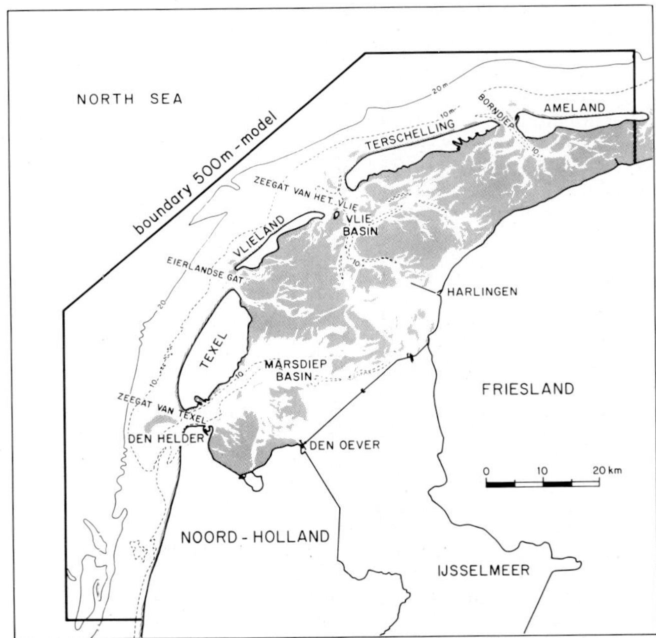

FIGURE 1. Map showing the boundaries and some isobaths (10, 20 m) of the 500 m gridsize numerical model of the western Dutch Wadden Sea. Shaded areas represent tidal flats. The drawn line in the North Sea gives the open boundary of the model at which the tidal elevations from a larger gridsize (3200 m) model of the North Sea are prescribed.

The model is driven by imposing on its open boundary, located inside the adjacent North Sea (see figure 1), the tidal elevations as they emerge from a large scale 3200m grid-size model of the North Sea area north of the Wadden Sea. The present model has its open boundary sufficiently far away from the inlets in order to reproduce the tidal current patterns at the seaward side of the inlets with sufficient precision. No noslip condition at the coastal perimeter was applied, which excludes the formation of frictional boundary layers and the associated rectification processes. However the sloping bottom near the coast will in most cases be a substitute for this process as it will provide the necessary vorticity production by means of differential bottom friction. The coefficient of horizontal viscosity, n, which was set at a constant value of 7 m^2/s in all runs discussed here, therefore does not play an important role in the dynamics.

The model was calibrated by adjusting the bottom friction coefficient such that at specific places the vertical tide and the tidal discharge through the inlets agreed with observations. Of course this does not mean that also the spatial structure of the currents is properly reproduced but it will be shown that the residual eddy field comes very near to the observational evidence. The residual field was obtained by averaging dynamical variables over the tidal period, which gives: the mean water level, $<\zeta>$, the (Eulerian) mean velocity $<u,v>$ and the mean transport $<U,V>=<(u,v)(H+\zeta)>$, where the brackets denote the time-averaging operator. The latter quantity defines a residual streamfunction by

$$<U> = \frac{\partial\psi}{\partial y}, \qquad <V> = -\frac{\partial\psi}{\partial x} \quad . \tag{4}$$

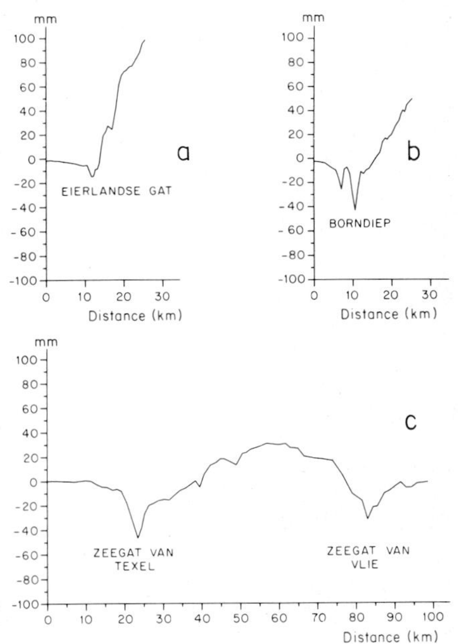

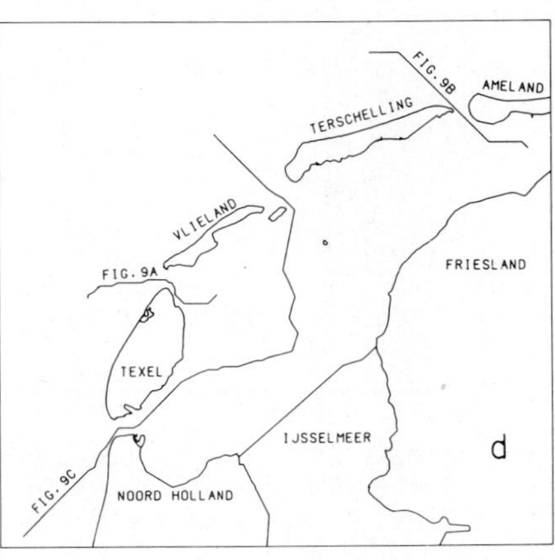

FIGURE 2. Residual water elevations along the axes of the main basin channels (mm) as a function of the distance from the seaward end of the inflow region. The locations of these axes are indicated in (d), (a) Eierlandse Gat, (b) Borndiep, (c) Marsdiep and Vlie.

The residual waterlevel is shown in figure 2 for some specific stretches in the area. A common feature of all these stretches is the sharp dip of the residual sea level in the inlets and a tidal set-up of the waterlevel in the direction of the tidal watersheds between drainage basins, or the coast. The effects are typically of the order of several centimeters.

The residual streamfunction is shown in figure 3 for the area inside the inlets. For the inlet area of the Vlie figure 4 gives a detailed view of the streamline pattern. Figure 3 clearly shows that the whole pattern can be subdivided in two parts. First there appears to exist a troughflow in the area from the Vlie basin to the Marsdiep basin, crossing the watershed between the two. The magnitude of this residual transport is about 900 m^3/s, which therefore is an important advective process for transport of dissolved and suspended matter in the area.

Hardly any cross-watershed transport exists between other inlets. This is due to the fact that the watershed between the Marsdiep basin and the Vlie basin is relatively deep, large parts not falling dry at low tide, which in turn is due to a morphological delay in adaptation to closure of the Zuiderzee (Ysselmeer).

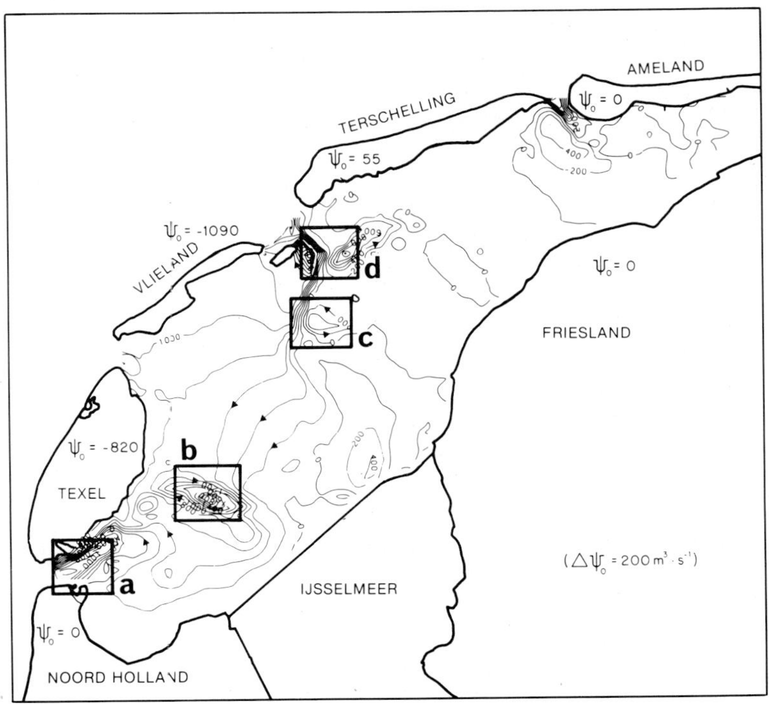

FIGURE 3. The residual streamfunction of the inner Wadden Sea. Intervals between streamlines are 200 m^3/s. The drawn rectangles give the areas of residual current observations of figure 5.

Superimposed on this unidirectional throughflow are smaller scale residual eddies. The weaker of these are obscured by the throughflow, but the stronger are quite outstanding, particularly near the inlets (figure 4), but also at some positions in the inner parts of the Wadden Sea. It is interesting to observe the shape and rotation sense of these eddies with the inferred patterns of Zimmerman (1976) (figure 5). The latter were constructed by using the scarce tidal current measurements in the area with some aid of the morphology in drawing the streamlines. Surprisingly the agreement between these inferences and the model is very good, particularly for the rotation sense of the eddies, the major difference being that Zimmerman (1976) poses at some places two eddies of the same sign whereas the model predicts one. Also the residual velocities of the eddies agree with the observations, their usual amplitude being between .10 and .15 m/s. The residual flow pattern around the Vlie inlet (figure 4) shows a structure that is typical of all inlets of the Wadden Sea: a highly distorted vorticity quadrupole with on the seaward side a large gyre to the northeast and a smaller counterrotating one to the southwest. Such quadrupoles are to be expected around inlets (Zimmerman, 1981) but their strong distortion which, as shall be shown in Ridderinkhof and Zimmerman, (1989), is systematic and must bear a connection to the morphology of the outer delta, is as yet unexplained. All in all the residual flow pattern seems to consist of two components of different origin, the throughflow and the eddies, each of which has its own dynamics. That of the former is most conveniently discussed in terms of a one-dimensional analytical model of two connected basins, which is the subject of the next chapter. The latter is more easily discussed in terms of tidal vorticity balance, which is the subject of Ridderinkhof and Zimmerman, (1989).

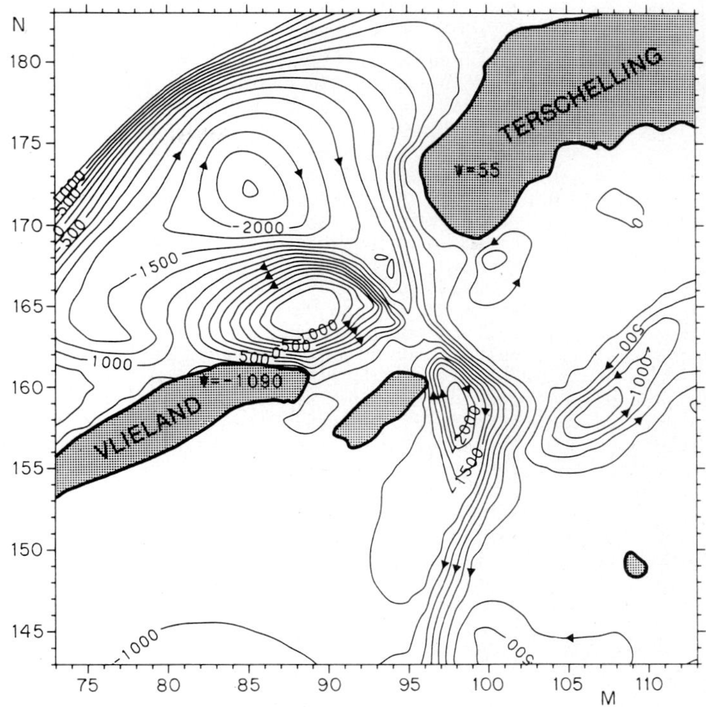

FIGURE 4. The residual streamfunction near the Vlie inlet (interval 250 m³/s).

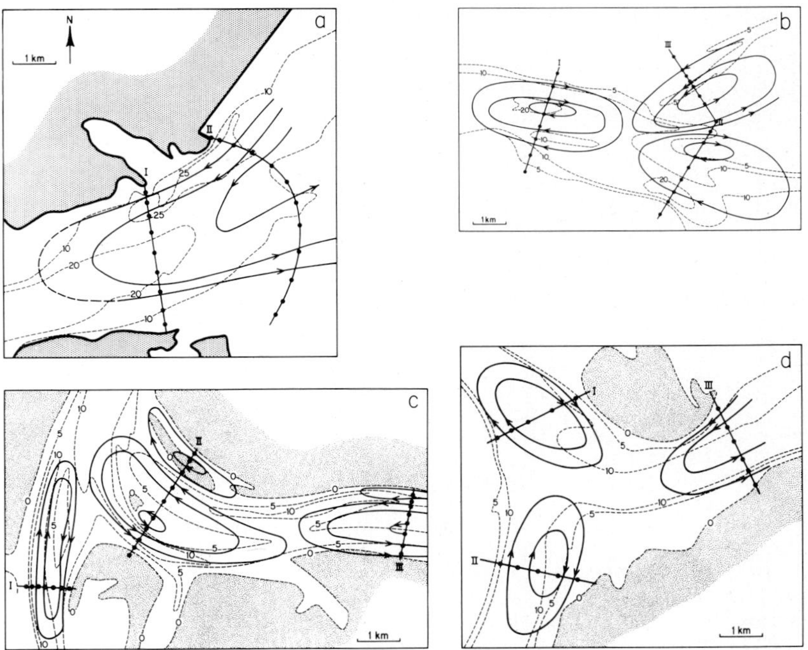

FIGURE 5. Observations of residual circulation in the areas marked by a - d in figure 3 (after Zimmerman, 1976).

III. The Throughflow

The existence of a tidal average water transport from one inlet to another in a tidal inland sea connected by two inlets to the continental shelf, as we may regard the combined Vlie and Marsdiep tidal basins in the western Dutch Wadden Sea, unveils an asymmetry in the forcing of the tides in the area. The asymmetry can be either of external or of internal origin. The first is the case if either the sea level amplitudes at both inlets differ in magnitude or the amplitudes of the discharge. The latter, for equal amplitudes of the sealevel, is equivalent to a phase difference in the vertical tide of both inlets. Internal asymmetries are due to different response characteristics of the two connected tidal basins. The response characteristics will be different in the case the depth, the length or the friction coefficient are different. Ridderinkhof (1989a) has investigated all possibilities by extending an analysis of Van de Kreeke and Dean (1975) for the throughflow of a simple tidal channel with uniform depth, forced at both open ends by a prescribed fluctuation of the sealevel. Ridderinkhof's extension consists of 1) a simple solution to the open boundary problem for the tidal mean sea level at the inlets, 2) an allowance for depth and length differences of the two connected basins and 3) a slightly different parametrization of linearized bottom friction for the tidal and the tidal mean flow. His analysis of the open boundary problem for the mean sealevel in the inlets shows that to a good approximation the drop in mean sea level at an inlet is equal to the tidal mean Bernoulli term in the dynamic pressure gradient. That is, at an inlet we have approximately $\langle\zeta\rangle = -\langle U^2\rangle/(2gH^2)$ or $\langle\zeta\rangle/H$ = -Froude number. For U/H = 1 m/s this amounts to a drop of about 2.5 cm in the inlets of the Wadden Sea. This is indeed the order of magnitude of the dip observed in the plots of figure 2. These dips enter the equation for the tidal mean discharge as boundary conditions for $\langle\zeta\rangle$.

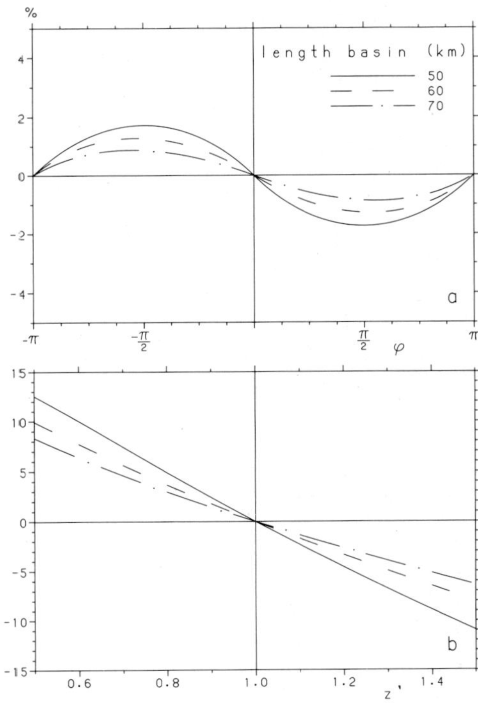

FIGURE 6. Ratio between residual discharge and tidal discharge amplitude at one of the inlets as a function of (a) phase differences between the vertical tide at both inlets and (b) of the amplitude ratio of the vertical tide.

In the absence of geometric differences between the two basins, in which case one deals with a single channel, both Van de Kreeke and Dean (1975) as Ridderinkhof (1989a) arrive at the following expression for the transport of the throughflow:

$$<U> = \lambda^{-1} \left[\left\{ cF(k,\lambda) - 2k^{-2} \right\} \left\{ Z^2 - 1 \right\} + c\, G(k,\lambda)\, Z \sin\phi \right] \tag{5}$$

Here $F(k,\lambda)$ and $G(k,\lambda)$ are complicated functions connected with the response characteristics of the channel, were k is the tidal wave length scaled by the basin length and λ a dimensionless friction parameter. Z is the ratio of the amplitudes of the vertical tide in both inlets and ϕ the phase difference of the vertical tide between both inlets. The numerical constant, c, equals 3/4 if allowance is made for externally forced residual waterlevel displacements at the inlets (Ridderinkhof, 1989a) or 1 if such an allowance is not made (Van de Kreeke and Dean, 1975). For the Vlie and Marsdiep inlets we have $Z = 1.23$ and $\phi = -.63$. The expression clearly shows that the throughflow vanishes only in the case of equal amplitudes at both inlets and either a forcing that is exactly in phase or in antiphase at both inlets. That is, the forcing must be exactly symmetric or antisymmetric as a function of time. Neither is the case in the Wadden Sea. Figure 6 shows the residual discharge relative to the tidal discharge at one of the inlets for varying ϕ and Z. The above given values for Z and ϕ as they apply to the western Dutch Wadden Sea give +1% and -3% for the effects of ϕ and Z respectively, thus a dominance of the effect of differences in vertical amplitude at the inlets. As a negative sign here means a tranport directed from the Vlie inlet to the Marsdiep inlet, this is the correct direction compared to the prototype and also the right order of magnitude although too weak compared to the value observed in the numerical model. A further analysis of the internal asymmetries of the system by Ridderinkhof (1989a) has shown that the shallower depth and shorter length of the Vlie basin may at most add another -1% to the throughflow, thus a slight reinforcement of the effect due to the external asymmetries. The primary factor that drives the residual discharge from inlet to inlet therefore appears to be the difference in vertical amplitude.

IV. Residual Eddies and the Vorticity Balance

The occurrence of residual eddies in a tidal area, such as those shown in figures 3 and 4, implies the existence of areas which experience a net gain or loss of relative vorticity over the tidal cycle (Zimmerman, 1981). The generation of these eddies can therefore best be studied by means of the vorticity balance. The relevant vorticity equation has been discussed extensively by Zimmerman (1978, 1980) and Robinson (1981, 1983). It is obtained by taking the curl of the shallow water equations of motion, Eqs.(1 and 2). Then we have:

$$\underset{(a)}{\frac{\partial\omega}{\partial t}} + u\frac{\partial\omega}{\partial x} \underset{(b)}{+} v\frac{\partial\omega}{\partial y} + \omega\left(\frac{\partial y}{\partial x} \underset{(c)}{+} \frac{\partial u}{\partial y}\right) = -f\left(\frac{\partial u}{\partial x} \underset{(d)}{+} \frac{\partial v}{\partial y}\right)$$

$$+ \frac{C}{(H+\zeta)^{4/3}} \underset{(e)}{\left(u\frac{\partial\sqrt{(u^2+v^2)}}{\partial y} - v\frac{\partial\sqrt{(u^2+v^2)}}{\partial x}\right)} +$$

$$+ \underset{(f)}{C\sqrt{(u^2+v^2)}\left(u\frac{\partial(H+\zeta)^{-4/3}}{\partial y} - v\frac{\partial(H+\zeta)^{-4/3}}{\partial x}\right)} + \underset{(g)}{\frac{C\sqrt{(u^2+v^2)}}{(H+\zeta)^{4/3}}\omega} + \underset{(h)}{\nu\nabla^2\omega} \tag{6}$$

where $w = \frac{\partial v}{\partial x} - \frac{\partial u}{\partial y}$.

The terms denoted by (a) - (h) stand, respectively, for: the local rate of change of the vorticity (a), vorticity advection (b), relative vortex stretching (c), planetary vortex stretching (d), differential bottom friction by differences in tidal velocities (e), differential bottom friction by the slope of the bottom perpendicular to the direction of the tidal current (f), damping of vorticity by bottom friction (g) and lateral diffusion of vorticity (h). Ridderinkhof (1989b) discusses the balance Eq.(6) in detail for various areas in the western Dutch Wadden Sea.

Here we concentrate on the area of the Vlie inlet (figure 4) for which we show: a) the behavior of the different terms in Eq.(6) during the tidal cycle (figure 7) for two different positions at opposite sides of the Vlie channel, and b) the current velocity field together with the distribution of vorticity during the ebb and flood stage of the tide (figures 8a,b) as well as the residual current pattern and the residual vorticity distribution (figure 8c).

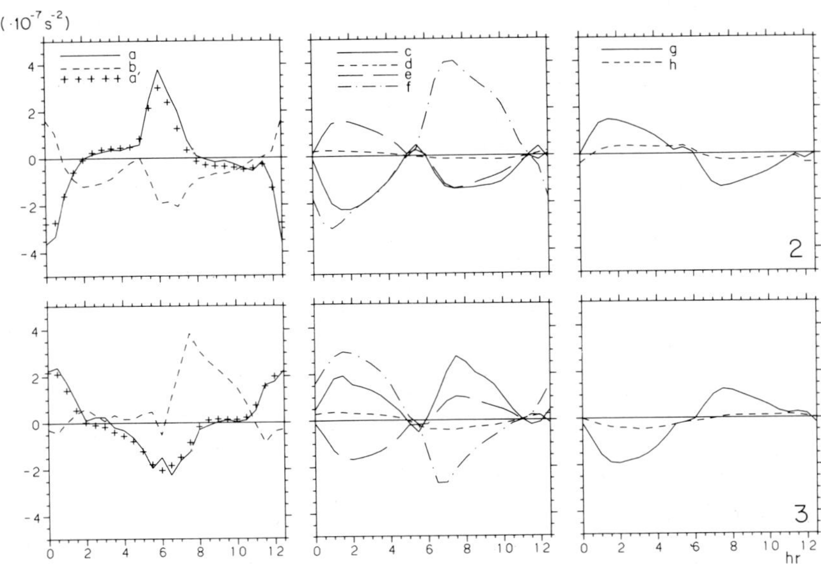

FIGURE 7. Variation with time of the various terms in Eq.(6) for two different positions at opposite sites of the Vlie channel, marked by 2 and 3 in figure 9b. The curves are grouped and labeled according to their labeling and position in Eq.(6). The curve labeled a' gives the sum of all terms (b) to (h).

In order to transfer vorticity from the time-varying field to the mean field it must be produced first. This is either achieved by planetary vortex stretching when the current flows in the direction of the bottom slope or by differential bottom friction when the current is along the isobaths. Once produced vorticity can be rectified by either advection or relative vortex stretching. The resulting residual vorticity must then be balanced by bottom friction. The curves labeled by d, e and f in the middle section of figures 7a,b show the variation over the tidal cycle of vorticity production by differential bottom friction due to velocity variations (d) and to the bottom slope (e), as well as by planetary vortex stretching (f). Evidently the latter is unimportant when compared to the other two processes. Differential bottom friction along the sides of the channel is by far the most important vorticity production mechanism. Looking at figures 8a,b this is reflected in the strong bends of tidal vorticity of opposite sign along both sides of the tidal channels. At maximum current speed tidal vorticity may reach values of the order of 10^{-3}/s, which is much larger than the planetary vorticity of 10^{-4}/s. This implies that relative vortex stretching will dominate planetary vortex stretching as can also be seen in figures 7a,b. The two components of differential bottom friction (d) and (e) are always of opposite sign. Actually term (d), differential bottom friction by velocity differences, is a negative feedback to term (e), differential bottom friction by the bottom slope, which is the primary vorticity source. Tidal vorticity produced that way now leads to residual vorticity by the two nonlinear terms in the vorticity equation, vorticity advection (b) and relative vortex stretching (c). Looking at the curves with those labels in figure 7a,b we see indeed that over a tidal cycle they have approximately the same sign during ebb and flood. Both are equally important as rectification processes. Particularly the vortex stretching is a mechanism that is typical for a shallow tidal sea with strong bottom slopes and adjacent tidal flats. The fact that vortex stretching is a production agency for residual vorticity can be understood from an asymmetry in the vorticity dynamics at the sides of the channels. Whereas differential bottom friction produces strong shear layers with vorticity of alternating sign during the tidal cycle, stretching changes sign too. During the flood, when differential bottom friction creates relative vorticity along the sides of the channel, the associated relative vortex columns are compressed when they move out of the main channel towards the flats, which thus is a negative feedback. During the ebb, however, the ebb current creates strong vorticity along the channel sides with an opposite sign compared to the flood but in principle of the same order of magnitude.

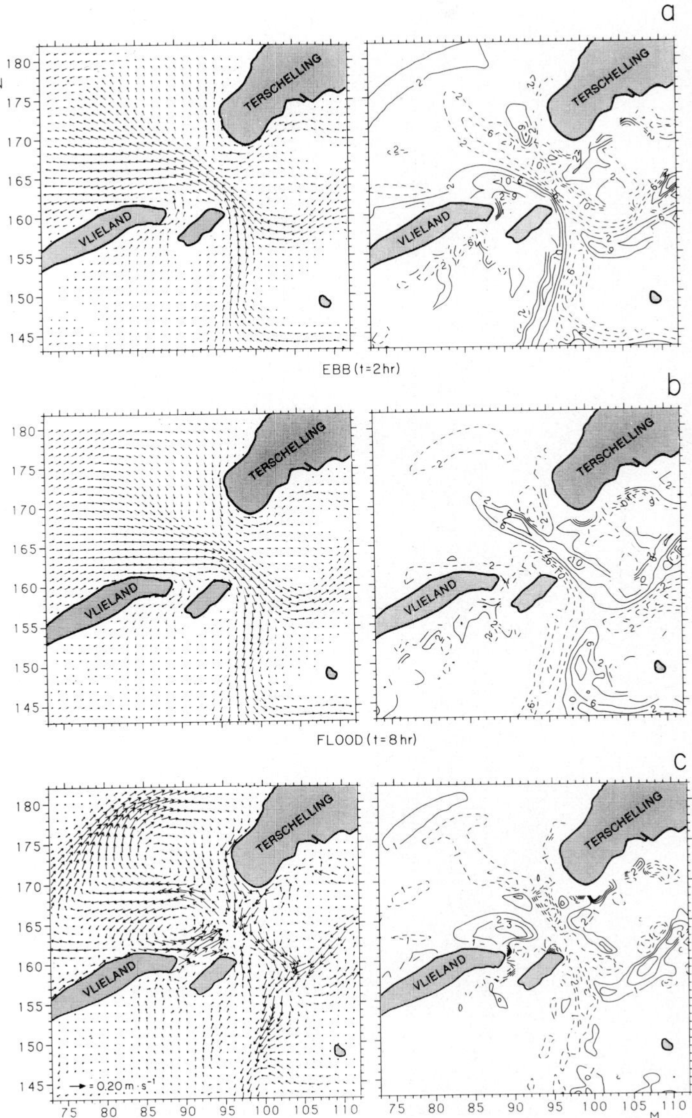

FIGURE 8. The distribution of tidal vorticity ($-10^{-4}s^{-1}$) during maximum ebb (a) and flood (b) currentspeed in the area of the Vlie inlet. Dashed lines denote negative vorticities, drawn lines denote positive vorticity. Figure 8c gives the residual circulation pattern as well as the residual vorticity distribution.

Now, concurrently the vortex columns are stretched when the water leaves the flats. This means a positive feedback. Hence averaged over the tidal cycle the vorticity produced during the ebb tide is stronger and thus there is a residual production mechanism. The same mechanism may operate at the center of the channel near the inlet since the bottom has a strong along-channel slope at that position. Finally the produced vorticity will be transported away from the production areas by the tidal current and be dumped at a place in the downstream direction. The advective rectification process is thus essentially a smearing agency which redistributes relative vorticity in contrast to relative vortex stretching which is a local rectification process acting on the position of large bottom slopes with cross-

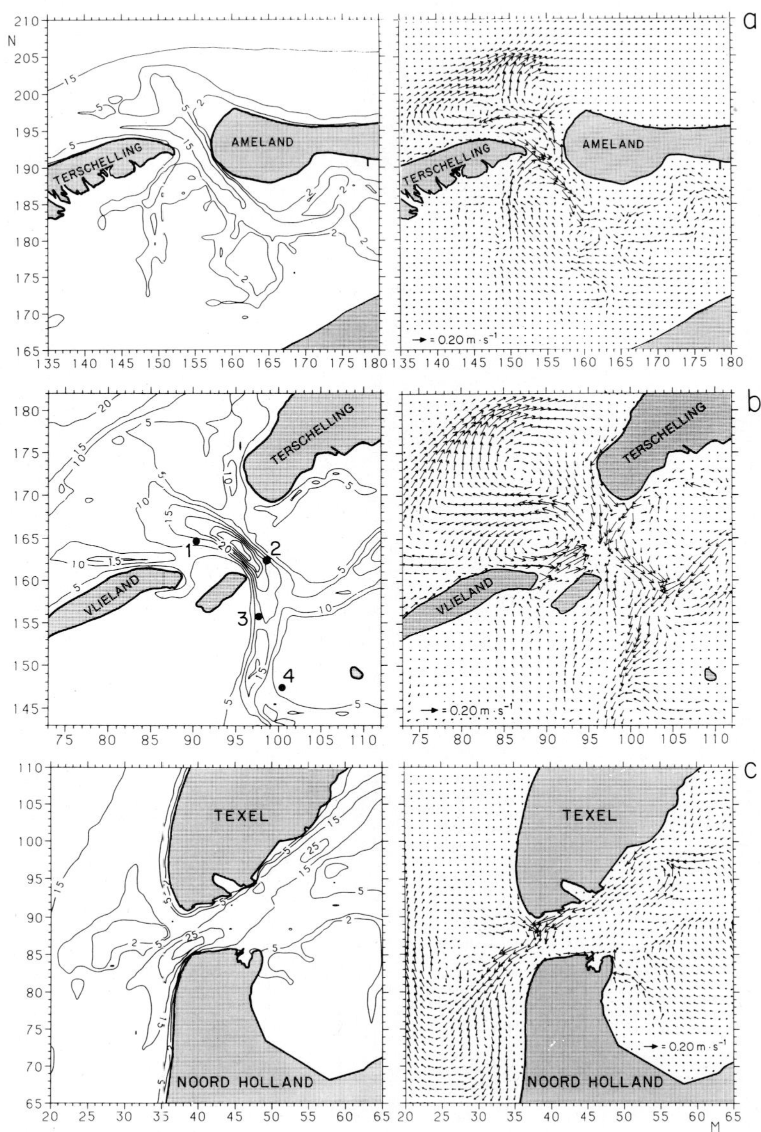

FIGURE 9. The bathymetry and the residual circulation pattern of the inlets Borndiep (a), Vlie (b) and Marsdiep (c) showing the strongly distorted quadrupoles.

isobath tidal currents. This is clearly shown by figure 8c when compared to figure 8a,b. The dominant residual vorticity balance thus reads:

$$\frac{\partial\langle u\omega\rangle}{\partial x} + \frac{\partial\langle v\omega\rangle}{\partial y} = -C\left[\frac{\partial}{\partial x}\left\langle\frac{\sqrt{(u^2+v^2)}\,v}{(H+\zeta)^{4/3}}\right\rangle - \frac{\partial}{\partial y}\left\langle\frac{\sqrt{(u^2+v^2)}\,u}{(H+\zeta)^{4/3}}\right\rangle\right]$$

Particularly for the outer delta of the inlet we see that the residual vorticity occurs at the downstream end of the tidal vorticity production areas; i.e. the vorticity produced at ebb-tide in the inner parts of the inlet is carried to the outer delta, but due to strong frictional damping, not carried back to its origin. This creates residual vorticity at the downstream end of the generation area. For the inner part of the inlet a similar mechanism exists but due to the complicated Pattern of bifurcating channels the picture is less obvious. Nonetheless we see that the basic mechanism of residual eddy formation around the inlet is production by differential bottom friction and subsequently by vortex stretching along the sides of the channel near to the inlet and in turn transport of the vorticity in the downstream direction, such that averaged over the tidal cycle there appears a residual vorticity quadrupole (figure 4 and 8c). Due to the complicated morphology of the area the quadrupole is strongly distorted. Particularly in the outer delta of the Vlie inlet we see that the cell with clockwise vorticity to the south west is smaller than the cell of anticlockwise vorticity to the north east, which seems to surround the smaller cell completely.

That this is a general behavior of all inlets of the Wadden sea is shown in figure 9 which gives the residual current patttern of the inlets Marsdiep and Borndiep as well. The broken symmetry of the quadrupole near these inlets is even more manifest than for the Vlie inlet, but always of the same sense, the cell to the north east being much more outstanding than the one on the soutwest side of the inlet. Actually this can be explained by the morphology of the inlets which is also asymmetric in having a large shoal in front of the inlets with two channels surrounding the shoal, the channel to the north east always being of minor importance relative to the one on the south west side. The broken symmetry therefore is an interesting and as yet unexplained geomorphological phenomenon that must be due to an inherent asymmetry in the tide-topography interactions around an inlet.

V. Conclusion

We have described in a relatively qualitative way the basic properties of the residual current pattern in the western Dutch Wadden Sea. The pattern is probably exemplary for other tidal inland seas and lagoons that open to the adjacent shelf sea by narrow inlets and that have a complicated internal morphology. The existence of more than one inlet will in general create residual discharges from inlet to inlet, unless the watersheds dividing the basin areas drained by those inlets are very shallow as is the usual case in the Wadden Sea except for the watershed between the Vlie and Marsdiep basins. The complicated morphology of ebb- and flood gullies, bifurcating channels and tidal flats, creates an environment that leads to strong vorticity poduction primarily by differential bottom friction. Vorticity thus produced is rectified by stretching and advection such that all over the area residual vortices are present. Around the headlands of the islands near the inlets this gives rise to the occurrence of highly distorted residual quadrupoles that seem to point to a hidden asymmetry in the morphodynamics of the inlet area.

Acknowledgements

We are much indebted to Rijkswaterstaat, Dienst Informatieverwerking, for the use of their computer and for the support in employing the extended system of computer programs.

This paper is also publication no. 35 of the project "Ecological Research of the North Sea and the Wadden Sea".

VI. References

Leendertse, J. J., 1967: Aspects of a computational model for long period water wave propagation, Memorandum RM-5294-PR, Rand Corporation, 1-116.

Ridderinkhof, H., 1988: Tidal and residual flows in the western Dutch Wadden Sea, I: numerical model results, Neth J. Sea Res., **22**, 1-21.

Ridderinkhof, H., 1989a: Tidal and residual flows in the western Dutch Wadden Sea, II: an analytical model to study the constant flow between connected basins, Neth. J. Sea Res., **22**, 185-198.

Ridderinkhof, H., 1989b: Tidal and residual flows in the western Dutch Wadden Sea III: vorticity balances, Neth. J. Sea Res. (submitted).

Ridderinkhof, H. and J. T. F. Zimmerman, 1989: Mixing processes in the western Dutch Wadden Sea, (this volume).

Robinson, I. S., 1981: Tidal vorticity and residual circulation, Deep-Sea Res., **28**, 195-212.

Robinson, I. S., 1983: Tidally induced residual flow in: B.Johns (ed.), Physical Oceanography of coastal and shelf seas, Elsevier, 321-357.

Stelling, G. S., 1984: On the construction of computational methods for shallow water flow problems, Communications Rijkswaterstaat, **35**, 1-226.

Van de Kreeke, J. and R. G. Dean, 1975: Tide-induced mass transport in lagoons, J. Waterw. Harb. Coast. Eng. Div. ASCE, **101**, WW4, 393-404.

Zimmerman, J. T. F., 1976: Mixing and flushing of tidal embayments in the western Dutch Wadden Sea, II: analysis of mixing processes, Neth. J. Sea Res., **10**, 397-439.

Zimmerman, J. T. F., 1978: Topographic generation of residual circulation by oscillatory (tidal) currents, Geophys. Astrophys. Fluid. Dyn., **11**, 35-47.

Zimmerman, J. T. F., 1980: Vorticity transfer by tidal currents over an irregular topography, J. Mar. Res., **38**, 601-630.

Zimmerman, J. T. F., 1981: Dynamics, diffusion and geomorphological significance of tidal residual eddies, Nature, **290**, 549-555.

Impacts of Relative Sea-level Rise on Evolution of Shallow Estuaries

Carl T. Friedrichs[1], David G. Aubrey[1], and Paul E. Speer[2]

[1]Woods Hole Oceanographic Institution
Woods Hole, MA 02543

[2]Center for Naval Analyses
Alexandria, VA (temporarily at[1])

ABSTRACT

The present study investigates the impact of relative sea-level rise on tidal propagation in shallow, well-mixed estuaries. Distortion of the co-oscillating estuarine tide is a composite of two principal non-linear effects: frictional interaction between the tide and channel (reflected in the scale a/h, or ratio of tidal amplitude to mean channel depth) and intertidal storage on tidal flats and marshes (measured by Vs/Vc, or ratio of volume of intertidal storage to volume of channels at mean sea level). Estuarine hypsometry (distribution of estuary surface area with height) defines a/h and Vs/Vc, and indicates whether these parameters will increase or decrease with increasing water level. The potential impact of sea-level rise is investigated utilizing both one-dimensional numerical modeling and seasonal fluctuations in mean sea levels at six shallow estuaries along the U.S. Atlantic Coast. These fluctuations are used as analogues to interannual trends in mean sea level rise. Numerical results and analysis of tidal propagation at the selected estuaries indicate that in flood dominant estuaries having relatively constant bank slope (rectangular or trapezoidal cross-section), a/h and degree of tidal distortion decrease with increased sea level. In estuaries having highly curved bank slope, a/h and Vs/Vc increase in flood and ebb dominant systems, increasing the degree of distortion and reinforcing existing tidal asymmetries. These findings have implications for the evolution of shallow estuaries as global sea level rises (an anticipated consequence of increased atmospheric trace gas loading). Whether estuaries import or export increased amounts of sediments as global sea-level rises depends on local estuarine geometry. Some systems will infill faster as sea level rises, while some will flush more efficiently. These contrasting responses to mean sea-level rise mandate a careful assessment of how any individual estuary may respond to rising water levels.

I. Introduction

Increased atmospheric trace gas loading and resulting climate change, may cause global mean sea level to rise on the order of one to several meters over the next century (NRC 1979; Hoffman et al., 1983). The position of the shoreline will change, and to some degree, the dynamics of coastal processes will be altered (Giese and Aubrey, 1987; NRC, 1987). Among the environments to show clear effects of encroaching seas along passive and barrier-type coastlines will be tidal lagoons and estuaries. The hydrodynamics of tidal estuaries are important to navigation, fisheries, flooding, and general water quality, as well as to the geological evolution of the coastline. One can use directly observable seasonal sea-level fluctuations in tidal inlets, channels and bays to infer the initial impact of

Coastal and Estuarine Studies, Vol. 38
R.T. Cheng (Ed.)
Residual Currents and Long-term Transport

relative sea-level rise on tidal dynamics (Aubrey and Friedrichs, 1988). In temperate zones, monthly mean sea level varies annually on the order of 30 cm due to variations in wind set-up, atmospheric pressure and water density. Continuous sea surface elevation data spanning a year or more in duration enable the examination of tidal variations under a range of oceanographic and climatological conditions analogous to the sea-level rise anticipated during coming decades.

One particularly relevant measure of estuarine hydrodynamics pertains to the distortion of the tide as it propagates through these shallow systems. Non-linear tidal distortion is a useful parameterization of tidal hydrodynamics, important to sediment transport and consequently to the evolution of shallow tidal estuaries (e.g., Postma, 1967; Boon and Byrne, 1981; Aubrey, 1986). "Flood dominant" lagoons and estuaries (having shorter duration, higher velocity floods) tend to infill their channels with coarse sediment. "Ebb dominant" systems (having shorter, higher velocity ebbs) tend to flush near-bed sediment seaward more effectively, and may represent more stable geometries (Speer and Aubrey, 1985; Aubrey, 1986). Observations taken from existing estuarine systems combined with numerical modeling results indicate that non-linear tidal distortion is primarily a result of frictional interaction with the channels and intertidal storage in tidal flats and marshes. Friction in shallow channels slows the propagation of low water through the inner estuary, shortening the duration of flood, whereas extensive intertidal storage slows the propagation of high water, shortening the duration of ebb. Strongly flood dominant systems are typically shallow (tidal amplitude/mean depth $> \sim 0.3$), having only small-to-moderate storage capacity in intertidal flats and marshes. Strongly ebb dominant estuaries are relatively deeper ($a/h < \sim 0.25$) and exhibit greater intertidal water storage (Aubrey and Speer, 1985; Friedrichs and Aubrey, 1988).

If shallow, flood dominant estuaries tend to shoal and deeper, ebb dominant channel systems tend to scour, then flood and ebb dominant systems should engender continued flood and ebb dominant behavior, respectively. Neglecting such important processes as storm overwash and transport of suspended sediment, this line of reasoning suggests that tidally dominant estuaries evolve into two geometrically and hydrodynamically divergent groups. However if present trends continue, rising global sea level has the potential of perturbing this divergence. If rising mean sea level causes channel depth to increase significantly or alters the volume of water stored intertidally, then the sense of tidal distortion and resulting sediment transport patterns may be altered fundamentally.

The distortion of a tide in bays and estuaries can be modeled as a non-linear growth of compound constituents and harmonics of the principal astronomical tidal components (e.g., Dronkers, 1964; Boon and Byrne, 1981; Uncles, 1981; Speer and Aubrey, 1985). Transfer of energy to even harmonics can produce asymmetric tidal velocities which result in net transport of coarse near-bed sediment and of suspended load. Along coastlines where the dominant astronomical constituent is M_2, the most significant overtide formed in well-mixed estuaries is M_4, the first harmonic of M_2. If estuarine tidal elevation is modeled as

$$A = a_{M2}\cos(\omega t-\theta_{M2}) + a_{M4}\cos(2\omega t-\theta_{M4}) \tag{1}$$

where t is time, ω is M_2 tidal frequency, a is amplitude and θ is phase, then the amplitude ratio and relative phase of M_4 to M_2 are defined as

$$M_4/M_2 = a_{M4}/a_{M2} \tag{2}$$

$$2M_2\text{-}M_4 = 2\theta_{M2}-\theta_{M4} \tag{3}$$

Flood and ebb dominant systems have relative sea surface phases of 0-180 and 180-360 degrees, respectively. For a fixed relative phase, the larger the M_4/M_2 ratio, the greater the magnitude of distortion and the more strongly flood or ebb dominant the system becomes.

II. Numerical Modeling

The physics of non-linear tidal distortion are investigated with a one-dimensional numerical model (Speer and Aubrey, 1985). One-dimensional approximations assume a small horizontal aspect ratio (channel depth/width<<1), long narrow channels (width/length<<1), and a well-mixed water column. The governing equations are conservation of mass and linear momentum:

$$\frac{\partial \zeta}{\partial t} + \frac{1}{b}\frac{\partial U}{\partial x} = 0 \tag{4}$$

$$\frac{\partial U}{\partial t} + \frac{\partial}{\partial x}\frac{U^2}{A} = -g\,A\,\frac{\partial \zeta}{\partial x} - \frac{\tau_b}{\rho}\,P \tag{5}$$

where $\zeta(x,t)$ = sea surface elevation, $U(x,t)$ = volume flux, τ_b = average shear stress on channel solid boundaries, ρ = density of water, P = channel wetted perimeter, A = channel cross-sectional area, and b = width of channel and flats. Friction is formulated as:

$$\tau_b = \rho f \frac{|U|}{A}\frac{U}{A} \tag{6}$$

where f is a dimensionless friction factor. The model approximates an ideal shallow estuary having two distinct elements: (1) a trapezoidal channel that transports all the momentum of the system and (2) shallow, sloping intertidal flats that act in a storage capacity only (Figure 1a). The boundary conditions on the model are M_2 surface forcing at the ocean end and no flux at the inland end. For systems modeled in this study, principal non-linear effects enter through friction in the momentum equation and through tidal interactions with estuarine geometry in the continuity equation (Speer and Aubrey, 1985).

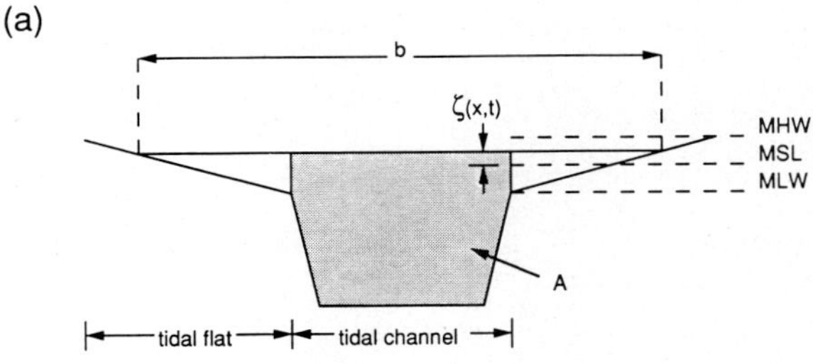

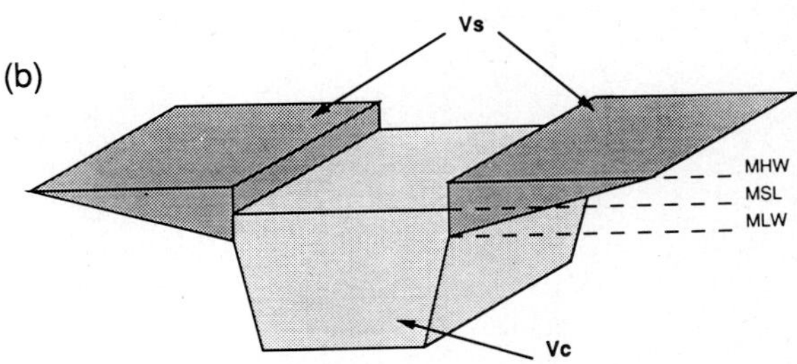

FIGURE 1. Schematic representations of the model system: (a) The numerical model is drawn in cross-section, showing the division of the model into (i) a momentum carrying channel and (ii) tidal flats which act in a storage capacity only. b = model width; $\zeta(x,t)$ = sea surface height; A = channel cross-sectional area, MHW, MSL and MLW = mean high water, mean sea level and mean low water. (b) A block diagram illustrates Vs = volume of storage in intertidal areas and Vc = mean volume in channels.

To quantify those morphological features of shallow estuaries relevant to the generation of non-linear tidal distortion, geometric parameters must be defined. The non-dimensional ratio a/h is defined as offshore M_2 amplitude divided by mean estuarine channel depth. In the shallow systems considered in this study, a/h scales the depth effect on bottom friction. A second non-dimensional ratio, Vs/Vc (Figure 1b), is defined as the volume of water stored, Vs, between mean high and low water in tidal flats and marshes divided by the volume of water contained in channels, Vc, at mean sea level (channels are defined as those areas submerged at mean low water). Vs and Vc represent the two dynamically distinct regions modeled numerically (flats which act in a storage capacity only and a momentum transporting channel). Vs/Vc scales the tidal interactions with estuarine geometry in the continuity equation (Friedrichs and Aubrey, 1988).

Numerical calculations of non-linear tidal distortion as a function of the geometric parameters (Friedrichs and Aubrey, 1988) are used to infer changes in distortion with sea-level rise. Once Vs/Vc and a/h are determined as functions of mean sea level for any given tidal estuary, probable trends in distortion can be determined. The distribution of non-linear distortion forms two fields, flood and ebb dominance, separated by the 180° contour in the $2M_2$-M_4 plot (Figure 2). This line corresponds to a trough in the contour plot of M_4/M_2. Among flood dominant systems, M_4/M_2 is primarily a function of a/h, whereas among ebb dominant systems it is primarily a function of Vs/Vc (Figure 2a). Relative phase is a more complicated function of both Vs/Vc and a/h (Figure 2b). Friedrichs and Aubrey (1988) confirmed trends in the numerical results by comparison to tide gauge records from 22 shallow tidal estuaries, and by examination of the non-linear response to variations in semi-diurnal tidal amplitude during the spring neap cycle. Numerical results represent average distortion of a symmetric ocean tide measured at four stations along a 7 km long estuary channel. The plots do not include the effects of system length or existing offshore tidal asymmetries, either of which can have a significant impact on non-linear tidal distortion (Dronkers, 1986; Friedrichs and Aubrey, 1988). However, initial increases in ocean height (tens of centimeters) will likely have a more significant impact on a/h and Vs/Vc than on channel length or distortion of the offshore tide.

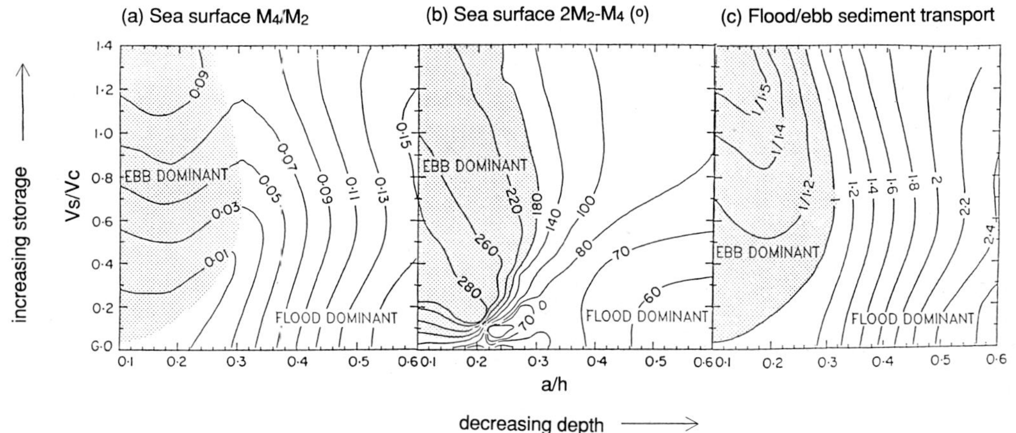

FIGURE 2. Contour plots of the parameters which determine non-linear tidal distortion as a function of a/h and Vs/Vc, resulting from 84 model systems (after Friedrichs and Aubrey, 1988): (a) surface M_4/M_2 amplitude ratio; (b) surface $2M_2$-M_4 relative phase; (c) flood-to-ebb near-bed sediment transport ratio. The distribution of non-linear distortion forms two fields, flood and ebb dominance, separated by the 180° contour in $2M_2$-M_4. Numerical results for coarse sediment transport are similar to trends in M_4/M_2 sea surface distortion.

Calculations of asymmetry in near-bed, coarse sediment transport (Figure 2c) produce trends that are similar to those in sea surface M_4/M_2 (Figure 2a). Therefore variations in sea surface distortion with sea-level rise can be used as reasonable indicators of probable changes in near-bed transport patterns (Friedrichs and Aubrey, 1988). Following Aubrey (1986) theoretical flood-to-ebb near-bed transport ratios are produced utilizing cross-sectionally averaged velocities obtained from numerical modeling and the transport formula of Meyer-Peter and Muller (1948). Results of numerical modeling suggest that changes in near-bed sediment transport with sea-level rise will be determined primarily by changes in a/h or Vs/Vc in flood or ebb dominant estuaries, respectively.

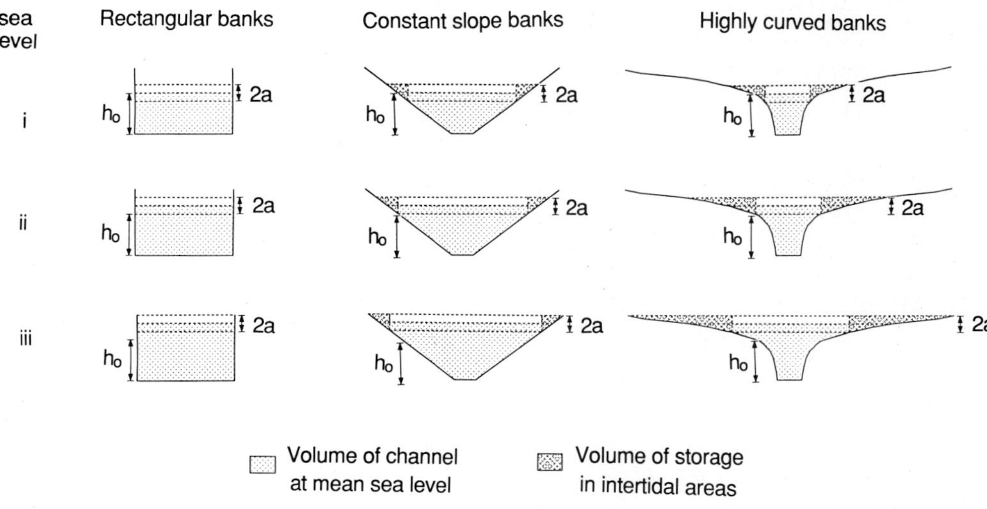

FIGURE 3. Cross-sections of conceptual rectangular, constant bank slope and highly curved bank slope channels at three sea-levels. h_o = reference depth and 2a = tidal range. Geometric parameters for each channel, measured at each of the three sea levels are displayed in Figure 4.

III. Geometry and Sea Level

To apply the numerical and empirical relationships between geometry and distortion to changes in global ocean height, the dependence of a/h and Vs/Vc on relative sea level must be determined. For a rectangular channel (Figure 3), the surface area of the system remains constant with depth. Because the channel banks are vertical, cross-sectionally averaged depth is equal to depth at the channel mid-point, and both increase by an amount equal to sea-level rise (Figure 4). Changes in geometry due to sea-level rise can be described fully by changes in a/h. If the cross-section is modeled with banks of constant slope (Figure 3), mean channel depth is less than that at the mid-point, and mean depth increases by an amount less than sea-level rise (Figure 4). In the constant slope model the Vs/Vc ratio is no longer zero. Vs remains constant with sea-level rise, so Vs/Vc decreases as Vc increases. If the cross-section of a shallow estuary has highly curved bank slopes (Figure 3), the response of a/h and Vs/Vc to sea-level rise can be markedly different. As sea level rises, shallows which were once beyond the edge of the channel are flooded and incorporated into the now wider channel. Including these newly flooded shallows into the channel can cause mean channel depth to decrease. In contrast to the rectangular and constant bank slope cases, a/h will then increase with mean sea-level rise (Figure 4). Also, if the bank slope is strongly curved, as sea level rises the areas submerged intertidally can increase dramatically, causing the Vs/Vc ratio to increase. Since the geometric parameters determine the type of distortion, changes in distortion and therefore near-bed transport with sea-level rise will be strongly dependent on bank slope.

The different bank slopes illustrated in Figures 3 and 4 provide insight into non-linear tidal response with sea-level rise at existing shallow tidal estuaries. Figure 5 shows hypsometric curves (profiles of surface area versus depth) at six tidally dominated estuaries. Bank slope can be derived directly from the hypsometric curve if the slope is averaged along the estuary. Main Creek, SC, and Little River, SC, (Figure 5a, b) exhibit only a slight decrease in bank slope with elevation, whereas Price, SC, and Wachapreague, VA, (Figure 5c, d) display highly curved bank slopes, with highest curvature occurring near the level of low tide. Manasquan, NJ, and Oaks Creek, SC, (Figure 5e, f) have more complicated geometries. The two extreme hypsometric groupings (relatively constant or highly curved bank slope) are analogous to the differences in the conceptual estuarine cross sections in Figure 3. With higher sea level, these basic differences in hypsometry cause opposite trends in non-linear tidal response.

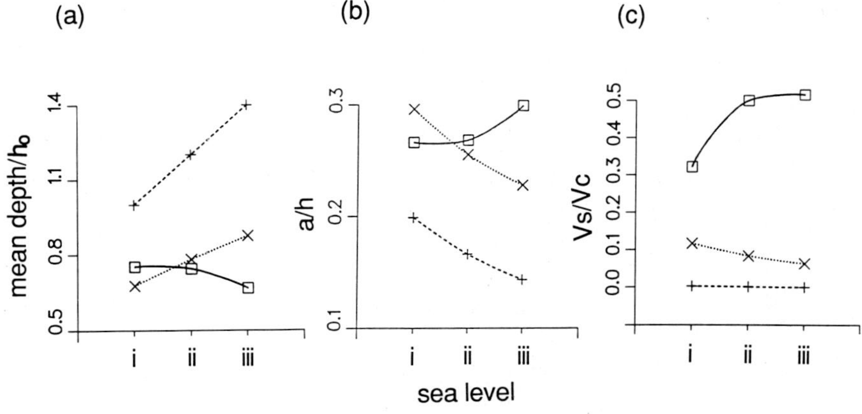

FIGURE 4. Changes in geometric parameters with sea-level rise in three conceptual model channels: effect of sea level on (a) mean channel depth, (b) a/h and (c) Vs/Vc. Response of geometric parameters to sea level is strongly dependent on system geometry.

Map views of these systems illustrate the differences in hypsometry. Main Creek and Little River (Figure 6) are shallow (average channel depth 2-2.5 m at MSL) and have marsh widths of 2 km or less measured perpendicular to the shoreline. In contrast, Price and Wachapreague (Figure 7) are significantly deeper (mean channel depths of 3.5-4 m) and have much wider marshes (5-10 km). Extensive intertidal mud flats contribute to highly curved bank slope. The surface area of flats at Price and Wachapreague is 4-5 times that of channels (FitzGerald and Nummedal, 1983; Byrne et al., 1975) but at Main Creek and Little River flats cover only 0.7 or less times the area of channels (Aubrey and Friedrichs, 1988; NOAA chart 11534). With limited marsh, narrow flats and shallow channels, the bank slopes of Main Creek and Little River are relatively constant; whereas with extensive marsh, large mud flats and deep channels, the bank slopes of Price and Wachapreague are strongly curved. Manasquan and Oaks Creek (Figure 8) contain features of each end member: they are at least as shallow as Main Creek or Little River, but they also exhibit sharp discontinuities in hypsometry near low tide, like Price and Wachapreague.

Quantitative changes in the geometric parameters with changes in sea level can be determined directly from the hypsometric curves. Mean sea level, low and high water, and the hypsometric curve define areas which can be integrated to determine a/h and Vs/Vc for any given sea level (Figure 9). Published hypsometries and

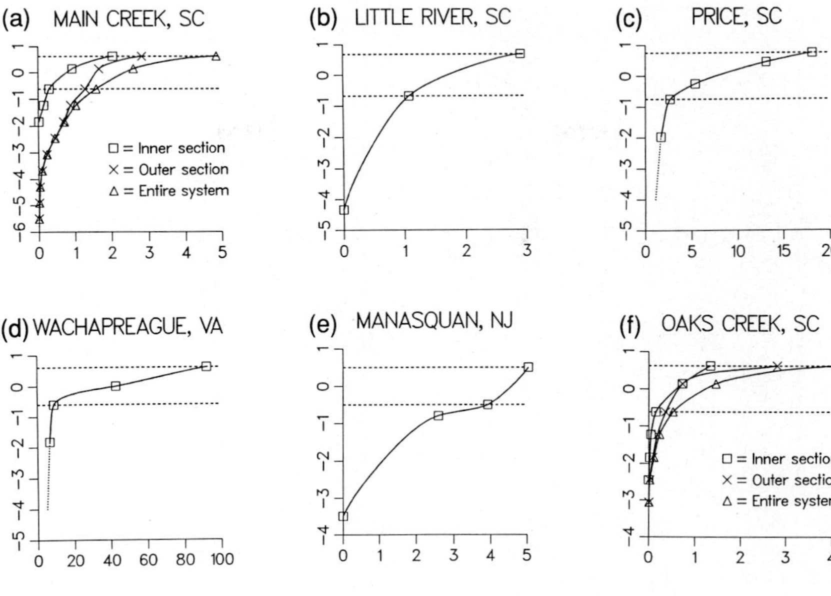

FIGURE 5. Hypsometric curves (estuarine surface area vs. elevation) for six tidally dominated estuaries: (a) Main Creek, SC; (b) Little River, SC; (c) Price, SC; (d) Wachapreague, VA; (e) Manasquan, NJ; (f) Oaks Creek, SC. Dashed lines show mean levels of high and low water within the estuaries.

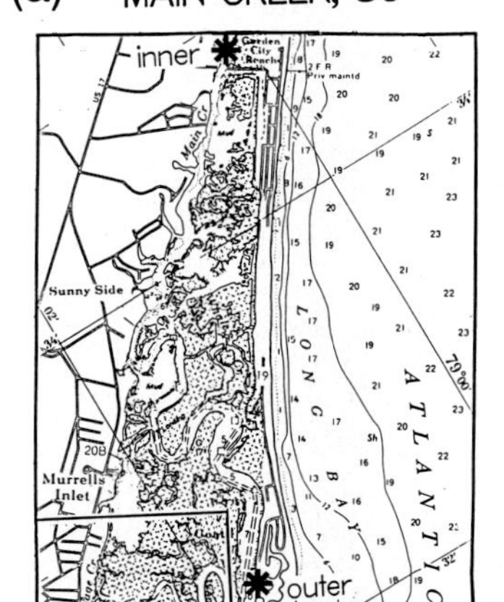

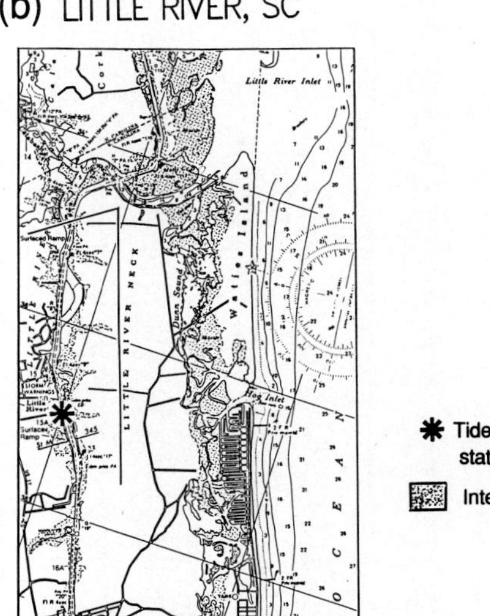

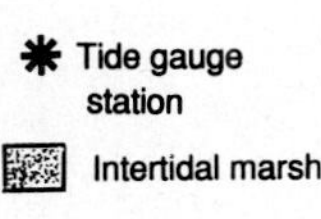

FIGURE 6. Bathymetric maps of flood dominant systems with relatively constant bank slope (after NOAA chart 11534): (a) Main Creek, SC; (b) Little River, SC. Heavy black lines crossing tidal channels and marsh indicate boundaries for purposes of geometric parameters. Soundings are depth in feet at mean low water (1 ft. = .305m). Scale is given by lines of latitude (two minutes = 3,704m).

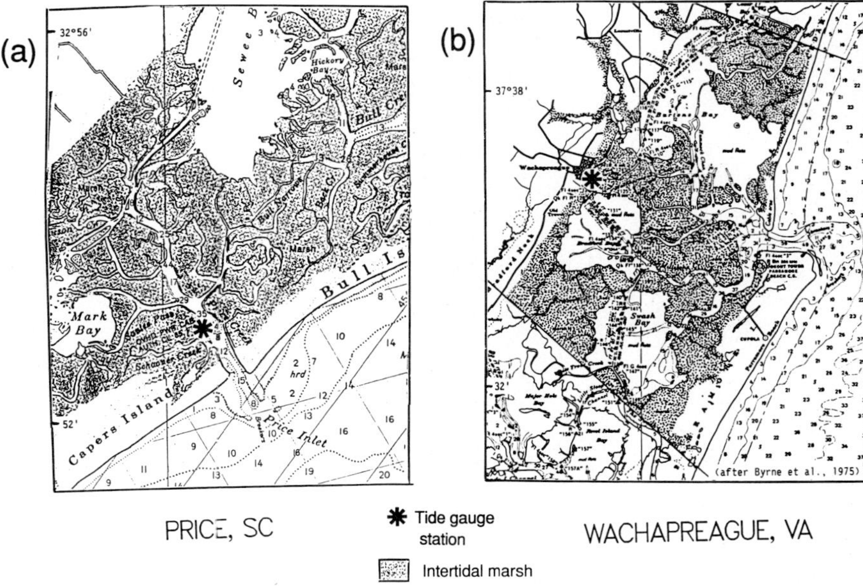

FIGURE 7. Bathymetric maps of ebb dominant systems with highly curved bank slopes: (a) Price Creek, SC (after NOAA chart 11531); (b) Wachapreague, VA (after Byrne et al., 1975). The entire tidal drainage area for each inlet is considered in calculating geometric parameters. Soundings are depth in feet at mean low water.

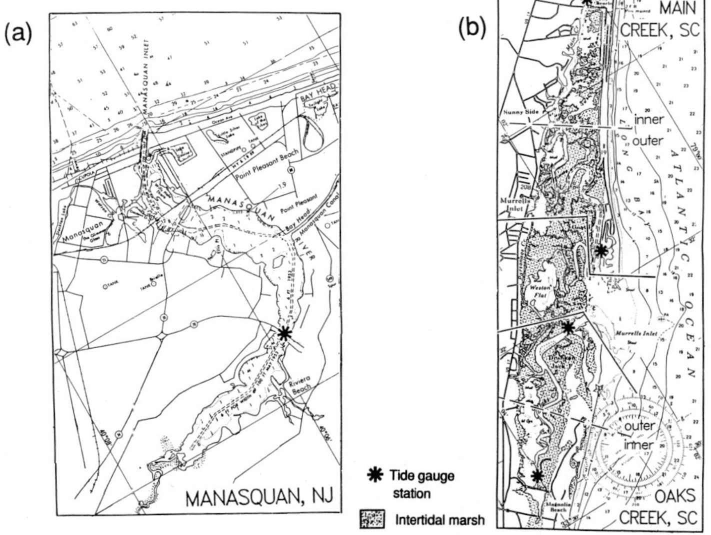

FIGURE 8. Bathymetric map of flood dominant systems with highly curved and banked slopes: (a) Manasquan, NJ (after NOAA chart 12324); (b) Oaks Creek and neighboring Main Creek, SC (after NOAA chart 11534). Heavy black lines crossing tidal channels and marsh indicate boundaries for purposes of geometric parameters. Dashed lines separate Oaks Creek and Main Creek into inner and outer sections for analyzing local changes in distortion and geometry. The entire tidal drainage area of the inlet is considered in calculating geometric parameters in Manasquan. Soundings are depth in feet at mean low water.

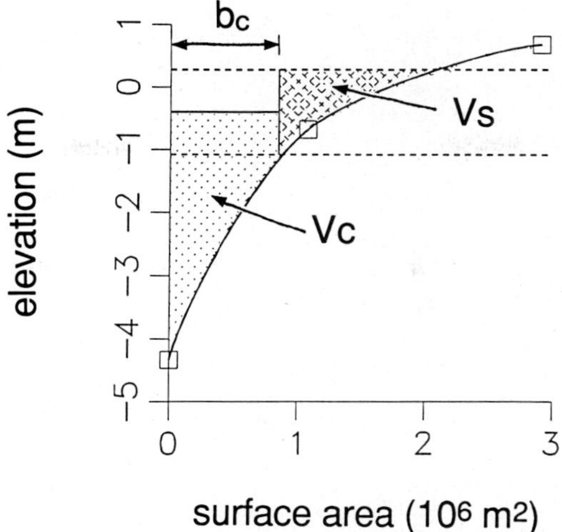

FIGURE 9. Methodology for determining geometric parameters from Little River, SC hypsometric curve. Vc (volume of channels) is the region between the hypsometric curve and low water plus the product of estuarine tidal amplitude and b_c (low water surface area). V_s is the region with surface area greater than b_c, bounded by the hypsometric curve and the high and low water lines. Mean channel depth is given by Vc divided by b_c.

topographic charts (Table 1) commonly depict the level of mean high water, but not the detailed information needed to determine bank slope for the first tens of centimeters above high water. Therefore, it is necessary to compare the geometric parameter values at present sea level with those at slightly reduced sea level to evaluate trends in Vs/Vc due to marginal changes in sea level (Table 2).

Table 1. Sources of hypsometric and sea height data used in calculation of geometric and non-linear distortion parameters:

Estuary	Geometry source	Sea surface record	Sea height source
Main Creek, SC	Perry <u>et al.</u>, 1978	10/01/74-01/31/76	NOS 1985; Aubrey and Friedrichs 1988
Little River, SC	NOAA chart 11534	07/30/74-09/28/74 11/21/74-05/13/75	NOS 1985
Price, SC	Fitzgerald and Nummedal 1983 (low to high water) USGS topos Sewee Bay/ Capers, SC (subtidal)	09/01/75-02/18/76; 06/08/76-09/16/76; 12/01/77-12/31/77	NOS 1985
Wachapregue, VA	Byrne <u>et al.</u>, 1975	01/01/83-12/31/83	Boon, personal communication
Manasquan, NJ	USGS topo Point Pleasant, NJ	04/12/76-06/30/77	NOS 1985
Oaks Creek, SC	Perry <u>et al.</u>, 1978	10/01/74-01/31/76	NOS 1985; Aubrey and Friedrichs 1988

Values determined for Vs/Vc at present sea level disagree with Friedrichs and Aubrey (1988) by 30% and 65% for Wachapreague, Va. and Main Creek, S.C., respectively. This discrepancy is mainly due to measurement of Vs in the previous study using estimates of tidal prism rather than changes in surface area with height. In the case of Main Creek, Friedrichs and Aubrey determined tidal prism by the empirical equation of Jarrett (1976) relating tidal prism to inlet cross-sectional area. Disagreement between values for Vs based on marsh outlines depicted on navigation maps and those based on empirical relationships to inlet size is not unexpected. It is not clear which of these methods is more accurate in the absence of detailed field studies (with goals specifically including measurements of tidal prism or intertidal hypsometry). Friedrichs and Aubrey (1988) examined the geometries of 22 systems to overcome uncertainties in individual measurements. The potential impact of errors inherent in estimates of intertidal storage (and the other geometric parameters) can also be minimized by examining responses to spring-neap amplitudes (Boon, 1988; Friedrichs and Aubrey, 1988) or seasonal mean sea level (Aubrey and Friedrichs, 1988). Trends in tidal distortion due to spring-neap and seasonal cycles in individual systems are determined by fewer variables than the more extreme differences in distortion observed between systems.

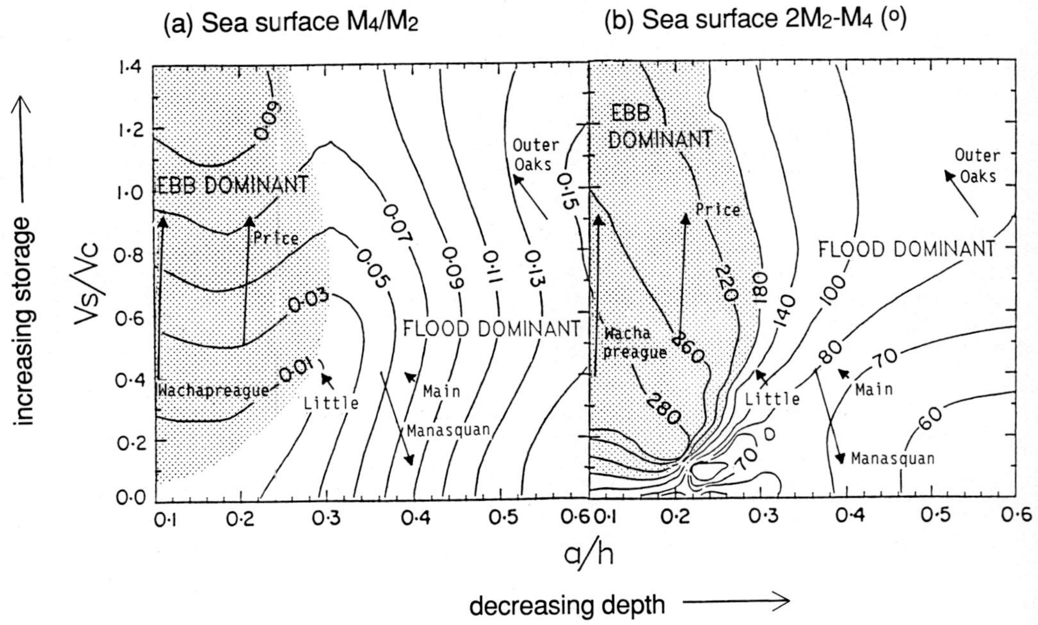

FIGURE 10. Changes in geometric parameters with sea-level rise at six tidally dominated estuaries, superimposed upon the results of numerical modeling: (a) surface M_4/M_2 amplitude ratio; (b) surface $2M_2$-M_4 relative phase. Arrows depict paths of a/h and Vs/Vc from relative mean sea level of -40cm to present mean water level. Paths indicate predicted changes in non-linear tidal distortion with sea level based upon hypsometry and numerical modeling.

IV. Distortion and Sea Level

Once changes in geometric parameters are determined as a function of sea level, initial trends in tidal distortion with global sea-level rise can be predicted. Paths illustrating variations in a/h and Vs/Vc with sea level are super-imposed upon contour plots of distortion versus geometry (Figure 10) to indicate predicted changes in M_4/M_2 and $2M_2$-M_4. Main Creek and Little River are relatively shallow ($0.3 < a/h < 0.4$) having bank slopes of low curvature, whereas Price and Wachapreague are deeper ($0.1 < a/h < 0.2$) and have highly curved banks. Each pair of systems displays a common trend for the predicted path of non-linear distortion in response to changes in ocean height (Figure 10). Manasquan and Oaks Creek have different geometric features and correspondingly distinct predicted paths of non-linear response to relative sea level.

Table 2. Values for the parameters which describe system geometry, calculated for present sea level and for -40 cm.

	Relative sea level	a/h	Vs/Vc
Entire Main Creek, SC	Present	0.39	0.41
	-40 cm	0.40	0.39
Inner Main Creek, SC	Present	0.59	2.0
	-40 cm	0.67	1.7
Outer Main Creek, SC	Present	0.37	0.20
	-40 cm	0.36	0.24
Little River, SC	Present	0.29	0.41
	-40 cm	0.31	0.36
Price, SC	Present	0.21	0.92
	-40 cm	0.20	0.50
Wachapregue, VA	Present	0.11	0.91
	-40 cm	0.10	0.39
Manasquan, NJ	Present	0.40	0.11
	-40 cm	0.37	0.40
Entire Oaks Creek, SC	Present	0.55	1.5
	-40 cm	0.57	1.7
Inner Oaks Creek, SC	Present	0.62	3.2
	-40 cm	0.63	3.4
Outer Oaks Creek, SC	Present	0.52	1.05
	-40 cm	0.55	0.90

Harmonic analyses of tide gauge records are compared to variations in seasonal mean sea level to evaluate trends predicted by hypsometry and numerical modeling. Hourly sea heights are obtained from two sources: existing observational studies and the National Ocean Survey data banks. The method of least squares harmonic analysis is used to extract tidal components from the sea-surface data (Boon and Kiley, 1978). Data analysed at each station range in length from eight to fifteen months (~6,000-10,000 hrs), depending on available record. Analyses at each station are of subsamples of 697 hourly observations, spaced every 15 days (~1/2 subsample length) for sufficiently continuous sea height records. At each station linear regression is used to relate the parameters which quantify non-linear tidal distortion (M_4/M_2 and $2M_2$-M_4) to monthly mean sea level. The probability that the best-fit coefficient (β) has the same sign as the statistically "true" coefficient is determined using the t-test (Table 3). For this study, results of linear regression are defined as statistically significant for t-test coefficients greater than 0.9.

Trends in non-linear distortion produced by harmonic analyses of tide gauge data at Main Creek and Little River are consistent with predictions based upon hypsometry and numerical modeling. Geometric parameter paths (Figure 10) and results of linear regression (Table 3; Figure 11) suggest that for flood dominant systems with relatively constant bank slope, M_4/M_2 decreases and $2M_2$-M_4 increases with an increase in relative sea level. For these systems, the decreased amplitude ratio along with increased relative phase indicate reduced flood dominance (and therefore reduced flood-to-ebb near-bed transport ratio). For flood dominant systems, magnitude of distortion (as measured by M_4/M_2) is primarily a function of a/h. Because the bank slopes of Main Creek and Little River are relatively constant, increased sea level causes mean channel depth to increase, a/h to decrease, and frictional effects on the tide to decrease. Thus observational and numerical results suggest that for flood dominant systems having relatively constant bank slope, the rate of channel shoaling is reduced with relative sea-level rise. If the initial values of a/h and Vs/Vc are near the 180° contour in $2M_2$-M_4, a large enough increase in mean sea level could conceivably alter the sense of tidal distortion from flood to ebb dominant, reversing the direction of net near-bed transport.

Table 3. Results of t-test relating monthly sea level to parameters describing non-linear tidal distortion. The equations are in the form $Y = \beta_0 + \beta_1 X$, where X = monthly sea level (with mean zero). [...] = probability that statistically 'true' coefficient is of same sign a β_1; N = number of observations.

	Y	β_0	β_1	N
Inner Main Creek, SC	M_4/M_2	0.094 [>.9999]	-0.076 [>.9999]	29
	$2M_2$-M_4(°)	90 [>.9999]	64 [>.9999]	
Outer Main Creek, SC	M_4/M_2	0.092 [>.9999]	-0.034 [.95]	23
	$2M_2$-M_4(°)	83 [>.9999]	-5 [.77]	
Little River, SC	M_4/M_2	0.054 [>.9999]	-0.079 [.9999]	12
	$2M_2$-M_4(°)	87 [>.9999]	-7 [.56]	
Price, SC	M_4/M_2	0.038 [>.9999]	0.031 [.995]	16
	$2M_2$-M_4(°)	244 [>.9999]	-16 [.99]	
Wachapregue, VA	M_4/M_2	0.042 [>.9999]	0.054 [>.9999]	22
	$2M_2$-M_4(°)	201 [>.9999]	20 [.82]	
Manasquan, NJ	M_4/M_2	0.092 [>.9999]	0.032 [.92]	26
	$2M_2$-M_4(°)	52 [>.9999]	20 [.93]	
Inner Oaks Creek, SC	M_4/M_2	0.082 [>.9999]	0.038 [.9992]	31
	$2M_2$-M_4(°)	112 [>.9999]	-72 [>.9999]	
Outer Oaks Creek, SC	M_4/M_2	0.074 [>.9999]	-0.104 [>.9999]	30
	$2M_2$-M_4(°)	77 [>.9999]	-24 [.97]	

Trends in non-linear distortion produced by harmonic analyses of tide gauge data at Price and Wachapreague are also consistent with predictions based upon hypsometry

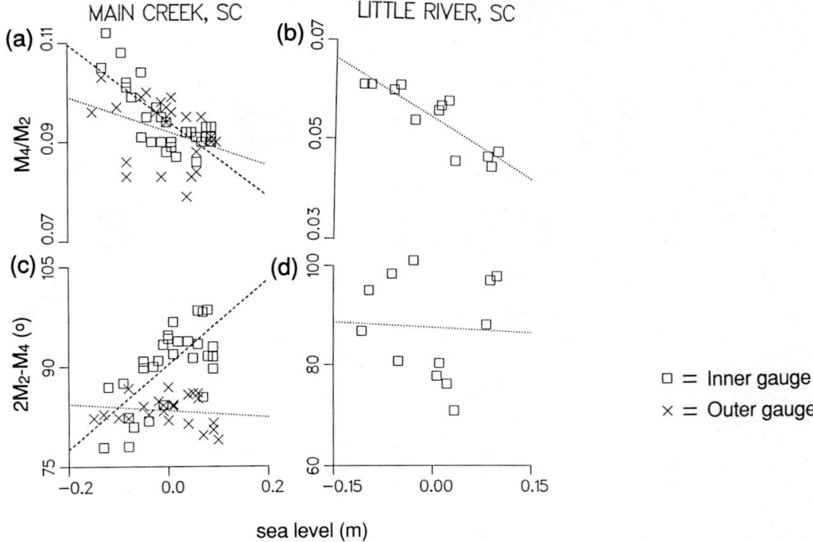

FIGURE 11. Parameters which determine non-linear tidal distortion at two flood dominant estuaries with relatively constant bank slope: surface M_4/M_2 amplitude ratio at (a) Main Creek, SC and (b) Little River, SC; surface $2M_2$-M_4 relative phase at (c) Main Creek and (d) Little River. Best-fit lines of linear regression are super-imposed upon results of a series of 697-hour harmonic analyses. Statistically significant regressions indicate that degree of flood dominance decreases with initial sea-level rise in this type of system.

and numerical modeling. Geometric parameter paths (Figure 10) and results of linear regression (Table 3; Figure 12) suggest that for ebb dominant systems having highly curved bank slopes, M_4/M_2 increases and $2M_2$-M_4 decreases with an increase in relative sea level. For ebb dominant systems, the increased amplitude ratio and decreased relative phase indicate increased ebb dominance (and therefore a greater ebb-to-flood near-bed transport ratio). For these systems, the magnitude of distortion is primarily a function of Vs/Vc. Because the bank slopes of Price and Wachapreague are highly curved, the volume of water stored intertidally and Vs/Vc increase dramatically with marginal increases in ocean height. Thus channel flushing is enhanced, the existing sense of tidal distortion is reinforced, and ebb dominant morphology may remain dynamically stable with higher mean sea level.

The trend in magnitude of non-linear distortion (M_4/M_2) produced by harmonic analyses of tide gauge data at Manasquan is consistent with the prediction of hypsometry and numerical modeling. The geometric parameter path (Figure 10) and results of linear regression (Table 3; Figure 13a) indicate that M_4/M_2 increases at Manasquan with an increase in relative sea level. However, predicted and observed trends for $2M_2$-M_4 relative phase are in disagreement, illustrating a limitation of this analysis technique. Manasquan is a shallow (mean channel depth 1.5m) lagoon with a dredged channel maintained along its axis (Figure 8a). Along much of the estuary the relatively deep dredged channel (low tide depth 2-3m) is bordered by broad shallows only 30-60cm deep at mean low water, which results in a highly curved bank slope. With seasonal sea-level rise at Manasquan, increasingly large sections of these shallows are submerged and become part of the momentum carrying central channel. For flood dominant systems, the magnitude of distortion (M_4/M_2) is primarily a function of a/h. Addition of the shallows to the main channel reduces average channel depth, increases a/h and increases the frictional effects on the tide. Thus landward near-bed transport is enhanced, the existing sense of tidal distortion is reinforced, and the need for dredging to maintain navigable channels may increase initially with higher mean sea level.

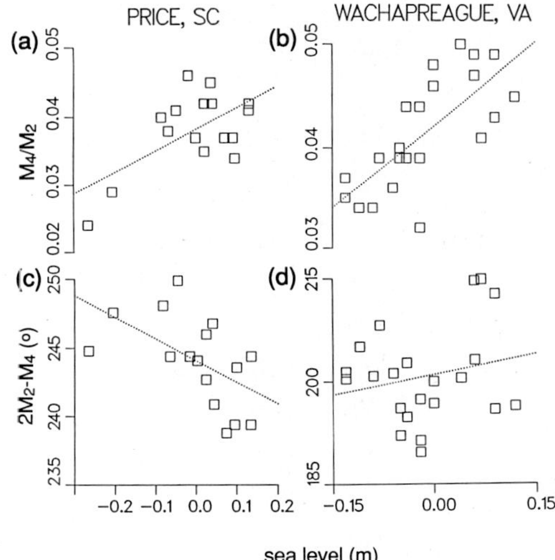

FIGURE 12. Parameters which determine non-linear tidal distortion at two ebb dominant estuaries with highly curved bank slope: surface M_4/M_2 amplitude ratio at (a) Price, SC and (b) Wachapregue, VA; surface $2M_2$-M_4 relative phase at (c) Price and (d) Wachapregue. Best-fit lines of linear regression are superimposed upon results of a series of 697-hour harmonic analyses. Statistically significant regressions indicate that degree of ebb dominance increases with initial sea level rise in this type of system.

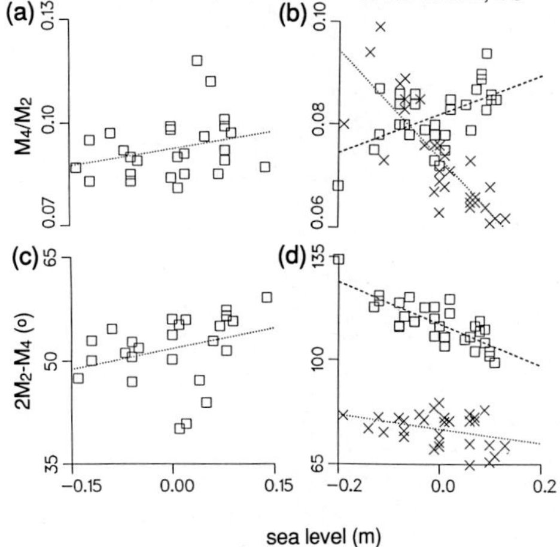

FIGURE 13. Parameters which determine non-linear tidal distortion at two flood dominant estuaries with regions of highly curved bank slope: surface M_4/M_2 amplitude ratio at (a) Manasquan, NJ and (b) Oaks Creek, SC; surface $2M_2$-M_4 relative phase at (c) Manasquan and (d) Oaks Creek. Best-fit lines of linear regression are super-imposed upon results of a series of 697-hour harmonic analyses. The magnitude of tidal distortion (M_4/M_2) increases with sea level in regions of highly curved bank slope.

Changes in the magnitude of non-linear distortion in response to seasonal sea-level fluctuations is opposite at the two tide gauges at Oaks Creek (Figure 13b): At the inner gauge M_4/M_2 increases with sea-level rise whereas at the outer gauge M_4/M_2 decreases; $2M_2$-M_4 decreases with sea-level rise at both tide gauges. Oaks Creek is flood dominant, and for sea-level rise from -40cm to the present level a/h decreases (Table 2). Thus previous analysis techniques suggest M_4/M_2 should decrease and $2M_2$-M_4 should increase at both gauges. Aubrey and Friedrichs (1988) attributed the trends in non-linear distortion with sea level at the inner gauge to the proximity of locally extensive tidal flats. To test local geometric effects on changes in distortion with sea level rise, Oaks Creek is divided into inner and outer sections (Figure 8b). As a control, a similar division is made for the two gauges at Main Creek, which is also connected to the Atlantic Ocean via Murrells Inlet. Hypsometric curves for Main Creek and Oaks Creek (Figure 5a, f) are based on a bathymetric map of two foot (0.61m) contour interval, which includes the distribution and elevation of intertidal flats (Perry et al., 1978). This detail allows the determination of a/h values with a reasonable degree of confidence for a range of ocean heights, both below and above mean sea level (Figure 14). Because calculation of a/h is based upon depths of areas flooded at low water, it is not affected by uncertainties in marsh extent at high water.

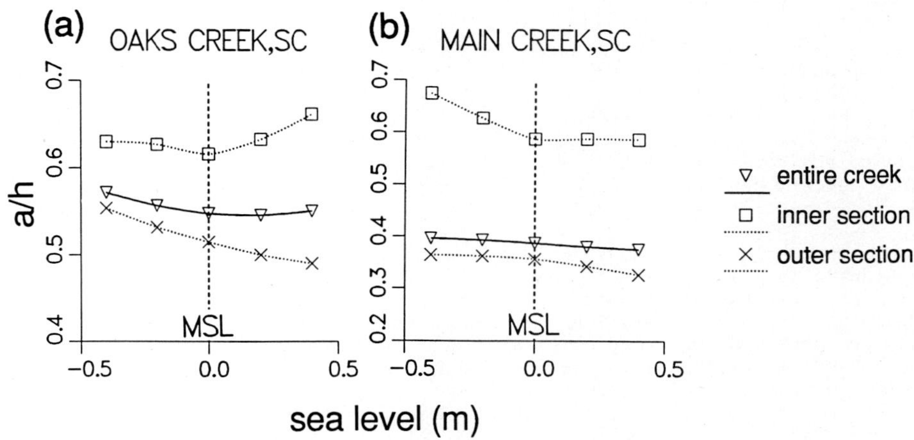

FIGURE 14. The a/h ratio as a function of relative sea level: (a) Oaks Creek, SC; (b) Main Creek, SC. Distinct trends in a/h with sea level in the inner and outer sections of Oaks Creek may account for opposite trends in local non-linear distortion with sea level in the two sections.

With sea-level rise, a/h is constant or continually decreasing for both sections of Main Creek and for the outer section of Oaks Creek (Figure 14). The a/h ratio increases with sea-level rise only for the inner section of Oaks Creek. The inner gauge at Oaks Creek also displays the only increase in M_4/M_2 with sea-level rise. Contrasts in bank slope in the two sections of Oaks Creek, as reflected in changes in a/h with sea-level rise, provide insight into the different trends in non-linear tidal distortion. Because bank slope in the outer section of Oaks Creek is relatively constant, increased sea level causes mean channel depth to increase and frictional effects on the tide to decrease. In contrast, locally extensive intertidal flats in the inner section of Oaks Creek contribute to a highly curved bank slope. A reduction in mean channel depth occurs as rising sea level incorporates former flats into the channel margins. Enhanced flood dominance in the inner section is reflected by the increase in M_4/M_2 and decrease in $2M_2$-M_4. The decrease in $2M_2$-M_4 at the outer gauge may reflect influences of the hypsometry of the inner section. In flood dominant systems having high bank curvature (such as the inner section of Oaks Creek), higher mean sea level reinforces the existing sense of tidal distortion. In such systems flood dominant morphology may remain dynamically stable with higher mean sea level.

V. Discussion and Conclusions

Shallow tidally dominated estuaries will be among the environments to show clear effects of rising global mean sea level. In this study a methodology is developed to predict the initial impact of increased ocean height on tidal distortion in these systems. Measurements of the effects of seasonal mean sea-level fluctuations on estuarine tidal distortion combined with numerical modeling results are used to infer future near-bed sediment transport patterns and general trends in estuarine evolution.

Different estuarine responses to sea-level rise occur because non-linear tidal distortion is a composite of two principle effects: Frictional interaction between the tide and channels (reflected in a/h = offshore M_2 amplitude/mean channel depth) and intertidal storage in flats and marshes (measured by Vs/Vc = volume of intertidal storage/volume of channels at mean sea level). For strongly flood dominant systems distortion is primarily a function of a/h; for ebb dominant systems it is primarily a function of Vs/Vc.

Estuarine geometries, from which a/h and Vs/Vc are derived as functions of ocean height, are grouped into two categories: relatively constant and highly curved bank slope. In flood dominant systems having relatively constant bank slope, higher sea level causes mean channel depth to increase. The a/h ratio is decreased, and therefore the degree of flood dominance is reduced. If a/h is initially near the transition to ebb dominance, then a large enough increase in mean sea level can conceivably alter the sense of tidal distortion from flood to ebb dominant, reversing the direction of net near bed transport.

In ebb dominant systems having highly curved bank slope, marginally higher sea level substantially increases the volume of water stored intertidally. Vs/Vc increases, and therefore ebb dominance is enhanced. This reinforcement may result in ebb dominant morphology remaining dynamically stable as mean sea-level rises. The existing sense of distortion is also enhanced with rising sea level in flood dominant systems having highly curved bank slope. A reduction in mean channel depth occurs as higher water level incorporates former shallows and flats into the channel margins. The a/h ratio and the degree of flood dominance are increased. Thus for tidally dominated estuaries with banks of high curvature, sea-level rise can reinforce existing near-bed sediment transport patterns whether flood or ebb dominant. This may contribute to the stable evolution of passive margins during periods of transgression.

There are limitations to the application of this method to long range estuarine evolution. In discussing sedimentation patterns this analysis does not consider fine-grained suspended material, complex channel geometry, or episodic events such as storms. Furthermore, complicated non-linear processes are analysed through relatively simple physical and conceptual models. For instance, coupled interactions between hydrodynamics and hypsometry are ignored; given sufficient time, the hypsometry would respond to higher sea-levels, causing a change in hydrodynamics. The change in hypsometry is not included here. Nonetheless, significant relationships are demonstrated among estuarine geometry, sea-level rise, and trends in non-linear distortion. From these relationships trends for estuarine evolution in response to initial sea-level rise may be inferred. The method successfully predicts changes in non-linear tidal distortion due to small increases in sea level at a number of systems. Finally, the technique is not limited to the few geometries examined here, but can be extended to any tidally dominated estuary.

Acknowledgements

This work was supported by the Office of Naval Research through the American Society for Engineering Education, the NOAA National Office of Sea Grant under grant NA80AA-D-0007, Woods Hole Oceanographic Institution Sea Grant Project Number R/B-2, the Waterways Experiment Station's Coastal Engineering Research Center, and the Woods Hole Oceanographic Institution's Coastal Research Center.

VI. References

Aubrey, D. G., 1986: Hydrodynamic controls on sediment transport in well-mixed bays and estuaries. Physics of Shallow Estuaries and Bays, J. van de Kreeke, Ed., Springer-Verlag, New York, 245-258.

Aubrey, D. G., and C. T. Friedrichs, 1988: Seasonal climatology of tidal non-linearities in a shallow estuary. Hydrodynamics and Sediment Dynamics of Tidal Inlets, D. G. Aubrey and L. Weishar, Eds., Springer-Verlag, New York, 103-124.

Aubrey, D. G., and P. E. Speer, 1985: A study of non-linear tidal propagation in shallow inlet/estuarine systems. Part I: Observations. Estuarine, Coastal and Shelf Science, **21,** 185-205.

Boon, J. D. III, 1988: Temporal variation of shallow-water tides in basin-inlet systems. Hydrodynamics and Sediment Dynamics of Tidal Inlets, D. G. Aubrey and L. Weishar, Eds., Springer-Verlag, New York, 125-136.

Boon, J. D. III, and R. J. Byrne, 1981: On basin hypsometry and the morphodynamic response of coastal inlet systems. Marine Geology, **40,** 27-48.

Boon, J. D. III, and K. P. Kiley, 1978: Harmonic analysis and tidal prediction by the method of least squares. Spec. Report No. 186, Virginia Institute of Marine Science, Gloucester Pt., VA. 49 pp.

Byrne, R. J., P. Bullock and D. G. Tyler, 1975: Response Characteristics of a tidal inlet: a case study. Estuarine Research, Vol. 2: Geology and Engineering, L. E. Cronin, Ed., Academic Press, New York, 201-216.

Dronkers, J. J., 1964: Tidal Computations in Rivers and Coastal Waters, North Holland Publishing, Amsterdam. 516 pp.

Dronkers, J., 1986: Tidal asymmetry and estuarine morphology. Netherlands Journal of Sea Research, **20,** 117-131.

Fitzgerald, D. M., and D. Nummedal, 1983: Response characteristics of an ebb-dominated tidal inlet channel. Journal of Sedimentary Petrology, **53,** 833-845.

Friedrichs, C. T., and D. G. Aubrey, 1988: Non-linear tidal distortion in shallow well-mixed estuaries: a synthesis. Estuarine, Coastal and Shelf Science, **26,** 521-545.

Giese, G. S., and D. G. Aubrey, 1987: Losing coastal upland to relative sea-level rise: 3 scenarios for Massachusetts. Oceanus, **30,** 16-22.

Hoffman, J. S., D. Keyes and J. G. Titus, 1983: Projecting future sea level rise: methodology, estimates to the year 2100, and research needs. Report 230-09-007, U.S. Environmental Protection Agency, Washington, DC. 121 pp.

Jarrett, J. T., 1976: Tidal prism-inlet area relationships. G.I.T.I. Report 3, US Army Coastal Engr. Res. Cent., 55 pp.

Meyer-Peter, E., and R. Muller, 1948: Formulas for bedload transport. Proc. Second Meeting, International Association of Hydraulic Research, Stockholm, Sweden. pp. 39-64.

National Research Council (NRC), 1979: Carbon Dioxide and Climate: A Scientific Assessment, National Academy Press, Washington, DC. 496 pp.

National Research Council (NRC), 1987: Responding to Changes in Sea Level, National Academy Press, Washington, DC. 148 pp.

National Ocean Service (NOS), 1985: Index of Tide Stations: United States of America and Miscellaneous Other Locations. National Oceanic and Atmospheric Administration, Washington, DC. 143 pp.

Perry, F. C., W. C. Seabergh and E. F. Lane, 1978: Improvements for Murrells Inlet, South Carolina. Technical Report H-78-4, US Army Eng. Waterways Exp. Sta., Vicksburg, MI. 339 pp.

Postma, H. 1967: Sediment transport and sedimentation in the marine environment. Estuaries, G. H. Lauff, Ed., Amer. Assoc. Adv. Sci., Publ. 83, Washington, DC, 158-179.

Speer, P. E., and D.G. Aubrey, 1985: A study of non-linear propagation in shallow inlet/estuarine systems. Part II: theory. Estuarine, Coastal and Shelf Science, **21**, 207-224.

Uncles, R. J. 1981: A note on tidal asymmetry in the Severn Estuary. Estuarine, Coastal and Shelf Science, **13**, 419-432.

Wind Domination of Residual Tidal Transport in a Coastal Lagoon

Ned P. Smith

Harbor Branch Oceanographic Institution
5600 Old Dixie Highway
Fort Pierce, Florida 34946 USA

ABSTRACT

A 4-layer, two-dimensional finite difference model is used to quantify residual motion in response to six tidal constituents and windstress. The model is applied to Indian River lagoon, on the Atlantic coast of central Florida. Wind, water level and near-bottom current measurements were made hourly during a 65-day study period in the summer of 1981. Tidal motion in the southern segment of the lagoon is a response to interacting tidal waves moving northward and southward from St. Lucie and Ft. Pierce Inlets, respectively. M_2 amplitudes of 12 cm and 31 cm/s were computed from water level and current data collected approximately midway between the inlets. In the absence of wind forcing, the depth-averaged mass transport residual current is 0.9 cm/s to the SSE; the depth-averaged Stokes drift is 0.5 cm/s to the SSE. The model is run with steady wind speeds of -4 to +5 m/s (NNWly headings defined positive) paralleling the axis of the lagoon. For NNWly winds of +2.4 m/s, a northward residual flow first appears in model layer 1. Winds of approximately +2.5 m/s and -2.5 m/s negate and double, respectively, the mass transport residual current calculated for calm conditions. Wind data from the 1981 study period are used in a final simulation. Results show that during relatively quiescent summer conditions wind forcing reversed and dominated the residual transport. The study suggests that wind forcing dominates residual tidal transport in the lagoon in all seasons.

I. Introduction

One characteristic commonly used to distinguish coastal lagoons from other types of estuaries is the restricted exchange of water between the lagoon and the adjacent inner continental shelf (Kjerfve 1986). As a result of constricting inlets, tidal motions within the lagoon are diminished, and local forcing assumes a more prominant role. In addition, most lagoons are shallow, and this amplifies further the response to local meteorological forcing. Smith (1983) compared nontidal variance to total variance in four 32-day current meter records from Florida's Indian River lagoon. Nontidal variance increased from 0.6% of the total near Fort Pierce Inlet to 27% at a study site 30 km from the inlet. Numerical simulations of local wind-driven currents (Smith 1988) investigated frictional damping in water 2-3 m deep. Once a steady-state was achieved under normal wind conditions, wind forcing was stopped abruptly. Current speeds decreased by 50% in 4 hours. Friction similarly damps the ebb and flood of the tide arising from exchanges through an inlet, and local windstress forcing becomes relatively dominant in the interior of the lagoon.

The ebb and flood of the tide in a coastal lagoon provide a baseline level of transport, regardless of the importance of local wind forcing. Thus, it is important to quantify tidal circulation. To obtain a proper perspective of either tidal or wind-driven transport however, both must be quantified in the same study.

Coastal and Estuarine Studies, Vol. 38
R. T. Cheng (Ed.)
Residual Currents and Long-term Transport

A field study was conducted during midsummer months in 1981 to investigate the relative importance of windstress and tidal forcing under relatively quiescent meteorological conditions. The study was conducted in the southern segment of Indian River lagoon, a microtidal lagoon on the Atlantic coast of central Florida [Figure (1)]. Indian River lagoon is 195 km long and 2-4 km wide. The southern segment, defined by Fort Pierce and St. Lucie Inlets, is 37 km long and 1-3 m deep. Earlier analyses of current and water level data (Sternberger 1983) described in a general way the relative importance of tidal and low-frequency variability in longitudinal flow. More recent work (Smith in press) isolated the six principal semi-diurnal and diurnal tidal constituents and found that the ebb and flood of the tide was responsible for a depth-averaged mass transport residual current of 0.9 cm/s to the SSE. While this represents a significant movement of water through the study area over time scales on the order of days to weeks, it appeared probable that even relatively light midsummer wind conditions could dominate the local circulation.

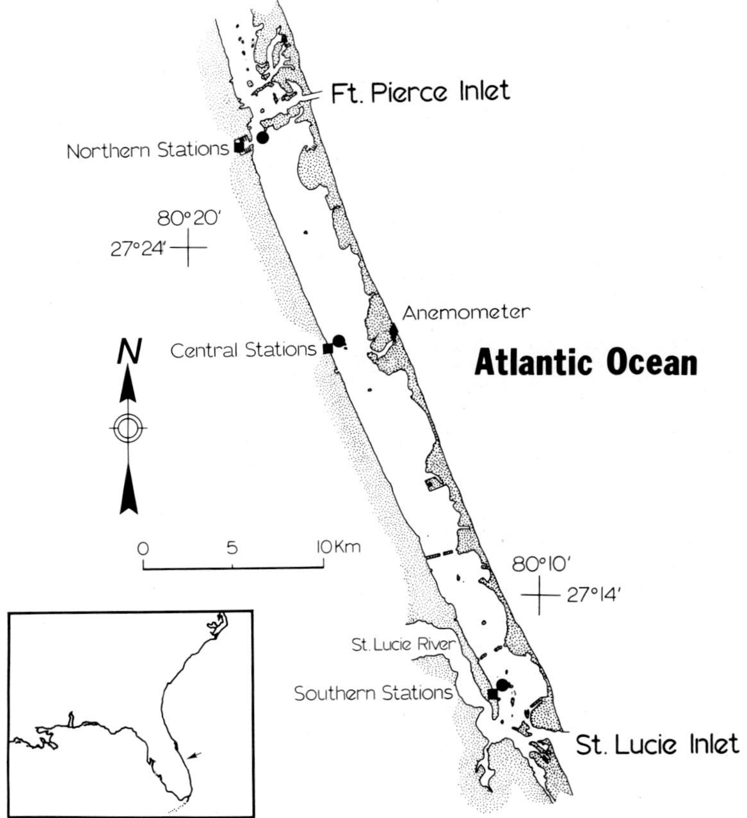

FIGURE 1. Map of northern, central and southern study sites in the southern segment of Florida's Indian River lagoon. Note location of anemometer 3 km east of central stations. Insert shows study area in the southeastern United States.

The purpose of this follow up study is to incorporate windstress into the model, and to compare several measures of transport to determine the relative importance of the wind-driven component of the circulation over time scales in excess of a few days. When the model is run with wind data recorded during the field study, simulated currents approximate measured values closely. Simulations with steady winds representative of quiescent summer conditions indicate that local wind forcing is dominant at that time of year, and probably throughout the year.

II. The Data

Time series of current speed and direction, water level and wind speed and direction were collected during a 65-day period of time from June 24 to August 28, 1981. The study period was anomalous in the sense that it coincided with a midsummer drought. Freshwater inflow to the lagoon through drainage canals and groundwater seepage was negligible, and as a result both horizontal and vertical salinity gradients were insignificant.

Water levels were read to the nearest 0.1 cm; tide staff measurements made weekly throughout the study indicated that values were accurate to within 0.5 cm. Harmonic analysis of 12 overlapping 29-day segments of the water level record (Dennis and Long 1971) provided harmonic constants (amplitudes and phase angles) for each of six principal semi-diurnal and diurnal tidal constituents (M_2, S_2, N_2, K_1, O_1, and P_1). Harmonic constants calculated for the northern and southern stations were vector-averaged (Haurwitz and Cowley 1975) and used to provide boundary conditions for model simulations. Vector-averaged amplitudes and phase angles calculated for the central station were compared with harmonic constants computed from simulated water levels to calibrate of the model and verify simulations.

Current meter data from the central station were also used to verify model simulations. Half-hourly measurements were made with a General Oceanics, Inc. Model 2010 film recording inclinometer. Currents were recorded 1 m above the bottom in 3 m of water at the edge of the Intracoastal Waterway. The natural depth of the lagoon in the immediate vicinity of the study site is 0.25-0.50 m shallower than the depth of the channel. Inclination angles and magnetic headings were read to the nearest degree. Instrument specifications state that the directional accuracy is ±5° for inclinations greater than 10°. Within the speed range characteristic of the study site, a ±0.5° precision in inclination is equivalent to a precision of approximately ±1.5 cm/s in current speed, according to the manufacturer's calibration curve. Current vectors were rotated, then decomposed into Cartesian coordinates to form along-channel and cross-channel components. Along-channel component currents were used for comparison with model simulations. Harmonic analysis of along-channel current components provided amplitudes and phase angles for the six tidal constituents listed above, and harmonic constants were vector-averaged for comparison with model results.

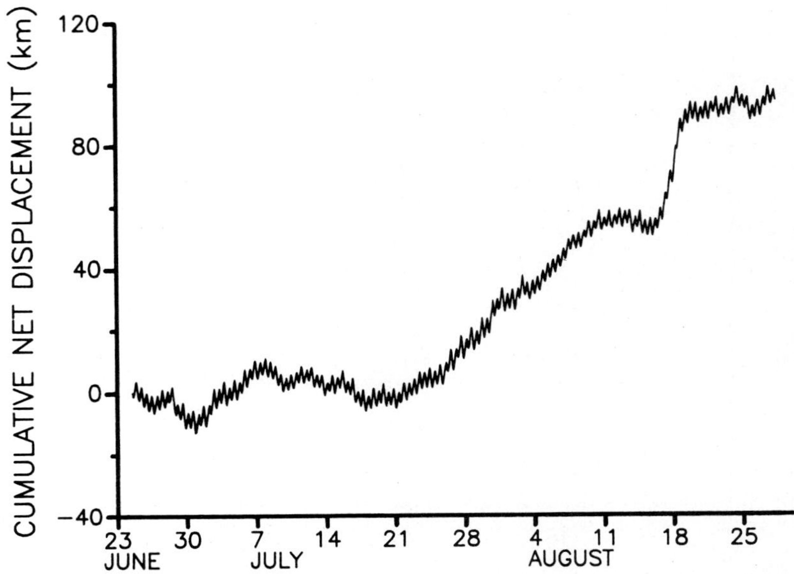

FIGURE 2. Cumulative net displacements, in km, computed from currents measured at the central study site, June 24 through August 28, 1981. Positive displacements indicate flow toward the NNW.

Currents recorded at the central study site are shown in Figure (2) in the form of cumulative net displacements. Displacement along the longitudinal axis of the lagoon is given by the product of the along-channel current component and the time interval it represents. Net displacement results from the cancelation of NNWly flow by SSEly flow; cumulative net displacement is achieved by summing products from the start of the study. Net displacement occurs over two distinct time scales. The ebb and flood of the tide appears over time scales on the order of several hours; low-frequency variations occur over time scales of from several days to several weeks. It is this latter component of the total that is of particular interest in this study.

Cumulative net displacements are less than 10 km in either direction during the first month of the study, then displacements begin to increase by just over 2 km/day in late July. A very rapid increase occurs in mid August; additional displacement in either direction is negligible through the end of the study. Comparing the endpoints of the cumulative net displacement diagram with the time interval separating them, the mean longitudinal flow past the current meter during the study period was 1.7 cm/s toward the NNW.

Wind data were obtained from a weather station on the barrier island and 3 km from the central study site. Winds were measured 58 m above the surface of the lagoon, and wind speeds were converted to 10-meter level values, assuming a logarithmic profile. Windstress was computed using the method suggested by Wu (1980). Windstress vectors were decomposed into Cartesian coordinates, and the along-channel component was used in the model to simulate transport.

Figure (3) shows winds recorded during the study in the form of cumulative net displacements computed from along-channel components of hourly wind vectors. The slope of the line connecting the end points of the plot indicate a mean longitudinal component toward the NNW of 1.25 m/s. Major deviations from the mean include a dip into negative values during the first week of the study, a rapid increase in positive values in mid August and a decrease in accumulation during the final week of the study--features seen in Figure (2).

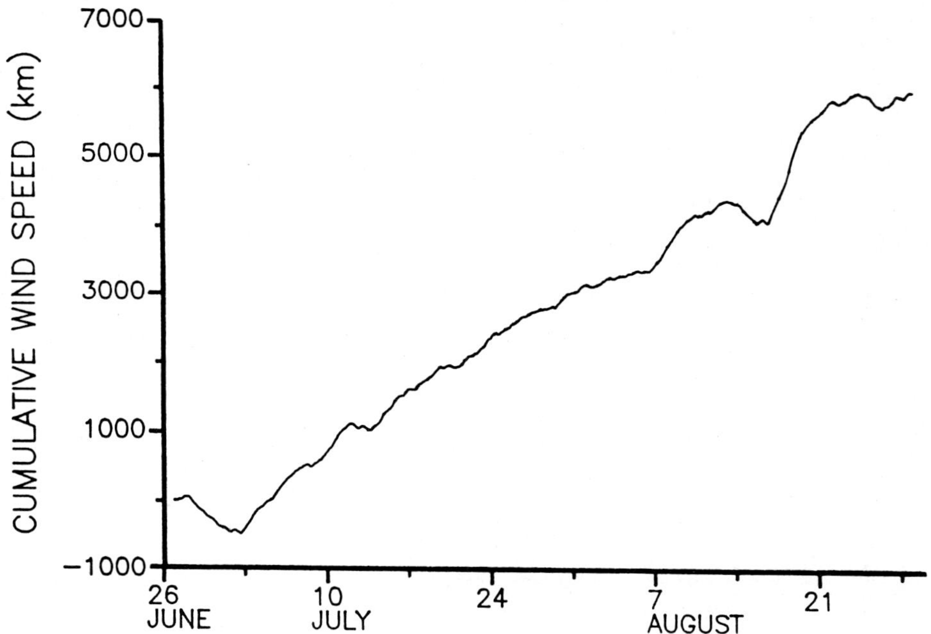

FIGURE 3. Cumulative net displacements, in km, computed from winds measured 3 km east of the central study sites, June 26 through August 28, 1981. Positive displacements indicate wind headings toward the NNW.

III. The Model

The model developed for this study quantified and compared the Eulerian residual current, u_E, the mass transport Stokes drift, u_S and the mass transport residual current, u_L, defined by Uncles et al. (1986) as

$$u_E = \langle U_m \rangle, \tag{1}$$

$$u_S = \langle \tilde{H}_m \tilde{U}_m \rangle / \langle H_m \rangle, \tag{2}$$

and

$$u_L = u_E + u_S, \tag{3}$$

where < > indicates a time averaging over the study period, H_m and U_m are the depth and the along-channel component current speed for the mth layer at time t, and

$$\tilde{H}_m = H_m - \langle H_m \rangle \tag{4}$$

and

$$\tilde{U}_m = U_m - \langle U_m \rangle. \tag{5}$$

Each of these measures of mass transport has particular advantages, and the selection of one over the others depends upon the nature of the study. The mass transport Stokes drift is best suited for quantifying transport associated with oscillatory motions involving coherent variations in both water level and current speed. The Eulerian residual current is best suited for quantifying wind-driven motion which is independent of, and thus incoherent with changes in water level. The mass transport residual current is preferred when the total current--measured or modeled--contains both quasi-steady and oscillatory components.

III.1 Model Formulation

Water level at the study site was the sum of northward and southward moving tidal waves from St. Lucie and Ft. Pierce Inlets, respectively. Northerly flow was defined positive; thus, flood currents from Ft. Pierce Inlet are directed along the negative x-axis, while flood currents from St. Lucie Inlets are along the positive axis. As noted in the previous section, six constituents were used. The model assumed an exponential damping of amplitude. For the ith constituent arriving from the jth inlet, water level at the study site was given by

$$\eta_{ij}(t) = A_{ij}\exp(-\lambda_{ij}X_j)\cos(\omega_i t + \alpha_{ij} - \kappa_{ij}X_j) \tag{6}$$

where A is amplitude, λ_{ij} is the damping coefficient, X_j is the distance from the jth inlet to the study site, ω_i is the angular frequency, α_{ij} is the difference between the Greenwich equilibrium argument of the ith constituent at the jth inlet and the modified epoch as defined by Schureman (1958), and κ_{ij} is the wave number.

Local water level at the study site, H_T, can be obtained from Eq.(6) by summing contributions of all six constituents and from both inlets:

$$H_T = \bar{H} + \sum_i \sum_j \eta_{ij} \tag{7}$$

where η is the water level relative to the mean depth, $\bar{H}$, of 300 cm. The expansion and contraction of the water column, or of individual layers, can be determined from the continuity equation

$$\partial u/\partial x + \partial w/\partial z = 0, \tag{8}$$

once spatial gradients in longitudinal flow have been evaluated.

Longitudinal current speed is obtained from the momentum equation for a wave in a uniform channel,

$$\partial u/\partial t + u\partial u/\partial x = -g\partial\eta/\partial x + N_z\partial^2 u/\partial z^2, \tag{9}$$

where g is gravity and N_z is the vertical eddy viscosity coefficient. The model simulates along-channel flow as a function of depth. The water column was divided into four equal layers which expanded and contracted with the rise and fall of the tide, and current speed was computed for each layer. Simulations are for the study site only; there is no shore-to-shore, lateral averaging to obtain a three-dimensional volume transport along the longitudinal axis of the lagoon.

Spatial gradients of water level and current speed needed for the pressure gradient and advective acceleration terms, respectively, in Eq.(9) must be evaluated from the x-derivative of Eq.(6). The value for $\partial\eta/\partial x$ is obtained directly; the value for $\partial u/\partial x$ is obtained by assuming that $\partial\eta/\partial x$ and $\partial u/\partial x$ are similar, although they may be phase shifted. In the mth layer of the model, the along-channel gradient of the current for the ith constituent from the jth inlet is

$$\partial u_{ijm}/\partial x = A_{ij}\exp(-\lambda_{ij}X_j)C_{ijm}[\kappa_{ij}\sin(\zeta + \phi_{ijm}) - \lambda_{ij}\cos(\zeta + \phi_{ijm})] \quad (10)$$

where ζ is the argument of the cosine term in Eq.(6), and ϕ_{ijm} is the phase lead of strongest flood currents at the mth level over highest water levels for the ith tidal constituent arriving from the jth inlet. C_{ijm} is a conversion factor defined by the ratio A'_{ijm}/A_{ij}, where A' is the amplitude of the tidal current at the mth level, and A is the amplitude of the water level change for the same constituent. At each level, a net value for $\partial u/\partial x$ is obtained by summing contributions of each constituent and from both inlets. The net value is used to evaluate the advective acceleration term. A similar summation provides a net surface slope, $\partial\eta/\partial x$, needed for the pressure gradient term.

The vertical eddy viscosity coefficient is given by

$$N_z = N_{zmax}N(z), \quad (11)$$

where $N_{zmax} = 0.08\ H_T|\bar{u}|$, following the general form suggested by Bowden and Hamilton (1975), and

$$N(z) = [(2z - H_T)/H_T]^2, \quad (12)$$

where z is the depth below the surface.

Frictional deceleration of the lowest layer is quantified using a quadratic term which assumes fully turbulent flow and a logarithmic current profile:

$$\tau_b = \rho\ k^2\ u\ |u|\ /\ [\ln(h_b/z_0)]^2 \quad (13)$$

where ρ is density, k is von Kaarman's constant, h_b is the height above the bottom at the midpoint of the lowest layer, and z_0 is the roughness length. A roughness length of 0.25 was used--slightly less than the 0.30 cm value suggested by Heathershaw and Langhorne (1988) for loosely packed gravel. At the Indian River study site, the roughness of the sandy mud bottom is increased by numerous clam shells.

The model was calibrated by adjusting C_{ijm} and ϕ_{ijm} until water level changes calculated from Eq.(8) matched those obtained from Eq.(7). Further tuning involved adjusting λ_{ij} and κ_{ij} in Eq.(6) until water level changes calculated from Eq.(7) and currents calculated from Eq.(9) matched observations. Simulations were made with 15-second time steps over 696 hours to quantify the perturbing effect of wind forcing during a complete synodic month.

III.2 Model Verification

Model verification was based upon its ability to reproduce harmonic constants computed from measured currents and water levels, and upon its ability to reproduce nontidal currents measured at the central station. Harmonic constants computed from measured and simulated currents are listed in Table 1. Simulated currents used for these calculations included a wind-driven component to make them directly comparable with calculations using measured values. In both cases, amplitudes and phase angles are vector-averages computed from 12 29-day harmonic analyses, as discussed above. M_2 constituent harmonic constants are within 1 cm/s in amplitude and 4° in local phase angle. Other constituents are generally within 1 cm/s of the amplitudes calculated from measurements. Phase angles are in error by as much as 33°, but it is difficult to determine precisely the phase of very low amplitude tidal waves.

Table 1. Harmonic constants of six principal tidal constituents computed from measured and simulated currents for the central station. Amplitudes are in cm/s; local phase angles are in degrees.

	Tidal Constituent					
	M_2	S_2	N_2	K_1	O_1	P_1
a. Measured currents						
1. amplitude	30.9	3.8	5.7	3.9	4.8	1.3
2. local phase angle	69	96	37	321	309	321
b. Simulated currents						
1. amplitude	31.9	4.8	4.7	5.1	3.7	1.7
2. local phase angle	65	90	30	288	296	288

The model's ability to simulate wind-driven currents was tested by comparing measured currents with currents simulated using winds recorded during the study period. Figure (4) shows cumulative net displacements, calculated from currents simulated for model layer 3. This layer represents a portion of the water column 0.75-1.50 m above the bottom which was occupied by the inclinometer. The pattern is very similar to that constructed from measured currents [Figure (2)]. Tidal co-oscillations match observations very closely, with tidal excursions of approximately 5-6 km. The nontidal component of the total contains both the appropriate general trend and the features superimposed onto the trend--with one exception. During the second and third weeks of July, simulated along-channel flow is positive, while measured currents produce a net negative displacement.

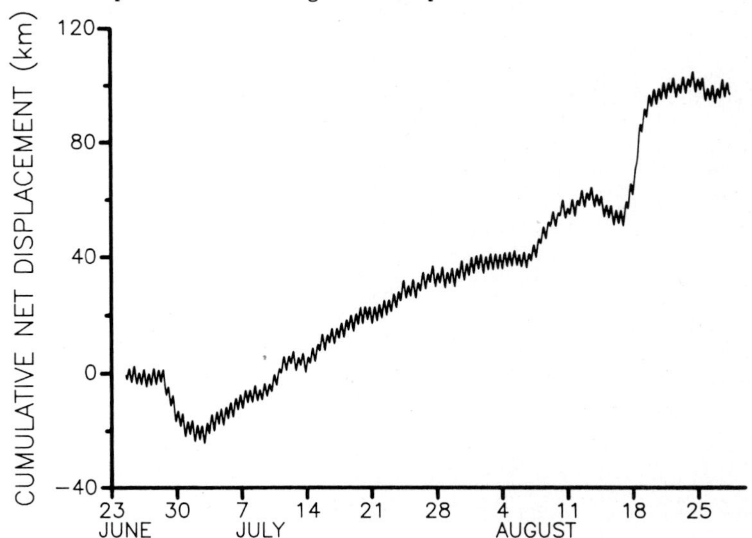

FIGURE 4. Cumulative net displacements, in km, computed using currents simulated for model layer 3, June 24 through August 28, 1981. Positive displacements indicate flow toward the NNW.

IV. Results

Stokes and Eulerian transport values simulated with steady winds are listed in Table 2. Figure (5) summarizes results by combining Stokes drift and Eulerian residual currents into the depth-integrated mass transport residual currents. Wind speed is shown along the horizontal axis. For calm conditions, residual currents are negative--toward the SSE--at 0.9 cm/s. The model indicates that just over half of the total flow is explained by the Stokes drift transport; the rest is contributed by the Eulerian residual current. This balance applies to the depth-integrated values, as well as to individual model layers.

Negative along-channel wind components reinforce the residual tidal flow. The model predicts that residual motion is doubled with a mean wind of approximately -2.5 m/s. Conversely, model results indicate that the mass transport residual current will be negated by an average wind of +2.5 m/s. Stronger along-channel component wind speeds in either sense will increase the wind domination of purely tidal processes. As wind speeds increase, the relative importance of Stokes drift decreases throughout the water column, and especially in the surface layer.

Table 2. Mass transport Stokes drift and Eulerian residual current simulated with steady wind speeds over a 29-day synodic month for the southern segment of Indian River lagoon. Negative wind speeds indicate southerly headings. The mass transport residual current for any model layer, or the mean for the water column, is given by the sum of the corresponding Stokes drift and Eulerian residual current.

Model	Stokes	Eulerian Transport Wind Speed, m/s									
Layer	Transport	-4	-3	-2	-1	0	1	2	3	4	5
1	-0.64	-3.88	-2.26	-1.20	-0.63	-0.45	-0.28	+0.30	+1.36	+2.97	+5.24
2	-0.58	-2.67	-1.61	-0.91	-0.53	-0.41	-0.30	+0.08	+0.79	+1.86	+3.37
3	-0.43	-1.90	-1.17	-0.69	-0.43	-0.35	-0.27	-0.01	+0.47	+1.20	+2.23
4	-0.34	-1.26	-0.79	-0.48	-0.32	-0.27	-0.22	-0.05	+0.26	+0.73	+1.39

Figure (6) shows the mass transport residual current, in profile form, for each of the 4 model layers, and for each of the ten wind speeds considered. Under calm conditions, the model predicts negative and relatively uniform residual flow of 0.5-1.1 cm/s throughout the water column. As wind speeds increase in a negative sense, or for positive winds above 2 m/s, the curve becomes increasingly distorted, and it becomes increasingly a response to wind forcing alone. Under the highest wind speeds considered, residual current speeds in the lowest layer are 1-2 cm/s in the downwind direction. Logically, transport in the near-surface layers is responsible for most of the depth-integrated total. The response to stronger winds was not investigated, because the dominance of wind forcing was established within this limited range of wind speeds.

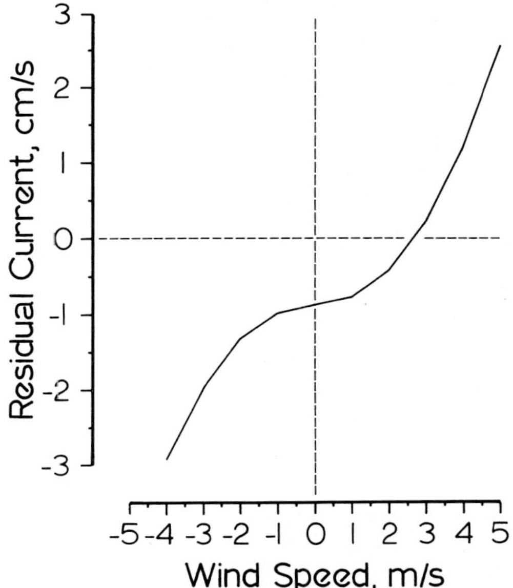

FIGURE 5. Mass transport residual current speeds, in cm/s, computed in response to tidal and steady wind forcing at the central study site. Current speeds are mean values for a 3-meter water column for 29-day simulations.

As a final simulation, the model was run with the time-varying winds which had been recorded during the study period and which had been used to calibrate the model. Results (not shown) indicate a Stokes drift identical to that given in Table 2. The Eulerian residual current was positive in all four layers, decreasing from 1.4 cm/s in the top layer to 0.3 cm/s in the lowest layer. The mass transport residual current--the sum of these two components--reversed at mid depth. Values of +0.7 and +0.2 cm/s were calculated for layers 1 and 2, respectively; a value of -0.1 cm/s was obtained for layers 3 and 4. For comparison, a steady flow of 0.1 cm/s over a 65-day period of time would results in a cumulative displacement of 5.6 km, or about 15% of the length of the southern segment of Indian River lagoon. The mean northward flow of +0.4 cm/s for the upper half of the water column would result in a displacement of 24 km, or about 67% of the length of the southern segment.

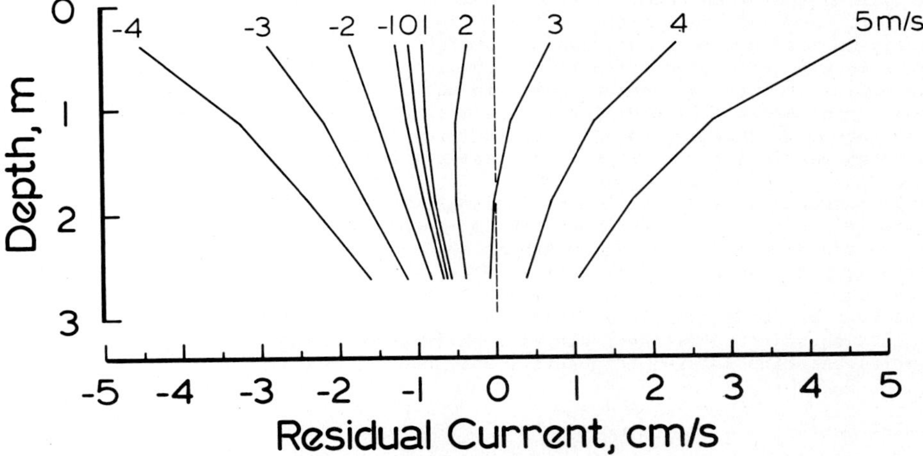

FIGURE 6. Mass transport residual current speeds, in cm/s, computed in response to tidal and steady wind forcing at the central study site. Mean wind speeds are indicated along the top of the plot. Current profiles represent mean conditions over 29-day simulations.

V. Discussion

The dominance of wind-forced residual transport in Indian River lagoon during relatively quiescent summer weather suggests that wind forcing plays the lead role throughout the year. Climatological data from the Vero Beach Municipal Airport, about 30 km northwest of the study site, shows relatively little annual variation in scalar mean monthly wind speeds. What variation there is, however, includes a minimum in midsummer months. The dominance of wind forcing in Indian River lagoon is not surprising, because this result is consistent with the relative importance of tidal and nontidal processes--over time scales in excess of several days--in most coastal lagoons with restricted exchanges. The importance of local meteorological forcing, however, justifies the more general approach taken here in which both tidal and wind forcing are considered together. While the magnitude and direction of residual flow over time scales on the order of days will be critically dependent upon the local wind history, results demonstrate that in Indian River lagoon tidal transport plays a secondary role under both steady-state and time-varying wind conditions.

The relatively close match of simulated flow [Figure (4)] to measured flow at the same level [Figure (2)] is encouraging, and it suggests that the terms contained in Eq.(9) are the necessary ones. Apparently, along the longitudinal axis of the lagoon, vertical momentum fluxes are dominant. Momentum fluxes from a vertical advective acceleration term, $w\partial u/\partial z$, do not seem to be playing a significant role, however. This term appeared in an earlier version of Eq.(9), but it was removed when it was found to be 3-5 orders of magnitude smaller than the other terms. When model layers expand and contract with the passage of tidal waves, little momentum is transferred vertically between adjacent layers.

While lateral gradients in along-channel flow (the $\partial u/\partial y$ term) appear to be small along the longitudinal axis of the lagoon, a logical follow-up to the present study would involve a model that quantifies flow though a transverse cross-section. Although lateral fluxes of momentum are minimal at the study site, this may not be the case closer to shore. In addition, an investigation of flow normal to a transverse cross-section would quantify total longitudinal transport, rather than the transport per unit width obtained in this study.

A multi-layered model such as the one described here is well suited for investigating the long-term transport of material that is concentrated in a particular part of the water column--regardless of whether that transport is primarily in response to windstress or tidal forcing. Suspended sediment, for example, would be concentrated in the lowest part of the water column; conversely, larvae and drift algae are concentrated in surface layers. The profiles shown in Figure 6, and especially the simulations made with time-varying winds, suggest that near-surface and near-bottom transport are quite different, and neither is represented particularly well by a vertically-integrated water column mean.

The dominance of wind forcing in Indian River lagoon is due in large part to its microtidal condition, which results in turn from relatively low-amplitude shelf tides and restricted exchanges through inlets. In another setting, a combination of higher-amplitude shelf tides and greater shelf-lagoon exchanges could tip the balance significantly in favor of tidal transport processes. For Indian River lagoon, however, and probably for most coastal lagoons with constricting inlets, any investigation of the baseline residual tidal transport must be supplemented by a study of local wind forcing to quantify and understand transport fully.

Acknowledgments

Mark Sternberger spearheaded the 1981 field study, assembled the current and water level data, digitized the data and performed the preliminary analysis. Patrick Pitts plotted the data used in Figs. (2-4), after assisting with the proofreading of the current meter data--a task as onerous as it is important. Wind data used in this study were provided by Robert Fisher, Florida Power and Light Company, Juno Beach, Florida.

Contribution Number 744, Harbor Branch Oceanographic Institution, Inc.

VI. References

Bowden, K. and P. Hamilton, 1975: Some experiments with a numerical model of circulation and mixing in a tidal estuary. Estuarine and Coastal Marine Science, **3**, 281-301.

Dennis, R. E. and E. E. Long, 1971: A user's guide to a computer program for harmonic analysis of data at tidal frequencies. NOAA Tech. Rept. NOS41, U.S. Dept. of Comm., 31 pp.

Haurwitz, B. and A. D. Cowley, 1975: The barometric tides at Zurich and on the summit of Santis. Pageoph., **113**, 355-364.

Heathershaw, A. D. and D. N. Langhorne, 1988: Observations of near-bed velocity profiles and seabed roughness in tidal currents flowing over sandy gravels. Estuarine, Coastal and Shelf Science, **26**, 459-482.

Kjerfve, B. J., 1986: Comparative oceanography of coastal lagoons. Pp. 63-81, In: Estuarine Variability (D.A. Wolfe, Ed.), Acad. Press, New York.

Schureman, P., 1958: Manual of harmonic analysis and prediction of tides. Special Publ. No. 98, U.S. Dept. Comm., Coast and Geodetic Survey. U.S. Govt. Printing Off., Washington, 317 pp.

Smith, N. P., 1983: Tidal and low-frequency net displacement in a coastal lagoon. Estuaries, **6**, 180-183.

Smith, N. P., 1988: Computer simulation of wind-driven circulation in a coastal lagoon. Pp. 113-131, In: Estuarine Circulation (B. Neilson, J. Brubaker and A. Kuo, ed.), Humana Press, Clifton, N.J.

Smith, N. P., (in press) Computer simulation of residual tidal transport in a coastal lagoon. J. Geophys. Research.

Sternberger, M. S., 1983: A physical description of long-period net displacement within the southern Indian River lagoon. Florida Scientist, **46(3/4)**, 396-406.

Uncles, R. J., R. C. A. Elliott, S. A. Weston, D. A. Pilgrim, D. R. Ackroyd, D. J. McMillan and N. M. Lynn, 1986: Synoptic observations of salinity, suspended sediment and vertical current structure in a partly mixed estuary. Pp. 58-70, In: Physics of shallow bays and estuaries (J. van de Kreeke, ed.). Vol. 16, Lecture Notes on Coastal and Estuarine Studies. Springer-Verlag, New York.

Wu, J., 1980: Wind-stress coefficients over sea surface near neutral conditions--a revisit. J. Phys. Oceanogr., **10**, 727-740.

Part III

Salt Mixing and Transport

Salinity Stratification and Vertical Shear Transport in an Estuary

R.J. Uncles and J.A. Stephens

Plymouth Marine Laboratory
Prospect Place, The Hoe
Plymouth, PL1 3DH, U.K.

ABSTRACT

The intratidal behaviour of the vertical shear dispersion transport for salt over three cross-sections in the upper reaches of the Tamar Estuary is presented for spring and neap tides. A variable, ψ, is defined which is proportional to the vertical shear salt flux at each instant and is equal to the negative salt flux due to vertical shear divided by the tidally-averaged longitudinal salinity gradient. The tidal average of ψ can be compared with dispersion coefficients deduced from tidally-averaged or long-term data from other estuaries. Temporal variability in ψ is discussed in terms of the root-mean-square deviations (about the depth-mean values) of the vertical profiles of salinity and longitudinal current over each tidal cycle. During spring tides ψ maximizes on the ebb within 3 h of high water and, depending on the section, this maximum lies in the range of 100-310 m^2/s (compared with 140-470 m^2/s at neap tides), with large values occurring for 2-3 h periods during the ebb. At the end of the ebb and during the flood ψ is very small and can be negative. Similar behaviour is exhibited by ψ during neap tides at the cross-section nearest the head. However, at the most seaward cross-section the velocity shear, stratification and ψ maximize at low water rather than the early ebb. A one-dimensional model of the vertical current structure which uses prescribed salinity fields and which calculates ψ is able to reproduce this behaviour. The model is also used to provide insight into the effects of stratification and its ebb-flood asymmetry on the intratidal and tidally-averaged (residual) currents.

I. Introduction

This paper has two objectives. First, to present data on the intratidal salt transport due to vertical shear dispersion in a partly mixed estuary and the associated vertical structure in salinity and longitudinal current throughout a tidal cycle. Second, to theoretically investigate the effect of stratification on the vertical distribution of longitudinal current and thereby explain the intratidal behaviour of the vertical current shear and shear dispersion of salt.

A knowledge of the various dispersion mechanisms for salt and other solutes is necessary for depth-averaged and cross-sectionally averaged water quality and salt budget modelling of estuaries (e.g. Lepage and Ingram, 1988; Fischer et al., 1979). The major unknown parameter in cross-sectionally averaged solute-budget models is the dispersion coefficient. Early measurements of salt transport indicated that transverse contributions to the shear dispersion (due to transverse shear in residual and oscillatory tidal currents and associated salinity variations) were small in a highly stratified estuary, whereas in partly mixed estuaries the transverse and vertical shear contributions were comparable (Dyer, 1974). However, according to Rattray and Dworski (1980) the role of vertical shear dispersion may have been underestimated in some of these studies.

Coastal and Estuarine Studies, Vol. 38
R.T. Cheng (Ed.)
Residual Currents and Long-term Transport

In the upper and central reaches of the macrotidal, partly-mixed Tamar Estuary, Southwest England (Fig. 1) vertical shear dispersion dominates transverse shear dispersion and is an important component of the total salt transport (Uncles et al., 1985). Therefore, it is of fundamental and practical interest to investigate the intratidal variations in vertical shear dispersion transport and to relate these to properties of the tidal flow in order to isolate the mechanisms which lead to large dispersive fluxes. Tides in the Tamar are semidiurnal with mean neap and spring ranges of 2.2 and 4.7 m, respectively. Variations in total water depth within a tidal cycle are therefore considerable, and tidal averaging, while providing a useful summary of data, conceals much of the physics involved in the transport processes.

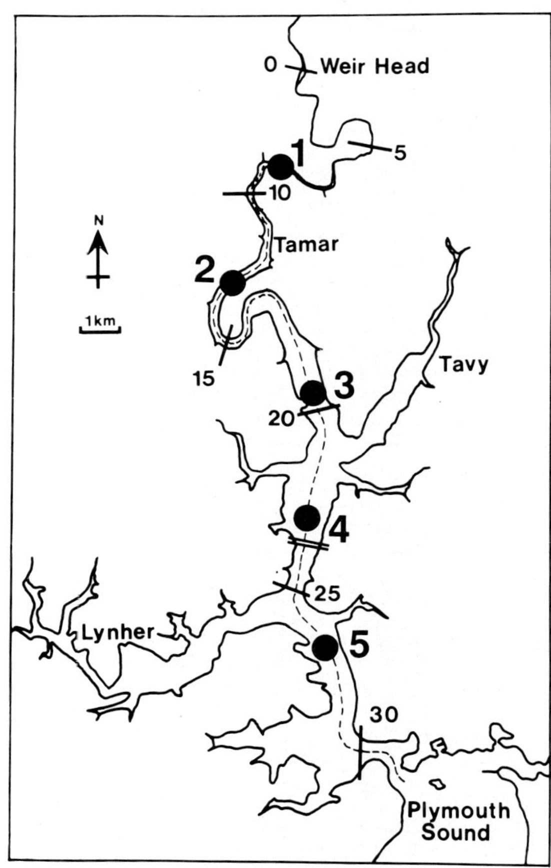

FIGURE 1. The Tamar Estuary, showing distances every 5 km along its axis and the locations of cross-sections 1-3 and synoptic, single-point anchor stations 1, 3, 4 and 5. Measurements were made on separate occasions over complete cross-sections at 1, 2, and 3. Measurements at single-point stations were made in the deep channel at different times to the cross-sectional surveys.

The experiments were of two types. In the first, measurements of salt fluxes over complete cross-sections were undertaken in the shallow upper and central reaches of the Tamar Estuary (less than 7m depth at spring-tide, high water). These cross-sections are numbered 1, 2, 3 in Fig. 1. In the second, synoptic measurements of salt fluxes were made in the deep channel at single-point Stations 1, 3, 4, 5.

II. Observations

Three cross-sections (1, 2 and 3 in Fig. 1) were worked on separate occasions for complete spring and neap tidal cycles (six survey periods). The measurements included vertical profiles of longitudinal current and salinity at either four or five observation points over each cross-section. A seatruck was anchored in the central part of a cross-section and an inflatable dinghy was used to travel between, and take measurements at, temporary moorings laid out over the estuary's width (Uncles *et al.*, 1985). Background data for cross-sections 1 to 3 are given in Table 1.

TABLE 1. Background data for cross-sections 1-3, springs (S) and neaps (N) residual flows due to daily-averaged run-off, diurnal inequality and their sum, Q, are shown (negative up-estuary).

Section	Date	Tidal Range (m)	Run-off (m^3/s)	Inequality (m^3/s)	Q (m^3/s)
1(N)	05/18/82	2.3	3.7	-0.5	3.2
2(N)	03/19/82	1.6	41.8	-1.8	40.0
3(N)	02/18/82	1.8	21.3	3.8	25.1
1(S)	05/25/82	4.9	3.6	0.0	3.6
2(S)	04/27/82	4.6	5.9	0.0	5.9
3(S)	02/26/82	4.9	25.9	14.3	40.2

The transverse topography at cross-sections 1, 2 and 3 is shown in Fig. 2. High and low water lines for the spring and neap-tide surveys are shown, together with the positions of observation points over the cross-sections. The tidal-mean, depth-averaged absolute current speeds are shown in Fig. 2.

Similar profiling measurements at single-point stations in the deep channel were made at Stations 1, 3, 4 and 5 during a spring tide survey on 7 July 1982 (Fig. 1). The tidal range was 3.9 m and the freshwater run-off was low (Uncles *et al.*, 1986). We will first investigate vertical shear dispersion of salt, stratification and vertical current structure at cross-sections 1-3. Subsequently, we will investigate theoretically the observed features using data at the deep channel, single-point Stations 1 and 5.

III. Calculation of Transport

The analytical method has been reported by Murray and Siripong (1978). At any instant of time each section is replaced by a lattice of points (i,j). Values of velocity, U, and salinity, S, at these points are denoted by U_{ij} and S_{ij}, where i is row subscript (vertical co-ordinate) and j is column subscript (traverse coordinate). Each lattice point has a rectangular sub-area associated with it, which is the same at all points, so that the instantaneous cross-sectional area, A, is the sum of all the sub-areas and the instantaneous salt transport is:

$$F = A\,\overline{U_{ij}S_{ij}} \tag{1}$$

where the overbar is a cross-sectional average. Considering shear dispersion processes we define the primary vertical and transverse deviations to be (for velocity):

$$U'_{i\bullet} = U_{i\bullet} - \overline{U} \text{ and } U'_{\bullet j} = U_{\bullet j} - \overline{U} \tag{2}$$

where $U_{i\bullet}$ is a width average at any depth and $U_{\bullet j}$ is a depth average at any point over the width. With these definitions for velocity and salinity the rate of salt transport through a cross-section is:

$$F = A\,\overline{U}\,\overline{S} + A\,\overline{U'_{i\bullet}S'_{i\bullet}} + A\,\overline{U'_{\bullet j}S'_{\bullet j}} + F_s^* \tag{3}$$

where the terms on the right hand side are, respectively, the transport due to (a) advection, (b) vertical shear, (c) transverse shear and (d) 'interaction' shear.

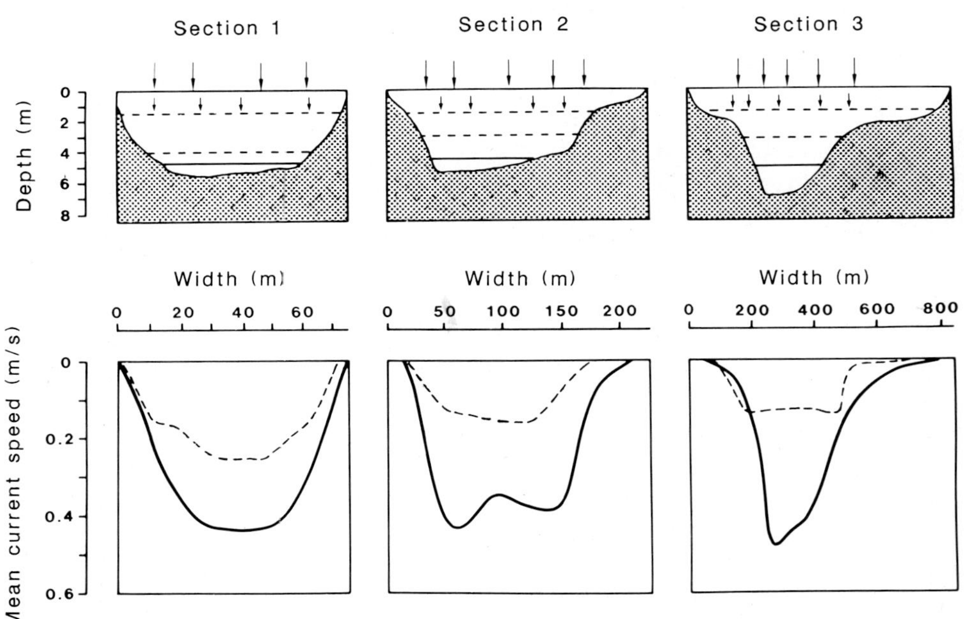

FIGURE 2. Topography and observation positions on the high-water lines at cross-sections 1-3 and the mean, absolute, depth-averaged tidal current speeds over the width for spring (continuous lines) and neap (dashed lines) tides.

The 'interaction' shear consists of seven terms each consisting of a spatial correlation between vertical and lateral deviations in salinity and longitudinal velocity [see Eq.(5) of Uncles et al., (1985) or Eq.(11) of Murray and Siripong (1978)]. There are no physical reasons to expect these correlations to be significant and the 'interaction' shear transport is indeed observed to be very small for both the partially-mixed Tamar (Uncles et al., 1985) and the well-mixed Rio Guayas Estuary (Murray and Siripong, 1978).

The advective salt transport arises from advection of the instantaneous, cross-sectionally averaged salinity by the instantaneous, cross-sectionally averaged currents. If this transport is tidally averaged it results in an advective transport due to the tidally-averaged water flow (the freshwater discharge in a steady state) and a tidal 'pumping' or 'non-local' transport [precise mathematical definitions are given in Uncles et al., (1985) and Dronkers and van de Kreeke (1986)].

The transverse shear dispersion transport is associated with correlated spatial variations in salinity and longitudinal velocity across the width of the estuary. This shear transport is due to both steady residual currents and oscillatory tidal currents.

The vertical shear dispersion transport is associated with correlated spatial variations in salinity and longitudinal velocity through the water column. This shear transport is due to both steady residual currents (the gravitational circulation being particularly important) and oscillatory tidal currents. In tidally-averaged, depth-axis models such as that of Hansen and Rattray (1965) the salt transport due to gravitational and wind-driven circulation is calculated explicitly and other mechanisms, such as the vertical shear transport due solely to tidal currents, is treated as a tidal mixing contribution to the transport.

Deep-channel measurements at single-point stations in the Tamar show that the longitudinal salt flux at spring tides in the deep channel is largely controlled by tidal pumping, whereas vertical shear transport is dominant at neap tides (Uncles et al., 1986). Tidal pumping also appears to be the dominant transport mechanism for salt in the Thames and Gironde Estuaries at spring tides (Dyer, 1978) and in the seaward reaches of the Volkerak Estuary (Dronkers and van de Kreeke, 1986). Measurements over cross-sections in the upper and central reaches of the Tamar, however, show that vertical shear transport can be comparable with or even exceed tidal pumping (Uncles et al., 1985, Table 3, p. 263) at both spring and neap tides. Transverse shear salt transport is small compared with vertical shear transport in the upper reaches but is significant in the wider central reaches. Vertical shear transport is the dominant salt transport mechanism in the upper reaches of the Volkerak Estuary (Dronkers and van de Kreeke, 1986) and in the Hudson (Hunkins, 1981).

The importance of vertical shear dispersion of salt in the macrotidal Tamar Estuary requires some explanation and the subsequent analysis is concerned with the physical behaviour of the longitudinal velocity and salinity stratification within a tidal cycle in order to isolate those mechanisms which lead to large dispersive fluxes.

We define a measure of the instantaneous vertical shear salt flux to be (with $\partial_x = \partial/\partial x$):

$$\psi = - \overline{U'_{i\cdot} \, S'_{i\cdot}}/\partial_x \langle\bar{S}\rangle \tag{4}$$

where $\partial_x\langle\bar{S}\rangle$ is the time-average (over that portion of the tidal cycle for which the salinity is non-zero) of the longitudinal gradient of the cross-sectionally averaged salinity. It is stressed that ψ is directly proportional to the vertical shear salt flux at any instant within a tidal cycle because $-\partial \langle\bar{S}\rangle$ is constant (by definition) throughout the cycle.

Negative values of ψ imply that the salt transport is up the salinity gradient, i.e., down-estuary. The equation for ψ also applies at single-point stations within the deep channel, except that the overbar now represents a depth-average and the width-average represents an average over a metre-wide section of water column. This quantity is more useful for presentation purposes than the basic salt-flux data because our measured values of ψ can be time-averaged over the tidal cycle to yield an experimental estimate of the dispersion coefficient due to vertical shear which can be compared with long-term dispersion coefficients for the Tamar and other estuaries. Some typical values for the total, long-term dispersion coefficient are 5000 m^2/s for the Columbia River, 24 m^2/s for the James River, 161-300 m^2/s for the Mersey, 158 m^2/s for Southampton Water, 50-300 m^2/s for the Tay and 53-328 m^2/s for the Thames (Dyer, 1974).

IV. Vertical Shear Transport of Salt

Time-series data for ψ (proportional to the salt flux due to vertical shear dispersion) and the cross-sectionally averaged salinity, $\bar{S}$, for spring tides at cross-sections 1-3 are shown in the left panel of Fig. 3. Values of ψ maximize in the range of about 100 - 310 m^2/s and show a consistent pattern of behaviour throughout the tidal cycle. Maximum values are reached on the ebb, before maximum currents and within 1-3 h (depending on section) of high water (H in Fig. 3). Values of ψ then fall rapidly to very small values before low water (L in Fig. 3)

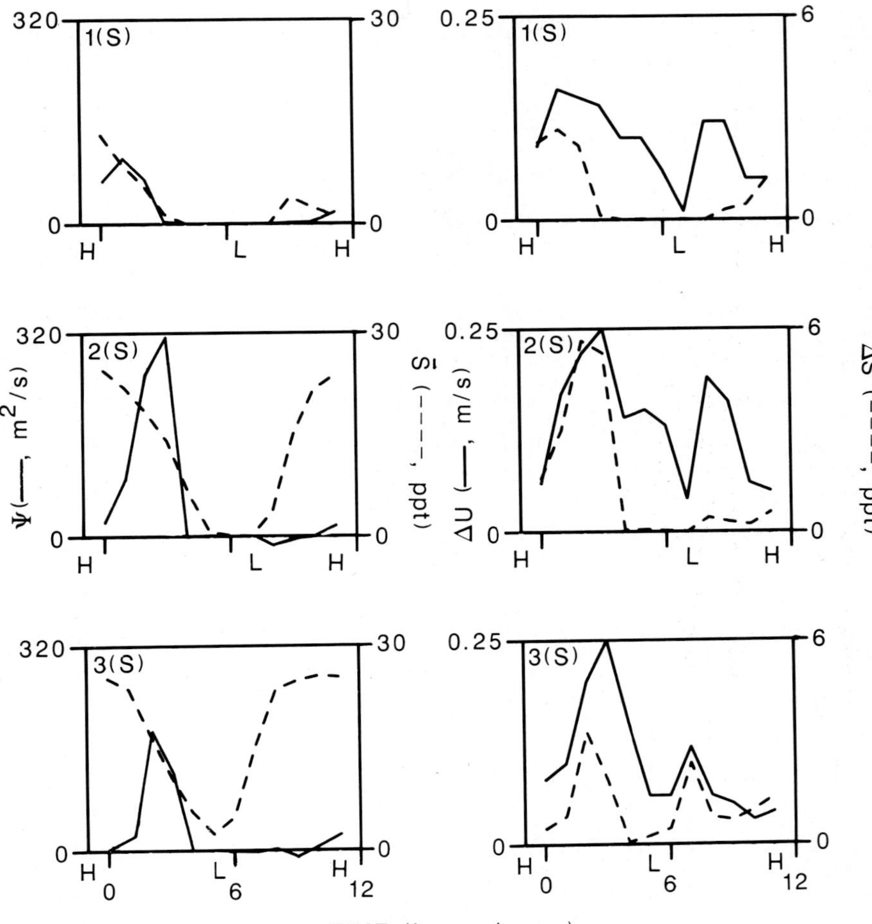

FIGURE 3. Spring-tide(S) data versus time for each tidal cycle at cross-sections 1-3 with the times of high and low-water beingdenoted H and L. Left panel: ψ (continuous lines), and cross-sectionally averaged salinity, $\bar{S}$ (dashed lines). Right panel: RMS deviations of the vertical profiles of longitudinal current (continuous lines) and salinity (dashed lines).

even when the salinity is non-zero, such as at section 3. During the flood, ψ isvery small and is occasionally negative (down-estuary transport) until shortly before high water, when it reaches significant positive values. Because ψ is proportional to the salt flux it must be zero when salinity vanishes. In addition, because salinity is not a passive tracer, the vertical gradient must usually be such that salinity increases with depth, which can lead to negative values (down-estuary vertical-shear salt transport) during the flood if faster, fresher surface water is advected over slower, more saline bottom waters. The salt-flux data can be derived from ψ at neap tides for cross-sections 1-3 using the following values for $\partial_x\langle\bar{S}\rangle$: 3.8, 0.6 and 2.4 ppt/km, respectively; at spring tides for cross-sections 1-3 the values are: 3.5, 3.8 and 3.2 ppt/km, respectively.

Winterwerp (1983) has presented data on vertical shear dispersion transport of salt from flume experiments and from single-point anchor stations in the Rotterdam

Waterway. Winterwerp's data displays similar intratidal behaviour to that shown here. More insight into the behaviour of ψ can be derived by plotting for each tidal cycle the root-mean-square velocity and salinity deviations through the column, ΔU and ΔS (right panel of Fig. 3):

$$\Delta U = \text{RMS}\ (U'_{i\cdot}) \text{ and } \Delta S = \text{RMS}\ (S'_{i\cdot}) \tag{5}$$

The RMS velocity deviations peak at around mid-flood and within 1-2 hours before mid-ebb, when currents are strong, and have minimum values at around high and low water when current speeds are slow. The largest RMS velocity deviation occurs on the ebb and implies that considerable vertical shear occurs in the longitudinal currents then. This is generated by additive summation of the vertical profiles of oscillatory tidal and steady gravitational circulation on the ebb, while vertical shear is reduced by their summation on the flood [see Bowden (1981) and Simpson _et al_., (in press) for theoretical analyses of this phenomenon and Officer (1976) for a discussion and review of estuarine tidal and gravitational currents].

The RMS salinity deviation maximizes between high water and maximum ebb, usually before the maximum velocity deviation occurs, and tends to be small elsewhere (Fig. 3). This is because the fast, ebb-flowing fresher water from up-estuary is advected over slower, more saline water, which generates a pronounced vertical salinity gradient and reduces vertical mixing; this in turn enables higher vertical shear to develop in the longitudinal currents and greater stratification. This process is halted when the currents become sufficiently fast and the depth sufficiently small for strong turbulence to develop and erode the vertical salinity gradient.

The temporal behaviour of ψ for the neap tide data is shown in the left panel of Fig. 4. Maximum values are in the range of about 140-470 m^2/s, which are significantly larger than those at spring tides. Data for cross-section 2 at neap tides cannot be considered representative of typical conditions because of the sudden onset of strong run-off during this period (Table 1) which led to salinity decreasing throughout the tidal cycle as a salt-wedge migrated down estuary. These data are presented for completeness only.

At cross-section 1, ψ has a very similar behaviour to that at spring tides. This is also true for the RMS salinity and velocity deviations at cross-section 1 (right panel in Fig. 4). The pattern is different at cross-section 3. The ψ values there maximize at low water, as do the RMS velocity and salinity deviations. At neap tides it appears that in the somewhat deeper, seaward part of the estuary the turbulent mixing in the water column at maximum ebb currents is unable to erode the vertical salinity gradient, which continues to increase until low water is reached.

V. Calculation of Currents

We consider a simplified one-dimensional model of the vertical structure of longitudinal currents which ignores lateral effects over a cross-section. The longitudinal acceleration is caused by the imbalance between vertical eddy viscosity and barotropic and baroclinic horizontal pressure gradients. The Munk and Anderson (1948) formulation is used for the eddy viscosity. There is zero horizontal shear stress at the surface and a quadratic friction law applies at the bed. Horizontal barotropic and baroclinic forcing is determined from a combination of synoptic measurements at single-point Stations 1, 3, 4 and 5 (Fig. 1) and a one-dimensional hydrodynamical, tidal model of the estuary. The residual surface gradient forcing is determined by iteration; solutions are recomputed until the calculated depth-averaged residual current corresponds to the observed value. The salinity field is prescribed and takes a highly schematic form; vertical salinity and longitudinal density gradient profiles are constrained to be linear throughout the water column and remain constant, but different, during the flood and ebb. Although we prescribe the density field the currents must satisfy the following equations of motion (with, for example $\partial_t = \partial/\partial t$):

Momentum:

$$\partial_t U = -(1/\rho)\partial_x P + \partial_z(N_z \partial_z U) \tag{6}$$

Eddy viscosity is constant with depth, and is a function of the bulk Richardson number, Ri:

$$N_z = 2.3\text{x}10^{-3} h|\bar{U}|(1 + 10Ri)^{-0.5} \tag{7}$$

Hydrostatic pressure:

$$0 = -\partial_z P - \rho g \tag{8}$$

with the boundary conditions at the surface:

$$P = \text{constant and } \partial_z U = 0 \tag{9}$$

and at the bed:

$$N_z \partial_z U = C_d U|U|, \text{ with } C_d = 3\text{x}10^{-3} \ . \tag{10}$$

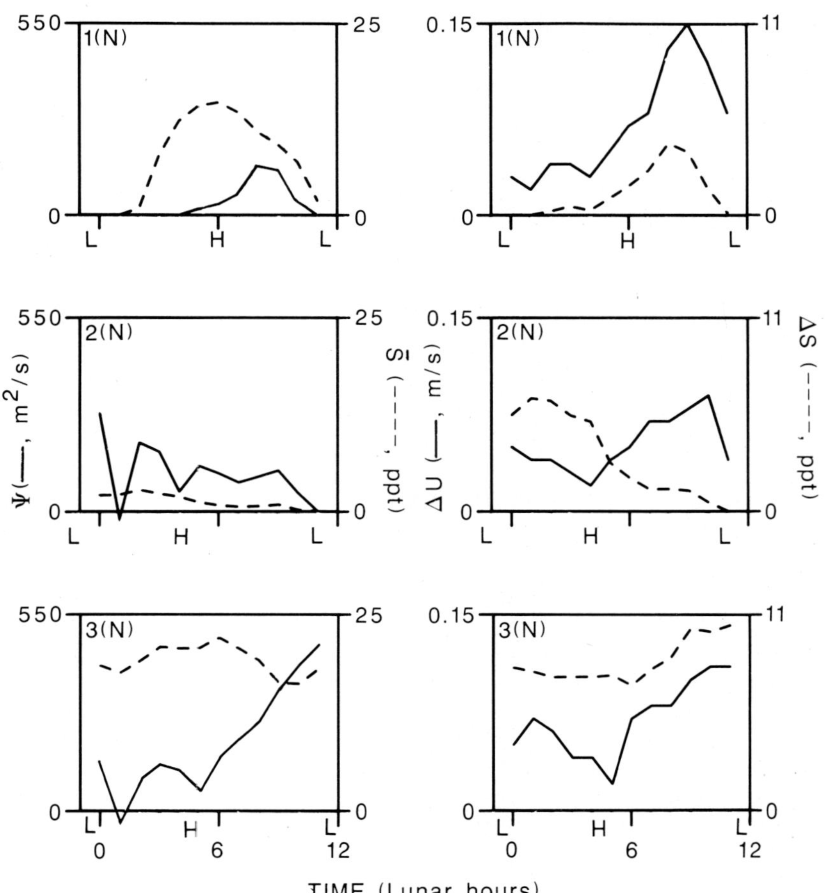

FIGURE 4. Neap-tide (N) data versus time for each tidal cycle at cross-sections 1-3 with the times of high and low water being denoted H and L. Left panel : ψ (continuous lines), and cross-sectionally averaged salinity, $\bar{S}$ (dashed lines). Right panel: RMS deviations of vertical profiles of longitudinal current (continuous lines) and salinity (dashed lines).

where U is cross-sectionally averaged velocity, ρ is density, P is pressure, and h is total depth. The drag coefficient, C_d, in the friction term is appropriate for the Tamar Estuary and the coefficient in the eddy viscosity formulation is derived from measurements in the vertically-mixed Bristol Channel (Uncles and Joint, 1983). The equations are solved to yield the depth distributions of tidal and residual longitudinal currents throughout a tidal cycle. The simulations are run until solutions do not vary between tidal cycles.

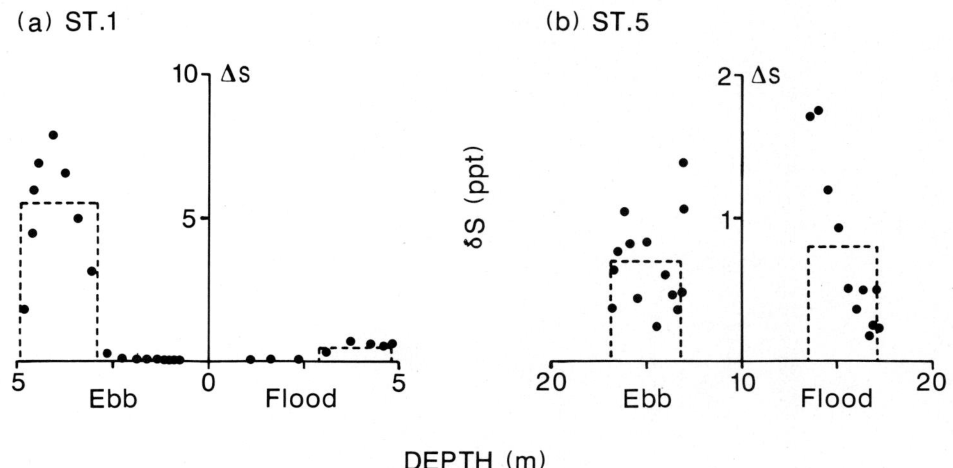

FIGURE 5. Salinity differences (ppt) between bed and surface (vertical axis) against water depth (m) during the ebb and flood for single-point (deep channel) Stations 1 and 5. Also shown are average values for the ebb and flood (dashed lines) which are used in the model (see text). Zero salinity differences occur at Station 1 when salinity is zero (water depth falls below 2.9 m).

We focus on single-point Stations 1 and 5 (Fig. 1) as being representative end-member sites. The measured surface to bed salinity differences at Stations 1 and Station 5 (denoted here by δS) are shown in Fig. 5 (a,b). Values of δS used in the model are averaged separately over the ebb and flood portions of the tidal cycle. Zero values of δS at Station 1 correspond to depths less than 2.9 m, when the salinity was zero. At the shallow Station 1 the average ebb stratification was 5.5 ppt, whereas the flood stratification was only 0.5 ppt. At Station 5 the average ebb and flood stratifications are 0.7 and 0.8 ppt, respectively. Results of calculations for residual currents made using actual values of δS are very similar to those made using the idealized forms, so that the solutions are not sensitive to rapid temporal variations in δS.

The computed and observed depth profiles of residual current at single-point Stations 1 and 5 are shown in Fig. 6. Depths are relative to high water (HW). Beneath the low water (LW) mark, the residual current is defined as the tidal average of the velocity at a fixed height above the bed. At a fixed height above the low water (LW) mark, the residual current is defined as the average velocity of the water at that height multiplied by the fraction of the tidal cycle for which water is present there.

The computed circulation at Station 1 shows some agreement with measurements, except near the bed. A slow up-estuary flow occurs in the upper 0.5 m which is due to slack water occurring after high water (e.g. Uncles and Jordan, 1980; Ianniello, 1977; Huang _et al_., 1986). Further down the water column the baroclinic pressure gradient generates a slow gravitational circulation which is confined above the low water mark. The computed residual current is ebb-directed in the bottom 1 m because of the fast ebb currents in the shallow water preceding low water.

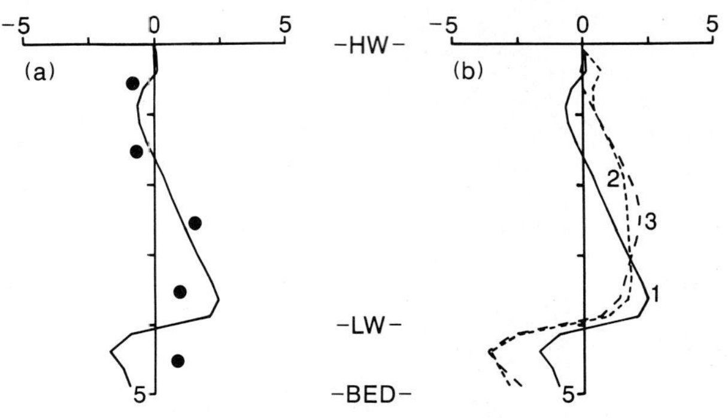

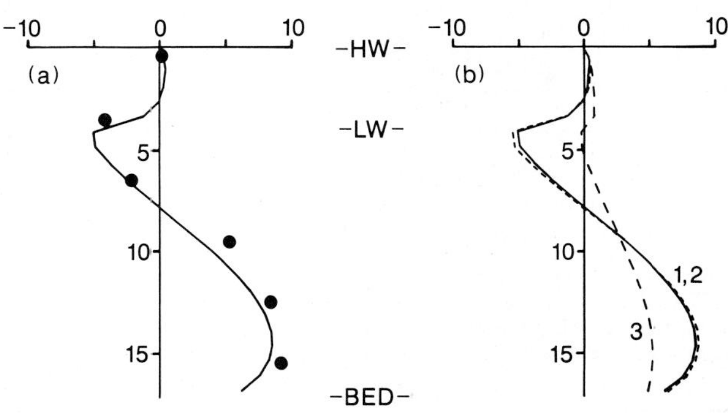

FIGURE 6. Residual currents (cm/s) versus depth (m) below high water level. Up-estuary velocities are counted positive. Upper panel: single-point Station 1. Lower panel: Station 5. (a) Computed and measured (•) profiles; (b) Computed profiles: 1) continuous line, computed profile with actual flood and ebb stratification; 2) short dashes, computed profile with equal flood and ebb stratification - the higher stratification value being used; 3) long dashes, computed profile with zero stratification.

Computed profiles of residual current at Station 1 are also presented for cases where the flood stratification is put equal to the large ebb stratification (short dashes) and when there is no stratification (long dashes). Clearly, both stratification and its flood-ebb asymmetry, acting together with the baroclinic pressure gradient, are required to generate a significant gravitational circulation in the shallow, upper reaches of the estuary.

The same set of calculations for single-point Station 5 is also shown in Fig. 6. A slow up-estuary current again flows in the upper layers, but is replaced by a gravitational circulation which now extends to the bed at this deeper site. Putting the ebb stratification (0.7 ppt) equal to the slightly larger flood stratification (0.8 ppt) has little effect on the circulation (short dashes) whereas putting the stratification equal to zero effectively suppresses the gravitational circulation (long dashes).

The computed intratidal velocity profiles at single-point Stations 1 and 5 are shown in Fig. 7. Velocities are plotted relative to their depth-mean values. The zero for each current profile is fixed at the time (in lunar hours) of its calculation during the tidal cycle.

At Station 1 there is considerable shear in the tidal currents at, and just

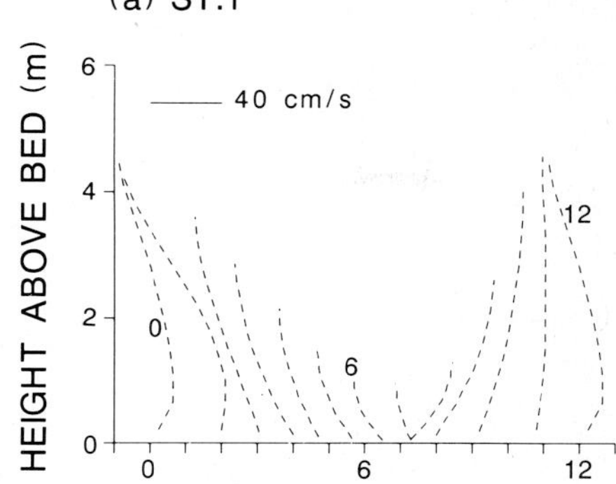

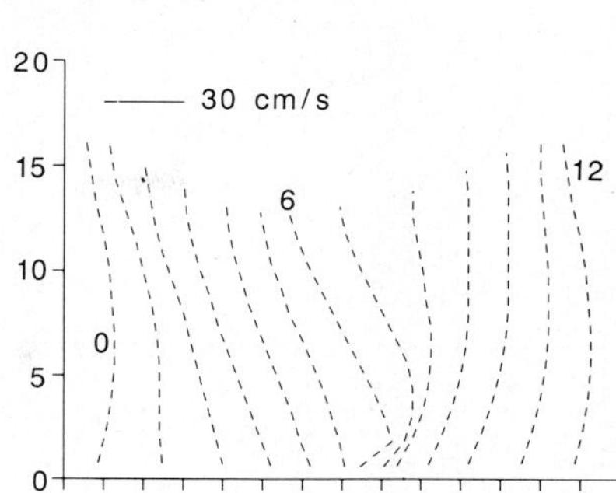

TIME (LUNAR HOURS)

FIGURE 7. Longitudinal currents versus height above the bed every lunar hour over a tidal cycle at single-point (deep channel) Stations 1 (a) and 5 (b). Depth-mean currents are subtracted from the actual currents to yield profiles of the velocity deviations. The zero for each deviation profile corresponds to the lunar-hour time for which the profile was computed.

after high water. This is a consequence of small vertical mixing due to slow currents enhancing the tendency of the longitudinal density gradient to drive a two-layer flow; soon after high water the surface current is ebbing while the near-bed current is still flooding. This behaviour at a time of small vertical mixing leads to fresher waters overlying denser saline waters and produces the large ebb-tide stratification shown in Fig. 5(a). The resultant stability enhances the velocity shear during the early ebb (Fig. 7a) and favours the formation of an ebb-directed, residual current in the upper layers (Fig. 6). The salinity is zero when depths are less than 2.9m, so that there is no tendency for a two-layer flow to develop then. By the time the salinity is non-zero on the flood the currents are fast and vertical mixing is large; in addition, the longitudinal density gradient tends to reduce the surface flood currents and enhance bottom currents, thereby further reducing vertical shear in the velocity and any tendency towards stratification.

Similar behaviour occurs at single-point Station 5 (Fig. 7b). However, in this case the salinity and its associated longitudinal density gradient exist throughout the tidal cycle. Strong vertical shear now occurs throughout the ebb and maximizes around low water (compounded by slow currents and low vertical mixing). At low water the near-bed currents have started to flood while the surface currents are still ebbing. Maximum stratification now occurs around low water (Fig. 5b).

VI. Calculation of Salt Transport

The qualitative behaviour of the salt flux due to vertical shear dispersion can be calculated at single-point Stations 1 and 5 using the numerical model of currents and the prescribed salinity profiles. We have not implicitly prescribed the salt flux by using a prescribed salinity field because the flux is dependent on the magnitude and vertical structure of the longitudinal currents, which are determined from the equations of motion. However, it does mean that the calculations are not self-consistent because the velocity field cannot influence the salinity distribution. The salt flux is again interpreted in terms of the variable ψ, which is plotted as a function of time in Fig. 8. The depth of water, h, is also plotted.

The computed behaviour of ψ at the single-point Station 1 is very similar to that observed over cross-sections at spring tides in the upper reaches (Fig. 3). At single-point Station 5 the behaviour is very similar to that observed over cross-section 3 (Figs. 1 and 4) during neap tides. Clearly, spring-tide conditions at the deep, single-point Station 5 in the lower reaches produce qualitatively similar

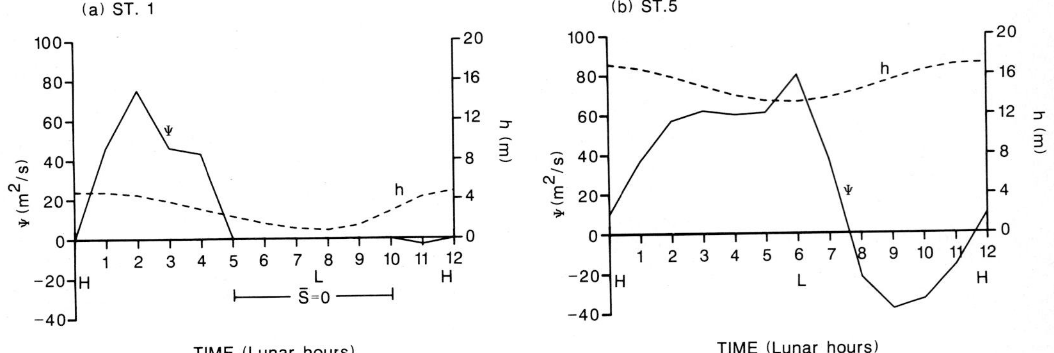

FIGURE 8. Modeled values of ψ (continuous lines), and water depth, h (dashed lines) versus time over a tidal cycle at single-point (deep channel) Stations 1 and 5 (a and b). The depth-averaged salinity, $\bar{S}$, is zero when depths are less than 2.9 m at Station 1. H and L /enote high and low water. Note that low water occurs in the middle of the cycle in (b) whereas it occurs roughly at the beginning and end in Fig. 4.

behaviour to neap-tide conditions over shallow, cross-section 3 in the central reaches. Maximum values of ψ occur around low water and minimum (negative) values of ψ occur during the flood in Figure 8 (b) and for section 3 (N) in Fig. 4 (note that low water occurs near the beginning and end of the tidal cycle in Fig. 4, whereas it occurs half-way through the cycle in Fig. 8).

VII. Conclusions

In the upper and central reaches of the estuary during spring tides, values of ψ (proportional to the salt flux due to vertical shear dispersion over a cross-section) maximize at around 100-310 m^2/s and show a consistent pattern of behaviour through each tidal cycle. Significant positive values are reached for 2-3 h during the ebb with peak values occurring on the ebb and within 1-3 h of high water (depending on cross-section). The ψ values are very small and can occasionally be negative near the end of the ebb and during the flood. Similar behaviour has been observed by Winterwerp (1983) and is a consequence of the vertical velocity structure and stratification being much higher on the ebb than the flood. During the ebb, the addition of gravitational and oscillatory tidal circulation leads to strong vertical shear and to the advection of fresher, up-estuary waters over much slower, saline waters near the bed (see Bowden (1981) and Simpson et al., (in press) for theoretical analyses). However, as currents increase and depths decrease around mid-ebb, turbulence erodes the stratification so that ψ rapidly decreases.

At neap tides, similar behaviour is exhibited by ψ over the cross-section nearest the head. However, in the deeper, central reaches of the estuary there is insufficient vertical mixing at maximum currents to erode the stratification and vertical structure in velocity, stratification and ψ then maximize at approximately low water. Maximum values of ψ are significantly higher at neap tides and lie in the range 140-470 m^2/s (depending on cross-section).

This observed temporal behaviour in both ψ and the velocity profiles can be reproduced qualitatively using a numerical model of the vertical structure in which salinity is prescribed. The one-dimensional model specifies the horizontal driving force as the imbalance between vertical eddy viscosity and horizontal barotropic and baroclinic pressure gradients. Calculations show that there is considerable vertical structure in the horizontal currents in the upper estuary during early to mid-ebb. In the deep water near the mouth of the estuary, strong vertical structure in the horizontal currents occurs throughout the ebb and maximizes around low water as a result of the slow currents, relatively deep water and weak vertical mixing at that time.

The vertical structure in horizontal currents is a consequence of vertical variations in the amplitudes and phases of the oscillatory tidal currents through the water column, as well as vertical variations in the residual currents. The model shows that both stratification (through its effect on stability) and its flood-ebb asymmetry (ebb stratification being much larger than flood stratification in the upper reaches) acting together with the horizontal baroclinic pressure gradient, are required to generate a significant gravitational circulation in the shallow, upper reaches of the estuary. The model also shows that stratification (through its effect on stability) acting together with the horizontal baroclinic pressure gradient, is required to generate a significant gravitational circulation in the deeper, lower reaches of the estuary. The flood-ebb asymmetry in stratification is, on average, negligible there.

VIII. References

Bowden, K. F., 1981: Turbulent mixing in estuaries. Ocean Management, **6**, 117-135.

Dronkers, J. and J. van de Kreeke, 1986: Experimental determination of salt intrusion mechanisms in the Volkerak Estuary. Netherlands J. Sea Research, 20, 1-19.

Dyer, K. R., 1974: The salt balance in stratified estuaries. Estuarine Coastal Marine Science, **2**, 273-281.

Dyer, K. R., 1978: The balance of suspended sediment in the Gironde and Thames Estuaries. In: Estuarine transport processes, B. Kjerfve (ed.). The Belle W. Baruch library in Marine Science, No. 7, 331pp.

Fischer, H. B., E. J. List, R. C. Y. Koh, J. Imberger and N. H. Brooks, 1979: Mixing in inland and coastal waters. Academic Press, N.Y., 483pp.

Hansen, D. V. and Rattray, M., 1965: Gravitational circulation in straits and estuaries. J. Marine Research, **23**, 104-122.

Huang, P. -S., D. -P. Wang and T. O. Najarian, 1986: Analysis of residual currents using a two-dimensional model. Lecture notes on coastal and estuarine studies, **16**, J. van. de Kreeke, Ed., Springer Verlag, New York, 71-80.

Ianniello, J. P. 1977: Tidally-induced residual currents in estuaries of constant breadth and depth. Journal of Marine Research, **35**, 755-78.

Lepage, S. and R. G. Ingram, 1988: Estuarine response to a freshwater pulse. Estuarine Coastal and Shelf Science, **26**, 657-667.

Munk, W. H. and E. R. Anderson, 1948: Note on the theory of the thermocline. Journal of Marine Research, **7**, 276-295.

Murray, S. P. and A. Siripong, 1978: Role of lateral gradients and longitudinal dispersion in the salt balance of a shallow well mixed estuary. Estuarine Transport Processes, B. Kjerfve, Ed., University of South Carolina Press, 113-124.

Officer, C. B., 1976: Physical oceanography of estuaries (and associated coastal waters). John Wiley and Sons, 465pp.

Rattray, M. and J. G. Dworski, 1980: Comparison of methods for analysis of the transverse and vertical circulation contributions to the longitudinal advective salt flux in estuaries. Estuarine Coastal Marine Science, **11**, 515-536.

Simpson, J. H., J. Brown, J. Mathews and G. Allen (in press): Tidal straining, density currents and stirring in the control of estuarine stratification. Estuaries.

Uncles, R. J. and M. B. Jordan, 1980: A one-dimensional representation of residual currents in the Severn Estuary and associated observations. Estuarine Coastal Marine Science, **10**, 39-60.

Uncles, R. J. and Joint, I. R. 1983: Vertical mixing and its effects on phytoplankton growth in a turbid estuary. Canadian J. Fisheries and aquatic Sciences, **40**, 221-228.

Uncles, R. J., R. C. A. Elliott, and S. A. Weston, 1985: Dispersion of salt and suspended sediment in a partly mixed estuary. Estuaries, **8**, 256-269.

Uncles, R. J., R. C. A. Elliott, S. A. Weston, D. A. Pilgrim, D. R. Ackroyd, D. J. McMillan and N. M. Lynn, 1986: Synoptic observations of salinity, suspended sediment and vertical current structure in a partly mixed estuary. Lecture notes on coastal and estuarine studies, **16**, J. van de Kreeke, Ed., Springer Verlag, New York, 58-70.

Winterwerp, J. C. 1983: Decomposition of the mass transport in narrow estuaries. Estuarine, Coastal and Shelf Science, **16**, 627-638.

Longitudinal Dispersion of Salt in the Volkerak Estuary

J. van de Kreeke

Division of Applied Marine Physics
Rosenstiel School of Marine and Atmospheric Science
University of Miami
4600 Rickenbacker Causeway
Miami, FL 33149

ABSTRACT

Values of longitudinal dispersion coefficients are calculated from observed steady longitudinal salinity distributions in the Volkerak estuary. The tidal regime and freshwater discharge for the estuary are controlled by respectively a storm surge barrier and discharge sluices. In the different experiments freshwater discharges are 25 m^3/s, 50 m^3/s and 100 m^3/s and the tidal ranges at the mouth are 2.70 m and 3.50 m. In the seaward part of the estuary values of the longitudinal dispersion coefficients are relatively uniform and for the different experiments vary between 175 m^2/s and 300 m^2/s. Values of the longitudinal dispersion coefficients decrease with decreasing tidal range. Variations in freshwater discharge hardly affect the value of the longitudinal dispersion coefficient. Values in the landward part have a slight tendency to increase in the seaward direction for the lower (25 m^3/s and 50 m^3/s) freshwater discharges and to decrease in the seaward direction for the large (100 m^3/s) freshwater discharge. Values in the different experiments range between 150 m^2/s and 325 m^2/s. Values of the dispersion coefficient increase with increasing freshwater inflow. Because the lower tidal range was accompanied by a reduction in freshwater discharge, the effect of a change in tidal range could not be determined. Changes in the values of the longitudinal dispersion coefficients are explained in terms of changes in the velocity field.

I. Introduction

The Volkerak Estuary is a partially mixed estuary. For this class of estuaries salinity differences in the cross-section are small compared to the cross-sectionally averaged value. In addition where the estuary has a well-defined longitudinal axis, tidal time variations in salinity exhibit a distinct pattern as the longitudinal salinity distribution oscillates with the tidal velocity. It follows that for most practical purposes and in particular when dealing with inputs to biological studies it is sufficient to know the cross-sectionally and tidally averaged salinities.

The temporal and spatial variations in the time- and cross-sectionally averaged salinitiy salinity $<\bar{\bar{s}}(x,t)>$ are described by the one-dimensional conservation of salt equation,

$$\frac{\partial \int_0^x <A\ \bar{\bar{s}}>\ dx}{\partial\ t} + Q_f <\bar{\bar{s}}> = D\ A_o\ \frac{\partial <\bar{\bar{s}}>}{\partial x} \tag{1}$$

Coastal and Estuarine Studies, Vol. 38
R.T. Cheng (Ed.)
Residual Currents and Long-term Transport

In this equation, $A_o(x)$ is the tidally averaged value of the cross-sectional area $A(x,t)$; x is the longitudinal coordinate, positive in the down estuary direction, x = 0 at the limit of salt intrusion; s is salinity; t is time; Q_f is the freshwater discharge and D is the longitudinal dispersion coefficient. Angle brackets denote averaging over an integral number of tidal periods.

The first term on the left hand side of Eq.(1) represents the time rate of change of the salt content upstream of the location x. The second term represents the downestuary advective flux. The term on the right hand side represents the upestuary salt flux; this flux is associated with temporal and transverse (i.e., in the cross-section) variations of the longitudinal velocity in combination with cross-sectional mixing. In Eq.(1) the upestuary salt flux is written as a gradient type flux. As will be discussed later, this does impose a number of restrictions on the salinity-and velocity field and the salt intrusion mechanisms. In the expression for the gradient type flux the longitudinal dispersion coefficient, D, is a measure for the combined efficiency of the exchange processes in transporting salt In principle, it's value solely depends on the longitudinal velocities, or more precisely the longitudinal velocities relative to a plane oscillating with the tidal velocity, and cross-sectional mixing.

Assuming stationary salinity distributions and known values of Q_f, $\langle\bar{\bar{s}}\rangle(x)$ and $A_o(x)$, values of longitudinal dispersion coefficients have been calculated from Eq.(1), with the time rate of change term taken equal to zero, for the Mersey Narrows (Hughes, 1958), the Severn and Thames estuaries (Bowden, 1963), the Delaware Estuary (Paulson, 1970), the Tay Estuary (West and Williams, 1972) and Southampton Water (Officer, 1976). Using a multiple regression technique, (Uncles and Radford, 1980) determined values of longitudinal dispersion coefficients for the Severn Estuary from non-stationary longitudinal salinity distributions. Instead of salt, Wilson and Okubo (1978) observed the spreading of a slug of dye in the seaward part of the York River. Using a two-dimensional conservation equation, they calculated values of longitudinal dispersion coefficients by matching observed and calculated dye distributions.

Values of longitudinal dispersion coefficients reported in the referenced papers range from 10 m^2/s to 1000 m^2/s. Values increase with increasing freshwater inflow. However, it seems obvious that the increase with increasing freshwater inflow cannot continue ad infinitum. In parts of the estuary that become fresh, longitudinal dispersion coefficients decrease because of the absence of gravitational circulation (West and Williams, 1972). Except for the Severn Estuary, values of of the longitudinal dispersion coefficients increase in a seaward direction.

When comparing the values of the longitudinal dispersion coefficients presented in the different studies, part of the variation can be explained by differences in bathymetry, freshwater inflow, tide and wind. However, in some of the studies in which salinity was used as a tracer, values are undoubtedly affected by the assumption of steady state. For estuaries in which the salinity distribution is not adjusted to the freshwater discharge, values of longitudinal dispersion coefficients are strongly influenced by the time history of the discharge. Also, in many of the referenced studies, the accuracy of the tidally- and cross-sectionally averaged salinities was limited as most of the values are based on measurements in the axis of the estuary.

II. Objectives

In this paper, values of longitudinal dispersion coefficients are calculated for four experiments in the Volkerak estuary. In each experiment, a true steady salt distribution is attained by keeping the freshwater discharge constant for periods of 6 months. The response time or decorrelation time scale for the estuary is approximately 2-3 months. Freshwater discharges range from 25 m^3/s to 100 m^3/s. In one of the experiments, the tidal range in the estuary is decreased by partly closing the storm surge barrier in the Eastern Scheldt, the body of water to which the Volkerak estuary is connected. Using the observed longitudinal cross-sectionally- and tidally-averaged salinity distributions and the known freshwater discharge, values of longitudinal dispersion coefficients are calculated from the steady state form of Eq.(1) The variation in the values of the dispersion coefficients for the different experiments is discussed in terms of the dominant salt intrusion mechanisms and the variations in freshwater discharge tide and wind.

III. The Volkerak Estuary

The Volkerak estuary is located in the Rhine-Meuse Delta in the southwestern part of the Netherlands; see figure (1). The landward part is characterized by a relatively uniform channel with few tidal flats. In the seaward part the bathymetry is irregular with many tidal flats and channels. A typical depth for the entire estuary is 10 m with local maxima of 35 m in the seaward part. Freshwater enters the estuary via the Volkerak locks (a combination of navigation locks and discharge sluice) located at the head of the estuary. The estuary via the Eastern Scheldt connects to the North Sea. The tide is dominated by the M_2 component. The average tidal range is approximately 3.5 m. The ratio between spring- and mean tidal range is 1.15. Maximum tidal velocities in the seaward part are 1.25 m/s corresponding to a tidal excursion of 15,000 m. In the landward part maximum tidal velocities are 0.25 m/s corresponding to a tidal excursion of 3,000 m.

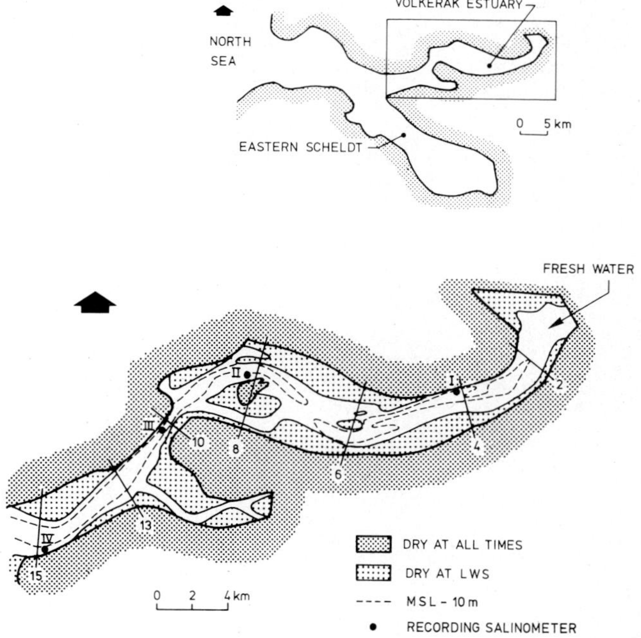

FIGURE 1. Volkerak Estuary; location of measuring sections and recording salinometers.

IV. Measurements

In the analysis presented in this paper, use is made of four observed steady state salinity distributions. The time periods, freshwater discharges and tidal ranges pertaining to these distributions are presented in Table I.

The salinity distributions for August 1977 and March 1980 are based on 13 hour measurements. Using a Beckman (induction type) salinometer, salinities are measured in seven cross-sections. For locations see Figure 1. The number of stations in each cross-section is 6 for sections 8 and 15, 5 for sections 2, 4, 6 and 13 and 1 for section 10. To verify the salinity readings water samples are taken at regular intervals and analyzed in the laboratory. Observations are made during both the downcast and the upcast. The first reading of the downcast and the last reading of the upcast are always 0.5 m below the water surface. The last reading of the downcast (= first reading of the upcast) is about 0.5 m above the bottom. The number of readings in a cast is such that the cast can be accomplished in 15 minutes.

Table I. Review of Experiments

Period	Freshwater m^3/s	Tidal Range m
August 1977	50	3.50
March 1980	100	3.50
August 1983	50	3.50
June 1986	25	2.70

The longitudinal salinity distributions for August 1983 and June 1986 are determined from observation in the permanent recording stations I-IV; for location see figure (1). In these stations conductivity and temperature is measured every 16 minutes at 1 m below the surface and at 1 m above the bottom. Except for station I, on a tidally averaged basis (using the M_2 period), the average of the two salinities corresponds closely to the cross-sectionally averaged salinity at the location of the permanent station (Robaczewska, 1986).

The tidally- and cross-sectionally averaged salinities for each of the experiments are plotted in Figure 2. Values are normalized with respect to the salinities at the mouth of the estuary. To arrive at cross-sectionally averaged values for August 1983 and June 1986, a correction is applied to the salinities measured in Station I.

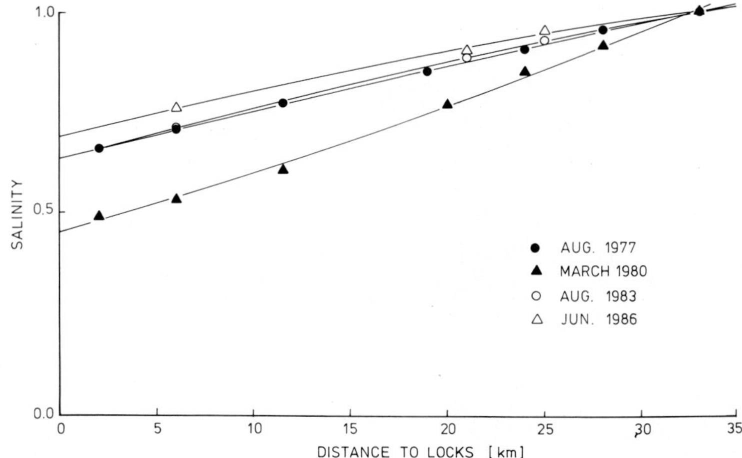

FIGURE 2. Longitudinal distributions of tidally and cross-sectionally averaged salinities. Salinities are normalized with respect to the salinities at the mouth; Aug. 1977, $\langle\bar{\bar{s}}\rangle_0$= 30.0 ppt, Aug. 1983, $\langle\bar{\bar{s}}\rangle_0$ = 29.3 ppt, June 1986, $\langle\bar{\bar{s}}\rangle_0$ = 29.3 ppt, March 1980, $\langle\bar{\bar{s}}\rangle_0$ = 27.2 ppt.

V. Salt Intrusion Mechanisms

For the measurements carried out in August 1977, a thorough analysis of the salt intrusion mechanisms is presented in Dronkers and van de Kreeke (1986). In their paper, the tidally averaged salt flux K in a fixed cross-section at a location x is written as

$$K = \langle Q \rangle \langle \overline{\overline{s}} \rangle + \langle Q_1 s_1 \rangle + \langle A\ \overline{\overline{u_2 s_2}} \rangle \tag{2}$$

(Note that for steady state conditions and assuming a simple harmonic tide, averaging over n tidal periods is equivalent to averaging over one tidal period.) The parameters on the right hand side of Eq.(2) follow from the decomposition of the volume flux $Q(t)$, the velocity $u(y,z,t)$ and the salinity $s(y,z,t)$, respectively.

$$Q = \langle Q \rangle + Q_1(t) \tag{3}$$

$$u = \langle \overline{\overline{u}} \rangle + u_1(t) + u_2(y,z,t) \tag{4}$$

$$s = \langle \overline{\overline{s}} \rangle + s_1(t) + s_2\ (y,z,t) \tag{5}$$

In these equations y is the lateral coordinate and z is the vertical coordinate. For steady state conditions, $\langle Q \rangle = Q_f$ is the freshwater discharge and $\langle \overline{\overline{u}} \rangle = u_f$ is the freshwater discharge velocity. Terms with subscript 1 represent tidal variations in velocity and salinity and terms with subscripts 2 are velocity and salinity deficits relative to the cross-sectional mean. In particular, u_2 is associated with oscillatory shear and longitudinal residual currents including gravitational circulation. For steady state conditions ($K = 0$), $\langle Q \rangle \langle \overline{\overline{s}} \rangle$ represents the downestuary directed advective flux associated with the freshwater inflow and ($\langle Q_1 s_1 \rangle + \langle A\overline{\overline{u_2 s_2}} \rangle$) is the upestuary flux. For reasons to be explained in the next paragraph, $\langle Q_1 s_1 \rangle$ is referred to as the non-local dispersive flux and $\langle A\overline{\overline{u_2 s_2}} \rangle$ is referred to as the local dispersive flux.

To delineate the processes responsible for $\langle Q_1 s_1 \rangle$ and to derive conditions that allow writing the upestuary flux as a gradient type transport, the salt flux, $\tilde{K}$, in a cross-section oscillating about a position x with the tidal velocity is decomposed in a similar fashion as K. Here K is the salt flux in the fixed cross-section at the position x. More precisely, at $t = 0$ the moving cross-section is at the position x of the fixed cross-section and oscillates in a fashion that during the tidal cycle the volume of water upestuary of the moving cross-section remains constant. Therefore, $\langle \tilde{Q} \rangle = Q_f$ and $\tilde{Q}_1 = 0$. The decomposition of $\tilde{K}$ can be written as

$$\tilde{K} = \langle \tilde{Q} \rangle \langle \overline{\overline{\tilde{s}}} \rangle + \langle \tilde{A}\ \overline{\overline{\tilde{u}_2 \tilde{s}_2}} \rangle \tag{6}$$

The tilde refers to quantities in the moving plane. In Eq.(6), the first and second term on the right hand side, respectively, are the downestuary and upestuary salt flux. For steady state conditions, implying

$$\langle \tilde{Q} \rangle = \langle Q \rangle = Q_f \tag{7}$$

and

$$\tilde{K} = K = 0 \tag{8}$$

and selecting $t = 0$ such that

$$\langle \overline{\overline{\tilde{s}}} \rangle = \langle \overline{\overline{s}} \rangle \tag{9}$$

it follows from Eqs.(2) and (6) that $\langle Q_1 s_1 \rangle$ is a result of differences in u_2 and s_2 in the fixed and oscillating cross-sections. Hence, the designation non-local salt flux. The flux $\langle A\overline{\overline{u_2 s_2}} \rangle$ is referred to as the local salt flux. The non-local salt flux is expected to be large where differences between $\tilde{u}_2$ and u_2 and differences between $\tilde{s}_2$ and s_2 are large, i.e., in areas with strong longitudinal variations in the bathymetry; for example, at the confluence of channels, where tidal flats are present and within one tidal excursion of the mouth. This agrees with the calculated salt fluxes for August 1977, presented in Figure 3. In the seaward part, with its many tidal flats and channels and the close proximity to the mouth, $\langle Q_1 s_1 \rangle$ is large whereas in the landward part its value is small. In a qualitative sense the same pattern is expected to hold for the other experiments.

VI. The Gradient Type Flux

By definition the volume transport through the moving cross-section associated with the velocity $\tilde{u}_2$ is zero. However, in combination with cross-sectional mixing,

$\tilde{u}_2$ causes a random flux of particles and an upestuary flux of salt. Provided certain conditions are satisfied, the salt flux when averaged over a suitable time period can be written as a gradient type flux, i.e.,

$$\overline{\overline{<\tilde{A}\tilde{u}_2\tilde{s}_2>}} = - \tilde{A}_o \tilde{D} \frac{\partial <\overline{\overline{\tilde{s}}}>}{\partial \zeta} \tag{10}$$

In Eq.(10), ζ is the distance measured from the oscillating plane in the positive x direction. $\tilde{A}_o$ is the cross-sectional area averaged over the path of the moving plane.

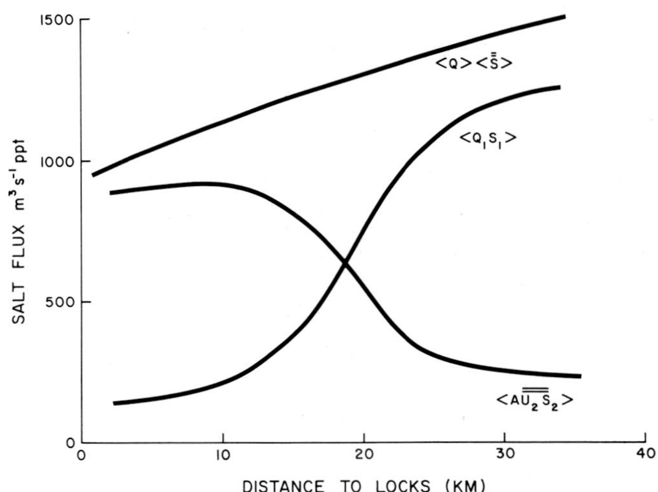

FIGURE 3. Values of salt fluxes in Eq.(2) for August 1977.

In part based on Dronkers (1982), conditions for a gradient type flux of salt in the oscillating cross-section are presented in Appendix A. They can be summarized as follows.

- The averaging period is equal or larger than the time scale for cross-sectional mixing, i.e.,

$$nT \geq \tau \tag{11}$$

For steady state conditions and assuming $\tau \geq T$, Eq.(11) may be replaced by

$$nT = T \quad \text{or} \quad n = 1$$

- The freshwater discharge velocity is much smaller than the tidal velocity, i.e.,

$$u_f \ll u_1 \tag{12}$$

- Variations of salinity over the cross-section are very small compared to the cross-sectionally averaged values, i.e.,

$$\tilde{s}_2 \ll \overline{\overline{\tilde{s}}} \tag{13}$$

- $\overline{\overline{\tilde{s}_2}}$ is sufficiently linear over a distance $\ell = \tilde{u}_2\tau$ (for a more precise definition of ℓ, see Eqs.(2), (3) and (4) in Appendix A), i.e.,

$$\frac{\partial^2 \overline{\overline{\tilde{s}}}}{\partial \zeta^2} \ll \frac{1}{\ell} \frac{\partial \overline{\overline{\tilde{s}}}}{\partial \zeta} \tag{14}$$

- and $\tilde{\bar{\bar{s}}}$ is sufficiently constant over the time period τ, i.e.,

$$\frac{\partial \tilde{\bar{\bar{s}}}}{\partial t} << \frac{\tilde{\bar{\bar{s}}}}{\tau} \tag{15}$$

Eq.(10) pertains to a cross-section oscillating with the tidal velocity u_1 about a position $x = x_o$. To arrive at a similar expression for the upestuary flux in a fixed cross-section located at $x = x_o$ it is assumed that

$$<Q_1 s_1> \quad << \quad <\overline{\overline{Au_2 s_2}}> \tag{16}$$

For steady state conditions it then follows from Eqs.(2) and (6) that the upestuary salt flux in the fixed cross-section is approximately equal to the upestuary flux in the moving cross-section, i.e.,

$$<\overline{\overline{\tilde{A}\tilde{u}_2 \tilde{s}_2}}> \approx <\overline{\overline{Au_2 s_2}}> \tag{17}$$

If in addition, it is assumed that over the length of the tidal excursion, λ, $<\bar{\bar{s}}>$ is sufficiently linear, i.e.

$$\frac{\partial^2 <\bar{\bar{s}}>}{\partial x^2} << \frac{1}{\lambda} \frac{\partial <\bar{\bar{s}}>}{\partial x} \tag{18}$$

then,

$$\frac{\partial <\tilde{\bar{\bar{s}}}>}{\partial \varsigma}\Big|_{x(t)} \approx \frac{\partial <\bar{\bar{s}}>}{\partial x}\Big|_{x(o)} \tag{19}$$

where x(t) is the position of the moving cross-section. In Appendix B it is suggested that if Eq.(16) holds than also Eq.(18) is satisfied. In general, $\ell > \lambda$ and therefore if Eq.(18) is satisfied, Eq.(14) is satisfied. Making use of Eqs.(10), (17) and (19) and provided the conditions expressed by Eqs.(11), (12), (13), (15) and (16) are met, it follows,

$$<A \, \overline{\overline{u_2 s_2}}> = - A_0 D \frac{d<\bar{\bar{s}}>}{dx} \tag{20}$$

Eq.(20) implies that the local salt flux when averaged over the tidal period can be expressed as a gradient type flux, i.e., the salt flux is proportional to the local salinity gradient.

For the landward and seaward part of the Volkerak Estuary, typical values of the parameters that play a role in the various conditions for a gradient type salt flux are presented in Table II. It can be deduced that conditions for a gradient flux are met in the landward part. Therefore, for this part of the Volkerak estuary and assuming steady state conditions, the advection dispersion equation

$$Q_f <\bar{\bar{s}}> + A_o \, D \frac{d<\bar{\bar{s}}>}{dx} = 0 \tag{21}$$

is a valid description for the salt balance and the dispersion coefficient, D, is a physically meaningful parameter.

In the seaward part some conditions are satisfied and others are not. In particular, the condition expressed by Eq.(16) is not valid. However, this does not necessarily imply that the salt flux in the seaward part cannot be of the gradient type. The conditions expressed by Eqs. (11), (12), (13), (15) and (16) should be looked upon as sufficient conditions rather than necessary conditions. For example, as shown in Dronkers (1978), in parts of the estuary were trapping is important and therefore $<Q_1 s_1>$ large, the upestuary salt flux can still be written as a gradient type flux provided the longitudinal salinity gradient is sufficiently uniform over a distance equal to the tidal excursion. In that case, in the absence of density gradients

$$D :: \hat{u}_1^2 T \tag{22}$$

where $\hat{u}_1$ is the amplitude of the tidal velocity u_1. In the seaward part of the Volkerak Estuary the condition of the a uniform longitudinal salinity gradient is only marginally satisfied. Accordingly, for this part of the estuary, Eqs.(21) and (22) should be treated as working formulae and D be regarded as an empirical coefficient.

Table II. Order of magnitude values of characteristics of the velocity and salinity field in the landward and seaward part of the Volkerak Estuary during the experiments

Landward Part			Seaward Part
5×10^{-3}	u_f	(m/s)	2×10^{-3}
1.5×10^{-1}	u_2	(m/s)	3×10^{-1}
2.5×10^{-1}	u_1	(m/s)	1.25
3×10^{3}	λ	(m)	1.5×10^{4}
2×10^{3}	ℓ	(m)	4×10^{3}
2×10^{1}	$\langle\bar{\bar{s}}\rangle$	(ppt)	2.5×10^{1}
1×10^{1}	s_2	(ppt)	3×10^{-1}
4×10^{-4}	$d\langle\bar{\bar{s}}\rangle/dx$	(ppt/m)	3×10^{-4}
5×10^{-9}	$d^2\langle\bar{\bar{s}}\rangle/dx^2$	(ppt/m²)	5×10^{-9}
1×10^{2}	$\langle Q_1 s_1\rangle$	(m³/s ppt)	1.2×10^{3}
9×10^{2}	$\langle \overline{\overline{A u_2 s_2}}\rangle$	(m³/s ppt)	1×10^{2}
4.5×10^{4}	τ	(s)	4.5×10^{4}

VII. Relationship Between Dispersion Coefficient and Velocity Field

As shown in the literature and summarized by Dronkers and Zimmerman (1982), the longitudinal dispersion coefficient D can be related to a characteristic value of the velocity u_2 and a time scale for mixing. The velocity u_2 is associated with the steady lateral and vertical circulations resulting from longitudinal salinity gradients and/or wind in combination with depth variations in the cross-section and the unsteady lateral and vertical oscillatory shear. For lateral circulation

$$D_\ell \approx u_\ell^2\, \tau_\ell \tag{23}$$

and for vertical circulation

$$D_v \approx u_v^2\, \tau_v \tag{24}$$

In Eqs. (23) and (24), u_ℓ and u_v are characteristic values for respectively lateral and vertical circulation. τ_ℓ is the time scale for lateral mixing and τ_v is the time scale for vertical mixing. For lateral and vertical oscillatory shear and assuming τ_ℓ and τ_v to be large compared to the tidal period, T, the expressions for the dispersion coefficients are respectively

$$D'_\ell \approx (u'_\ell)^2\, T\, \frac{T}{\tau_\ell} \tag{25}$$

and

$$D'_v \approx (u'_v)^2\, T\, \frac{T}{\tau_v} \tag{26}$$

In Eqs.(25) and (26), u'_ℓ and u'_v are measures for respectively the velocity variations associated with lateral and vertical shear.

VIII. Values of the Longitudinal Dispersion Coefficients

To determine the values of the longitudinal dispersion coefficients from Eq.(21), second degree polynomials are fitted through the data [see Figure (2)]. Derivatives $d\langle \bar{\bar{s}} \rangle/dx$ are determined by differentiating the polynomials. With the known values of $\langle \bar{\bar{s}} \rangle(x)$, $d\langle \bar{\bar{s}} \rangle/dx$, $\langle Q \rangle$ and $A_o(x)$, values of longitudinal dispersion coefficient D(x) are calculated. The results are presented in Figure 4. In the calculations values of $A_o(x)$ are smoothed by taking a running average with a length scale equal to the tidal excursion.

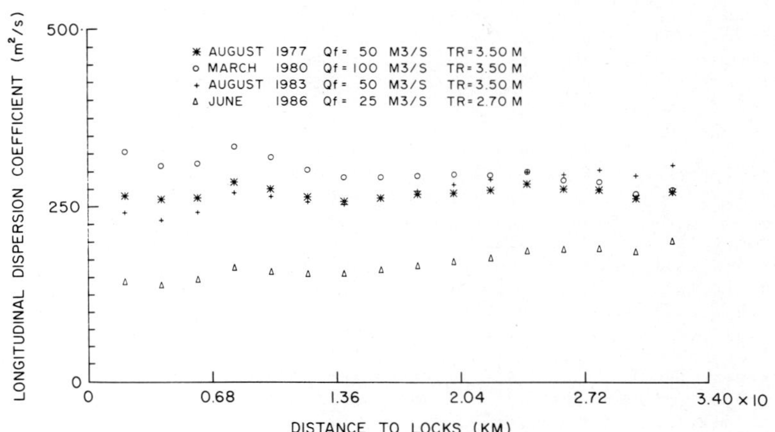

FIGURE 4. Values of longitudinal dispersion coefficient.

Typical values of the longitudinal dispersion coefficients in the seaward part of the estuary are 280 m^2/s for August '77, March '80 and August '83 , and 175 m^2/s for June '86. Values are relatively uniform along this part of the estuary. Typical values in the landward part are 250 m^2/s for August '77 and August '83, 325 m^2/s for March '80 and 150 m^2/s for June '86. In June '86, August '77 and August '83, there is a weak trend for the values to increase and in March '80 a weak trend for the values of the dispersion coefficient to decrease in a seaward direction. In comparing with other estuaries it should be kept in mind that the length of the Volkerak Estuary is smaller than the salt intrusion length scale resulting in non-zero values of the salinity at the head.

IX. Discussion

In discussing the variation in the values of the longitudinal dispersion coefficients of the Volkerak estuary a distinction is made between the landward part (km 1 to km 10) and the seaward part (km 20 - km 34) of the estuary. In the landward part the upestuary salt flux is most strongly reflected in the local salt flux $\langle \underline{\underline{A u_2 s_2}} \rangle$; see Figure (2). As shown in Dronkers and van de Kreeke (1986) for the period Aug. '77, the local salt flux in this part of the estuary is associated with steady (=tidal mean) lateral and vertical variations in the longitudinal velocity and unsteady vertical variations in the longitudinal velocity. Unsteady lateral variations play a minor role in the upestuary flux. Although a detailed analysis was not carried out it seems safe to assume that the foregoing results also hold for the other experiments.

Expressions for the contributions to the longitudinal dispersion coefficient by steady lateral, steady vertical and unsteady vertical variations of the longitudinal velocity in a cross-section are given by Eqs. (23), (24) and (26), respectively. An estimate of the value of the time scale, τ_v, follows from (Fischer, 1979)

$$\tau_v \approx 0.35 \frac{h^2}{K_z} \tag{27}$$

in which h is mean depth and K_z is an effective mixing coefficient. For the landward part of the Volkerak estuary with a value of the Richardson number $\langle Ri \rangle \approx 4$, $K_z \approx 5$ cm²/s (van de Kreeke and Robaczewska, 1989). With h = 10 m it follows $\tau_v \simeq 0.7 \times 10^5$ s. Most of the lateral mixing is expected to take place during times of current reversal and therefore the order of the time scale for lateral mixing τ_ℓ equals the order of τ_v equals the tidal period T.

During each period of constant freshwater inflow, the longitudinal distribution of the tidally and cross-sectionally averaged salinities varied little in spite of fortnightly variations in the tide and variations in vertical circulation and stratification resulting from variations in wind speed and direction (Dronkers and van de Kreeke, 1986), (van de Kreeke and Robaczewska, 1989). This is also confirmed by the longitudinal salinity distributions for Aug. '77 and Aug '83 and presented in Figure (2). The salinity distributions are the same for both periods whereas wind conditions differed.

To explain the differences in the values of the longitudinal dispersion coefficients for the different experiments in the landward part of the estuary it is reasoned that for the same tidal range, an increase in freshwater discharge results in an increase in stratification and a decrease in the turbulent exchange of mass and momentum. As a result, values of u_v and τ_v increase. To the extent that u_ℓ is determined by density effects its value is also expected to increase. The velocity u'_v and the time scale for lateral mixing are mainly determined by the tide and therefore are little affected by the increase in freshwater inflow. Making use of Eqs.(23), (24) and (26) and taking into account the increase in freshwater inflow it follows that compared to Aug. '77/Aug. '83, for March 80 in the landward part of the estuary values of D_v, and D_ℓ increase and values of D'_v decrease. Provided the increase in $D_v + D_\ell$ is larger than the decrease in D'_v this could explain the observed increase in the value of the overall dispersion coefficient D, see Figure 4.

For constant freshwater inflow, a decrease in tidal range and therefore in tidal velocities leads to a reduction in the vertical turbulent exchange of mass and momentum. This is reinforced by an increase in stratification. Therefore, values of u_v, τ_v and u_ℓ are expected to increase. Because the tidal period remains the same, the time scale for lateral mixing is not expected to change. The velocity u'_v decreases. Making use of Eqs.(23), (24) and (26) it follows that D_v and D_ℓ are expected to increase for a decrease in tidal range. The value of D'_v can go either way depending on the relative change in $(u'_v)^2$ and τ_v. Therefore, the decrease in tidal range most likely leads to an increase in the value of the longitudinal dispersion coefficient D. Comparing the results of June '86 and Aug '77/Aug '83 it follows that this increase is more than offset by the decrease in the longitudinal dispersion coefficient resulting from the decrease in freshwater discharge.

In the seaward part of the Volkerak estuary, salt intrusion is caused by large scale exchanges of water bodies and salt between the estuary and the Eastern Scheldt, between channels at channel junctions and between channels and flats. These processes are reflected in the non-local salt flux $\langle Q_1 s_1 \rangle$. Because density gradients are small and waters are well-mixed in this part of the estuary, density currents hardly play a role in the exchange of salt. Therefore values of longitudinal dispersion coefficients vary little with a change in freshwater discharge; compare values of Aug. '77/Aug. '83 and March '80. On the contrary, the reduction in tidal range from 3.5 m to 2.7 m leads to a significant decrease in the values of the dispersion coefficient; compare values of Aug. '77/Aug. 83 and June '86. As stated earlier, in the absence of density gradients, the longitudinal dispersion coefficient is proportional to the square of the tidal velocity. Starting with a value of D = 280 m²/s for Aug. '77/Aug '83 and assuming the tidal velocity to be proportional to the tidal range this would lead to a value of $((2.7/3.5)^2 \times 280 =)$ 166 m²/s for June '86. This is close to the observed value of 175 m²/s.

X. Conclusions

For the range of freshwater discharges and tidal ranges encountered in the experiments, values of longitudinal dispersion coefficients determined from steady salinity distributions range from 150 to 250 m²/s in the landward part and from 175 to 280 m^2/s in the seaward part. In a qualitative sense variations in the values of the longitudinal dispersion coefficients for the different experiments can be explained in terms of variations in freshwater discharge and tide. Predicting values of longitudinal dispersion coefficients in estuaries in terms of tide and freshwater discharge is still not possible. In part this is due to the large spectrum of processes that is responsible for the upestuary salt flux and the strict conditions that need to be satisfied to justify a gradient type transport. Short of field measurements, for engineering purposes values of longitudinal dispersion coefficients can best be estimated from known values of estuaries with a similar geometry, tide and freshwater inflow.

Acknowledgement

Measurements were carried out by the hydrographic survey of the Rykswaterstaat, The Netherlands. Most of the calculations were carried out by Ms. K. Robaczewska.

XI. References

Bowden, K. F., 1963: The mixing processes in a tidal estuary. International Journal of Air and Water Pollution, **7**, 343-356.

Dronkers, J., 1978: Longitudinal dispersion in shallow well mixed estuaries. Proceedings of the 16th Coastal Engineering Conference, Hamburg, **3**, 2761-2777.

Dronkers, J., 1982: Conditions for gradient-type dispersive transport in one dimensional, tidally averaged transport models. Estuarine, Coastal, and Shelf Science, **14**, 599-621.

Dronkers, J. and J. T. F. Zimmerman, 1982: Some principles of mixing in tidal lagoons. Oceanologica Acta no sp.

Dronkers, J. and J. van de Kreeke, 1986: Experimental determination of salt intrusion mechanisms in the Volkerak Estuary. Netherlands Journal of Sea Research, **20(1)**, 1-19.

Fischer, H. B., E. J. List, R. C. Y. Koh, J. Imberger and N. H. Brooks, 1979: Mixing in Inland and Coastal Waters. Academic Press, New York.

Hughes, P., 1958: Tidal mixing in the narrows of the Mersey Estuary. Geophysical Journal of the Royal Astronomical Society, **1**, 271-283.

Officer, C. B., 1976: Physical oceanography of estuaries and associated coastal waters. John Wiley and Sons, New York, p. 465.

Paulson, R. W., 1970: Variation of the longitudinal dispersion coefficient in the Delaware River Estuary as a function of freshwater inflow. Water Resources Research, Vol. 6, No. 2, pp. 516-526.

Robaczewska, K., 1986: Relationship between dispersion coefficients, tidal velocity and freshwater discharge. Rijkswaterstaat, Tidal Hydraulics Division. Notitie GWAO-86.436, p. 10 (Dutch text).

Uncles, R. J. and P. J. Radford, 1980: Seasonal and spring-neap tidal dependence of axial dispersion coefficients in the Severn - a wide, vertically mixed estuary. J. Fluid Mech., Vol. 98, part 4, pp. 703-726.

van de Kreeke, J. and K. Robaczewska, 1989: Vertical circulation and stratification in the partially mixed part of the Volkerak Estuary. Netherlands Journal of Sea Research, **23(3)**, 239-253.

Wilson, R. E. and A. Okubo, 1978: Longitudinal dispersion in a partially mixed estuary. Journal of Marine Research, Vol. 36, No. 3.

West, J. R. and D. J. A. Williams, 1972: An evaluation of mixing in the Tay Estuary. American Society of Civil Engineers. Proceedings of the Thirteenth Conference on Coastal Engineering, pp. 2153-2169.

Appendix A

Conditions for a Gradient Type Flux in a Moving Frame

Consider the path of a fluid particle in a two dimensional flow field relative to a frame moving with the depth averaged velocity. Assuming the freshwater velocity, u_f, to be very small compared to the tidal velocity u_1, i.e.,

$$u_f \ll u_1 \tag{A.1}$$

and defining τ as the time scale for cross-sectional mixing, the particle describes an eddy type motion in the interval τ by the combined action of residual currents and cross-sectional mixing; see Figure (A-1). Here residual currents are defined as currents observed relative to the moving frame. The residual currents can either be steady, e.g., gravitational circulation or unsteady, e.g., currents associated with oscillatory shear. Defining $\tilde{u}_2$ as the longitudinal velocity relative to the moving frame, for steady residual currents the length scale ℓ of the eddy is

$$\ell \approx \tilde{u}_2 \tau \tag{A.2}$$

For unsteady oscillatory motions

$$\ell \approx \tilde{u}_2 \tau \qquad \tau \ll T \tag{A.3}$$

$$\ell \approx \tilde{u}_2 T \qquad \tau \gg T \tag{A.4}$$

where T is the tidal period. Except for the last case the length scale of the eddy and the time scale for cross-sectional mixing are related. Note that the time scale for cross-sectional mixing is the equivalent of the Lagrangian decorrelation time scale when dealing with turbulent mixing (Fischer et al., 1979).

To derive an expression for the salt flux associated with the eddy type motion a simple box model is used; see Figure (A.1). The length of each box equals ℓ and each box contains an equal number of particles n. Each particle carries a mass m. It is assumed that the variation of m in the cross-section is relatively small, i.e.

$$\tilde{m}_2 \ll \tilde{\bar{\bar{m}}} \tag{A.5}$$

Therefore $m \approx m(\zeta)$, where ζ is the distance of the particle to the moving frame. Furthermore, it is assumed that $m(\zeta)$ is a monotone increasing function and is sufficiently linear over the distance ℓ, i.e.

$$\frac{\partial^2 \tilde{m}}{\partial \zeta^2} \ll \frac{1}{\ell} \frac{\partial \tilde{m}}{\partial \zeta} \tag{A.6}$$

and is sufficiently constant over the time interval τ, i.e.

$$\frac{\partial \tilde{m}}{\partial t} \ll \frac{\tilde{\bar{\bar{m}}}}{\tau} \tag{A.7}$$

It is postulated that in view of the eddy type motion over a time period τ a particle has a probability k to transfer from one box to the other. The dimension of k is s^{-1}. Making use of Eq.(A.7), it follows that when averaging over the time period, τ, (the condition expressed by Eq.(11) in the main text) the mass flux across the dividing line of the boxes associated with particles located originally between $-\zeta$ and $-(\zeta + d\zeta)$ and between ζ and $(\zeta + d\zeta)$ is

$$dq = K \frac{d\zeta}{\ell} n \, [m(-\zeta) - m(+\zeta)] \tag{A.8}$$

Introducing the concentration

$$\overline{\overline{\tilde{s}}}(\varsigma) = \frac{n}{\ell}\, m(\varsigma) \tag{A.9}$$

and making use of Eq.(A.6), it follows

$$dq = -\, k\, 2\varsigma \frac{d\overline{\overline{\tilde{s}}}}{d\varsigma}\, d\varsigma \tag{A.10}$$

The average flux per unit time over the time interval ς when accounting for all the particles in the entire area covered by the eddy is

$$K = -\, k \int_0^{\ell} 2\varsigma d\varsigma \frac{d\overline{\overline{\tilde{s}}}}{d\varsigma} \tag{A.11}$$

and therefore,

$$K = -\, k\ell^2 \frac{d\overline{\overline{\tilde{s}}}}{d\varsigma} \tag{A.12}$$

It can be reasoned that in a time τ the probability of a particle in the boxes crossing the dividing line is close to one and therefore,

$$k \approx \frac{1}{\tau} \tag{A.13}$$

It follows

$$K = -\, \frac{\ell^2}{\tau} \frac{d\overline{\overline{\tilde{s}}}}{d\varsigma} \tag{A.14}$$

with ℓ given by Eqs. (A.2)-(A.4). Note that when making use of Eq.(A.9), Eqs. (A.1), (A.5), (A.6) and (A.7) correspond to respectively Eqs. (12), (13), (14) and (15) of the main text.

The foregoing simplified model demonstrates that the longitudinal salt flux in a moving frame when averaged over the time scale for cross-sectional mixing can be written as a gradient type flux provided certain conditions, i.e., Eqs. (11)-(15) in the main text are satisfied. For a more detailed and theoretical treatment of this problem the reader is referred to Dronkers (1982).

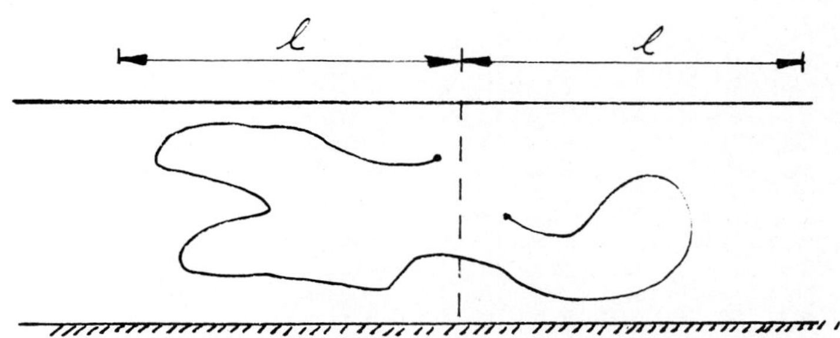

FIGURE A-1. Particle path. 2 ℓ is length scale of eddy.

Appendix B

The Covariance of Tidal Variations in Velocity and Salinity as an Indicator for a Gradient Type Flux in a Fixed Frame

Assuming the salt flux in the oscillating cross-section to be of the gradient type, additional conditions for a gradient type flux in the fixed cross-section are that $\partial\langle\overline{\overline{s}}\rangle/\partial x$ is sufficiently uniform over a distance on the order of the tidal excursion and $<Q_1 s_1> \ll <\overline{\overline{Au_2 s_2}}>$. The following suggests that if $<Q_1 s_1> \lesssim\!< <\overline{\overline{Au_2 s_2}}>$ than $\partial<\overline{\overline{s}}>/\partial x$ is sufficiently linear. We start with the assumption that $\partial\overline{\overline{s}}/\partial x$, and the tidal velocity $u_1(x,t)$ are sufficiently uniform over a distance equal to the tidal excursion, λ. Therefore,

$$\frac{\partial^2 \overline{\overline{s}}}{\partial x^2} \ll \frac{1}{\lambda}\frac{\partial \overline{\overline{s}}}{\partial x} \tag{B.1}$$

$$\frac{\partial u_1}{\partial x} \ll \frac{u_1}{\lambda} \tag{B.2}$$

Furthermore, we assume that the longitudinal distribution of the cross-sectionally averaged salinity translates with the tidal velocity u_1 and therefore,

$$\frac{\partial \overline{\overline{s}}}{\partial t} + u_1 \frac{\partial \overline{\overline{s}}}{\partial x} = 0 \tag{B.3}$$

When differentiating Eq.(B.3) with respect to x and making use of Eqs.(B.1) and (B.2) it follows that to a first approximation $\partial\overline{\overline{s}}/\partial x$ is independent of time. Integrating Eq.(B.3) with respect to time yields

$$\overline{\overline{s}} = \frac{d\overline{\overline{s}}}{dx}\int_0^t u_1 dt + \text{const} \tag{B.4}$$

and therefore when assuming u_1 to be periodic in time

$$<u_1 \overline{\overline{s}}> = < u_1 \int u_1 dt > \frac{d\overline{\overline{s}}}{dx} \tag{B.5}$$

Writing

$$\overline{\overline{s}} = <\overline{\overline{s}}> + s_1 \tag{B.6}$$

it follows

$$<u_1 s_1> = 0 \tag{B.7}$$

When neglecting time variations in the cross-sectional area

$$Q_1 = A_o u_1 \tag{B.8}$$

and therefore

$$<Q_1 s_1> = 0 \tag{B.9}$$

In that case,

$$<Q_1 s_1> \ll <\overline{\overline{Au_2 s_2}}> \tag{B.10}$$

Although not necessarily true from a strict mathematical point of view, if Eq.(B.9) and therefore Eq.(B.10) holds then the inequalities (1) and (2) are valid.

Mixing and Trapping in Australian Tropical Coastal Waters

Eric Wolanski and Peter Ridd

Australian Institute of Marine Science
P.M.B. No. 3, Townsville, M.C., Queensland, 4810
Australia

ABSTRACT

Evidence is presented indicating the existence in tropical Australia of a coastal boundary layer, often visible as a nearshore band of turbid water. The coastal boundary layer inhibits mixing between the nearshore and offshore water by two effects, firstly by trapping in mangrove swamps, and secondly due to frictional effects which inhibit inertial jets. Contrast was found between coastal boundary layer dynamics in the Gulf of Carpentaria, which has shallow inshore water and a smooth coastline, and the continental shelf of the Great Barrier Reef which has deeper inshore water and a more rugged coastline. In the Gulf, the coastal boundary layer maintains its integrity for much longer time scales than in the Great Barrier Reef continental shelf.

I. Introduction

The coastal boundary layer (hereafter CBL) is a zone of water with a particular physical, chemical or biological property that is different from the offshore water. A CBL may trap freshwater runoff, nutrients, suspended sediments, water of temperature different from estuarine and offshore waters, etc... When the CBL is well defined, it acts as a barrier to direct mixing between the estuarine and the offshore water, mixing occurring instead between estuarine and CBL water, and between CBL and offshore water. An obvious example is when steady river plumes that extend to the sea floor prevail in coastal waters (Csanady, 1984). Some well documented attention has been given to CBL dynamics in European and American temperate estuaries, principally as river plumes (e.g. Bowman, 1988). The CBL may not be important if the coastal waters are deep enough, and the freshwater discharge important all year round, that the tidal currents and (or) the wind are generally unable to break down the vertical salinity stratification in coastal waters. In this case, sea water moves up-river and controls the hydrographic properties up-river, (e.g. the analytical model of Zhang et al. 1987) and mixing between estuarine and offshore water occurs in the estuary (e.g. the pioneer field and model studies of Pritchard, 1952; Rattray and Hansen, 1966, and others).

Mixing between estuarine and offshore water also occurs at tidal frequency at the river mouth. A tidal jet may exist at ebb flood tide, and radial flow at flood tide (e.g. Fischer et al., 1979; Joshi and Taylor, 1983; Ozsoy and Unluata, 1982; Ozsoy, 1986). This asymmetry helps determine the value of the return coefficient, R, of an estuary, i.e. the fraction of the volume of water leaving an estuary at ebb tide that returns at flood tide. Pioneer studies of the return coefficient were undertaken by Pritchard (1960). However, the value of R has been measured for few estuaries. This value was found to range typically from 0.24 to 0.6 (van de Kreeke, 1988). The value of R is difficult to predict. Simple tidal jets or steady river plume dynamics do not always prevail, and the value of R may be more controlled in some cases by the dynamics of unsteady river plumes (e.g. a review in Bowman, 1988). Numerical models, as opposed to simpler analytical models available for steady river plumes, are needed to estimate the fate of unsteady river plumes (e.g. Chao, 1988; Chao and Boicourt, 1986; Royer and Emery, 1985; Wolanski and Banner, 1978). These

Coastal and Estuarine Studies, Vol. 38
R. T. Cheng (Ed.)
Residual Currents and Long-term Transport

models are not always able to predict the value of R because the assumptions behind these models are too simplistic, and to remove them requires further detailed field studies.

Nearshore dynamics are even more complex, and predictions of the value of R are even less amendable to modelling in systems comprising many estuaries closely spaced along a smooth coastline. In that case, water leaving an estuary may simply be carried to the next estuary at tidal frequency, estuarine water leaving an estuary at ebb tide and entering the same or nearby estuaries at flood tide. In that case, a CBL may exist though the water may be located in the estuaries at high tide and nearshore along the shoreline at low tide. Recirculation from one estuary to the other may prevail also at low-frequencies, when atmospheric-driven sea level fluctuations in coastal waters alternatively withdraw estuarine water from estuaries, or inject coastal water into estuaries (e.g. Wong, 1986; Smith, 1988). Presumably, estuarine water may be circulated from estuary to estuary, though presumably it may be somewhat diluted with offshore water.

Mixing rates between CBL and offshore waters can be expected to decrease with increasing uncoupling of coastal and offshore dynamics. Bottom friction would lead to such uncoupling if coastal waters are shallow even in the absence of buoyancy effects, provided the sea bottom has a slope (Csanady,1982; Hopkins and Swoboda, 1986).

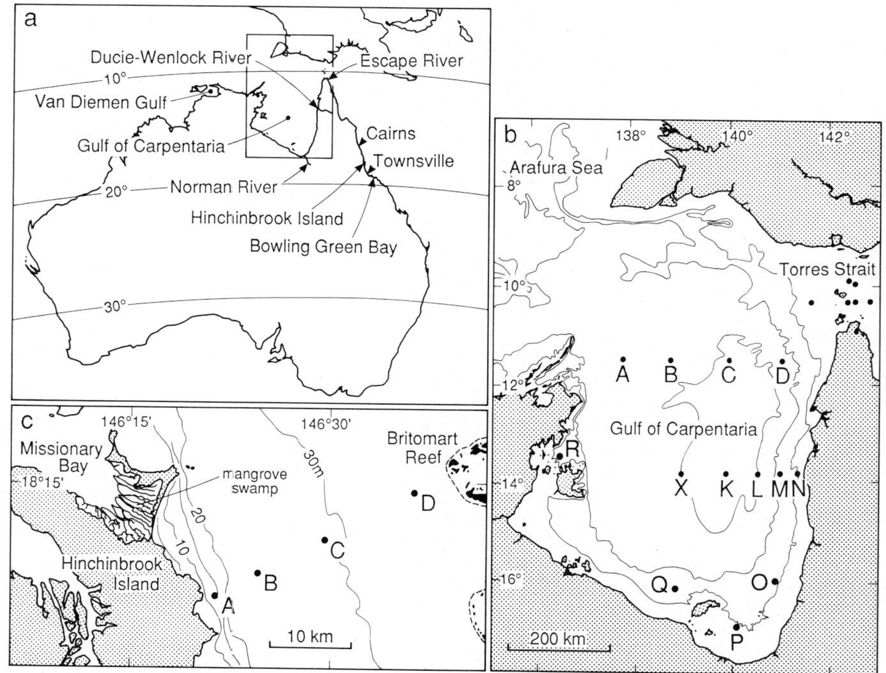

FIGURE 1. Location map of our study sites in tropical Australia (a), of Missionary Bay and the inner shelf of the Great Barrier Reef (b), and of the Gulf of Carpentaria (c). The bathymetry (in m) is shown in (b) and (c). The letters and solid circles indicate the location of current meter moorings.

In this study, it is demonstrated that CBL processes inhibit the mixing between estuarine and offshore waters in tropical Australia. We begin by describing the situation in the Gulf of Carpentaria where we present evidence of long-term (several months) trapping of brackish water in the nearshore zone. This trapping is attributed to the existence of a barotropic friction-driven CBL, and we find evidence for this layer in satellite images, CTD and current meter data, and numerical predictions. Data from the Great Barrier Reef continental shelf are then presented, showing that the coastal boundary layer is only short-lived (a few days) and patchy. This finding is attributed to the inshore waters being much deeper than those of the Gulf of Carpentaria, and to the presence of numerous capes and headlands on the Great Barrier Reef continental shelf but not in the Gulf of Carpentaria. This finding implies rapid cross-shelf mixing, on the Great Barrier Reef inner shelf, between inshore and offshore waters. However, for the mangrove-fringed estuaries in the bays protected by these capes and headlands, both Lagrangian data and two-dimensional numerical simulations suggest high values of the return coefficient R. Salinity data following short-lived river floods also suggest trapping in such embayments for up to three weeks. These findings suggest the existence of a barotropic friction-driven CBL occurring in these bays. At the end of the dry season, mixing rate between estuarine and offshore waters is further decreased by the presence of an evapotranspiration-driven salinity maximum zone in estuaries. Nutrient-enriched CBL water is periodically ejected into offshore water as a tidal jet peeling off from the capes and headlands, and this water, enriched by outwelling from mangrove swamps, may fertilize offshore waters.

II. The Gulf of Carpentaria

The Gulf of Carpentaria (Figs. 1a and 1b) is large (900 by 550 km) with extensive areas (typically 20 km wide) of shallow water along the coast, (shallow is defined here as depth <20m). The basic hydrological conditions were described by Rochford (1966) and Newell (1973). The coastline is generally smooth, and in the nearshore zone, the isobaths are parallel with the coast. The Gulf waters are usually well-mixed, especially in the nearshore zone, with the exception of short periods after heavy rainfall and runoff, and the presence of a thermocline after prolonged periods of calm weather in the (austral) summer (Church and Forbes, 1983). Continuous measurements of temperature throughout the summer (October 1986-March 1987) during the Australian Monsoon Experiment (AMEX) revealed that the temperature difference was never larger than 1.5°C top to bottom in the deepest parts of the Gulf. This thermocline disappeared after the first tropical cyclone, in January 1987, and vertical homogeneity thereafter prevailed.

Dozens of estuaries, many of them tidal creeks only a few km long, are spread along the coast of the Gulf, and these are generally surrounded by extensive areas of mangrove swamps and salt pans. The nearshore waters of the Gulf support commercially important prawn fisheries. Both the estuaries and the sea grass beds in nearshore water are important habitats for prawns (Rothlisberg et al.,1985).

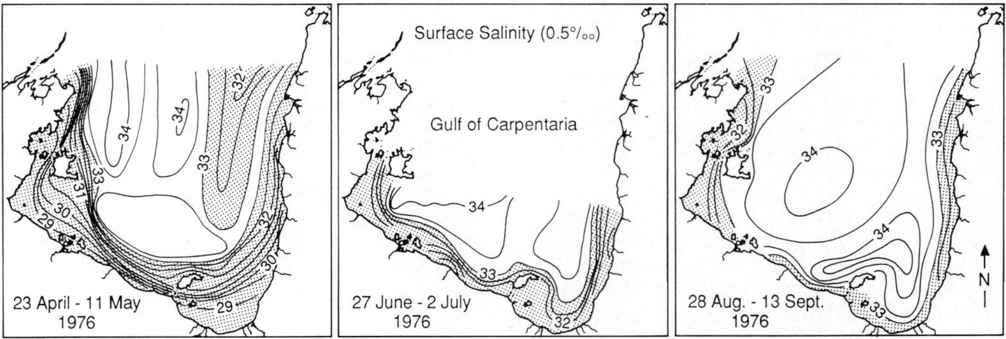

FIGURE 2. Synoptic maps of the surface salinity in the Gulf of Carpentaria at about two-monthly interval in 1976. These data were kindly provided by P. Rothlisberg.

The estuaries are usually very shallow, typically only a few m deep at most during low tide. Rainfall is restricted to the northwest monsoon in the (austral) summer, and can be intense, but is usually short-lived. River floods result, and for most rivers, usually last only a few days. Salt water is usually flushed entirely out of these estuaries during river floods. The recovery of the estuaries, following cessation of the freshwater inflow, has not been studied in detail, though some lateral trapping of freshwater in mangrove swamps may occur when salt water creeps back in the main channels (Wolanski and Ridd, 1986). From July to December, river runoff usually vanishes, and an evaporation-driven salinity maximum zone exists in the estuaries (Wolanski, 1986; Ridd et al., 1988).

Plots of the horizontal distribution of salinity in the Gulf at three-monthly intervals show the presence of a brackish water mass in the nearshore zone and spread all along the shore of the Gulf (Figure 2). The survey in April-May was taken just after the wet season. The survey in August-September was taken in the dry season. The salinity contours in April-May show that river plumes did not propagate outside the nearshore zone and suggest the presence of a coastally-trapped baroclinic CBL (i.e. a river plume). Four months later, in an August-September survey, when river runoff was by then negligible, brackish water was still found in the nearshore zone, in water typically less than 15 m deep (Fig. 2). This again suggests the presence of a CBL, and implies that cross-shore mixing between CBL and offshore water is proceeding very slowly indeed.

The dynamics of this brackish water mass in the dry season, remaining nearshore long after freshwater runoff has ceased, are unusual in as much as baroclinic effects may be negligible, as shown below. Strong tidal currents maintain vertical homogeneity and the isohalines are vertical (Forbes and Church, 1983). In these conditions, as proposed by Forbes and Church (1983) following Heaps (1972), the cross-shore density-induced gradients of buoyancy, b, may, by geostrophy, generate a longshore current v,

$$f\ v = 0.5\ H \frac{db}{dx} \tag{1}$$

where f is the Coriolis parameter, H the depth, and the x axis is pointing cross-shore. The prediction from equation (1) of a steady current v is inconsistent with observations. Indeed, a steady longshore current in nearshore waters results in a longshore friction-driven sea-level slope (Csanady, 1978; Murray and Young, 1985). However, when the freshwater inflow has stopped, there is no mechanism to maintain this slope. Indeed, any sea level set-up generated by freshwater inflow, once the inflow ceases, disperses in about a day to deep water because this dispersion occurs at a speed of $\sqrt{(gH)}$, where g is the acceleration due to gravity. Instead, as suggested by Csanady (1978), it is more likely that the only effect the cross-shore salinity gradient has is to raise the sea level at the coast by an amount h,

$$\frac{dh}{dx} = -(H\ db/dx)/2g \tag{2}$$

No longshore currents result (v=0). This finding also results from the analytical model of Murray and Young (1985) for shallow waters and zero freshwater inflow. Further, we will present below current meter data in the Great Barrier Reef continental shelf, also suggesting that once river runoff ceases, brackish water plumes in shallow coastal waters ride passively on the prevailing current. These findings suggest the presence of a dry-season CBL trapping brackish water after cessation of the freshwater inflow. There is also some evidence for the existence of a dry-season CBL, in NOAA satellite images in the visible bands which invariably show a mass of turbid water in the nearshore zone (Fig. 3). The front separating inshore and offshore waters is sharp and runs parallel to the coast. There are no apparent eddies or jets streaming off from the front, and this suggests a lack of intense cross-shore mixing between these two water masses. Fig. (3) was taken during an AMEX cruise in the hot dry season, when there was no river runoff and CTD data show negligible vertical or horizontal buoyancy gradients. This implies that the coastal boundary layer is barotropic.

To explain the likely presence of a barotropic CBL in the Gulf, we modelled the depth-averaged barotropic water circulation. There have been some previous attempts to model the depth-averaged water circulation in the interior of the Gulf, as forced by the tides of the Arafura Sea only (Rienecker, 1979; Webb, 1981) and both the Arafura Sea and the Torres Strait (Church and Forbes, 1981). Forbes and Church (1983) also added the effect of a uniform steady wind, by modelling the influence of the two prevailing winds (in the absence of tropical cyclones), namely a

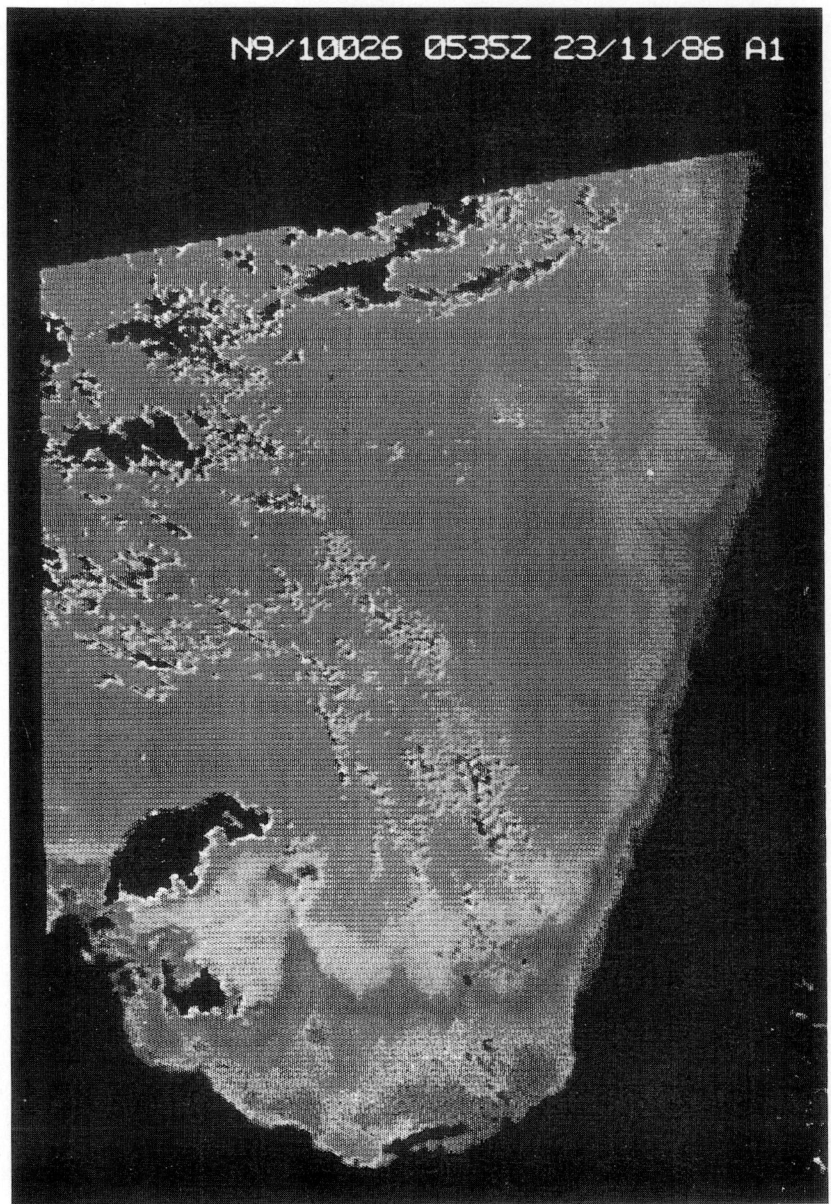

FIGURE 3. NOAA satellite view in the visible band of the east and south coast of the Gulf of Carpentaria on 23 November 1986, during AMEX, kindly provided by Mr. C. Steinberg.

southeasterly trade wind and a northwesterly monsoon wind. These models could not be verified in-depth for three reasons. Firstly because the coastal tide data are very sparse. Secondly, only current meter data from sites O, P and Q (Fig. 1b), in the southernmost part of the Gulf were used. Thirdly, because current meter data at Site R (Radok, 1978), in the western side of the Gulf (Fig. 1b), were ignored. AMEX resulted in additional current meter data from sites A, B, C, D, K, L, M and N, as well as two tidal stations within the Gulf at sites A and C. All these data together provide an excellent spatial coverage for a reliable verification of the model.

We have used the depth-averaged two-dimensional, implicit, centered, infinite-difference, numerical model described by Falconer et al., (1986) to study the tide and wind-driven circulation. The model necessitates the specification of the Manning roughness coefficient, which was set equal to 0.025. The approximate model boundaries are shown in Fig. (1b). For the open boundaries, we used a tidal forcing by the Arafura Sea similar to that proposed by Church and Forbes (1981), with however, based on an examination of the AMEX data, a smaller phase lag difference for semi-diurnal tides between Australia and Irian Jaya. We applied a tidal forcing by Torres Strait, using AMEX data as detailed in Wolanski et al., (1988b). The mesh size was 18.53 km. As shown in Fig. (4), the model yields a very encouraging agreement between observed and computed tidal ellipses at all sites, for the K_1 and M_2 tides. Comparison (not shown) for other dominant diurnal and semi-diurnal constituents is also encouraging, as is the comparison between observed and predicted tidal constants at the (few) coastal ports.

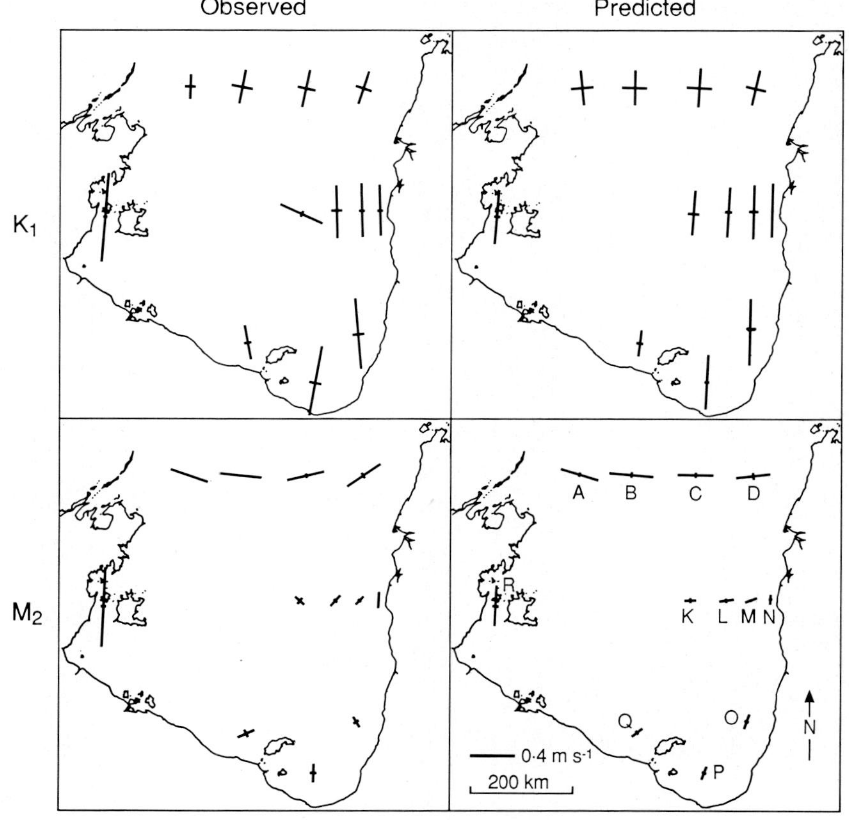

FIGURE 4. Observed (left) and predicted (right) tidal ellipses at mooring sites in the Gulf of Carpentaria for the dominant K_1 and M_2 tides. Similar good agreement is found for the other important diurnal and semi-diurnal tides (principally O_1 and S_2). The field data from site K are shown though they are suspicious and may indicate an instrument problem.

An examination of hourly tidal currents maps (not shown) reveals no evidence of large shear near the coast that may result in a tidally-driven barotropic coastal boundary layer. The plot of the tidal-averaged water circulation (Fig. 5a) shows only weak currents and no evidence for coastal boundary layers. We then proceeded to include a steady and uniform wind in the model. We found that, in the absence of tides, a south-easterly trade wind results in topographic gyres consisting of wide coastal jets along all shores, fed by a weaker circulation in deeper water (Fig. 5b). The coastal jets result from a simple balance in shallow water between bottom friction and wind stress. In deep water the along-Gulf momentum balance is between and along-Gulf sea level slope and the wind stress, the water being at rest except for what is needed to feed the coastal jet. For a northwesterly wind, we found (not shown) a flow basically in the opposite direction, water in the coastal jet flowing southward and being released offshore in the southern part of the Gulf. When both winds and tides are applied simultaneously in the model, as they occur simultaneously in nature, non-linear effects result in a decrease of the width of the CBL and the currents within it (Fig. 5c). Forbes and Church (1983) failed to report the presence of coastal jets, maybe because their mesh size was too large to resolve shallow water effects. The finding of thin barotropic coastal jets is not an artifact of our model, but simply a result of our smaller mesh size. Indeed, Sodusta (1988) recently used the Chiu and van de Kreeke (1980) explicit depth-averaged numerical model, and for a similar mesh size as ours, he also found barotropic wind-driven coastal jets.

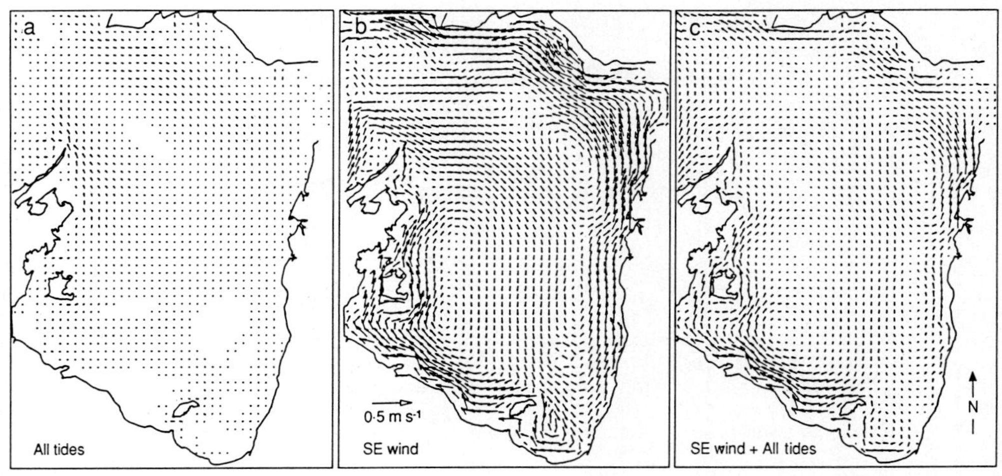

FIGURE 5. Predicted tidal-averaged Eulerian water circulation in the Gulf of Carpentaria under the action of all the dominant diurnal and semi-diurnal tides (a), a steady and uniform 15 m/s southeasterly trade wind (b), and under the combined influence of both all tides and this wind (c). The length of the arrows is directly proportional to the water current.

A barotropic CBL thus probably exists in the Gulf of Carpentaria and is restricted to shallow coastal nearshore waters no more than 15 to 20 m deep. The existence of this layer is suggested by three independent means, namely the numerical simulations (Fig. 5c), the satellite observations (Fig. 3), and the dry-season salinity data (Fig. 2).

Current meter data from the cross-shore section at 14°S (sites K,L,M,N) also suggest the existence of a wind-driven CBL. As shown in Fig. 6, the current meter sites K, L and M are in deep water, while site N, in water 20 m deep, may just be at the outer edge of the CBL. In Fig. 6 is shown a time series plot of the tidal residual longshore currents at all sites, the tidal influence having been removed from the data using Foreman (1978) technique. Also shown in Fig. 6 is the time series plot of the longshore wind component as measured at the AMEX weather buoy at site X in the middle of the Gulf. The correlation between the longshore wind component and the tidal residual longshore current is best at site N and decreases rapidly offshore.

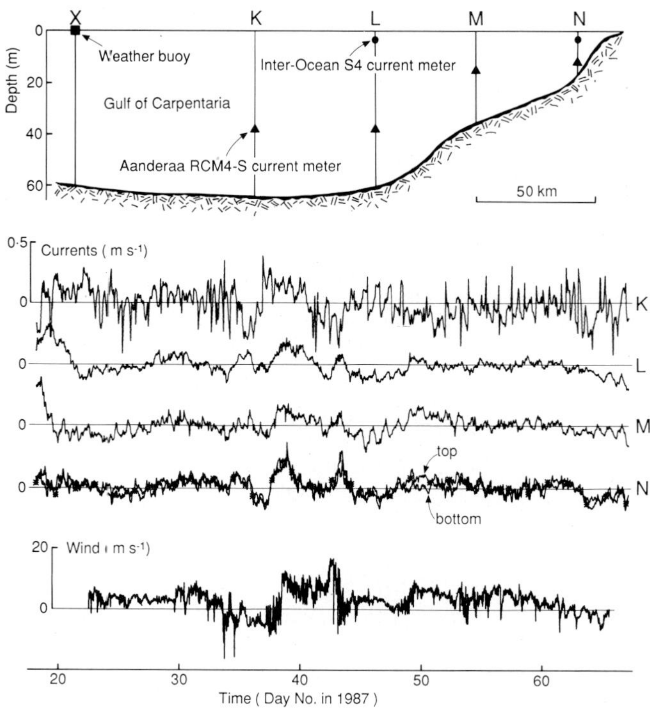

FIGURE 6. Cross-shore distribution of moorings in the Gulf of Carpentaria during AMEX (a; see a location map in Fig. 1c), and time series plot of the longshore wind and tidal residual current at the mooring sites (b). Data from site K are included though they are suspicious and may indicate an instrument problem.

An explanation of this has been proposed by Franklin and Jones (personal communication). They proposed that the wind was unsteady with a significant curl, particularly for the two tropical cyclone events during AMEX. Under such forcing the leading terms in the momentum balance are wind stress, sea surface slope and Coriolis force, and in inshore water, bottom friction. The resulting currents in the interior of the Gulf are stronger than those under steady uniform winds.

A significant cross-shore shear of the low-frequency longshore current thus existed in coastal waters. At site N, the cross-shore currents were too small to be measurable. One may expect in such conditions the formation of a front (e.g. Bowman, 1988) separating CBL and offshore waters, even in the absence of baroclinic effects. Hopkins and Swoboda (1978) also found, from current meter observations, a nearshore friction-driven regime partially uncoupled from a midshelf regime off Long Island. In their case, the boundary was between the 25 and 30 m isobaths. In the Gulf of Carpentaria, this boundary appears to be between the 15 and 20 m isobaths.

The existence of a CBL in the Gulf, trapping nearshore water for months, implies that the bulk of the Gulf water entering and leaving an estuary at tidal frequencies is in fact CBL water, not offshore water. As a result, the return coefficient of the ensemble of closely-spaced estuaries in the Gulf is probably high, and the estuaries are thus largely isolated from offshore influences.

This inhibition of mixing between offshore and estuarine waters in the Gulf of Carpentaria is reinforced at the end of the dry season by the presence of an evaporation- (or evapotranspiration-) driven salinity maximum zone (Wolanski, 1986; Ridd et al., 1988). This salinity maximum zone has been observed in all tropical estuaries that have been studied, namely the van Diemen Gulf, the Ducie-Wenlock River, the Norman River, the Escape River, Bowling Green Bay and Coral Creek (Fig. 1a).

The trapping effect of the coastal boundary layer in the Gulf of Carpentaria may have profound biological consequences, with relevance to commercial fisheries, which deserve investigations.

III. The Great Barrier Reef Continental Shelf

Contrasting with the Gulf of Carpentaria, the Great Barrier Reef continental shelf has numerous headlands, reefs and islands (Fig. 7). The shallow water area is much narrower (the 20 m isobath being located only 3 km from the coast; see a sea floor profile in Fig. 8a) than in the Gulf of Carpentaria (where the 20 m isobath is located 20 km from the coast). The inner shelf waters are usually well-mixed in temperature and salinity, except occasionally after river floods (Wolanski and Jones, 1980; Wolanski and van Senden, 1983). The latter are, like in the Gulf of Carpentaria, very short-lived, lasting typically only a few days in the (austral) summer, river runoff being negligible the rest of the year. Salt water is usually completely flushed out of the estuaries during river floods.

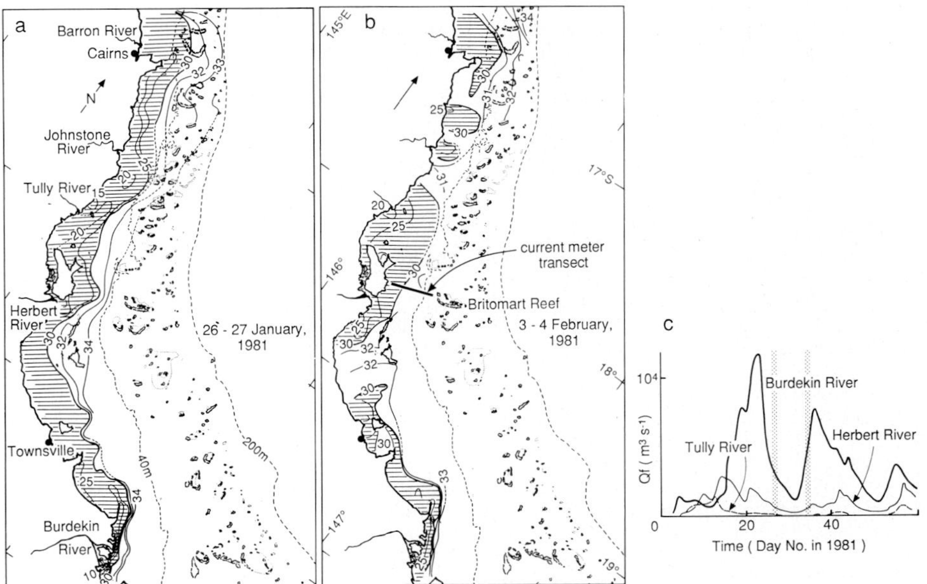

FIGURE 7. In (c) is shown a time series plot of the freshwater dis- charge Qf (m^3/s) from the Burdekin River and other local rivers during the 1981 wet season. See Fig. (1a) for a location map. The stipples indicate the time when the salinity distribution was measured. The synoptic distribution of the surface salinity is shown, in (a) on 26-27 January 1981, and in (b) on 3-4 February (adapted from Wolanski and van Senden, 1983).

A time series plot of the freshwater inflow from the Burdekin River in the summer of 1981 illustrates two river floods, each rising and falling in a matter of days (Fig. 7c). Also shown are the hydrographs from other rivers between Townsville and Cairns. The Burdekin river discharge was typically ten times that of the other nearby rivers individually, and twice the total of the other rivers taken together. We also show in Figs. 7a and 7b the surface salinity distribution at weekly interval. Just after the Burdekin River flood, on 26-27 January, a distinct river plume was formed, originating at the mouth of the Burdekin River and moving northward against a weak prevailing southward current (Wolanski and van Senden, 1983). The plume touched the coast on its left-hand side, as Coriolis forces dictated, and the nearshore waters were vertically well-mixed. The plume width increased with distance from the Burdekin River mouth, and this indicated either that the river plume was mixing with the offshore waters and spreading offshore by buoyancy, or that Burdekin River water, moving northward, was pushed further offshore by the additional freshwater discharge of the Herbert, Tully and Johnstone Rivers. However, a week later, on 3-4 February, the plume was found to be broken up in patches. Similar surveys in 1980 showed that the plume broke up in patches in a few days after cessation of the freshwater inflow, leaving brackish water that remained for several weeks in the wide bays sheltered by headlands (Wolanski and Jones, 1980).

These findings suggest that a long-lasting CBL does not exist on the Great Barrier Reef continental shelf, contrary to the situation in the Gulf of Carpentaria, but that some brackish water is trapped for several weeks in embayments after cessation of the inflow. The rapid break-up of a river plume following the cessation of freshwater inflow may be attributed to the presence of headlands and capes and the resulting complex coastal flows and mixing processes (e.g. Crepon _et al._, 1984; Narimousa and Maxworthy, 1987; Pingree _et al._, 1985; Wolanski _et al._, 1984). Nof (pers. comm.) has suggested that, in deep water with no coast and no bottom friction, a long buoyant plume suddenly cut off from its source would break up in patches within a time scale of f^{-1}, i.e. 1 to 2 days. The fact that these patches do not appear to form in the Gulf of Carpentaria suggests that river plumes confined to shallow waters may be stabilized by bottom friction.

This absence of a long-lasting CBL is also apparent in current meter data from four moorings that were deployed across the inner shelf, with current meters deployed throughout the water column (Fig. 8a). The transect of current meters cuts across the expected path of river plumes (Figs. 1c and 7). Time series plots of longshore currents (Fig. 8b) show the tidal and low-frequency currents of equal amplitude, as is typical of this area, but with a barely measurable vertical and horizontal shear (Fig. 8c), suggesting the absence of a CBL. The data cover the period (roughly days 25 to 33) where tropical cyclone Winifred, coming from offshore, crossed the shelf north of the transect at a distance from the transect of about twice the radius of maximum wind speed. Winifred made landfall on day 32. This was followed by a short-lived flood of the Herbert, Johnstone and Tully Rivers. CTD data were collected on day 40, by which time winds were light, and river runoff had ceased for practical purposes. The CTD data (Fig. 8b) show the presence of a warm river plume in coastal waters, with a weak (top to bottom temperature differences of order 0.4°C) summer thermocline in deeper water. Even in the presence of this river plume on day 40, the vertical and horizontal shear of longshore current was too small to be reliably measured (Fig. 8c). This suggests, as was argued earlier when discussing the situation in the Gulf of Carpentaria, that once river floods cease, baroclinic currents, driven by the brackish water trapped inshore, are probably negligible.

The rapid break-up of river plumes implies that estuarine and offshore waters mix rapidly at least for those estuaries not located in the wide and shallow bays sheltered by headlands and capes. However, as discussed earlier, salinity data in these bays suggest that some brackish water may be trapped (with an e-folding time of three weeks) after cessation of the freshwater inflow, three times longer than can be expected from tidal prisms calculations. This suggests that a CBL may exist in these bays.

To investigate this, we studied the dry-weather nearshore circulation in two such bays, Missionary Bay and Bowling Green Bay (see location maps in Fig. 1) using radar-tracked curtain and sail drogues. Both bays are shallow, mean water depth 2 km offshore being only typically 2 m.

In Missionary Bay, a southward nearshore residual current exists (Wolanski _et al._, 1980) and river runoff is negligible in the dry season. Drogues released at

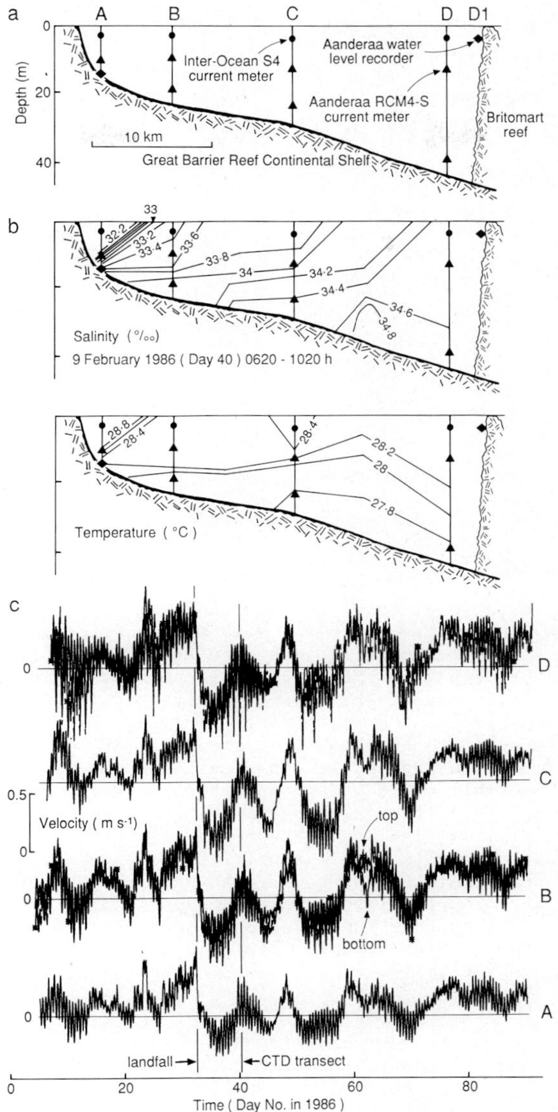

FIGURE 8. Cross-shore distribution of moorings across the inner Great Barrier Reef continental shelf (a; see a location map in Fig. 1b). Cross-shored distribution of salinity and temperature along the current meter transect on day 40 (b). Time series plot of the raw longshore current (positive if northward) at various sites (c). 'Landfall' indicates the time when tropical cyclone Winifred landed, and 'CTD transect' when CTD data were collected along the current meter transect (on day 40).

ebb tide at the mouth of Coral Creek, a mangrove-fringed tidal creek at the northern end of the bay, moved offshore with ebb tidal currents before turning back southward and entering other mangrove-fringed estuaries at the flood tide (Figure 9). This demonstrates considerable exchange of water at tidal frequencies between closely-spaced estuaries. Indeed water entering the mangrove-fringed estuaries in the southern part of the bay at flood tide was nearly entirely water that had circulated in other mangrove-fringed estuaries at previous tidal cycles.

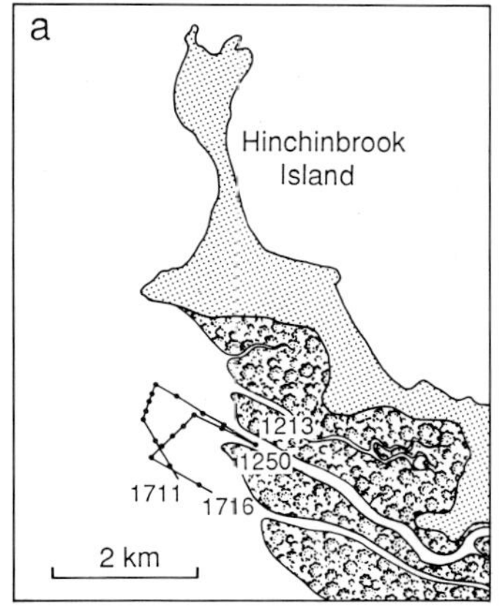

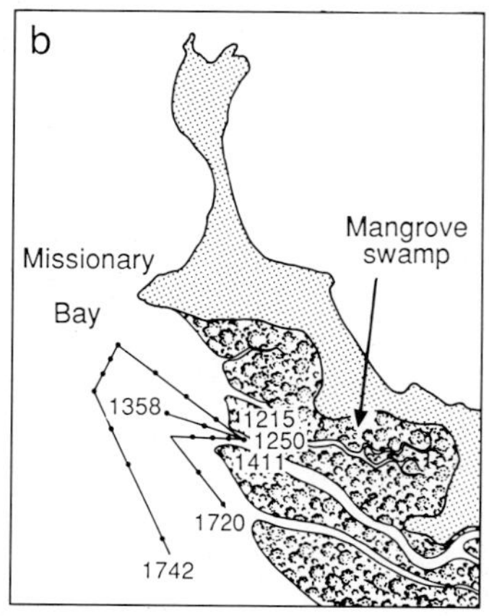

FIGURE 9. Drogue trajectories in Missionary Bay (see a location map in Fig. 1) on 15 September 1979. The drogues were released at ebb tide at the centre of Coral Creek, a tidal creek fringed with mangrove swamps. At the following flood tide all the drogues entered mangrove creeks further downstream in the bay.

Because these bays are very shallow (a few m deep), bottom friction and sea level slope are the dominant terms in the momentum equation at the mouth of these tidal creeks. As a result, no tidal jets form at ebb tides and instead radial flow prevails. This can be seen from the drogue trajectories throughout a spring ebb tide at the mouth of Barrata Creek, a mangrove-fringed tidal creek, with negligible runoff in the dry season, located in Bowling Green Bay (Fig. 10a). Though the tidal currents at the mouth of Barrata Creek are large (up to 0.6 m/s) because of the tidal prism of the mangrove swamps, flow separation and jets did not exist. The longshore current was small on that day, and radial outflow appeared to prevail. At flood tide, a similar radial flow also prevails near such shallow inlets (Wolanski and Imberger, 1987). Numerical depth-averaged simulations, using Falconer et al., (1986) numerical model and with model boundaries shown in Fig. (10a) also suggest (Figs.10 b and c) that radial flow prevails at both flood and ebb tides, implying a high value of the return coefficient. Other drogue studies (not shown) at the mouth of Barrata Creek revealed that, under strong longshore winds, the tidal outflow is deflected longshore to enter the mangrove-fringed estuaries downwind at flood tide. In that case, downwind estuaries receive at flood tide water that has circulated in upwind estuaries at previous tidal cycles.

This exchange of water between closely-spaced estuaries demonstrates nearshore trapping, and explains the observations of long-term (3 weeks) trapping of brackish waters in such bays following short-lived river floods.

IV. Leaking of Coastal Boundary Layers

CBL water in embayments may be further isolated from offshore water by the presence of barotropic, friction-driven front separating bay water from offshore water. Using the same numerical model as described above, we have studied the water circulation in Bowling Green Bay (see a bathymetric map in Fig. 11). The bay is very shallow (a few m deep), and surrounded by extensive areas of tidal wetlands (mangrove swamps and salt pans). We have forced the open boundaries (as shown in Fig. 11) by tides and a net longshore current in open waters as in the model of Kelly and Andrews (1985). However, we chose not to neglect the influence of the tidal wetlands. To do this we incorporated in the model, at the shoreline, the water flux at tidal frequencies between the bay and the wetlands, the tidal average of this flux being zero (no freshwater inflow). The tidally-averaged residual velocity field in the absence of wind is shown in Fig. 11.

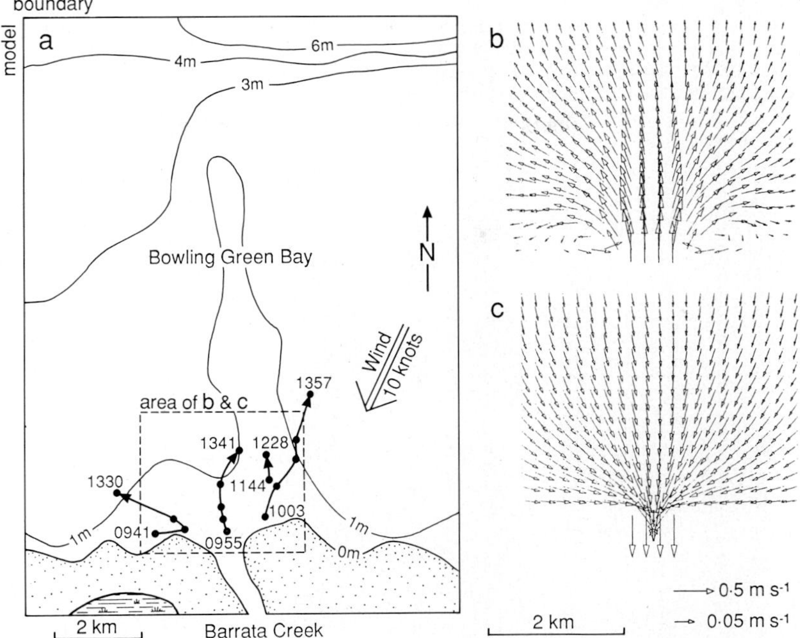

FIGURE 10. Bathymetry (in m) near the mouth of Barrata Creek, a tidal creek fringed with mangrove swamps in Missionary Bay (a; see a location map in Fig. 1a). Also the trajectories at ebb tides of drogues released near slack high tide, on 15 November 1985 and the location of the boundaries of the numerical model. Synoptic distribution of the numerically-predicted depth-averaged currents at ebb (b) and flood tides (c). The velocity scale for the currents is logarithmic, i.e. the length of the arrows is proportional to the logarithm of the water current.

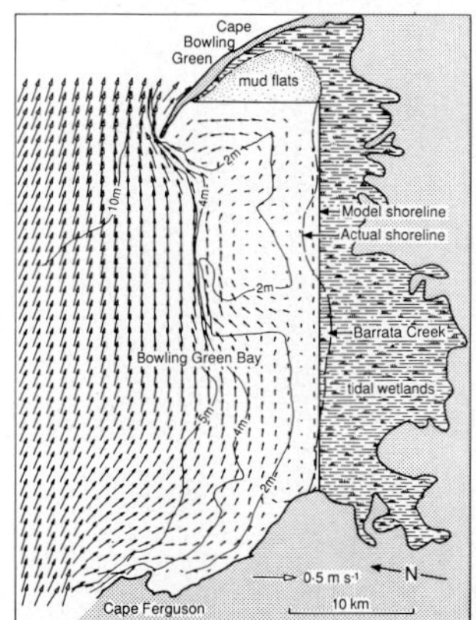

FIGURE 11. Bathymetry (in m) of Bowling Green Bay (see a location map in Fig. 1a) showing also the extensive tidal wetland area. The open water boundaries of the numerical model are those shown in the figure. The figure also shows, using a logarithmic scale as in Fig. 10, the predicted depth-averaged residual currents in the absence of wind.

This figure suggests that there exists a large shear zone extending into the bay about southwestward from Cape Bowling Green. This velocity shear separates bay water from offshore water. This shear can be expected to result in a front. Indeed, LANDSAT images (e.g. Fig. 12) in such calm weather conditions also show the presence of this front running about southwestward from Cape Bowling Green. The front in Fig. 12 crosses isobaths, suggesting that the LANDSAT image does not only reveal the bathymetry, as is apparent in this image in offshore water, but also turbidity differences (particularly near Cape Bowling Green).

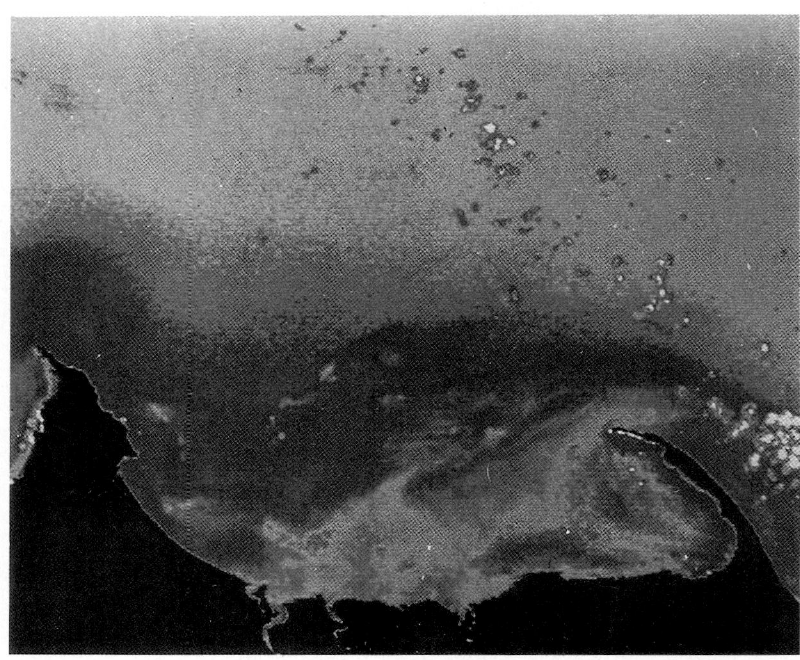

FIGURE 12. Computer-enhanced LANDSAT image of Bowling Green Bay on 26 December, 1984.

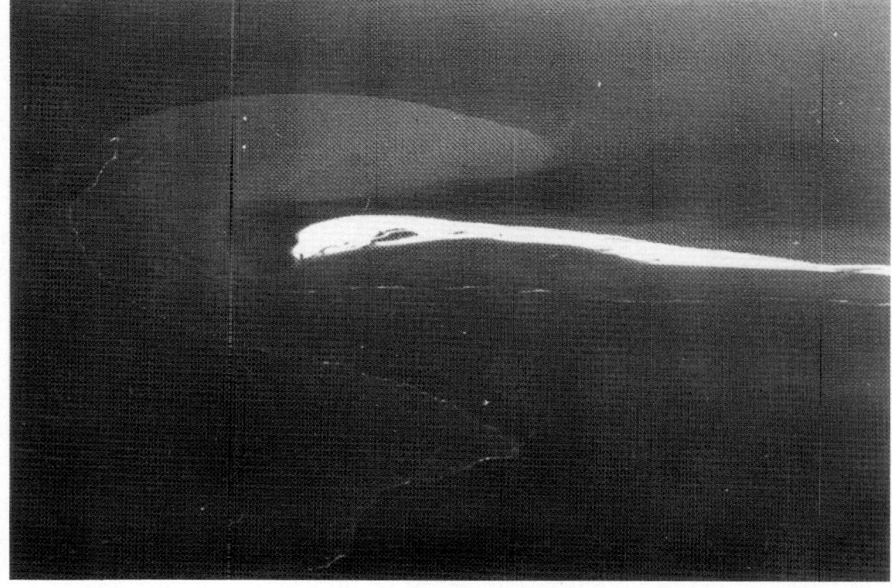

FIGURE 13. Aerial view of an ebb tidal jet peeling off from Cape Bowling Green and propagating offshore.

Note that both Figs. 11 and 12 suggest that water from the bay leaves the bay in a thin outflow zone (a jet) near Cape Bowling Green. This is numerically predicted to occur on ebb tide. When this occurs, a pulse of coastal boundary water is ejected as a tidal jet peeling off from Cape Bowling Green and propagating offshore. This jet is readily visible in aerial observations (Fig. 13). Note from Fig. 13 the clear front in the bay, separating bay water from CBL water. CTD surveys across such a front in 1984 show that these fronts are essentially turbidity fronts generated by the topography, with baroclinic effects being negligible.

At spring ebb tide, the predicted currents near Cape Bowling Green are measurably smaller when the tidal prism effect of the wetlands is not taken into account (Fig. 14a) than when they are (Fig. 14b). These higher currents are important in the dynamics of the tidal jets because they permit flow separation to occur (Pingree and Mardell, 1986; Wolanski et al., 1988a). The mangrove swamps may thus be important as their tidal prism effect helps maintain deep channels in mangrove swamps (Wolanski et al., 1980) and facilitate flow separation and formation of jets off Cape Bowling Green.

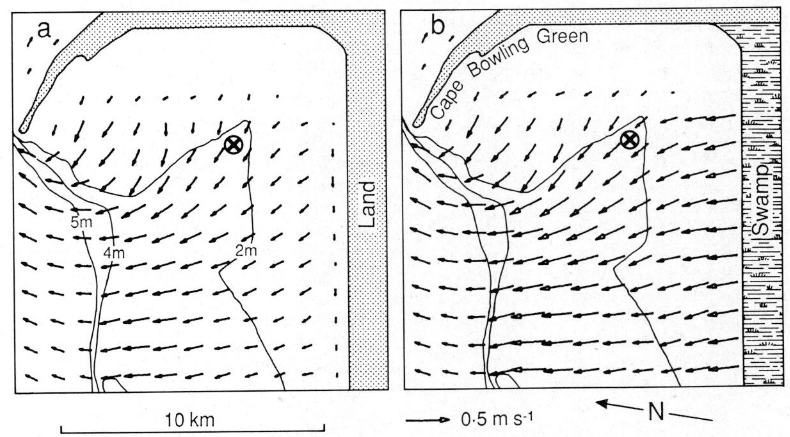

FIGURE 14. Predicted depth-averaged peak ebb tidal currents near Cape Bowling Green in the absence (a) and presence (b) of coastal wetlands. The length of the arrows is directly proportional to the water current.

These jets of CBL water, presumably enriched by outwelling from tidal wetlands (Boto and Bunt 1981; Robertson et al., 1988), may fertilize the phytoplankton in offshore waters. In this manner, the mangrove swamps of Bowling Green Bay, through periodic ejection of coastal boundary layer water, may help support the baitfish and billfish schools found a few km offshore from Cape Bowling Green (Williams, 1988). These schools support thriving sport fisheries. Bowman (1988) argued for a similar enrichment of offshore water near Crane Neck, a headland in Long Island Sound. We have frequently observed a similar ebb tidal jet peeling off other headlands along this coast, suggesting that many mangrove-fringed bays may, under the right wind and tide conditions, locally enrich a patch of offshore water on the inner Great Barrier Reef continental shelf. Wolanski and Hamner (1988) showed that plankton and water-borne eggs and larvae are aggregated by secondary currents in such jets, and this aggregation influences the distribution and density of benthic assemblages and of pelagic secondary and tertiary predators. This may explain the presence of baitfish and billfish fisheries off Cape Bowling Green.

The long-term trapping of water in a CBL in such mangrove-fringed embayments may have important ecological implications that need further studies.

The recirculation at tidal frequency water from one mangrove creek to the other is important in selecting a mangrove creek in outwelling studies. When there is a prevailing longshore current estuaries in the 'upstream' end of the bay will have a lower 'bulk' return coefficient than one at the 'downstream' end. This may greatly affect outwelling budget studies. Coral Creek, the principal northern Australian mangrove creek used in outwelling studies (Boto and Bunt, 1981; Robertson et al., 1988) is on the 'upstream' side of Missionary Bay. This made possible an outwelling budget study. If another mangrove creek further 'downstream' had been selected, water entering the creek at flood tide would have originated from another mangrove creek at the previous ebb tide, the chemical contrast between ebb and flood tide water would become blurry, and calculations of an outwelling budget less reliable.

V. Discussion

There is a distinct seasonality in estuarine processes in tropical Australia. In the wet season, river plumes are formed which are very short-lived. Once the freshwater inflow ceases, the brackish water remaining in shallow coastal waters is submitted to the dominant influence of bottom friction, and provides a fairly passive tracer whose distribution can be used to evaluate the estuarine flushing rates. In the case of the Gulf of Carpentaria, with its long smooth coastline, parallel isobaths, and no capes and headlands, some brackish water is found to remain trapped several months in a thin coastal strip. This is explained in terms of barotropic wind-driven coastal jets in shallow coastal waters preventing mixing between inshore and offshore waters. In such conditions, the many mangrove-fringed estuaries along the Gulf of Carpentaria may exchange water with each other, but estuarine waters probably mix very little with offshore waters. In the case of the Great Barrier Reef continental shelf, the river plume is rapidly (in a matter of days) broken up in patches. Mixing between inshore and offshore waters is thus very rapid. However, in the dry season, in the absence of any freshwater inflow, a CBL is formed in the shallow coastal waters of the wide bays sheltered by the capes and headlands. These wide bays offer a habitat for closely-spaced mangrove-fringed tidal creeks. This CBL inhibits mixing between offshore and estuarine waters, but allows recirculation of water from one estuary to the other. Water from this CBL is presumably enriched by outwelling from the mangrove swamps. This water moves slowly longshore towards the headland sheltering the bays. Under certain conditions of topography and weather, it may be periodically ejected as an ebb tidal jet peeling off from the headlands, and presumably enriching in nutrient the offshore waters. This may explain the presence of juvenile billfish fisheries a few kilometers offshore of Cape Bowling Green. Also, in such conditions, the bulk return coefficient of the mangrove-fringed estuarine waters is very large, and these tidal creeks are, taken as an ensemble, very poorly flushed in the dry season. This may be important in planning development projects around mangrove creeks in such wide bays. This also suggests that little of the outwelled nutrient will be rapidly 'outwelled' from the estuaries into coastal waters, with the bulk of it staying initially trapped in the CBL.

VI. Conclusion

Through a variety of dynamical processes in the nearshore zone, a coastal boundary layer effect may prevail that traps water in the nearshore zone and inhibits mixing between estuarine and offshore waters, particularly on a long smooth coastline like that of the Gulf of Carpentaria. This mixing is further reduced in the dry season by the presence of an evaporation-driven salinity maximum zone. When capes and headlands are present, such as on the continental shelf of the Great Barrier Reef, a wet-season coastal boundary layer is formed but is rapidly (in a few days) broken down in patches, and this increases the estuarine flushing rates as tidal recirculation is minimized. No dry-weather coastal boundary layer is observed on the continental shelf of the Great Barrier Reef. However, a mini-coastal boundary layer exists in the wide shallow bays in the lee of headlands. The existence of this mini-coastal boundary layer is due to the presence of extensive mangrove swamps bordering the estuaries. Coastal boundary layer water is presumably enriched by outwelling from mangrove swamps and, being periodically ejected as tidal jets peeling off from capes and headlands, may locally enrich offshore waters.

Although this study has been restricted to tropical environments with mangrove swamps, the same conclusions on the existence of a coastal boundary layer should be applicable to temperate regions with shallow shelves and tidal wetlands.

VII. References

Boto, K. and Bunt, J. S., 1981: Tidal export of particulate organic matter from a northern Australian mangrove system. Estuarine Coastal Shelf Sci., **13**, 247-255.

Bowman, M. J., 1988: Estuarine fronts. In B. Kjerfve (Ed.) "Hydrodynamics of Estuaries", Vol. I, CRC Press, Boca Raton, Florida, 85-132.

Chao, S. -Y., 1988: River-forced estuarine plumes. J. Phys. Oceanogr., **18**, 72-88.

Chao, S. -Y. and W. Boicourt, 1986: Onset of estuarine plumes. J.Phys.Oceanogr., **16**, 2137-2149.

Chiu, S. and J. van de Kreeke, 1980: A finite difference solution for the two dimensional (horizontal) long wave equation. Rosenstiel School of Marine and Atmospheric Science, University of Miami, Tech. Rep. NSF Grant No. Eng -76-08288.

Church, J. A. and A. M. G. Forbes, 1981: A non-linear model of the tides in the Gulf of Carpentaria. Aust. J. Marine Freshwater Res., **32**, 685-697.

Church, J. A. and A. M. G. Forbes, 1983: Circulation in the Gulf of Carpentaria. I: Direct observations of currents in the South-East corner of the Gulf of Carpentaria. Aust. J. Marine Freshwater Res., **34**, 1-10.

Crepon, M., C. Richew and M. Chartier, 1984: Effects of coastline geometry on upwellings. J. Phys. Oceanogr., **14**, 1365-1383.

Csanady, G. T., 1978: The arrested topographic wave. J. Phys. Oceanogr., **8**, 47-62.

Csanady, G. T., 1982: Circulation in the Coastal Ocean, Reidel, Dordrecht, 279 pp.

Csanady, G. T., 1984: The influence of wind stress and river runoff on a shelf-sea front. J. Phys. Oceanogr., **14**, 1383-1392.

Falconer, R. A., E. Wolanski and L. Mardapitta-Hadjipandeli, 1986: Modelling tidal circulation in an island's wake. J. Waterway, Port, Coastal and Ocean Engineering, ASCE, **112**, 234-254.

Fischer, H. B., E. J. List, R. C. Y. Koh, J. Imberger and N. H. Brooks, 1979: Mixing in inland and coastal waters. Academic Press, New York, 484 pp.

Forbes, A. M. G. and J. A. Church, 1983: Circulation in the Gulf of Carpentaria. II: residual currents and mean sea level. Aust. J. Marine Freshwater Res., **34**, 11-22.

Foreman, M. G. B., 1978: Manual for tidal current analysis and prediction. Pacific Science Rep. 78-6. Inst. of Ocean Sciences, Sidney, B.C., Canada.

Heaps, N. S., 1972: Estimation of density currents in the Liverpool Bay area of the Irish Sea. Geophys. J.R. Astron. Soc., **30**, 415-432.

Hopkins, T. A. and A. L. Swoboda, 1986: The nearshore circulation off Long Island, August 1978. Continental Shelf Res., **5**, 431-473.

Joshi, P. B. and R. B. Taylor, 1983: Circulation induced by tidal jets. J. Waterway, Port, Coastal Land Ocean Engineering, ASCE, **109**, 445-465.

Kelly, L. R. and J. C. Andrews, 1985: Numerical models for planning coastal circulation studies. Proc. 1985 Australasian Conf. Coastal & Ocean Engineering, Christchurch, New Zealand, vol.2, 445-454.

Murray, S. P. and M. Young, 1985: The nearshore current along a high-rainfall, trade-wind coast- Nicaragua. Estuarine, Coastal and Shelf Sci., **21**, 687-699.

Narimousa, S. and T. Maxworthy, 1987: On the effects of coastline perturbations on coastal currents and fronts. J. Phys. Oceanogr., **17**, 1297-1303.

Newell, B. S., 1973: Hydrology of the Gulf of Carpentaria. CSIRO Div. Fish. Oceanogr., Australia, Tech. Paper No.7.

Ozsoy, E., 1986: Ebb-tidal jets: a model of suspended sediment and mass transport at tidal inlets. Estuarine, Coastal and Shelf Sci., **22**, 45-62.

Ozsoy, E. and U. Unluata, 1982: Ebb-tidal flow characteristics near inlets. Estuarine, Coastal and Shelf Sci., **14**, 251-263.

Pingree, R. D., G. T. Mardell and L. Maddock, 1985: Tidal mixing in the channel Aisles region derived from the results of remote sensing and measurements at sea. Estuarine, Coastal and Shelf Sci., **20**, 1-18.

Pingree, R. D. and G. T. Mardell, 1986: Coastal tidal jets and tidal fringe development around the isles of Scilly. Estuarine, Coastal and Shelf Sci., **23**, 581-594.

Pritchard, D. W., 1952: Salinity distribution and circulation in the Chesapeake Bay estuarine system. J. Mar. Res., **11**, 106-123.

Pritchard, D. W., 1960: Salt balance and exchange rate for Chincoteague Bay. Chesapeake Sci., **1**, 48.

Radok, R., 1978: Tidal height and stream prediction at Groote Eylandt. BHP Technical Bulletin, **22(2)**, 46-60.

Rattray, M., Jr. and D. V. Hansen, 1966: A similarity solution for circulation in an estuary. J. Mar. Res., **20**, 121-133.

Ridd, P., M. W. Sandstrom and E. Wolanski, 1988: Outwelling from tropical tidal salt flats. Estu. Coastal Shelf Sci., **26**, 243-253.

Rienecker, M. M. 1979: Tidal propagation in the Gulf of Carpentaria. Ph.D. Thesis, University of Adelaide. 143 pp.

Robertson, A. I., D. M. Alongi, P. Daniel and K. G. Boto, 1988: How much mangrove detritus enters the Great Barrier Reef lagoon. Proc. 6th Int. Symp. Corals Reefs, Townsville (in press).

Rochford, D. J., 1966: Some hydrological features of the eastern Arafura Sea and the Gulf of Carpentaria in August 1964. Aust. J. Mar. Freshwater Res., **17**, 31-60.

Rothlisberg, P. C., B. J. Hill, and D. J. Staples, (Editors), 1984: Second Australian National Prawn Seminar, NPS, Cleveland, Australia, 361 pp.

Royer, L. and W. J. Emery, 1985: Computer simulations of the Fraser River plume. J. Mar. Res., **43**, 289-306.

Smith, N. P., 1988: The Laguna Madre of Texas: Hydrography of a hypersaline lagoon. In B, Kjerfve (Ed.) "Hydrodynamics of estuaries", Vol. II, CRC Press, Boca Raton, Florida, 31-40.

Sodousta, J. A., 1988: Non-linear interaction of wind and tide driven currents in the Gulf of Carpentaria. M.Sc. Thesis, Sydney University, 123 pp.

Van de Kreeke, J., 1985: Stability of tidal inlets - Pass Cavallo, Texas. Estuarine, Coastal and Shelf Sci., **21**, 33-43.

Van de Kreeke, J., 1988: Dispersion in shallow estuaries. In B. Kjerfve (Editor): "Hydrodynamics of Estuaries", CRC Press, Boca Raton, Florida, Vol.I, 27-39.

Webb, D. J., 1981: A numerical model of the tides in the Gulf of Carpentaria and the Arafura Sea. Aust. J. Marine Freshwater Res., **32**, 31-44.

Williams, D. McB., 1988: Significance of coastal resources to sailfish and blackmarlin in NE Australia: An ongoing research program. Proc. International Billfish Symposium II, Kailua-Kona, Hawaii, August 1-5, 1988, (in press).

Wolanski, E., 1986: An evaporation-driven salinity maximum zone in Australian tropical estuaries. Estuarine, Coastal and Shelf Sci., **22**, 415-424.

Wolanski, E. and M. L. Banner, 1978: Buoyant surface jet in tidal longshore current. J. Hydraulics Division, ASCE, **104**, 1505-1519.

Wolanski, E. and M. Jones, 1980: Physical properties of Great Barrier Reef lagoon water near Townsville. I. Effects of Burdekin River floods. Aust. J. Mar. Freshwater Res., **32**, 305-319.

Wolanski, E., M. Jones and J. S. Bunt, 1980: Hydrodynamics of a tidal creek-mangrove swamp system. Aust. J. Marine Freshwater Res., **31**, 431-450.

Wolanski, E. and D. van Senden, 1983: Mixing of Burdekin River flood waters in the Great Barrier Reef. Aust. J. Marine Freshwater Res., **34**, 49-63.

Wolanski, E., J. Imberger and M. L. Heron, 1984: Island wakes in shallow coastal waters. J. Geophys. Res., **89**, 10553-10569.

Wolanski, E. and P. Ridd, 1986: Tidal mixing and trapping in mangrove swamps. Estuarine, Coastal Shelf Sci., **23**, 759-771.

Wolanski, E. and J. Imberger, 1987: Friction-controlled selective withdrawal near inlets. Estuarine, Coastal and Shelf Sci., **24**, 327-333.

Wolanski, E. and W. M. Hamner, 1988: Topographically controlled fronts in the ocean and their biological influence. Science, **241**, 177-181.

Wolanski, E., E. Drew, K. Abel and J. O'Brien, 1988a: Tidal jets, nutrient upwelling and their influence on the productivity of the alga Halimeda in the Ribbon Reefs, Great Barrier Reef. Estuarine, Coastal and Shelf Sci., **26**, 169-201.

Wolanski, E., P. Ridd and M. Inoue, 1988b: Currents through Torres Strait. J. Phys. Oceanogr., **18**, 1535-1545.

Wong, K. -C., 1986: Sea-level fluctuations in a coastal lagoon. Estuarine, Coastal and Shelf Sci., **22**, 739-752.

Zhang, Q. H., G. S. Janowitz and L. J. Pietrafesa, 1987: The interaction of estuarine and shelf waters: a model and applications. J. Phys. Oceanogr., **17**, 455-469.

Control of Estuarine Salinities by Coastal Ocean Salinity

William J. Wiseman, Jr.[1], E.M. Swenson[2], and F.J. Kelly[3]

[1]Coastal Studies Institute and
Department of Geology and Geophysics
Louisiana State University
Baton Rouge, LA 70803

[2]Coastal Ecology Institute
Center for Wetland Resources
Louisiana State University
Baton Rouge, LA 70803

[3]Department of Environmental Engineering
Texas A&M University
College Station, TX 77843

ABSTRACT

Shelf-estuarine exchange along the Louisiana coast is known to be dominated by barotropic, tidal and sub-tidal exchanges. Simple regression of the estuarine salinities against Mississippi River discharge and cross-spectrum analysis do not produce a successful hindcast of the observations. Parsimonious auto-regressive, moving-average predictive models of the weekly-averaged salinity data are quite skillful in hindcasting the measurements. These models are generally consistent with a conceptual model in which Mississippi River discharge alters coastal salinities and this salinity signal propagates downcoast and then up-estuary.

I. Introduction

The shallow estuaries and adjacent marshes of coastal Louisiana are highly productive nursery grounds for the abundant fisheries of the northern Gulf of Mexico (Shaw et al., 1985; Turner, 1982). Subsidence, global sea-level rise, and anthropogenic modification to the watersheds may be contributing to a gradual alteration in the salinity regime within these estuarine environments (Turner and Cahoon, 1987). Such variation could not only alter the area of acceptable nursery ground for fisheries, but also affect the health of the surrounding marsh and, consequently, influence the rate of land loss, since dead plants cannot stabilize marsh soil against wave attack. Louisiana coastal land loss is presently estimated at 12,700 ha/yr. and seems to be increasing (Turner and Cahoon, 1987). Changes in mean salinity and salinity variance have been detected in long-term records from Louisiana estuaries, although the level of most of these changes has not been demonstrated to be harmful to the adjacent marsh (Wiseman and Swenson, in press; Mendelssohn and McKee, 1987). Most of the documented rates of change are, at best, tentative because of the low signal-to-noise ratio in the data sets.

In a previous study (Wiseman and Swenson, in press), it was realized that the estuarine salinities exhibited not only a strong annual cycle, but also variability on time-scales of many years, which was, presumably, related to climatic variations. Furthermore, these variations were inversely related to Mississippi River discharge (Figure 1), the dominant source of fresh water for the northern Gulf of Mexico shelf. The estuarine drainage basins were, however, isolated from that of the Mississippi River system.

Coastal and Estuarine Studies, Vol. 38
R. T. Cheng (Ed.)
Residual Currents and Long-term Transport

Cross-spectrum analysis produced high coherence estimates between estuarine salinity and Mississippi discharge at frequencies of 0.06 to 0.10 cpm (cycles per month), the 'annual' cycle, and at frequencies lower than 0.025 cpm. In both bands, the phase was near 180°. If this coherence truly reflected a cause and effect relationship, it could arise only because the Mississippi River discharge was altering the salinity of the nearshore zone and, consequently, that of the adjacent estuaries.

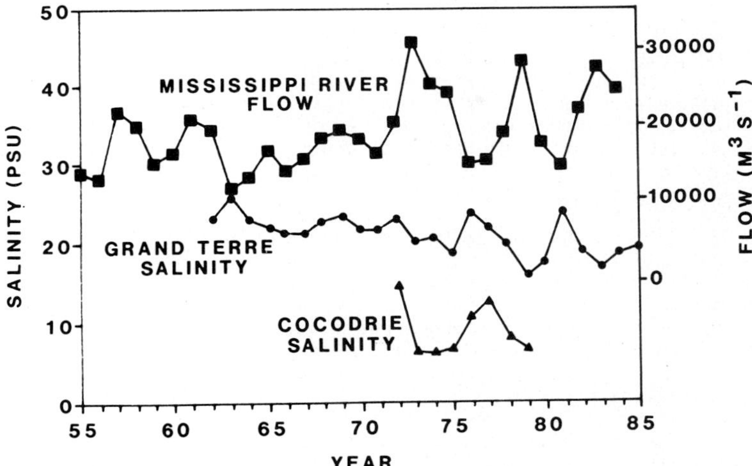

FIGURE 1. Time series of annual mean Mississippi River discharge, salinity at Grand Terre, and salinity at Cocodrie. Note the inverse relationship between salinity and runoff on multi-year time scales.

Previous studies of the fate of fresh water on the Louisiana-Texas shelf have shown the Mississippi River discharge to have a residence time on the shelf of the order of 1 year or less (Dinnel and Wiseman, 1986). The distribution of this water over the shelf varies with season. The Mississippi flood is felt within 1 to 2 months all the way to the Mexican border as a distinct reduction in coastal salinity. A band of low-salinity water is found year-round against the shore from the Mississippi River Delta to the Texas border. These are the waters that actively exchange with those of the estuaries proper.

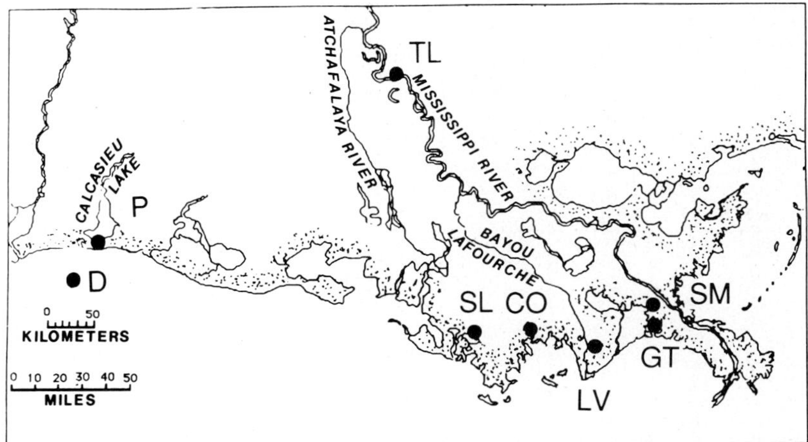

FIGURE 2. Location map showing the salinity measurement sites at Grand Terre (GT), St. Mary's Point (SM), Leeville (LV), Cocodrie (CO), and Sister Lake (SL), the Mississippi River gauging station at Tarbert Landing (TL), and velocity and salinity measurement sites at Calcasieu Pass (P) and the adjacent nearshore (D).

Time-series of nearshore and estuarine currents and salinities near the mouth of Calcasieu Lake in west Louisiana (Figure 2) are available. The two salinity time-series are highly coherent. Analysis of the records shows that the nearshore salinities are also coherent with alongshore flow. When the river feeding the head of the estuary is in flood, large pulses of fresh water leave the estuary on the ebb tide and are replaced by coastal water on the flood tide. The coastal salinities remain controlled by downcoast advection, not local effluent (Kelly _et al_., 1983).

Analysis of long-term current measurements from the mouth of the same estuary have recently shown the flow regime to be predominantly barotropic at tidal frequencies and at the time-scales of the synoptic weather systems (Lee, 1988). The other bar-built estuaries of south Louisiana may be assumed to have similar circulation patterns. This suggests that one-dimensional advection-diffusion of salt from the coastal ocean into the estuary may be a reasonable model for the processes that dominate the salt balance in these estuaries.

Runoff to most Louisiana estuaries is small. This is one reason for the success of a simple barotropic model in explaining shelf-estuarine exchange. Refill times for the estuaries immediately west of the Mississippi River Delta, estimated as estuarine volume divided by upland runoff, generally range from approximately 1 month to 10 months or longer, depending upon the estuary and season. Excess runoff (rainfall minus potential evapotranspiration) is generally small and occasionally negative during the summer months (Byrne _et al_., 1976). In most estuaries, salinity changes are driven by variations in river flow into the head of the estuary, and the salinity at the mouth of an estuary is often considered to be relatively stable. In these Louisiana estuaries, the river flow at the head of the estuary undergoes small absolute changes. The salinity of the coastal ocean, though, can change by 10 to 20 psu during the year.

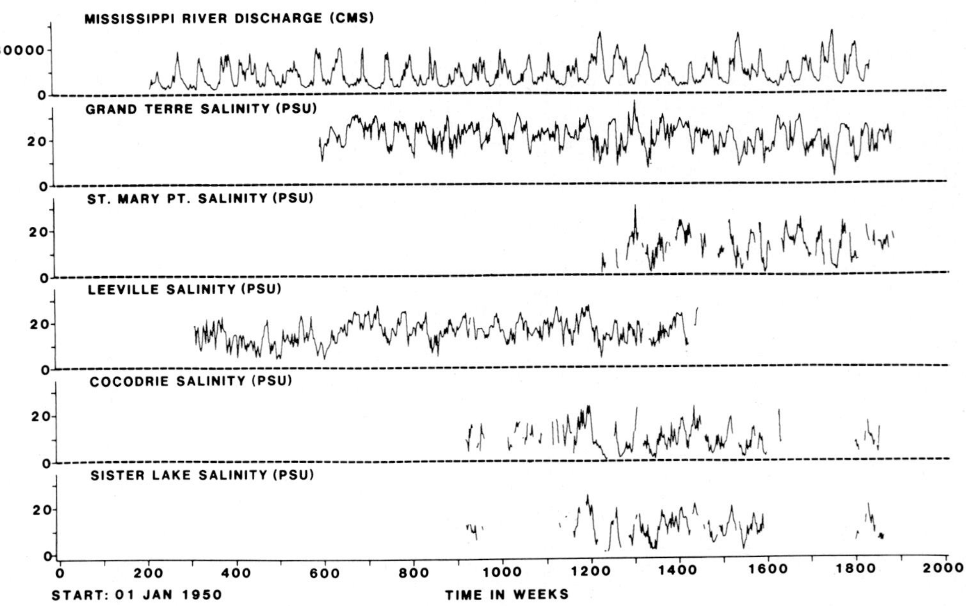

FIGURE 3. Time series of weekly mean salinities and discharge used in the analyses.

In light of these observations, we attempted to predict salinities at a selected set of estuarine monitoring sites as a function of past salinity at these sites and Mississippi River discharge. A number of different statistical models were attempted, with varying degrees of success. Models were selected if they were parsimonious and physically reasonable.

II. The Data and The Analysis Techniques

Two sources of data were utilized. The Louisiana Department of Wildlife and Fisheries provided us with hourly samples of salinity from the surface waters at four estuarine sites: Grand Terre, St. Mary's Point, Cocodrie, and Sister Lake (Figure 2). The U.S. Army Corps of Engineers provided daily samples, nominally collected at 0800 each day, from Leeville (Figure 2). These data were averaged to provide time-series of weekly and monthly mean salinity. Small gaps in the resultant series were filled by linear interpolation. All data records exceeded 9 years in length, and the longest record was 24 years long (Figure 3). Concurrent Mississippi River discharge data at Tarbert Landing were also obtained from the U.S. Army Corps of Engineers (Figures 2, 3).

We wished to develop time-series models to describe the salinity variations at these locations. We hoped that the structure of these models and their changes between measurement locations would lend insight into the dynamics controlling estuarine salinity. Three modeling techniques were employed in this analysis: simple regression (Neter and Wasserman, 1974), Fourier analysis, and auto-regressive, moving-average modeling (Box and Jenkins, 1970). For the first and third techniques, we used programs from a commercially available statistical analysis package (SAS, 1984). The Fourier analysis was performed using programs that we wrote and based on the Fast Fourier Transform (Cooley and Tukey, 1965).

III. Regression Models

The simplest prediction scheme attempted was an ordinary regression of both weekly and monthly salinity means against Mississippi River discharge. Although the initial indication that Mississippi River discharge might provide a good predictor for estuarine salinities came from the annual mean data, when the annual means were regressed against one another the R-squared values were lower than expected (Table 1). We had initially anticipated that river discharge alone would account for most of the salinity variance.

Best fits of the monthly mean salinities using a single lag were obtained for a 1-month lag of the river discharge. This is partially due to the time it takes for water to flow from the gauging station, north of Baton Rouge, Louisiana, down the river to the coast and along the coast to the mouths of the estuaries. It is, though, also due to the finite flushing times of the estuaries. At least two important time-scales influence the data! The R-squared for these models was less than 0.5 (Table 1). This was actually less than the R-squared values found in the

Table 1. Results of regression analyses on the annual and monthly mean data

REGRESSION OF ANNUAL MEAN SALINITY AGAINST RIVER DISCHARGE

Station	R-squared	P > F*
Grand Terre	0.57	0.0001
St. Mary's Pt.	0.59	0.0033
Leeville	0.22	0.0241
Cocodrie	0.31	0.0238
Sister Lake	0.36	0.0296

REGRESSION OF MONTHLY MEAN SALINITY AGAINST RIVER DISCHARGE LAGGED BY ONE MONTH

Station	R-squared	P > F*
Grand Terre	0.48	0.0001
St. Mary's Pt.	0.46	0.0001
Leeville	0.11	0.0001
Cocodrie	0.35	0.0001
Sister Lake	0.48	0.0001

*Probability that the calculated F-value will be exceeded under the null hypothesis.

analysis of the annual data. The time-series of monthly means contained significantly more data points than the time-series of annual means. Furthermore, they contained much greater variability. The models of the monthly mean salinities produced acceptable hindcasts of the annual signal and its phase, but they were unable to reproduce the extremes of the data accurately. The R-squared values decreased with distance of the site from the Mississippi River and from the mouth of the estuary. This suggests the increasing importance of small-scale local circulation patterns and short-lived upland runoff events in the upper reaches of the estuaries.

Numerous alternate regression models were attempted, including forward and backward step-wise regression. Using the concurrent river flow and the first four monthly lagged series, we could improve the fit somewhat. It is not clear that the improvement in fit warrants the use of the additional regression parameters. When only the most significant inputs to the model were retained, the resulting model was often physically unreasonable. For example, the best model might involve runoff lagged by one and three months, but not by two months. There is no physical justification for the lack of two-month lagged runoff in the model.

Similarly poor results were obtained when we used weekly time-series rather than monthly time-series. The annual cycle was, again, well modeled. The extremes were slightly better modeled. Even with eight lags in the model, though, river flow alone was unable to hindcast the details of the salinity records. The higher frequency fluctuations remained poorly represented.

Using lagged series of both runoff and salinity as inputs, the regression analysis was more successful. The resulting models were still physically unreasonable.

IV. Fourier Series Analysis

The available time-series of monthly values were, generally, too short to allow a reliable cross-spectrum analysis to be performed. The weekly series, however, contained enough data points to allow meaningful resolution in a frequency analysis. Cross-spectrum analysis determined those time-scales at which a linear model might have some hope of making an accurate prediction of one series from the other.

All the spectra were red (Figure 4), with significant energy at the annual frequency, 0.019 cycles per week (cpw). The decay rate varied between f^{-1} and f^{-3}. Most of the statistically significant coherence between runoff and estuarine salinity was generally limited to frequencies below 0.03 cpw. Even though the energy in both signals is concentrated at low frequencies, much of the salinity variance is not accounted for by the model. Almost none of the signal at time scales shorter than 7 months can be reliably explained. It is clear that a simple linear transfer model based on these frequency domain results could not efficiently relate estuarine salinities to river runoff. The problem is partially related to the statistical reliability of the cross-spectrum analyses, which is limited by the record lengths. With the records available, statistical stability of the estimates had to be traded for frequency resolution.

V. ARMA Modeling

Auto-regressive, moving average (ARMA) modeling assumes that the present state of a system depends on the past history of the system and the present and past values of its inputs (Box and Jenkins, 1970). The residuals are required to be white noise. Because of the expected absence of well-defined harmonic oscillators in the data, it is not surprising that periodogram analysis, such as presented in the previous section, is not particularly successful or enlightening. In cases such as these, it has been suggested that auto-regressive, moving-average models may be superior to direct Fourier analyses. Furthermore, the results of ARMA analyses may ultimately be translated into the frequency domain. We attempted a series of ARMA models, first on the monthly mean time-series and then on the weekly mean time-series. Our goal was to derive a parsimonious model with minimum variance in the residuals. We used the Akaike Information Criterion (Akaike, 1974) to select a model with the minimum number of free parameters. Furthermore, the model had to make physical sense.

The selected models for the monthly mean time-series are presented in Table 2. The R-squared values are slightly greater than those achieved with simple regression. We can now account for more than half the variance in the observed signal by allowing for an auto-regressive portion in the model. The hindcast, though, still misses the extrema in the data. We may simply be undersampling the important processes with a sampling interval of 1 month.

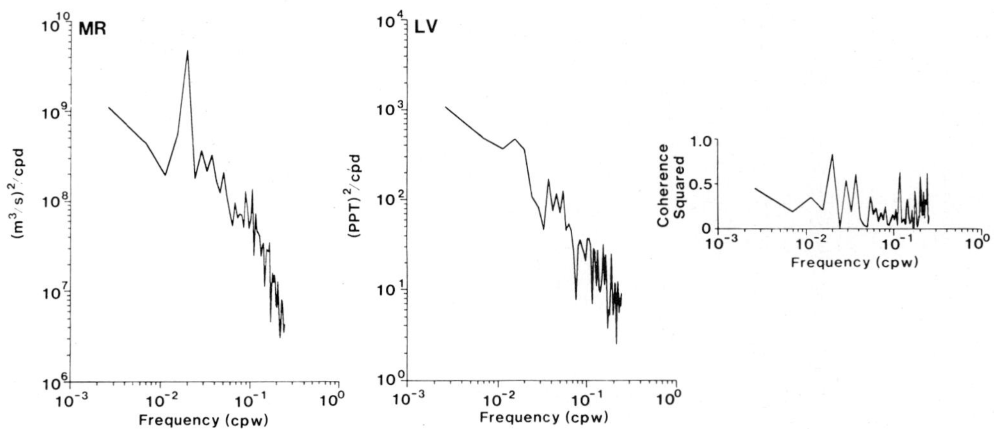

FIGURE 4. Left: Spectrum of weekly mean Mississippi River discharge for period concurrent with Cocodrie salinity records. Middle: Spectrum of Cocodrie weekly mean salinities. Right: Coherence between Cocodrie salinity and Mississippi River discharge estimated from available records. All estimates have 10 degrees of freedom. The 95% confidence interval for the coherence is 0.53.

The selected models of the weekly mean time series have an impressive ability to account for the observed variability (Table 3, Figure (5)). Note that the Akaike Information Criterion selects a more complex model for the weekly data than for the monthly data. The models have the general form

$$S(t) = \langle S \rangle - a_0 B^n (1 - a_1 B) R(t) + N(t)/(1 + b_1 B + b_2 B^2) \quad (1)$$

where $S(t)$ is the salinity time series, $R(t)$ is the discharge time series, $N(t)$ is a white noise time series, $\langle S \rangle$ is the mean salinity at the station, and a_0, a_1, b_1, b_2, and n are constants. B is the backshift operator defined such that

$$B^k F(t) = F(t-kT) \quad (2)$$

where F is an arbitrary function of time and T is the sampling interval. The lag, n, associated with the river discharge time-series in the model increases for those estuaries lying farther downcoast from the delta, as expected. It takes the river discharge plume a finite period of time to disperse downstream from the river mouth. The implied alongshore speed of the freshwater signal is 10-20 cm/s, which is of the correct order of magnitude. The moving-average coefficient, a_1, of the discharge time series is small for the Grand Terre and Sister Lake monitoring sites, which are located near the mouths of their respective estuaries. This coefficient increases at the stations farther up-estuary. If we assume that the discharge signal propagates through the system with an apparent velocity (which may be time varying), then a time change can be converted to a distance change. The term $(1+a_1B)R(t)$ may be written as $a_1(B-1)R(t) + (1+a_1)R(t)$. The portion $a_1(B-1)R(t)$ may tentatively be identified with a longitudinal salinity gradient. Thus, one possible interpretation of the increased values of a_1 at up-estuary sites is that they reflect the importance of longitudinal gradients to either mixing processes or gravitational circulation.

Table 2. ARMA models for the monthly mean data

Model	
$S_{GT}(t) = 26.907 - 2.9x10^{-4}B(1 + 0.485B)R(t) + N(t)/(1 - 0.431B)$	$R^2 = 0.57$
$S_{SM}(t) = 17.966 - 2.5x10^{-4}B(1 + 0.377B) R(t) + N(t)/(1 - 0.571B)$	$R^2 = 0.59$
$S_{LV}(t) = 16.470 - 5.5x10^{-5}B(1 + 0.415B) R(t) + N(t)/(1 - 0.630B)$	$R^2 = 0.53$
$S_{CO}(t) = 14.025 - 2.0x10^{-4}B(1 + 0.680B) R(t) + N(t)/(1 - 0.630B)$	$R^2 = 0.61$
$S_{SL}(t) = 17.102 - 2.4x10^{-4}B(1 + 0.742B) R(t) + N(t)/(1 - 0.637B)$	$R^2 = 0.70$

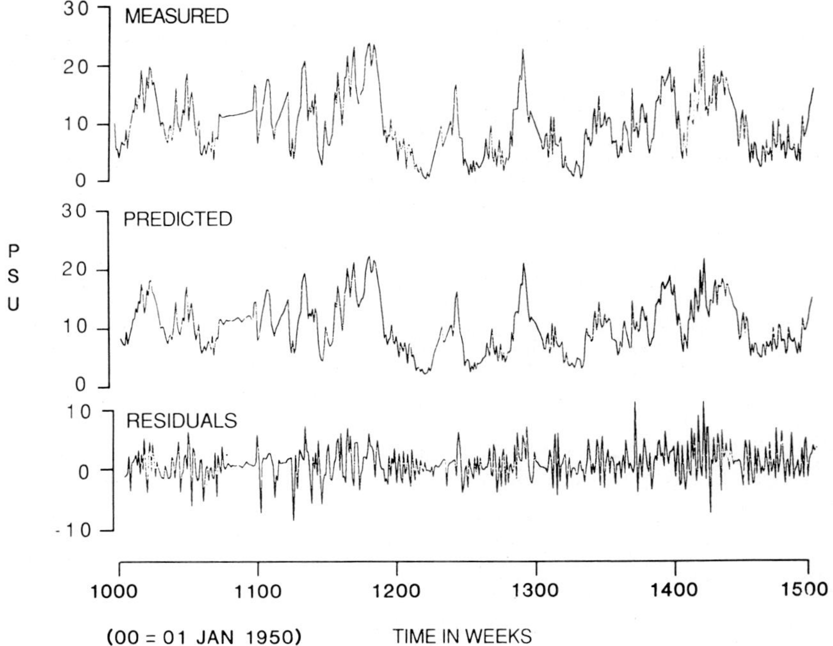

FIGURE 5. Measured salinities at Cocodrie compared with the hindcast salinities from an ARMA model. Residuals are plotted in the lower panel.

The auto-regressive portion of the model describes dynamics of the system that are not directly driven by the river. These include such things as the tidal and wind-driven circulations and the resultant mixing characteristics of the system, although the response caused by any individual process cannot be isolated. At all five stations, the operator

$$\frac{1}{1 + b_1 B + b_2 B^2} \qquad (3)$$

is similar (Figure 6). Low amplitudes prevail down to frequencies of about 0.05 to 0.1 cpw. Below these frequencies, the spectrum rapidly increases. Since the energy in this spectrum is confined to low frequencies, there must exist processes with long time scales (>10-20 weeks) that are incoherent with Mississippi River discharge and important controls on estuarine salinity. Two candidates are local evaporation/precipitation and wind-driven shelf-estuarine exchange. The present analysis, though, cannot determine the validity of this conjecture.

It should be pointed out that the auto-regressive portion of the model can be related to a continuous second-order differential equation (Bartlett, 1946). Our attempts to interpret the coefficients of these equations in terms of a physical model of the estuarine salt balance have proven futile.

Table 3. ARMA results for the weekly mean data.

$$S_{GT}(t) = 25.213 - 2.8\text{x}10^{-4}B^2(1+0.086B)\ R(t) + N(t)/(1-0.686B-0.104B^2);\ R^2 = 0.73$$

$$S_{SM}(t) = 15.629 - 1.5\text{x}10^{-4}B^2(1+0.213B)\ R(t) + N(t)/(1-0.729B-0.116B^2);\ R^2 = 0.76$$

$$S_{LV}(t) = 16.416 - 5.0\text{x}10^{-5}B^3(1+0.239B)\ R(t) + N(t)/(1-0.722B-0.131B^2);\ R^2 = 0.70$$

$$S_{CO}(t) = 11.795 - 1.1\text{x}10^{-4}B^4(1+0.526B)\ R(t) + N(t)/(1-0.721B-0.124B^2);\ R^2 = 0.75$$

$$S_{SL}(t) = 14.228 - 1.9\text{x}10^{-4}B^4(1+0.057B)\ R(t) + N(t)/(1-0.953B+0.065B^2);\ R^2 = 0.86$$

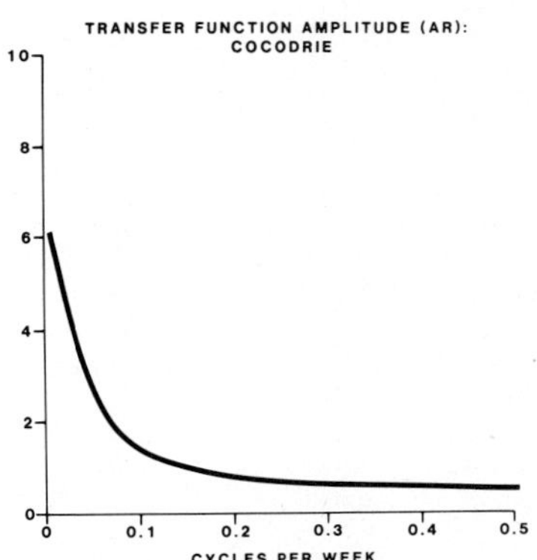

FIGURE 6. Spectrum of the auto-regressive model for Cocodrie salinity driven white noise.

VI. Discussion

Long-term records of estuarine salinity from coastal Louisiana exhibit inter-annual variability, which appears to be inversely related to Mississippi River discharge. Since Mississippi River runoff does not flow directly into the heads of these estuaries, such a relationship must arise from alteration of the coastal ocean's salinity structure. The salinity signal must subsequently propagate upstream, not downstream, within the estuaries. That such a process could control estuarine salinities is not a new idea. Pritchard and Bunce (1959) proposed a similar mechanism for the control of circulation and salinity inside the Magothy River estuary. The flood of the Susquehanna River and its alteration of salinities along the western shore of Chesapeake Bay played the same role in their model as we suggest for the Mississippi River discharge. Barrett (1971), from analysis of a single year of data, pointed out that, during certain months of the year, massive Mississippi River discharge flowed westward along the coast and low-salinity water entered the estuaries through the tidal passes to alter the salinity of the stations near the mouth of an estuary. Others have suggested that this is a mechanism for larval recruitment from the shelf to the estuarine nursery grounds (J. H. Stone, personal communication).

Auto-regressive, moving-average models of the salinity records were able to hindcast the observations with acceptable accuracy. The models were consistent with a downstream dispersion of Mississippi River water along the coast and upstream dispersion within the estuary. In the case of the ARMA models of weekly mean salinity, the portion of the model resulting from the Mississippi River discharge accounted for between 30 and 50% of the explained variance at all sites except Leeville. At Leeville, the river flow accounted for only 6% of the explained variance. This appears to be due to the fact that Leeville lies along Bayou Lafourche, an abandoned distributary of the Mississippi River. Circulation is restricted within the bayou, and moderate freshwater inflow at the head of the bayou tends to push salt from the estuary. River flow at the head of the bayou is large and undoubtedly is the dominant control on salinity.

The river discharge portions of the models of monthly mean salinity account for much higher percentages of the explained variance. We believe that this merely reflects the ability of the river flow to act as a surrogate for an annual cycle in the analysis.

The auto-regressive portion of the models describe processes not directly related to Mississippi River discharge. These may include tidal dispersion, wind-driven estuarine-shelf exchange, and flow driven by the small upland runoff to these estuaries. In all cases, the response characteristics of these portions of the models were weak at periods shorter than 10 to 20 weeks, but increased greatly at longer periods. Much of this variability is probably associated with low-frequency variations in mean water level in the northwestern Gulf of Mexico (Blaha and Sturges, 1981) and seasonal variations in the meteorological forcing (DiMego et al., 1976).

The models we have described are purely statistical. No physics is incorporated in their derivation. They are, though, consistent with a qualitative physical description of large-scale features of the system. Because of their simplicity, it is to be expected that they do not account for all the observed salinity variability. What is surprising is that they do so well with constant coefficients. The local meteorology is highly seasonal (DiMego et al., 1976). Wind-driven shelf-estuarine exchange is expected to be strong in winter and much weaker in summer (Schroeder and Wiseman, 1986). It is, perhaps, possible to discern a hint of such a seasonal pattern in the residuals from the models. Further resolution of the questions concerning the dynamics of these shallow systems will require carefully designed field studies, not further analysis of data monitored for other purposes.

Acknowledgments

This work was funded under Contract 14-12-0001-30252 with the U.S. Minerals Management Service. Particular thanks go to Mr. B. Barrett of the Louisiana Department of Wildlife and Fisheries for releasing data to us. This manuscript is Contribution No. 125 of the School of Geoscience, Louisiana State University.

VII. References

Akaike, H., 1974: A new look at the statistical model identification. IEEE Trans. Autom. Control, AC-**19**, 716-723.

Barrett, B. B., 1971: Cooperative Gulf of Mexico estuarine inventory and study, Louisiana, phase II, hydrology, and phase III, sedimentology. Louisiana Wild Life and Fisheries Commission, New Orleans, LA, 191 pp.

Bartlett, M. S., 1946: On the theoretical specification and sampling properties of autocorrelated time series. J. Roy. Statist. Soc., **B8**, 27-41.

Blaha, J., and W. Sturges, 1981: Evidence for wind-forced circulation in the Gulf of Mexico. J. Mar. Res., **39**, 711-734.

Box, G. E. P., and G. M. Jenkins, 1970: Time Series Analysis: Forecasting and Control. Holden-Day.

Byrne, P., M. Borengasser, A. Grew, R. A. Miller, B. L. Smith, Jr., and C. Wax, 1976: Barataria Basin: hydrologic and climatologic processes. Sea Grant Publ. LSU-T-76-010, Center for Wetland Resources, Louisiana State University, Baton Rouge, LA, 176 pp.

Cooley, J. W., and J. W. Tukey, 1965: An algorithm for the machine calculation of complex Fourier series. Math. of Comput., **19**, 297-301.

DiMego, G. J., L. F. Bosart, and G. W. Enderson, 1976: An examination of the frequency and mean conditions surrounding frontal incursions into the Gulf of Mexico and Caribbean Sea. Mon. Wea. Rev., **104**, 709-718.

Dinnel, S. P, and Wm. J. Wiseman, Jr., 1986: Fresh water on the Louisiana and Texas shelf. Continental Shelf Research, **6**(6), 765-784.

Kelly, F. J., J. D. Cochrane, F. E. Randall, and J. E. Schmitz, 1983: Chapter 2, Physical Oceanography, in (L. R. DeRouen, R. W. Hann, D. M. Casserly, C. Giammona, and V. J. Lascara, eds.) West Hackberry Strategic Petroleum Reserve Site Brine Disposal Monitoring, Year I Report, Final Report, Vol. II, Physical and Chemical Oceanography, submitted to Department of Energy, Strategic Petroleum Reserve Project Management Office, under DoE Contract No. DE-AC96-80P010288 by McNeese State University, Lake Charles, La., and Texas A & M University, College Station, Tex.

Lee, J. M., 1988: Subtidal Circulation at Calcasieu Pass. M.S. thesis, Louisiana State Univ., Baton Rouge, LA. 61 pp.

Mendelssohn, I. A., and K. L. McKee, 1987: Experimental field and greenhouse verification of the influence of saltwater intrusion and submergence on marsh deterioration: Mechanisms of action. Causes of Wetland Loss in the Coastal Central Gulf of Mexico, Volume II: Technical Narrative, R. E. Turner, and D. R. Cahoon, Eds., final report to Minerals Management Service, New Orleans, LA, Contract No. 14-12-0001-30252, OCS Study/MMS 87-0120, 145-180.

Neter, John, and W. Wasserman, 1974: Applied Linear Statistical Models, Richard D. Irwin, 842 pp.

Pritchard, D. W. and R. E. Bunce, 1959: Physical and chemical hydrography of the Magothy River. Tech. Rep. XVII, Ref. 59-2, Chesapeake Bay Institute, The Johns Hopkins University, Baltimore, MD.

SAS Institute, Inc., 1984: SAS/ETS User's Guide, Version 5. Cary, NC, 738 pp.

Schroeder, W. W., and Wm. J. Wiseman, Jr., 1986: Low-frequency shelf-estuarine exchange processes in Mobile Bay and other estuarine systems on the northern Gulf of Mexico. Estuarine Variability, D. A. Wolfe, Ed., Academic Press, 355-367.

Shaw, R. F., Wm. J. Wiseman, Jr., R. E. Turner, L. J. Rouse, R. E. Condrey, and F. J. Kelly, 1985: Transport of larval gulf menhaden Brevoortia patronus in continental shelf waters of western Louisiana: a hypothesis. Trans. Am. Fish. Soc., **114**, 452-460.

Turner, R. E., 1982: Wetland losses and coastal fisheries: an enigmatic and economically significant dependency. Proc. Conf. on Coastal Erosion and Wetland Modification in Louisiana: Causes, Consequences and Options, D. F. Boesch, Ed., U.S. Fish and Wildlife Serv., Biol. Serv. Program, FWS/OBS-82/59, Washington, D.C., pp. 112-120.

Turner, R. E., and D. R. Cahoon, 1987: Introduction to the study. Causes of Wetland Loss in the Coastal Central Gulf of Mexico, Volume II: Technical Narrative, R. E. Turner and D. R. Cahoon, Eds., final report to Minerals Management Service, New Orleans, LA, Contract No. 14-12-0001-30252, OCS Study/MMS 87-0120, 1-12.

Wiseman, Wm. J., Jr., and E. M. Swenson, in press: Salinity trends in Louisiana estuaries. Estuaries.

Mixing Processes in a Numerical Model of the Western Dutch Wadden Sea

H. Ridderinkhof[1] and J.T.F. Zimmerman[1] & [2]

[1]Netherlands Institute for Sea Research
P. O. Box 59, 1790 AB Texel
The Netherlands

[2]Institute for Meteorology and Oceanography
Buys Ballot Laboratory
University of Utrecht
The Netherlands

ABSTRACT

Particle trajectories in a numerical model of the Western Dutch Wadden Sea are used to study the horizontal mixing processes in this area. The Lagrangian displacement of watermasses appears to be very sensitive to the time of release, the initial position and the initial volume of the watermass considered. The advective component of this displacement agrees with the Eulerian residual currents on relatively large scales. In a Lagrangian sense the small scale Eulerian residual currents only have no physical meaning. With respect to tidal dispersion our results are in conflict with the classical shear dispersion concept in that dispersion occurs without the presence of random motions (turbulence). In our model dispersion is caused by the interaction of longitudinal and lateral velocity variations. These spatial variations are caused by the complexity of the bathymetry that varies over a length that is short compared with the tidal excursion length. Regarding the tidal distortion in more detail, similarities with results from studies on the chaotic aspects of fluid mixing can be recognized, e.g. the extremely fast growth of the perimeter of a watermass and the strong sensitivity of the Lagrangian trajectory of a particle to its initial position.

I. Introduction

The Western Dutch Wadden Sea is a typical shallow tidal area that consists of some more or less separated tidal basins. In previous studies (Postma, 1954 and Zimmerman, 1976) one dimensional tidally averaged advection-diffusion models were used to describe observed salinity distributions in this part of the Wadden Sea. This approach, which uses "gradient type" fluxes for salt transport, has also been applied to the Ems Dollart estuary (Helder & Ruardy, 1982), the Eastern Scheldt (Dronkers et al., 1981) and the Bay of Fundy (Holloway,1981). In these one-dimensional tidally averaged box models the advective horizontal velocity is induced by the fresh water input. The horizontal dispersion coefficient is either derived from equating salt exchange (dispersion) and residual salt transport (advection) during stationary periods or is chosen such that observed and computed salinity distributions agree reasonably well. In two dimensional transport models for the North Sea (Prandle, 1984 and De Ruyter et al., 1987) basically the same concept has been applied. The advective horizontal velocity results from locally averaging the time dependent velocity vector, and the parametrisation of the horizontal dispersion coefficient is chosen such that the outcome of the model agrees reasonably with tracer observations. The dispersion coefficient can be

Coastal and Estuarine Studies, Vol. 38
R. T. Cheng (Ed.)
Residual Currents and Long-term Transport

interpreted as a bulk parameter in these models. Its value lies in the range of 50 -500 m^2/s. The physical mechanism responsible for these relatively large effective dispersion coefficients is often assumed to be a cascade of shear dispersion processes (Young et al., 1982 and Zimmerman, 1986). In estuaries tidal trapping and pumping (Fischer et al., 1979) can also play an important role. In the classical shear dispersion concept (Taylor, 1954), dispersion due to turbulence proper is increased by the interaction of turbulence with vertical and/or horizontal shear in the velocity field. This concept was originally developed for steady turbulent flows (Taylor, 1954) and later extended to periodic turbulent flows by Okubo (1967). An important aspect of this concept is that random motions need to be present for an effective dispersion of watermasses.

However, experiments with a numerical model of the Seto Sea in Japan (Imasato et al., 1980, Imasato, 1983 and Awaji, 1982) have shown that mixing of watermasses can also occur if random motions are absent. They used a Lagrangian approach by simulating the track of free flowing particles in the neighbourhood of straits and showed that the resulting mixing or dispersion after a tidal period does not change significantly by adding small scale turbulence to their simulation. These results are in conflict with the shear dispersion concept in that the horizontal dispersion after a tidal period does not seem to rely on the presence of turbulence.

More recently, studies on the chaotic aspects of mixing of fluids in two dimensional time-periodic flows have been presented (e.g. Aref, 1984; Ottino et al., 1988 and Pasmanter, 1988). These laboratory and numerical studies indicate that relatively simple periodic flow fields can give rise to spatially complicated particle behavior. This complex behavior, in which, on relatively small distances apart, so called islands with practically no mixing are surrounded by areas with very strong mixing, is ascribed to the nonlinear aspects of the Euler-Lagrange transformation. These theoretical studies result in Lagrangian dispersion without the presence of random components in the underlying velocity field.

In the present study we also apply the Lagrangian approach to examine tidal dispersion. Particle trajectories in a detailed two-dimensional hydrodynamical model of the western Dutch Wadden Sea are determined on the basis of the deterministic velocity field in which we do not incorporate the effect of random motions. After a short description of our model and the method applied, we shall treat separately the advective and dispersive component of the Lagrangian displacement of watermasses.

With respect to the advective component the discussion will be focussed on the relation between Eulerian and Lagrangian residual velocities. For the dispersive component most attention is paid to the basic mechanisms responsible for tidal dispersion in our model.

II. Numerical Model and Method of Calculation

To compute trajectories of free flowing particles we use a detailed two dimensional vertically averaged numerical model of the tidal flows in the Western Dutch Wadden Sea. The model covers four tidal basins and the open boundaries are located in the adjacent North Sea. The grid size is 500 m and the time step in a numerical simulation is 150 s. Results concerning the computed tidal and residual water levels and velocity field have been discussed extensively by Ridderinkhof (1988, 1989) and have been summarized in a companion paper (Ridderinkhof and Zimmerman, 1989). Here we only give some characteristics of the velocity field that are of interest for the present study. A characteristic value of the amplitude of the periodic tidal currents is about 1.25 m/s in the tidal inlets and about 1 m/s in the center of the main channels. The Eulerian residual flow field can be interpreted as a composition of a small scale part, the residual eddies, and a large scale part, the throughgoing flow. In most regions residual velocities associated with the small scale partare an order of magnitude [O(0.1 m/s)] larger than the velocities associated with the large scale flow [O(0.01 m/s)]. For the present study at each time step the Lagrangian velocity of a particle is bilinearly interpolated from the velocities in the surrounding gridpoints. For the integration scheme a second order midpoint rule is employed (Heemink, 1988, Heemink and van Stijn, 1989). It is assumed that the numerical error introduced by this method, which has a random character, can be ignored.

In our vertically averaged model each particle can be interpreted as representative for a water column equal to the local waterdepth at the time of release (H). The center of mass of a particle group is defined as:

$$(X_c, Y_c) = \left[\frac{\sum_{i=1}^{n} X_i H_i}{\sum_{i=1}^{n} H_i}, \frac{\sum_{i=1}^{n} Y_i H_i}{\sum_{i=1}^{n} H_i} \right] \tag{1}$$

in which (X_c, Y_c) is the position of the center of mass, (X_i, Y_i) the position and H_i the water depth at the time of release of particle i and n the number of particles in a specific group. The displacement of this center of mass is used to examine the advective component of the transport.

The distortion of a particle group can be described by the displacement covariance tensor(σ). The components of this tensor define the mean square distances about the center of mass:

$$\sigma^2_{xx} = \frac{\sum_{i=1}^{n} (X_i - X_c)^2 H_i}{\sum_{i=1}^{n} H_i}, \qquad \sigma^2_{yy} = \frac{\sum_{i=1}^{n} (Y_i - Y_c)^2 H_i}{\sum_{i=1}^{n} H_i}$$

$$\sigma_{xy} = \frac{\sum_{i=1}^{n} (X_i - X_c)(Y_i - Y_c) H_i}{\sum_{i=1}^{n} H_i} \tag{2}$$

The Lagrangian dispersion coefficient is calculated from the distortion after one tidal period (T):

$$D_{xx} = \frac{\sigma^2_{xx}\big|_{t_0+T} - \sigma^2_{xx}\big|_{t_0}}{2T}, \qquad D_{yy} = \frac{\sigma^2_{yy}\big|_{t_0+T} - \sigma^2_{yy}\big|_{t_0}}{2T}$$

$$D_{xy} = \frac{\sigma^2_{xy}\big|_{t_0+T} - \sigma^2_{xy}\big|_{t_0}}{2T} \tag{3}$$

Tensor transformation rules can subsequently be used to determine the orientation of the dominant axes of distortion [along these dominant axes the cross terms in Eqs.(2-3) equal zero].

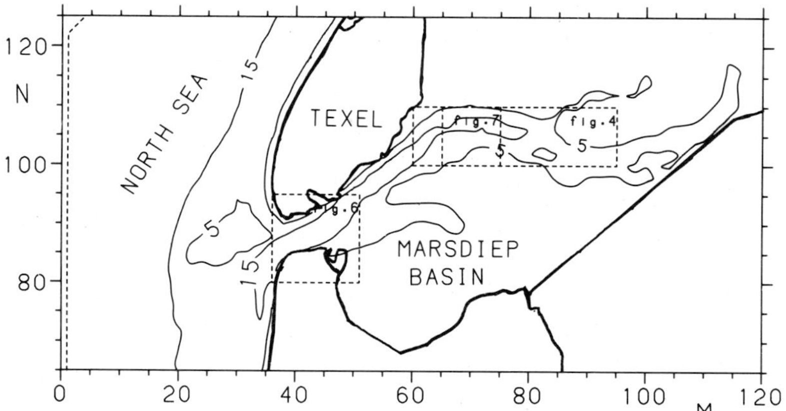

FIGURE 1. Isobaths in a representative area of the two dimensional numerical model for the western Dutch Wadden Sea that is used for particle computations. Grid numbers of this numerical model are indicated on the m- and n-axes. The open boundary is located at m = 2 (dashed line) in the adjacent North Sea. Dashed rectangles mark sub regions as used in figures 4, 6 and 7.

The above defined measures are discussed for particle displacements in a part of our numerical model. Figure 1 displays isobaths in this representative area which covers the tidal inlet region and the inner part of the Marsdiep basin.

III. Lagrangian Displacement of Water Masses

III.1 Advection

In this section we will treat the advective component of particle spreading by regarding the displacement of the center of mass of groups of particles after a tidal period. This Lagrangian residual displacement appears to be a function of the time of release (initial tidal phase), the initial position and of the initial dimensions of the water mass considered.

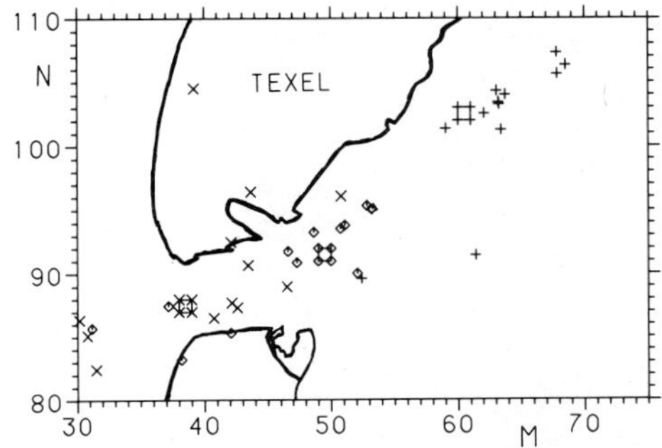

FIGURE 2. Positions of the center of mass after one tidal period of particle groups initially released inside the dashed grid units. The 12 different final positions for each particle group depend on the time of release, which was varied with an interval of 1 hour. Marks at the final position correspond with marks that indicate the initial angular points of a particle group.

The dependence on the initial phase is shown in figure 2. This figure gives the final positions of the center of mass of 16 particles which are initially released inside the three marked grid units. The 12 different final positions stem from successive computations in which the time of release was varied with an interval of 1 hour. It clearly shows that the Lagrangian residual displacement is strongly dependent on the time of release. This initial phase dependence stems from the relatively large spatial variations in the velocity field of our numerical model. The time of release determines the velocity field in which a particle group travels and the Lagrangian residual displacement is a result of the velocity field

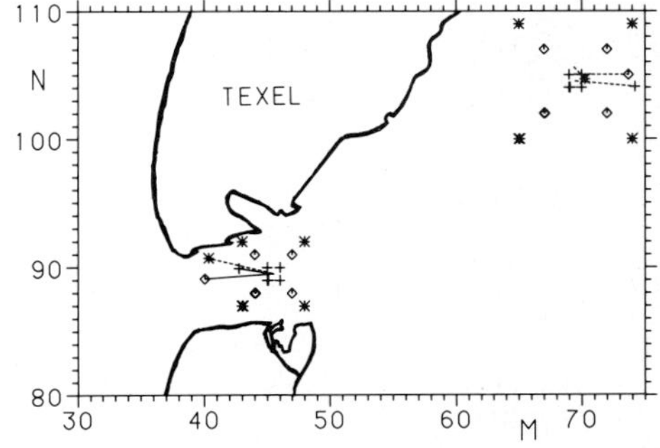

FIGURE 3. Displacement of the center of mass after one tidal period depending on the initial dimensions of the square considered. Corresponding marks indicate the initial angular points of these squares as well as the final position of the center of mass after one tidal period.

experienced. Thus the initial phase strongly influences the residual displacement. The dependence of the displacement of the center of mass on the initial position is also clear from figure 2. The initial position also determines the velocity field in which a watermass travels. Thus the same arguments as for the phase dependence can be used to explain the dependence on the initial position.

The influence of the initial dimensions of a water mass on its residual displacement is shown in figure 3. This figure gives for two different regions the residual displacement of the center of mass depending on the volume represented by the particles at the time of release, during ebb tide. From this figure it is clear that the initial dimensions of a watermass strongly influence its residual displacement. Enlarging the initial volume results in averaging over the displacement of separate subvolumes. This influences the results because the residual displacements are very sensitive to the initial position.

III.1.1. Eulerian - Lagrangian residual velocities

Computations on the basis of the time dependent velocity field require relatively large computer storage and computer time. Therefore it is of interest to study the agreement between the Lagrangian residual displacement of the center of mass of a grid unit and the Eulerian residual velocities. If both agree, the advective component of spreading can be computed relatively simply on the basis of the Eulerian residual velocity field.

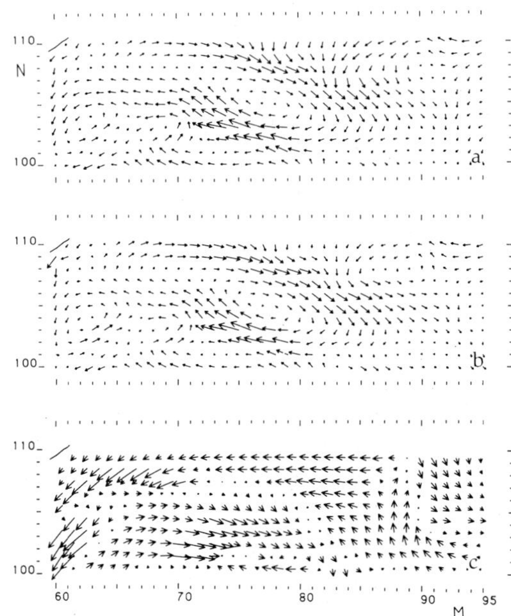

FIGURE 4. Eulerian and Lagrangian residual velocities in a representative region of the Marsdiep basin (see figure 1), a) Eulerian residual velocity field ($\underline{u}_E$), b) Eulerian residual transport velocity field ($\underline{u}_T$), c) Lagrangian residual velocity field ($\underline{u}_L$). One grid unit corresponds with 0.10 m/s. See text for a definition of these velocities.

The Eulerian residual velocity field can be computed by time averaging the computed time dependent velocity in each specific grid unit:

$$\overline{\underline{u}_E} = \frac{1}{T} \int_0^T \underline{u}(t)\, dt \tag{4}$$

Averaging the vertically integrated mass transport gives the so called residual transport velocity:

$$\overline{\underline{u}_T} = \frac{1}{T} \frac{\int_0^T \underline{u}(t)[H+\zeta(t)]\, dt}{\int_0^T [H+\zeta(t)]\, dt} \tag{5}$$

It should be noted that both the residual velocity as well as the residual transport velocity are Eulerian quantities.

The Lagrangian residual velocity is defined as:

$$\overline{\underline{u}_L} = \frac{X(T) - X(0)}{T} \tag{6}$$

in which X(0) and X(T) are the position of the center of mass of a watermass at the start of the computation and after one tidal period respectively. In the figures shown the Lagrangian residual velocity is attributed to the starting point X(0).

Figure 4 shows the three different quantities for a representative region (see figure 1) in the inner part of the Marsdiep basin. Figure 4a,b show the Eulerian residual velocity field ($\underline{u}_E$) and the Eulerian residual transport velocity field ($\underline{u}_T$) respectively. The difference between both velocity fields is small. Both show the dominant presence of tidal residual eddies in this area. However, figure 4c shows that Lagrangian residual velocities ($\underline{u}_L$) differ appreciably from the Eulerian residual velocity field. For these Lagrangian computations each grid unit was represented by 16 particles, evenly distributed over the grid unit at the time of release, at maximum ebb tide. As discussed before, the initial tidal phase strongly influences the Lagrangian residual velocity. Although not shown here, averaging the results from computations in which the initial phase is varied gives the same conclusion that Lagrangian residual velocities differ substantially from the Eulerian residual velocity field. Thus the detailed Eulerian residual velocity field cannot be used to predict the advective component of the transport of mass in our numerical model.

III.2 Dispersion

The dispersive component of the displacement of a watermass can be described by the displacement covariance tensor (defined in section 2).

Figure 5 marks some grid units in which 16 particles were released at maximum ebb tide and gives the residual displacement of the center of mass and the

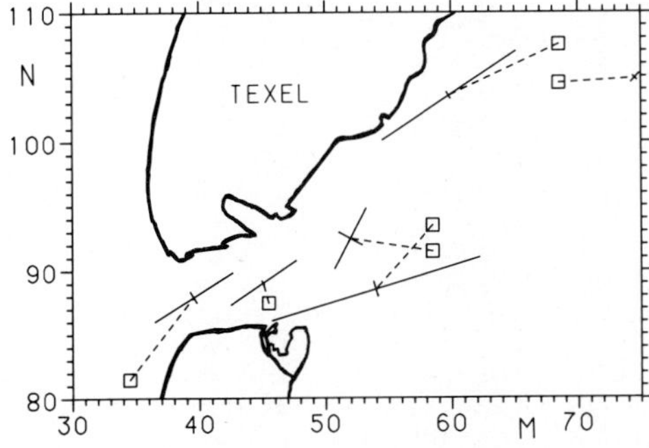

FIGURE 5. The tidal distortion of different particle groups initially released inside the dashed squares. Dashed lines represent the Lagrangian residual displacement of the center of mass. Solid lines give the orientation of the principal axes of distortion and the root-mean square displacement along these axes after one tidal period.

dimensions and orientation of the principal axes of the distortion tensor after one tidal period. Again, most striking are the large differences between the various particle groups. Some particle groups remain relatively close together while others show a remarkable distortion and growth. The distortion is anisotropic with the main axes tending to be oriented in the dominant current direction, parallel to the local isobaths. The length of the axis perpendicular to the dominant current direction remains small. In general, the direction of the residual displacement of the center of mass does not agree with the main axis of the covariance tensor. Thus a decomposition of the Lagrangian motion in an advective and dispersive component would require the introduction of cross terms in the dispersion tensor to be able to describe the dispersive component correctly.

Figure 5 only shows the influence of the initial position on particle dispersion. Although not shown, the dispersion is also strongly dependent on the initial tidal phase. The same arguments as in section 3.1 can be used to explain these dependencies.

As described before, the distortion can be used to compute the dispersion coefficient. Figure (6) gives isopleths for this coefficient along the dominant axes of distortion for the region near the Zeegat van Texel. Particle computations were started during ebb tide with 16 particles evenly distributed over each grid unit. In this region the dispersion coefficient reaches values up to 700 m^2/s. Dispersion coefficients perpendicular to this main axis are much smaller.

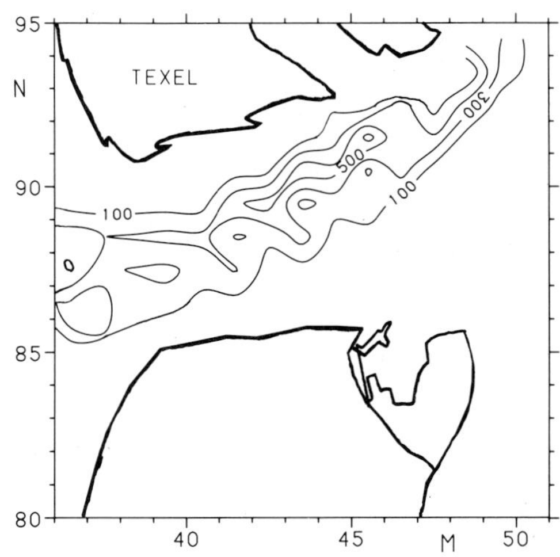

FIGURE 6. Isopleths of the dispersion coefficient (m^2/s) along the dominant axis of dispersion in the tidal inlet region (see figure 1).

Figure 7a-d show the dispersion coefficient in a part of the inner basin when the initial dimensions of each separate watermass are subsequently increased. figure (7a), which corresponds with figure (6) with respect to the initial dimensions of each sub volume, shows that the dispersion coefficients in the inner basin are much smaller than in the tidal inlet region. In this area the maximum value is about 100 m^2/s. However, figure (7b-7d) show that this value increases with increasing dimensions of the subvolumes considered. In figure (7d) the dispersion coefficient is even larger than the maximum value in figure (7a). A greater part of particle spreading is attributed to dispersion instead of advection if a larger water mass is used to make a distinction between the advective and dispersive component. Thus also the ratio between advection and dispersion strongly depends on the initial dimensions of a particle group. In general this ratio decreases with increasing initial dimensions.

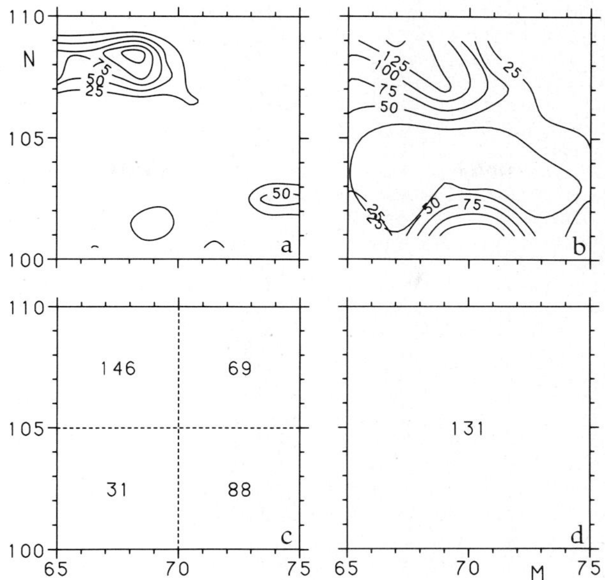

FIGURE 7. Dispersion coefficients (m^2/s) along the dominant axis of distortion in a part of the inner basin (see figure 1) dependent on the initial dimensions of each watermass considered. The dimension of each watermass is 1, 4, 25 and 100 grid units in figures (7a-7d) respectively.

III.3 Basic Mechanisms

Previous examples of particle behavior have clearly shown that a particle does not return to its initial position after one tidal period. Furthermore the position after one tidal period is very sensitive to the initial phase of release and the initial position of the particle considered. As a consequence, a watermass, marked by numerous particles, is displaced and distorted by the flow field in our numerical model. The fact that distortion occurs in our model is in contrast to the theoretical shear-dispersion concept for oscillating flow in which, without the presence of random motion, a particle would return to its original position. This makes it interesting to examine which mechanism causes distortion in our numerical model. For this we first look more closely to the Lagrangian distortion within a tidal period. Then we will relate our results with (theoretical) studies that predict the occurrence of chaos as a result of the Euler-Lagrange transformation.

III.3.1. Dispersion Within a Tidal Period

The dispersion within a tidal period is discussed by following the deformation of two originally straight lines. These lines initially extend over about 6 grid units and each line is labelled with about 2500 particles with an initial distance of 0.0025 grid unit between successive particles. Both lines cover a cross section of the major tidal channel in the southern basin, one in the tidal inlet (a) and the other in the inner basin (b). The computation was started during ebb tide.

Figure (8a) displays isobaths in this part of our model and both particle lines at the time of release. figure (8b-8g) display successive particle positions after 2, 4, 6, 8, 10 and 12.5 hours (one tidal period). Two hours after the time of release both lines have been stretched by lateral shear. At the seaside of the inlet the particle line is divided in two separate groups by the branching of the main channel. Next picture (figure 8c) shows that two hours later, around the turn of the tide, stretching by lateral shear has further increased the length of both lines. Six hours after the time of release (figure 8d) the tide has been reversed, which strongly influences the position and shape of both lines. At the seaside of the inlet a part of the particle group at the northern side remains outside the inflow area of the tidal inlet and continues to flow in a northward direction, even after the reversal of the tide. This illustrates that a part of the outflowing watermass is influenced by two different current systems, which enlarges the stretching. At the southern side of the inlet region, most of the particles remain in the sphere of influence of the tidal inlet. The relatively regular shape of this particle group during ebb has been distorted by lateral phase and amplitude differences. The most striking of the shape of the other line at this tidal phase is that after the reversal of the tide an important part of this line has not travelled along the center of the main channel, but follows a more southward route. Besides this, lateral phase differences in the main channel, in which slack tide in the shallower parts preceeds slack tide in the deeper parts, causes an increased stretching of this line. Lateral shear in the velocity field further stretches both lines during the flood period, see figures (8e-8f). A small part of the particles that left the inlet at the southern side during ebb appears to travel around the emersed sandbank at the seaside of the inlet during flood. However, most particles from the southern part of the inlet return to the inner basin during flood. In the northern part most particles remain in the adjacent sea. In the inner basin a part of the line of particles appears to branch off to another tidal channel where it is stretched further by lateral shear. The final positions after one tidal period (and one more slack tide) are shown in figure (8g).

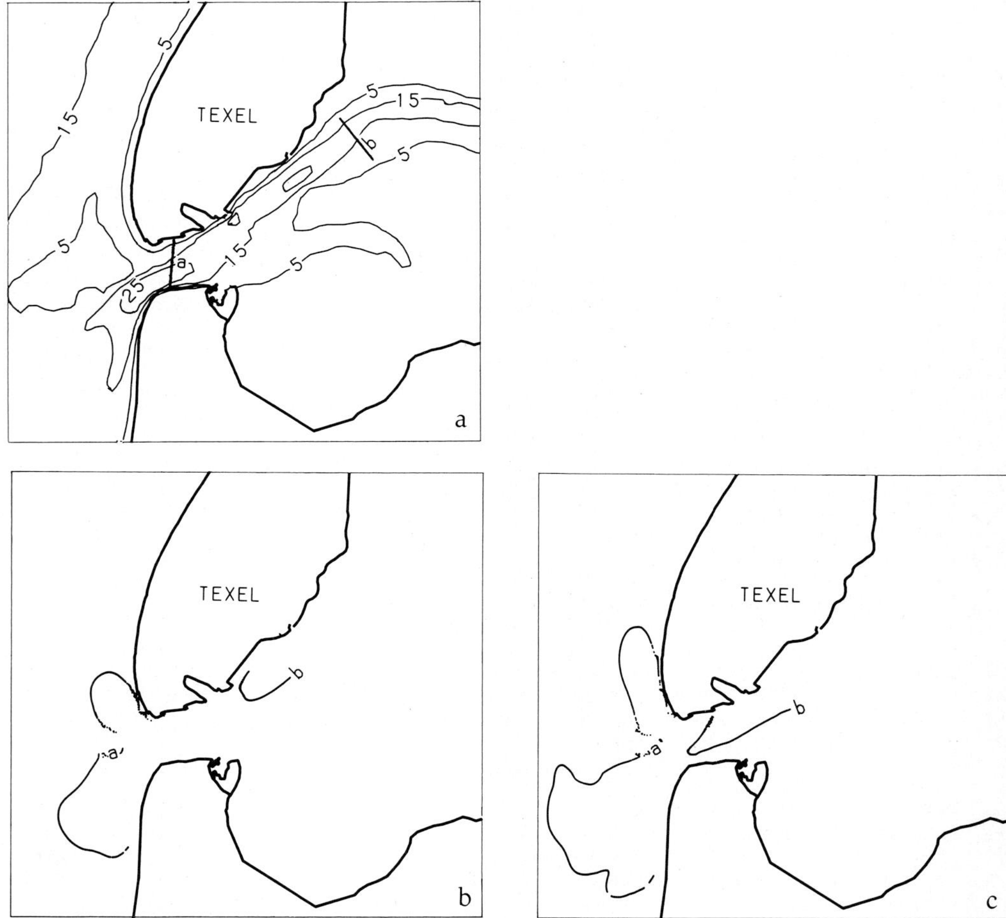

FIGURE 8. (Caption on the next page).

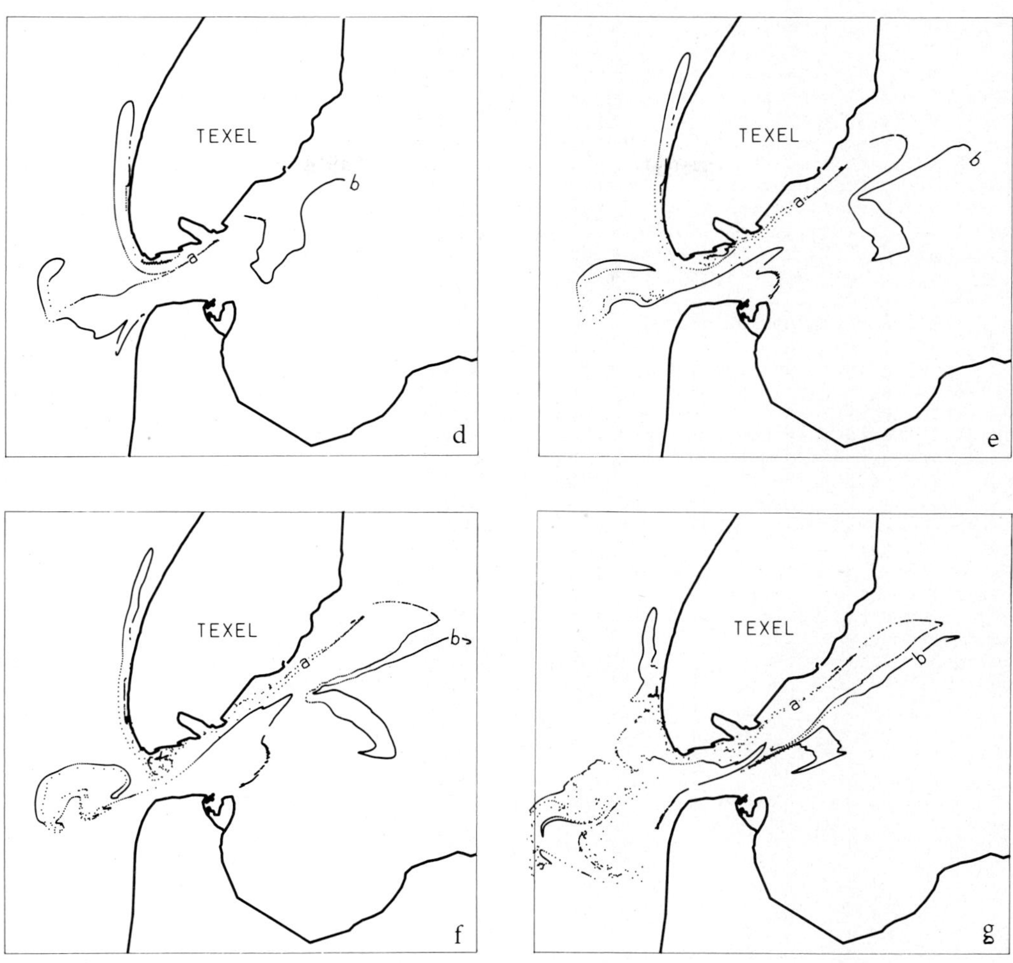

FIGURE 8. Distortion of two separate lines of particles within a tidal period. Figure (8a) gives the positions at the time of release and some isobaths (5, 15 and 25 m) in this part of the model. Figure (8b-8g) give successive particle positions after 2, 4, 6, 8, 10 and 12.5 hours (one tidal period).

It is hard to reconstruct the shape of the originally straight line in the tidal inlet. Although somewhat less pronounced, the same can be said about the shape of the line in the inner basin.

From these particle distributions dispersion coefficients can be derived. Decomposed along and perpendicular to the main axis of deformation these coefficients are 340 m^2/s and 100 m^2/s respectively for the line in the inlet region and 185 m^2/s and 0 m^2/s for the line in the inner basin.

From this sequence of figures it is clear that dispersion results from the spatial variations in amplitude, direction and phase of the oscillatory currents. At the seaside of the inlet the spatial variations in dominant current directions and tidal phase of the currents are extremely large. This is the transition zone of two current regimes, one belonging to the tidal basin and the other to the adjacent open sea. A watermass travelling through both regimes is therefore strongly deformed. In the tidal basin itself the interaction between the complicated bottom topography and the tidal wave creates large spatial variations in amplitude, phase and direction of the oscillatory currents. Here the most important cause for Lagrangian dispersion appears to be that the bottom topography varies over a length that is short compared to the tidal excursion length. Thus a watermass meets substantial different velocity fields during its tidal excursion.

A more detailed study of the particle distribution after one tidal period shows that the stretching within both lines is rather irregular. There appear to be large differences in the growth of segments of both lines. This strong variability in line stretching has also been recognized in studies of the chaotic aspects of fluid mixing. In the next section we will relate the particle behavior in our numerical model to these studies.

III.3.2. Lagrangian chaos?

Some recent publications on mixing in periodic velocity fields (Aref, 1984 & Ottino et al., 1988) have shown the possibility of chaotic particle behavior in that particle positions may become random functions of time even if the underlying velocity field does not contain a stochastic component. Laboratory studies of flows with very low Reynold numbers as well as numerical studies with relatively simple periodic velocity fields have shown such Lagrangian chaotic behavior. Most fascinating is that these velocity fields, which are very simple in an Eulerian sense, can cause very complicated Lagrangian trajectories. The velocity field used by Pasmanter (1986) comes close to the velocity field in our numerical model of the Wadden Sea. He shows that depending on the initial positions, particles can stick together, like (patchiness) as well as showing the distance between particles can grow exponentially in the same velocity field. This kind of particle behavior is demonstrated by showing maps of particle positions after relatively many (tidal) periods. Thus particles remain in a specific velocity field over a relatively long period. In our numerical model we cannot simulate particle trajectories over long

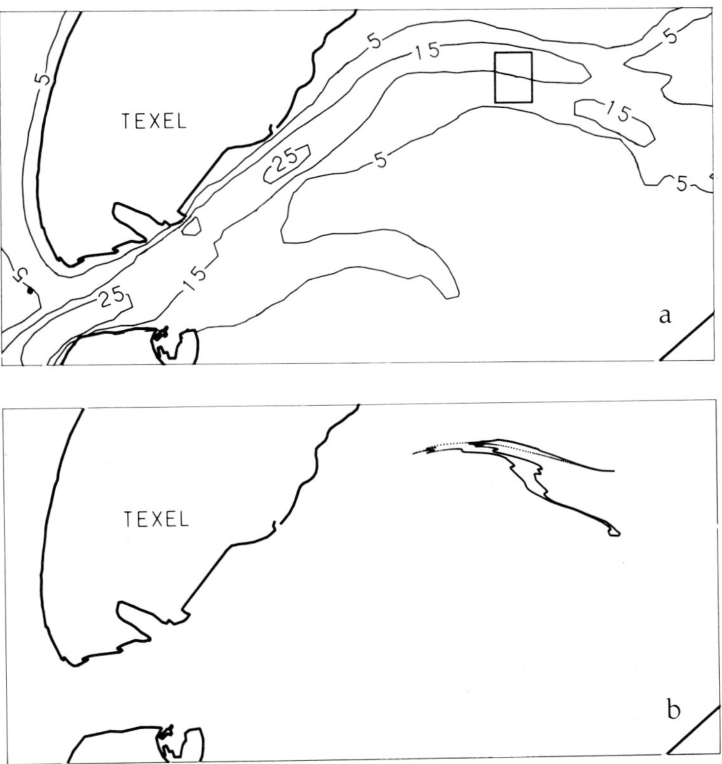

FIGURE 9. (Caption on the next page).

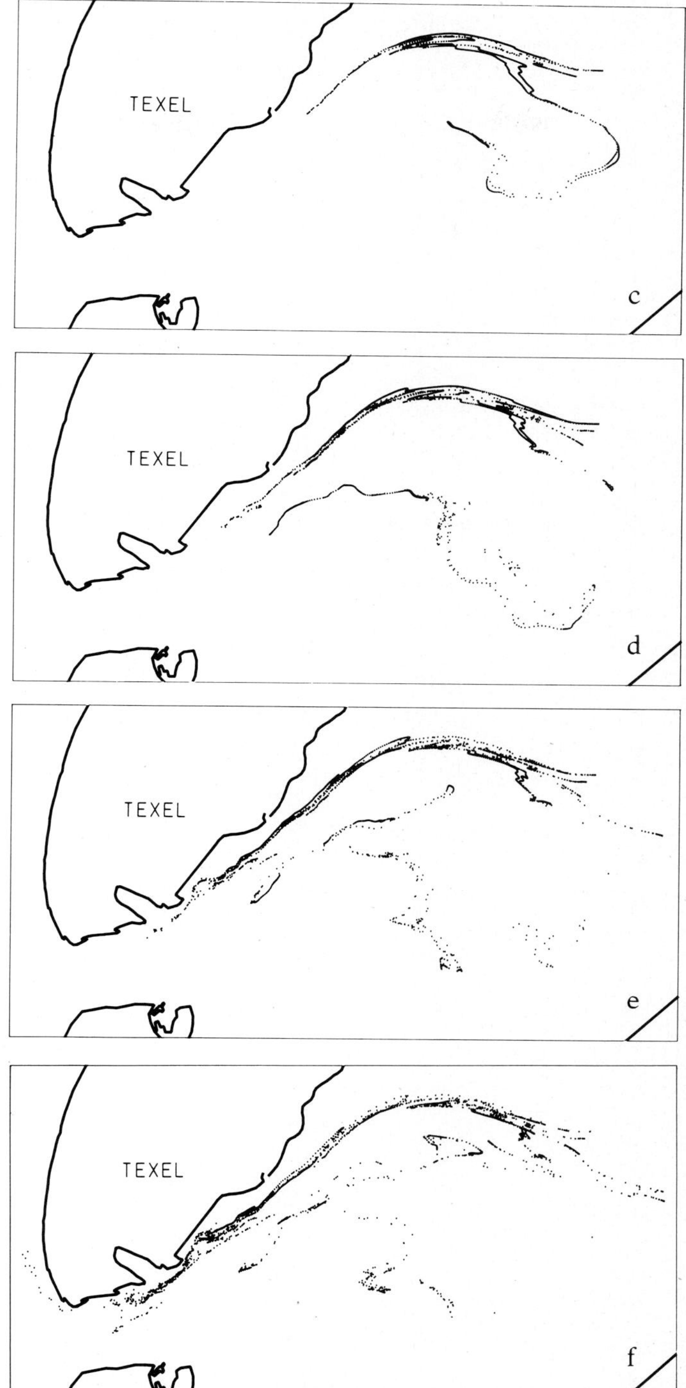

FIGURE 9. Tidal deformation of a watercolumn initially released in the middle of the Marsdiep basin. Figure (9a) gives particle positions at the time of release and some isobaths. Figure (9b-9f) give the particle positions after 1, 2, 3, 4 and 5 tidal periods.

periods in the same velocity field. Most particles of a specific group would have left such a velocity field within a few tidal periods. Even if we regard the velocity field in a complete tidal basin as our velocity field of interest only a few tidal periods can be used if we do not want our results to be influenced by the closed boundaries of the basin or by substantially different flow fields belonging to the tidal inlet region or the adjacent sea.

However, some general features of this anomalous particle behavior can also be recognized in our numerical model. To demonstrate this we show the deformation of a water column initially released in the middle of the southern basin. On the perimeter of this column we labelled 200 particles per grid, equivalent to an initial distance of 2.5 m between subsequent particles. Figure (9) shows the position of these particles at the start of the computation and after 1, 2, 3, 4 and 5 tidal periods successively. From these maps we see that the watermass considered deforms strongly and that its perimeter grows extremely fast. Already after one tidal period the perimeter is strongly stretched and the deformation is so strong that it is hard to recognize the shape of the perimeter in the following pictures.

A characteristic feature of chaotic behavior is that the distance between particles that are initially close to each other grows exponentially in time. Thus the length of the perimeter of the watermass should grow exponentially in a chaotic regime. Figure (10a) shows the length of the perimeter as a function of time. Although only five tidal periods have been shown it is clear that the growth is stronger than linear. The computed length should be interpreted as a minimum value,

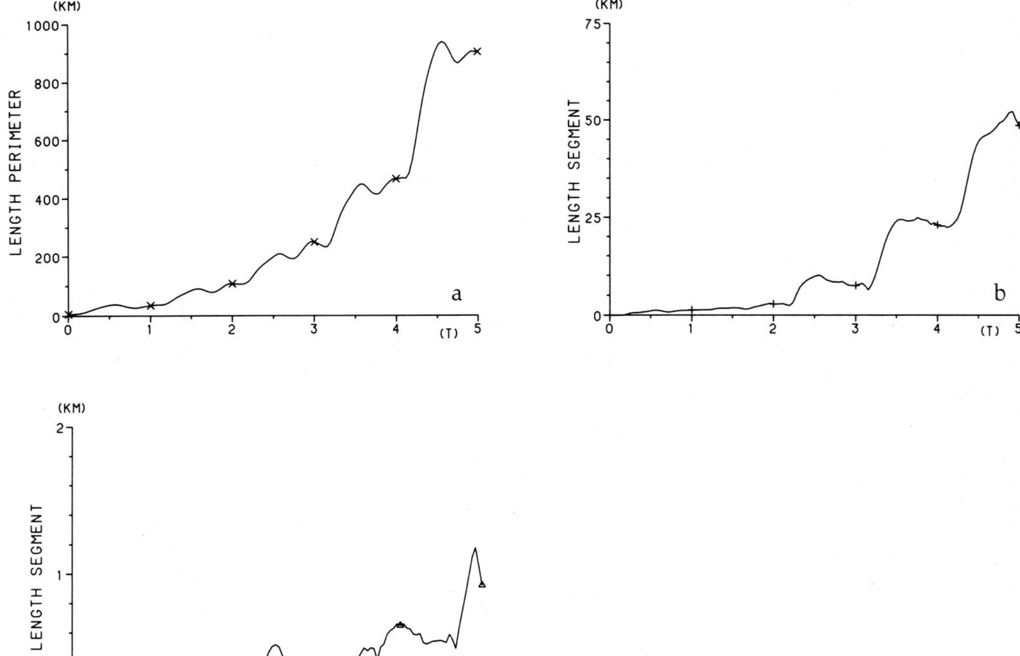

FIGURE 10. a) Length of the perimeter of a watermass (see figure 9) as a function of tidal period, b) and c) Length of two segments of 20 particles as a function of tidal period. Initially both segments are one grid unit apart. Note the different scales of the vertical axes.

because in comparison with this computation decreasing the initial distance between successive particles would, after a few tidal periods, certainly give an increased length .

In some parts of the maps in figure (9) one can recognize that there is a large difference in the stretching of different segments of this perimeter. This phenomenon has much in common with the joint occurrence of patchiness as well as chaos, depending on the initial positions of the particles considered in Pasmanter's study. This strong spatial variability of line stretching is illustrated in figure (10b) and figure (10c), which give the length of a segment of 20 particles as a function of time for two different segments initially being about one grid unit apart. In spite of the small initial distance between these segments the growth of their length differs extremely (note the large difference in vertical scale in both figures). Thus large variations of line stretching and mixing, can occur on relatively small distances apart.

IV. Discussion and Conclusions

With respect to the advective part of water mass displacement our results confirm previous studies (Zimmerman, 1979, Cheng & Casuli, 1982) in that the small scale Eulerian residual velocity field or transport velocity field does not agree with Lagrangian residual velocities. Only in those regions where the Eulerian residual current as well as the oscillatory tidal current is uniform over the watermass considered during its complete tidal excursion, the Eulerian residual velocity is equal to the Lagrangian residual velocity. This is certainly not the case in estuaries like the Wadden Sea. Therefore we have to be very careful in using Eulerian residual quantities in explaining or even predicting residual transport of material which is a Lagrangian process. Although Lagrangian and Eulerian residual velocities are never identical in an estuary, the difference between both decreases if the initial dimensions of the watermass are large. This can be explained from a decomposition of the Eulerian residual flow field in a large scale part, the throughgoing flow, and a small scale part, the residual eddies. If the dimensions of a water mass are such that it covers the cross section of the estuary (or tidal channel) and that its length exceeds the length scale of the residual eddies, spatial averaging results in the relatively uniform large scale residual velocity. Then the Lagrangian residual velocity equals approximately the Eulerian residual velocity. Thus, only if the inital volume covers the cross section of a channel and its length exceeds the tidal excursion length, the difference between its Lagrangian and Eulerian residual velocity will be small.

With respect to the dispersive part of watermass displacement, our results show that in a complex bathymetry like the western Wadden Sea relatively large horizontal dispersion coefficients stem from spatial variations in the periodic tidal current. Thus dispersion does not depend exclusively on the presence of random motion. In the classical shear dispersion concept the effective dispersion coefficient is a function of lateral variations in velocity amplitude and of the magnitude of turbulence. Applying this concept to tidal flow basically gives the same results. The only difference is that the time scale ratio between the tidal period and the cross-sectional (turbulent) mixing time scale now enters the expression for the effective horizontal dispersion coefficient. In essence turbulence is indispensible for horizontal dispersion in this concept. Our particle computations show that in an estuary like the Wadden Sea, longitudinal variations in the velocity field play an important role in particle spreading. Due to these longitudinal variations a particle does not return to its initial position and, considering the displacement of a watermass, horizontal dispersion is the result. Thus horizontal dispersion stems from the strong spatial variability of the oscillatory currents and the effective horizontal dispersion strongly depends on the combined effect of longitudinal and lateral velocity variations.

Regarding the tidal distortion in more detail, large spatial variations in stretching can be recognized. Extremely fast as well as extremely slow stretching of fluid segments initially released close together, is a feature of particle behavior that is similar to results from theoretical studies on chaotic mixing in periodic flow fields. The overall effect is that the perimeter of a watermass grows stronger than linear in time. This fast growth of the perimeter of a water mass is of importance if one considers the mixing with the surrounding water, as this is decisive for the interfacial area over which turbulent diffusion can operate effectively.

Summarizing, tidal dispersion in an estuary is mainly caused by the irregularity of the flow field which stems from the interaction of the tidal flow with the complex bathymetry. The tidal excursion length far exceeds the length over which the bathymetry and associated velocity field varies. Thus the non-uniformity of the flow field in a longitudinal direction can be seen as the essential mechanism for tidal dispersion. The actual value of the dispersion coefficient primarily depends on the complexity of the bathymetry.

Acknowledgements

We are much indebted to Rijkswaterstaat, Dienst Informatie Verwerking, for the use of their computer and the support in employing their extended system of computer programs for simulating two dimensional tidal flows (WAQUA system).

IX. References

Aref, H., 1984: Stirring by chaotic advection. J. Fluid. Mech., **143**, 1-21.

Awaji, T., 1982: Water mixing in a tidal current and the effect of turbulence on tidal exchange through a strait. J. Phys. Oceanogr., **12**, 501-514.

Cheng, R. T. & V. Casulli, 1982: On Lagrangian residual currents with applications in South San Francisco Bay, California. Wat. Res. Res., **18**, 1652-1662.

Dronkers, J., A. G. van Os and J. J. Leendertse, 1981: Predictive salinity modelling of the Oosterschelde with hydraulic and mathematical models. In: Fischer, H.B. (ed.). Transport models for inland and coastal water. Acad. Press, New York: 451-482.

Fischer, H. B., E. J. List, R. C. Y. Koh, J. Imberger and N.H. Brooks, 1979: Mixing in inland and coastal waters. Acad. Press, New York: 1-483.

Heemink, A. W., 1988: Stochastic modelling of dispersion in shallow water. Stochastic Hydrology and Hydraulics (submitted)

Heemink, A. W. and Th. L. van Stijn, 1989: Numerical and stochastic modelling of advection and diffusion of water pollutants. To appear in Proc. VIth Int. Conf. on Numerical Methods in Laminar and Turbulent Flow. C.Taylor (ed), 1989

Helder, W. and P. Ruardij, 1982: A one-dimensional mixing and flushing model of the Ems-Dollard estuary: Calculation of time scales at different river discharges. Neth. J. Sea Res., **15**, 293-312.

Holloway, P. E., 1981: Longitudinal mixing in the upper reaches of the Bay of Fundy. Est. Coast. Shelf Sci., **13**, 495-515.

Imasato, N., 1983: A numerical experiment on water and salt exchange through the Akashi and Naruto Straits. J. Phys. Oceanogr., **13**, 1526-1533.

Imasato, N., T. Awaji and H. Kunishi, 1980: Tidal exchange through Naruto, Akashi and Kitan Straits. J. Oceanogr. Soc. Japan, **36**, 151-162.

Okubo, A., 1967: The effect of shear in an oscillatory current on horizontal diffusion from an instantancous source. Int. J. Oceanol. Limnol., **1**, 194-204.

Ottino, J. M., C. W. Leong, H. Rising and P. D. Swanson, 1988: Morphological structures produced by mixing in chaotic flows. Nature, **333**, 419-425.

Pasmanter, R., 1986: Dynamical systems, deterministic chaos and dispersion in shallow tidal flow. Preprint.

Postma, H., 1954: Hydrography of the Dutch Wadden Sea. Archs. Neerl. Zool., **10**, 405-511.

Prandle, D., 1984: A modelling study of the mixing of Cs^{137} in the seas of the European continental shelf. Phil. Trans. R. Soc., (A) **310**, 407-436.

Ridderinkhof, H., 1988: Tidal and residual flows in the Western Dutch Wadden Sea, I: Numerical model results. Neth. J. Sea Res., **22**, 1-22

Ridderinkhof, H., 1989: Tidal and residual flows in the Western Dutch Wadden Sea, III: Vorticity balances. Accepted for publication in Neth. J. Sea Res.

Ridderinkhof, H. and J. T. F. Zimmerman, 1989: Residual currents in the Western Dutch Wadden Sea. (this volume)

Ruijter, W. P. M. de, L. Postma and J. M. de Kok, 1987: Transport Atlas of the Southern North Sea. Mooren, Nijmegen: 1-33.

Taylor, G. I., 1954: The dispersal of matter in turbulent flow through a pipe. Proc. Roy. Soc., (A) **223**, 446-468.

Young, W. R., P. B. Rhones and C. J. R. Garret, 1982: Shear flow dispersion, internal waves and horizontal mixing in the ocean. J. Phys. Oceanogr., **12**, 515-527.

Zimmerman, J. T. F., 1976: Mixing and flushing of tidal embayments in the western Dutch Wadden Sea, I: Distribution of salinity and calculation of mixing time scales. Neth. J. Sea Res., **10**, 149-191.

Zimmerman, J. T. F., 1979: On the Euler-Lagrange transformation and the Stokes drift in the presence of oscillatory and residual currents. Deep Sea Res., **26**, 505-520.

Zimmerman, J. T. F., 1986: The tidal whirlpool: A review of horizontal dispersion by tidal and residual currents. Neth. J. Sea Res., **20**, 133-154.

Numerical Simulation of Tidal Dispersion Around a Coastal Headland

Richard P. Signell and W. Rockwell Geyer

Woods Hole Oceanographic Institution
Woods Hole, MA 02543

ABSTRACT

Tidal flows around headlands can exhibit strong spatial gradients in the Eulerian currents, resulting in complex Lagrangian trajectories and dispersion of the vertically integrated flow. This typically occurs when the horizontal length scale of the headland is comparable to or smaller than the tidal excursion. The effects of these headlands on dispersion are investigated using a depth-averaged hydrodynamic model combined with a particle tracking model. The dispersion of patches of fluid is found to vary by more than an order of magnitude, depending both on position and tidal phase at the time of release. This is due to the infrequent interaction of material with the strongly sheared flow at the tip of the headland, where flow separation occurs during times of maximum tidal flow. Spreading of these patches over many tidal cycles is not Gaussian, but rather shows a patchy, streaky structure.

I. Introduction

In many shallow estuaries and embayments, tidal dispersion plays an important role in the horizontal transport of pollutants, larvae and other suspended or dissolved material. Zimmerman (1976) found that in the Dutch Wadden Sea, the large effective diffusivities calculated from salinity observations (100-1000 m^2/s) could be explained by tide-induced dispersion based on an analytic model which considered interactions of a purely oscillatory tidal flow with a spatially varying tide-induced residual eddy field. Awaji et al. (1980) and Awaji (1982) conducted numerical experiments of mixing around a tidally dominated strait and found that even without considering the tide-induced residual effects, the rapid spatial variation in the oscillatory current could give rise to large residual particle displacements and to large values of dispersion (800 m^2/s) for material released in the vicinity of the strait.

Insight into this type of dispersion was obtained by Aref (1984), who showed that a completely deterministic, periodic flow of simple form could give rise to rapid mixing due to chaotic particle trajectories. The efficiency of the mixing is controlled by the strength of a small region of high strain and the frequency with which material interacts with this region of high strain. Building on the work of Aref (1984), Pasmanter (1988) examined particle trajectories in a kinematic tidal flow consisting of a mean and oscillatory current that vary harmonically in space. For some values of the flow parameters, he found that the area of released patches grows linearly with time, directly analogous to Brownian motion or turbulent diffusion. For other values, however, Pasmanter (1988) found "islands" of unmixed material embedded in the well-mixed regions, and discovered that the area of patches can grow faster or slower than linearly with time, a phenomenon he termed "anomalous diffusion".

Aref (1984) and Pasmanter (1988) specified simple tidal flow fields, studying the nature of mixing over many tidal cycles, and over a range of flow parameters.

Coastal and Estuarine Studies, Vol. 38
R. T. Cheng (Ed.)
Residual Currents and Long-term Transport

Awaji et al.(1980) and Awaji (1982) used a numerical model to generate a flow field, and primarily studied the nature of mixing over a single tidal cycle. In this paper, the technique of Awaji et al.(1980) is used to investigate the tidal dispersion around an idealized headland, but the simulations are extended over multiple tidal cycles to better understand the longer-term mixing behavior. In the case presented here, the tidal excursion is comparable to the scale of the headland, and the dynamics around the headland are strongly nonlinear, evidenced by flow separation and transient eddy formation. The objectives are to first give a clear description of the basic mixing mechanism around this headland and then to illustrate the dramatic spatial and temporal variation in dispersion characteristics that result.

II. Methods

The basic physics of how vertically well-mixed material is horizontally dispersed around the headland can be understood by neglecting vertical structure and considering only depth-averaged dynamics (Signell, 1989). In this case, the depth-averaged equations governing Eulerian flow in a homogeneous fluid without wind stress can be expressed as:

$$\frac{\partial \mathbf{u}}{\partial t} = \mathbf{u}\cdot\nabla\mathbf{u} + f(\mathbf{k} \times \mathbf{u}) = -g\nabla\eta - \frac{C_D\mathbf{u}|\mathbf{u}|}{h + \eta} + \nabla\cdot(A_H\nabla\mathbf{u}), \qquad (1)$$

$$\frac{\partial \eta}{\partial t} + \nabla\cdot[\mathbf{u}(h + \eta)] = 0, \qquad (2)$$

where $\mathbf{u}$ denotes the depth averaged velocity vector, η the free-surface elevation, h the water depth measured from the at-rest sea surface z=0, f the Coriolis parameter, g the gravitational acceleration, C_D the bottom drag coefficient, A_H the horizontal eddy viscosity, ∇ the horizontal gradient operator, and $\mathbf{k}$ the vertical unit vector acting upward.

The numerical scheme to solve these equations is derived from Flather and Heaps (1975), but modified for a curvilinear orthogonal grid (Signell, 1989). The curvilinear grid was chosen to allow increased resolution near the tip of the headland and give a smooth representation of the coastline. The scheme is second-order explicit, using a forward-backward scheme for the gravity waves (e.g., Mesinger and Arakawa, 1976) and an angled-derivative scheme (Roberts and Weiss, 1966) for the advective terms. Spatial variations close to the scale of the grid (2-4 gridlengths) which are treated poorly by the numerical scheme are removed by Shapiro filtering (Shapiro, 1975).

Figure 1 illustrates the geometry of the case study. The channel is 50 km long and 15 km wide and is discretized with a 60 x 30 grid. With the curvilinear grid, this results in a minimum distance between grid points of 53 m at the tip of the headland. Experimentation with grid resolution showed that dispersion results obtained with this grid were negligibly different than results obtained with a grid of twice the resolution. The headland is represented by a Gaussian profile

$$y(x) = A \exp\left[-\frac{1}{2}\left(\frac{x}{\sigma}\right)^2\right] \qquad (4)$$

with amplitude A = 7 km and width specified by σ = 2 km. The depth increases linearly northward away from the southern boundary, reaching a depth of 20 m at a distance of 3 km from the boundary. The basin is a constant depth of 20 m elsewhere. The western boundary is forced by a normal depth-averaged velocity that varied sinusoidally with an amplitude of 0.5 m/s and a M_2 tidal period of 12.42 hours (12 lunar hours). The eastern boundary has a gravity wave radiation condition, the southern boundary has a no-slip condition, and the northern boundary has a free-slip condition. Parameters values are C_D= .0025, A_h = 1.0 m^2/s and f = 1.0 x 10^{-4}/s. The specifications of this problem correspond roughly to Gay Head, Massachusetts, a headland at which the detailed spatial structure of the tidal flow has been mapped with an acoustic Doppler current profiler (Geyer and Signell, 1989).

With the explicit solution technique and fine grid spacing, the time step was restricted by the gravity wave Courant condition

$$\Delta t < \left[\frac{s_1 s_2}{\sqrt{gh(s_1^2 + s_2^2}} \right] \quad , \tag{3}$$

where s_1 and s_2 are the metric factors of the orthogonal transformation. Satisfying this criterion in the vicinity of the headland resulted in a time step of 9.0 s. The efficiency of the scheme was improved by using a time-splitting technique

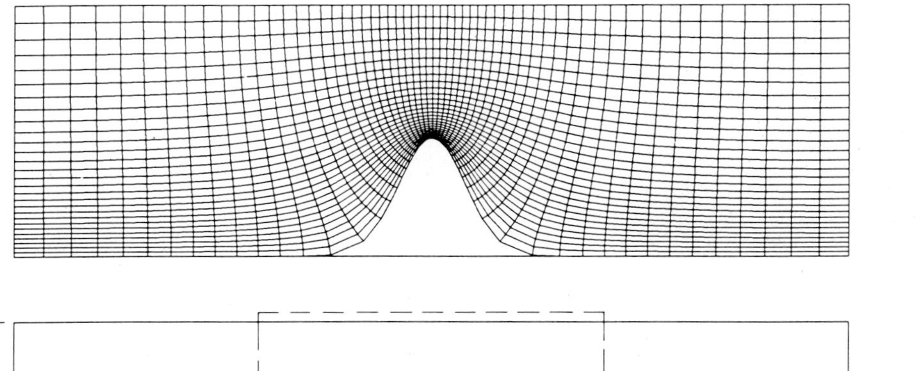

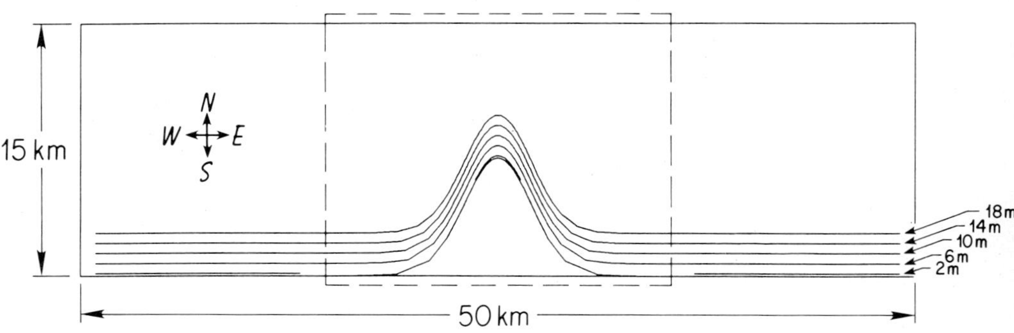

FIGURE 1. Model grid and bathymetry. The top panel shows the 60 X 30 curvilinear orthogonal grid used in the study. The basin is 50 km long, 15 km wide and the minimum spacing between grid points is 53 m at the tip of the headland. The bottom panel shows contours of the model bathymetry at intervals of 2 m. The depth along the southren boundary increases linearly to the north, reaching a depth of 20 m in 3 km. The interior of the basin has a constant depth of 20 m. The dashed line indicates the sub-domain that will be presented in succeeding plots of results.

(Marchuk, 1974), whereby momentum is propagated, advected, and diffused in separate steps. The propagation and diffusion steps both vectorize, resulting in a significant speed-up over the scheme which solves the entire momentum equation at once. Since the gravity wave speed is much larger than the maximum advection speed, the advection step, which does not vectorize, is computed every 5 timesteps. The overall speed-up of the time-split scheme relative to a combined scheme is nearly a factor of five, with a negligible change in accuracy.

Particle trajectories were calculated by numerical integration of the velocity field. The velocity field used for the particle tracking was obtained by running the model for five tidal cycles until all transients had disappeared. The sixth cycle was then used repeatedly in the dispersion calculations. First the velocity at the particle location was determined by bilinearly interpolating the velocity field to the particle location. Bicubic interpolation was also tried, but did not result in significant differences in particle trajectories. With the velocity fieldspecified in space by interpolation, the particle positions were advanced in time using a 4th order Runge-Kutta integration scheme. The time step for particle tracking was 240 s.

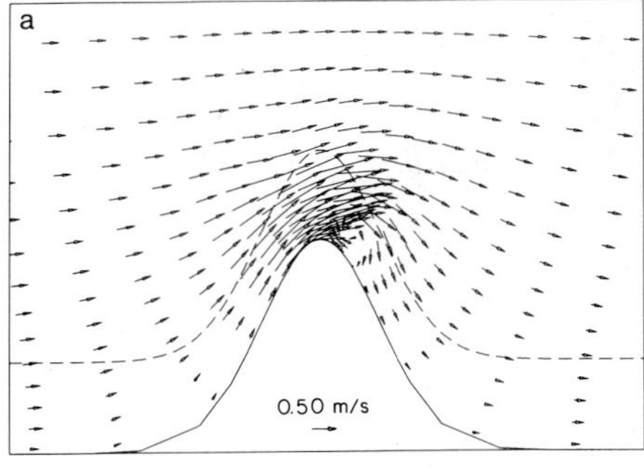

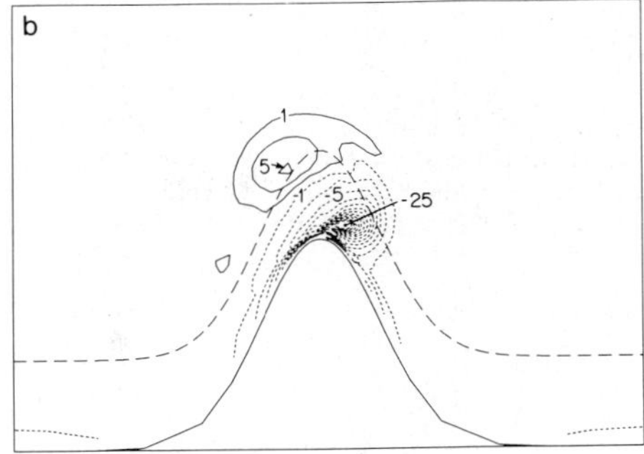

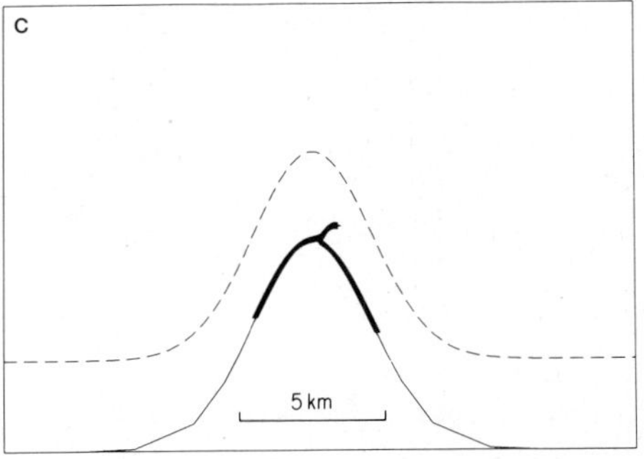

FIGURE 2. The modeled tide at 2 lunar hours. (a) The depth-averaged velocity vectors (plotted every other grid point). The offshore extent of the sloping bottom is indicated by the dashed line. (b) The vertical component of the vorticity: the curl of the depth-averaged flow. Values are scaled by 10^{-4}/s. (c) The positions of particles released in a line parallel to the coast 1/2 grid space away from the southern boundary at the beginning of the tidal cycle.

The effects of sub-grid scale turbulence on particle motion were simulated by using a random-walk Markov-chain model whereby the particle velocity is perturbed at each time step by a random fluctuation corresponding to a specified level of diffusivity. In this way, a level of diffusivity that parameterizes the effects of three-dimensional sub-gridscale turbulence or vertical shear dispersion can be represented in the motion of the particles. Details of the Markov-chain model can be found in Awaji (1982).

III. Results

By examining the flow field over a tidal period, the basic mixing process around the headland is illuminated. In the following discussion, three different aspects of the flow are presented: (1) depth-averaged current vectors; (2) contour maps of the vertical component of vorticity; (3) the position of particles released in a line parallel to the coast 1/2 grid space away from the southern boundary at the beginning of the tidal cycle. The beginning of the tidal cycle (0 lunar hours) is defined as the beginning of eastward flow.

At 2 lunar hours, roughly one hour before maximum eastward flow, the upstream flow is smoothly increasing from less than 0.4 m/s west of the headland to nearly 1.5 m/s off the headland tip (Figure 2a). Along the upstream southern boundary, the current is sheared by bottom friction over the sloping bottom. At the tip of the headland, the flow separates from the coast and a lee eddy begins to form, resulting in large spatial velocity gradients. The rotational part of the velocity is represented in the vorticity field (Figure 2b), which shows a narrow band of negative vorticity along the leading edge of the headland. The magnitude in this band reaches a maximum value of 25×10^{-4}/s at the tip of the headland. At the point of flow separation, this band is stripped away from the boundary and injected into the interior. The vorticity is concentrated at the center of the developing lee eddy with magnitude of about 25×10^{-4}/s. The separation can be clearly seen in the deformation of the particle line (Figure 2c), which shows that particles are converging toward the point of separation, where they are pulled away from the boundary. The strong deformation of the line at the point of separation indicates a large degree of strain.

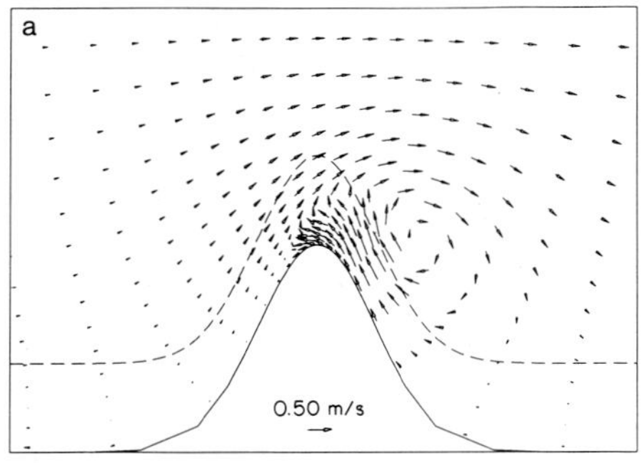

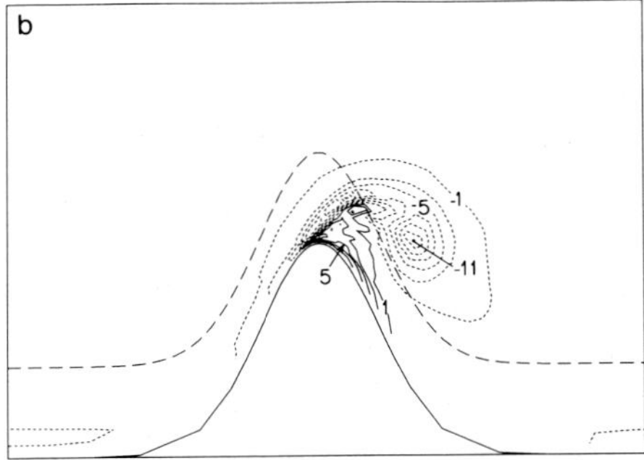

FIGURE 3. The modeled tide at 6 lunar hours. Panels as in Figure 2.

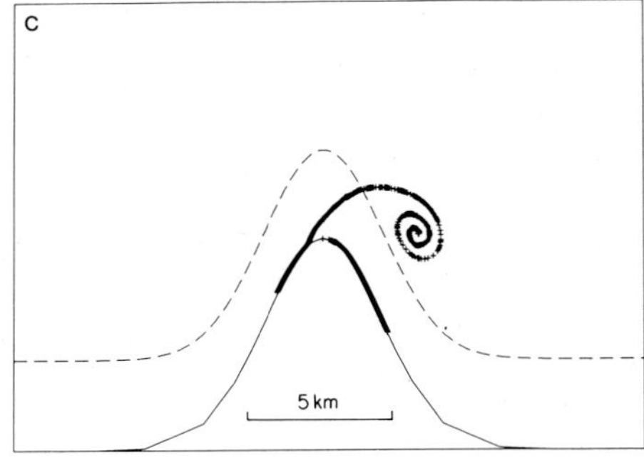

At 6 lunar hours, the tide is beginning to turn. The flow at the headland tip has already turned back toward the west, while the eastward flow offshore of the headland is rapidly approaching slack (Figure 3a). The lee eddy has grown in size and has advected to the east. The vorticity plot (Figure 3b) shows that as the size of the eddy increases, the vorticity magnitude at the eddy core decreases. Positive vorticity is already being produced along the southern boundary by the effects of the sloping bottom on the westward flow. The particle line (Figure 3c) shows that particles that were injected into the interior by the flow separation have been stretched and wrapped around the eddy core. Strain associated with the flow separation and the vortex roll-up has led to large deformation of water parcels in the vicinity of the headland tip.

At 8 lunar hours, roughly one hour before maximum westward flow (Figure 4a), the flow is beginning to separate on the western side, forming nearly a mirror image of the flow at two lunar hours. The vorticity plot (Figure 4b) shows that the counter-clockwise eddy is growing, while the clockwise eddy formed of the eastward flow is rapidly disappearing. As the clockwise eddy weakens, Figure 4c shows that particles associated with this eddy cease to wrap and are now mostly advected toward the west, while the newly formed flow separation results in another injection of particles into the interior.

At 12 lunar hours, the end of westward flow (Figure 5a), we see nearly a mirror image of the current at the end of the eastward phase (Figure 3a). There is an isolated vorticity maximum associated with the counter-clockwise eddy (Figure 5b), and the vorticity signature of the clockwise eddy has completely disappeared.

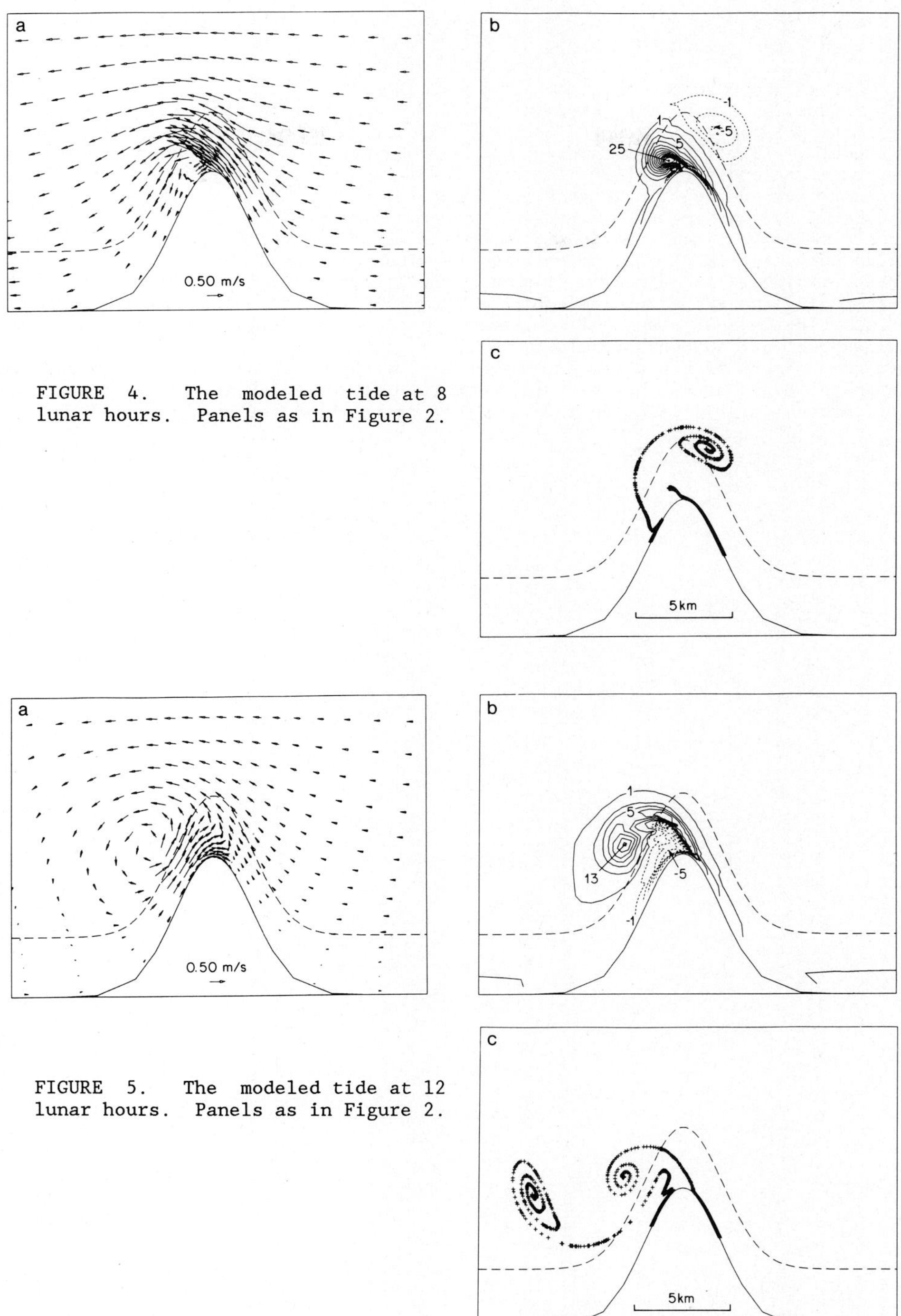

FIGURE 4. The modeled tide at 8 lunar hours. Panels as in Figure 2.

FIGURE 5. The modeled tide at 12 lunar hours. Panels as in Figure 2.

Particles drawn away from the boundary on the westward flow separation have wrapped around the counter-clockwise eddy (Figure 5c). The particle distribution shows that over the course of one tidal cycle the initial particle line has been strongly deformed by the strain associated with the flow separation, and particles are spread over a much larger distance.

To illustrate that the modeled flow is similar in nature to the flow at Gay Head, Figure 6 shows the observed tidal flow at maximum flood and at slack high water, interpolated from repeated acoustic Doppler current profiler tracks (Geyer and Signell, 1989). It is clear that the observed flow at Gay Head separates and forms transient eddies, similar to the idealized numerical model. A detailed comparison between the observations and a more realistic model of Gay Head using the actual geometry, bathymetry and tidal forcing is currently underway. The Eulerian residual velocity from the idealized model was calculated by averaging the depth-averaged current at each grid point over the tidal cycle. The structure of the Eulerian residual is a nearly symmetric double-gyre circulation that results from time-averaging the transient lee eddies (Figure 7). It is sometimes argued that the Eulerian tide-induced residual current is an important transport mechanism, since even weak residual currents result in large displacements over time. In the case where the length of the tidal excursion is comparable to the length over which the Eulerian residual varies, however, the structure of the Eulerian residual has very little physical significance to the transport of material (e.g. Imasato, 1983).

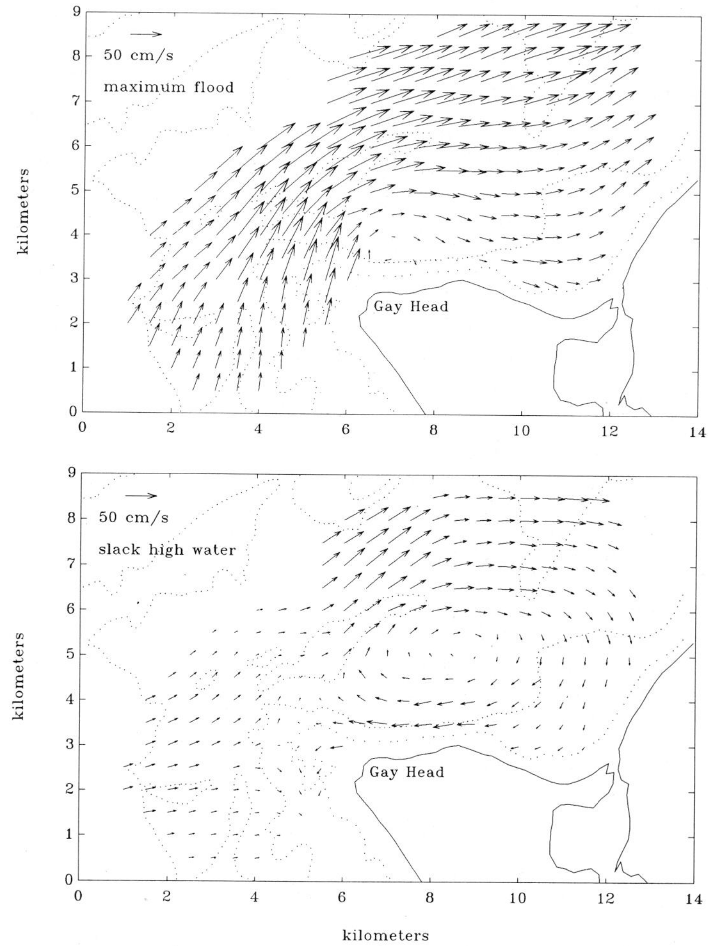

FIGURE 6. The observed tide at Gay Hea, Massachusetts at maximum flood and slack high water. The flow vectors were interpolated from repeated Doppler cruise tracks using a technique discussed in (Geyer and Signell, 1989).

The field most relevant to the transport of material is the "Lagrangian residual velocity," a quantity which indicates the average velocity of a particular particle over a tidal cycle (Zimmerman, 1979). This can be defined as the vector quantity formed by dividing the net particle displacements over a tidal cycle by the tidal period:

$$\vec{u}_L(\vec{x}_o, t_o) \equiv \frac{\vec{x}(\vec{x}_o,\ t_o + T) - \vec{x}_o}{T} \tag{5}$$

where T is the tidal period, $\vec{x}_o$ is the location of the particle at t_o, and $\vec{x}(\vec{x}_o + T)$ is the location of the particle one tidal cycle later. If the Lagrangian residual velocities are scaled by the tidal period, then this field expresses the transformation which maps particle positions at a specified phase of the tide onto new positions one tidal cycle later. Since the flow is periodic, this map (if calculated at infinite spatial resolution) contains all the information to calculate the subsequent evolution of particles released at a given phase of the tide. The path traveled by a released particle varies dramatically depending upon the phase of the tide at the time of release. The Lagrangian residual velocity, therefore, also varies as a function of release phase.

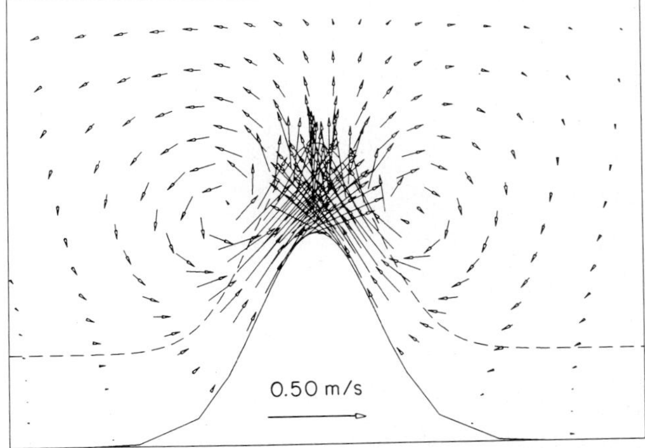

FIGURE 7. The Eulerian tide-induced residual flow, obtained by averaging the depth-averaged flow at each grid point over a tidal cycle.

The map of the Lagrangian residual velocity corresponding to the particle release at the beginning of the tidal cycle was constructed and is shown in Figure 8. In contrast to the Eulerian residual velocity (Figure 7), far away from the headland the Lagrangian residual velocity field is relatively smooth, but near the headland there is a region of rapid variation which is unresolved by the plotted vectors. If the vectors are plotted with higher resolution, the structure has more detail, but is still not fully resolved. In fact, particles arbitrarily close together can become widely separated over a tidal cycle, indicating chaotic behavior of the particle trajectories in this region. The implication of this map for dispersion is that a patch of material released in this region will be strongly stretched and deformed, while a patch released far away from the headland will exhibit some residual displacement, but will not be strongly deformed. Maps of the Lagrangian residual velocity at other phases of the tide, while displaying much different local particle displacements, share this common global feature: they all reveal that material in a certain region near the headland will be strongly deformed, while material far away from the headland will not.

To illustrate the key aspects of the dispersion process and how it evolves with time, two rectangular patches of 900 particles each were released at 0 lunar hours (the start of eastward flow) and tracked for 24 tidal cycles. The first patch was 200 m wide and released 3.5 km north of the tip, while the second patch was 100 m wide and released 0.2 km north (Figure 9a). After 6 tidal cycles (Figure 9b), the outer patch, released in a region of low strain, has migrated a few kilometers eastward, but has spread very little, while the inner patch, released in a region of high strain associated with the flow separation, has been distributed over a large area. Along the east edge of the inner patch distribution, multiple bands of material indicate successive injections of material from the flow separation, which deform relatively little after reaching the interior. After 12 tidal cycles, the outer patch and the bands of inner patch material have advected to the southeast, slowing recirculating back into the region of strong deformation (Figure 9c). This recirculation pattern is also seen in the map of the Lagrangian residual velocity (Figure 8). After 24 tidal cycles, part of the outer patch has finally made its way back into the region of strong deformation and has been injected into the interior (Figure 9d). Outer patch material can clearly be seen being stretched by the flow separation and wrapped around the transient eddy. The inner patch material is continuing to evolve, with a tongue protruding to the west along the northern boundary and the bands along the southeast folding back into the interior.

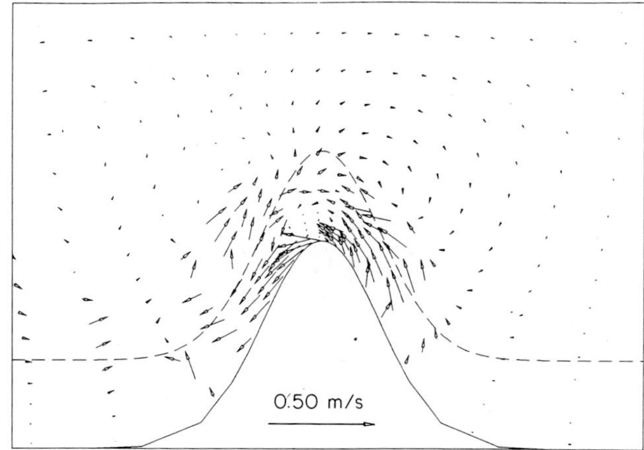

FIGURE 8. The Lagrangian residual flow calculated at 0 lunar hours. This quantity is obtained by tracking a particle released at each grid point at a particular phase of the tide over a complete cycle. The net displacement of the particle over the cycle is then divided by the tidal period, forming a vector quantity, and is assigned to the release point. This field, if scaled by the tidal period, expresses the transformation that maps particle positions at a specified phase of the tide onto new positions one tidal cycle later.

The patch spreading over the 24 tidal cycles can by quantified by calculating the change in variance of the particle locations over a tidal cycle. In an infinite fluid of uniform diffusivity, the variance of particle position evolves according to $\sigma^2 = 2Dt$ where D is the diffusivity, and t is the elapsed time since release. In a similar manner, we can quantify the spreading by deterministic motions by defining a dispersion coefficient

$$K(t) \equiv \frac{1}{2}[K_x(t) + K_y(t)] = \frac{1}{4}\left(\frac{\sigma_x^2(t+T) - \sigma_x^2(t)}{T} + \frac{\sigma_y^2(t+T) - \sigma_y^2(t)}{T}\right) \tag{9}$$

where σ_x and σ_y are defined by

$$\sigma_x^2(t) \equiv \frac{1}{N}\sum_{i=1}^{N}[x_i(t) - \bar{x}(t)]^2 \quad ,$$

$$\sigma_y^2(t) \equiv \frac{1}{N}\sum_{i=1}^{N}[y_i(t) - \bar{y}(t)]^2 \quad . \tag{10}$$

Here T is the tidal period, (x_i, y_i), i = 1,..., n are the particle locations at time t, and $(\bar{x},\bar{y})$ are the coordinates of the particles' center of mass. In the deterministic mixing process, K only describes the growth in the average variance of the patch: it does not completely describe the distribution, which requires higher-order moments to represent attributes such as streakiness.

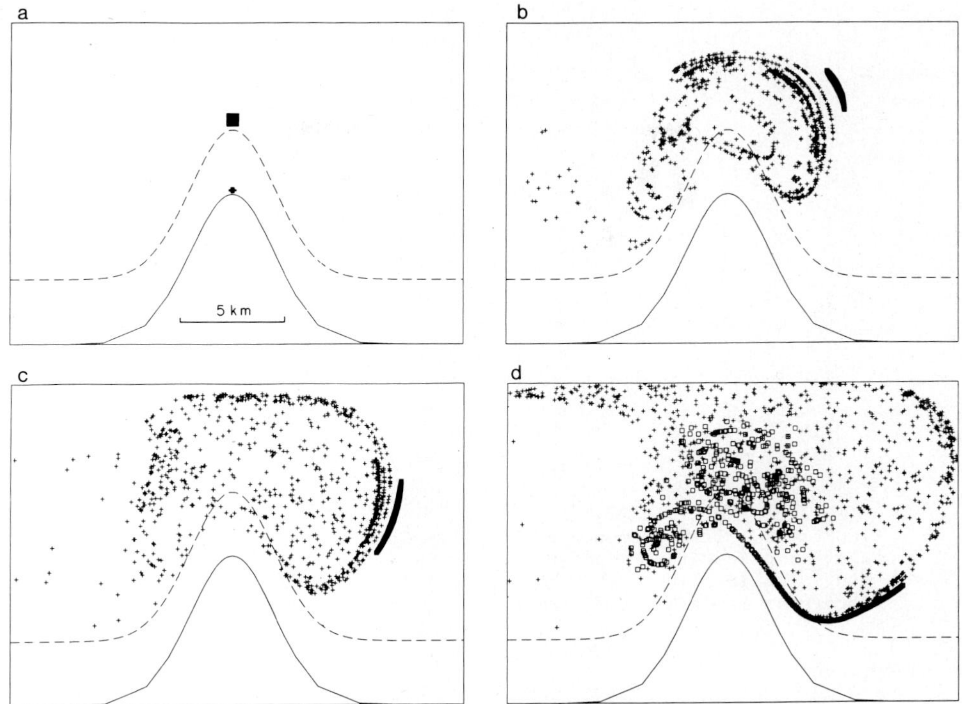

FIGURE 9. Dispersion of two patches of material released at 0 lunar hours. (a) Particle positions at 0 lunar hours (release configuration). The outer patch is initially 200 m wide and is released 3.5 km north of the tip (open boxes). The second patch is 100 m wide and released at 0.2 km north of the tip (pluses). (b) Particle positions 6 tidal cycles after release. (c) Particle positions 12 tidal cycles after release. (d) Particle positions 24 tidal cycles after release.

Figure 10 shows the time sequence of the dispersion coefficient K for the inner and outer patches at intervals of 1 tidal cycle after release. For the inner patch, K varies rapidly from less than 6 to more than 27 m^2/s over the first eight tidal cycles, then fluctuates from 7 to 13 m^2/s for the duration of the simulation. For the outer patch, K increases slowly from less than 1 to 3 m^2/s over 18 tidal cycles, suddenly leaps to 20 m^2/s at 20 cycles, then falls back to around 7 m^2/s at 24 cycles. These values of dispersion are not as large as the values found by Zimmerman (1976) and Awaji (1982), probably due in part to the increased strength of the tide in their studies. The values are high enough to represent an important mixing mechanism, however, since dispersion due to three-dimensional turbulence and vertical shear effects should be on the order 1-5 m^2/s for these values of velocity and water depth.

Awaji (1982) found that the inclusion of small levels of turbulence in the tracking of particles led to large changes in local mixing characteristics. To assess the impact of diffusion on our dispersion results, the patch dispersion simulation was repeated including the Markov-chain model with 1 m^2/s diffusion. Comparing the turbulent case after 6 tidal cycles (Figure 11) to the non-turbulent case (Figure 9b), the outer patch has spread laterally in a gradient-diffusive manner. The effect of turbulence on the inner patch, however, has not only smeared out the banded structure, but has resulted in the formation of a new lobe of the distribution to the west of the headland.

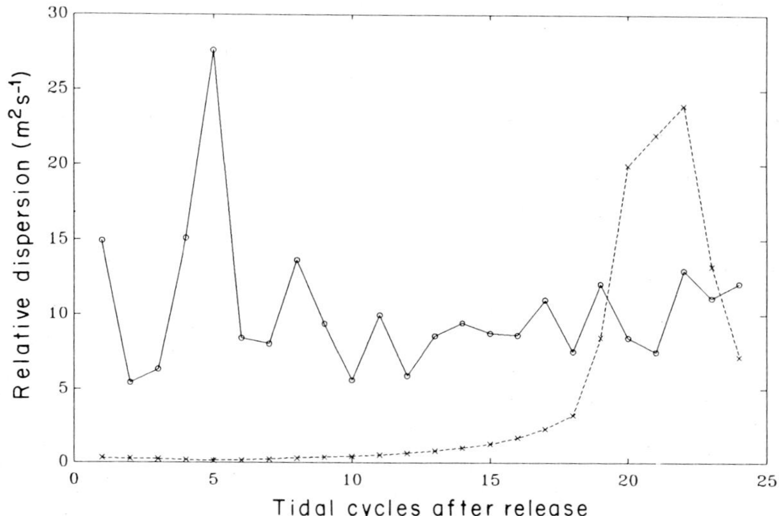

FIGURE 10. Time sequence of relative dispersion. The effective relative dispersion K (defined in text) is shown for the inner patch (solid line) and the outer patch (dashed line). The dispersion is calculated between successive tidal cycles only.

IV. Discussion

In the absence of stochastic processes, mixing occurs as the result of material being repeatedly stretched and folded (e.g. Ottino, 1988), like the action of an egg beater blending a viscous batter. Material is stretched when it passes through regions of high strain. If the stretching is exponential in time (as surrounding a point of pure strain), then reversal of the stretching process is unstable to infinitesimal perturbations, and deformed material cannot be restored to its original shape. Thus "unmixing" is not possible.

If the stretching is linear in time (as in parallel shear flow) then the stretching process is reversible and mixing may not occur. The mixing in a given area will depend on the character of the strain field, and also on the large-scale structure of the flow which determines the frequency with which material interacts with the regions of high strain.

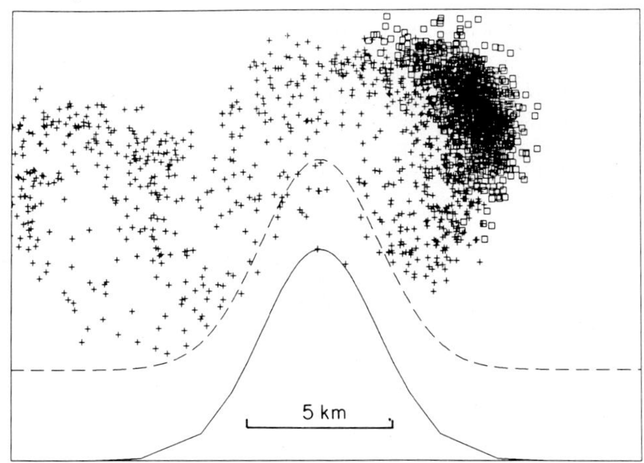

FIGURE 11. Same as figure 9b, but with 1 m^2/s diffusion included using a random walk Markov-chain model.

In our study of tidal dispersion around a headland, the snapshots of the flow field over the tidal cycle illustrated that material is rapidly stretched in the vicinity of the flow separation at the headland tip. As a tongue of material is drawn away from the tip, it begins to wrap around the core of the forming lee eddy, and it continues to wrap until the vorticity of the eddy decays to zero. The wrapped tongue of material is not subsequently unwrapped by the flow, but it does not soon return to the vicinity of the flow separation either. In fact, by examining the Lagrangian residual velocity (Figure 8) and the recirculation time of outer patch material in the dispersion experiment (Figure 9), the period between successive interactions with the region of high strain for this particular flow field appears to be 10-20 tidal cycles. Since it takes many stretching and folding operations for a fluid to become well mixed, the time scale for material around the headland to achieve maximum variance, in this case study, may be hundreds of tidal cycles. The implication is that over several days the spreading of a patch in the vicinity of the headland will not have reached its asymptotic value. The distribution, therefore, will depend crucially on the initial release location and phase of the tide, since this will determine how soon the patch will interact with the region of high strain. The distribution of particles in the patch over several days must also be expected to have a streaky, patchy nature due to the limited amount of stretching and folding that has taken place.

The inclusion of turbulent mixing has basically two effects on the dispersion of material by deterministic processes. The first effect is to allow a particle traveling along a deterministic trajectory to "jump" to another trajectory through the random motions of the turbulence. Since even a small changes in position can lead to large particle separations in a region of high strain, the distribution of particles in the turbulent case may look quite different from the non-turbulent case. This was first discussed by Awaji (1982). The second effect of turbulent mixing is to smear out the streakiness of the deterministic particle distribution, since diffusive processes rapidly decrease the amplitude of short wavelength structure. Of course, if the turbulent fluctuations are large enough, i.e. if the time scale required to diffuse material over the region influenced by nonlinear tidal effects is short compared to the time scale over which mixing is of interest, then the mixing due to deterministic processes will be of minor importance.

In this idealized numerical study, we have shown that a tidal flow field around an isolated headland can cause effective mixing of material. In the real ocean, however, there will be low-frequency wind and density induced currents, as well as the possibility of multiple bathymetric and coastline features. Clearly these effects will modify the results presented here. Low-frequency motions would cause a broader area to be influenced by the headland effects. Material released in the vicinity of the headland would be advected out of the area of nonlinear tidal effects after being stretched once or twice, if at all. In our case study, for example, an along-shore flow of 0.05 m/s would replace the water in the region of the headland in 3 days (6 semi-diurnal tidal cycles). A substance released near the headland would not be homogenized in that period of time, and it may not even be mixed appreciably if it doesn't encounter the flow separation.

Multiple headlands or bottom bumps, of course, could allow further spreading of material once it left the direct influence of a particular feature. For a sufficient number of features, one may enter the regime discussed by Zimmerman (1976) and Pasmanter (1988), where fields of tide-induced residual eddies interact with the oscillatory flow to act diffusively over large scales.

In conclusion, the deterministic motions of a tidal flow around a headland can be an effective dispersive mechanism. The basic mixing mechanism is the stretching of material in a high strain region associated with the flow separation, followed by wrapping of this material around the transient eddies. Because the dispersion depends on the infrequent interaction of material with a small region of high strain, the spreading of an individual patch is extremely sensitive to the initial release location and phase of the tide. This implies that the distribution of particles from a released patch may be expected to be streaky and non-Gaussian over many tidal cycles.

Acknowledgments

This work was supported by NSF grant OCE-87-11031 and the National Center for Atmospheric Research. Woods Hole Oceanographic Institution contribution number 7164.

V. References

Aref, H., 1984: Stirring by chaotic advection. J. Fluid Mech., **143**, 1-21.

Awaji, T. A., 1982: Water mixing in a tidal current and the effect of turbulence on tidal exchange through a straight. J. Phys. Oceanogr., **12**, 501-514.

Awaji, T. A., N. Imasato, and H. Kunishi, 1980: Tidal exchange through a strait: A numerical experiment using a simple model basin. J. Phys. Oceanogr., **10**, 1499-1508.

Flather, R. A. and N. S. Heaps, 1975: Tidal computations for Morecambe Bay, Geophys. J. R. Astr. Soc., 42, 489-517.

Geyer, W. R. and R. P. Signell, 1989: Measurements of tidal flow around a headland with a shipboard acoustic Doppler current profiler, in Proceedings of the International Conference on Tidal Hydrodynamics., edited by B. Parker.

Imasato, N., 1983: What is tide-induced residual current?. J. Phys. Oceanogr., **13**, 1307-1317.

Marchuk, G. I., 1974: Numerical methods in weather prediction. Academic Press, 277 pp.

Mesinger, F. and A. Arakawa, 1976: Numerical methods used in atmospheric models, GARP Publication series 17, WMO-ICSU Joint Organizing Committee, 64 pp.

Ottino, J., C. Leong, H. Rising, and P. Swanson, 1988: Morphological structures produced by mixing in chaotic flows. Nature, **333**, 419-425.

Pasmanter, R. A., 1988: Deterministic diffusion, effective shear and patchiness in shallow tidal flows, in Physical Processes in Estuaries, edited by J. Dronkers and W. van Leussen, pp. 42-52, Springer-Verlag, Berlin.

Roberts, K. V. and N. O. Weiss, 1966: Convective difference schemes. Maths. Comput., **20**, 272-299.

Shapiro, R., 1975: Linear filtering, Mathematics of Computation, **29**, 1094--1097.

Signell, R. P., 1989: Tidal dynamics and dispersion around a headland, PhD thesis, Massachusetts Institute of Technology, In preparation.

Zimmerman, J. T. F., 1976: Mixing and flushing of tidal embayments in the Western Dutch Wadden Sea, part II: analysis of mixing processes. Neth. J. Sea Res., **10(4)**, 397-439.

Zimmerman, J. T. F., 1979: On the Euler-Lagrange transformation and the Stokes drift in the presence of oscillatory and residual currents. Deep Sea Res., **26A**, 505-520.

Circulation in the Salt Wedge Estuary

Gerhard H. Jirka

DeFrees Hydraulics Laboratory
Cornell University
Ithaca, NY 14853

ABSTRACT

The dynamics of the salt wedge estuary are studied in detailed steady-state laboratory experiments with Reynolds numbers of the order of 10^4. Flow observations show an interfacial region that is in a laminar-turbulent state of marginal instability. Results are presented from the application of a two-layer model for salt wedge dynamics that has been developed on the basis of the experimentally observed flow structure and on the basis of asymptotic similarity theory for stratified shear flows. The model results are in good agreement with the limited available large scale field observations on overall salt wedge circulation and geometry.

I. Introduction

The salt wedge estuary is an important class of estuarine flow configurations. The salt wedge is formed under weakly energetic conditions, characterized primarily by a small tidal range and secondly by the absence of strong wind shear as in well-channelized narrow estuaries. In the saline wedge flow the ocean salt water intrudes upstream against the direction of the overlying freshwater flow. A sharp interface separates the fresh and salt water layers. In its ideal steady-state form, the so-called arrested saline wedge, a force equilibrium is maintained between internal buoyant pressure gradients, interfacial shear stress, and convective accelerations.

Reasonably reliable classification criteria have been developed that separate the salt wedge estuary from the partially mixed and the well mixed estuary types, respectively, on the basis of the tidal forcing function (Fischer et al., 1979; Thatcher and Harlemen, 1981; Jay, 1987). At present, however, there is no equivalent classification on the basis of wind stress induced mixing which is clearly much more difficult to quantify in a general manner as it is highly dependent on estuarine morphology.

Understanding the fluid mechanics of the salt wedge estuary is of great ecological and economic importance, as it effects decisions on freshwater supply withdrawals, flow diversions, and river navigation (e.g. channel deepening). Furthermore, extended droughts with reduced freshwater runoff may have long-term consequences on the salt wedge estuarine behavior. The salt wedge is also a fundamentally interesting geophysical flow phenomenon. The flow has significant variations in the vertical direction with strong gradients of velocity and density in the interface zone. As a consequence the fluid turbulence is strongly suppressed in this zone and limited to intermittent mixing events leading to very small mass and momentum transfer rates (per unit area) across the interface. The analysis and prediction of these local transfer rates is difficult - due to the laminar-turbulent transition phases - but crucial since these small rates apply over the large horizontal extent of such intrusions and therefore may lead to significant total exchanges of mass and momentum. This feature is common to a larger class of geophysical flows including exchange flows, surface plumes, inclined density

Coastal and Estuarine Studies, Vol. 38
R.T. Cheng (Ed.)
Residual Currents and Long-term Transport

currents, and wind-driven mixed layers.

In an ideal well-channelized estuary of constant cross-section the overall flow dynamics of the salt wedge are governed by two global parameters: the channel densimetric Froude number $F_o = q_f/(g_o'H^3)^{1/2}$, and the channel Reynolds number $Re = q_f/\nu$, in which q_f = freshwater flow per unit width, g_o' = fresh/saltwater buoyant acceleration, H = channel depth, and ν = kinematic viscosity. The necessary condition for the existence of a salt wedge intrusion, $F_o < 1$, is provided by shallow water wave theory.

In field conditions, the Froude number can vary over a wide range down to values of 0.1 and less. The Reynolds numbers are typically very large, $Re = 10^5$ to 10^8, which - on the basis of mechanical turbulence considerations - might lead one to expect certain invariant behavior. However, as will be shown in the following, in this strongly stratified flow situation with its delicate balance of turbulent energy production at the interface and simultaneous buoyant damping the Reynolds number remains an important parameter in its own right and no invariance occurs. Under field conditions, relative intrusion lengths, L_o/H, where L_o = wedge length, can be very large (of order of 10^4). Unfortunately, very limited field data exist with few observations on overall salt wedge properties, and even fewer data are available on the internal flow and density distributions. Some measurements on the Mississippi River (Arita and Jirka, 1987b) do indicate, however, that the induced salt circulation can be very high, of the same order or even larger than the freshwater flow. Furthermore, appreciable mixing and net upward entrainment seem to take place across the density interface as evidenced by longitudinal changes in density in both upper and lower layers.

In the absence of comprehensive field data, most of the available information on salt wedge mechanics stems from laboratory experiments that, for equipment reasons, are necessarily limited to a different parameter range: the smallest achievable Froude numbers are of the order of 0.2 and the Reynolds number is in a much lower range, $Re = 500$ to 10^4 [for a review of past experiments see Sargent and Jirka (1987)]. Given a critical open channel flow Reynolds number, $Re_c \approx 500$, such laboratory conditions provide what one might consider a reasonably turbulent flow condition in the freshwater overflow. However, the dynamic situation at the density interface can only be qualitatively similar to the high Reynolds number field conditions; it must be expected to be quite different in quantitative (suitably normalized) terms. Relative intrusion lengths, L_o/H, in the laboratory are limited to a maximum of several hundreds.

The purpose of this paper is to demonstrate that the overall properties (i.e. force balance, intrusion length, and induced circulation) of the nominally highly turbulent (high Re) salt wedge are intimately controlled by the laminarizing effect of the density interface. The strategy is two-pronged: First, detailed experimental insight from a laboratory salt wedge (Re from 4,000 to 10,000) is presented. The flow observations show an interfacial region that is in a laminar-turbulent transition state of marginal instability. Second, a two-layer model for the salt wedge dynamics is developed on the basis of the experimentally observed flow structure and on the basis of asymptotic similarity theory for stratified shear flows. The model prediction when applied to the few available field data gives satisfactory predictions and gives support to the hypothesis that field conditions are governed by a similar marginal instability state albeit with quantitatively different rates.

II. Laboratory Experiments

The experiments were conducted in the stratified flow flume in the DeFrees Hydraulics Laboratory with a channel 730 cm long, 46 cm wide and 25 cm deep. The flume has an upstream stilling basin and a large downstream reservoir for the establishment of a control section and for separation of fresh and salt water flow by means of a selective withdrawal device. Experiments were conducted in two modes, as either "open boundary" experiments (similar to actual estuarine conditions) or using a "barrier device" with a specified injection rate to supply the salt circulation and hence establish the wedge. In either case, a steady-state wedge was established in which the buoyant pressure force of the intrusion is balanced in first order by the interfacial shear along the wedge and by convective forces. It should be noted in this context that the salt wedge is only one of two possible steady-state force balances for such intrusions. The other one leads to a flow configuration in form of a density current ("front") in which the buoyant pressure is balanced by a form drag around the enlarged head of the intrusion front. The detailed flow conditions governing these diverse intrusion possibilities have recently been investigated by Jirka and Arita (1987).

Wedge measurements included flow visualization, detailed velocity profiles produced by a falling "dye-rake" (with thin tubes which injected several dye traces across the flume with time-lapse photographic observation), and by a vertically traversing conductivity micro-probe. A total of 10 wedge experiments were conducted with a Froude number range from about 0.4 to 0.5 and a Reynolds number range from about 4,000 to 10,000. Details on these experiments are given in Sargent and Jirka (1983, 1987).

The entire salt wedge can be divided into three regions (see Fig. 1): the tip region, the quasi-equilibrium region, and the exit region. The quasi-equilibrium region occupies the major portion of the wedge. The flow is in a near-equilibrium condition between convective inertial forces, buoyant pressure forces and shear forces with only gradual changes in the longitudinal direction. (This situation is quite analogous to a gradually varied flow region in free surface flow). As a consequence, velocity and density distributions appear similar, independent of position, although strict self-similarity, of course, cannot exist due to the multitude of scales involved. For example, as shown in Fig. 2 the density interface and the velocity interface (defined by the point where the velocity is 50% of the maximum freshwater velocity) coincide in this region and the height of the zero velocity line is a constant proportion (about 0.6) of the interface height, h_2.

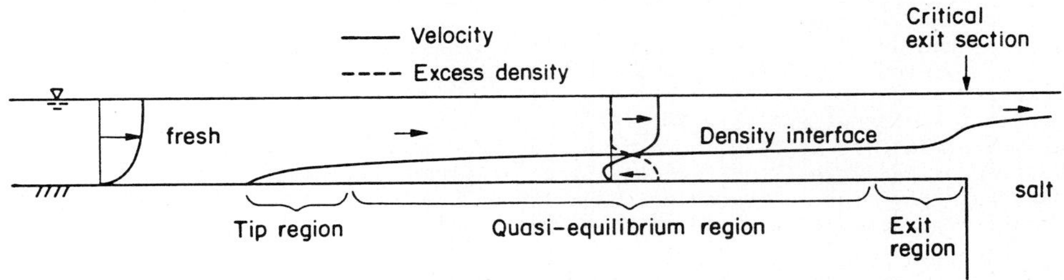

FIGURE 1. Schematic structure of salt wedge with three major hydrodynamic regions.

In the tip region the ambient flow bottom boundary layer undergoes a transformation into a mixing layer. Using estimates for a laminar mixing layer, $\delta \approx 2\sqrt{\nu x/U}$, in which U = velocity difference across the mixing layer, the distance x required to achieve a mixing layer depth of about 0.1H, is about 40 cm. This is consistent with observations of the interface stability from flow visualization as well as indicated by the vertical profiles: In the tip region, the density and velocity interfaces approach each other, with the velocity interface being clearly higher. In general, it can be expected that the tip region will be at most several water depths long, even for the high Reynolds number field environment in which turbulent growth rate estimates, $\delta \approx 0.1\,x$, would be appropriate for the mixing layer.

The exit region is another non-equilibrium reach governed by vertical accelerations and non-hydrostatic pressure distributions quite similar to the flow in the vicinity of any control section in free-surface flow. Flow profiles show the velocity interface falling well below the density interface. The extent of this region is again a few (2 to 3) multiples of the channel depth using free surface flow over an overfall as a simple analogy.

The discussion of local wedge properties in the main quasi-equilibrium region, i.e. mean velocity and density profiles, entrainment velocities etc., is complicated by the fact that individual measurement are always influenced by turbulent fluctuations, wave activity and experimental errors. To eliminate this variability, data from different wedge experiments have been assembled into groups with similar relative interface height, h_2/H, and averaged. This procedure seems justified given the gradual evolution along each wedge and the relative -though not exact- affinity among wedges.

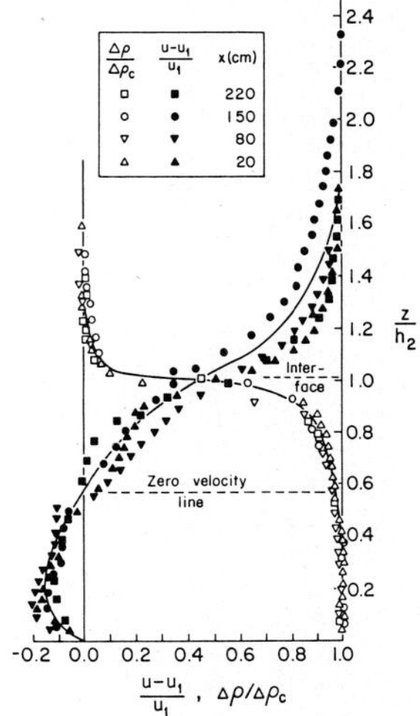

FIGURE 2. Normalized velocity and density distributions in quasi-equilibrium region of salt wedge. Data are from an experiment with total wedge length of 400cm in which the quasi-equilibrium region extends from x = 20cm to 350cm, approximately.

Detailed normalized data for five wedge sections, labeled W1 to W5, with h_2/H ranging from 0.15 to 0.41, have been presented by Sargent and Jirka (1983). Within the present parameter restrictions, these sections should provide a precise data base for any future modeling effort on saline wedges. As an example, normalized data for the wedge section W2, $h_2/H = 0.245$, are given in Figs. 3 and 4. Fig. 3

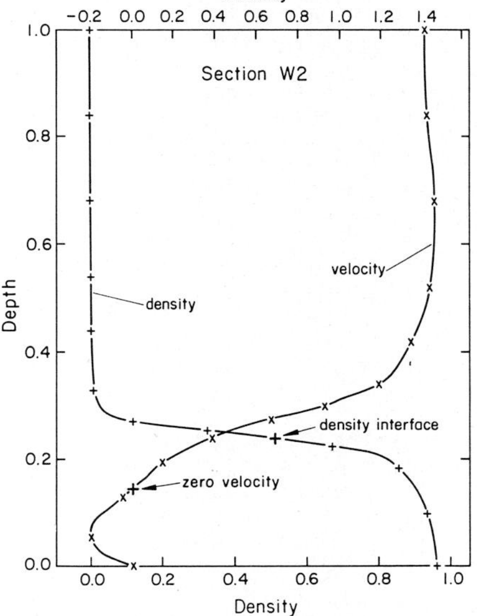

FIGURE 3. Normalized velocity and density profiles in main region of salt wedge. Data for $h_2/H = 0.245$.

presents the mean velocity profile, normalized by the mean ambient approach velocity U, and the mean normalized density profile. The density gradient is considerably steeper (factor of about 2) than the velocity gradient. The gradient Richardson number profile, $Ri = -(g/\rho)(d\rho/dz)/(du/dz)^2$, obtained from a finite differencing procedure, is plotted in Fig. 4. The data show considerable scatter. Nevertheless, a zone at the interface with a value of $Ri_{max} \approx 2$ and with a strong drop-off on either side is clearly distinguishable. These conditions will be further examined in the light of flow visualization results.

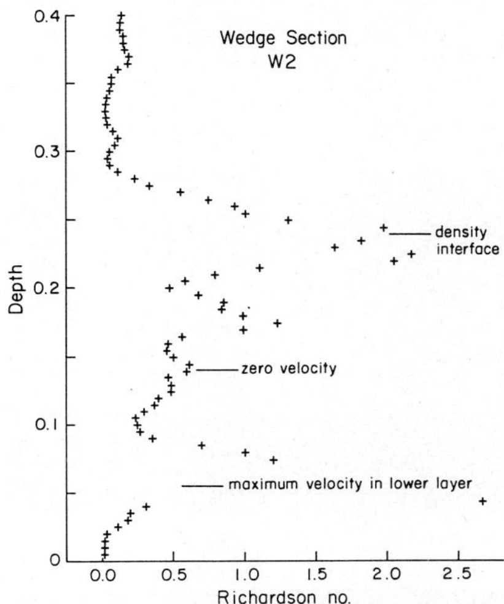

FIGURE 4. Gradient Richardson numbers profile in main region of salt wedge. Data for $h_2/H = 0.245$.

While the wedge tip region had a predominantly laminar interface in the experiments, interfacial waves were first observed at the end of the tip region and then persisted over the main region of the wedge. The waves were seen to move in both the upstream and downstream directions. The predominant waves were highly three-dimensional with three or four wave crests/troughs typical across any wedge section (46 cm wide flume) and somewhat irregular in shape with a cusp-like structure (Fig. 5a). Typical wave lengths were observed to be 5 to 10 cm with no apparent correlation over the wedge length. The internal waves were seen to break intermittently with apparently equal amounts being mixed in both directions across the density interface. The appearance of these waves which appear to be the dominant agent for interfacial mixing is similar to Holmboe waves as described by Browand and Winant (1973) in a study of the buoyant mixing layer. Similar wave observations for the salt wedge were made by Grubert (1980) and Hino (1980). Occasionally, and much less frequently, two-dimensional vortical wave structures of the Kelvin-Helmholtz type were observed extending across the entire flume width (Fig. 5b). These were centered at the density interface and slowly propagated downstream.

The observations of interfacial processes as well as the measured gradient Richardson number seem consistent with Hazel's (1972) analysis of buoyant shear flow instabilities for cases in which the normalized density gradient is considerably stronger than the normalized velocity gradient. Under these conditions (see also Browand and Winant, 1973) the primary vortical mode of instability becomes progressively damped while the secondary instability mode in form of Holmboe waves takes over. The marginal stability boundary controlling the Holmboe wave formation is given in the present notation by

$$Ri_{max} = \alpha R , \tag{1}$$

in which R = ratio of normalized density and velocity gradients, and $\alpha = 2\pi h/\lambda$ = non-dimensional wave number, with h = shear layer half-thickness and λ = wave length. Under the observed conditions (see Figs. 3 and 4a) $h \approx 1$ to 2cm, $\lambda \approx 10$cm, and $R \approx 2$, thus $\alpha \approx 1$, one obtains $Ri_{max} \approx 2$ in agreement with the measurements. Note that, for the symmetrical profiles considered by Hazel (1972), the gradient Richardson number is related to the bulk shear layer Richardson number $Ri^* = g(\Delta\rho/\rho)\delta_u/\Delta U^2$ in which δ_u = shear layer (velocity) thickness and ΔU = velocity difference, $Ri^* = Ri_{max}/R$. Thus, the bulk shear layer Richardson number, $Ri^* \approx 1$, is in rough agreement with Browand and Winant's measurements, but somewhat larger than usually quoted values of $Ri^* \approx 0.3$ for stratified shear layers [for example, see the review paper of Hopfinger (1987)]. This larger value may be caused by the fact that the outside (freshwater) flow above the shear layer was strongly turbulent in the present experiments unlike most other shear layer studies that have been conducted at a smaller scale.

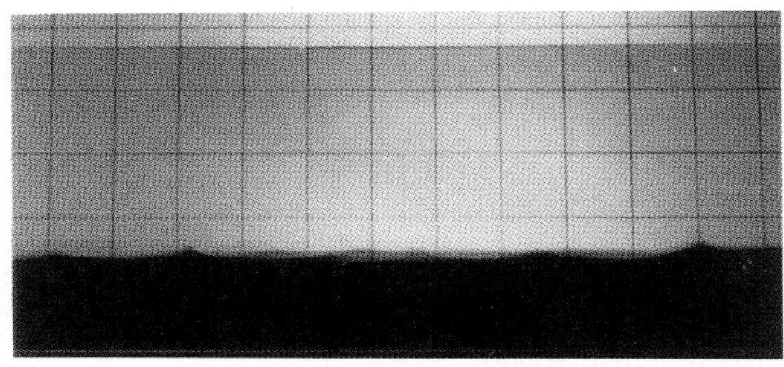

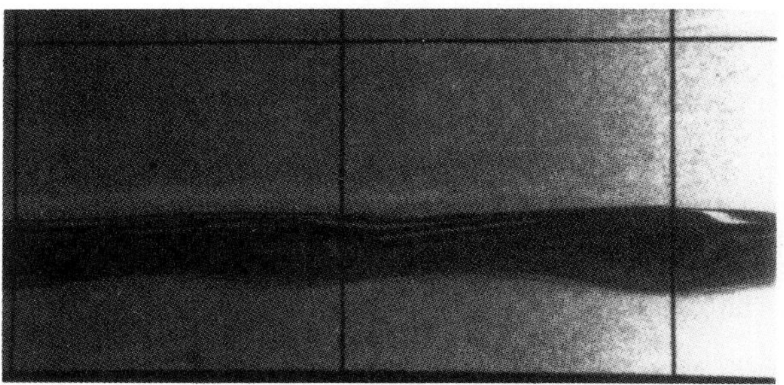

FIGURE 5. Interfacial wave observations in quasi-equilibrium wedge region: a) Three-dimensional waves of the Holmbee type as the dominant mechanism. b) Occasional two-dimensional waves of the K-H type shown here after breaking. Grid size 5 cm x 5 cm.

Another indicator of the laminar/turbulent transition state of the interface is given by Gibson's (1982, 1987) measure for the fossilization of stratified flow turbulence due to viscous action. Fossilization occurs when

$$\epsilon \approx (15 \text{ to } 30)\ \nu N^2 \tag{2}$$

in which ϵ = turbulent energy dissipation rate and N = buoyancy frequency. For the salt wedge condition the dissipation rate at the interface (assuming for a moment that there would not be simultaneous buoyant destruction) is given as $\epsilon = u_*/\ell$ in which $u*$ = shear velocity evaluated from the overall force balance in the wedge, and $\ell = \kappa h$ = turbulent length scale with $\kappa = 0.4$. For the conditions of the Sargent and Jirka experiment, $u_* \approx 0.2$ cm/s and $\ell \approx 0.4$ cm, the left hand side of Eq.(2) is evaluated as $\epsilon = 0.02\ cm^2/s^3$. While, with $N \approx 1.0/s$ and $\nu = 0.01\ cm^2/s$, the right hand side yields 0.015 to $0.03\ cm^2/s^3$. Thus, as mechanical turbulence is generated due to shearing at the interface it is simultaneously modulated by the action of buoyancy and destroyed by viscous damping.

The following internal flow dynamics in the quasi-equilibrium region of the present experiments are suggested by the data (see Fig. 6): The entire flow is governed by an external pressure gradient that drives the turbulent freshwater flow which, in turn, exerts a high shear effect over the underlying salt layer. This destabilizing shear is counteracted by a very strong density gradient. In total, the interface is in a marginal stability condition with laminar flow but intermittent wave breaking and mixing. As a result, viscous and turbulent shear transfer is taking place at the interface as well as some net entrainment into the upper layer (as is shown further below). These transfer processes provide energy for a weakly turbulent middle layer, bounded by the density interface and the zero velocity line. Decreased stability (due to smaller density gradients) in this return flow zone allows weak, yet persistent, turbulent activity. The turbulent middle layer, in turn, causes net entrainment of the non-turbulent salt water flow and up-lifting across the zero velocity line. In summary, a spatial cascading process is sketched in Fig. 6 in which an active upper layer transmits its energy downward and, ultimately, drives a passive lower layer whose motion is further counteracted by bottom friction (purely laminar friction in the present experimental range).

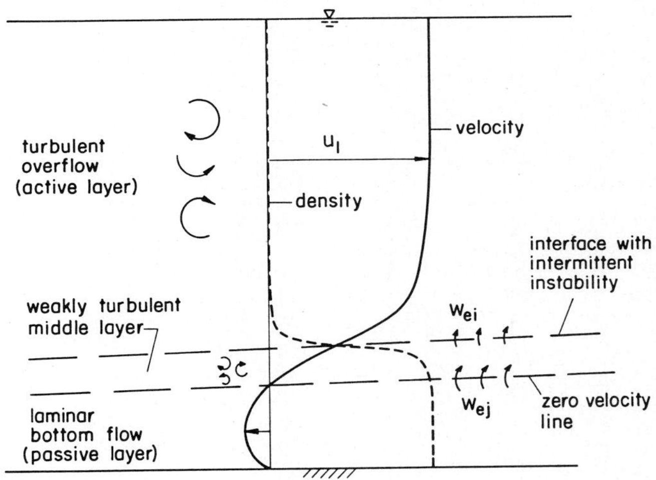

FIGURE 6. Schematics of the internal flow structure of the salt wedge for the experimental Reynolds number range ($Re \approx 10^4$).

Estimates of the net entrainment velocities w_{ei} at the density interface and w_{ej} at the zero velocity line have been obtained by measuring the change with distance of both volume flux as well as salt flux quantities. The normalized values of local entrainment velocities at the density interface w_{ei}/u_1 and at the zero velocity line w_{ej}/u_1 in which u_1 = local fresh water velocity are plotted in Fig. 7 as a function of local bulk Richardson number $\overline{Ri} = (g'h_2)/u_1^2$ in which g'= buoyant acceleration due to local fresh/salt water difference, and h_2 = interface height. Both w_{ei} and w_{ej} show a decrease with increasing $\overline{Ri}$ as would be expected from stability considerations. Neglecting this decrease, however, over the limited $\overline{Ri}$ range, the average value at the interface, $w_{ei}/u_1 = 0.26 \times 10^{-3}$, is only about 20% of that at the zero velocity line, $w_{ej}/u_1 = 1.3 \times 10^{-3}$, reflecting the strong stabilization at the interface which inhibits turbulent entrainment.

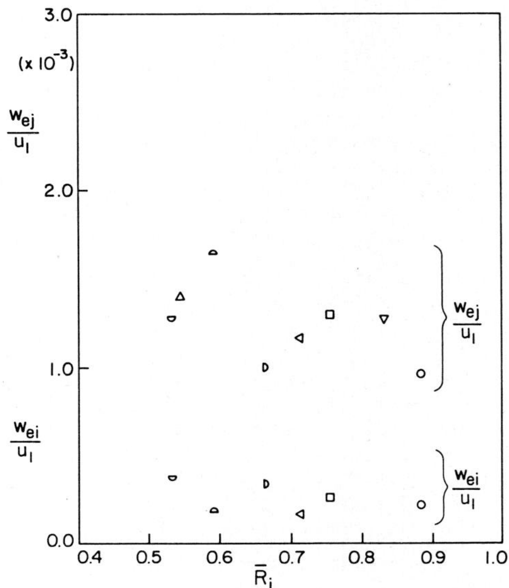

FIGURE 7. Normalized entrainment velocities in a laboratory salt wedge at density interface, wei, and at zero velocity here, wej, as a function of local bulk Richardson number, $\overline{Ri}$.

The internal circulation in the wedge is generated by the entrainment flux at the zero velocity line, w_{ej}. The salt water volume flux, q_s, was taken as that portion of flow in the wedge advancing upstream, and calculated by summing the velocity profile from the flume bottom to the zero velocity line. The local discharge, q_s, is normalized by the fresh water discharge, q_f(Fig. 8). The internal flux varies in a nearly linear fashion over the major portion of the wedge. For the present F_o range the maximum circulation is $q_{so}/q_f \simeq 4.0\%$ and occurs at the downstream critical point, $x^*/L_o = 1.0$ where x^* = distance from wedge tip.

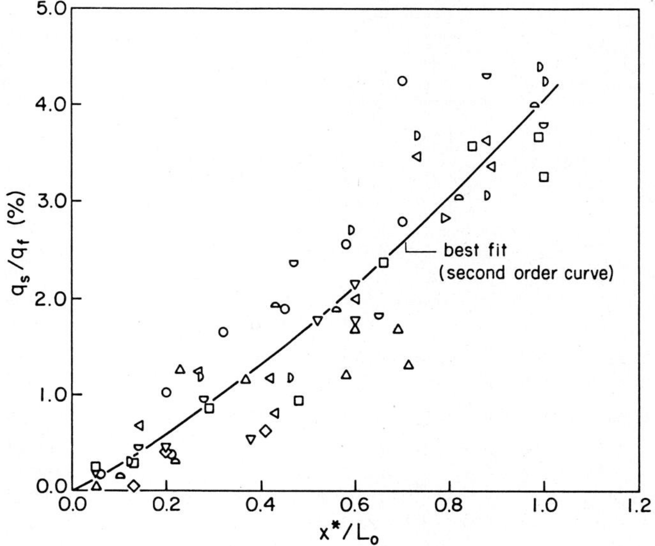

FIGURE 8. Normalized salt circulation flux qs/qf in salt wedge as a function of distance.

A final point on the dynamics of the laboratory salt wedge concerns the shear stresses τ_i at the interface and τ_j at the zero velocity line. As a typical example, carefully evaluated local force balances yield an actual interfacial shear stress $\tau_i/\rho = \lambda_i u_1^2 = 0.038$ cm²/s² (with $u_1 = 4.0$ cm/s) in one of Sargent and Jirka's experiments in which λ_i = local shear stress coefficient. This shear stress is transferred through viscous stresses, turbulent Reynolds stresses (including internal wave radiation) and through net entrainment of lower layer fluid momentum. The viscous stresses at the interface are obtained directly from the measured velocity profiles, $\nu \frac{du}{dz}\Big|_{max} = 0.018$ cm /s² or 47% of the total. The entrainment drag is given by $w_{ei} u_1$ in which the interface velocity is taken as $u_i = u_1/2$. From Fig. 6, $w_{ei}/u_1 = 0.0002$, and the contribution $w_{ei} u_1 = 0.0016$ cm²/s², or only 4% of the total. The remainder is therefore made up by the Reynolds stress, $\overline{-u'w'} = 0.018$ cm²/s².

The shear stress τ_j at the zero velocity line (ZVL) is primarily carried by viscous stresses, which are of order $\nu\frac{du}{dz}\Big|_{ZVL} = 0.005$ cm²/s² from the measurements. Turbulent stresses are negligible as the upstream salt water flow always appeared laminar, both from visual observations and from the conductivity records. Also the entrainment drag is negligible as the entrained salt flow has very small velocities relative to the freshwater flow. (This holds even though the volumetric entrainment rate w_{ej} is much larger than w_{ei} at the density interface).

The internal dynamics for the present experiments (typical channel Reynolds number of order 10^4) can be summarized as follows: The <u>density interface</u> is governed by large interfacial stresses but small net entrainment. Furthermore, viscous stresses and turbulent Reynolds stresses are of about equal magnitude. [A similarly large role of viscous stresses in a laboratory salt wedge was found by Dermissis and Partheniades (1985) using hot-film anemometry]. The <u>zero velocity line</u> has much smaller stresses (primarily viscous) but very high entrainment rates.

III. Two-Layer Model of Salt Wedge

As mentioned earlier, the results of the laboratory salt wedge should be a qualitative indication of the dynamic structure of real salt wedge estuaries but cannot be expected to apply quantitatively. For quantitative predictions of field conditions over a wide Reynolds number range, a two-layer model of the salt wedge mechanics has been formulated.

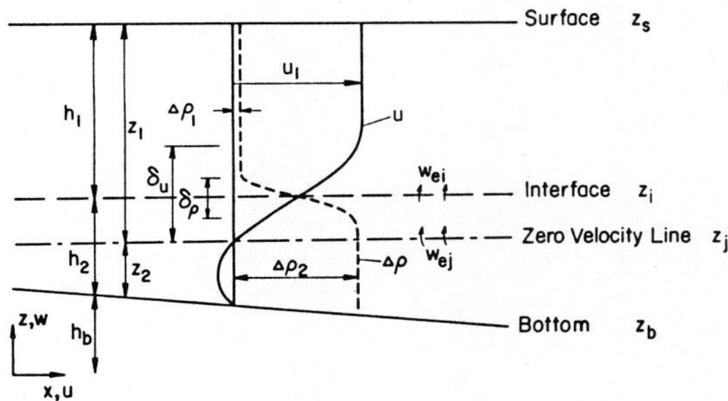

FIGURE 9. Definition diagram for flow distributions in two-layer model of salt wedge.

A definition diagram for the flow in the quasi-equilibrium region (the deviations within the tip and exit regions are neglected as unimportant in the overall evolution of a long wedge!) is given in Fig. 9. Unlike earlier two-layer models of the salt wedge (Schijf and Schonfeld, 1953) the present formulation uses the zero velocity line (ZVL) as the demarcation between the two flow layers. This is suggested by the energetics of the flow (Fig. 6) with its active upper layer which controls the lower passive zone primarily through entrainment at the ZVL. Use of this two-layer schematization has several advantages: (1) It allows a direct prediction of the saltwater circulation. (2) In deriving a model for the entrainment rate, it is sufficient to consider in first order the energetics of the active zone alone. The passive-layer dynamics can be neglected in this stage, but can be incorporated at a layer stage for a more complete model. (3) The entrainment model can be used to predict the shear stress at the density interface (a dynamically related quantity).

With the hydrostatic and boundary-layer approximations for the predominantly horizontal flow, the governing equations are vertically integrated to yield conservation equations for volume flux, horizontal momentum flux, mechanical energy flux, and buoyancy flux. Using the energy equation as a dependent equation (in addition to the continuity and momentum equations) serves to elicit the relationship between the entrainment velocity and the internal energy balance. Following an earlier development by Jirka (1982) for entrainment in a horizontal buoyant surface jet, Arita and Jirka (1987a) propose an entrainment formulation for the salt wedge. For the case of high Reynolds number flow, the model consists of an asymptotic matching between the entrainment conditions in a shear flow perturbed by slight buoyancy (jet-like behavior) and a strongly stratified shear flow near its marginal instability state. An additional transition is included for the effect of low Reynolds number flows in which the shear layer distributions (for both velocity and density) and the associated viscous entrainment rate are controlled by laminar flow conditions (with Schmidt number dependence). In total, the closure relation developed by Arita and Jirka predicts not only the entrainment rate $\alpha_j = w_{ei}/u_1$ as a function of the Re_1 - Ri_1 domain in which $Re_1 = u_1 z_1/\nu$ = local Reynolds number and $Ri_1 = g' z_1/u_1^2$ = local Richardson number, but also, as a related measure, the relative velocity shear layer thickness, δ_u/z_1.

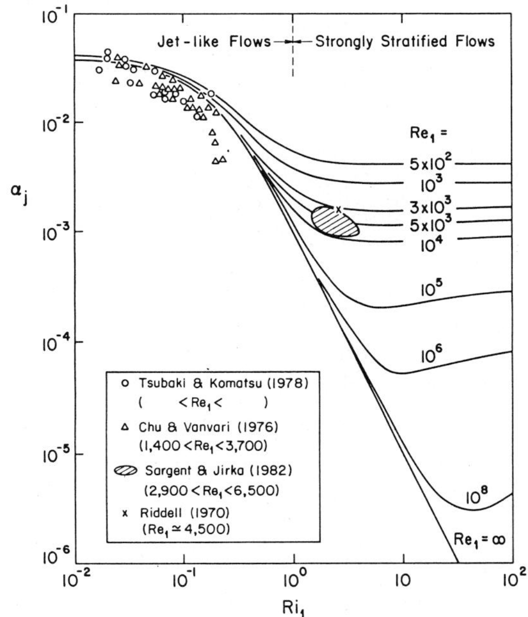

FIGURE 10. Entrainment coefficient α_j at edge of active zone in two-layer stratified flows. Comparison of Eq.(3) with data for salt wedge and buoyant surface jets.

The entrainment relationship is

$$\alpha_j = 0.038 \left(1 - \frac{Ri_1}{\sqrt{Ri_1^2 + Ri_*^2}}\right) + \frac{2}{Re_1}\left(\frac{z_1}{\delta_u}\right) \tag{3}$$

in which Ri_* = initial shear layer Richardson number (= 1/4 from theory) and the shear layer thickness is

$$\frac{\delta_u}{z_1} = \begin{cases} 1 & ; Re_1 < 500 \\ \left(\frac{500}{Re_1}\right)^n + \frac{Ri_*}{\sqrt{Ri_1^2 + Ri_1^2}}\left[1 - \left(\frac{500}{Re_1}\right)^n \right] & ; Re_1 > 500 \end{cases} \tag{4}$$

in which the exponent n = 1/2 appears reasonable for this empirical transition. The first term on the right hand side of Eq.(3) describes entrainment under high Reynolds number conditions, with the value 0.038 appropriate for a purely non-buoyant shear flow. The second term represents purely laminar entrainment dependent on the local shear layer thickness. Fig. 10 displays Eq.(3) and gives a comparison, with the few available data for the salt wedge that include direct entrainment rate measurements as well as for related data on weakly buoyant surface jets. The agreement in this intermediate range is excellent - though the data are limited - suggesting that the entrainment model, indeed, includes a realistic transition between its theoretically based asymptotic limits.

With Eqs.(3) and (4) as the closure equations Arita and Jirka (1987b) have numerically integrated (using a simple Runge-Kutta routine) the two-layer equations to provide prediction of several global wedge properties which then can be compared to other salt wedge data theory providing another - albeit secondary - validation of the proposed salt wedge dynamics.

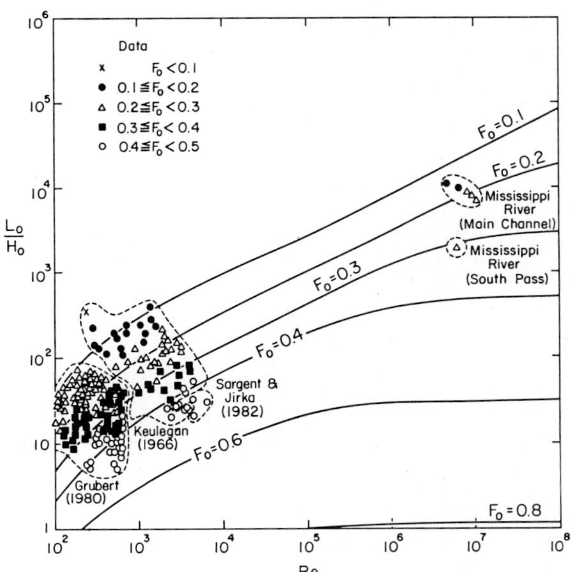

FIGURE 11. Comparison of predicted wedge lengths L_o/H and observations in field and laboratory a function of Reynolds number Re and Froude number F_o.

Fig. 11 shows the predicted wedge length L_o/H for constant depth H and constant width estuaries as a function of the governing bulk parameters, the channel Reynolds number Re and Froude number F_o, respectively. Several laboratory and field data are available for comparison and show good agreement (for a detailed data discussion, see Arita and Jirka, 1987b). The magnitude of the Froude number F_o is, of course, a measure of the buoyant gravitational effect, with smaller F_o (stronger stratification or weaker freshwater flow) leading to longer intrusions. The most striking aspect of Fig. 11 is the strong dependence of intrusion length on Reynolds number. For long intrusions (small F_o), the two-layer model does not indicate any Reynolds number invariance in the plotted range, $R < 10^8$. Only for short intrusions is there an invariant region (e.g., for $F_o = 0.6$ and $R \geq 10^6$) as might be expected from other types of high Reynolds number flows. As was discussed earlier, this behavior is due to the laminarizing effect of strongly stratified flow as represented in the present entrainment parameterization.

Furthermore, the limited role of physical models for salt intrusion is clearly evident from Fig. 11. Taking a moderate $F_o = 0.3$, a physical model at $R = 10^4$ (a large value for laboratory conditions) would underpredict an intrusion in the field at $R = 10^6$ by a factor of 10. Thus, the use of laboratory models for salt intrusion (and other strongly stratified flow phenomena) is limited to qualitative insights. They are unreliable for quantitative predictions.

The longitudinal distribution of the normalized salt water circulation, q_s/q_f, is plotted in Fig. 12 ($S_b = 0$) as a function F_o and, in each case, for the extreme Reynolds numbers $R = 10^2$ and 10^8. The only detailed data available by Sargent and Jirka show excellent agreement, which is to be expected since the same data already compared well for the measured α_j as discussed earlier. Of major interest are the high salt fluxes, q_s/q_f of order 1^j or larger, for low $F_o \leq 0.1$, which encompasses field situations such as the Mississippi River. The maximum induced salt discharge, q_{so}/q_f, at the channel exit is shown in Fig. 13 as a function of Fo. Data for the laboratory and two field measurements are in good quantitative agreement. These data points have also been included in Fig. 12.

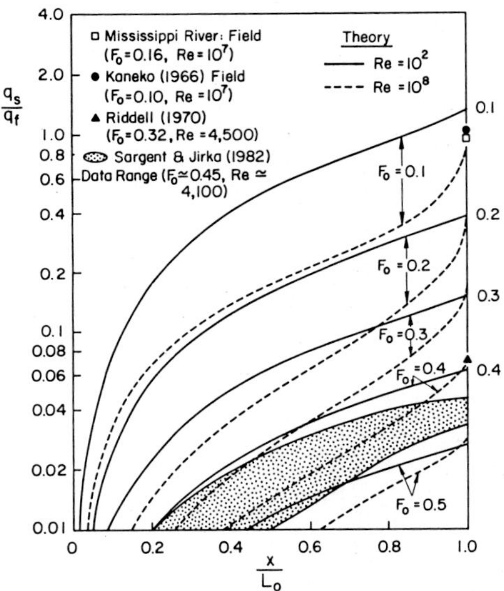

FIGURE 12. Distribution of relative salt circulation q_s/q_f at channel exit as function of Froude number F_o.

Further predictions on wedge shape and on the effect of a bottom slope are given by Arita and Jirka (1987b). A final point concerns the effective interfacial friction coefficient at the density interface, which has also been calculated by the two-layer model. By comparison, Schijf and Schonfeld (1953) - in the earliest, and lowest order, model of salt wedge intrusion - have specified an "apparent" shear stress $\overline{\tau}_i/\rho = \overline{\lambda}_i(q_f/h_1)^2$ at the interface while neglecting altogether the lower layer dynamics, i.e. any entrainment, circulation and bottom friction. The analytical integration of the Schijf and Schonfeld equations gives the well established result for wedge length

$$\frac{L_o}{H} = \frac{1}{\overline{\lambda}_i}\left(\frac{1}{5F_o^2} - 2 + 3F_o^{2/3} - \frac{6}{5}F_o^{4/3}\right) \tag{5}$$

A value for the "apparent" interfacial friction coefficient that is constant along the wedge, $\overline{\lambda}_i$, can be predicted by setting the wedge length from Eq.(5) equal to the detailed model predictions given in Fig. 11. This is shown in Fig. 14 as a function of F_o and Re. Also given are laboratory and field data, which have been reduced on the basis of Eq.(5). The agreement is excellent, and Fig. 14 can be considered as the appropriate friction diagram whenever a simple calculation of a Schijf and Schonfeld-type model is desired for estuarine wedge calculations.

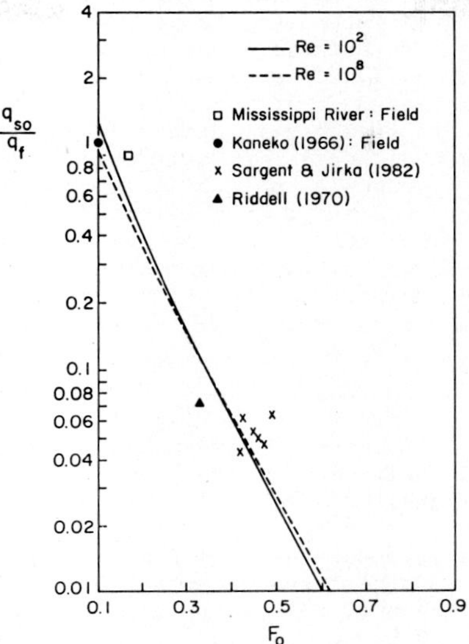

FIGURE 13. Maximum relative salt circulation q_{so}/q_f at channel exit as function of Froude number F_o.

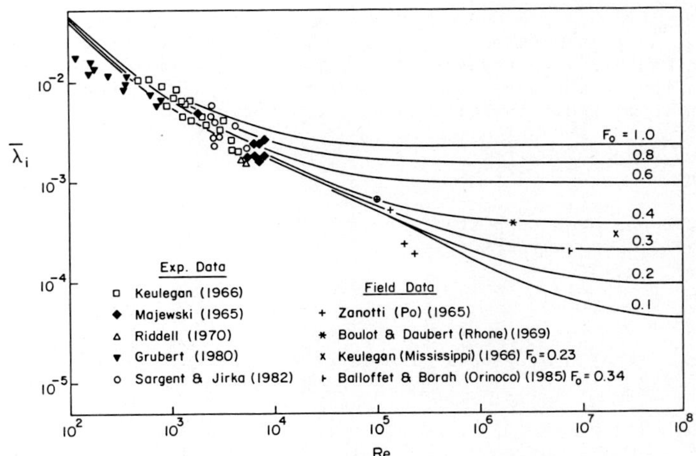

FIGURE 14. Friction diagram for wedge-averaged "apparent" interfacial shear stress coefficient $\overline{\lambda}_i$ used in Schijf and Schonfeld (1953) formulation.

IV. Conclusions

Given the conventional measure of the generation of mechanical turbulence, i.e. a bulk Reynolds number, the salt wedge intrusions observed in some of the world's largest rivers (with Re up to 10^8!) must surely represent some of the largest scale stratified shear flows occurring on Earth. Yet in their internal properties, such as turbulent entrainment, interfacial shear and induced circulations, these flows do not exhibit any significant Reynolds number invariances. Undoubtedly, this behavior is related to the laminarizing effects due to the buoyancy at the fresh-salt water interface.

Detailed experimental data at moderately high Reynolds numbers (up to 10^4) in the laboratory indicate an interface in a marginal state of laminar/turbulent transition with intermittent interfacial wave activity. No equivalent detailed data do currently exist for the field. However, predictions for higher Reynolds field conditions with a two-layer model that includes an entrainment parameterization valid for the limiting state of a nearly damped buoyant shear flow are in excellent agreement with available global data on salt wedge properties. This provides indirect support for the hypothesis that the salt wedge estuarine dynamics are over their entire parameter range dominated by a laminar/turbulent transition state at its density interface. Incidently, yet another circumstantial evidence for this state of affairs is the surprisingly low value for the interfacial friction coefficient in strongly buoyant (low F_o) salt wedges: for Re = 10^8 and F_o = 0.1, the apparent friction coefficient $\overline{\lambda}_i$ (see Fig. 13) is at least an order of magnitude lower than any friction factor obtained from a Moody diagram in purely mechanical turbulence with smooth wall boundaries! Thus, in essence, the laminarizing effect at the density interface reduces the "interfacial skin friction" and makes the salt wedge flow "slip under" the oncoming high Reynolds number river flow leading to intrusion lengths considerably larger than would be estimated on the basis of conventional friction estimates.

In the quest for a better understanding of salt wedge estuaries - and of related geophysical flows - it seems urgent to conduct similarly detailed measurements at yet larger scales: For example, laboratory experiments in a flume with a depth of order of 1m seem readily possible and would allow to investigate the complete range of Re up to 10^5. Also, detailed field observations with modern profiling instruments and tracer methods would yield further insight into the process dynamics.

Finally, the evidence presented herein - along with data presented by others - should provide a considerable challenge to the mathematical modeling of the turbulent field equations in stratified estuary flows. Clearly, any closure scheme must describe the complete transition to a locally fully laminar state and high Reynolds number invariances theories cannot be invoked!

Acknowledgments

This research has been supported, in part, by a contract from the U.S. Army Corps of Engineers, Waterways Experiment Station.

V. References

Arita, M. and Jirka, G.H., 1987a: Two-layer model of saline wedge, I: Entrainment and interfacial friction, J. of Hydr. Engrg., ASCE, **113** (10), 1229-1243.

Arita, M., and Jirka, G.H., (1987b): Two-layer model of saline wedge, II: Prediction of mean properties, J. of Hydr. Engrg., ASCE, **113**(10), 1249-1263.

Browand, F.K. and Winant, C.D., 1973: Laboratory observations of shear-layer instability in stratified fluid, Boundary-Layer Meteorology, **5**, 67-77.

Dermissis, V. and Partheniades, E., 1985: Dominant shear stresses in arrested saline wedges, J. Waterway, Port, Coastal and Ocean Eng., **111**, 733-752.

Fischer, H.B., et al., 1979: "Mixing in Island and Coastal Waters," Academic Press.

Gibson, C.H., 1982: Alternative interpretation for microstructure patches in the thermocline, J. Phys. Oceanogr., **12**, 374-383.

Gibson, C.H., 1987: Fossil turbulence and intermittency in sampling oceanic mixing processes, J. Geophys. Res., **92**, (C5), 5383-5404.

Grubert, J.P., 1980: Experiments on arrested saline wedge, J. Hydr. Div., Proc. ASCE, **106**, 945-960.

Hazel, P., 1972: Numerical studies of the stability of inviscid stratified shear flows, J. Fluid Mech., **51**, 39-61.

Hino, M., 1980: Entrainment and friction at the interface of salt wedge, in Stratified Flows, T. Carstens and T. McClimans, Ed.s, Tapir, Trondheim, Norway.

Hopfinger, J., 1987: Turbulence in stratified fluids: A review, J. Geophys. Res., **92**, (C5), 5287-5303.

Jay, D.A., 1988: in Physical Processes in Estuaries, J. Dronkers and W. Van Leussen, Ed.s, Springer-Verlag.

Jirka, G.H., 1982: Turbulent buoyant jets in shallow fluid layers, Turbulent Buoyant Jets and Plumes, W. Rodi, Ed., Pergamon Press, New York, NY., 69-119.

Jirka, G.H., and M. Arita, 1987: Density currents or density wedges: Boundary-layer influence and control methods, J. Fluid Mech., **177**, 187-206.

Sargent, F.E. and G.H. Jirka, 1983: A comparative study of density current and density wedges, Tech. report, School of Civil and Environmental Engineering, Cornell University, Ithaca, New York.

Sargent, F.E. and Jirka, G.H., 1987: Experiments on the saline wedge, J. Hydr. Engrg., ASCE, **113** (10), 1307-1324.

Schijf, J.B. and Schonfeld, J.C., 1953: Theoretical considerations on the motion of salt and fresh water, Proc. Minnesota Int. Hydr. Conv., ASCE and IAHR.

Thatcher, M.L. and Harleman, D.R.F., 1981: Long-term salinity calculations in Delaware estuary, J. Env. Engrg Div., ASCE, **107** (EE1), 11-27.

Part IV

Sediment Transport

Laboratory Experiments on Sedimentation of Fine-grained Sediments: A State-of-the-Art Review in the Light of Experiments with the Delft Tidal Flume

Wim van Leussen[1] and Johan C. Winterwerp[2]

[1]Rijkswaterstaat, Tidal Waters Division
P.O. Box 20907, 2500 EX The Hague
The Netherlands

[2]Delft Hydraulics, Estuaries and Seas Division
P.O. Box 177, 2600 MH Delft
The Netherlands

ABSTRACT

A review is given of laboratory experiments on sedimentation of fine-grained sediments, with emphasis on the aspect of not disrupting the fragile aggregates. A comparison between sedimentary processes in a tidal flume and in nature shows the importance of various time scales: time scale of settling, time scale of vertical mixing and time scale of flocculation. A concise description of the Delft Tidal Flume is included. The paper concludes with experiments under tidal conditions. They show the importance of the tidal slack period, sufficient time for the development of the large aggregates which results in large settling velocities, as well as the effect of tidal distortion and asymmetric concentration curves on the longitudinal fine sediment transport. The development of the concentration curves during the successive tidal cycles, starting with a consolidated bed, shows the dominant role of the structure and properties of the top layer of the bed, and illustrates the difficulty of predicting the suspended sediment concentration during a tidal cycle from results of erosion experiments under steady flow conditions. The development of this top layer under unsteady flow conditions needs further research.

I. Introduction

In the past decades much effort has been given to develop mathematical models of the transport of mud in estuaries and coastal waters. Although both numerical and physical problems have yet to be solved, presently the main task is to obtain a full understanding of the dominant physical processes, and to develop reliable mathematical relations for these processes. Moreover, it is difficult to get experimental data on the integrated effects, which are needed to verify the calculated results and to verify the predictive capability of mathematical models.

With the objective to resolve such problems and difficulties, a research programme is launched in The Netherlands, in which field studies, mathematical modelling and laboratory experiments are performed simultaneously. Each phase of research has its advantages and disadvantages. Therefore interactions and continuous feedback among these phases are needed to understand and describe all important aspects of fine cohesive sediment dynamics.

A central piece of equipment in this research programme is the Delft Tidal Flume. It was built in 1986 for basic experimental research on stratified (tidal)

Coastal and Estuarine Studies, Vol. 38
R. T. Cheng (Ed.)
Residual Currents and Long-term Transport

flows and mud transport phenomena. It is a schematized estuary with a length of 130 m, in which processes like mixing, salt intrusion and fine sediment transport can be studied. A description of this facility is given in section II.

This paper concentrates on laboratory experiments on the sedimentation of fine-grained sediments. Section III compares sedimentary processes in the tidal flume with those in nature. Special attention is given to the various time scales in the laboratory facility with respect to the real processes in nature. It will be shown that the tidal flume is not a scale model of a real estuary, but a facility in which basic processes that happen in nature can be studied systematically under controlled conditions.

A literature review on sedimentation research in laboratory experimental systems is given in section IV. To obtain reliable functional relationships between the deposition rate and the sediment properties and flow parameters, various experimental systems were developed. The formation of aggregates is most important in the deposition of the fine sediment particles. However, because these aggregates are very fragile, special attention should be given to avoid the disruption of these aggregates. An example is the flow in a recirculating flume, in which the aggregates are disrupted by the high shear stresses in the pump and pipes, so that the settling velocities of the aggregates are not the same as in nature. Flows in annular flumes and tidal flumes are relatively free of such aggregate-disrupting mechanisms, but other problems can arise as will be shown. Section IV concentrates on experiments under steady flow conditions. Laboratory research on sedimentation under tidal conditions has just begun. Some preliminary results are given in section V.

II. The Delft Tidal Flume

A schematic view of the Delft Tidal Flume is given in figure 1. It was designed as a tool for fundamental studies on turbulence and mixing processes in stratified flows and on fine sediment transport in estuaries. It may be considered as a schematized estuary. Tides, water depth, salinity and sediment concentration can be controlled over a wide range at its sea boundary. Fresh water discharge and fine sediment concentration can be varied at the river boundary. The wide range of boundary conditions permits the generation of flow situations from strongly stratified to well-mixed. The boundary conditions are summarized in table 1.

Table 1. Possible boundary conditions of the Delft Tidal Flume

sea boundary	
• mean water level	0.10 to 0.90 m
• tidal range	0 to 0.15 m
• tidal period	30 to 1800 s
• density (salinity)	1000 to 1030 kg/m^3
• fine sediment concentration	0 to 3000 mg/l
river boundary	
• fresh water discharge	0.3 to 30 l/s
• fine sediment concentration	0 to 1000 mg/l
• tidal flow	-300 to +300 l/s

The flume has a width and depth of 1.00 m and a length of 130 m. The length of the flume is of great importance for studies on fine sediment transport under tidal conditions, as will be shown in section III. It permits the study of sedimentation processes without the effect of floc-disrupting elements. A longer length can be simulated by injecting and substracting water at the river boundary. A numerical model is used to compute these flow rates at the river boundary to mimic the required "virtual" length of the flume.

The sea level is controlled by a cylinder shaped overflow weir. The water-sediment mixture, discharged at the sea boundary, is stored in two tanks with a total volume of 150 m^3. In these tanks the sediment is kept in suspension by a rotating propeller system. Sea water density and fine sediment concentration are regulated by brine and mud injection systems. The vertical density distribution in the model sea is additionally controlled by a fresh water skimmer at the water surface and a jet-mixing system. Photographs of the tidal flume are given in figure 2. The tidal flume can also be applied as a recirculating flume with a discharge up to 500 l/s.

Plan view

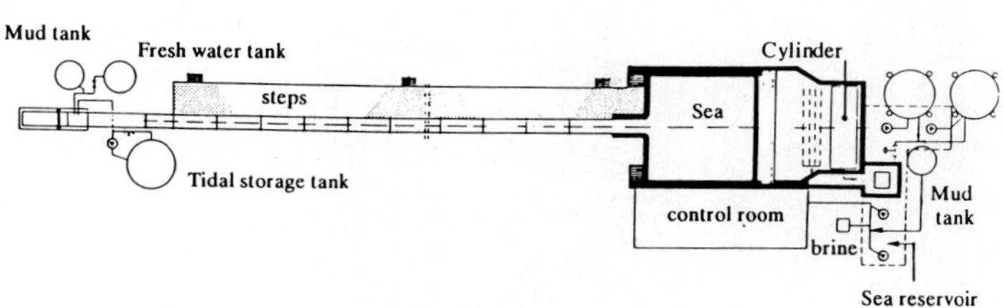

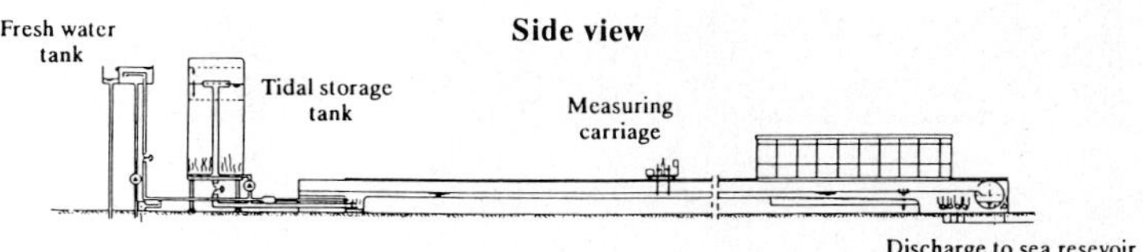

FIGURE 1. Schematic view of Delft Tidal Flume

A special Information System was developed to control the sea and river boundary conditions and to sample, store and process data, using a VAX 730 minicomputer as a central facility. Setpoints for the boundary controls are updated 10 times per second using a feed-back system. If the required experimental conditions are achieved, data collection is started. Final data storage contains only those data selected by the user for further processing on a VAX 750 minicomputer.

Measuring instruments are mounted on fixed places along the flume (water levels), as well as on a position controlled carriage (flow velocity, salinity, temperature, fine sediment concentration, position of the bed, etc.). The fine sediment concentration is detected by optical methods. The probe consists of 6 transducers, to measure the concentration at 6 points in the vertical simultaneously. A special bottom layer device has been developed, equipped with conductivity electrodes (dimensions 200 to 400 μm), to measure the position and the density profile of the bed.

The flume was funded by The Netherlands' Ministries of Transport and Public Works and of Education and Science, expressing the public importance of long-term research with applications in solving practical problems such as the siltation of harbors and channels and the transport of pollutants. Besides gaining the necessary insight into the basic processes in fine sediment transport, the flume has also to provide data for verifying the predictive capability of the numerical models for estuarine cohesive sediment transport.

III. Sedimentary Processes in the Tidal Flume and in Nature

The transport of fine cohesive sediments in estuaries is influenced by a number of factors, such as tidal flows, river discharge, wind-driven flows, waves, salinity and the properties of the fine-grained sediment. In nature these factors can vary enormously. Because various processes occur simultaneously at different time scales, good correlations between the measured sediment concentrations and relevant parameters representing tidal range, river discharge, wind, etc. cannot be obtained in general.

Hydraulic experiments in the laboratory offer the possibility to perform tests under controlled boundary conditions. Therefore, the Delft Tidal flume, as

FIGURE 2a. The Delft Tidal Flume (L = 130 m) with a view on the sea boundary (100 m^2) and the control room.

FIGURE 2b. The Delft Tidal Flume, seen in the direction of the sea. Within the flume, also unidirectional, recirculating flow can be maintained. Furthermore, facilities are available to establish a steady two-layer flow system. The necessary pipes and pumps are shown in the foreground. In the background, tanks for storing and keeping in suspension the water sediment mixture (160 m^3) are seen.

described in section II, is particularly suited to investigate various estuarine processes systematically. However, an important question may arise as to whether there are scale effects on the processes to be investigated.

The tidal flume is not a scale model of any real estuary, but designed to generate a unique prototype flow in the laboratory environment. It is a very schematic estuary, especially designed to gain insight into the various processes. Experiments can be performed in two ways:

1) Tests on cohesive sediments with the same properties as in nature.

2) Tests on artificial sediments with properties on model scale. In the past years some materials were developed to simulate the various processes in a scale model, such as treated natural sediment (by coatings), pulverized synthetic materials, addition of some chemicals to the water (flocculents and deflocculents). It has to be realized that it is almost impossible to simulate all the processes on the correct scale at the same time.

In this paper we will only focus our attention on research in the tidal flume with natural sediments and kaolinite. The latter one has been chosen because of its reproducible properties.

The fine sediment transports are dominantly influenced by the exchange of sediments at the bed (i.e., the erosion and deposition processes). This means that besides the properties of the sediment the shear stresses are of special importance. Applying the same sediment as in nature, comparable bed shear stresses have to be realized in the tidal flume.

A review of the literature reveals that the critical shear stress for deposition τ_d is 0.04 to 0.20 Pa, and that for erosion τ_e varies between 0.1 and 0.5 Pa. Steady flow tests in the tidal flume with a smooth mud bottom showed Chezy-coefficients of about 70 $m^{1/2}/s$ at a water depth of 0.20 m. This means critical flow velocities for deposition of 0.1 to 0.3 m/s and for erosion of 0.2 to 0.5 m/s from which it can be concluded that mean flow velocities up to 0.50 m/s should be obtainable in the flume.

The maximum flow velocities are essentially influenced by the tidal range and the length of the estuary. Starting from a infinite long smooth flume, the maximum flow velocity can be approximated by $u_{max} = \frac{a_o}{h} \sqrt{gh}$, in which a_o is the amplitude of the tide and h the water depth. When the length of the estuary is approximately a quarter of the wave length, higher flow velocities will occur at the same sea boundary conditions. A maximum tidal amplitude a = 0.075 m at h = 0.20 m then gives the maximum flow velocity of 0.53 m/s in the case of an infinite long flume. The flow velocities during an experiment in the tidal flume with a bed of kaolinite are given in figure 16a. The tidal amplitude was 0.03 m and the estuarine length 636 m. Because this length approximates a quarter of the wave length (90%), the maximum flow velocities are larger than the estimates above for an infinite long smooth flume.

The following formulae are generally applied for the exchange processes at the bottom:

for erosion, $$\frac{dm}{dt} = M \left(\frac{\tau_b}{\tau_e} - 1\right) \qquad \text{for } \tau_b \geq \tau_e, \qquad (1)$$

and for deposition, $$\frac{dm}{dt} = w_s C_b \left(1 - \frac{\tau_b}{\tau_e}\right) \qquad \text{for } \tau_b \geq \tau_d, \qquad (2)$$

where τ_b = bottom shear stress

τ_d = critical shear stress for deposition

τ_e = critical shear stress for erosion

M = empirical constant

w_s = settling velocity

C_b = sediment concentration at the bed

m = mass of exchanged sediments per unit area

There will be erosion nor deposition in the case of $\tau_d < \tau_b < \tau_e$.

Although the tidal flume is not a scale model of any real estuary, it is essential to make comparisons between results from the flume and from the field. When the settling velocity w_s and the erosion parameter M would be the same in the field as in the flume, then the erosion and deposition of the fine-grained material would be fully governed by the bed shear stresses. Then for equal bed shear stresses the concentrations of suspended sediment in the flume will be equal to that in nature when the time scale (n_t) is equal to the depth scale (n_h):

$$n_t = n_h \tag{3}$$

For example, when the water depth in nature is 10 m and that in the flume is 0.20 m to 0.40 m, this results in a tidal period T = 447 to 894 sec in the flume, assuming a semi-diurnal tide in nature. For a sinusoidal velocity cycle, the tidal excursion length is $U_{max} T/\pi$. This means for a maximum flow velocity u_{max} = 0.50 m/s an excursion length of 71 to 142 m.

The tidal flume can be applied for homogeneous and nonhomogeneous tests. Because of the required tidal excursion length, the length of the flume is an important dimension of the tidal flume in both cases:

a. nonhomogeneous tests:

Tests with a saline sea and fresh river water discharge. The experiments should be conducted in such a way that the maximum salt intrusion length is less than the physical length of the flume. During the tide the point of maximum salt intrusion will move forward and backward over a distance more or less equal to the tidal excursion length. Therefore sufficient length of the flume will be needed.

b. homogeneous tests:

In this case the water has the same salinity at both the sea and river boundaries. The water in the flume is flowing to and from under the given desired flow conditions. Therefore, the fine sediment transport can be studied under unsteady flow conditions without influencing the suspended flocs by pumps or pipes. The water mass with suspended sediment at a location under investigation should remain at sufficient distance from both ends of the flume. This again stresses the importance of the sufficient length of the flume.

For these two types of experiments various calculations and estimates were made of the needed length. Eventually a length of 130 m was chosen. The total length of the flume, including the sea area, is about 150 m.

It has to be realized that the parameter M and bed shear strength in Eq.(1) for erosion is strongly dependent on the degree of consolidation (Partheniades, 1962; Postma, 1967; Thorn et al., 1976; Mehta, 1988). Shorter tidal periods result in shorter periods of consolidation and thus in increasing erosion rates. Also in the sedimentation process various time scales are of importance. These are discussed below.

The sedimentation process in flowing water can be imagined as a continuous process of aggregation and breakup of mud flocs (see figure 3). The deposition of the mud flocs is controlled by the turbulent processes in a zone near the bed. Only flocs that are strong enough to resist the bottom shear stresses will settle to the bed. Flocs which strength is too low, will be broken up and re-entrained into suspension by the hydrodynamic lift forces. The broken flocs will again participate in the aggregation process in the water column. In this continuous process the following time scales are of importance:

The time scale of vertical mixing : $T_\epsilon = \frac{h^2}{\epsilon_v}$

The time scale of vertical settling: $T_s = \frac{h}{w_s}$,

where ϵ_v is the vertical exchange coefficient.

To give some ideas of the relative magnitude between these two time scales, some of them are calculated in the following table for a water depth of 10 m (characteristic for natural conditions) and a water depth of 0.20 m (characteristic for laboratory conditions). In the case of the flume it is assumed that $\epsilon_v \propto u_* h$ and w_s has the same value as in nature.

T_ϵ	h = 10 m (nature)	h = 0.20 m (flume)
$\epsilon_v = 400\ cm^2/s$ (strong tidal flow in homogeneous water)	42 hrs	50 sec
$\epsilon_v = 40\ cm^2/s$ (estuary with slight stratification)	7 hrs	8.4 min
$\epsilon_v = 4\ cm^2/s$ (strongly stratified estuary)	70 hrs	84 min

T_s	h = 10 m (nature)	h = 0.20 m (flume)
$w_s = 1$ mm/s	3 hrs	3.3 min
$w_s = 0.1$ mm/s	28 hrs	33 min
$w_s = 0.01$ mm/s	278 hrs	5.6 hrs
$w_s = 0.001$ mm/s	2780 hrs	56 hrs

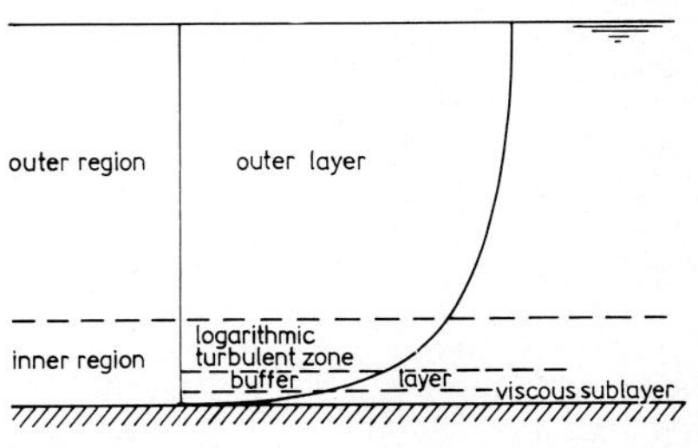

a. Velocity profile with zones in turbulent flow above a smooth wall.

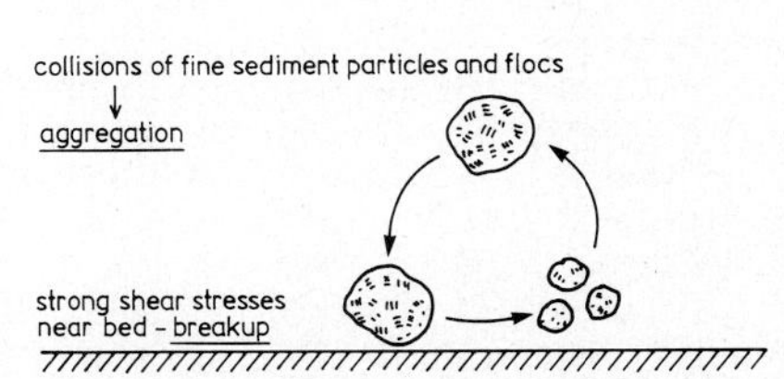

b. Schematic picture of disruption and aggregation of mud flocs.

FIGURE 3. Schematic picture of the continous process of aggregation and breakup of mud flocs in a turbulent flow (figures not to scale).

From a comparison of the calculated time scales, the following remarks can be made:

- In the case of homogeneous flow, generally $T_\epsilon \ll T_s$. Therefore, the fine sediment particles are rapidly mixed over the water depth, especially in the flume. Well-mixing occurs in a very short period of time in the flume; therefore, attention is usually limited to depth-averaged concentrations.

• The time scale of vertical mixing is significantly influenced by stratification. Nonhomogeneous tests in the tidal flume should give more insight into the effect of stratification on concentration profiles and sediment transport.

• Because $\epsilon_v \sim u_* h$, both time scales are proportional to the water depth. This corresponds to the time scale in Eq.(3).

An important parameter is the ratio between T_s and T_ϵ:

$$\frac{T_s}{T_\epsilon} = \frac{\epsilon_v}{w_s h} \tag{4}$$

To get the same concentration profiles in the flume as in nature the value of this dimensionless parameter should be the same. Assuming that $\epsilon_v \propto u_* h$, it can be shown from Eq.(4) that the settling velocity in the flume must be the same as in nature in the case of equal bed shear stresses.

However, the formation of aggregates can be the result of several processes: Brownian motion, turbulent shear, and differential settling. Therefore there are several time scales for flocculation. Brownian motion is only of importance for very fine particles <1 μm and can be neglected here. For turbulent shear, the half-time t_{50}, in which 50% of the particles is flocculated, can be estimated as $t_{50} = \pi \ln 2/(8C \frac{\rho_w du}{\rho_d dy})$ (Einstein and Krone, 1961; Allersma, 1971). C = sediment concentration, ρ_w = density water ρ_d = density particle, $\frac{du}{dy}$ = velocity gradient. This means for equal suspended sediment concentrations that $n_t = n_h/n_u$, or $n_t \approx n_h$ when the velocities in the flume and in nature are of the same order. However, as was shown by Van Leussen (1988) even in flowing water, differential settling may be expected to be dominant. In that case only differences in settling velocity are of importance. When these are the same in the flume as in nature, the same times will be needed to produce the same flocs.

The last requirement is that the water depth should be large enough. Previous experiments in settling columns (McLaughlin, 1961; Van Leussen, 1986) showed that for water depths smaller than 1.00 m there is a strong effect of the water depth on the ultimate settling velocity because of the needed detention time. It can therefore be expected that, starting at the quasi-equilibrium processes as depicted in figure 3, the settling velocities in a flume with a water depth of approximately 0.20 m will be significantly smaller than in nature. Comparisons between results from flumes without floc-disrupting elements (e.g. annular flumes and tidal flumes), settling columns, and field measurements should answer this question.

To avoid this problem it could be suggested to conduct flume tests at higher suspended sediment concentrations than in nature. This increases strongly the settling velocity because of an increase in the number of collisions between the particles per unit time and because the sedimentation rate also increases. It does not affect the erosion rate.

As mentioned previously, because of the various time scales, the Delft Tidal Flume is not a scale model of any real estuary to reproduce its fine sediment transport under various circumstances. The tidal flume is an estuary in itself, by which the basic processes of prototype fine sediment transport can be studied. These experimental studies should be performed in a continuous feedback with field measurements. Numerical models thus play the intermediary role linking between experimental and field research areas.

IV. Laboratory Experiments on Sedimentation

The aggregation of fine sediment is an essential mechanism in the deposition of these sediments. The aggregates are generally a few orders of magnitude larger than the individual particles, resulting in much larger settling velocities. The aggregation process is affected by a large number of factors, such as the physico-chemical surface properties of the particles, organic compounds in the water and sediment, detention times, turbulence in the water, etc. Especially the turbulence structure in a zone near the bed is a dominant factor. The dependence of the properties of the settling sediment (size distribution, strength and density of the aggregates and their settling velocities) and of the ultimate deposition both on the flow conditions and the type of sediment makes the deposition process rather complex. Therefore systematic studies on these processes have mainly been concentrated to experimental research in laboratory flumes.

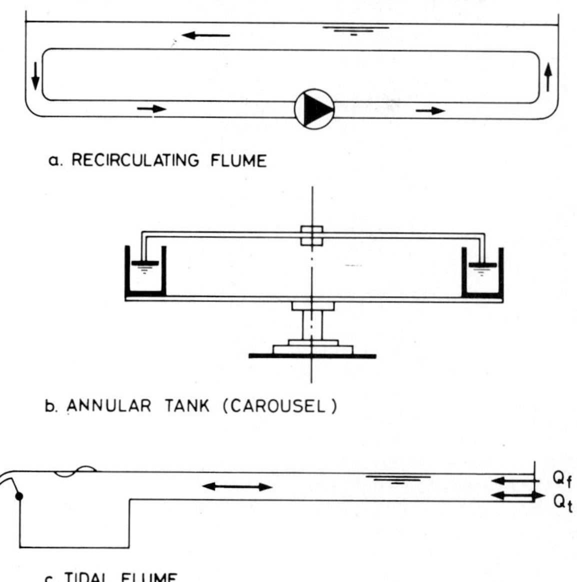

FIGURE 4. Laboratory facilities for sedimentation research.

The first laboratory studies on deposition were conducted in recirculating flumes (cf. figure 4a). A series of tests were reported by Krone (1962) more than 25 years ago. The water and sediment were recirculated with high velocity so that all the sediment remained in suspension. Then a lower velocity was chosen and the deposition of the sediment was studied by measuring the decrease of the concentration as a function of time. The tests were conducted on San Francisco Bay mud with initial concentrations of up to 17,000 mg/l. The length of the flume was 30 m, its height 0.30 m and its width 0.90 m. Krone's results are reproduced in figure 5 for tests with low and high initial concentrations. The smaller the flow velocity, the larger is the decrease in the concentration as a function of time and thus the larger the deposition rate. In the case of the high initial concentration (20,000 mg/l), the concentration remains nearly constant for a while before it starts to decrease slowly. When the concentration is below a value of about 10,000 mg/l the decrease is much faster (figure 5b). The low initial deposition rate is due to the effects of hindered settling.

Krone defined a critical shear stress τ_d, below which fine sediment particles can settle on the bed. He proposed that the probability Pr for settling flocs to remain on the bed varies from one to zero as the bed shear stress increases from zero to τ_d. The deposition rate is then described by:

$$\frac{dm}{dt} = w_s C_b \left(1 - \frac{\tau_b}{\tau_d}\right), \quad \tau_b \leq \tau_d \tag{2}$$

The San Francisco Bay mud showed a critical bed shear stress for deposition τ_d = 0.06 Pa. The above formula is not valid for $\tau_b > \tau_d$. Experience has indicated that this method works well for modelling the fine sediment transport in real estuaries (Odd, 1988).

An apparent problem in performing deposition tests in recirculating flumes is the effect of the pump and return pipes on the distribution of floc sizes and settling velocities. The flocs are disturbed because of the high shear stresses in the pump and return pipes. Therefore, it can be expected that the settling velocities in the flume are strongly influenced by the properties of various disrupting elements in the facility.

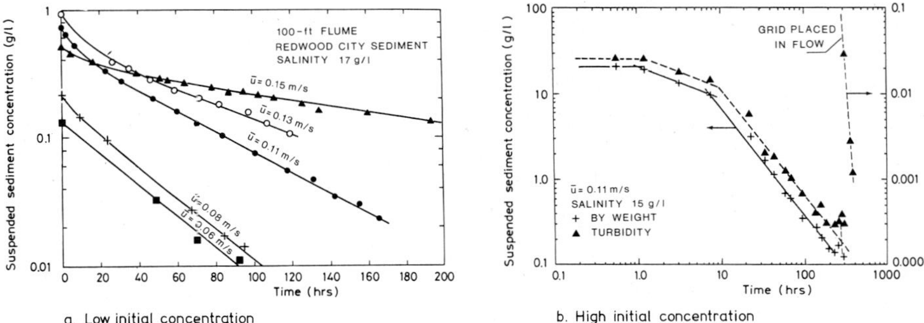

FIGURE 5. Results of depostion tests in a recirculating flume at low and high initial concentrations (Krone, 1962).

In order to overcome this problem, annular flumes (carousels) were constructed (Partheniades et al., 1966), consisting of an annular channel which contains the water-sediment mixture, and an annular ring in contact with the water surface (Figure 4b). A simultaneous rotation of the ring and the channel in opposite direction generates the flow in the flume. Because of its infinite length, there are no floc-disrupting elements in this facility. A problem of circular flumes is the generation of secondary flows (spiral flows), but by proper selection of the rotation speeds of the channel and ring, this problem can be minimized.

Results of deposition tests conducted by Mehta and Partheniades (1975) in an annular flume are given in figure 6. The experiments were conducted in a similar way as in the recirculating flume. The mixture of water and sediment was first brought into suspension by rotating the flume at high speed during one hour. Then the flume speed was lowered to obtain that at which a particular "depositing" bed shear stress. Figure 6 shows that above a certain value $\tau_{b\,min}$ an equilibrium concentration C_{eq} is reached, which depends on the bed shear stress τ_b, but not on the initial concentration C_o. Figure 7 illustrates these equilibrium concentrations as a function of the bed shear stress, showing a well-defined minimum, below which ultimately no material can remain in suspension. The recirculating flume experiments done by Krone and mentioned above were conducted at situations with $\tau_b < \tau_{b\,min}$, so that no equilibrium concentration were obtained. The results also show that in the case of $\tau_b > \tau_{b\,min}$ deposition can occur up to a certain equilibrium concentration.

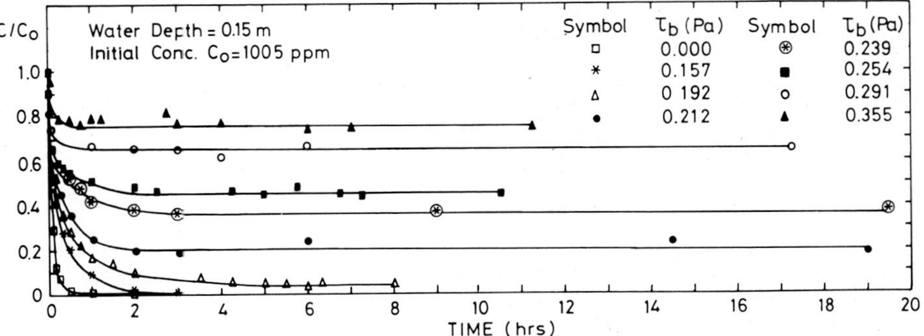

FIGURE 6. Results of deposition tests in an annular flume (Mehta and Partheniades, 1975).

For $\tau_b < \tau_{b\,min}$ the annular flume tests also show that the deposition rate can be described by a formula as given by Eq.(2) for recirculating flume tests. For $\tau_b < \tau_{b\,min}$ the concentration seems to decrease according to a logarithmic-normal relationship:

$$C^{**} = 1/2 \text{ erfc } \left(\frac{T}{\sqrt{2}}\right), \tag{5}$$

where

$T = \frac{1}{\sigma_2}(\log\frac{t}{t_{50}})$; erfc = the complementary error function;

$C^{**} = \frac{C - C_{eq}}{C_o - C_{eq}}$ = the fraction of the depositable sediment in suspension;

t_{50} = the time required for a 50 percent deposition of the depositable part of the sediment. (i.e. when $C^{**} = 0.5$); and

σ_2 = standard deviation.

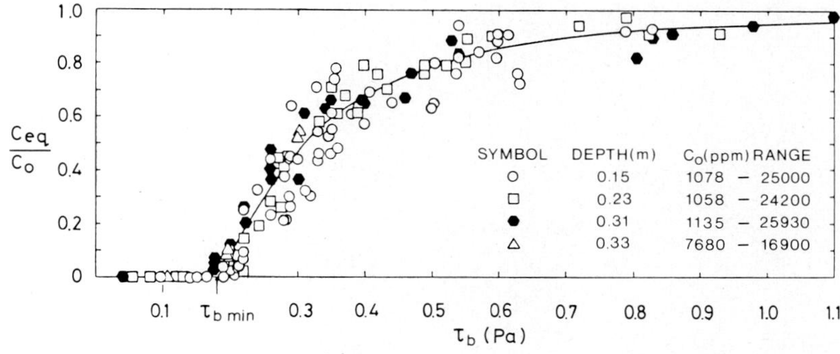

FIGURE 7. The dimensionless equilibrium concentration C_{eq}/C_o as a function of the bed shear stress τ_b (Mehta and Partheniades, 1975).

As an example, Eq.(5) is plotted in Figure 8. Differentiation of Eq.(5) with respect to time and then multiplying it by $(C_o - C_{eq})$ and the depth h gives the deposition rate:

$$\frac{dm}{dt} = (C_o - C_{eq}) \frac{h}{t_{50}} \frac{0.434}{\sqrt{(2\pi)}\,\sigma_2} \frac{e^{-\frac{T^2}{2}}}{t/t_{50}}$$

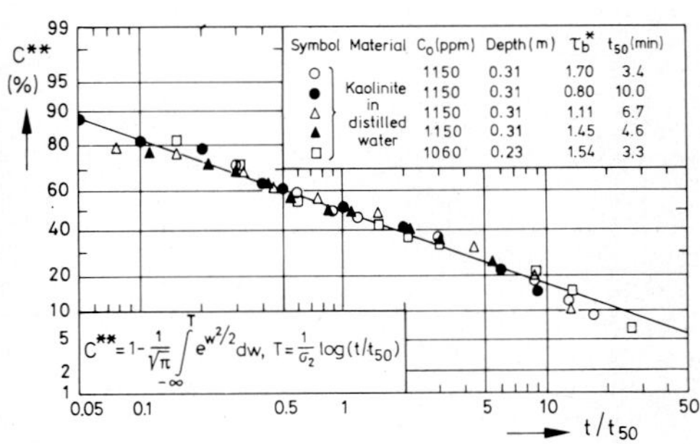

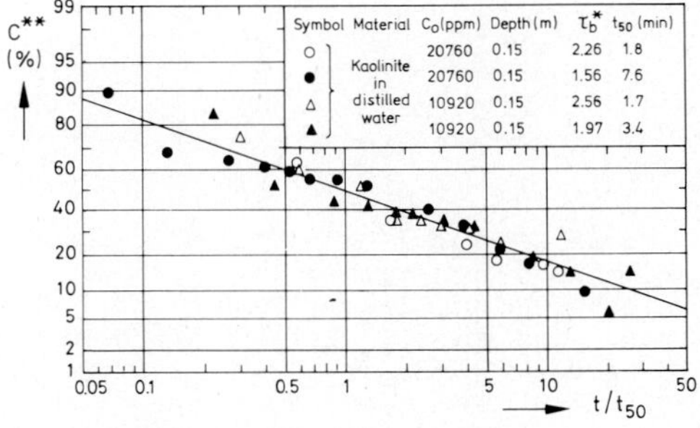

FIGURE 8. Typical examples of log-normal relationship between C^{**} and t/t_{50} for kaolinite in salt water at sea from experiments in an annular flume (after Mehta and Partheniades, 1975).

Mehta and Partheniades (1975) concluded that for predicting the true rates of concentration change only two parameters, t_{50} and σ_2, are needed, provided that the initial and equilibrium concentration are known. Experiments have shown, however, that the parameters t_{50} and σ_2 depends on h, τ_b and C_o, as shown in figure 9.

A tidal flume, in which the flocs are not disturbed by floc-disrupting elements such as pumps or return-pipes, is especially designed for studies on fine-grained sedimentation. Under the influence of a tide, special attention should be given to the various time scales as discussed in section III.

It has been stressed by Partheniades that the above-mentioned formulas from experiments in straight and annular flumes are only valid for non-dispersive systems, i.e. systems with no concentration gradients in the direction of the flow. Therefore their application to real flow systems with considerable transport and longitudinal concentration gradients requires modification (Partheniades, 1986). Experiments with the Delft Tidal Flume being conducted with the assistance of numerical modelling and field measurements are intended to fill the gap between the study of the isolated processes in the laboratory and the integrated processes in nature.

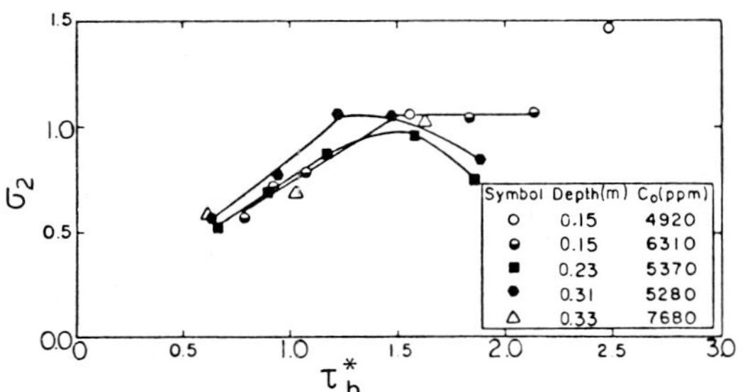

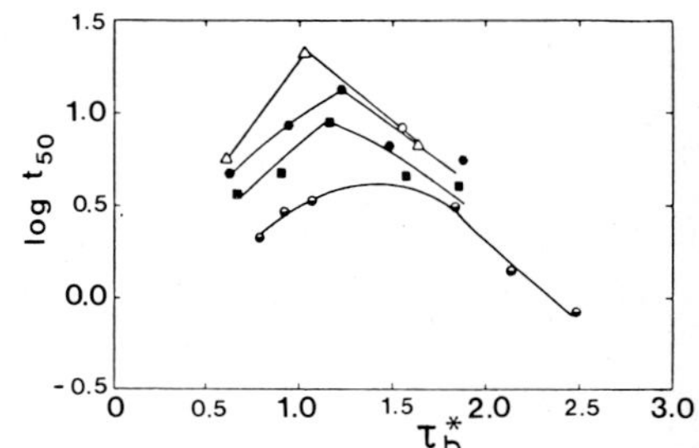

FIGURE 9. Dependence of σ_2 and t_{50} on the dimensionless bed shear stress $\tau^*_b = \tau_b/\tau_{bmin}$ (Mehta and Partheniades, 1975).

Table 2. Test conditions in the Delft Tidal Flume (recirculating tests).

test	u [m/s]	τ_b [Pa]	C_o [mg/l]	Ceq [mg/l]	t50 [hours]
SP1	0.31	0.22	820	490	3.5
SP2	0.22	0.12	730	150	5.0
SP3	0.16	0.06	690	0	6.5
SP4	0.23	0.12	688	180	4.5
SP5	0.22	0.11	1867	470	4.5

Besides tidal conditions (to be described in section V), the Delft Tidal Flume can be used as a recirculating flume. Some tests were performed using commercially available kaolinite. During the tests the initial concentration and bed shear stress were varied. Results are summarized in table 2. The time-deposition curves are given in figure 10. It follows the log-normal relationship obtained for tests in annular plumes. However, the values of t_{50} in the case of the recirculating flume are 3 to 6 hours. For comparison, in the case of the annular flume t_{50} was found to be smaller than 10 minutes in general. This difference in t_{50} may be attributed to the difference in aggregation because the growth in the recirculating flume was limited by floc-disrupting elements such as pumps and return pipe, thus resulting in much lower settling velocities.

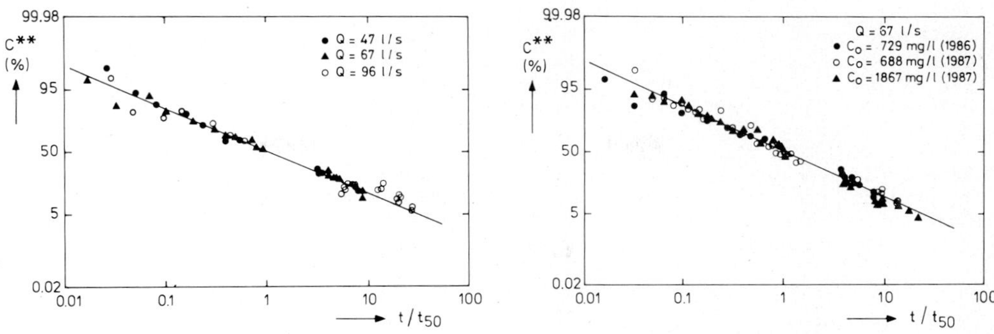

FIGURE 10. Log-normal relationship between C^{**} and t_{50} for recirculating flume type tests in the Delft Tidal Flume with kaolinite in salt water.

V. Sedimentation Under Tidal Conditions

In estuaries, the sedimentation of fine-grained sediment is largely governed by the accelerating and decelerating tidal flows. The processes are further complicated by river inflow, density gradients, waves and wind stresses. Thus far almost all research on sedimentation has been conducted under steady flow conditions. Generally the results so obtained are applicable to real estuaries if one can divide the tidal period into a number of quasi-steady periods. Laboratory experiments under unsteady flow conditions have been conducted only very recently. Therefore, our understanding of the complex sedimentation process in a continuously varying flow, such as in tidal flows, is still in its infancy.

Results of unsteady-flow sedimentation experiments were published by Hayter (1983) and Umita _et al_. (1984). Both conducted experiments in annular flumes. The tidal period was equal to that in nature: $T \approx 12$ hours (semi-diurnal tide).

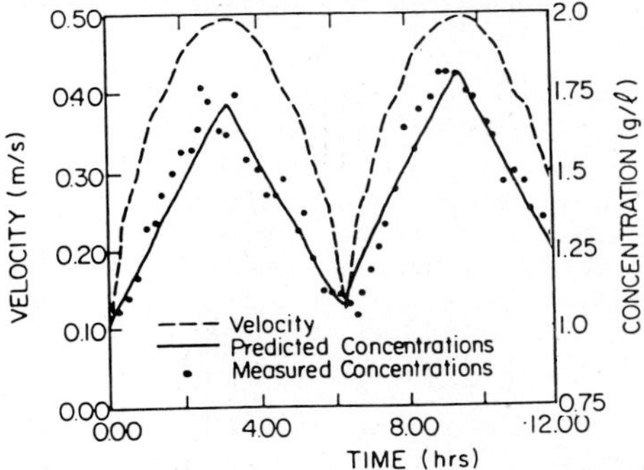

FIGURE 11. Flow velocity and concentration curves during an unsteady flow experiment in an annular flume. Sediment from Lake Francis in salt water. Consolidation time = 40 hrs (Hayter, 1983).

Hayter performed tests with sediment from Lake Francis in salt water, measuring the variation in suspended sediment concentration over the tidal cycle. The flow was unidirectional and the flow velocity varied sinusoidal between 0.1 and 0.5 m/s. The bed was formed from a suspension and was allowed to consolidate for 40 hours. After 4 cycles (25 hours) a dynamic equilibrium was reported to be obtained. The suspended sediment concentration varied almost linear with time, with a phase shift only of a few minutes with the velocity curve (figure 11). The experimental results were mainly used to verify a numerical model. No analysis was made to study the specific features for tidal flow.

Umita et al., (1984) performed two test series with a representative Japanese mud (River Chicugo) in salt water. The flow was unidirectional again. In test series 1, the bed shear stress was varied sinusoidally between 0 and 0.2 Pa and in test series 2 between 0 and 0.4 Pa. They observed a significant sorting effect during the tests and in particular discussed the role of flocculation processes. From the shear stress curves, such as shown in figure 12, three major phases were distinguished. In phase 1 the concentration increases due to erosion of the bed, in phase 2 the concentration remains constant or decreases slightly due to the settling of primary particles or of some flocs, and in phase 3 flocculation processes are important and the concentration drops rapidly because of the large settling velocity of the large flocs. This latter process took about 30 to 60 minutes.

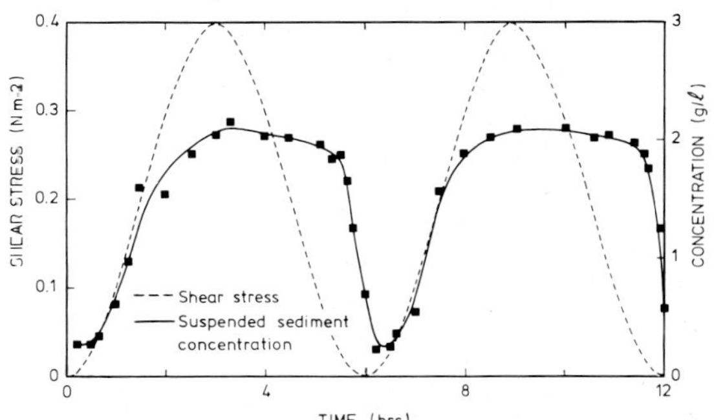

FIGURE 12. Bed shear stress concentration curves during an unsteady flow in an annular flume. Sediment from River Chicugo in salt water. Consolidation time = 24 hrs. (Umita et al., 1984).

Similar experiments were also conducted in the Delft annular flume under tidal conditions. However, the flow direction was changed after each slack tide. The external diameter of this carousel is 2.10 m. The width and depth of the flume are 0.20 m and 0.30 m, respectively. Secondary flows are minimized by rotating the flume and surface ring in the opposite direction. The tests were performed on kaolinite in salt water for tidal periods of 420 sec., 1800 sec., 6 hours and 12 hours.

Results of these tests are shown in figure 13, exhibiting the time variation of the flow velocity, bed shear stress, and suspended sediment concentration. A comparison of figure 13c and d with figure 11 and 12 reveals that both results are qualitatively similar, namely an increase in the suspended sediment concentration during the erosion phase under increasing shear stresses, followed by a phase of almost constant concentration and finally a phase of a relatively rapid deposition around slack tide, some time after the bed shear stress decreases below a critical level.

For tidal periods of 1800 sec, 6 hrs, and 12 hrs, this rapid decrease in the suspended sediment concentration started 2.7 min. 4.7 min, and 4.1 min, respectively, after decreasing of τ_b below a critical shear stress τ_d = 0.06 Pa (value of τ_d from tidal flume tests in section IV). However in the tests with tidal periods of 6 and 12 hours, a slow decrease in suspended sediment concentration started already at 0.64 and 1.5 hours, respectively, before the point of rapid decrease. These times correspond to times needed for flocculation of the same material in a settling column test (H = 4 m, D = 0.30 m), as shown in figure 14.

The time periods of steep, almost linear, decreasing concentrations (see Figure 13) are 28 sec., 8 minutes and 9 minutes, respectively, for T = 1800 sec., 6 hours and 12 hours. In the case of short tidal periods (T<6 hrs), the concentration increased again before full deposition was realized because of the reviving erosion process. This results in lags of 2, 12 and 22 minutes between the minimum suspended sediment concentration and the minimum flow velocity for T = 1800 sec., 6 hours, and 12 hours, respectively, as is shown in figure 13. In the test with T = 420 sec, the slack tide period is too short to reveal any decrease in the suspended sediment concentration (cf. fig. 13a).

The rates of decreasing suspended sediment concentration are summarized in table 3. Settling velocities were calculated from $\frac{dC}{dt} = - \frac{w_s}{h} C$. The higher settling velocity at T = 1800 s may be attributed to the higher suspended sediment concentration. These values of w_s are of the order of magnitude as expected, though they are somewhat lower than those obtained from experiments in settling columns if the effect of the concentration is taken into account (cf. fig. 14). The higher suspended sediment concentrations for smaller tidal periods may be attributed to the short tidal slack time. Thus insufficient time is available to complete the flocculation process and the deposition of the flocculated sediment.

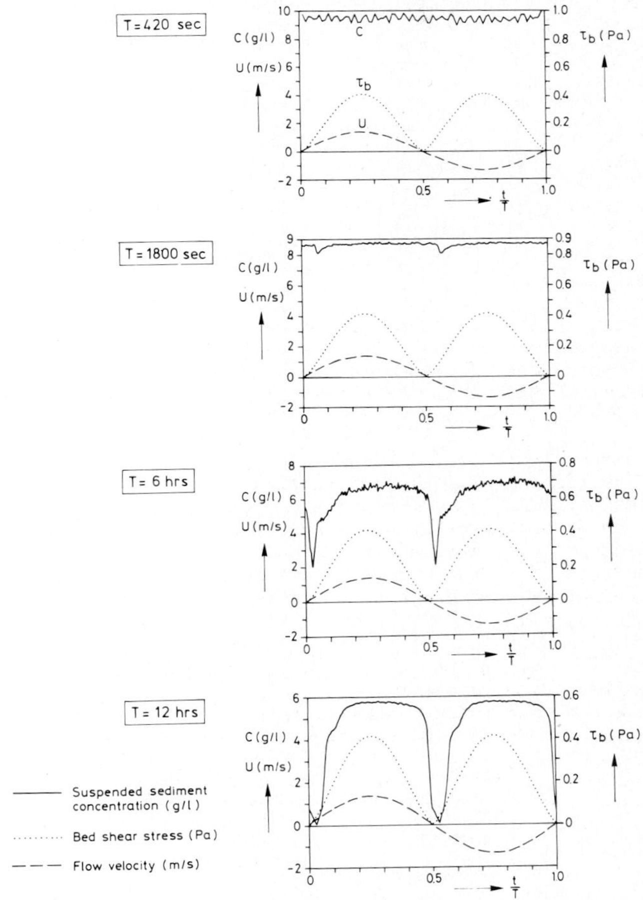

FIGURE 13. Unsteady flow experiments in the Delft annular flume on kaolinite in salt water at tidal periods of 420 s, 1800 s, 6 hrs and 12 hrs.

Table 3. Results from tidal tests in the Delft annular flume.

	$\frac{dC}{dt}\Big\vert_{max}$ (mg/ls)	w_s (mm/s)
T = 1800 s	33.8	0.85
T = 6 hrs	9.67	0.48
T = 12 hrs	7.47	0.50

These results support the earlier remarks on the time scales namely that the detention time should be large enough to realize the flocculation process and hence the relatively large settling velocities. In other words, the slack tide period should be long enough for the flocs to settle on the bed. (Note that $w_s = 0.50$ mm/s is a characteristic value in these tests, and hence $T = h/w_s = 500$ sec.). Therefore, it can be concluded that the slack tide period needs special attention in laboratory experiments under tidal conditions.

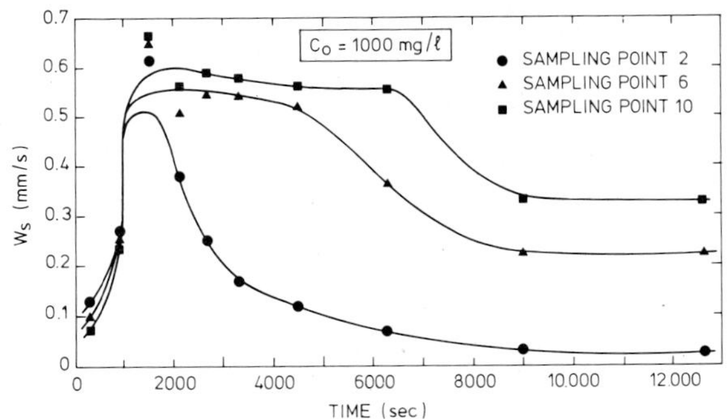

FIGURE 14. Settling velocities as a function of time starting from a homogeneous suspension in a settling column (H = 4.00 m). Sampling points no. 2, 6 and 10 are at z = 3.50, 1.50 and 0.25 m respectively. (Kaolinite, C_o = 1000 mg/l, salt water). (Van Leussen, 1986).

For the test with T = 6 hours, a comparison of the concentration curves between the first and 27th tidal cycles is made in figure 15. It shows that the suspended sediment concentration during the 27th tidal cycle is much higher than that during the first tidal cycle although the bed shear stresses are equal in both cases. Apparently a weakening process is going on in the top layer of the bed, so that after an increasing number of tidal cycles more sediment will be eroded in one tidal cycle, notwithstanding the same bed shear stresses in each tidal cycle. It may be concluded that the development of the top layer of the consolidated bed (T_{cons} = 1 week) under the varying bed shear stresses is essential. The figure also shows that the concentration in the first tidal cycle continues to increase over a longer period than in the 27th tidal cycle. The rate of erosion during the 27th tidal cycle is much larger and stops already before the bed shear stress reaches its maximum. (Fig. 15 shows a decreasing erosion rate after t ≈ 2000 sec in the 27th

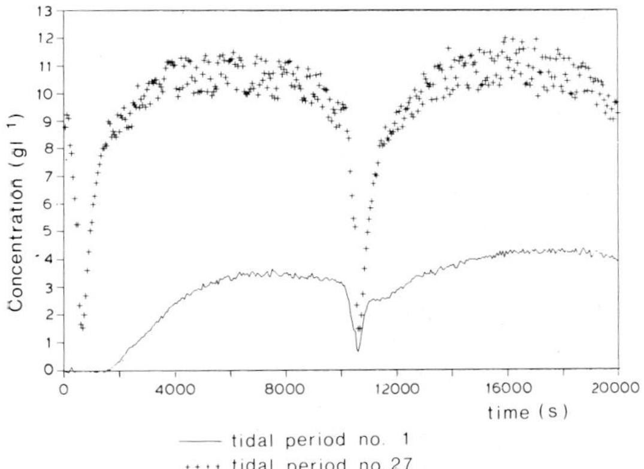

FIGURE 15. Concentration curve during the 1st and 27th tidal cycle in an unsteady flow experiment on kaolinite in salt water in the Delft annualar flume (Tidal period 6 hours).

tidal cycle, while the bed shear stresses are still increasing up to t = 5400 sec). This stresses the conclusion that the resulting suspended sediment concentration during a tidal cycle is also governed by the vertical structure and properties of the top layer of the bed. [cf., the discussion on the role of this top layer in the fine sediment transport in the Rotterdam Waterway by Van Leussen and Van Velzen (1989)]. The figure illustrates the difficulty of predicting the suspended sediment concentration during a tidal cycle from the results of erosion experiments under steady flow conditions of a consolidated bed.

Results of the first experiment on fine sediment transport under unsteady flow conditions in the Delft Tidal Flume are given in figure 16. The tidal amplitude was 0.03 m, waterdepth 0.25 m, tidal period 1800 sec, and the simulated estuary length 636 m. The experiment was performed with kaolinite in homogeneous salt water (no fresh water discharge). Results are given at x = 12 m and x = 60 m (distance from the sea boundary).

Under the sinusoidal water level variations at the sea boundary, there is an asymmetry in the curve of flow velocities and bed shear stresses. The larger bed shear stresses in the ebb period, especially at x = 12 m, resulted in a higher suspended sediment concentration and ultimately in a tide averaged fine sediment transport in the direction of the sea. Neglecting the dispersive terms, the tide-averaged fine sediment transports were estimated as about 90 and 60 kg per tide at x = 12 m and x = 60 m, respectively. It must be stressed that the cumulative transports per tide are larger than the quantity of remaining suspended sediment in the flume.

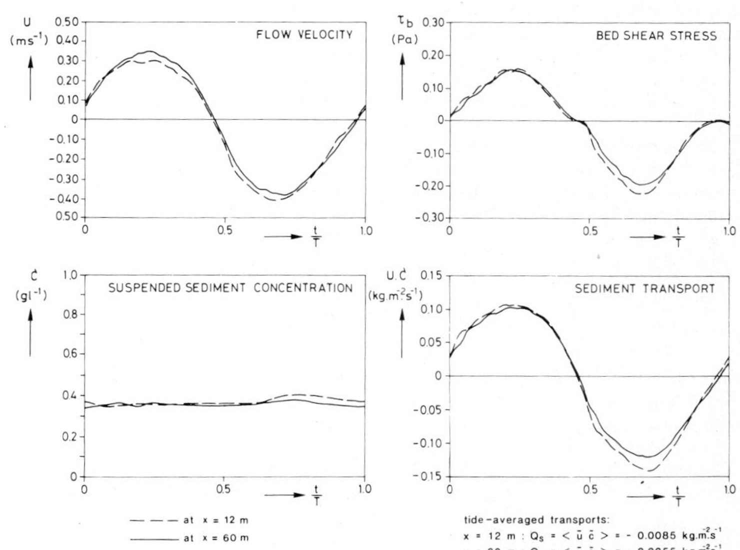

FIGURE 16. Results of tidal flume experiments on kaolinite in salt water. Tidal range 0.03 m, water depth 0.25 m, simulated flume length 636 m, tidal period 1800 s. No fresh water discharge.

The results of this tidal flume experiment are as follows:

a. Longitudinal transport can enormously affect the suspended sediment concentration distribution.

b. Some difference between $\tau_{b,flood}$ and $\tau_{b,ebb}$ results in large horizontal transports.

It can be concluded that the form of the tidal curve (asymmetry, tidal slack period, etc.) and its distribution over the estuary length deserve more attention in fine sediment transport studies.

VI. Conculsions and Recommendations

A comparison between the sedimentary processes in a tidal flume and in nature shows the importance of various time scales: time scale of settling, time scale of vertical mixing, and the time scale of flocculation. This leads to the conclusion that a tidal flume will never be a scale model of any real estuary, but that it is a facility, in which the basic processes that occur in real estuaries can be studied.

A review has been given of laboratory systems to investigate the sedimentation rate of cohesive sediments, with emphasis on experimental systems, in which the fragile aggregates are not destroyed by floc-disrupting elements as pumps or return-pipes. Such systems are annular flumes and tidal flumes with sufficient length. A comparison of times of 50% deposition illustrated the dominant effect of floc-disrupting elements in a recirculating flume on the deposition rates.

The results of experiments under tidal conditions showed the importance of the tidal slack period, sufficient time for the development of the large aggregates which results in large settling velocities, and the effect of tidal distortion and asymmetric concentration curves on the longitudinal fine sediment transport.

The development of the suspended sediment concentration curves during the successive tidal cycles, starting with a consolidated bed, showed the dominant role of the structure and properties of the top layer of the bed. This illustrates the difficulty of predicting the suspended sediment concentration during a tidal cycle from results of erosion experiments under steady flow conditions. The development of the top layer of the bed under unsteady flow conditions needs further research.

Although this paper concentrates on laboratory experiments, it is stressed that a basic research programme on fine sediment transport in estuaries can only be made balanced when laboratory experiments, field studies and numerical modelling are all performed concurrently with a continuous feedback. Only such a balanced programme will lead to a better understanding of the various aspects of cohesive sediment dynamics, reliable descriptions of the processes and judgement of the predictive capability of the numerical models for fine sediment transport.

Acknowledgements

The experiments on fine sediment transport were conducted at Delft Hydraulics. Thankfully we mention the contribution to this paper by our colleagues John Cornelisse (annular flume experiments) and Cees Kuijper (Tidal Flume experiments). The cohesive sediment research programme was commissioned by Rijkswaterstaat, Ministry of Transport and Public Works in The Netherlands. Its permission to publish results from this research programme is gratefully acknowledged.

VII. References

Allersma, E., 1971: Observations on the simulation of estuarine silt transports in hydraulics models. Proc. 14th Congress of the IAHr, Paris, Subject D, 51-58.

Einstein, H. A. and Krone, R. B., 1961: Estuarial sediment transport patterns. Journal of Hydraulics Division, American Society of Civil Engineers, Vol. 87, No.HY 7, 51-59.

Hayter, E. J., 1983: Prediction of cohesive sediment transport in estuarial waters. Ph.D. Diss., University of Florida, Gainesville, Florida, 348 pp.

Krone, R. B., 1962: Flume studies of the transport of sediment in estuarial shoaling processes. Final Report Hydraulic Engr. Lab. and Sanitary Engr. Res. Lab., University of California, Berkeley, 110 pp.

Mehta, A. J., 1988: Laboratory studies on cohesive sediment deposition and erosion. In: J. Dronkers and W. van Leussen (eds.): Physical Processes in Estuaries. Springer Verlag, Berlin Heidelberg New York, 427-445.

Mehta, A. J. and Partheniades. E., 1975: An investigation of the depositional properties of flocculated fine sediment. Journal of Hydraulic Research, **13**, No.4, 361-381.

McLaughlin, R. T., 1961: Settling proporties of suspensions. Trans. Am. Soc. Civ. Eng., **126**, 1734-1767.

Odd, N. V. M., 1988: Mathematical modelling of mud transport in estuaries. In: J. Dronkers and W. van Leussen (eds.): Physical Processes in Estuaries. Springer Verlag, Berlin Heidelberg New York, 503-531.

Partheniades, E., 1962: A study of erosion and deposition of cohesive soils in salt water. Ph.D. Thesis, University of California, Berkeley, CA, 192 pp.

Partheniades. E., 1986. A fundamental framework for cohesive sediment dynamics. In: A.J. Mehta (ed.): Estuarine Cohesive Sediment Dynamics, Springer Verlag, Berlin Heidelberg New York, 219-250.

Partheniades, E., Kennedy. J. F., Etter, R. J. and Hayer, R. P., 1966: Investigations of the depositional behavior of fine cohesive sediments in an annular rotating channel. Rep. No. 96. Hydrodynamics Lab., MIT, Cambridge, Mass.

Postma, H., 1967: Sediment transport and sedimentation in the marine environment. In: Lauff. G.H. (ed.): Estuaries Assoc. Am. Sci. Publ., Washington DC, **83**, 158-179.

Thorn, M. F. C. and Parsons, J. G., 1980: Erosion of cohesive sediments in estuaries: an engineering guide. Proceedings of the Third International Symposium on Dredging Technology, Bordeaux, March 5-7, 1980, 349-358.

Umita, T., Kusuda, T., Awaya, Y., Onuma, M. and Futawateri, T., 1984: The behaviour of suspended sediments and muds in an estuary. Wat. Sci. Tech., **17**, 915-927.

Van Leussen, W., 1986: Laboratory experiments on the settling velocity of mud flocs. In: S.Y. Wang, H.W. Shen and L.Z. Ding (eds.): Proceedings of the Third Int. Symp. on River Sedimentation, School of Engineering, University of Mississippi, 1803-1812.

Van Leussen, W., 1988: Aggregation of particles, settling velocity of mud flocs - a review. In: J. Dronkers and W. van Leussen (eds.): Physical Processes in Estuaries, Springer Verlag, Berlin Heidelberg New York, 347-403.

Van Leussen, W., and Van Velzen, E., 1989: High concentration suspensions: their origin and importance in Dutch estuaries and coastal waters. Journal of Coastal Research (in press).

Estuarine Boundary Layers and Sediment Transport

W.H. McAnally, Jr.[1] and E.J. Hayter[2]

U.S. Army Corps of Engineers
Waterways Experiment Station
Vicksburg, MS 39180

College of Engineering
Clemson University
Clemson, SC 29634

I. Introduction

Calculation of sediment transport in estuarine areas is used to evaluate deposition rates (typically in channels and fishery areas) and to predict removal rates (typically dredged material disposal sites). These calculations require accurate determination of stresses exerted on the bed by the flow and of velocity profiles near the bed.

Methods commonly in use to calculate these quantities, based on simple hydrodynamic conditions, are poorly suited to estuarine conditions, i.e. strong density gradients, highly unsteady and nonuniform flow, high sediment concentrations, bed interaction with the flow, and poorly defined bed-water interfaces. Application of inappropriate expressions can lead to large errors in calculated stresses and even directions of flow (both instantaneous and residual) opposite to that occuring in nature. The resulting errors in sediment transport calculations can be enormous in some circumstances.

II. Objectives

The objective of the work described here is to provide appropriate improved expressions for estuarine bed stresses and velocity profiles such that sediment transport is more accurately calculated.

The purpose of this paper is to state the problem, present some options, formulate appropriate questions, and generate discussion that will provide guidance on suitable research to reach the objectives. In that regard this paper is somewhat different from the usual presentation of research results.

III. Scope

The scope is limited to free surface flows typical of estuaries (flow affected strongly by tides, river flows, density gradients, winds, and geometry) acting on both cohesive and noncohesive sediment beds. Effects of wind waves are excluded for the present.

Methods of calculation generated by this work must be of practical use to engineers engaged in production level work. The problem of most direct interest is movement of sediment out of open water dredged material placement sites.

IV. Character of Estuarine Flows and Boundary Stresses

Two aspects of estuaries require us to evaluate estuarine currents differently from the usual steady, uniform case. First, their oscillatory nature creates time-

Coastal and Estuarine Studies, Vol. 38
R. T. Cheng (Ed.)
Residual Currents and Long-term Transport

dependent accelerations that alter the thickness and velocity profile of the bottom boundary layer. Second, freshwater runoff meeting saline ocean water creates longitudinal, and usually lateral and vertical, density gradients that cause the ambient, vertical velocity profile to be substantially different from the classic forms. In order to compute bottom shear stresses in estuarine flows we must consider the boundary layer.

Unlike unidirectional flows, in which the entire depth of flow may be occupied by the boundary layer, estuarine boundary layers grow between successive slack water periods, and thus their thickness may be small with respect to the water depth, even though conventional wisdom holds that it is large. This has significant implications with regard to use of standard velocity profile equations for computing stresses.

To illustrate how the nature of estuarine flows can affect boundary stresses, consider the total effective boundary stress for open channel flows. Total bed shear stress in two-dimensional open channel flow can be expressed as

$$\tau = \rho g h S e = \rho u_*^2 \tag{1}$$

where

ρ = density of fluid
g = acceleration of gravity
h = water depth
Se = slope of the energy grade line
u_* = shear velocity.

For steady, uniform flow, Se becomes the water surface slope, but for estuarine flows the variation of Se (and the other terms) in space and time must be considered. Consider the Reynolds form of the conservation of momentum equation in the x direction for the case shown in Figure 1,

$$\frac{\partial u}{\partial t} + u\frac{\partial u}{\partial x} + v\frac{\partial u}{\partial y} + w\frac{\partial u}{\partial z} = -g\frac{\partial \eta}{\partial x} - \frac{g(y-\eta)}{\rho}\frac{\partial \rho}{\partial x} + \frac{1}{\rho}\left(\frac{\partial \tau_{xy}}{\partial y} + \frac{\partial \tau_{xz}}{\partial z}\right) \tag{2}$$

where

u = velocity in x direction
v = velocity in y (vertical) direction
w = velocity in the z direction
η = displacement of the water surface from mean water level
τ_{xy} = shear stress in the x direction acting on the top and bottom of an elemental particle of fluid
τ_{xz} = shear stress in the x direction acting on the sides of an elemental particle of fluid.

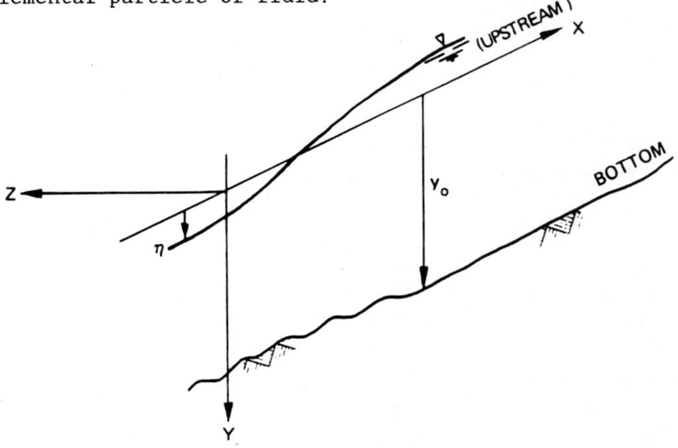

FIGURE 1. Definition Sketch for Equations

This form of the equation is fairly general. The primary assumptions are

1) gravity acts in the y direction,
2) pressure varies hydrostatically,
3) normal-direction viscous stresses are negligible,
4) Coriolis effect is negligible.

Eq.(2) must be integrated over depth to obtain an expression for the total shear. To make the integration tractable, it is usually assumed that the variation of ρ with respect to y is negligible in comparison with the other terms. This is not a particularly satisfying assumption since we are seeking to establish the importance of the various terms, not to exclude them automatically. Therefore, we will not make this assumption right away. Integrating over depth and rearranging the terms results in

$$\tau = (\tau_x)_{y_o} = A + B + C + D + E - F \tag{3}$$

where $(\tau_x)_{y_o}$ = shear stress in the x direction at the bed ($y = y_o$), and terms A-F are shown in column 1 of Table 1.

TABLE 1. Shear Stress Terms in Equation 3

Term	Effect	Form 1	Form 2
A	Unsteadiness	$\int_{\eta}^{y_o} \rho \frac{\partial u}{\partial t} dy$	$\rho \int_{\eta}^{y_o} \frac{\partial u}{\partial t} dy$
B	Nonuniformity	$\int_{\eta}^{y_o} \rho \left(u \frac{\partial u}{\partial x} + v \frac{\partial u}{\partial y} + w \frac{\partial u}{\partial z}\right) dy$	$\rho \int_{\eta}^{y_o} \left(u\frac{\partial u}{\partial x} + v\frac{\partial u}{\partial y} + w\frac{\partial u}{\partial z}\right) dy$
C	Surface slope	$g \frac{\partial \eta}{\partial x} \int_{h}^{y_o} \rho \, dy$	$\rho g \frac{\partial \eta}{\partial x} (y_o - \eta)$
D	Density gradient	$g \int_{\eta}^{y_o} (y - \eta) \frac{\partial \rho}{\partial x} dy$	$\frac{g}{2} (y_o - \eta)^2 \frac{\partial \rho}{\partial x}$
E	Wind stress	$\tau_{xy,\eta}$	$\tau_{xy,\eta}$
F	Side wall stress	$\int_{\eta}^{y_o} \frac{\partial \tau_{xz}}{\partial z} dy$	$\frac{\partial \tau_{xz}}{\partial z} (y_o - \eta)$

From this point several more simplifications are possible, most requiring an assumption that some variable or function is nearly constant with respect to y. Adopting the assumptions that $\partial\tau_{xz}/\partial z$, ρ, and $\partial\rho/\partial x$ are not significantly dependent on y (with respect to the other terms) results in the forms shown in Table 1. Still further simplification can be obtained by assuming that velocities and gradients are not functions of y.

An experimental evaluation of the terms in Eq.(3) was performed using data from the Waterways Experiment Station (WES) salinity flume test 14, in which a 10 min tide of 3 cm amplitude was induced in the ocean headbay (salinity 30 ppt) while a fresh water flow of 0.0002 m^3/s was introduced at the upstream end of the 100-m-long, 0.23-m-wide flume. The vertical salinity gradient in this test was moderate, with about a 30 percent difference between salinities at the bottom and surface at a point 12 m from the ocean end of the flume.

Using a three-point piecewise linear fit to data points in the vertical and difference forms for the derivatives, terms A-D in column 2 of Table 1 were evaluated at one station in salinity flume test 14. Figure 2 shows the individual term's variation in magnitude plus their sum over a complete cycle. At the bottom of the figure the depth average current velocity and water surface elevation histories at that station are shown. Terms A-D were nondimensionalized by dividing by $\rho g y_o$, where y_o is mean water depth.

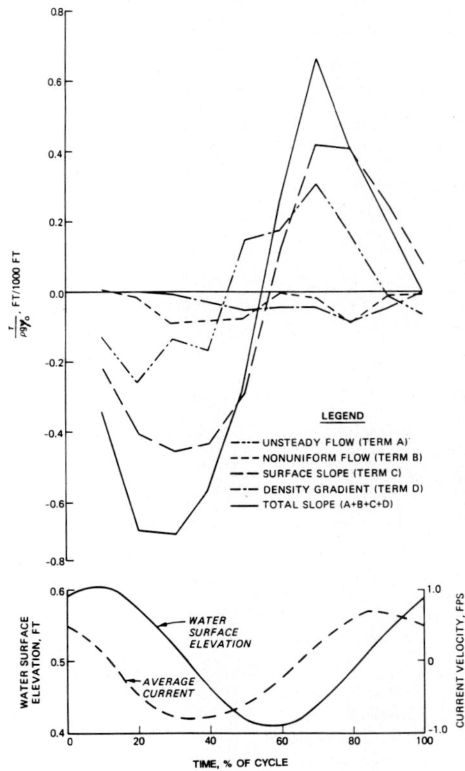

FIGURE 2. Flume Boundary Stress Contribution by Terms A-D

The surface slope and unsteady terms were the largest, but at times during the cycle the density and nonuniform flow terms became significant in comparison. The nonuniform flow term should not be expected to be large since the flume is prismatic. Term F , the sidewall stress, was not evaluated, but will be quite large in the flume because of sidewall roughness strips. Using the three-point piecewise fit and not neglecting variation of ρ with respect to y resulted in about a 5% reduction in the magnitude of the surface slope terms for the flume data. This suggests that for similar conditions that can be adequately described by linear approximations, the variation of ρ with y may be neglected for individual terms and the total shear except when the direction of flow is reversing. However, the above analyses showed that it may not be appropriate to automatically assume further simplifications to column 1 of Table 1 for all cases. Rather, for each application an evaluation must be made as to the proper simplifying assumptions.

Figure 3 shows how a typical steady flow shear stress equation of the form shown below compares to the total stress.

$$\tau = \frac{1}{2} f_c \rho u_1^2 \tag{4}$$

where u_1 = a flow velocity at a specified distance above the bed, f_c = a constant stress coefficient, with an empirically derived value of 0.02 for the salinity flume. The phase shift between peak stress and peak velocity is clearly demonstrated by Figure 3.

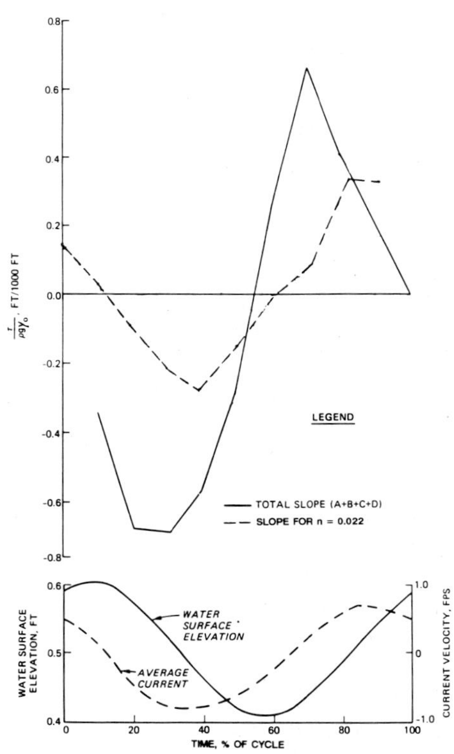

FIGURE 3. Flume Boundary Stress by Equation of Motion and by Constant Friction Factor

Figures 2 and 3 show that, for the fairly simple case of stratified flow in a prismatic flume, using the surface slope or a simple velocity squared equation for calculating the boundary stresses can lead to significantly low magnitude estimates. At small values, even the direction of stress can be in error. In more complicated flows, characteristic of natural environments, the errors can be even larger.

A perhaps more difficult problem is that estuarine flow is not steady and uniform, calling into doubt application of most experimentally determined constant friction coefficients. Several researchers have observed that shear stresses are dependent on the pressure gradient as well as the flow speed; and following from the variation demonstrated in Figure 3, it seems inappropriate to assume that f_c in Eq.(4) is a constant.

IV.1 Soft sediment boundaries

The bed elevation in open channels is usually defined as the location at which the flow velocity is zero (although some profile definitions put it at a distance z_o below that point). Where beds are rigid this definition is appropriate, but for mobile sand beds the definition lacks precision and for soft muds it may be substantially erroneous.

In many estuarine situations, sediment concentration and bulk density increase rapidly near the bed zone, but without a clearly defined interface between water and mud. In the words of W. R. Parker, as the bed is approached, the suspension changes gradually from muddy water to watery mud to bed without a definite interface. Under these conditions defining the bed elevation becomes a matter of selecting an arbitrary definition, usually based on density. Even with a density-defined bed interface, the zero velocity criterion may be inapplicable, since at any given moment the high density suspension may flow to a significant depth or it may possess sufficient structure (stationary suspension) to hold it apparently motionless. In either case, ordinary definition of the boundary and boundary layer may become inappropriate.

V. Steady, Uniform Flows

For steady, constant density, uniform flows the coefficient f_c in Eq. (4) is equal to 1/4 the Darcy pipe flow friction factor, f, and when the flow is also rough turbulent ($u_* k_s/\nu > 74$) it can be related to the Chezy, C, Manning, n, (English Units) and effective roughness, k_s, coefficients by

$$f_c = \frac{2g}{C^2} = \frac{2gn^2}{\left[1.49R^{1/6}\right]^2} = 0.028\left[\frac{k_s}{R}\right]^{1/3} = \frac{f}{4} \tag{5}$$

where R = hydraulic radius.

VI. Unsteady, Nonuniform Flows

VI.1 Use of steady uniform resistance equations in tidal flows

Researchers working with coastal and estuarine currents have employed a variety of methods to compute bed shear stresses. Mehta and Christensen (1977) applied the modified von Karman-Prandtl equation to flow in tidal inlets using several velocity measurements within 1 m of the bed. They reported a satisfactory fit to the data points by the theoretical velocity profile. Ludwick (1973), assuming the boundary layer thickness to be equal to the flow depth, applied the velocity defect concept to Chesapeake Bay flows.

Sternberg (1968) measured Puget Sound velocities at three points within 1 m of the bed and, using a log profile, calculated values of f_c for a characteristic velocity 1 m above the bed. He found a mean value of $f_c = 0.0062$, with 95 percent confidence limits of 0.0017 and 0.022. For Reynolds numbers greater than 1.5 x 10^5, results were less variable and the mean value of f_c was 0.0060.

Bowden et al. (1959) found that data from tidal channels showed good results for $f_c = 0.007$ and a characteristic velocity measured at a distance above the bed equal to 5% of the water depth. Teche (1975) reported favorable results in the Savannah estuary using Bowden's f_c with a velocity 50 cm above the bed.

Though several of the above cited researchers have found that velocity profiles can often be well approximated by a logarithmic profile, it does not necessarily follow that assumption of a log profile yields a satisfactory estimate of shear stress. There is usually a great deal of noise in velocity measurements, and any continuous (in space) expression is generated on a best-fit basis to that data. Thus, a theoretical velocity profile may "look" like a satisfactory approximation to the measured velocity points without correctly predicting the shear stress at the bed.

As previously mentioned, part of the difficulty in defining f_c from field data is that it is time dependent, in that flow may be changing from hydraulically smooth to hydraulically rough, the bed itself may be changing with time and space, and the pressure gradient dependency of friction (Ludwick 1973; Smith and McLean, 1977) is not included. Under these conditions average shear stresses may be computed from suitably time- and space-averaged velocity profiles, but instantaneous local shear particularly for the noncohesive bed cases. Values of f_c for the flat bed case may be as much as an order of magnitude less than field-derived values. In the absence of better information, one of the forms of Eq.(5) can be used for a constant friction coefficient, and an estimate of a time varying bed surface shear stress can be obtained by multiplying the values from Eq.(5) by the cube root of the ratio of a k_s based on grain size to an effective total k_s.

VI.2 Eddy viscosity model

Next in complexity after the steady flow equations come the simple turbulence closure models. They are typically based on a turbulent eddy viscosity coefficient, ν_t, that is analogous to the kinematic viscosity, ν , and is used in an expression for bed shear stress

$$\tau_o = \nu_t \frac{\partial u}{\partial y} \tag{6}$$

which parallels the expression for laminar shear stress. Knight (1978), and Sleath (1984) provide excellent reviews of these models.

The difficulty with this approach is defining an appropriate value for the eddy viscosity. Since it is a function of the flow, instead of the fluid (like kinematic viscosity) eddy viscosity is not constant in space or time. In the simplest case it is assumed constant. In progressively more realistic (and computationally more complex) approaches it is assumed to be constant over two or three layers. Only one such model is presented here.

VI.3 Three layer model

Kajiura (1968) proposed a widely cited three layer model that defines an inner layer next to the bed, an overlap layer, and an outer layer. The eddy viscosity varies simply across each layer and is given by

$$\nu_{t,inner} \equiv \nu_{t1} = \begin{cases} \nu \text{ for smooth flow} \\ 2.71\ \kappa u_* \delta_i \end{cases} \tag{7}$$

$$\nu_{t,overlap} \equiv \nu_{t2} = \kappa u_* y \tag{8}$$

$$\nu_{t,outer} \equiv \nu_{t3} = \kappa u_* \delta_o \tag{9}$$

where κ = von Karman constant
δ_i = inner layer thickness
δ_o = thickness of all 3 layers

and

$$\delta_o = 0.05\ u*/\sigma \tag{10}$$

$$\delta_i = \begin{cases} \dfrac{12\nu}{u_*} & \text{smooth beds} \\ \dfrac{k_s}{2} & \text{rough beds} \end{cases} \tag{11}$$

δ = frequency of flow oscillation.

The overlap layer disappears if $\delta_i \geq \delta_o$. For an oscillatory free stream velocity of the form

$$u = U_o \cos\sigma t, \tag{12}$$

the equation of motion yields for the boundary layer velocity

$$u = \text{REAL}\left[U_o\ (1 - e^{\alpha_1 y} + A_1 \sinh \alpha_1 y)e^{i\sigma t}\right] \tag{13}$$

where

$$A_1 = \frac{(\alpha_1 + \alpha_2)e^{\alpha_1 \delta_o}}{\alpha_1 \cosh\ (\alpha_1 \delta_i) + \alpha_2 \sinh\ (\alpha_1 \delta_i)} \tag{14}$$

$$\alpha_1 = \begin{cases} \frac{i\sigma}{\nu} \text{ for smooth beds} \\ \frac{i\sigma}{0.185\ \kappa u_* k_s} \text{ for rough beds} \end{cases} \tag{15}$$

$$\alpha_2 = \frac{i\sigma^2}{0.05\ \kappa u_*^2} \tag{16}$$

From these equations it follows that the bed shear stress is

$$\tau = \nu_t \text{REAL} \left[\alpha_1\ U_o\ (A_1 \cosh \alpha_1 y - e^{\alpha_1 y}) e^{i\sigma t}\right] \tag{17}$$

Computation of velocity profiles and stresses using Eqs. (7) thru (17) requires knowledge of the free stream velocity and period and effective roughness size. Since the equations also include the shear velocity term, an iterative solution is required.

VI.4 Turbulence transport models

Recently, unsteady flow research has shifted away from analytic expressions and constants for eddy viscosity/mixing length and concentrated on various higher order approaches that consider the turbulence field in more detail. The aim of these approaches is to develop additional expressions based on constants that are truly constant among a given class of problems (e.g., open channel flow) and not flow dependent. A variety of approaches are given in the literature under a variety of titles.

Reviews of the several turbulence transport approaches are available (e.g. Markatos 1986, ASCE 1988) and only two are presented here.

Turbulence transport models begin with the Eulerian conservation of mass and momentum equations and an equation of state, which are shown below.

$$\frac{D\rho}{Dx_i} = \frac{\partial \rho}{\partial t} + \frac{\partial}{\partial x_i} (\rho u_i) = 0 \tag{18}$$

$$\frac{D(\rho u_i)}{Dt} = - \frac{\partial \rho}{\partial x_1} + \rho g_i - e_{ijk}\ \Omega_j \rho u_k + \frac{\partial \tau_{ij}}{\partial x_j} \tag{19}$$

$$\frac{DC_n}{Dt} = \frac{\frac{\partial\ D_{in}\ \partial C_n}{\partial x_i}}{\partial x_i} + K_n \tag{20}$$

$$\rho = f(C_n) \tag{21}$$

$$\tau_{ij} = \nu_t \frac{\partial u_i}{\partial x_j} \tag{22}$$

where the usual Cartesian index notation is used with repeated index summation and

ρ = density,
P = pressure,
u_i = time mean velocity in the ith direction,
g = gravitational acceleration,
e = alternating tensor,
Ω = Coriolis effect parameter,
C_n = temperature or concentration of dissolved or suspended substance affecting water density,
K_n = sum of sinks and sources of C_n,
D_{in} = Diffusion coefficient of C_n in ith direction.
τ_{ij} = stress on the jth surface in the ith direction

In turbulence transport models, the turbulent stress terms are expressed in further equations that involve production, transport, and dissipation of turbulent kinetic energy. Two prominent approaches are the k-ϵ model and the level 2 1/2 model, both described below.

In k-ϵ models the turbulent kinetic energy, k, is produced at rate P and dissipated at rate ϵ. They have been widely described and applied (e.g. see Rodi, 1982; Chapman and Kuo, 1986). The conservation of energy equation is expressed as

$$\frac{Dk}{Dt} = \frac{\partial}{\partial x_i}\left(C_{k1}\nu_t \frac{\partial k}{\partial x_i}\right) + P - \epsilon \tag{23}$$

$$P = \nu_t\left(\frac{\partial u_i}{\partial x_j} + \frac{\partial u_j}{\partial x_i}\right)\frac{\partial u_i}{\partial x_j} \tag{24}$$

$$\frac{D\epsilon}{Dt} = \frac{\partial}{\partial x_i}\left(C_{\epsilon 1}\nu_t \frac{\partial \epsilon}{\partial x_i}\right) + C_{\epsilon 2}\frac{\epsilon}{k}P - C_{\epsilon 3}\frac{\epsilon^2}{k} \tag{25a}$$

$$\nu_t = C_\nu \frac{k^2}{\epsilon} \tag{26}$$

where if the equations are truly general, the five C's will actually be constant. They are quoted as being (Bradshaw et al. 1981)

$C_\nu = 0.09$

$C_{k1} = 1.$

$C_{\epsilon 1} = 1/1.3$

$C_{\epsilon 2} = 1.45$

$C_{\epsilon 3} = 1.9$

Takemitsu (1988) has proposed a modification to the k-ϵ model that replaces Eq.(25a) with

$$\begin{aligned}\frac{D\epsilon}{Dt} = {} & \frac{\partial}{\partial x_i}\left(C_{\epsilon 1}\nu_t \frac{\partial \epsilon}{\partial x_i}\right) + C_{\epsilon 2}\frac{\epsilon}{k}P - C_{\epsilon 3}\frac{\epsilon^2}{k} \\ & - \frac{\epsilon}{k}\frac{\partial}{\partial x_i}\left(C_{\epsilon 4}\nu_t \frac{\partial k}{\partial x_i}\right) - C_{\epsilon 5}\frac{\nu_t}{\epsilon}\left(\frac{\partial \epsilon}{\partial x_i}\right)^2 \\ & + C_{\epsilon 6}\frac{\nu_t}{k}\left(\frac{\partial k}{\partial x_i}\right)\left(\frac{\partial \epsilon}{\partial x_i}\right) - C_\nu C_{\epsilon 7}\left(\frac{\partial k}{\partial x_i}\right)^2\end{aligned} \tag{25b}$$

which increases the number of constants to 9. He specifies the range of constant values as a minimum, maximum, and most probable value.

Eq.(25b) is said by Takemitsu (1988) to improve the k-ϵ formulation since it includes the effect of diffusion on ϵ. As evidence of its superiority, he shows that in a trial application to open channel flow (using the probable constant values) the resulting eddy viscosity is a minimum at the bed and surface, plus it exhibits a third minimum at middepth, a reasonable distribution that is not obtained with some approaches.

A level 2-1/2 model as an alternate approach to higher order turbulence transport modeling is presented by Mellor and Yamada (1982) and endorsed for numerical flow models by Blumberg and Kantha (1986). The level 2-1/2 model describes turbulent kinetic energy by

$$\frac{Dk}{Dt} = \frac{\partial}{\partial x_j}\left(\ell S_q \sqrt{2k}\,\frac{\partial k}{\partial x_j}\right) + P_s + P_B - \epsilon \tag{27}$$

where P_s is the shear production of turbulence,

$$P_s = \nu_t\left(\left(\frac{\partial u_i}{\partial x_j}\right)^2 + \left(\frac{\partial u_k}{\partial x_j}\right)^2\right) \tag{28}$$

and P_B is the buoyancy production of turbulence

$$P_B = \frac{g}{\rho_o} D_{in} \frac{\partial \rho}{\partial x_j} \tag{29}$$

and ϵ is again dissipation, expressed as

$$\epsilon = \frac{(2k)^{3/2}}{\Lambda_1} \tag{30}$$

and Λ_1 is a length scale.

The level 2-1/2 model also includes an equation for the master length scale, ℓ, to which all other length scales are proportional.

$$\frac{D(2k\ell)}{Dt} = \frac{\partial}{\partial x_j}\left(\ell S_\ell \sqrt{2k}\,\frac{\partial}{\partial x_j}(2k\ell)\right) + \ell E_1 (P_s + P_B) - \frac{(2k)^{3/2}}{B_1}\left[1 + E_2\left(\frac{1}{\kappa(2b - x_j)}\right)^2\right] \tag{32}$$

$$\nu_t = \sqrt{2k}\,\ell\, S_m \tag{32}$$

$$D_{im} = \sqrt{2k}\,\ell\, S_H \tag{33}$$

where

$$S_m = A_2 \frac{\left[1 - \frac{6A_1}{B_1}\frac{P_s + P_B}{\epsilon}\right]}{1 - (3A_2B_2 + 18A_1A_2)G_H} \tag{34}$$

$$S_H = A_1 \frac{\left[1 - 3C_1 - \frac{6A_1}{B_1}\frac{P_s + P_B}{\epsilon}\right] + S_m\left[1 - 9A_1 A_2 G_H\right]}{18A_1^2 + 9A_1 A_2 G_H} \tag{35}$$

$$G_m = \frac{-\ell^2}{2k}\frac{g}{\rho_o}\frac{\partial \rho}{\partial x_j} \tag{36}$$

where

k = turbulent kinetic energy,

ℓ = master length scale,

ρ_o = reference density,

ρ = density,

Λ_1 = length scale,

κ = von Karman constant,

b = channel width,

$A_1, A_2, B_1, B_2, C_1, E_1, E_2, S_q, S_\ell$ are constants.

The complexity of the level 2-1/2 model is daunting; however, if we are using a computer calculation process the complexity is a problem only to the extent that numerical solution methods may be non-straightforward and/or that the computations may be too slow (or too expensive) for the purpose. A significantly simplified level 2 model described by Mellor and Yamada (1982) may be appropriate for many applications.

The attractiveness of the k-ϵ and level 2-1/2 models are that they incorporate implicitly many of the processes we wish to capture in our calculation of boundary layer flows. One contributing factor that they do not capture is the effect that suspended sediment may have on turbulence damping.

VI.5 Intermittence

These calculation methods do not explicitly include the effects of intermittent pulses of turbulence that may be the major factor in boundary stresses in decelerating flows. Such processes have been detected empirically and represent a significant challenge to use of steady flow laboratory experimental results.

VII. Stratification Effects

Methods of correcting for density stratification (e.g., due to temperature, salinity, and/or suspended sediment) in determination of boundary shear and velocity profiles in surface water flows are necessary if the turbulence calculation method does not explicity account for it.

VII.1 Salinity stratification

McCutcheon (1981), started with the Prandtl-von Karman law of the wall, incorporated the similarity analysis of Monin and Obukhov, and developed the following formula for the vertical velocity distribution for uni-directional, hydraulically rough, continuously stratified open channel flow:

$$\frac{u(y)}{u_*} = \frac{1}{\kappa} \int_{y_o}^{h} \frac{dy}{y[1+\alpha g \kappa^2 y^2 f(y)/\rho_m u_*^2]} \tag{37}$$

where $f(y) = \partial\rho/\partial y$; $\alpha = c\alpha_f$ with c = ratio of mass mixing length to momentum mixing length; and ρ_m = depth-averaged density. Eq.(32) can be substituted into the equation for Chézy coefficient. Eqs.(37-38) are solved iteratively for C and u_*. McCutcheon (1981) performed several laboratory tests in which salt was the stratifying agent in order to verify Eq. (37). One limitation of the method is that they both assume the law of the wall to be valid over the entire depth of flow. In addition, if the vertical density profile is not known a priori, these equations for the velocity profile would have to be solved coupled with the conservation of mass equation for salt or sediment.

VII.2 Suspended sediment stratification

Coleman (1981) derived an expression for u(y) in uniform, suspended sediment stratified open channel flow using the vertical velocity profile given by Scottron (1967):

$$\frac{u}{u_*} = \left[\frac{1}{\kappa} \ln \frac{u_* y}{\nu'} + A\right] - \frac{\Delta u}{u_*} + - \frac{\Pi}{\kappa} \omega \left(\frac{y}{\delta}\right) \tag{38}$$

in which A = integration constant; δ = boundary layer thickness; $\Delta u/u_*$ = channel roughness velocity reduction function (Hinze, 1959); and $(\Pi/\kappa)\omega(y/\delta)$ = wake region velocity augmentation function (Coles, 1956), where ω is a function of y/δ, and Π = wake strength coefficient. In Eq. 38 ν' is the kinematic viscosity of a solid-liquid two phase mixture at an elevation z above the bed, given by

$$\nu' = \frac{(1 + 2.5c + 6.25c^2 + 15.62c^3)}{\rho + (\rho_s - \rho)c} \tag{39}$$

where c = sediment volume concentration; ρ = water density; and ρ_s= sediment density. Thus, the effect of suspended sediment on the boundary layer structure is explicit in Eq. (38).

The expression in square brackets in Eq. (38) is the Prandtl-von Karman law of the wall, while the last two terms have been added as a result of subsequent developments in boundary layer theory and constitute the law of the wake. Coles (1956) stated that this latter law is characterized by a velocity profile at the point of separation (or reattachment), at which the flow is locally a pure wake flow. The wake region in turbulent, fully-developed boundary layers is an indication of large-scale eddy structure which is constrained primarily by inertia rather than by viscosity (as is flow in the viscous sublayer), and therefore is similar to the flow region in a wake. Coles (1956) found the following empirical relationship for ω:

$$\omega(y/h) = 2\sin^2\left(\frac{\Pi}{2}\frac{y}{h}\right) \tag{40}$$

Evaluating Eq. (38) at $y/h = 1$ gives the expression for u_{max}, from which the velocity defect law is found to be

$$\frac{u_{max} - u}{u_*} \quad \left[\frac{1}{\kappa}\ln\frac{y}{h} + \frac{2\Pi}{\kappa}\right] - \frac{\Pi}{\kappa}\,\omega\left(\frac{y}{h}\right) \tag{41}$$

Coleman (1981) stated that this equation describes the velocity distribution for the entire flow depth except for the viscous sublayer. The terms inside the square brackets are the logarithmic part of the Prandtl-von Karman velocity defect law and an intercept term, respectively. The value of the wake strength coefficient, Π, can be determined by applying a straight line fit to the measured data points from the asymptotically logarithmic part of the profile to the high values of y/h, giving

$$\Pi = \frac{\kappa}{2}\left[\frac{u_{max} - u}{u_*}\right]_{y/h=1} \tag{42}$$

Coleman (1981) found a relationship between Π and the following form of a gross flow Richardson number (Monin and Yaglom, 1971):

$$Ri = \frac{gh(\rho_h - \rho_o)}{\rho_m u_*^2} \tag{43}$$

where ρ_h, ρ_o and ρ_m are the densities of the sediment-water mixture at the water surface, channel bed, and depth-averaged value, respectively. Thus, for the velocity defect profile given by Eq. (41) the suspended sediment distribution (and therefore the effect of sediment-induced stratification) is only implicitly accounted for through the relationship between Π and Ri. Unfortunately, Coleman (1981) was not able to verify this empirical relationship found from laboratory tests he conducted using any other data sets. In addition, the validity of applying the empirical relationship found for ω, Eq. (40), in all bounded shear flows has not been ascertained.

With these limitations in mind, to use Coleman's method for determining u(y) and u_*, given a known suspended sediment profile c(y), Eqs. (40-42) would have to be iteratively solved along with a digitized form of the relationship between Ri and Π.

Gust (1976) and Gust and Walger (1976) reported that measurements of mean velocity and suspended cohesive sediment profiles in a tidal channel in the North

Sea yielded deviations from the law of the wall in a dilute sea water/clay mineral suspension. The same result was obtained from laboratory experiments in a recirculating flume in which tidal flow of a sea water/clay mixture was simulated. Specifically, deviations from the logarithmic law of the wall were found for $10<y'$ 30, where $y' = yu_*/\nu$. Gust (1976) concluded that the law of the wall cannot be used to compute bed shear stress in clay suspension flows. Yet he did not present an alternate method for determining the velocity profile, and therefore u_*, in flows with clay suspensions.

Gust (1976) also found that the measured velocity profiles were very similar to those for drag-reducing polymer flows. Thus, he further concluded that drag reduction was occurring in the clay suspension. He hypothesized that drag reduction was at least partially caused by aggregates of clay particles which change shape under the action of shear or vorticity in a turbulent flow.

Gust and Walger (1976) noted the following differences between clear water flows (Newtonian) and clay suspension flows (non-Newtonian): 1) the viscous sublayer thickness for clay suspensions was about five times larger than that for clear water flows under the same flow conditions; 2) the values of κ obtained from the slopes of measured clay suspension velocity profiles varied between 0.3 to 0.4; 3) the velocities in the logarithmic layer were the same for the two types of flow for "comparable Reynolds numbers"; and 4) u_* values for clay suspension flows were approximately 40% less than those obtained for clear water flows.

VIII. Summary and Conclusions

Conventional steady flow equations for boundary layer velocity profiles and bed stresses can be seriously misleading when applied to sediment transport problems in estuarine flows. They have been used with considerable success in providing estimates of water levels and general flow patterns in manual calculations and two-dimensional depth-integrated numerical models. They have even been used in sediment transport calculations, but considerable skill and knowledge is required. Better descriptions are needed for detailed (high temporal and spatial resolution) analysis of dredged material placed in open water.

VIII.1 Choices

Multiple alternatives are available that will lead to an improvement in computed sediment transport. Evaluation of the choices should consider physical realism and usefulness, but also satisfy proportionality - a balance of realism, usefulness, effort required, input data required, and accuracy of the sediment transport equations in which they will be used now or in the near future.

Some choices of basic approaches that are available include the following:

a. minor or no change from present practice (steady flow equations adapted for estuarine use)

b. unsteady flow equations based on eddy viscosity that is constant over one or several layers.

c. higher order models that account for transport and production/dissipation of turbulence.

After choosing from among these or others we must then decide how to add effects of acceleration-deceleration, variable density, sediment concentration, bed interaction (porous medium flow or bed deformation) and poorly defined bed interface, if they are not already included. It may also be more appropriate to work in a stochastic rather than a deterministic form.

Some work has been done on the effect of flow through porous beds, but virtually none has been done on the effect that deformable soft sediment beds have on the flow. Bed interface definition problems occur when the transition from sediment suspension to settled bed is so gradual that ordinary definitions of what is the bed surface become inoperative. What appears to be the bed may move at a speed approaching flow velocity and may either remain in a fluid mud state, consolidate into fixed bed, or suddenly resuspend.

VIII.2 Proposed Approach

One attractive approach is to select an array of options of various level of complexity and sophistication and provide guidelines that steer the user toward the most appropriate application for a given situation. For example, the equations based on steady flow could be used in hand calculations and rapid computer applications when data are sparse and for well developed flows where density gradients are small and peak values (near strength of flood or ebb) are of interest. For cases where more rigor is required by virtue of computational need and quality of input data, microcomputer-based routines that employ numerical methods short of full 3D modeling could be used. For cases where time and resources are available, fully 3D numerical models with rigorous higher order turbulence closures could be used. All could be modified, if appropriate, for other significant contributing phenomena.

The following conclusions are drawn from this review:

a. For cases where limited resources prevent the most rigorous boundary layer descriptions, Eqs. (6-17) may be sufficient for rapid estimates of sediment transport.

b. For 3-dimensional modeling and highly detailed sediment transport calculations, turbulence transport modeling [Eqs. (18) and (23-26) or (27-36)] is most appropriate and should, depending on resources, be used.

c. To take into account salinity-induced stratification in computation bed shear (or u_*) and velocity profiles in estuarine flows, the method developed by McCutcheon (1981) is recommended. For suspended sediment stratified flows, it is recommended that the values for bed shear obtained using the method presented by Coleman (1981) be compared to those obtained using McCutcheon's method. Both methods require iterative solution of the governing equations for vertically non-uniform density profiles. However, finding of Gust (1976) regarding occurrence of dray reduction in sediment suspensions (which causes alterations in the velocity profile in particular in the near-bed zone) should be kept in mind.

The work described here was performed under the Corps of Engineers Dredging Research Program. The Office, Chief of Engineers, has given permission to publish this paper.

IX. References

ASCE, 1988, "Turbulence Modeling of Surface Water Flow and Transport": Parts I-V, Journal of Hydraulic Engineering, Vol 114, No. 9, September; American Society of Civil Engineers, New York, pp 970-1073.

Blumberg, A. F. and Kantha, L. H., 1986: "Computing Storm-Induced Currents Using Second Moment Turbulence Closure Draft chapter for ASCE publication, "Physics-Based Environmental Modeling of Estuaries".

Bowden, K. F., Fairbairn, L. A., and Hughes, P., 1959: "The Distribution of Shearing Stresses in a Tidal Current," Geophysical Journal of the Royal Astronomical Society, Vol 2, 288-305.

Bradshaw, P., Cebeci, T., and Whitelaw, J. H., "Engineering Calculation Methods for Turbulence," Academic Press, 1981.

Chapman, R. S. and Kuo, C. Y., 1982: "Application of the Two-Equation (k-e) Turbulence Model to a Two-Dimension," Proceeding of the Symposium -- Refined Modeling of Flows International Association for Hydraulic Research.

Coleman, N. L., 1981: "Velocity Profiles with Suspended Sediment," Journal of Hydraulic Research, Vol 19, No. 3.

Coles, D., 1956: "The Law of the Wake in the Turbulent Boundary Layer," Journal of Fluid Mechanics, Vol 1.

Gust, G., 1976: "Observation of Turbulent-drag Reduction in a Dilute Suspension of Clay in Sea-water," Journal of Fluid Mechanics, Vol 75, Part 1.

Gust, G. and Walger, E., 1976: "The Influence of Suspended Cohesive Sediments on Boundary-Layer Structure and Erosive Activity of Turbulent Seawater Flow," Marine Geology, Vol 22.

Hinze, J. O., 1959: "Turbulence," McGraw-Hill Book Co., New York.

Kajiura, K., 1968: "A Model of the Bottom Boundary Layer in Water Waves," Bulletin of the Earthquake Research Institute University of Tokyo, No. 46, Tokyo, Japan.

Knight, D. W., 1978: "Review of Oscillatory Boundary Layer Flow," Journal of the Hydraulics Division, ASCE, HY6 Vol. 104, June.

Ludwick, J. C., 1973: "Tidal Currents and Zig-Zag Sand Shoals in a Wide Estuary Entrance," TR No. 7, Institute of Oceanography, Old Dominion University, Norfolk, Virginia.

Markatos, N. C., 1986: "The Mathematical Modelling of Turbulent Flows," Appli. Math Modeling, Vol. 10, June.

McCutcheon, S. C., 1981: "Vertical Velocity Profiles in Stratified Flows," Journal of the Hydraulics Division, ASCE, Vol 107, No. 8.

Mehta, A. J., and Christensen, B. A., 1977: "Incipient Sediment Motion in Entrances with Shell Beds," Coastal Sediments' 77, Proceedings of the Conference, American Society of Civil Engineers, New York.

Mellor, G. L. and Yamada, T., 1982: "Development of a Turbulence Closure Model for Geophysical Fluid Problems," Reviews of Geophysics and Space Physics, Vol 20, No. 4.

Monin, A. S. and Yaglom, A. M., 1971: "Statistical Fluid Mechanics," (English translation of: Statisicheskaya gidromekhanika--Mekhhanika Turbulentnosti), M.I.T. Press, Cambridge, Massachusetts.

Rodi, W., 1982: "The Modelling of Turbulent Flows Using Transport Equation Models", Proceedings of the Symposium-IAHR: Refined Modeling of Flows. Presses de l'Ecole Natl des Ponts et chaussees, Paris.

Scottron, V. E., 1967: "Turbulent Boundary Layer Characteristics Over a Rough Surface in an Adverse Pressure Gradient," Report No. 2659, Naval Ship Research and Development center.

Sleath, J. F. A., 1984: Sea Bed Mechanics, John Wiley & Sons, New York.

Smith, J. D., and McLean, S. R., 1977: "Spatially Averaged Flow over a Wavy Surface," Journal of Geophysical Research, Vol 82, No. 12.

Soulsby, R. L. and Dyer, K. R., 1981: "The Form of the Near-Bed Velocity Profile in a Tidally Accelerating Flow," Journal of Geophysical Research, Vol 86, No. C9.

Sternberg, R. W., 1968: "Friction factors in Tidal Channels with Differing Bed Roughness," Marine Geology, Vol 6, No. 3.

Takemitsu, N., 1988: "Revised k-e Model (A Turbulence Model)," Computational Methods in Flow Analysis, Proceedings of the International Conference on

Computational Method in Flow, Okayama University of Science, 5-8 September 1988.

Tesche, T. W., 1975: "Temporal Estuarine Shear Stress Distributions," ASCE National Convention, Denver, Colorado, American Society of Civil Engineers, New York.

Morphological Processes in Shallow Tidal Seas

H.J. de Vriend

Delft Hydraulics
P.O. Box 152, 8300 AD Emmeloord
The Netherlands

ABSTRACT

The formation and the elementary behaviour (e.g. propagation, growth and decay, dispersion) of linear sand banks in shallow tidal seas are analyzed. The principal tool is a linearized mathematical model of three-dimensional topographical changes of a sandy bottom in a tidal environment.

The resulting insight is used to explain this class of sand banks, in general, and the sand bank system off the coast of Holland, The Netherlands, in particular.

I. Introduction

The large coherent sand bank systems known as "linear sand banks" are among the most striking morphological phenomena in shallow tidal seas. In spite of their regular and systematic features, there are still various essentially different explanations around (for instance, see: Pattiaratchi and Collins, 1987), none of which is generally accepted.

One of the analyses, first given by Huthnance (1982a,b), considers the inherent stability the morphodynamic system far away from lateral boundaries. This seems to explain much of the observed behaviour of linear sand bank systems far offshore, qualitatively and to some extent also quantitatively. Based on an essentially local analysis, however, this stability approach does not explain the longitudinal coherence of the banks, which can extend over many times their wave length. Making use of another technique to analyze the same system of constitutive equations, the present paper addresses this point.

Huthnance's analysis applies to shallow seas, where the water is still deep enough for the sediment transport to be determined almost exclusively by the tidal current. This is certainly not true in shallower water closer to the coast. The present paper uses the same technique, applied to a more extensive mathematical model, to analyze the behaviour (especially the different orientation; see Figure (1) of linear sand bank systems closer to the shore.

The basic element of the approach is a consistent mathematical model of the relevant morphodynamic process. In principle, such a model can be solved numerically (Lepetit and and Hauguel, 1978; McAnally _et al_., 1984). However, the real-time simulation approach taken by these models, with the tidal period as a basic time scale, limits their practical applicability to rather short time spans in terms of the morphological time scale. Besides, numerical model simulations like these hardly make sense without a thorough understanding of the phenomena and the mechanisms to be simulated.

For linear sand banks, the morphological time scale is very much larger than the tidal period (centuries, or even more). Also are the phenomena and the mechanisms still insufficiently understood. Therefore, an analysis of the inherent behaviour

Coastal and Estuarine Studies, Vol. 38
R. T. Cheng (Ed.)
Residual Currents and Long-term Transport

of the mathematical system is more appropriate, at this moment, than a numerical model simulation.

II. Basic Mathematical Model

II.1 Model concept

The mathematical model concept (Figure 2) is based on "quasi-stationarity". Within each morphological time step, the water and sediment motion is computed as if the bed topography were fixed, and subsequently the topographical changes are computed while keeping the water and sediment motion invariant to morphological changes. Thus the influence of the rate of bed level change on the wate motion is left out of consideration. Since the bed level changes per tidal cycle are very small, this rate will be very small compared with e.g. the rate of change of the water surface elevation. Hence, the quasi-stationary approach is acceptable here.

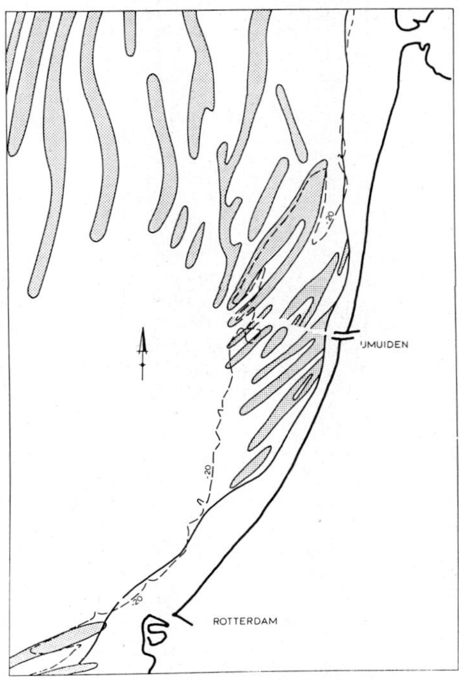

FIGURE 1. Sand bank system in front of the Holland coast. (Courtesy Rijkswaterstaat)

The constituent models in the system are all fairly standard in a general sense, whereas details are irrelevant to the present discussion. Therefore, these constituent models will only be described in words here. The full mathematical formulation, after linearization, will be given in a next section.

II.2 Wind waves

In contrast with the model underlying Huthnance's (1982a,b) stability analysis, the water motion module includes both wind waves and (tidal) currents. Wave propagation is described using the mild-slope equation for progressive waves, which includes refraction, shoaling and diffraction, but ignores reflection, non-linearity and randomness. Besides, wave-current interaction and breaking are left out of consideration here.

The latter simplifications seem to be justified, since the current in the area of interest will exhibit only small spatial gradients and depth-induced wave breaking will hardly occur at these depths (15 m and more). Using the mild-slope equation in its proposed form, however, implies the a priori exclusion of a number of wave-related morphodynamic interaction phenomena, like "infragravity wave" effects (Holman and Bowen, 1982; Boczar-Karakiewicz and Bona, 1986), bragg-scattering (Mei, 1985) and resonance due to non-linearity effects (Boczar-Karakiewicz and Davidson-Arnot, 1987). As far as locally generated wind waves (sea) are concerned, most of these phenomena will be concentrated in the nearshore, but the same effects associated with swell may also be relevant in the area of interest.

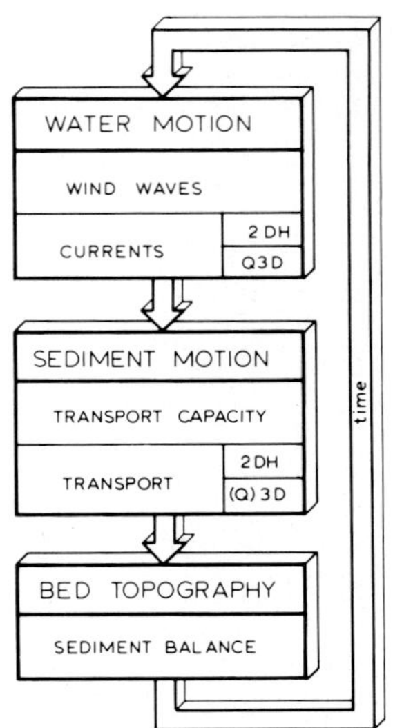

FIGURE 2. Aggregate flow chart of basic mathematical model.

The reason to exclude these wave-related interaction phenomena here is not their expected unimportance, but rather the fact, that their analysis requires quite a different basic model. Therefore, the present analysis cannot be claimed to give a full explanation of the observed behaviour of the sand bank systems. The role of wave-related interaction phenomena remains to be considered, in addition.

II.3 Currents

The current model is two-dimensional depth-integrated, with an implicitly assumed vertical distribution function. It can easily be extended to a "quasi-3D" model, including secondary circulations in the vertical plane (for instance, see: Davies, 1988). There are no wave-driven currents included.

Omitting 3D-effects, especially the geostrophic ones, clearly restricts the ability of the model to reproduce the water motion in detail. However, Huthnance's (1982a,b) analysis of linear sand banks on the open shelf (depths of 30 m and more) showed, that these banks can be explained very well without taking the Coriolis-induced vertical circulation into account. Apparently, this effect has little influence on the morphodynamic stability of the sea bed, even at these depths. Only when the sand banks have grown to a finite size, it will play a part in their deformation (also see: Pattiaratchi and Collins, 1987). Since the intensity of the Coriolis-induced vertical circulation decreases with the water depth, these findings will stand at smaller water depths.

Currents driven by breaking waves will be concentrated in the surf zone, which is outside the area of interest. Wave-induced boundary layer streaming, certainly a rectifying effect that responds to water depth variations, can be taken into account through a local relationship with the wave parameters.

Effects of stratification, though important in many shelf seas, are not taken into account. This may restrict the validity of the analysis. However, in shallow seas like the southern bight of the North Sea, with its rather strong wind and wave action, vertical homogeneity seems plausible for the major part of the time.

II.4 Sediment transport

The morphologically relevant sediment transport is assumed to concern sand, in keeping with the observed composition of the banks. Besides, the sand is assumed to be of uniform grain size, as the bank generation mechanism is not expected to be primarily related to grain sorting. This assumption suffices for the present qualitative analysis, but it may lead to problems in the quantification of the results (what is the "representative" grain size?).

The availability of sand for transportation is assumed to be unlimited, at least for the time being. Since the observed behaviour of the banks on shallower water (Figure 1) is not likely to be due to limitations in sand availability, the incorporation of such limitations seems an unnecessary complication.

Two modes of sediment transport will be considered, viz.

- transport (bed load, total load) that can be described with a local transport formula,
- suspended load transport, with the sediment concentration described by an advection/diffusion equation.

Either of these descriptions is formulated in as general terms as possible, since sediment transport models are still far from being well-established. The formulation allows for the inclusion of effects of currents (stirring, transport), waves (stirring, transport by streaming, asymmetry-transport) and gravitation.

III. Linearized Mathematical Formulation

III.1 Linearization

Various mathematical techniques can be utilized to investigate the elementary behaviour of the results to be expected from the mathematical model, without knowing the solution in every detail. Some of these techniques are linear (characteristics analysis, harmonic analysis) and some are non-linear (perturbation methods, bifurcation analysis). Here attention will be focused on linear techniques.

Since these techniques apply to systems of linear differential equations only, the mathematical model is linearized using

$$F(\underline{f}) \approx F(\underline{f}_o) + (\underline{f} - \underline{f}_o)\cdot(\underline{\nabla}F)_o \tag{1}$$

in which:

F = a (non-linear) function relationship,
$\underline{f}$ = vector of dependent variables in the model,
$\underline{f}_o$ = reference solution of the model,
$\underline{\nabla}$ = gradient vector in the $\underline{f}$-space, and
$(..)_o$ = evaluation for the reference solution.

The result is a system of linear differential equations, with unknowns $\underline{f}' \approx \underline{f} - \underline{f}_o$ and parameters $(\underline{\nabla}F)_o$ and $F(\underline{f}_o)$. Although the applicability of the linear techniques does not require a simple reference solution, a spatially uniform reference solution with a horizontal sea bed is assumed here. This means that phenomena on a systematically sloping sea bed, such as the shoreface, are not considered.

III.2 Wave module

As stated before, the starting point for the wave module is the mild-slope equation (Berkhoff, 1976). Assuming progressive waves and ignoring current refraction and energy dissipation, the linearized equations for a small perturbation (a', θ') of the uniform wave field (a_o, θ_o) read (cf. De Vriend, 1986)

$$L_2(\theta') - \frac{C_{ph}}{h_o} L_1(h') + \frac{1}{2a_o k_o^2} L_1(\nabla^2 a') = 0 \tag{2}$$

$$-L_1(\theta') + \frac{C_{gh}}{h_o} L_2(h') + \frac{2}{a_o} L_2(a') + \frac{1}{2a_o k_o^2} L_2(\nabla^2 a') = 0 \tag{3}$$

in which:

$L_1 = \sin\theta_0 \frac{\partial}{\partial x} - \cos\theta_0 \frac{\partial}{\partial y}$;

$L_2 = \cos\theta_0 \frac{\partial}{\partial x} + \sin\theta_0 \frac{\partial}{\partial y}$;

x,y = horizontal cartesian co-ordinates;
θ = direction of wave propagation (reference: X-axis);
a = wave amplitude;
k = wave number;
h = water depth;
c_p = phase celerity of the waves;
c_g = group celerity of the waves;

$C_{ph} = \left(\frac{h}{c_p}\frac{dc_p}{dh}\right)_o$ = normalized depth-dependence of c_p;

$C_{gh} = \left(\frac{h}{c_g}\frac{dc_g}{dh}\right)_o$ = normalized depth-dependence of c_g.

Note that C_{ph} and C_{gh} are parameters of the type $(\underline{\nabla}F)_o$ in Eq.(1).

III.3 Flow module

The mean flow is described with the shallow water equations. As the perturbations of interest have rather large wave lengths, the horizontal dispersion terms are disregarded. Wind wave effects are restricted to the magnitude and the direction of the bottom shear stress:

$$\tau_{bx} = \tau_b(U,h,a,\theta-\alpha) \cos\{\alpha + \phi(U,h,a,\theta-\alpha)\};\ \gamma_{by} = \gamma_b \sin(\alpha + \phi) \tag{4}$$

in which U is the magnitude of the depth-averaged mean flow velocity and α its direction. Linearized for small perturbations of the unsteady uniform flow (u_o, v_o, h_o), the flow equations read

$$\frac{\partial u'}{\partial t} + u_o\frac{\partial u'}{\partial x} + v_o\frac{\partial u'}{\partial y} - f_c v' = -g\frac{\partial z'_s}{\partial x} - R'_x \tag{5}$$

$$\frac{\partial v'}{\partial t} + u_o\frac{\partial v'}{\partial x} + v_o\frac{\partial v'}{\partial y} + f_c u' = -g\frac{\partial z'_s}{\partial y} - R'_y \tag{6}$$

$$\frac{\partial h'}{\partial t} + u_o\frac{\partial h'}{\partial x} + v_o\frac{\partial h'}{\partial y} + h_o\frac{\partial u'}{\partial x} + h_o\frac{\partial v'}{\partial y} = 0 \tag{7}$$

in which:

t = time,
u,v = depth-averaged velocity components,
g = acceleration due to gravity,
f_c = Coriolis factor,
z_s = water level,
R_x, R_y = resistance terms, which can be written as

$$R'_x = \frac{\tau_{b_o}\cos(\alpha_o+\phi_o)}{\rho_m h_o}\left[T_u\left(\frac{u'u_o}{U_o^2} + \frac{v'v_o}{U_o^2}\right) + (T_h - 1)\frac{h'}{h_o} + T_a\frac{a'}{a_o} + T_o(\theta'-\alpha')\right]$$

$$- \frac{\tau_{b_o}\sin(\alpha_o+\phi_o)}{\rho_m h_o}\left[\Phi_u\left(\frac{u'u_o}{U_o^2} + \frac{v'v_o}{U_o^2}\right) + \Phi_h\frac{h'}{h h_o} + \Phi_a\frac{a'}{a a_o} + \Phi_\theta(\theta'-\alpha') + \alpha'\right] \tag{8}$$

and a similar expression for R'_y. Here ρ_m denotes the mass density of the fluid and the normalized dependence factors are defined by

$$T_u = \frac{U}{\tau_b}\frac{\partial\tau_b}{\partial U} \;;\; T_h = \frac{h}{\tau_b}\frac{\partial\tau_b}{\partial h} \;;\; T_a = \frac{a}{\tau_b}\frac{\partial\tau_b}{\partial a} \;;\; T_\theta = \frac{1}{\tau_b}\frac{\partial\tau_b}{\partial(\theta-\alpha)} \;; \tag{9}$$

$$\Phi_u = U\frac{\partial\phi}{\partial U} \;;\; \Phi_h = h\frac{\partial\phi}{\partial h} \;;\; \Phi_a = a\frac{\partial\phi}{\partial a} \;;\; \Phi_\theta = \frac{\partial\tau_b}{\partial(\theta-\alpha)} \;. \tag{10}$$

These factors, of the type $(\underline{\nabla}F)_o$, again, are invariant to small perturbations of (U, h, a, θ-α).

III.4 Sediment transport module

The bed-load (or total-load) transport mode is assumed to be described by a formula of the type

$$S_{bx} = S_b[\cos(\alpha+\psi) - \beta\frac{\partial z_b}{\partial x}] \;,$$
$$S_{by} = S_b[\sin(\alpha+\psi) - \beta\frac{\partial z_b}{\partial y}] \;, \tag{11}$$

in which:
S_{bx}, S_{by} = volumetric transport components,
$S_b(U,h,a,\theta-\alpha)$ = transport rate on a horizontal bed,
$\psi(U,h,a,\theta-\alpha)$ = transport deviation angle on hor. bed,
β = a constant,
z_b = bottom level.

The β-terms in Eq.(11) represent the adjustment for gravitational effects on a sloping bottom (cf. Bagnold, 1956; Huthnance, 1982b, Deigaard, 1983; also see De Vriend, 1986). This formulation of the downslope gravitational transport, which is fairly crucial in the description of the amplitude behaviour of morphological features (De Vriend, 1986), slightly differs from the one applied successfully in river bend morphology (Struiksma et al.,1985). It is still a point of discussion whether the downslope effect is basically isotropic, as reflected by Eq.(11), or anisotropic and strongest if the flow is directed exactly down- or up-slope, as suggested by Struiksma et al.

In the linearized system for small perturbations to the uniform transport field (S_{x_o}, S_{y_o}), the transport perturbations become

$$S'_{bx} = S_{bo}\cos(\alpha_o+\psi_o)[S_{bu}\frac{u_o u' + v_o v'}{U_o^2} + S_{bh}\frac{h'}{h_o} + S_{ba}\frac{a'}{a_o} + S_{b\theta}(\theta'-\alpha')]$$
$$-S_{bo}\sin(\alpha_o+\psi_o)[T_u\frac{u_o u' + v_o v'}{U_o^2} + T_h\frac{h'}{h_o} + T_a\frac{a'}{a_o}$$
$$- T_\theta(\theta'-\alpha') + \alpha'] - S_{bo}\,\beta\frac{\partial z'_b}{\partial x} \;. \tag{12}$$

and a similar expression for S'_{by}. The normalized dependence factors are defined in a similar way as in Eqs.(9) and (10), respectively. These factors are assumed invariant to small perturbations of (U, h, a, θ-α), again.

The suspended sediment concentration in the second transport mode is assumed to be described by a depth-integrated equation of the form (see Galappatti and Vreugdenhil, 1985; Wang and Ribberink, 1986)

$$\gamma_t h\frac{\partial c}{\partial t} + \gamma_s h(u\frac{\partial c}{\partial x} + v\frac{\partial c}{\partial y}) = -(c-c_e) \tag{13}$$

in which:
c = volumetric depth-averaged mean concentration,
$\gamma_t h$ = relaxation time,
$\gamma_s Uh$ = relaxation length,
c_e = equilibrium concentration, given by

$$c_e = \frac{S_{se}(U,\ h,\ a,\ \theta-\alpha)}{f_s Uh} \;. \tag{14}$$

Here S_{se} denotes the suspended-load transport rate under equilibrium conditions, given by a local transport relationship. The factor f_s is a constant representing the depth-averaged product of the concentration and velocity profile functions.

In the linearized system for small perturbations to the uniform concentration field (c_o, c_{eo}), and with invariant γ_t and γ_s, Eq.(13) becomes

$$\gamma_t(h'\frac{\partial c_o}{\partial t} + h_o\frac{\partial c'}{\partial t}) + \gamma_s(u_o h_o\frac{\partial c'}{\partial x} + v_o h_o\frac{\partial c'}{\partial y}) = -(c'-c'_e) \quad , \tag{15}$$

$$c'_e = c_{eo}[(S_{su}-1)\frac{u_o u' + v_o v'}{U_o^2} + (S_{sh}-1)\frac{h'}{h_o} + S_{sa}\frac{a'}{a_o} + S_{s\theta}(\theta'-a')] \quad . \tag{16}$$

The normalized dependence factors are defined in a similar way as in Eq.(9), are assumed invariant to perturbations of U, h, a and θ-α.

III.5 Bottom level changes

In principle, the bottom level changes follow from the conservation of sediment mass. In terms of perturbations

$$(1 - \epsilon_p)\frac{\partial z'_b}{\partial t} + \frac{\partial S'_{bx}}{\partial x} + \frac{\partial S'_{by}}{\partial y} + \frac{\partial S'_{sx}}{\partial x} + \frac{\partial S'_{sy}}{\partial y} + \frac{\partial(h_o c')}{\partial t} + \frac{\partial(h' c_o)}{\partial t} = 0 \quad , \tag{17}$$

in which ϵ_p denotes the porosity of the bottom and $S_s = f_s Uhc$. For the moment, the two transport modes are considered separately, which means that either the S_b-terms or the S_s-terms are retained.

In the S_b-case, substitution of Eq.(12) and its S_{by}-equivalent into Eq.(17) yields, with $\alpha_o \approx \text{atan}(v_o/u_o)$ and the equation of continuity Eq.(7),

$$\frac{\partial z'}{\partial t} + C_x\frac{\partial z'_b}{\partial x} + C_y\frac{\partial z'_b}{\partial y} - D(\frac{\partial^2 z'_b}{\partial x^2} + \frac{\partial^2 z'_b}{\partial y^2})$$

$$= P_1(u') + P_2(v') + P_3(z'_s) + P_4(a') + P_5(\theta') \tag{18}$$

with:

$$C_x = \frac{S_{bo}}{(1-\epsilon_p)h_o}[(S_{bu}\cos\psi_o - T_u\sin\psi_o)\frac{u_o}{U_o} - S_{bh}\cos(\psi_o + \alpha_o) + T_h\sin(\psi_o + \alpha_o)], \tag{19}$$

$$C_y = \frac{S_{bo}}{(1-\epsilon_p)h_o}[(S_{bu}\cos\psi_o - T_u\sin\psi_o)\frac{v_o}{U_o} - S_{bh}\sin(\psi_o + \alpha_o) - T_h\cos(\psi_o + \alpha_o)], \tag{20}$$

$$D = \beta S_{bo} \quad . \tag{21}$$

P_1 through P_5 denote complicated linear first-order differential operators, which will not be specified here (see De Vriend, 1986, for an example).

Eq.(18) could be considered as an advection/diffusion equation for z'_b, with the P-terms as source terms. This option may be of use in devising tide-integrated simulation models. In the following analysis, the system, Eqs.(2), (3), (5), (6), (7) and (18) will be considered in its full interaction.

In the suspended-load case, the sediment balance Eq.(17) can be elaborated using Eqs.(15) and (7), to yield

$$(1 - \epsilon_p)\frac{\partial z'_b}{\partial t} - \frac{f_s}{\gamma_s}(c' - c'_e) - (\frac{f_s \gamma_t}{\gamma_s} - 1)(h'\frac{\partial c_o}{\partial t} + h_o\frac{\partial c'}{\partial t}) = 0 \tag{22}$$

with c'_e as given by Eq.(16). In this case, the interaction of the system Eqs.(2), (3), (5), (6), (7), (15) and (22) will be considered.

IV. Linear Analysis Techniques

Two techniques of linear analysis are utilized to investigate the transient behaviour of the present model: the characteristics analysis and the harmonic analysis. The role of these two methods in morphological modelling, together with that of the harmonic analysis of the equilibrium bottom configuration, was discussed in an earlier publication by the author (1987a).

The characteristics analysis, described in further detail by Lin and Shen (1984) and by De Vriend (1987b), provides information on the mathematical character of the system. If this is hyperbolic, it yields the celerity of bottom disturbances. It shows, for instance, how information travels through the area of interest. This helps to identify where in a mathematical model boundary conditions are needed, but also to understand the elementary behaviour of the solution. It is this insight that will be used to explain the longitudinal coherence of linear sand banks.

The harmonic analysis describes the transient behaviour of harmonic perturbations to the basic solution. It proves very useful to determine the growth/decay rate and the celerity of each harmonic mode. Starting from the assumption that only the fastest growing mode will survive, the technique can be used to analyse large-scale harmonic features of the sea bed, such as linear sand banks (Huthnance, 1982a; Deigaard, 1983; De Vriend, 1988).

The two methods, which exhibit strong parallels in their elaboration, turn out to be more or less complementary. Both use algebraic elaboration, which provides the possibility to trace the physical mechanisms underlying specific features of the morphological interaction process.

V. Characteristics Analysis: Results and Discussion

The principal results of applying the characteristics analysis to a morphological system with a local transport formula as a sediment transport model have been given by De Vriend (1987b). All results of the characteristics analysis discussed hereafter were described in that paper, together with the underlying mathematics. Therefore, the method and its application will not be described here, again.

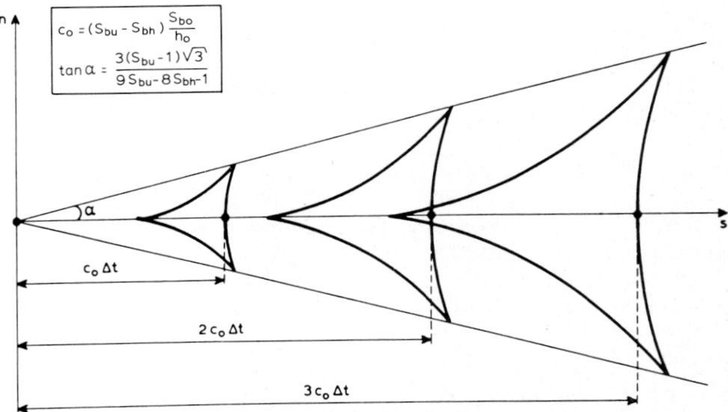

FIGURE 3. Evolution of a point disturbance in the bottom level, under steady flow with bed load transport. The s-direction coincides with the transport direction. (From: De Vriend, 1987a)

A typical result is shown in Figure 3, illustrating that the interaction between the water and sediment motion, on the one hand, and the bed topography changes, on the other, are essentially two-dimensional in the horizontal plane, rather than basically one-dimensional in the transport direction. This result was derived for a specific class of models, but an analysis of the interaction mechanism, outlined in Figure 4 shows that the two-dimensionality is primarily due to the two-dimensional response of the flow to a local bottom perturbation. Hence a two-dimensional morphological interaction will always be found if the transport is current-driven, irrespective of how the sediment is stirred.

In essence, the elements in Figure 4 are the same as in the analysis of the evolution of a small bump on a horizontal sea bed, given by Huthnance (1982b). There, too, the two-dimensional response of the flow is the crucial element.

Although Huthnance considers only initial scour and deposition, he also finds the typical lobed deformation pattern that is found from the characteristics analysis (see Figure 5).

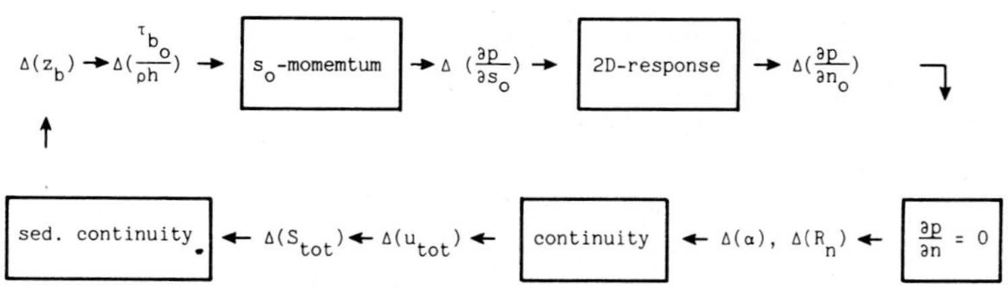

FIGURE 4. Morphological interaction process. (From: De Vriend, 1987b)

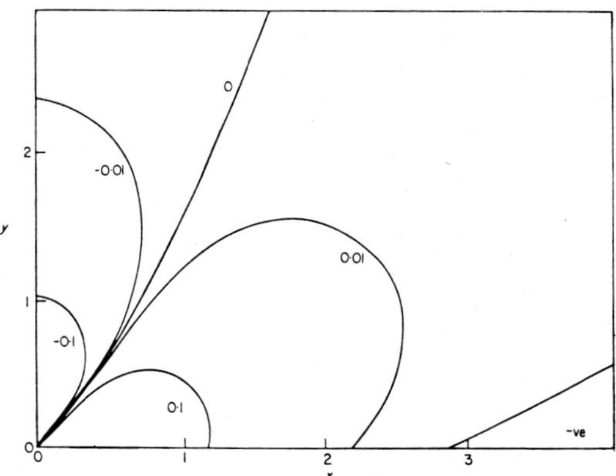

FIGURE 5. Contours of the scour function corresponding to a unit point-depression in the origin. Undisturbed flow in x-direction, scour function symmetric in x and y. (From: Huthnance, 1982b; courtesy Academic Press, London)

[Note that Figures 3 and 5, though indicating similar patterns, are not comparable. Figure 3 pertains to the dynamic morphological interaction, so the transient process, whereas Figure 5 concerns the initial situation].

When applied to a schematic tidal situation, the results of the characteristics analysis can explain the tendency of bottom perturbations to stretch to long sand banks under some 20° (cyclonic) with the flow direction (De Vriend, 1987b). This is illustrated in Figure 6, for a sinusoidal bump on a horizontal bottom. For simplicity and without loss of generality, the flow dynamics are reduced to the assumption of zero potential vorticity. Due to the coriolos effect, the flow exhibits a slight preference to pass the bump at the left side. This introduces a slight asymmetry in the development of the lobes at either side, and hence a stretching and a cyclonic rotation of some 20 to 30°. This corresponds well with observations in nature (see Pattiaratchi and Collins, 1987, for an extensive review).

Huthnance (1982b) describes a similar numerical experiment, starting from a somewhat different initial bump (an exponential ellipsoid) and extended over a larger time span (up to more than 10 times the morphological time scale; the 600 days in Figure 6 are less than once this time scale). If there is sufficient sand available, these computations indicate the same tendencies as observed in Figure 6, plus a distinct tendency of the bump to grow in volume and height.

Apparently, the mechanism outlined in Figure 4, in combination with the Coriolis-effect on the depth-averaged current, explains the onset and the longitudinal coherence of linear sand bank systems in shelf seas.

The topography-induced vertical circulation, with the near-bed current converging towards the crest, plays no part in this explanation. It is expected to have a reshaping effect on the banks (McCave, 1979; Caston, 1972), rather than essentially determining their onset or nature.

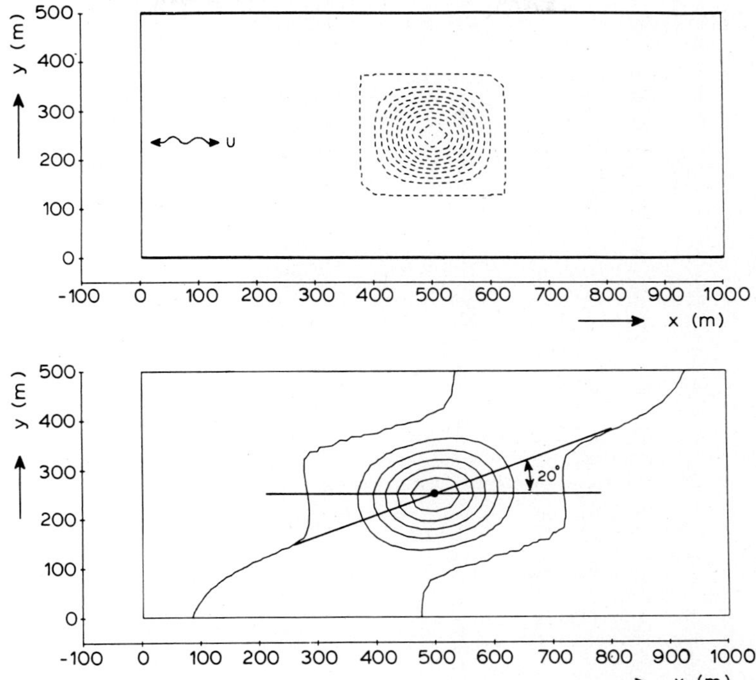

FIGURE 6. Evolution of a sinusoidal hump on a horizontal bottom in tidal flow. (From: De Vriend, 1987b)

VI. Harmonic Analysis (Currents only)

VI.1 Objectives

Extending Huthnance's (1982a) pioneering work on the formation of linear sand banks at deeper water, a harmonic analysis was carried out for sand bank systems under slightly different conditions: more complex tidal flows (free surface, open tidal ellipse), different transport modes (bed load, suspended load). This gives an impression of the model's sensitivity to its components and its input conditions.

Part of this work has been reported before (De Vriend, 1988). In order to have a complete reference for a similar analysis including wave effects, a brief summary of the result of this work is included in the discussion hereafter. Before going into this discussion, the procedure of the harmonic analysis in the case of unsteady flow is summarized.

VI.2 Procedure

Morphological evolutions in tidal areas involve at least two time scales, viz. the tidal period and the morphological time scale, with the latter usually much larger than the former. If suspended-load transport is involved, another time scale is introduced, viz. the relaxation time of the sediment concentration, for sand usually small compared with the tidal period.

If the morphological time scale is much larger than the tidal period, a quasi-stationary approach can be followed, meaning that:

- the flow can be computed as if the bottom topography were fixed, and

- the frequency of the water motion is so much higher than that of the morphological changes, that the following approximation holds well:

$$\frac{1}{T}\int_o^T f_1(\omega_1 t)f_2(\omega_2 t)dt \approx f_2(\omega_2 T)\frac{1}{T}\int_o^T f_1(\omega_1 t)dt \quad \text{with: } \omega_2 \ll \omega_1 \quad , \tag{23}$$

in which ω_1 denotes the frequency of the water motion, ω_2 is the frequency of the morphological changes, and the integration period T is long enough to average over the tide, but still much smaller than the morphological time-scale.

Since the system to be analysed is linear, every Fourier component of the perturbations can be considered separately. Hence it is sufficient to introduce bottom perturbations of the type

$$z_b'(x,y,t) = \hat{z}_b'\exp\{i(\rho t + k_x x + k_y y)\} \tag{24}$$

in which ρ denotes a complex frequency and (k_x, k_y) are real wave number components. For mathematical reasons, it is convenient to choose the x-axis parallel to the bottom wave crest, so that $k_x = 0$ and $k_y = k$, k denoting the total wave number, and

$$z_b'(x,y,t) = \hat{z}_b'\exp\{i(\rho t + ky)\} \tag{25}$$

Note that, in accordance with the assumption of quasi-stationarity, z_b' varies with t at the slow (morphological) time scale only. In order to allow for fast (tidal) variations of the velocity perturbations, they are written in the form

$$u'(x,y,t) = \hat{u}'(t)\ \exp\{iky\} \quad \text{with} \quad u'(t)\ \text{complex}. \tag{26}$$

Since the velocity perturbations are due to the slowly varying z_b', the fast variation of u', say in a tidal cycle around $t = t_o$, can be scaled with $z_b'(t_o)$. This is easily shown by substituting Eq.(25) and (26) into the flow perturbation Eqs.(5) through (7).

All fast varying perturbations, also for the waves and the sediment transport, can be written in the form, Eq.(26), and can be scaled with $\hat{z}_b'(t_o)$. Hence the latter factor can be eliminated from the equations for the fast varying quantities. So if $\tilde{.}$ denotes the time-dependent amplitude function $\hat{.}'(t)$ scaled by $\hat{z}_b'(t_o)$, the system describing ($\tilde{a}$, $\tilde{\theta}$, $\tilde{u}$, $\tilde{v}$, $\tilde{z}_s$, and possibly $\tilde{c}$ and $\tilde{c}_e$) is independent of $z_b'(t_o)$.

For tidal flow without waves and with a Chezy-type bottom shear stress in the flow direction, for example, the system reads

$$\frac{\partial\tilde{u}}{\partial t} + [ikv_o+(1+\frac{u_o^2}{U_o^2})\frac{\tau_{bo}}{\rho_m U}]\tilde{u} + [-f_c+\frac{u_o v_o \tau_{bo}}{U_o^2\ \rho_m U_o}]\tilde{v} - \frac{u_o\tau_{bo}}{U_o\rho_m h_o}\tilde{z}_s = -\frac{u_o\tau_{bo}}{U_o\rho_m h_o} \quad , \tag{27}$$

$$\frac{\partial\tilde{v}}{\partial t} + [f_c+\frac{u_o v_o}{U_o^2}\frac{\tau_{bo}}{\rho_m U_o}]\tilde{u} + [ikv_o+(1+\frac{v_o^2}{U_o^2})\frac{\tau_{bo}}{\rho_m U_o}]\tilde{v} + [ikg-\frac{v_o\tau_{bo}}{U_o\rho_m h_o}]\tilde{z}_s = -\frac{v_o\tau_{bo}}{U_o\rho_m h_o} \tag{28}$$

$$\frac{\partial\tilde{z}_s}{\partial t} + ikh_o\tilde{v} + ikv_o\tilde{z}_s = ikv_o \quad . \tag{29}$$

This system, or a more extensive form including suspended load transport, can be solved separately from the bottom evolution. Once the perturbations of the time-varying water motion are known, the sediment transport perturbation is determined from the y-equivalent of Eq.(12) in the case of bed load, or from Eqs.(15) and (16) in the case of suspended load. The result is substituted into the sediment balance equation and averaged over the tide, making use of Eq.(23). This yields an equation of the form

$$\frac{\partial\hat{z}_b'}{\partial t} - i\rho\hat{z}_b' = 0 \tag{30}$$

in which ρ is the complex amplification factor, again.

If, for example, the transport rate is a function of U_o only, and the transport is directed along with the flow (except for the gravitational effect), then the expression for ρ reads

$$\rho = \frac{1}{1-\epsilon_p}[i\beta k^2 \overline{S_{bo}} - kS_{bu} \overline{\frac{S_{bo}\tilde{v}}{U_o}} - k(S_{bu}-1)\{\overline{\left(\frac{u_o v_o S_{bo}\tilde{u}}{U_o^3}\right)} - \overline{\left(\frac{u_o^2 S_{bo}\tilde{v}}{U_o^3}\right)}\}] \tag{31}$$

in which the overbar denotes time-averaging. Note that $\tilde{u}$ and $\tilde{v}$ can be complex, so that the second and third terms can contribute to $Im(\rho)$.

Starting from these equations, the following procedure is used in the analysis:

(i) specify, in a given cartesian coordinate system, the undisturbed water motion as a function of (fast) time,

(ii) specify the wave number, k, and the crest orientation, α_o, of the harmonic bottom perturbation,

(iii) rotate the coordinate system, such that the x-axis is parallel to the bottom wave crests,

(iv) solve for the perturbation of the water motion, $(\sim, \sim, \sim_s)$, from Eqs.(27) through (29) or a simplified subset,

(v,a) in the case of bed load transport: determine the perturbation of the sediment transport rate from the y-equivalent of Eq.(12), or

(v,b) in the case of suspended load transport: solve for the perturbation of the sediment concentration from Eqs.(15) and (16),

(vi) determine the complex amplification factor from Eq.(31) or its suspended load equivalent, and

(vii) translate the complex amplification factor into physical quantities:

- wave length [L] : $\lambda_{bed} = \frac{2\pi}{k}$ (32)
- celerity $[T^{-1}]$: $c_{bed} = -\frac{Re(\rho)}{k}$ (33)
- wave period [T] : $T_{bed} = -2\pi Re^{-1}(\rho)$ (34)
- growth rate $[T^{-1}]$: $A_{bed} = -Im(\rho)$ (35)

Figure 7 illustrates how the results of step (vi) will be presented. To every point a in the $\vec{k}$-plane, a value ρ can be attributed. The distributions of $Re(\rho)$ and $Im(\rho)$ over the $\vec{k}$-plane will be given in the form of contour plots.

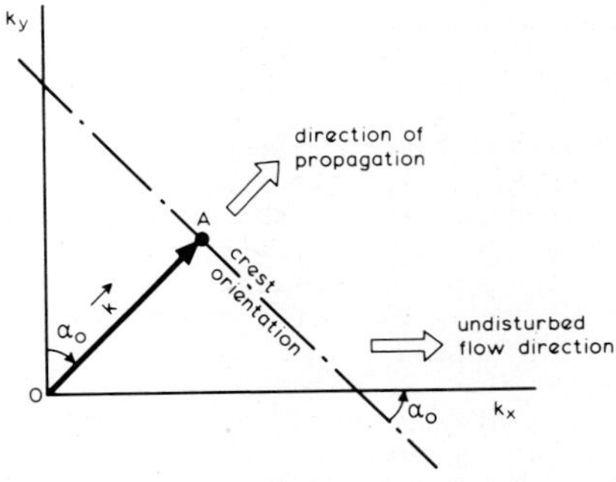

FIGURE 7. Presentation of results of the harmonic analysis.

VI.3 Reference case

In order to set a reference for further sensitivity analyses, the procedure was applied to a case similar to that considered by Huthnance (1982a). It concerns situations where a rigid-lid approximation can be used for the water surface, where the flow is purely oscillatory (i.e. the tidal ellipse is a single line), where the role of the Coriolis-effect can be disregarded and where the Chzy-formula can be used to describe the bottom shear stress. The sediment transport is assumed to be bed load, proportional to U^5 and directed along with the flow, and the gravitational parameter β is taken 1% of the inverse friction factor C^2/g.

Figure 8 shows the results of the steady-flow equivalent of this analysis. It gives the real and imaginary parts of the complex amplification factor as functions of the magnitude and the direction of the bottom wave number vector. The amplitude factor is scaled by

$$\rho = \frac{S_{bo}}{(1 - \epsilon_p)h_o\lambda_w}\,\zeta \quad \text{with} \quad \lambda_w = \frac{C^2}{2g}\,h_o \tag{36}$$

and the wave number components in the original coordinate system are given by

$$\kappa_x = \kappa\lambda_w \sin\alpha_o \quad \text{and} \quad \kappa_y = \kappa\lambda_w \cos\alpha_o \tag{37}$$

In that case, the only remaining parameter in the system is the ratio of the flow relaxation length λ_w and the tidal excursion scale. This ratio is assumed to be small enough to disregard its influence (cf. De Vriend, 1988). Figure 8 is then valid for any combination of tidal velocity amplitude, mean water depth, bottom roughness and sediment transport rate that satisfies the aforementioned conditions.

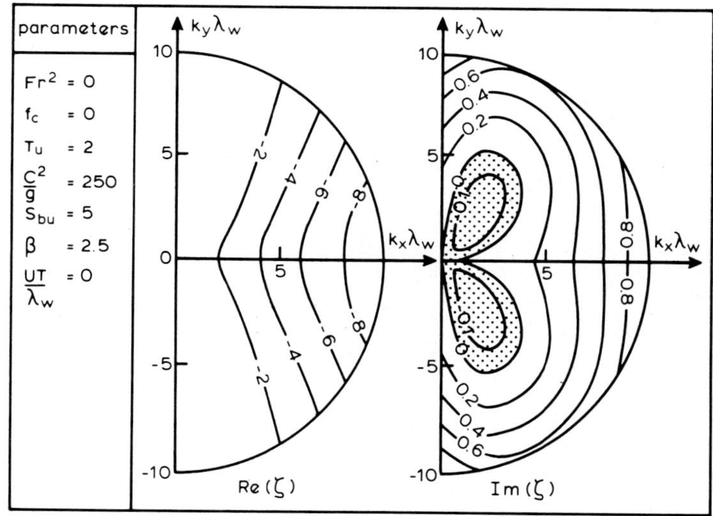

FIGURE 8. Amplification factor for the reference case.

Since there is no residual current, $Re(\zeta)$ is identically equal to zero, which means that the bottom waves are not propagating. The picture of $Im(\zeta)$ shows that certain bottom wave modes tend to grow, with the maximum growth rate at $\kappa_x \approx 1.6$ and $\kappa_y \approx 2$. This means that bottom waves with their crest under some 38° with the undisturbed flow direction will grow the fastest. For $S_o = 10^{-5}$ m²/s, $\epsilon_p = 0.5$, $C = 50$ $m^{1/2}/s$ and $h_o = 10$ m, the doubling period of the amplitude is about 22 years.

The result shown in Figure 8 agree qualitatively with those obtained by Huthnance (1982a), but quantitatively there are differences due to a different parameter setting (Huthnance: transport ~ U^3; $\beta g/C^2 = 0.005$). Since the present case is meant as a reference for a sensitivity analysis, these quantitative differences are considered immaterial.

VI.4 Sensitivity to the flow model.

The flow model is based on a variety of simplifying assumptions, the influence of which should be assessed in a sensitivity analysis. Although the present system provides the framework for such an analysis, the work remains to be done.

In view of earlier discussions in the literature (for a review, see: Pattiaratchi and Collins, 1987), the role of the Ekman veering of the velocity near the bottom deserves special attention. The argumentation in Section II.3 is rather heuristic and requires further substantiation. This can be done within the present system by introducing an Ekman-model for the near-bottom velocity and the bottom shear stress. These quantities, rather than the depth-averaged velocity, should then be used as an input into the sediment transport model.

Also the sensitivity of the results to the parameter setting in the flow model should be assessed. So far, this has only been done by switching physical phenomena on and off:

- Huthnance (1982a) showed that the Coriolis effect on the depth-averaged flow upsets the symmetry between cyclonic and anti-cyclonic orientation of the fastest growing banks with respect to the flow direction: the cyclonic ones tend to grow faster.
- De Vriend (1988) showed for steady flow that the free surface effects tend to reduce the angle of rotation of the critical banks with respect to the flow direction; this requires a rather high Froude number in order to be perceptible.

An interesting point of further study, especially for bank systems in shallow water, is the effect of the tidal water level variations. Even if they play a minor role in the mass conservation, they will be felt in the flow resistance, an important factor in shallow water cases.

VI.5 Sensitivity to the undisturbed flow conditions.

The reference case concerns purely oscillatory undisturbed flow, without residual currents and with the tidal ellipse reduced to a single line. In the context of linearization, tidal flow with a residual current can be constructed by adding a steady current to purely oscillatory flow.

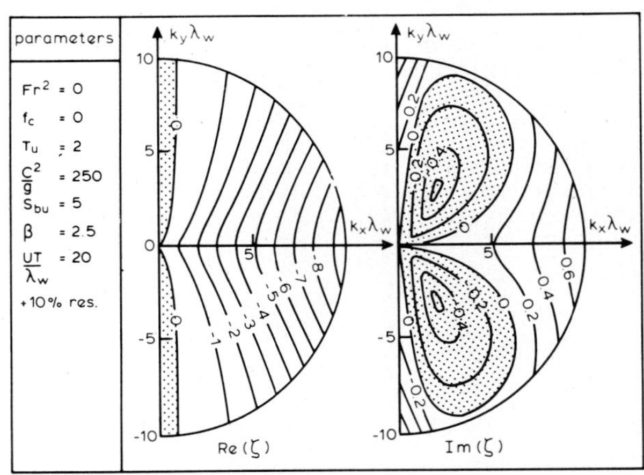

FIGURE 9. Amplification factor for tidal flow with a 10% residual current

Figure 9 shows the effect of adding a steady current of $0.1\ \hat{U}_o$ to an oscillatory flow of amplitude $\hat{U}_o$. Comparing Figures 8 and 9 leads to the conclusion that the imaginary part of the complex amplification factor is hardly affected, i.e. the wave length, crest direction and growth rate of the critical mode remain virtually the same. The real part of ζ is affected much more: the residual current makes the bottom wave propagate.

The occurrence of tidal sand banks (ridges) in front of the coast of Holland, The Netherlands (see Figure 1), seems to coincide with an area of almost purely oscillatory tidal flow. At other parts of this coast, further south and north, the tidal ellipses are more open and ridges are absent. The procedure described in the foregoing provides the possibility to check whether this correlation is substantiated by the present system.

The most extreme form of an open tidal ellipse is a circle, represented by

$$U_o = \hat{U}_o \cos(2\pi \frac{t}{T}) \quad \text{and} \quad V_o = \hat{U}_o \sin(2\pi \frac{t}{T}) \quad . \tag{38}$$

In that case, ζ is independent of the crest angle, which means that it is a function of κ alone. Figure 10 shows this function, for the same parameter setting as in Figures 8 and 9. Apparently, the growth rate and the wave length of the critical mode are not much different from what was found for a completely flat ellipse.

This means that the only effect of an open tidal ellipse, as compared with a flat one, is a weaker preference of the critical mode for a certain direction. So, in contrast with earlier suggestions (De Vriend, 1988), the system underlying Figures 8, 9, and 10 gives no explanation for the absence of ridges in the northern and southern parts of the Holland coast.

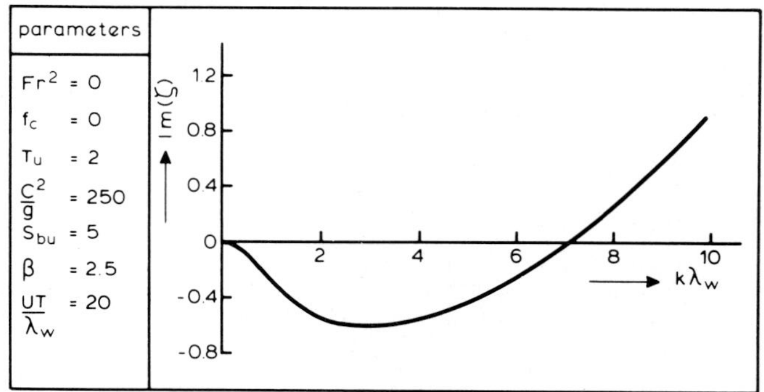

FIGURE 10. Amplification factor in case of a circular tidal ellipse (bed load transport)

VI.6 Sensitivity to the type of sediment transport

The influence of relaxation effects in the case of suspended load transport can be investigated by replacing the bed load formula by the following system, derived from Eqs.(15) and (16) with $h' = 0$ (rigid lid, negligible bottom level variations):

$$\tau_A \frac{\partial \tilde{c}}{\partial t} + [ik\lambda_A \frac{V_o}{U_o} + 1]\tilde{c} = \tilde{c}_e \quad , \tag{39}$$

$$\tilde{c}_e = \frac{S_{so}}{f_s U_o h_o}(S_{su}-1)\frac{u_o\tilde{c} + v_o\tilde{v}}{U_o^2} \quad , \tag{40}$$

in which:

τ_A = relaxation time (assumed constant),

λ_A = relaxation length (assumed constant),

$\tilde{c}$ = amplitude, scaled by $\tilde{z}_b'(t_o)$, of c',

$\tilde{c}_e$ = amplitude, scaled by $\tilde{z}_b'(t_o)$, of c_e',

S_{so} = suspended load transport capacity for steady flow with velocity U_o and water depth h_o.

Substituted into the sediment balance equation, this yields an equation of the form (30), but now with the amplification factor

$$\rho = \frac{1}{1-\epsilon_p}\,\frac{1}{\gamma t}\left[(\gamma_s - f_s\gamma_t)k\,\overline{v_o h_o \tilde{c}} - i\,\overline{(\tilde{c} - \tilde{c}_e)}\right] \quad . \tag{41}$$

This can be scaled by

$$\rho = \frac{S_{so}}{(1-\epsilon_p)\, f_s U_o h_o \tau_A}\,\zeta \quad . \tag{42}$$

As compared with the bed load case, this system contains two more parameters, viz. the relaxation length ratio λ_A/λ_w, and the ratio of what could be called the "concentration excursion" and λ_w:

$$\sigma \approx \frac{f_s\, U_o\, \tau_A}{\lambda_w} \tag{43}$$

Figure 11 shows the results of the analysis for the suspended-load variant of the reference case, with $\lambda_A/\lambda_w = 2$, $\tau_A/T = 0.1$ and $\sigma = 1$. Now the critical mode has a crest orientation much closer to the flow direction (deviation less than 10°), and its growth rate is much smaller (about a factor 5) than in the bed load case. Apparently, the relaxation of the suspended sediment concentration has a smoothing effect on the bottom.

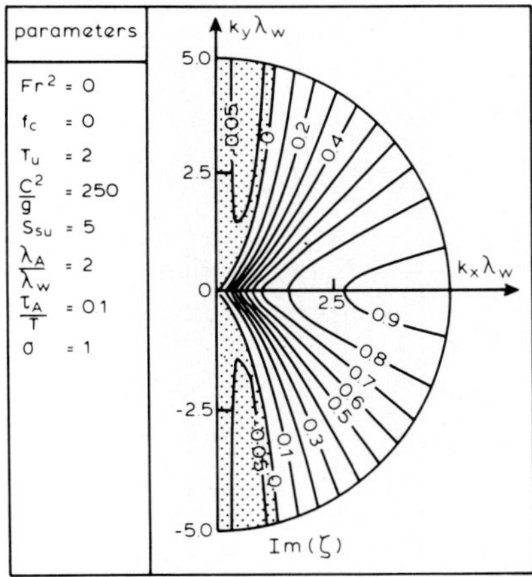

Figure 11. Amplification factor in case of tidal flow with suspended load transport

This becomes even more manifest in the case of an open tidal ellipse, as appears from Figure 12 for the purely circular path, Eq.(38). Figure (13) shows the corresponding result for the lemmiscate-type path

$$u_o = X\cos\alpha_o - Y\sin\alpha_o \;;\; v_o = X\sin\alpha_o + Y\cos\alpha_o \quad ; \tag{44}$$

$$X = \hat{U}_o \cos(2\pi \tfrac{t}{T}) \quad ; \; Y = \hat{U}_o \tan\phi_o \cos(4\pi \tfrac{t}{T}) \quad ;$$

in which ϕ_o was taken 30°. In the former case, none of the modes tends to grow, and also in the latter case growing modes have virtually vanished. Apparently, the combination of an open tidal ellipse and suspended load transport is very effective in suppressing tidal ridges. This may explain the absence of ridges at the northern and southern parts of the Holland coast.

VI.7 Sensitivity to additional transport rates.

So far, the transport direction was assumed to coincide with the flow direction, i.e. the angle ψ in Eq.(11) was taken equal to zero. In practice, this is not always true. Not only do the banks give rise to residual Coriolis-induced vertical circulations, also in the undisturbed situation the transport vector can deviate systematically from the current direction. Possible causes are upwelling effects due to wind or horizontal density gradients, wave-induced streaming (Roelvink and Stive, 1989), and wave-asymmetry effects (ibid.).

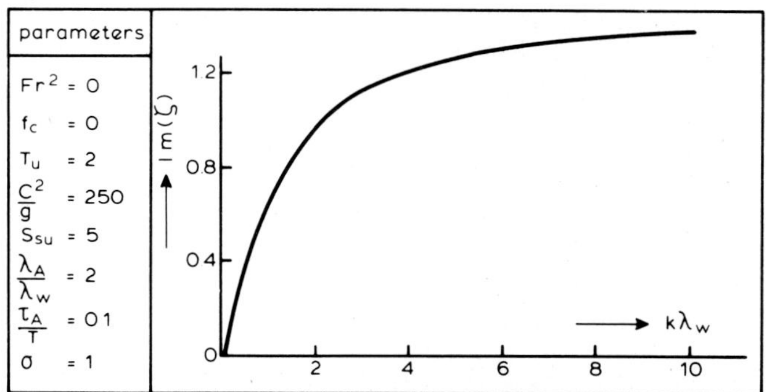

FIGURE 12. Amplification factor in case of a circular tidal ellipse (suspended load transport)

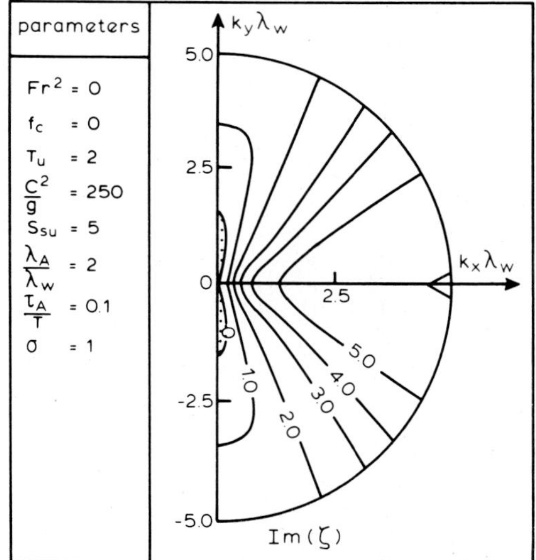

FIGURE 13. Amplification factor in case of a limniscate-type tidal ellipse (suspended load transport)

In order to assess the influence of a systematic deviation of the transport direction, the analysis was repeated with ψ invariant to bottom perturbations ($\psi = \psi_o$). This led to [cf. Eq.(31)],

$$\rho = \frac{1}{1-\epsilon_p}[i\beta k^2\ \overline{S_{bo}} - kS_{bu}\{\overline{\frac{S_{bo}\tilde{v}\cos\psi_o}{U_o}}\} - k\{\overline{\frac{S_{bo}\tilde{u}\sin\psi_o}{U_o}}\}$$

$$-k(S_{bu}-1)\{\overline{\frac{S_{bo}u_o v_o(\tilde{u}\cos\psi_o + \tilde{v}\sin\psi_o)}{U_o^3}}\}$$

$$+ k(S_{bu}-1)\{\overline{\frac{S_{bu}u_o^2(\tilde{v}\cos\psi_o - \tilde{u}\sin\psi_o)}{U_o^3}}\}] \qquad (45)$$

When applied to the case

$$U_o = \hat{U}_o\{\delta_1 + \cos(2\pi \frac{t}{T})\} \quad \text{and} \quad \psi_o = \text{atan}\{\delta_2, \delta_1 + \cos(2\pi \frac{t}{T})\} \qquad (46)$$

with $\delta_1 = 0$ or 0.1 and $\delta_2 = -0.2$, this yields the results shown in Figures 14a and 14b, respectively.

Apparently, the additional transport rate in these examples tends to favour the formation of banks with a cyclonic deviation of the crest direction from the current direction. This tendency is enhanced in the case of flood dominance.

VI.8 Conclusions on the "currents only" case

The behaviour of the sand bank was found to be rather insensitive to the investigated modifications of the flow model (simplifications, parameter setting, input conditions). Together with the heuristic arguments given in Section II.3, this indicates that the proposed flow model is adequate to explain the formation of linear sand banks in (well-mixed) tidal seas.

The observed sensitivity to the sediment transport model shows the transport formulation to be rather critical here. Not only should the transport mode (bedload/total load or suspended load) be reproduced appropriately, also rectification effects can be of significant influence. On the other hand, the absolute magnitude of the transport is only influencing the time scale of the process. Hence it can easily be eliminated by scaling the mathematical system. Apparently, the key point in sediment transport modelling is to establish the relationship with the other dependent variables in the system, rather than to sort out every detail on the absolute magnitude.

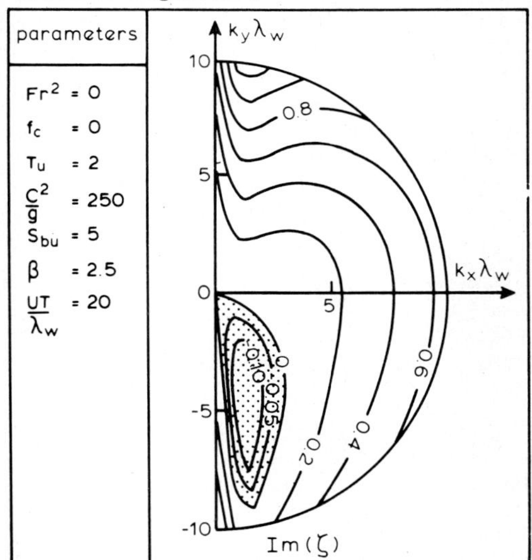

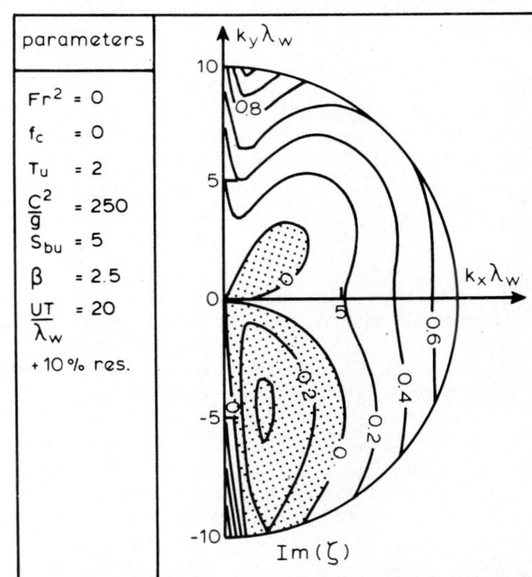

FIGURE 14. (a) Amplification factor in case of a systematic deviaton of the transport direction (symmetrical tide).
(b) (symm. tide + 10% residual current).

The present analysis also shows the role of interactions of phenomena. Even though the shape of the tidal ellipse hardly influences the sand bank behaviour in the reference case, it has a significant influence in combination with suspended load transport. These interaction effects, more of which will be shown in the next sections, significantly complicate the sensitivity analysis.

VII. Harmonic Analysis (Waves and Currents)

VII.1 Approach

Wind waves have a variety of effects on the sediment transport. Two of these will be considered here:

- the stirring of sediment, making it available for transport by the tidal current, and

- the net transport due to asymmetry of the wave orbital motion.

For simplicity reasons, the analysis was performed for steady flow, supposing the results to pertain to unsteady flow,as well (cf. De Vriend, 1988). Besides, the systems were simplified as much as possible, in order to focus attention on the phenomenon to be investigated. This means: no diffraction, rigid-lid, no wave effects in the bottom shear stress, no wave effects on the downslope gravitational transport (other than maybe a higher value of β), and only wave amplitude effects on the magnitude of the transport (S_{ba}, $S_{sa} \neq 0$).

VII.2 Influence of wave stirring

In the analysis of the effects of stirring, the influence of the waves on the direction of the transport is disregarded. Their influence on the magnitude is assumed to be exclusively due to amplitude effects.

The harmonic analysis, carried out for bed load as well as suspended load transport, shows that in either case the wave influence is proportional to the factor

$$F_w = C_{gh} - C_{ph} \frac{k_\nu^2}{k_\sigma^2} \tag{47}$$

in which k_σ and k_ν are the bottom wave number components parallel and perpendicular to the direction of surface wave propagation, respectively. With the dispersion relationship for linear gravity waves, this can be elaborated to

$$F_w = C_{gh}\left[1 - \frac{c_g}{c\left(1 - \frac{\omega^2 h}{g}\right)} \frac{k_\nu^2}{k_\sigma^2}\right] \tag{48}$$

This factor is positive for $h > g/\omega^2$ and for small angles between the bottom and surface wave crests. In those cases, the shoaling effect, represented by the C_{gh}-term in Eq.(47), is predominant. The factor F_w is negative if $h < g/\omega^2$ and the angles between the bottom and surface wave crests are large. In that case, the ray divergence (refraction) effect is predominant.

Note that for $k_\sigma \to 0$, so for surface waves propagating in the crest direction of the banks, the factor goes to infinity. In that range of conditions, however, the applied wave model is no longer valid: if the angle between the bank crests and the direction of propagation of the incoming surface waves lies below a certain critical value, the waves tend to be trapped in the troughs between the banks (LeBlond and Mysak, 1977; Liu *et al.*, 1988; also see: Mynett, 1983, who found this phenomenon when applying a wave ray model to the Norfolk Banks). In that case, the wave height variation, and hence the sediment stirring, is more or less in phase with the banks. The waves must therefore be expected to affect only the propagation of the banks, giving them a tendency to move upstream.

Without going into details of the analyses, the influence of wave stirring on sand bank behaviour is summarized in Table 1. The term "inversion" in this table means, that the growth rate becomes negative and the bottom waves are damped.

VII.3 Influence of wave-asymmetry

As was pointed out by Ribberink and De Vriend (1988), the effect of wave asymmetry on the sediment transport is influenced by the presence of a current. In formulae,

$$S_{bx} = S_c\cos\alpha + S_w\cos\theta - \beta S_b\frac{\partial z_b}{\partial x} \quad , \tag{49}$$

$$S_{by} = S_c\sin\alpha + S_w\sin\theta - \beta S_b\frac{\partial a_b}{\partial y} \quad , \tag{50}$$

in which:

$S_c(U, h, a, \theta\text{-}\alpha)$ = the "convective" transport rate, carried by the current,
$S_w(U, h, a, \theta 0\alpha)$ = the "asymmetry"-transport rate, and
$S_b = \{S_c^2 + S_w^2 + 2S_cS_w\cos(\theta-\alpha)\}^{1/2}$.

Table 1. Influence of wave stirring on bottom wave behaviour (valid for waves with their crest under not too large angles with the bank crests)

	bed load		suspended load	
	growth rate	celerity	growth rate	celerity
F_w<0 (shoaling)	increase	none	increase	decrease
F_w<0 (refraction)	decrease/ inversion	none	decrease/ inversion	increase

The analysis shows, that the waves influence the growth rate of the banks:

$$\mathrm{Im}(\zeta) = \beta h_o\lambda_w(k_x^2 + k_y^2) - P.X(k_x, k_y) \tag{51}$$

in which $X(k_x, k_y)$ is the destabilisation factor that was also found in the case without waves (De Vriend, 1986). There the factor P was equal to $(S_{bu}-1)$, here it is given by

$$P = \frac{S_{co}}{S_{bo}}(S_{cu} - 1) + \frac{S_{wo}}{S_{bo}}(S_{w\theta}\frac{k_\sigma}{k_y} + S_{wu}\frac{k_\sigma}{k_x}) \tag{52}$$

in which the suffix o refers to the undisturbed situation, again. The normalized dependence factors, S_{cu}, $S_{w\theta}$ and S_{wu}, are defined in the same way as those in Eq.(9). Apparently, the growth rate is influenced by the dependence of S_w on U and θ-α, but not by that on h and a. Besides, the effects are strongest if $k_\sigma^w = k$, so the wave crests are parallel to the bank crests.

VII.4 Conclusions on the "waves and currents" case

The analysis shows a sensitivity of the sand bank behaviour to the investigated wave effects. For the stirring effect, this is fairly straightforward, with the wave amplitude (or rather the near-bottom orbital velocity) as a linking pin. For the asymmetry-effect, however, the influence goes through an interaction with the current velocity, implying that pure wave-asymmetry transport is no generating agent for sand bank systems of this type.

Results from earlier work show that bottom diffraction is of minor influence on sand bank formation (De Vriend, 1986), and that rectification effects can have a significant effect on the sediment transport in a tide/wave environment (Roelvink and Stive, 1989). Together with the sensitivity shown by the present analysis, this suggests that wave modelling in this context should not be focused on a sophisticated description of linear wave propagation, but rather on the combined effects of waves, currents and gravitation via the sediment transport.

From that point of view, the present wave model must be considered as rather primitive. Hence the conclusions on the effects of waves can only be preliminary (cf. Section II.2).

VIII. The Holland Ridges

VIII.1 Observations

The fundamental question whether a sand bank system like the Holland ridges (Figure 1) is a morphodynamic feature, or rather a geological relict (e.g. ancient dune ridges, left after a rapid transgression), ought to be answered before going into a morphodynamic analysis of this system. A heuristic argument in favour of a morphodynamic feature is the regularity of the pattern: most probably, relicts of consecutive transgressions would have shown much more irregularity. Besides, recent measurements at one of the ridges connected to the Holland shoreface (Van de Meene, 1989, private communication) showed a continuous layer of significant thickness to be reworked during a single, not excessively severe storm. This means that a significant part of these ridges is exposed to morphodynamic action. So, after thousands of years and numerous severe storms, the whole system must have become a morphodynamic feature, even if the origin would be a geological relict.

Assuming the origin to be morphodynamic, the ridge system in front of the Holland coast exhibits some peculiarities, as appearing from Figure 1. Firstly, there is the absence of ridges in front of the southern and northern parts of this coast, which can be explained by the combined effect of suspended load transport and an open tidal ellipse. That sediment is brought into suspension, especially under storm conditions, becomes clear when flying over the ridge area just after a storm (Wiersma, 1988, private communication): streaks of high sediment concentration are observed, with their orientation and spacing in accordance with the ridges.

Secondly, the shallower the water, the stronger the anti-cyclonic rotation of the ridges with respect to the deep-water orientation. The ridges at the central part of the Holland coast even exhibit an anti-cyclonic rotation with respect to the tidal current. Since the rotation increases shorewards, it must be caused by one or more phenomena that gain importance on approaching the coast, such as (also see: Roelvink and Stive, 1989):

- wave-induced near-bottom water motion and sediment stirring,
- wave-asymmetry effects,
- wave-induced boundary layer streaming,
- near-bottom currents induced by wind (upwelling) or horizontal density gradients, and
- the tidal water level variation.

Figure 15 shows that, due to these effects, the onshore transport component increases significantly on approaching the shore. The analysis given in the previous section could help to explain the role of these phenomena in the orientation of the ridges.

VIII.2 Role of wave-induced stirring

In view of the assumption that only the fastest growing mode of the bottom perturbation will survive, a phenomenon can only influence the bank orientation if it affects the growth rate. Hence wave stirring in combination with bed load transport cannot explain the ridge orientation, but in combination with suspended load transport it possibly can.

The period of the most important sediment stirring waves is typically 6 to 8 s. For these waves, the water depth beyond which the refraction component in F_w [see Eq.(48)] is always positive amounts 9 to 16 m, whereas the ridges are found at depths of 15 m and more. This means that the sediment stirring waves will nearly always have a damping effect on the bottom perturbations.

The damping effect is weakest for $k_\nu = 0$, so for banks with their crests perpendicular to the predominant wave direction. At the central part of the Holland coast, this direction is approximately West, i.e. more or less perpendicular to the tidal flow. Therefore, wave stirring cannot explain the anti-cyclonic rotation of the shoreface-connected ridges with respect to the tidal flow.

VIII.3 Role of wave-asymmetry effects

Wave asymmetry effects, in contrast with wave-induced stirring, tend to enhance the growth rate, and hence will have a stronger impact on the bank orientation. However, the effect is strongest for $k_\nu = 0$, again. Therefore, the conclusion drawn for the effects of wave-induced stirring applies here, too: wave asymmetry effects alone do not explain the orientation of the shoreface-connected ridges. This refutes the suggestion of the opposite in De Vriend (1988).

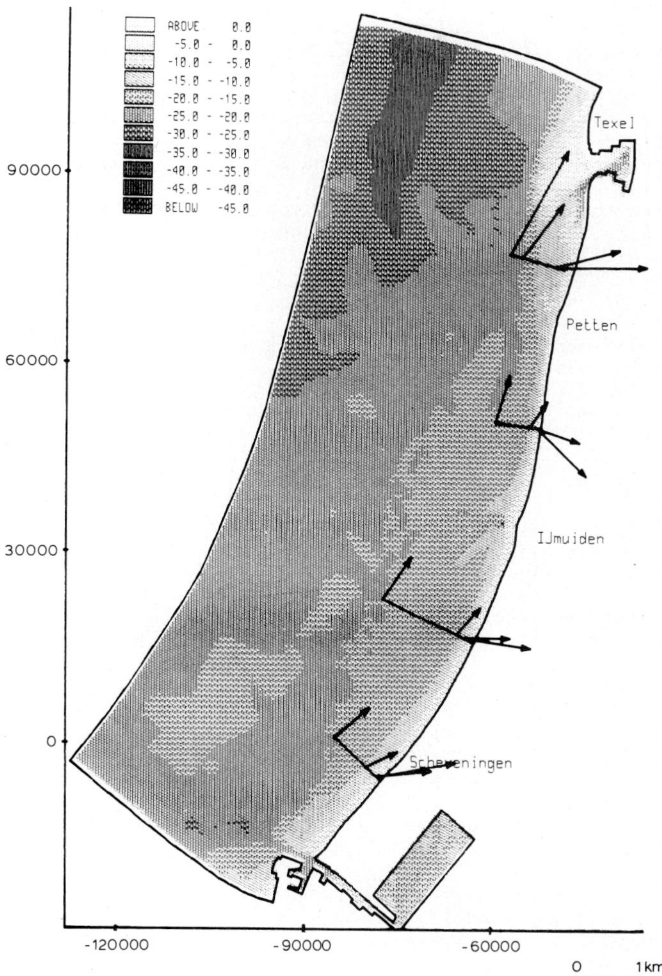

FIGURE 15. Estimated residual transports at the Holland coast. (From: Roelvink and Stive, 1989)

VIII.4 Role of wave-induced boundary layer streaming

The wave-induced boundary layer streaming is directed along with the waves and is driven by an effective stress arising from shear effects in the wave orbital motion (Longuet-Higgins, 1953). In the case of waves only, the bottom boundary layer is very thin and the induced streaming occurs only close to the bottom. In a turbulent environment, however, the wave orbital motion is subject to shear effects throughout the water colomn and the induced streaming is likely to extend up to the water surface (cf. Trowbridge *et al.*, 1986).

As this turbulence-dependence of the induced streaming is insufficiently known, only two extreme cases will be considered, viz.:

a) the induced streaming, and the attending sediment transport, vary along with the tidal current and transport, respectively, and

b) the induced streaming is independent of the tide, and the attending transport rate is determined by the tidal and wave-induced sediment stirring.

ad a) This means that the resultant transport direction is more or less constant and that the system behaves in a similar way as quasi-steady flow with wave-asymmetry. So the induced streaming, if of this nature, provides no explanation of the orientation of the shoreface-connected ridges.

ad b) This is a case with an additional transport rate depending on the wave height, the water depth and the tidal current velocity. Its steady-state equivalent is the wave-asymmetry case, again, with $S_{w\theta} = 0$ and $S_{wu} = S_{cu} - 1$ in Eq.(52).

If St denotes the amount of sediment mobilized by stirring, and V is the transport velocity associated with the induced streaming, then the transport rates S_{co} and S_{wo} in Eq.(52) can be written as

$$S_{co} = U_o St \quad \text{and} \quad S_{wo} = V\ St \tag{53}$$

Besides, the predominant wave direction at this part of the Holland coast is approximate perpendicular to the tidal current. As the flood current is north-going, $k_\sigma \approx -k_y$ during flood and $k_\sigma \approx k_y$ during ebb. If, in addition, $V \ll U_o$, Eq.(52) can be elaborated to

$$P_{flood} \approx (S_{cu} - 1)(1 - \frac{V}{U_o}\frac{k_y}{k_x}) \quad \text{and} \quad P_{ebb} \approx (S_{cu} - 1)(1 + \frac{V}{U_o}\frac{k_y}{k_x}) \tag{54}$$

This means that during flood banks with negative k_y, i.e. with cyclonic rotation, grow faster. During ebb, the banks with positive k_y, i.e. with anti-cyclonic rotation, grow faster. If the tide is symmetric, there is no preference. The flood dominance of the transport in this area (Roelvink and Stive, 1989) seems to be in favour of the cyclonic banks. In that case, the induced streaming provides no explanation for the orientation of the shoreface-connected ridges.

VIII.5 Role of systematic vertical circulations

The relatively fresh water from the Rhine-Meuse river system moves north in a gradually widening stream adjacent to the Holland coast. The corresponding horizontal density gradients, together with wind-induced upwelling, give rise to a shoreward residual near-bottom current of a few cm/s (Van der Giessen et al., 1988). This current determines a significant part of the onshore transport component in the area (Roelvink and Stive, 1989); also see Figure 15.

As the mathematical system is linearized, this small residual current can be added (in a vectorial sense) to the tidal current. Due to its smallness, the residual current will mainly give rise to a systematic shoreward deviation of the transport direction. This is a typical additional transport case, of the kind shown in Figures 14a and 14b. In the case of flood dominance, there is a tendency of cyclonic banks to grow faster, but the difference in growth rate with the anti-cyclonic banks is of the order $O(V.\Delta U/U_o^2)$, in which V and ΔU are the cross-shore and longshore residual velocities, respectively.

VIII.6 Role of water level variations

At various points in the previous analysis, the effects of ebb and flood are opposite and balancing, unless tidal asymmetry is involved. At shallower water, the water level variation can contribute substantially to this asymmetry. Therefore, its role is worth investigating.

A typical result of a steady flow analysis, e.g. for the wave-asymmetry case and the steady-state version of the additional transport case, is given by Eq.(51). In a dimensional form this reads

$$\mathrm{Im}(\rho) = \frac{S_{bo}}{1-\epsilon_p}[\beta k^2 - \frac{P}{h_o \lambda_w} X\ (k_x, k_y)] \tag{55}$$

So far, h_o was assumed constant through the tide and the factors P were compared for the ebb and flood situation. In shallower water tides, however, the mean water level during flood can be significantly higher than during ebb. This means that if the P-terms in Eq.(51) almost balance for ebb and flood, they will be ebb-dominated in Eq.(55).

If that is the case, the ebb-effect on the bank formation will prevail. Both for the wave-asymmetry effect and for the effects of cross-shore residual currents (wave-induced or otherwise), this implies that anti-cylonic banks will grow faster. This may explain the orientation of the shoreface-connected ridges at the central part of the Holland coast. Before being able to draw definitive conclusions, however, quantitative substantiation by field data is needed.

VIII.7 Role of infragravity waves

Non-linear infragravity waves can lead to the formation of bank systems of similar dimensions as the shoreface-connected ridges at the Holland coast (Boczar-Karakiewicz and Bona, 1986). By the formulation of the wave model this phenomenon was ignored a priori.

The swell at the North Sea, entering mainly from the North and refracting slowly towards the Holland coast, and also the bound long waves induced by wave groups, could lead to wave-induced bank formation at the shelf. If the banks would have been formed by these waves alone, however, their crest orientation should reflect the refraction towards the coast. Figure 9 suggests rather the contrary.

Another possibility is, that waves and tide together are forming the banks. At deeper water, where the tide is predominant, the bank orientation is typically tidal (20° cyclonic), whereas at shallower water the long waves gain importance and tend to rotate the banks towards their crest direction. Qualitatively, the observed bank pattern could be interpreted in this way, but here, again, quantitative substantiation is needed.

IX. Conclusion

Even though the quantitative description of a number of physical phenomena involved in morphological processes in shallow tidal seas is still subject to doubt, the linear analysis techniques presented herein appear to be useful tools in gaining qualitative insight into the morphological interaction phenomena. Both the characteristics analysis and the harmonic analysis could be used in early phases of a study, in order to judge hypotheses about mechanisms underlying observed phenomena.

When applied to the Holland coast, The Netherlands, the harmonic analysis makes it possible to test hypothese and find seemingly plausible explanations for the absence of bank systems at the northern and southern parts of this coast, and for the orientation of the shoreface-connected ridges at the central part.

Acknowledgements

Part of the work presented herein was done in the framework of the Rijkswaterstaat Coastal Research Programme, financed by the Ministry of Transport and Public Works (Rijkswaterstaat) and executed in co-operation with DELFT HYDRAULICS. The author wishes to thank Dr. J. Wiersma of Rijkswaterstaat for his useful suggestions, and Mr. J. van de Meene of Utrecht State University for providing information from his recent measurements at the shoreface-connected ridges at the Holland coast.

X. References

Bagnold, R. A., 1956: The flow of cohesionless grains in fluids. Phil. Trans. Royal Soc., **A249**, 235-297.

Berkhoff, J. C. W., 1976: Mathematical models for simple harmonic linear water waves; wave diffraction and refraction. Doct. thesis, Delft Univ. of Techn., 103 pp.

Boczar-Karakiewicz, and Bona, J. L., 1986: Wave-dominated shelves: a model of sand-ridge formation by progressive infragravity waves. In: Knight, R. J. and McLean, J. R. (eds.): "Shelf Sands and Sandstones." Can. Soc. Petr. Geol., Memoir II, 163-179.

Boczar-Karakiewicz, and Davidson-Arnot, R. G. D., 1987: Nearshore bar formation by non-linear wave processes - a comparison of model results and field data. Marine Geol., **77**, 287-304.

Caston, V. N. D., 1972: Linear sand banks in the southern North Sea. Sedimentology, **18**, no. 1 & 2, 63-78.

Davies, A. M., 1988: On formulating two-dimensional vertically integrated hydrodynamic numerical models with an enhanced represenation of bed stress. J. Geoph. Res., **93**, no. C2, 1241-1263.

Deigaard, R., 1983: Formation of large bed undulations under waves and currents - a stability analysis. Tech. Univ. Denmark, Inst. Hydrdyn. and Hydraulic Engrg., Prog. Rept., 59, 3-11.

De Vriend, H. J., 1986: 2DH computation of transient sea bed evolutions. Proc. 20th ICCE, Taipei, Taiwan, 1689-1712.

De Vriend, H. J., 1987a: 2DH mathematical modelling of morphological evolutions in shallow water, Coastal Engineering, **11**, no. 1, 1-27.

De Vriend, H. J., 1987b: Analysis of 2DH morphological evolutions in shallow water. J. Geoph. Res., **92**, no. C4, 3877-3893.

De Vriend, H. J., 1988: Inherent stability of depth-integrated mathematical models of coastal morphology. Proc. IAHR-Symp. "Math. Modelling of Sed. Transp. in the Coastal Zone," Copenhagen, Denmark, 320-329.

Galappatti, R. and Vreugdenhil, C. B., 1985: A depth-integrated model for suspended sediment transport. J. Hydr. Res., **23**, no. 4, 359-77.

Holman, R. A. and Bowen, A. J., 1982: Bars, bumps and holes: models of the generation of complex beach topography. J. Geoph. Res., **87**, no. C1, 475-468.

Huthnance, J. M., 1982a: On one mechanism forming linear sand banks. Est., Coastal and Shelf Sci., 14, 79-99.

Huthnance, J. M., 1982b: On the formation of sand banks of finite extent. Est., Coastal and Shelf Sci., **15**, 277-299.

Katopodi, I. and Ribberink, J. S., 1988: Quasi-3D modelling of suspended sediment transport by currents and waves. Delft Hydraulics, Rept. H460, 59 pp.

LeBlond, P. H. and Mysak, L. A., 1977: Trapped coastal waves and their role in shelf dynamics. In: E. D. Goldberg et al., (eds.). "The Sea, Vol. 6: Marine Modelling," Wiley-Interscience, New York, 459-495.

Lepetit, J. P. and Hauguel, A., 1978: A numerical model of sediment transport. Proc. 16th ICCE, Hamburg, Fed. Rep. Germany, 1715-1724.

Lin, P. N. and Shen, H. -W., 1984: Two-D flow with sediment by characteristics method. J. Hydr. Engrg., **110**, no. 5, 615-625.

Liu, P. L. -F., Kostense, J. K. and Dingemans, M. W., 1988: Long waves generated by short-wave groups over a step. Submitted for publ.

Longuet-Higgens, M. S., 1953: Mass transport in water waves. Phil. Trans. Royal Soc., **A 254**, 535-581.

McAnally, W. H., Letter, J. V., Stewart, J. P., Thomas, W. A. and Brogdon, N. J., 1984: Application of Columbia hybrid modelling system. J. Hydr. Engnr., **110**, no. 5, 627-642.

McCave, I. N., 1979: Tidal currents at the North Hinder lightship, Southern North Sea: flow directions and turbulence in relation to maintenance of sand banks. Mar. Geol., **31**, 101-114.

Mei, C. C., 1985: Resonant reflection of surface waves by periodic sand bars, J. Fluid Mech., **152**, 35-335.

Mynett, A., 1983: Wave refraction/shoaling analysis for the near-shore section of the S.E. Inde trunkline. Delft Hydr. Lab., Rept. R1296 (limited distribution).

Pattiaratchi, C. and Collins, M., 1987: Mechanisms for linear sandbank formation and maintenance in relation to dynamical oceanographical observations. Prog. Oceanog., **19**, 117-176.

Ribberink, J. S. and De Vriend, H. J., 1988: Sediment transport formulations for waves and currents as developed for and applied in the Voordelta area of the Dutch coast. Delft Hydraulics, Rept. H526-2, 42 pp.

Roelvink, J. A. and Stive, M. J. F., 1989: Initial sediment transport at the Holland coast. Delft Hydraulics, Rept. H825-IV, 23 pp. (in Dutch)

Struiksma, N., Olesen, K. W., Flokstra, C. and De Vriend, H. J., 1985: Bed deformation in curved alluvial channels, J. Hydr. Res., **23**, no. 1, 57-79.

Trowbridge, J. H., Kanetkar, C. N. and Wu, N. T., 1986: Numerical simulation of turbulent wave boundary layers. Proc. 20th ICCE, Taipei, Taiwan, 1623-1637.

Van der Giessen, A., de Ruyter, W. P. M., and Borst, J. C., 1988: Three-dimensional current structure in the Netherlands coastal zone. Neth. J. Sea Res. (in press).

Suspended Sediment Transport Processes in the Southern North Sea

J. Dronkers[1], J.S.L.J. Van Alphen[2], and J.C. Borst[1]

[1]Rijkswaterstaat, Tidal Waters Division
P.O. Box 20907, 2500 EX The Hague
The Netherlands

[2]Rijkswaterstaat, North Sea Directorate
P.O. Box 5807, 2280 HV Rijswijk
The Netherlands

ABSTRACT

The structure and the composition of the shoreface and adjacent shelf is permanently subject to changes. Particulate material and adsorbed substances are brought in from external and internal sources (rivers, atmosphere, sea bottom) and redistributed. Sediment transport is mainly confined to the coastal zone of the continental shelf. Many factors contribute to this transport, such as coastal drift, tidal currents, residual circulations and wave activity. These processes are examined and some theoretical estimates are presented. The suspended sediment balance of the Dutch coastal zone serves as an example, and provides experimental support. Major conclusions are:

1. The average residual water flux is only partly responsible for the average alongshore sediment transport.

2. Density- and tide-induced processes strongly influence the coastal sediment transport rate and the turbidity in the coastal zone.

3. Human activities currently involve changes in the coastal sediment budget capable of changing significantly sedimentation and turbidity.

I. Introduction

Continental shelves receive terrigeneous erosion products which are not retained in rivers and estuaries. The main input comes from rivers, the atmosphere contributes for a relatively small part. The erosion products are principally fine grained sediments ($\mu < 50$), which are redistributed on the continental shelf by currents and waves. The ocean inflow into the continental shelves carries little sediment (concentrations are far below 1 mg/l). Similarly there is little outflow to the ocean, except in cases where large, heavily loaded rivers enter a narrow shelf (for example, the Amazon outflow on the South American shelf, the Mississippi outflow in the Gulf of Mexico). Otherwise the net flux between the continental shelves and the open ocean is small (Postma 1980).

The continental seafloor seldom attains complete equilibrium. Natural causes are tectonic motions and sea level rise. Therefore, shelves are not just passive receivers of sediment, but exhibit an own dynamic behavior. Large parts of the continental shelves have been surf zones at some time in the past. Therefore they are mainly covered by course grained material (Swift 1976). Behind relict sand bars or morainic structures mud deposits have been formed. Under actual conditions the sheltering may have become less effective, allowing erosion by waves or tidal

Coastal and Estuarine Studies, Vol. 38
R. T. Cheng (Ed.)
Residual Currents and Long-term Transport

currents. On the contrary, other regions have become less affected by currents and waves, and new potential deposition areas are formed.

Thus in addition to the externally imported sediment also internally eroded sediment is redistributed on the shelf; most shelves are still adapting to actual hydrodynamic conditions. Besides, particulate material is formed by primary production. Although most of it is destroyed by mineralization, this organic matter contributes to turbidity and to coastal transport (Postma 1978). It even plays an essential role in the formation of particle aggregates and in binding dissolved trace elements (in particular organic contaminants, Van Leussen 1988).

In some cases the equilibrium of the continental shelf is effected by human interventions. An example is the Southern Bight of the North Sea, and especially the Dutch coastal zone. The coastal waters of the Netherlands have been strongly altered during the past 60 years by civil engineering works. Most of the existing shallow tidal basins, covering more than 4000 km^2, have been dammed up (closure of the Zuyderzee in 1932, closure of Lauwerszee, Haringvliet and Grevelingen during the past decades and partial closure of the Eastern Scheldt recently). At the same time the remaining estuaries, the Rotterdam Waterway and the Western Scheldt have been considerably deepened for navigational purposes. The tidal amplitude along the Dutch coast has increased by approximately 5% during this period (De Ronde 1982). The dissipation zone of tidal energy has been displaced from the former tidal basins to the coastal zone. Human interventions in the coastal morphology will continue in the future in order to assure protection of the Netherlands against flooding and to further improve harbour and navigation facilities.

In order to appreciate to what degree human activities influence the coastal sediment budget it is necessary to investigate sources and sinks and to analyse the transport processes. In this paper an analysis of these coastal sediment transport processes is presented. Attention is especially directed to the fine sediment fraction (μ < 50).

Sediment fluxes are governed by sedimentation and resuspension processes, which are subject to strong fluctuations in space and time. With present knowledge they are hardly predictable. From isolated time series or from incidental surveys it is not possible to reconstruct a reliable picture valid for different time and space scales. Therefore, theoretical concepts are necessary to interprete field data and to arrive at generally applicable conclusions concerning coastal sediment transport. Along this line the present study proceeds, with special attention to the Dutch coastal zone.

For the North Sea sediment sources and sinks have been inventorised by Mc Cave (1973) and Eisma (1981) and more recently by Kalf & Eisma (1988) and by Van Alphen (1989). Their estimates yield comparable orders of magnitude for the input and output rates of the major sources (Channel, Rivers, Flemish Banks, Atmosphere) and major sinks (German Bight, Norwegian Trench). The input through the Channel exceeds the output to the Atlantic Ocean along Norway. The North Sea therefore behaves as a sink of sediment. Van Alphen (1989) compared his source-sink balance qualitatively with transport rates which can be derived from the flow field and the suspended sediment distribution. In this study this comparison is elaborated with respect to physical processes.

The main driving forces for coastal circulation are examined, being the large scale and local wind fields, the tidal stresses and the coastal zone density distribution. A large portion of the Dutch coast is influenced by the outflow of the rivers Rhine, Meuse and Scheldt, causing a cross-shore salinity gradient. Due to earth rotation this density gradient drives a northward near surface residual flow. In the near coastal zone this northward residual flow appears to be one of the major components of the average alongshore drift. The northward drift is thus concentrated in the upper part of the vertical. In the lower portion of the vertical, where sediment concentrations are highest, the northward residual current is small or even absent. This observation implies that the alongshore sediment flux cannot be computed only from the depth integrated alongshore drift and the average suspended sediment load. A correction is necessary in order to take into account the vertical structure of velocity and sediment profiles.

A second correction appears necessary when considering the tidal influence on sediment transport. The ebb-flood asymmetry of the tidal current evoluates along the Dutch coast. The tidal wave is increasingly distorted when propagating northward along the coast. Along the southern coast maximum ebb currents are on the average larger than maximum flood currents, especially in the bottom region. Apart from the distortion of the tidal wave, the presence of estuaries plays a role here. The result is a net southward tidally induced sediment transport, counteracting the transport due to the residual northward water flux. On the contrary, the distortion of the tidal wave along the northern Dutch coast is such that flood sediment transport exceeds ebb sediment transport. Tidal influences provide an explanation for the differences observed between the computed sediment fluxes along the southern and the northern portions of the Dutch coast.

Finally it is argued that the role of wind waves should also be taken into consideration. When northerly winds prevail, higher wind waves with longer periods arrive at the dutch coast than for southerly winds. This implies that during periods of southward alongshore wind drift the suspended load will be on the average higher than during periods of northward alongshore wind drift.

In the following sections the different processes contributing to the residual alongshore sediment transport will be examined more in detail.

II. Sediment Budget and Average Transport

The North Sea shelf area presents certain particularities which characterize it with respect to other continental shelves. The depth is generally less than 50 m and the bottom relief shows rather pronounced structures, see Fig. 1. A shallow area with a size of approximately hundred km, the Doggerbank, protects a large part

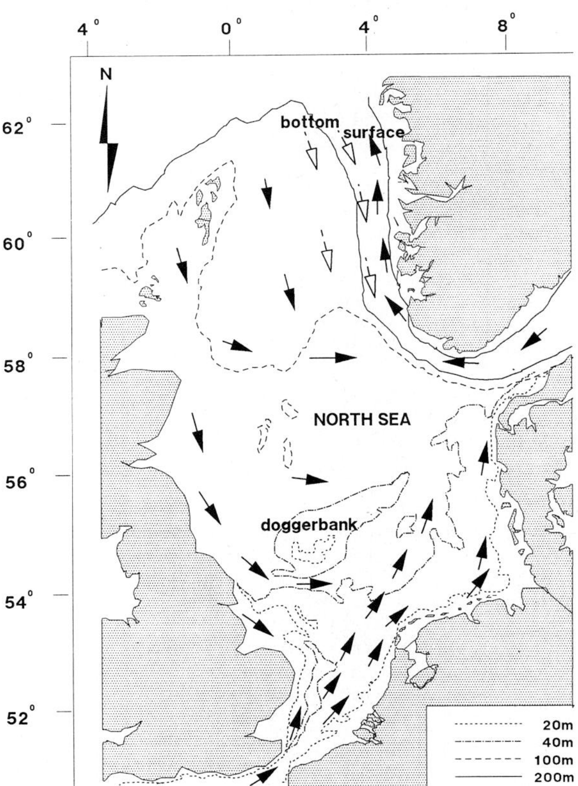

FIGURE 1. North Sea topography and average current system.

of the Southern North Sea from strong swell coming in from the Atlantic Ocean. Besides, sand ridges with length in the order of 10 km, locally offer protection from waves. In the northern part, along the Norwegian coast, a deep trench acts as a trap for the sediment flowing into this region.

In the coastal zone tidal motion is responsible for rather strong currents. Figure 1 also shows the general circulation pattern. Wind is the major driving force and therefore this circulation pattern is subject to temporal variations on a time scale of days to weeks.

Along the Dutch coast the average flow is in northern direction with an average drift velocity of 3 - 5 cm/s (Otto 1964 and 1983). Figure 2 shows the major erosion and sedimentation areas, with an estimate of their contribution to the sediment balance of the North Sea (after Eisma 1981, Irion et al. 1988, Van Alphen 1989, Zuo et al.1989).

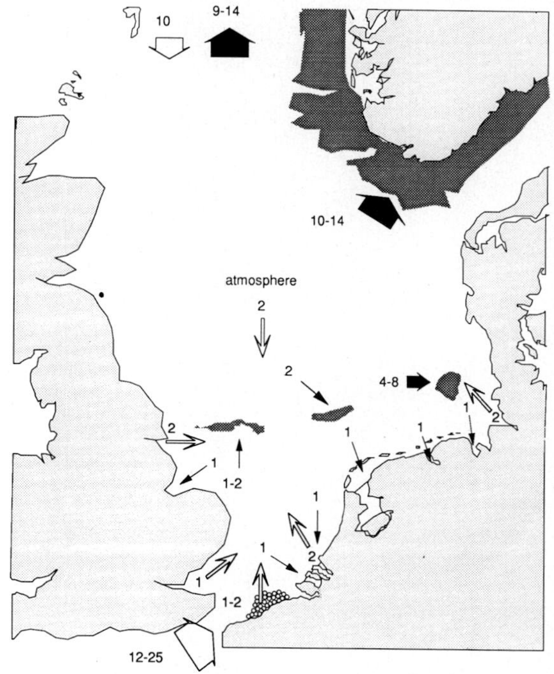

FIGURE 2. Sediment budget of the North Sea: major sources (white arrows, dotted areas) and sinks (dark arrows, dark areas) in 10^6 ton/y, after Eisma (1981), Irion et al. (1988), van Alphen (1989), Zuo et al. (1989).

Mud patches are present off rivers and harbours, mainly as a result of dredging and dumping activities. Sediment is not permanently stored at the dumping sites, but brought into transport in periods of strong waves and currents (Louisse 1986). Temporary storage of sediment on near shore areas also occurs at other locations along the Dutch coast. Deposits containing up to 50 % of mud ($\mu < 50$) have been detected in troughs between sand ridges at depths less than 10 m (Van Alphen 1989). A density driven circulation brings sediment shoreward along the bottom in a broad coastal area around the Rhine outflow (Van Alphen et al. 1988). For the resuspension of sediment in those near-shore locations the necessary energy is mainly produced by wind waves or swell (Creutzberg & Postma 1979).

The Dutch coastal zone is a rather turbid environment, especially during stormy conditions. An impression of the near-surface sediment concentrations is given in Figure 3. Measurements in the vertical indicate that concentrations increase near the bottom (Figure 4). Near the entrance of Rotterdam Waterway the formation of a fluid mud layer has been observed at several occasions. The presence of high concentration bottom layers is a common feature for continental shelves, as shown, for example, by Halper & McGrail (1988). In the entrance channels to estuaries or bays the tidal currents are generally strong enough to achieve almost complete mixing of sediment over the vertical. Along the coast, however, tidal currents are insufficient to assure vertical homogeneity, so that surface concentrations are not representative for the vertical average. As a rough estimate it is derived from the available vertical profiles that vertically averaged concentrations are approximately 1.5 times higher than the surface concentration. Using this figure, as well as estimates for the mean surface distribution of suspended sediment and for the mean alongshore circulation, the average alongshore sediment transport is computed and compared to the sediment input onto the shelf, see table 1. This table clearly shows a discrepancy between the computed sediment flux and the flux resulting from the source-sink balance. The computed values are larger, especially in the southern coastal range.

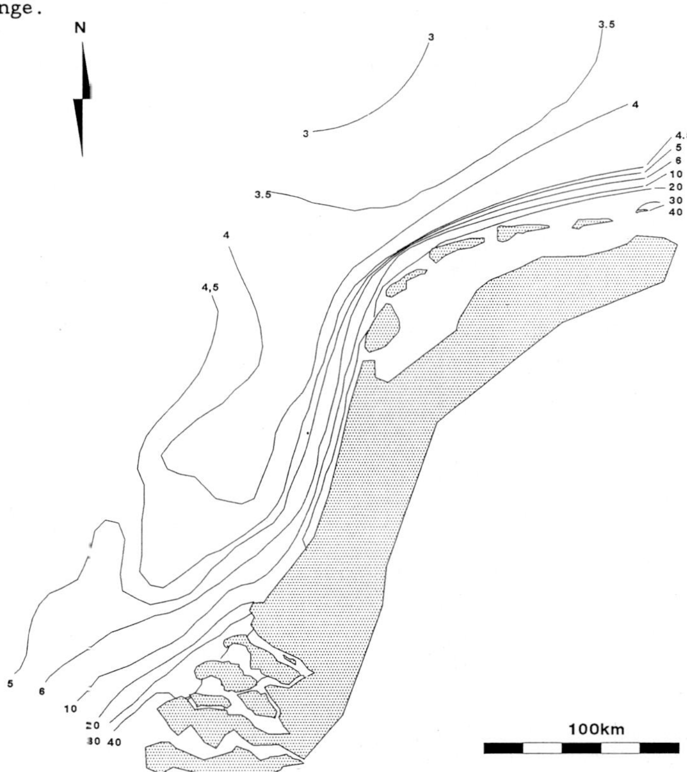

FIGURE 3. Average surface sediment distribution [mg/l] in the Dutch coastal zone

Eisma and Kalf (1979) relate the high turbidity off the Belgian coast to local erosion of mud banks, the so-called Flemish Banks. According to these authors the high turbidity is confined to this region due to the presence of a gyre in the residual flow field. The high turbidity zone extends along the Southern Dutch coast where the residual flow is northward. Therefore, one should expect that high turbidity is advected along the northern Dutch coast, but this appears not to be true.

The comparison of the alongshore drift, the average sediment load, the sediment sinks and sources and the average suspended sediment distribution leads to apparently conflicting results. In the next sections the discrepancies will be discussed considering space and time variations of the currents and the suspended sediment concentration.

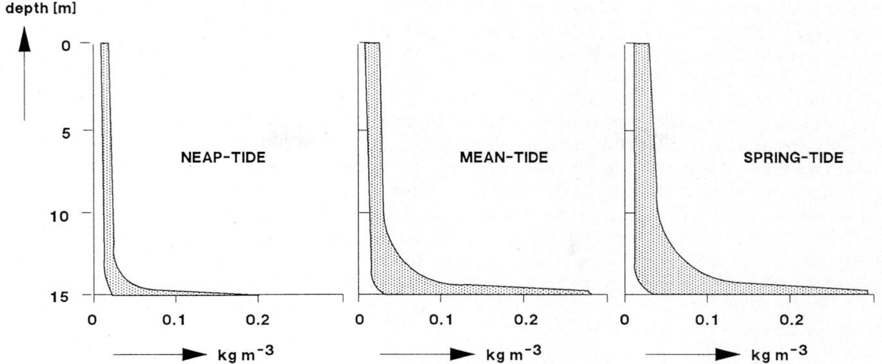

FIGURE 4. Vertical profiles of suspended sediment. Indicated are the ranges which fit most of the profiles measured during some surveys in the coastal zone.

Table 1. Alongshore transport

COASTAL ZONE	South (Schouwen)	North (Noordwjik)
Water flux [km³/y]	600	600
Residual current [m/s]	0.03-0.05	0.03-0.05
Depth section [m²]	20	20
Width coastal zone [m]	20,000-30,000	20,000-30,000
Average concentration surface [kg/m³]	0.017	0.008
Average concentration cross-section [kg/m³]	0.025	0.012
Computed alongshore sediment transport (flux*conc.) [ton/y]	$15x10^6$	$7x10^6$
Input Channel [ton/y]	$(2.5-5)x10^6$	$(2.5-5)x10^6$
Input Flemish banks + seafloor erosion [ton/y]	$(1-2)x10^6$	$(1-2)x10^6$
Sedimentation Western Scheldt [ton/y]	$-0.5x10^6$	$-0.5x10^6$
Net deposition coastal zone (Voordelta) [ton/y]		$-0.5x10^6$
Input Rhine [ton/y]		$1.5x10^6$
Deposition Loswal dumping site [ton/y]		$-0.5x10^6$
Flux of organic matter [ton/y]	$0.2x10^6$	$0.2x10^6$
Alongshore sediment transport from source-sink balance [ton/y]	$(3-6.7)x10^6$	$(3.5-7.2)x10^6$

III. Shelf Circulation

The continental shelf is a region of intensive water circulation. The exposure to wind fields causes important wind driven flows, especially in the alongshore direction. The essential characteristics have been described by Csanady (1978). A wind driven alongshore drift exists for almost any wind direction. For more or less coast-parallel winds the flow follows the coast in the direction of the wind. For all depths the flow direction is approximately alongshore, but the magnitude decreases downwards, especially when stratification is present. Besides alongshore flow, wind may also induce a cross-shore circulation, cross-shore winds being more effective than alongshore winds. For cross-shore winds an alongshore wind drift is directed to the right of the wind vector. Tides and density differences may effect the wind induced coastal circulation substantially. For a qualitative discussion the effect of tides and density differences will be treated separately.

The North Sea communicates with the Atlantic Ocean at the northern boundary via wide passes between Scotland, the Shetland Islands and Norway, and at the southern boundary via the relatively narrow Strait of Dover. A net flow through the North Sea is occasioned by a difference in water level and tidal conditions at these boundaries and by wind stress (Maier-Reimer 1977). A difference in wind set-up is related to the large scale structure of the wind field on the NW-European shelf. The strong amplification of the tide in the Channel builds at the southern boundary a tidal stress which acts as a virtual increase in water level

$$\Delta\zeta = u^2/4g$$

where u = amplitude tidal velocity, g = gravitational acceleration. The net result is a yearly inflow through the Strait of Dover of approximately 5000 km³ (Prandle 1978).

The dominant wind direction in the Southern North Sea is south-west, but strong fluctuations are present which are reflected in the direction of the alongshore drift. The average drift is northward and the average subsurface current velocity due to wind and tidal stress effects is in the order of 0.05 m/s.

Tides produce circulations on the coastal shelf mainly through interaction with topographical irregularities. In this respect "headland eddies" and circulations around sand ridges are well established features (Heathershaw & Hammond 1980, Zimmerman 1981). The interaction of estuarine cross-shore tidal flow and alongshore tidal currents also produces residual displacements of water parcels (see Fig. 5, Dronkers 1988). In the Dutch Coastal Zone tidally induced circulations act on a much smaller scale than the wind induced circulation. However, locally the effect is visible in the sediment distribution. If opposed to the large scale drift the tide-induced circulation locally causes retention of sediment, and a turbidity maximum appears (Figure 6).

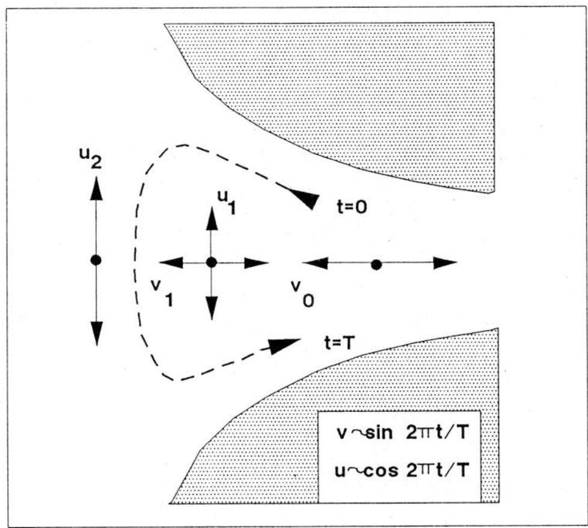

FIGURE 5. Schematic representation of the Lagrangian drift of a water parcel at the transition region of estuarine and coastal tidal flow. The net southward shift is due to 90° phase difference between the along-shore coastal currents and cross-shore estuarine currents.

In the previous section the alongshore coastal sediment transport has been estimated by assuming the alongshore residual current velocity approximately identical in every location of the cross-section. In fact, as shown by the previous considerations, this assumption is questionable. In the following the vertical current structure will be examined for the Dutch coastal zone.

Generally continental shelf waters are not homogeneous. In the deeper offshore parts temperature stratification occurs. In the shallow near coastal zone fresh water inflow causes a decrease in salinity. The density difference drives a cross-shore circulation, and also influences the alongshore drift. Density driven cross-shore currents are thought to play an important role in concentrating sediment in the near coastal zone. The flow convergence at the frontal outcrop is visualized by the appearance of a foam line (Garvine 1974). The downward fluid motion at this location causes a minimum in the suspended sediment concentration (Postma 1981), visible also in Figure 3.

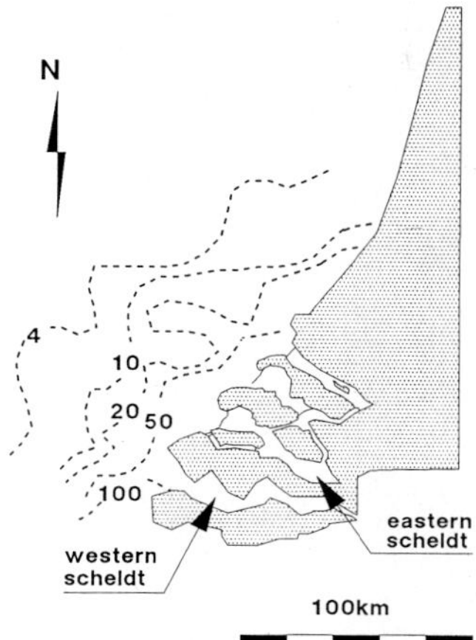

FIGURE 6. Mid-depth sediment distribution on 8-10 Jan 1975 at moderately rough sea conditions. Note the high concentrations in the Eastern and Western Scheldt outflow regions.

In the Dutch coastal zone salinity is decreased by the outflow of the rivers Rhine, Meuse and Scheldt. Horizontal and cross-sectional distributions representative for yearly average conditions are shown in Figures 7 and 8. The associated density differences influence the circulation in the coastal zone.

$$\frac{\partial p}{\partial x} - \rho f v - \frac{\partial \tau_x}{\partial z} = 0 \tag{1}$$

$$\frac{\partial p}{\partial y} + \rho f u - \frac{\partial \tau_y}{\partial z} = 0 \tag{2}$$

where p = pressure $\approx g \int_z^{\zeta(x,y)} \rho(x,y,z)dz$,

$x(y)$ = alongshore (cross-shore) coordinate,

$u(v)$ = alongshore (cross-shore) velocity,

ρ = density

f = Coriolis factor $\approx 10^{-4}/s$

τ_x (τ_y) = shear stress in x- (y-) direction

All quantities are long term averages and therefore time independent. Now consider a region in the vertical where the shear stress varies more or less linearly with z (assuming that such a region exists). Let z_1 be a position in the upper half of the vertical where the velocity equals the average value for the upper half of the vertical, and z_2 a corresponding position in the lower half of the vertical. Substracting Eq.(2) for $z = z_1$ from Eq.(2) for $z = z_2$ one finds approximately:

$$u(z_1) - u(z_2) \approx \frac{1}{2}\frac{gh}{f}\frac{1}{\rho}\frac{\partial p}{\partial y} \tag{3}$$

Now substitute as representative values $h = 10$ m, $\partial p/\partial y \approx 10^{-4}$ kg/m⁴, then $u(z_1) - u(z_2) \approx 0.05$ m/s.

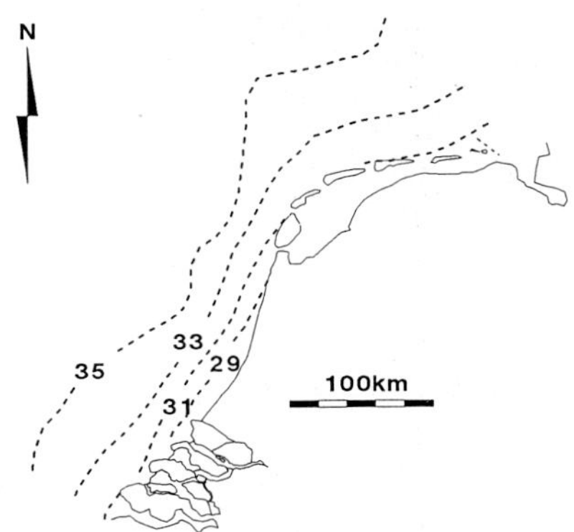

FIGURE 7. Surface salinity distribution [ppt] in the Dutch Coastal Zone

The experimentally observed average near surface drift is also in the order of .05 m/s. From the above result it follows that in the lower part of the vertical the northward drift is very small or even absent. An analytical solution of Eqs. (1) and (2) has been given by Heaps (1970), assuming a uniform cross-shore density gradient. This solution exhibits a rotation with depth of the residual current vector due to earth rotation. The result Eq.(3) yields a fair approximation of this more elaborate analytical expression.

Current velocity measurements by Van der Giessen et al. (1989) provide experimental support for a smaller alongshore drift in the lower part of the vertical compared to the upper part. Figure 9 shows that the wind force and direction also influence the vertical variation of the residual alongshore current. However, for almost all wind directions the difference between the alongshore residual currents at a mid-depth or near-bottom location and at a near-surface location is directed southward and has an order of magnitude of .05 m/s. As, in addition, the sediment concentration is highest in the lower part of the vertical,

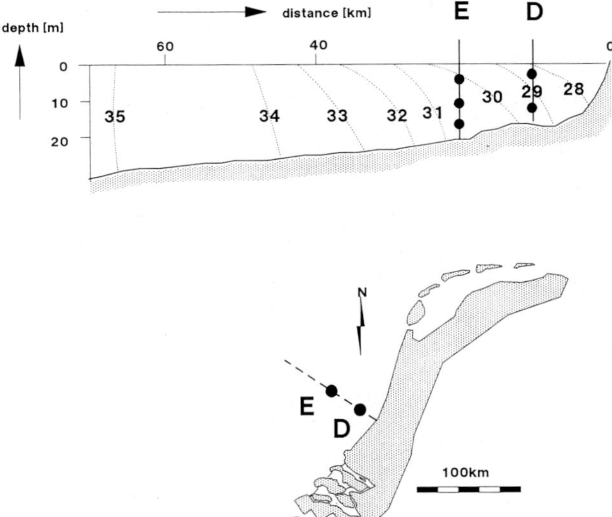

FIGURE 8. Cross-shore salinity distribution and current meter moorings.

it may be concluded that the sediment transport along the coast is smaller than the computed value presented in Table 1, assuming homogeneous distributions for the residual current and the suspended sediment. In fact, in the near bottom zone of those coastal regions which are under the influence of river outflow, sediment tends to be retained or to be transported much slower than in the upper part of the vertical. Taking this effect into account, the computed alongshore transport in Table 1 should be reduced roughly by a factor 2 for the northern coastal zone. In the southern coastal zone the cross-shore salinity gradient is smaller. There a reduction factor of 1.5 seems more appropriate as a first approximation. In this way a first correction on the average alongshore suspended sediment transport is obtained (Table 2).

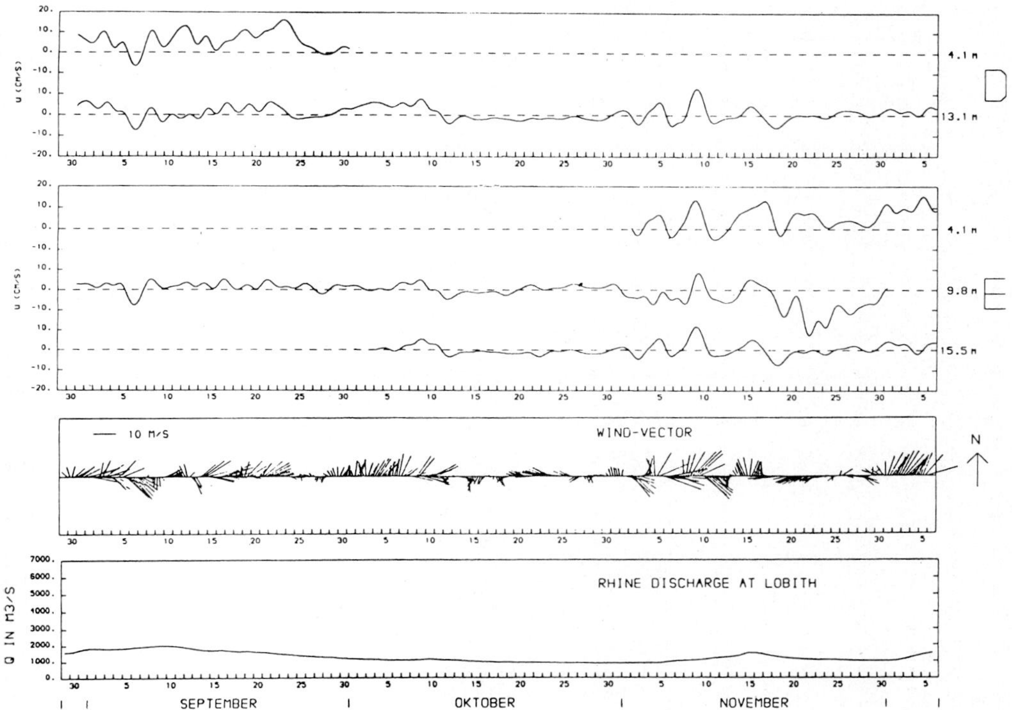

FIGURE 9. Low-pass filtered along-shore velocity component at the measuring locations of Fig. 8, wind vector and Rhine discharge (after Van der Giessen et al. 1988)

IV. Tide-Induced Alongshore Sediment Transport

The suspended sediment load exhibits strong fluctuations in the coastal zone. An important component of these fluctuations is associated with tidal motion. Wind waves constitute another factor influencing the resuspension of sediment, which will be discussed later.

The suspended load increases when tidal currents become strong. For complete suspension in the vertical generally a time lag exists, which can amount to a few hours. When the tidal currents become small the suspended load tends to decrease as a result of sedimentation. From this description it is clear that tides, if strong enough, will cause a residual transport of sediment. The magnitude of this transport depends on differences in tidal variation and in strength between ebb and flood currents. The effect of tidal asymmetry on sediment transport will be examined more in detail below.

Table 2. Processes Contributing to Alongshore Sediment Transport [10^6 ton/y]

	COASTAL ZONE South (Schouwen)	COASTAL ZONE North (Noordwijk)
Sediment transported by average alongshore water flux	15	7
Density influence on vertical residual flow profile	-5	-3.5
Tide-induced residual transport	-4	5
Correllation wind drift and wave induced resuspension	-1	-1
Fluid mud flow	Local influence at harbors and navigation channels	
Topographic circulations	Local influence at estuary entrances	
Net transport (sum)	5	7.5

The residual alongshore sediment transport can be written

$$M = \int_{\text{tidal period}} A\,u\,c\,dt \qquad (4)$$

where

A = cross-sectional area
u = cross-sectionally averaged tidal velocity
c = cross-sectionally averaged sediment concentration
< > = tide averaged value

The integration of Eq.(4) is effectuated in a frame moving with the average tidal velocity Au/<A>. It is supposed that no residual water flux is present: <Au> = 0. In the moving frame variations of the sediment concentration c are essentially related to local sedimentation or resuspension. The tidal period is divided in four intervals:

ΔHWS = period of settling preceding and around high water slack (dc/dt<0);
ΔHWR = period of resuspension after high water slack (dc/dt>0);
ΔLWS = period of settling preceding and around low water slack (dc/dt<0);
ΔLWR = period of resuspension after low water slack (dc/dt>0).

The sediment load per unit length is designated μ = Ac, μ^f = maximum flood load, μ^e = maximum ebb load. A straightforward calculation shows that the residually transported mass can be decomposed as follows:

$$M = \mu^+ \lambda^+ - \mu^- \lambda^- + \mu^f \lambda^f - \mu^e \lambda^e \qquad (5)$$

with:

$$\mu^+ = \frac{1}{\Delta HW} \left[\int_{\Delta HWS} (\mu^f - \mu)dt + \int_{\Delta HWR} (\mu^e - \mu)dt \right],$$

$$\mu^- = \frac{1}{\Delta LW} \left[\int_{\Delta LWS} (\mu^e - \mu)dt + \int_{\Delta LWR} (\mu^f - \mu)dt \right],$$

$$\lambda^{+} = \frac{1}{\mu^{+}} [\int_{\Delta HWS} u(\mu^{f}-\mu)dt + \int_{\Delta HWR} u(\mu^{e}-\mu)dt],$$

$$\lambda^{-} = \frac{1}{\mu^{-}} [\int_{\Delta LWS} u(\mu^{e}-\mu)dt + \int_{\Delta LWR} u(\mu^{f}-\mu)dt],$$

$$\lambda^{f} = \int_{\Delta LWR} udt + \int_{\Delta HWS} udt,$$

$$\text{and} \quad \lambda^{e} = \int_{\Delta HWR} udt + \int_{\Delta LWS} udt$$

These different quantities can be interpreted approximately as follows:

μ^{+} (μ^{-}) = average settled and resuspended mass during a period around high water slack (resp. low water slack).

λ^{+}(λ^{-})= average horizontal settling and scour lag around high water slack (resp.low water slack).

$\lambda^{f} \approx \lambda^{e}$ = tidal excursion.

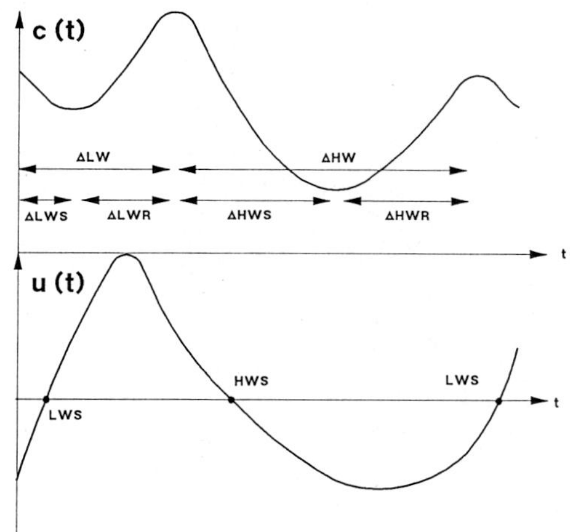

FIGURE 10. Possible tidal variation of c(t) and u(t) in a moving frame; definition of the intervals ΔHWS, ΔHWR, ΔLWS and ΔLWR

The first two terms in Eq.(5) represent the relative displacement of sediment which is withdrawn from the tidal motion during high and low water slack by settling. They correspond to the residual transport resulting from an ebb-flood asymmetry in settling and subsequent resuspension. The average displacements λ^{+} and λ^{-} are large if time lag effects are important, i.e. the tidal variation of c(t) is out of phase with $|u(t)|$. The last two terms in Eq.(5) represent the residual transport due to a difference in suspended load between ebb and flood currents, due, for example, to a larger magnitude of the flood or ebb current and a stronger uptake of sediment. A more detailed discussion of the interpretation of Eq.(5) has been presented in a previous study (Dronkers 1986a).

An asymmetry between ebb and flood currents results from the propagation of the tidal wave in shallow water. Non-linearities are related to bottom friction and to the influence of tide levels on the propagation speed of the tidal wave. The distortion of the tidal wave increases during the course of propagation. When a tidal wave in the absence of a residual water flux enters the coastal shelf from deep water the maximum ebb current is larger than the maximum flood current. This asymmetry compensates for the so-called Stokes drift (coincidence of flood current and high water level). The difference between maximum ebb and flood currents is in the order of a/2h, i.e. a few cm/s for Dutch coastal conditions.

After propagating some hundred kilometer along a shallow coast the tidal wave is distorted such that the the current asymmetry is reversed: the maximum flood current exceeds the maximum ebb current (see Figure 11), and therefore carries more sediment ($\mu^f > \mu^e$). Besides, the current variation around slack water before flood is faster than before ebb: The flood current carries a large amount of sediment which in the short period of slack water has not settled on the bottom ($\mu^+ > \mu^-$ and $\lambda^+ > \lambda^-$). This asymmetry also favours flood transport over ebb transport.

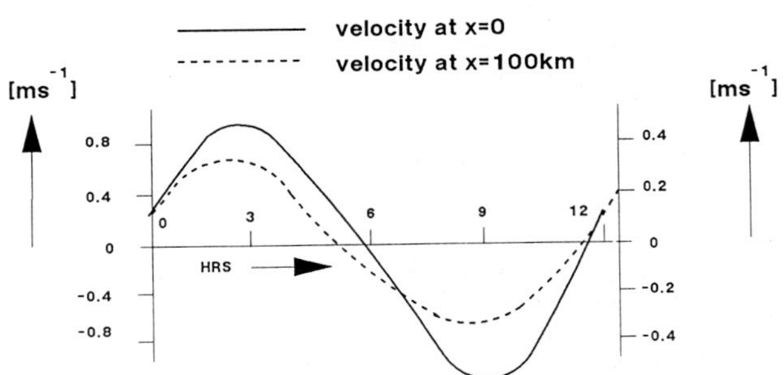

FIGURE 11. Current velocity variation of a damped non-linear tidal wave at x = 0 and at x = 100 km entering from deep water (x = 0) and propagating along a uniform shallow coast. The tidal amplitude a = 1 m, the depth h = 10 m, the coefficient for linearized friction F = 2.10^{-1} ms^{-3}. (from Dronkers 1986 b)

The tidal Kelvin wave travelling in northward direction along the Dutch coast is only slightly distorted in the south; the asymmetry of the tidal wave increases in northward direction (see Fig. 12). In the southern part of the Dutch coast ebb currents on the average will thus exceed flood currents, see Fig. 13. This asymmetry is even more pronounced in the near bottom zone because of the smallness of the residual velocity (see the previous section). Tide-induced sediment transport is directed southward. On the contrary, along the northern part of the Dutch coast the tide induced sediment transport enhances the northward transport caused by the residual longshore drift. Here the maximum flood current exceeds the maximum ebb current and, in addition, the settling and resuspension lag is greater before ebb than before flood.

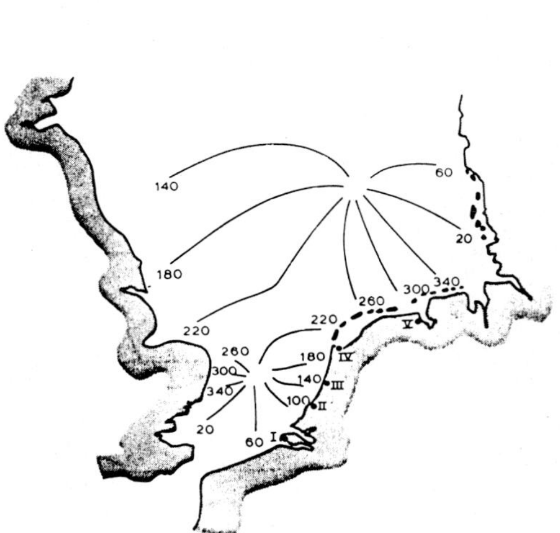

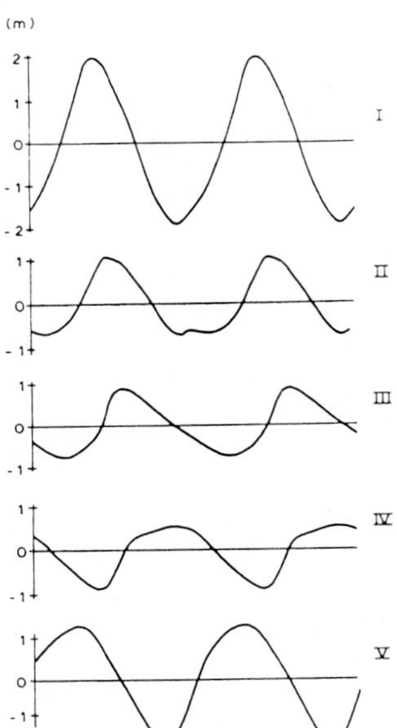

FIGURE 12. Distortion of the tidal wave travelling along the Dutch coast

Tide induced transport is not included in the computed value of table 1. The magnitude can only be roughly estimated from presently available data, see the appendix. As shown in table 2 tide induced transport provides the main explanation for the discrepancy between the sediment fluxes along the southern and northern parts of the Dutch Coast. Along the southern coast the tidally induced transport counteracts the average northward drift transport, while along the northern coast the drift transport is enhanced. As a result, along the northern coast sediment is transported more rapidly than along the southern coast. Sediment therefore tends to be retained in the southern coastal zone, giving rise to the high suspended concentrations which are observed there.

V. Discussion

Substances released on the continental shelf can be rapidly displaced over large distances, especially in alongshore direction. This is true for both dissolved and particulate constituents. The transport mode and the rate of transport may be very different, however. A few differences between dissolved and particulate alongshore transport in the North Sea are identified in the previous sections. A quantitative simulation of coastal transport processes is beyond the scope of the present study. One might ask, however, if sufficient insight in the transport physics is available to undertake such a quantitative modelling effort successfully. A tedious obstacle is the modelling of near bottom processes, in particular the formation of a high concentration layer in relation to settling of sediment on the bottom and to resuspension.

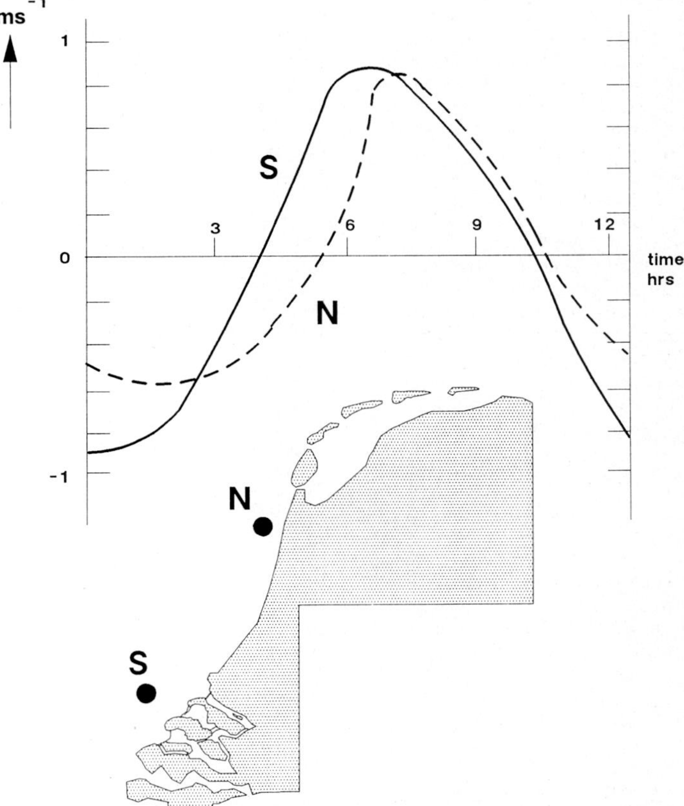

FIGURE 13. Alongshore tidal velocity variation at two locations along the Dutch coast. At S the maximum ebb velocity exceeds the maximum flood velocity. At N the inverse holds. Besides the period of current reversal is longer at slack water before flood than at slack water before ebb.

On certain continental shelves, where highly loaded rivers discharge, the transport is believed to proceed mainly from the motion of a fluid mud layer. A well known example is the Amazon discharge on the Atlantic Coastal Shelf of South America (Froidefond et al., 1988). The sediment load of rivers discharging into the North Sea is relatively low. The appearance of fluid mud layers here probably is an incidental and local phenomenon. Locally the contribution of fluid mud to the transport and sedimentation may be important, especially in navigation channels and harbours (van Leussen & van Velzen 1989). For the large scale sediment displacement the role of fluid mud is thought to be less important. However, even if fluid mud flow is not a major transport mechanism, the process of sedimentation and resuspension taking place in the transition zone between suspension and bottom are very important and not well understood. Moreover, organisms living in and on the bottom certainly play a role, as well as the presence of organic matter, providing "glue" for sediment particles and other substances present in the fluid (Van Leussen 1988). For the time being these processes only can be modelled by introducing bulk parameters which require calibration of the models for each particular field situation.

In most coastal sediment transport studies the important influence of wind waves is stressed. Wind waves may strongly increase the shear stress of coastal currents (Grant & Madsen 1979, Huntley & Hazen 1988). Due to this influence the bottom top layer can be fluidized (Mehta 1988) and sediment be brought in suspension. As a result of the asymmetry of shallow water waves a residual transport of sediment may occur. A necessary condition is that time lag effects do not exceed the wave period. In fact, wave asymmetry is only effective in transporting sediment if the wave period is in the same order or larger than settling and resuspension time lags (Groen 1967). Wind waves thus may occasion a residual transport of coarse sediment, while tides are most effective in transporting fine sediment. The wave spectrum in the Dutch coastal zone receives substantial contributions of waves with periods between 3 and 9 seconds; the significant wave height can reach 3 m. Wave activity is capable of resuspending fine sediment in the coastal zone where the depth is in the order of 30 m or less. Sedimentation occurs offshore at places where the depth of the coastal current exceeds 30 m and in the nearshore zone at places sheltered by breaker bars and tidal ridges. Wind waves from N and NW directions on the average possess longer periods and larger heights than waves from southern directions. Figure 14 shows that approximately the largest 10% of the northerly waves disturb the bottom more strongly than the largest southerly waves. In periods of strong southward winddrift the suspended sediment load should therefore be higher than in periods of equivalent northward winddrift. The existing data are too scarce to yield experimental confirmation. Nonetheless it is reasonable to believe that a difference in the intensity of southward and northward waves causes an additional residual sediment transport. A rough guess from the existing data is that the suspended sediment concentration in periods of strong winds is at least doubled with respect to average conditions. The additional wave induced transport then can be estimated at approximately 10% of the residual wind drift transport and proceeds in the opposite (southward) direction.

VI. Conclusions

Gathering the results of this study, it may be concluded that the alongshore sediment transport is essentially different from the value obtained when multiplying the average suspended sediment concentration with the average alongshore current velocity. Table 2 summarizes the processes which have to be taken into account together with an indication of their importance. These corrections bring the coastal sediment transport in better agreement with the source-sink balance established by Eisma en Van Alphen (1989).

Van Alphen (1989) deduces the amount of mud involved in sedimentation or resuspension due to human activities: the construction of the harbour of Zeebrugge between 1980 and 1985 resulted in an input of about 10^6 tons of fine sediment per year; sedimentation in front of the closed estuaries of the southern part of Holland involves annually $(1\text{-}2)\times10^6$ and 0.5×10^6 tons respectively; storage of contaminated dredgings in an artificial basin on land withdraws about 10^6 tons per year from the Rotterdam harbours while dumping of dredged material at sea amounts to several million tons per year. In addition fishery activities lead to resuspension of sea bed sediments once a year on average. In the future it is expected that additional resuspension will occur due to increased extraction of sea sand for civil engineering works. Comparing these figures with those in Table 2, it can be concluded that the influence of human activities on suspended sediment transport is of comparable order of magnitude as that of natural processes.

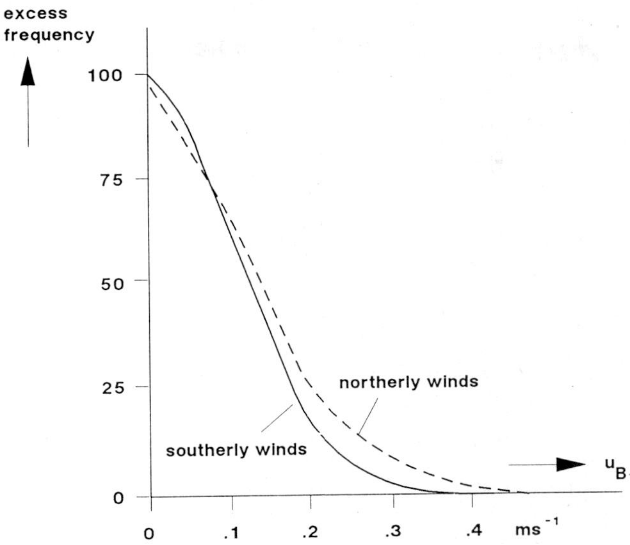

FIGURE 14. Excess frequency of maximum botton current u_B due to waves caused by northerly and southerly winds in the Dutch Coastal Zone

VI. References

Creutzberg, F. and H. Postma, 1979: An experimental approach to the distribution mud in the North Sea. Netherlands Journal of Sea Research, **13**, 99-116.

Csanady, G. T., 1984: Circulation in the coastal zone. D. Reidel Publ. Co.

Dronkers, J., 1986a: Tidal-induced residual transport of sediment. In: J. van de Kreeke. Physics of Shallow Estuaries and Bays, Miami 1984. Springer Verlag, 228-244.

Dronkers, J., 1986b: Tidal asymmetry and estuarine morphology. Netherlands Journal of Sea Research, **20**, 117-131.

Dronkers, J., 1988: Inshore/offshore water exchange in shallow coastal systems. In: B.O. Jansson. Coastal-offshore Ecosystem Interactions, Springer-Verlag, 3-39.

Eisma, D. and Kalf, J., 1987: Dispersal, concentration and deposition of suspended matter in the North Sea. Journal of the Ecological Society, London, **144**, 161-178.

Eisma, D., 1981: Supply and deposition of suspended matter in the North Sea. Spec. Publications International Association of Sedimentologists, **5**, 1-14.

Froidefond, J. H., Pujas, M. and X. Andr, 1988: Migration of mud banks and changing coastline in French Guiana. Marine Ecology, **84**, 19-30.

Garvine, W., 1977: Observations of the motion field of the Connecticut river plume. Journal of Geophysical Research, **82**, 441-454.

Grant, W. D. and O. S. Madsen, 1979: Combined wave and current interaction with a rough bottom. Journal of Geophysical Research, **84**, 1797-1808.

Groen, P., 1967: On the residual transport of suspended matter by an alternating tidal current. Netherlands Journal of Sea Research, **3**, 564-574.

Halper, F. B. and D. W. Mc. Grail, 1988: Long-term measurements of the near bottom currents and suspended sediment concentration on the outer Texas-Louisiana Continental Shelf. Continental Shelf Research, **8**, 23-36.

Heaps, N. S., 1972: Estimation of density currents in the Liverpool bay area of the Irish Sea. Geophysical Journal of the Royal Astronomical Society, **30**, 415-432.

Heathershaw, A. D. and F. D. C. Hammond, 1980: Secondary circulations near sandbanks and in coastal embayments. Deutsche Hydrographische Zeitschrift, **33**, 135-151.

Huntley, D. A. and D. G. Hazen, 1988: Seabed stresses in combined wave and steady flow conditions on the Nova Scotia Continental Shelf; Field measurements and predictions. Journal of Physical Oceanography, **18**, 347-362.

Irion, G., Wunderlich, F. and E. Schwadhelm, 1987: Transport of clay minerals and anthropogenic components into the German Bight and the procenance of fine-grained sediments SE of Helgoland. Journal of the Geological Society, **144**, 153-160.

Louisse, C., 1986: Private Communication.

Maier-Reimer, E., 1977: Residual circulation in the North Sea due to the M2-tide and mean annual windstress. Deutsche Hydrographische Zeitschrift, **30**, 69-80.

Mc. Cave, J. N., 1973: Mud in the North Sea. In: E. D. Goldberg. North Sea Science, MIT Press, 75-100.

Mehta, A. J., 1988: Laboratory studies on cohesive sediment deposition and erosion. In: J. Dronkers and W. van Leussen. Physical Processes in Estuaries, Springer Verlag, 427-445.

Otto, L., 1964: Results of current observations at the Netherlands light vessels over the period 1910-1939. Med. Ver. KNMI, 85.

Otto, L., 1983: Currents and water balance in the North Sea. In: J. Sndermann and W. Lenz. North Sea Dynamics. Springer Verlag, 26-43.

Postma, H., 1981: Exchange of materials between the North Sea and the Wadden Sea. Marine Geology, **40**, 199-213.

Postma, H., 1980: Sediment transport and sedimentation. In: E. Olausson and I. Cato. Chemistry and Biochemistry of Estuaries. John Wiley, 153-186.

Prandle, D., 1978: Residual flows and elevations in the southern North Sea. Proceedings of the Royal Society of London, A **359**, 189-228.

Swift, D. J. P., 1976: Continental shelf sedimentation. In: Stanley & Swift. Marine Sediment Transport and Environmental Management. John Wiley, 311-351.

Van Alphen, J. S. L. J., 1989: A mud balance for the Belgian-Dutch Coastal waters between 1969-1986. Netherlands Journal of Sea Research (in press).

Van Alphen, J. S. L. J., De Ruyter, W. P. M. and J. C. Borst, 1989: Outflow and three-dimensional spreading of Rhine river water in the Netherlands Coastal Zone. In: J. Dronkers & W. van Leussen. Physical Processes in Estuaries. Springer Verlag, 70-90.

Van der Giessen, A., De Ruyter, W. P. M. and J. C. Borst, 1988: Three-dimensional current structure in the Dutch Coastal Zone. Netherlands Journal of Sea Research, in press.

Van Leussen, W., 1988: Aggregation of particles, settling velocity of mud flows - a review. In: J. Dronkers & W. van Leussen. Physical Processes in Estuaries. Springer Verlag, 347-403.

Van Leussen, W. and E. Van Velzen, 1989: High concentration suspensions: their origin and importance in Dutch estuaries and coastal waters. Netherlands Journal of Sea Research, to be published.

Zuo, Z., Eisma, D. and G. W. Berger, 1989: Recent sediment deposition rates in the Oyster Grounds, North Sea. Netherlands Journal of Sea Research, **23**, 263-269.

Appendix -- Transport Estimates

1. Tidally induced sediment transport at S.

No long term measurement of suspended sediment is available. Therefore a representative tidal time series has to be constructed on the basis of theoretical arguments. It is assumed that the suspended sediment concentration $c(t)$ depends on the current velocity $|u(t)|$ with a time lag of at most a few hours, the time lag being shortest when the current velocity is at its maximum. When $|u|$ is approximately constant $c \sim |u|^n t$, with $n \approx 2$. (Erosion of a dense bed with constant erosion rate is approximately proportional to the shear stress, Mehta 1988).

At station S the behaviour of the current around SBE and SBF is similar, for $|u| < 0.5$ m/s (see Figure 13, showing an average spring tide). For this period $c(t - t_{SBE}) \approx c(t - t_{SBF})$, and there is no contribution to the mass transport integral

$$M = \int Aucdt.$$

For $u > 0.5$ m/s one should distinguish between periods of maximum ebb and maximum flood currents. The average max ebb current, $\langle|u|\rangle_{max\ ebb} \approx 0.79$ m/s, and the average max flood current, $\langle|u|\rangle_{max\ flood} \approx 0.75$ m/s.

Now it is assumed that:

$$\langle c\rangle_{max\ ebb} \approx \langle c\rangle \left[\frac{\langle|u|\rangle_{max\ ebb}}{\langle|u|\rangle}\right]^n ,$$

and a similar relationship for $\langle c\rangle_{max\ flood}$.

Substitution in the mass transport integral yields the following approximate expression:

$$M \simeq \frac{1}{2} \frac{\langle|u|\rangle^n_{max\ flood} - \langle|u|\rangle^n_{max\ ebb}}{\langle u \rangle^n} \lambda_{0.5} \cdot \langle u\rangle .$$

Here $\lambda_{0.5}$ is the distance travelled by the suspended sediment in the time interval $|u| > 0.5$ m/s. By substitution of numerical values one finds $M \approx -1.25 \times 10^4$ ton for an average spring tide. The yearly flux is estimated by assuming that the contribution of neap tides can be neglected and the contribution of mean tides is approximately half the contribution of spring tides. The result is $M \approx -4.5 \times 10^6$ ton/year.

2. Tidally induced sediment transport at N.

In this case the behaviour of $u(t)$ around SBE and SBF is not similar. However, there is such a strong difference between maximum ebb and flood currents, that the contribution of the periods around SBE and SBF (corresponding to $|u| < 0.4$ m/s) to the mass transport integral can safely be neglected.

As a consequence a similar formula can be used as for station S, with:

$$\langle|u|\rangle_{max\ ebb} = 0.5 \text{ m/s}, \quad \langle|u|\rangle_{max\ flood} = 0.6 \text{ m/s}.$$

The result is $M \approx 5 \times 10^6$ ton/year.

Some Observations on Erosion and Entrainment of Estuarine Fluid Muds

Panagiotis D. Scarlatos[1] and Ashish J. Mehta[2]

[1]South Florida Water Management District
P.O. Box 24680
West Palm Beach, FL 33416-4680.

[2]Coastal and Oceanographic Engineering
University of Florida
Gainesville, FL 32611

ABSTRACT

A matter of particular interest to estuarine sedimentation problems is the dynamic behavior of fluid muds. It is suggested that shear flow-induced interfacial instabilities leading to the erosion of fluid muds is a significant resuspension mechanism. Thus, the erosion process is controlled by the scale of near-bottom turbulence and the vertical velocity and density profile structures. Under strongly entraining conditions in which settling is small, the density profile microstructure of the entraining material is governed by buoyancy stabilized turbulent diffusion. A step-wise profile of the suspended sediment is theoretically developed under simplified physical conditions. Field and experimental data support the theoretical assumptions, while experience from other stratified flow phenomena provides valuable insight relative to studies required to quantify fluid mud erosion response to hydrodynamic forcing.

I. Introduction

Simulation of estuarine residual cohesive sediment transport leading to sedimentation is an important but also a complicated task. This complexity is largely defined by our level of understanding of a wide variety of near-bed flow-sediment interactions which characterize the overall transport problem. Due to the small particle size, cohesive sediment particle movement is controlled not only by inertial, gravitational, buoyancy and drag forces, but also by electro-chemical ones. Since however a microscopic description of the processes involved in cohesive sediment transport is not presently feasible, for simulation purposes a macroscopic approach using lumped parameters is usually applied (Scarlatos and Partheniades, 1986).

During a tidal cycle or an episodic event, spatial and temporal redistribution of the suspended sediment concentration occurs by virtue of the effects of the different phases of the sedimentation cycle. From field and laboratory observations it has been noticed that, generally, as shown in Figure 1 the vertical concentration (density) profile of fine sediments can be categorized into three zones: an upper zone of mobile, relatively low concentration suspension, a middle zone of higher concentration suspension or fluid mud, and the lower zone of settled bed (Kirby and Parker, 1983). The fluid mud zone is of particular practical importance because, due to its high concentration and weak internal structure, it can be easily entrained and can substantially contribute to turbidity even under relatively low energy inputs (Ross, 1988). Therefore, assuming an initially sharp interface between the overlying water and fluid mud during quiescent periods as occurs for example during slack water, it is important to understand the erosion behavior of the mud-water interface, and the evolution of the interface as the fluid mud entrains into the water column due to shear flow.

Coastal and Estuarine Studies, Vol. 38
R. T. Cheng (Ed.)
Residual Currents and Long-term Transport

Many past studies on cohesive sediment transport seem to have neglected the fluid mud zone, emphasizing instead the erosion of the deposited bed (Mehta, 1986). Such has been the case in spite of the fact that it has been long recognized that fluidized muds do not erode in the same manner as a settled bed (Parker and Kirby, 1982). Yet, because of inadequate knowledge of the underlying mechanism for fluid mud erosion, numerical models have had to rely on conjecture and empiricism (Ariathurai, 1974; Hayter, 1983). The purpose of this paper is to briefly describe the physical properties of fluid muds, to review erosion and entrainment processes of initially stationary high concentration suspensions, to present a simple model for simulation of vertical suspension density distribution which results from entrainment, and, thereby, to highlight some areas for further research.

II. Physical Properties of Fluid Muds

The suspended sediment concentration typically increases rather dramatically downwards from the surface to the bottom of the water column (Figure 1). The concentration profile may be smooth or step-wise depending on, among other factors, the concentration of total sediment mass in suspension (Kirby, 1986). The transition from a mobile suspension of low concentration (< 300-500 mg/l), to a stationary suspension of high concentration (> 10 g/l), usually occurs within a very narrow vertical distance, creating thus a steep density gradient. The zone of the steepest gradient within the water column is called the lutocline layer (Figure 1). The thickness of the various zones and the degree of vertical stratification depend on the magnitude of energy input by the tides, wave action, wind stress, geophysical motions, bioturbation or sometimes even human activity.

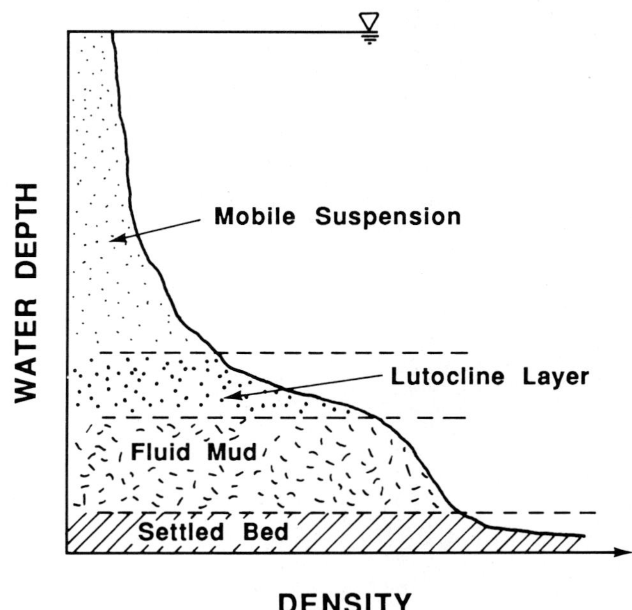

FIGURE 1. Schematic definition of fluid mud relative to mud density profile.

Fluid muds are usually contained between the lutocline and the settled bed. The thickness of the fluid mud layer may be from a few centimeters to several meters. Typical concentrations for fluid muds are usually within the range of 10 to 320 g/l, which correspond to bulk densities approximately between 1.03 to 1.2 g/cm^3, respectively (Bradley et al., 1988). However, as Been and Sills (1981) have noted from their laboratory work on quiescent settling, a weak solid structure is usually present when the concentration exceeds about 220 g/l.

Fluid muds exhibit a strong time-dependent, non-Newtonian fluid response (Bryant et al., 1980); therefore any realistic description of their dynamic behavior must account for this essential property. Various constitutive models have been used for simulation of their mechanics including kinematics and energy dissipation. These include: elastic, poro-plastic, viscous fluid, Bingham plastic, and viscoelastic models (Maa, 1986). Laboratory and field tests indicate that the rheological behavior of fluid muds, i.e., the relationship between stress and rate of strain, may change from pseudoplastic at relatively low concentrations, to Bingham plastic at high concentrations (Sills and Elder, 1986; Maa, 1986). The Bingham yield stress documented in high concentrations provides an indication of the work required to disrupt the floc structure in the cohesive suspension. Under very high shear stresses fluid muds approach Newtonian behavior, i.e., a linear relation between stress and rate of strain. This is due to the fact that increasing shear rates break down the flocs and reduce the particle's partial support by the cohesive structure.

Besides concentration, the rheological properties of fluid muds are greatly influenced by the physico-chemical characteristics of the supporting fluid and the composition of the sediment, which essentially indicates the importance of cohesion in defining mud behavior. Experimentally it has been observed that, for example, very weakly cohesive kaolinite-quartz mixtures tend to behave like a pseudoplastic material, while highly cohesive estuarine muds may apparently behave as Bingham plastics (Krone, 1963). Very few muds however seem to be truly Bingham plastics; most exhibit pseudoplastic character at very low rates of shear, and consequently creep under even mild forcing due to shear flow or gravity (Bryant et al., 1980; Kirby, 1986).

III. Erosion and Entrainment

III.1 Interfacial response

Under low energy conditions, such as during slack water, when the suspended sediment settles rapidly, fluid mud is formed near the bed, provided the rate of deposition is high enough to prevent dewatering of the underlying material. The thickness of the fluid mud-water interface depends on the concentration, the particle size distribution, and the time from the initiation of deposition. Based on these physical characteristics, Davis and Hassen (1988) estimated the spreading of the interface during deposition by considering hindered settling effect, particle interaction, and a slight polydispersity of the suspension. Hindered settling effect, caused by increasing sediment concentrations in the lower layers and very important in influencing sediment dynamics near the bed, was shown to lead to a decrease of the interface thickness.

Given a two-layer flow with a distinct interface, erosion of the fluid mud layer is believed to be controlled by the mechanisms of shear flow interfacial instability. This is unlike, for example, the erosion of a settled bed which erodes by dislodgement and entrainment of surface flocs due to applied stress. Liu (1957) was first to suggest the notion of interface instability in sediment transport studies, while attempting to explain ripple formation in moveable beds. Unfortunately his application may not have been entirely appropriate, because the motion of granular beds is more discrete than continuous, as identified by rolling and saltation of individual grains. On the other hand, fluid mud and the overlying water column can be treated in many respects as a system of two miscible continua with different densities and different rheological properties separated by an interface of a certain thickness.

In general, under shear flow conditions, the thickness of the density interface (d) affects the mode of instability that will erode the interface. The mode of instability is equally affected by the thickness (D) of the region of high velocity gradient (Figure 2). In case the mid-axis of the velocity and density gradients coincide, then when d is approximately equal to D, the prime mode of instability is that of the Kelvin-Helmholtz type; when d is smaller than D, the instability is of the Holmboe type. The main characteristic of the Kelvin-Helmholtz instability is the roll-up and pairing of the interfacial vortices (Delisi and Corcos, 1973). The

Holmboe mode of instability is recognized by the sharp crests along the interface which finger alternatively into both fluids (Browand and Wang, 1972).

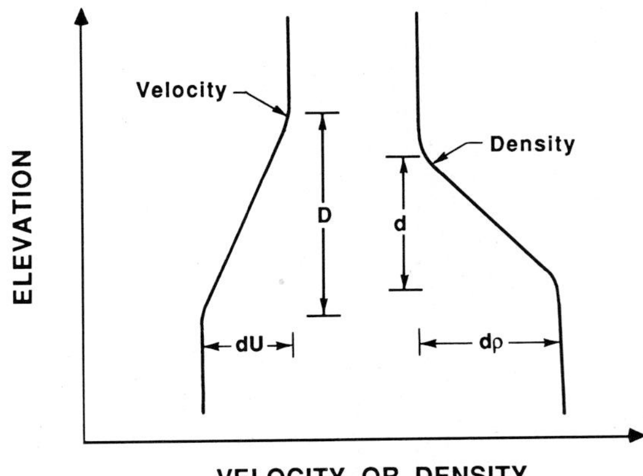

FIGURE 2. Characteristics of density and velocity profiles in the interfacial region.

Large scale internal wave instabilities over high concentration mud underflows have recently been recorded in the Bohai Gulf offshore near the mouth of the Yellow River, China (Wright et al., 1988). These features (Figure 3), thought to be ubiquitous, were of rather high frequency (2.5-5.0 x 10^{-3} Hz) close to the local Brunt-Vaisala frequency (3.3 0.8 x 10^{-3} Hz). They were generated by shear arising from the relative motion between the gravity underflow and the overlying water, and were believed to contribute to underflow deceleration. Undoubtedly, such large disturbances are likely to play a significant role in mixing the mud flow with the upper fluid in which another, more diffused, low concentration plume was observed.

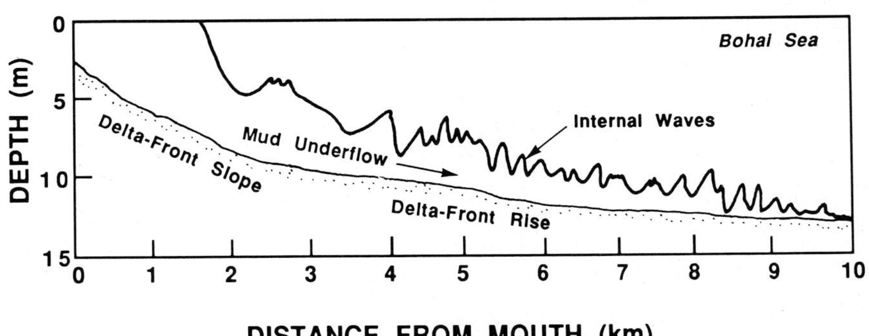

FIGURE 3. High concentration mud underflow over Yellow River delta (adapted from Wright _et al_., 1988).

Instability can be studied by superimposing small perturbations on the mean flow, and representing the continuity and Navier-Stokes momentum equations in terms of the perturbated quantities. By using the theory of normal modes, the linearized governing equation for stability is given by the well-known Orr-Sommerfeld equation (Betchov and Criminale, 1967). The flow is considered as stable if the perturbations decay in time. Analytical solutions of the Orr-Sommerfeld equation are feasible only when the two-phased fluid system under consideration is either inviscid-inviscid or inviscid-viscid, i.e., when there is a velocity discontinuity at the interface. The case of interfacial stability between viscous-viscous fluids requires numerical treatment of the governing stability equation. The case of instability, where the velocity and density gradient axes do not coincide, have been studied by Lawrence _et al_. (1987).

The analytical results obtained from the Orr-Sommerfeld equation provide an understanding of the conditions that lead to instability and breaking of the interface. These results are given in terms of the physical flow characteristics such as flow velocity, viscosity, surface tension, and density variations in the system. For inviscid-viscid flow, which can be considered in a very simplified treatment of the water-fluid mud type problem, the linear stability criterion is given as

$$\{2T^{3/2}/[\nu^2 \rho_2 (g\Delta\rho)^{1/2}]\}^{1/2} >> 1 \tag{1}$$

where T is the surface tension, ν is the kinematic viscosity of the upper fluid, g is the acceleration of gravity, $\Delta\rho = \rho_2 - \rho_1$, and ρ_1, ρ_2 are the densities of the upper and lower fluid layers, respectively. Eq.(1) was generated considering the fact that the time scale of the perturbation must be much shorter than vorticity diffusion in the mean flow. It is illustrative of the number of characteristic parameters which are involved even in this simplified approach. The nature of the assumption made concerning the physical properties of the two fluids critically influences stability. Thus, for example, the stability criterion for viscid-inviscid fluid systems is more restrictive than the criterion for inviscid-inviscid fluids (Hogan and Ayyaswamy, 1985).

A realistic assessment of the instability of the water-fluid mud interface, presently unavailable, requires a more rigorous approach which considers not only the viscous effects, but also the rheological properties of the fluid mud. The non-Newtonian nature of fluid mud complicates the instability problem by introducing variable viscosity, non-zero normal stress differences, and out of phase relationships between stress and rate of strain (Maa, 1986). Depending on the magnitude of the characteristic flow velocity, characteristic length and fluid relaxation time, the fluid mud system can be simplified and treated either as a non-Newtonian fluid with non-zero normal stresses and in-phase stress versus strain rate, or with zero normal stresses and out-of-phase stress versus strain relationship (Harris, 1977). Attention must also be given to the fact that Squire's theorem, which states that two-dimensional flows provide more strict instability conditions than three dimensional flows, may also not be valid for non-Newtonian fluid and, consequently, mud flow. These issues are the subject of ongoing research at the University of Florida, and will be examined in a subsequent publication.

III.2 Entrainment

Once the interface breaks, the sediment particles entrain into the overlying water column, and there is mixing between the fluid mud and the overlying water. Entrainment processes have been extensively studied for the cases of thermal and salinity stratification, but apparently not for fluid muds. However, the role of this mechanism in modeling fine sediment resuspension has been recognized by, among others, McLean (1985). Entrainment is usually quantified as a function of the bulk Richardson number, Ri, which expresses the balance between buoyancy and inertia forces. There are different laws pertaining to the rate of entrainment based on the degree of stratification. The entrainment velocity, u_e, can for example be normalized either by the mean flow velocity, U, the shear velocity, u_*, or the velocity jump, dU. Accordingly, depending on which of these is used as the characteristic velocity, the Richardson number is defined as Ri_U, Ri_*, or Ri_{dU}. Narimousa *et al*. (1986) suggested the shear velocity as the scaling velocity, and proposed the expression:

$$u_e/u_* = aRi_*^{\,b} \tag{2}$$

where a and b are numerical coefficients defined accordingly as: a = 0.65, b = -1/2 for $15 < Ri_* < 150$, a = 7.00, b = -1 for $150 < Ri_* < 800$, and a = 5.00, b = -3/2 for $800 < Ri_*$. These exponential laws can be explained on physical grounds as follows: for low Ri_*, shear is high and the density gradient is small, so that mixing is produced by the larger, energy carrying eddies. For intermediate values of Ri_*, the processes are less chaotic, so that entrainment is more ordered and caused by Kelvin-Helmholtz instability. Finally, for large Ri_*, shear is very weak and entrainment is due primarily to the kinetic energy of the turbulent layer. If the density gradient is extremely large, then entrainment is supressed and the only mixing is caused by molecular processes. In another publication, Narimousa and

Fernando (1987) indicated that the best scaling velocity is the velocity jump at the interface. The explanation of their suggestion was based on the fact that the energy for turbulent mixing of the interface is produced at this high shear zone.

III.3 Experimental observation

Five very preliminary experimental runs were performed at the University of Florida in order to monitor the interfacial instability of a two-layered flow of clear water over kaolinite-water suspensions. The tests were conducted in a 18.3 m long and 0.6 m wide recirculating flume. A description of this flume is given by Dixit (1982). Initially the kaolinite-water suspension was thoroughly mixed through high speed velocity, and then by reducing the velocity substantially, the sediment was allowed to settle. After 10-20 minutes when a distinct interface was formed, the velocity was increased to a predetermined magnitude and the interfacial activity was filmed for a duration of approximately 30 seconds. The initial density jump, $d\rho/\rho_1 = (\rho_2-\rho_1)/\rho_1$, was varied from 0.005 to 0.009.

Results from these experiments showed that for $Ri_U = 19.0$ there was a distinct wave at the interface but no breaking (no turbulent entrainment). For $Ri_U = 5.0$ (Figure 4) there was breaking of the wavy interface, while for $Ri_U = 1.7$ the wave breaking caused smoke-like wisps that were entraining into the upper layer, and resembled the Holmboe type of instability. Furthermore, the interfacial thickness, d, was found to be on the order of 5% of the depth of the mixed upper layer. The observed value of $Ri_U = 19.0$ for stability (non-entrainment) of a water-mud system coincides with the experimental data presented by Narimousa _et al_. (1986), for fresh water-salt water system. However, more data are necessary in order to establish confidence in the range of applicability of the existing state-of-the-art understanding of instability and entrainment to the case of fluid mud systems, as suggested by Odd and Cooper (1989).

Qualitative differences (e.g., in the meaning of coefficients a and b in Eq.(2)) are likely to arise in replacing salt water by fluid mud due to differences in the rate of dissipation of turbulent kinetic energy in these two fluids, as suggested by the recent laboratory results of Srinivas (1989) and Wolanski _et al_. (1989), in which sediment-induced collapse of turbulence was observed at the water-fluid mud interface.

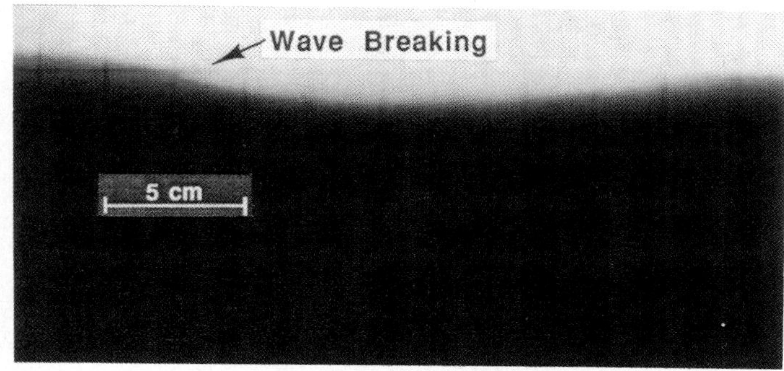

FIGURE 4. Interfacial disturbance at $Ri_U = 5$ ($d\rho/\rho_1 = 0.007$). Wave height ranged from 1.3 to 2.5 cm, period from 3 to 5 s. Grid squares are 2.54 cm².

IV. Microstructure and Density Profile

Continuous monitoring of the vertical density profile in stratified estuaries has indicated the occurrence of a step-wise profile structure. This microstructure has been identified in cases of salinity, temperature, and sediment stratification. Typical profiles of suspended cohesive sediment, recorded a few minutes apart in the Severn Estuary in England, are shown in Figure 5 (Kirby, 1986). The time scale of the microstructure development, in the differentiated (stratified) layer between the mixed upper layer and the main or primary lutocline near the bed, was on the order of 10 to 15 minutes. The general location of the primary lutocline is largely controlled by hindering effects during settling (Ross, 1988), while turbulent diffusion strongly influences mixed layer thickness. Posmentier (1977) presented data for salinity microstructure in the Hudson River with a longer representative time scale on the order of 90 minutes. Another difference between Kirby's and Posmentier's data is that the former documented a dynamic system where the number and position of secondary lutocline steps changed in time, while the latter presented a profile where the microstructure apparently stabilized after a certain time. Since the microstructure is indicative of the vertical stratification process, its understanding is essential for quantifying sediment mass exchange between stationary and mobile sediment suspensions.

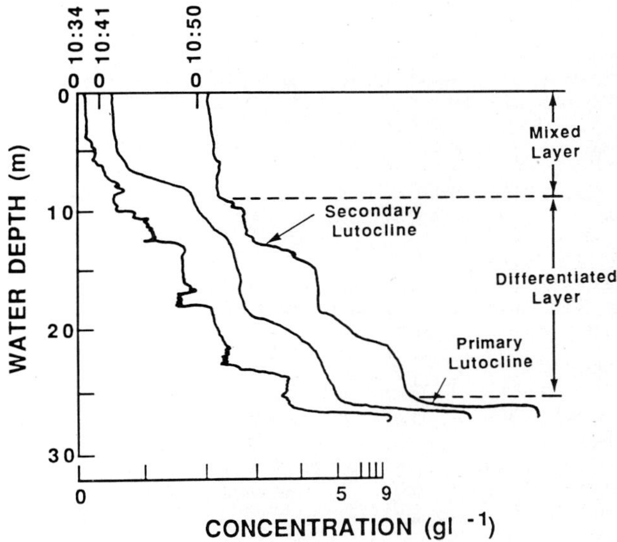

FIGURE 5. Microstructure of three sequential sediment vertical profiles in the Severn estuary, England, taken over a 16 min interval (adapted from Kirby, 1986).

A full understanding of the mechanisms of density microstructure have not as yet been established. There are a number of theories which attribute the step-wise density structure to different causes such as Kelvin-Helmholtz instability, breaking of internal waves, viscous overturning, or salt fingering. A common agreement among these theories is that the microstructure is due to turbulent energy patches resulting in vigorous and uniform mixing which creates sharp gradients with the rest of the density layers. Since the representative length scale of the phenomenon is the profile step, the turbulent eddies affecting the system are expected to be of the same magnitude (Monim and Ozmidov, 1986).

Theoretical analysis of the vertical suspension density distribution under general conditions is quite complicated. Assuming steady state flow conditions, homogeneity along the longitudinal axis, and considering only the entraining phase of flow when gravitational settling effects may be ignored for simplicity, the vertical concentration profile can be studied very approximately by means of the one-dimensional diffusion equation written as

$$C_{,t} = F_{,z} \tag{3}$$

where C is the sediment concentration, F is the sediment flux, z is the vertical axis, and t is time. A comma in front of a subscript denotes partial differentiation with respect to that subscript. Applying Fick's law for turbulence induced fluctuations the sediment flux reads

$$F = - KC_{,z} \tag{4}$$

where K is the coefficient of vertical diffusivity for mass. In sediment stratified systems turbulence is strongly affected by the stratification characteristics (McLean, 1985). Therefore, the buoyancy stabilized coefficient of diffusivity can be expressed by a semi-empirical form as

$$K = A\,(1 + sRi_u)^r \tag{5}$$

where A is the coefficient of vertical eddy diffusivity under neutral stability, and s, r are empirical coefficients (Okubo, 1970). The coefficient A is related to the velocity and density fluctuations in the vertical direction. Using the mixing-length theory, A can be approximated as

$$A \approx (l_o)^2\, |u_{,z}| \tag{6}$$

where l_o is the mixing length in a neutrally stratified flow, and u is the depth average horizontal velocity. For fluid muds the Richardson number can be expressed as (Scarlatos and Mehta, 1988)

$$Ri_u = \{g\, C_{,z}\, [(\rho_s - \rho_w)]/\rho_w\}/u^2_{,z} \tag{7}$$

where ρ_w, ρ_s are respectively the density of water and sediment. Assuming for simplicity a logarithmic profile for the velocity, the mixing length is given as

$$l_o = \{kz\, [1 - (z/H)]\}^{1/2} \tag{8}$$

where k is the von Karman's constant, and H is the depth of the boundary layer. Setting $X = C,_z$, Eqs.(3-4) yield

$$C_{,t} = - F_{,x}\, C_{,zz} \tag{9}$$

Due to the existence of the gradient $F_{,x}$, Eq.(9) is a nonlinear partial differential equation. Posmentier (1977) suggested that when the factor $F,_x$ is negative then the system becomes unstable. As a result of the instability, there is thorough mixing which creates steep density gradients and a more stable system. When stability increases the factor $F,_x$ becomes positive. Combining Eqs.(4-8) yields

$$F_{,x} = R\,[1 + (1 + r)BX]\,(1 + BX)^{r-1} \tag{10}$$

where R and B are defined as

$$R = kz\, [1 - (z/H)]\, |u_{,z}| \tag{11}$$

$$B = s\, g[(\rho_s - \rho_w)/\rho_w]/u^2_{,z} \tag{12}$$

Since R is always positive, the sign of the quantity $F,_x$ depends on the sign of the remaining right hand side of Eq.(10). From field observations it has been estimated that the values for the coefficients s and r are approximately between $3.33 < s < 10$ and $- 2.0 < r < - 0.5$, (Ross, 1988).

It is not known whether all the values between the upper and lower inequality limits of s and r are physically realistic. For $r = -1$, $F,_x$ is always positive, and therefore Eq.(9) is well-posed. In this case, it is not possible to generate instability and consequent microstructure. For $r = -2$ the stability requirement is given as

$$(BX)^2 = (sRi_u)^2 < 1 \tag{13}$$

Assuming $s = 4$, Eq.(4) yields $Ri_u < 1/4$, which is the well-known stability criterion for shear flows (Okubo, 1970). A similar expression can be derived for $r = -0.5$. Whenever there is violation of the stability criterion, internal waves break and a patch of weak turbulence is generated. This turbulent patch spreads horizontally and creates a layer of homogeneous water which causes the generation of microstructure.

Simulation of the microstructure development process can be achieved by solving Eqs.(9-12). The numerical scheme used for the solution was the Newton linearization iterative technique which applies to the governing equation in a form of:

$$C_{,zz} = G(z;\ t;\ C;\ C_{,z};\ C_{,t}) \tag{14}$$

Then, Eq.(14) is expanded as

$$C_{,zz}^{(n+1)} = G^{(n)} + [C^{(n+1)} - C^{(n)}]\ G_{,C} + [C_{,z}^{(n+1)} - C_{,z}^{(n)}]\ C_{,x} + [C_{,t}^{(n+1)} - C_{,t}^{(n)}]\ G_{,(C,t)} \tag{15}$$

where n is the number of iterations (Ames, 1977). The term C_{zz} is discretized by using a central difference scheme. For this numerical scheme' the solution matrix is tri-diagonal while the iteration converges quadratically. Boundary conditions for the system must be specified at the free water surface and at the consolidated bed. At the free surface the concentration is assumed as constant while at the bed a mass flux is defined to essentially account for net erosion of sediment (Mehta, 1986).

Simulated result for a hypothetical case in a 20 m deep unit width channel is presented in Figure 6. In this simulation a logarithmic velocity profile with a surface current of 0.2 m/s was used, and the values for the coefficients s and r were s = 4.0 and r = -2.0. These values for s and r are typical values estimated from field data (Ross, 1988). The time step used was 5 s while the flow depth was discretized into 80 segments for one run and into 50 segments for another. The microstructure generation was however not measurably affected by the number of segments over the depth. It can be clearly seen that starting from a smooth profile, a step-wise profile developed after 90 s of simulation time. The lutocline steps defining the differentiated layer are not however as dramatic as those recorded in nature (Figure 5). Neglect of horizontal advection may be one reason for this discrepancy. For any velocity profile other than logarithmic, e.g. for a bottom reverse current, the vertical mixing coefficient must be redefined since Eqs.(4-8) will not be valid. In general, the mechanics of generation of the differentiated layer is the same as that in a halocline, and is physically explained by the non-linear dependence of diffusion flux on the density gradient (Postmentier, 1977; Ross, 1988).

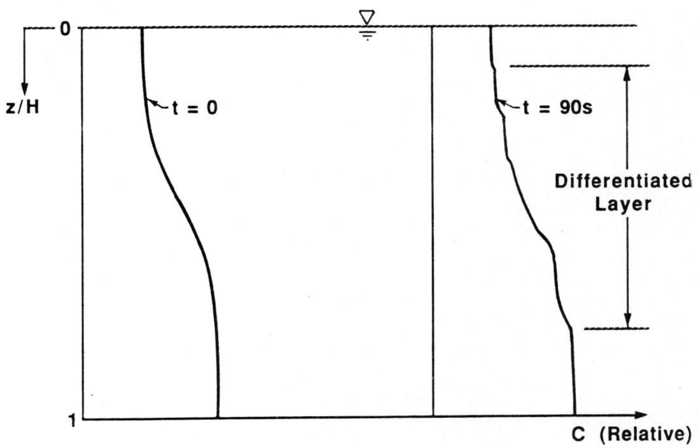

FIGURE 6. Simulated development of suspension density microstructure.

V. Concluding Remarks

It appears that fluid mud erosion and entrainment mechanisms can be simulated by approaches applied to salinity or thermal density flows. Erosion of fluidized mud-water interface depends strongly on the velocity and density gradients, and resembles interfacial instability phenomena typical to stratified flows characterized by haloclines and thermoclines. Laboratory experiments with water and kaolinite-water suspension seem to show a qualitative agreement with other existing instability data. However, due to differences in the rate of turbulence dissipation near the interface, entrainment rates of fluid muds, while undoubtedly Richardson number dependent, are likely to show qualitative differences from those for salt and heat.

The vertical profile of the mobile suspension at the upper water levels exhibits a step-like density microstructure. This microstructure can be simulated very approximately by using a diffusion type equation with an appropriate value for the vertical mixing coefficient. The simulated density "steps" are not however as dramatic as those observed in nature. The discrepancy may result from the neglect of horizontal advection. In addition, the use of the Fickian closure approach is inherently limited by empiricism, and a higher order closure scheme may be essential for quantitative simulation.

Due to the non-Newtonian nature of fluid muds, additional theoretical and experimental research is needed on the instability and entrainment mechanisms in shear flows arising from hydrodynamic forcing under generalized conditions of out-of-phase stress versus strain rates, variable viscosity, and non-zero normal stress differences.

Support from the U.S. Army Engineer Waterways Experiment Station, Vicksburg, MS through contract DACW39-89-K-0012 is acknowledged.

VI. References

Ames, W. F., 1977: Numerical Methods for Partial Differential Equations. Academic Press.

Ariathurai, R., 1974: A finite element model for sediment transport in estuaries. Ph.D. Thesis, University of California, Davis.

Been, K., and G. C. Sills, 1981: Self-weight consolidation of soft soils: an experimental and theoretical study. Geotechnique, **31(4)**, 519-535.

Betchov, R., and W. O. Criminale, Jr. 1967: Stability of Parallel Flows. Academic Press.

Bradley, J. B., R. C. MacArthur and B. J. Brown, 1988: Analytical and numerical modeling at high sediment concentrations. Proc. Hydraul. Eng. Meet., S.R. Abt and J. Gessler Eds., ASCE, Colorado Springs, 230-235.

Browand, F. K., and Y. H. Wang, 1972: An experiment on the growth of small disturbances at the interface between two streams of different densities and velocities. Proc. Int. Symp. Stratified Flows, Novosibirsk, USSR, 491-498.

Bryant, R., A. E. James and D. J. A. Williams, 1980: Rheology of cohesive suspensions. Industrialised Embayments and Their Environmental Problems, M.B. Collins, et al. Eds., Pergamon Press, 279-287.

Davis, R. H., and M. A. Hassen, 1988: Spreading of the interface at the top of a slightly polydispersed sedimentation suspension. J. Fluid Mech., **2**, 107-134.

Delisi, D., and G. M. Corcos, 1973: A study of internal waves in a wind tunnel. Boundary Layer Meteorol., **5**, 121-137.

Dixit, J. G., 1982: Resuspension potential of deposited kaolinite beds. M.S. Thesis, University of Florida, Gainesville.

Harris, J., 1977: Rheology and Non-Newtonian Flow. Longman.

Hayter, E. J., 1983: Prediction of cohesive sediment movement in estuarial waters. Ph.D. Thesis, University of Florida, Gainesville.

Hogan, J. M., and P. S. Ayyaswamy, 1985: Linear stability of viscous-inviscid interface. Phys. of Fluids, **28(9)**, 2709-2715.

Kirby, R., 1986: Suspended fine cohesive sediment in the Severn estuary and Inner Bristol channel. Rep. ETSU-STP-4042, U.K. Atomic Energy Authority, Harwell, United Kingdom.

Kirby, R., and W. R. Parker, 1983: The distribution and behaviour of fine sediment in the Severn Estuary and Bristol Channel. Can. J. Fish. and Aquat. Sci., **40(1)**, 83-95.

Krone, R. B., 1963: A study of rheologic properties of estuarial sediments. Rep. SERL 63-8, Hydraulic Engineering Laboratory, University of California, Berkeley.

Lawrence, G. A., J. C. Lasheras and F. K. Browand, 1987: Shear instabilities in stratified flow. Proc. 3rd Int. Symp. Stratified Flows, Session A.3, California Institute of Technology, Pasadena, 1-13.

Liu, H. -K., 1957: Mechanics of sediment-ripple formation. J. Hydraul. Div., ASCE, paper 1197, 83, 1-23.

Maa, P. -Y., 1986: Erosion of soft muds by waves. Ph.D. Thesis, University of Florida, Gainesville.

McLean, S. R., 1985: Theoretical modelling of deep ocean sediment transport. Mar. Geol., **66**, 243-265.

Mehta, A. J., 1986: Characterization of cohesive sediment properties and transport processes in estuaries. Estuarine Cohesive Sediment Dynamics, A. J. Mehta Ed., Springer-Verlag, 290-325.

Monin, A. S., and R. V. Ozmidov, 1986: Turbulence in Ocean. D. Reidel.

Narimousa, S., R. R. Long and S. A. Kitaigorodskii, 1986: Entrainment due to turbulent shear flow at the interface of a stably stratified fluid. Tellus, **38A**, 76-87.

Narimousa, S., and H. J. S. Fernando, 1987: On the sheared interface of an entraining stratified fluid. J. Fluid Mech., **174**, 1-22.

Odd, N. M. V., and A. J. Cooper, 1989: A two-dimensional model of the movement of fluid mud in a high energy turbid estuary. J. Coast. Res., Special Issue 5, in press.

Okubo, A., 1970: Oceanic Mixing. Management Oceanic Services, Detroit, Michigan.

Parker, W. R., and R. Kirby, 1982: Time-dependent properties of cohesive sediment relevant to sedimentation management. Estuarine Comparisons, V. S. Kennedy Ed., Academic Press, 573-589.

Posmentier, E. S., 1977: The generation of salinity finestructure by vertical diffusion. J. Phys. Oceanogr., **7**, 298-300.

Ross, M. A., 1988: Vertical structure of estuarine fine sediment suspensions. Ph.D. Thesis, University of Florida, Gainesville.

Scarlatos, P. D., and A. J. Mehta, 1988: Microstructure of cohesive sediment suspensions. Proc. Hydraul. Eng. Meet., ASCE, Colorado Springs, Colorado, 218-223.

Scarlatos, P. D., and E. Partheniades, 1986: Numerical simulation of fine sediment motion in estuaries. Proc. Int. Conf. Hydraul. Eng. Software, Southampton University, United Kingdom, 111-123.

Sills, G. C., and McG. Elder, 1986: The transition from sediment suspension to settled bed. Estuarine Cohesive Sediment Dynamics, A. J. Mehta Ed., Springer-Verlag, 192-205.

Srinivas, R., 1989: Response of fine sediment-water interface to shear flow. M.S. Thesis, University of Florida, Gainesville.

Wolanski, E., T. Asaeda and J. Imberger, 1989: Mixing across a lutocline. Limnol. and Oceanogr., (in press).

Wright, L. D., W. J. Wiseman, B. D. Bornhold, D. B. Prior, J. N. Suhayda, G. H. Keller, Z. -S. Yang and Y. G. Fan, 1988: Marine dispersal and deposition of yellow river silts by gravity-driven underflows. Nature, **332(6164)**, 629-632.

A Plume Front in Hangzhou Bay and its Role in Suspended Sediment Transport

Su Jilan, Wang Kangshan, and Li Yan

Second Institute of Oceanography, State Oceanic Administration
Hangzhou, Zhejiang, China

ABSTRACT

Based on hydrographic and anchored-station data it is shown that a persistent NE-SW front exists inside Hangzhou Bay year-round. This front is actually the merging of the secondary Changjiang plume front and the Qiantang plume front from the upstream side of the bay. At neap tides during the low runoff periods these two plumes can readily be identified. Our findings suggest that suspended sediment is transported into the bay through the northern end by the secondary Changjiang plume. The front inside the bay acts to concentrate suspended sediment and transport it southwestward into the bay.

I. Introduction

Hangzhou Bay is the outer reach of the Qiangtang estuary, the largest estuary along the China coast of East China Sea (figure 1). There have been constant accretion of land along the bay's southern shore throughout the Chinese history. Along a 40 km stretch of the south shore near Andong, the land has accreted steadily at an average rate of 20 m/y over the last 600 years. The accreted material consists mainly of silt and fine sand, believed to be derived from offshore sources (Sun and Huang, 1984). The Bay is situated immediately south of the mouth of Changjiang, the fourth largest river in the world in both water and sediment discharge. Therefore, Changjiang river could be an initial and important source of the accreted material inside Hangzhou Bay (Sun and Huang, 1984). However, how the material is transported into the bay is far from being understood.

Between 1981 and 1983 the local provincial coastal zone office conducted several general hydrographic surveys of the bay. We have done some analysis of this set of data (Su and Wang, 1989). The results suggest that a secondary Changjiang river plume front plays an important role in carrying suspended sediment from Changjiang into the bay. One of the shortcomings of these general surveys is that the transects did not cover the northern one-third of the bay area where the secondary plume enters the bay. Another shortcoming is that the anchored section was not placed across the front and thus not suitable to view the frontal structure. Furthermore, no observations were made during the neap tides.

Recently we have conducted observations in Hangzhou Bay to study this plume front. We find that after the secondary Changjiang plume enters the bay its front aligns with the Qiantang river plume front, at least during medium and spring tides. During neap tides in low runoff season the two plume fronts may be disconnected. The front(s) appears to transport suspended sediment southwestward into the bay.

II. The Oceanographic Setting of Hangzhou Bay

Hangzhou Bay is situated along a meso-tidal coast. The bay is shallow (10 m average depth) and relatively flat. The tidal range increases rapidly inside the bay owing to the sharp reduction of the bay width. West of the bay there is a

Coastal and Estuarine Studies, Vol. 38
R. T. Cheng (Ed.)
Residual Currents and Long-term Transport

strong tidal bore. Sediment in Hangzhou Bay is predominantly fine and medium silt which can be easily resuspended. However, in spite of the strong tidal currents, there has been only slight change of the bathymetry in the east half of the bay over the last 60 years.

Rivers upstream of the bay discharge altogether an average water flow of 42 km^3/y and an average suspended sediment load of 7.9 x 10^6 t/y. All but the finest portion of the sediment discharged upstream of the bay is trapped in a massive

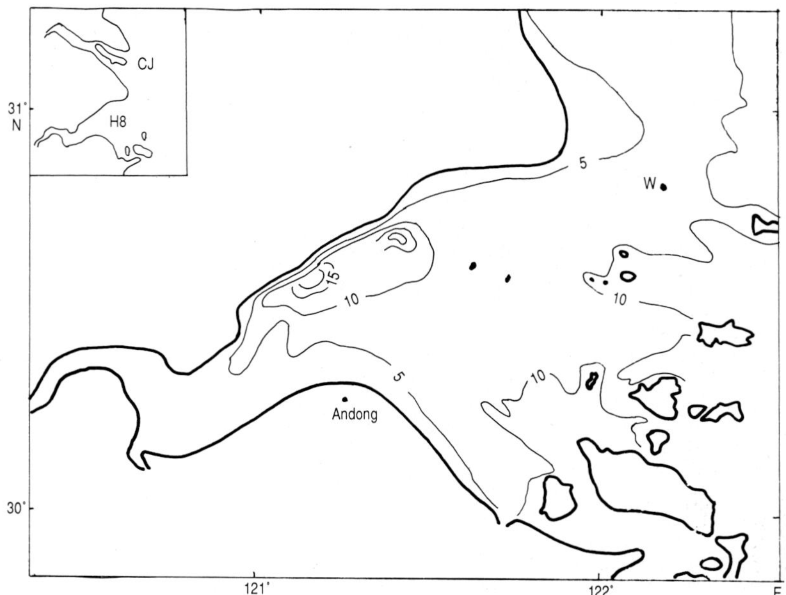

FIGURE 1. Bathymetry in Hangzhou Bay (units are in meters; W denotes position of the Dajishan wind station). In insert map: CJ-Changjiang; HB-Hangzhou Bay.

sandbar under the tidal bore. Immediately to the north of the bay Changjiang carries offshore average water and sediment discharges of, respectively, 925 km^3/y and 486 x 10^6 t/y. Consequently, lower salinity water is usually found along the northeastern part of the Hangzhou Bay (Hu et al., 1986). Mineral composition of the Hangzhou Bay sediment has also been shown to be similar to that of the Changjiang estuary sediment (Yan and Hu, 1986)

Offshore of the bay close to the 50 m isobath there is a strong coastal current, the Taiwan Warm Current, which flows northward all year around. This is believed to originate from the Kuroshio (e.g. see Su and Pan, 1987). During winter its water is significantly warmer (and saltier) than coastal water and hence it can readily be identified by a warm tongue at the sea surface (figure 2). Influence of Taiwan Warm Current can even reach inside the Hangzhou Bay, as is evidenced by the warmer water at the southeast part of the bay in winter (figure 7a). In summer the Taiwan Warm Current can be traced by its subsurface high-salinity and low-temperature characteristics, and consequently the southeast part of the bay has a slightly lower temperature than the rest of the bay (figure 8b).

In winter strong northerly winds drive the Changjiang river plume all the way south along the China coast. The primary Changjiang plume front is located about 50 km away off the bay mouth as determined by the position of the 30 ppt isohaline along the sea floor. Water and sediment discharges from Changjiang are much higher in summer and then the primary plume front is about 100 km seaward of the bay. However, combination of several factors, namely, mild southerly winds, stronger Taiwan Warm Current, and local topography, help to turn the plume water northeastward (Yu et al., 1983). As a consequence, the intruding offshore water at the southeast part of Hangzhou Bay is more saline and less turbid in summer than in winter. In addition to the primary plume front Changjiang also has a secondary plume front which appears to enter Hangzhou Bay near the northern end of the bay mouth year-round (Su and Wang, 1989).

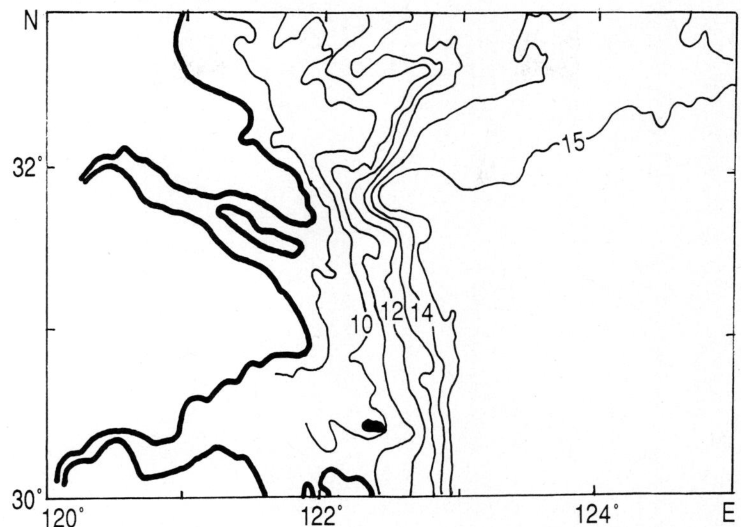

FIGURE 2. Sea surface temperature (°C) in coastal sea near Hangzhou Bay on 21-22 December 1987 based on infrared satellite imageries of NOAA-9 (Courtesy of Sha Xingwei).

III. The Data

Two observation programs were made by us in Hangzhou Bay, one in winter (13-24 December 1987) and the other in summer (22 July - 2 August 1988). Figure 3 shows the transects and the mooring station of both cruises. One current meter was moored at about 4 m above the bed throughout each observation period. The transects were traversed seven times by different vessels concurrently. Periods of the seven runs are indicated in figure 4, along with the corresponding current speed variation as recorded at the mooring. As a time reference, we mention that the tidal phase difference between the southeast end of Transect D and the southeast end of Transect H is about 10 minutes and that between either end of Transect H about 1 hour (see figure 3 of Su and Xu, 1984). Twenty-five-hour anchored measurements were also made during each cruise.

The locations of the winter neap-tide anchored stations are also marked in figure 3. At the winter spring-tide observation C1, C3 and C6 were about 8 km WSW, 1.5 km SSE, and 3.5 km SE of, respectively, the positions indicated in figure 1. At the summer observation C2, C3, C4, and C6 were all displaced between 3-5 km SE of the corresponding winter neap-tide locations. In addition, we also tracked floats equipped with a cross vane at 2.5-3 m depth below the sea surface. In winter the tracking was performed by one vessel and it lasted for four tidal periods (51 hours) at neap tides. Two vessels were used in summer to carry out trackings concurrently in two consecutive daytimes at neap tides. In addition, one daytime tracking was also done with one vessel during spring tides in summer. Tracking periods are marked in figure 5.

Current measurements were made at the anchored stations every hour at 1 m below the surface, six-tenth of the water depth level, and 1 m above the bottom. Water samples were taken at the same levels during both winter and summer cruises for salinity and total suspended matter (TSM) determinations. Along the transects only the salinity and TSM were measured at the same levels. During the summer cruise, in addition to the methods described above, one Partech TCDS-81 and two Partech

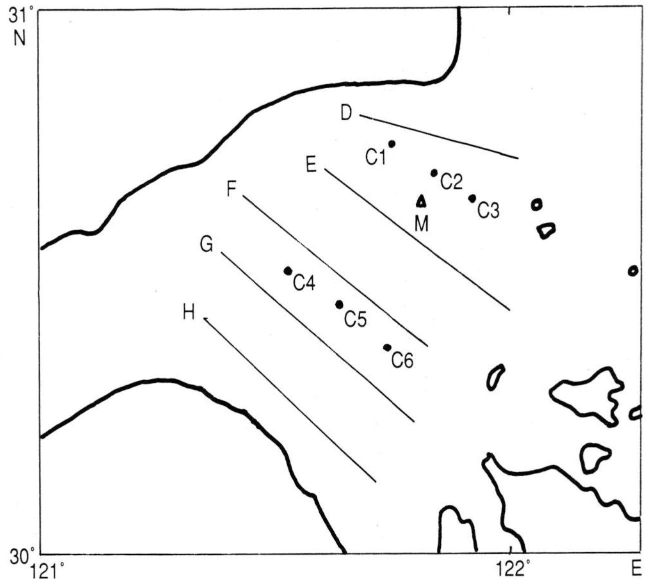

FIGURE 3. Locations of transect lines, mooring station M (triangle), and anchor stations in winter neap-tides C1-C6 (solid circles). Transect D is extended farther SE by 18 km in the summer observations.

7000-3RP were also used, respectively, for the salinity and turbidity measurements on the transect runs. The turbidity readings were converted to TSM data with empirical formulas determined in our laboratory.

The prevailing wind in Hangzhou Bay is from the south in summer and the north in winter. Figure 5 shows the wind vector stick diagrams for the two periods of observations. The wind station Dajishan is located off the northern end of the bay mouth (figure 1).

IV. Fronts in Hangzhou Bay

IV.1 Hydrographic structure

Figure 6 gives the bottom salinity distributions during the Transect-run Nos. 2 and 3, both traversed in the SE directions. In the winter neap-tide run there are evidently two plumes of low salinity water (figure 6a). The one near the bay mouth is the secondary Changjiang river plume mentioned before in Section II and the other is the plume of the Qiantang river discharge from the upstream side of Hangzhou Bay. Large tidal execusions during spring tides obviously result in the alignment of these two plume fronts into a single NE-SW oriented front (figure 6b). Previous transect observation in the southern two-thirds of the bay during medium and spring tides also showed similar frontal structure (Su and Wang, 1989). The transects are covered by different vessels concurrently over a period of 3-5 hours in a bay where the tidal currents are quite strong. To ascertain whether the salinity distributions in figure 6 give a reasonable representation of the frontal structures we choose two infrared satellite imageries (figure 7). Figure 7a was taken around the maximum ebb the next day after Transect-run No.2 was conducted. The surface isotherms show similar pattern as the bottom isohalines in figure 6a except for the area northeast of the south shore of the bay. One of the reasons for this discrepancy may be due to the fact that this area was observed in the early phase of the ebb during the Transect-run No. 2, whereas in figure 7a this area was likely to be in the late ebbing stage. Another possible explanation is because of the

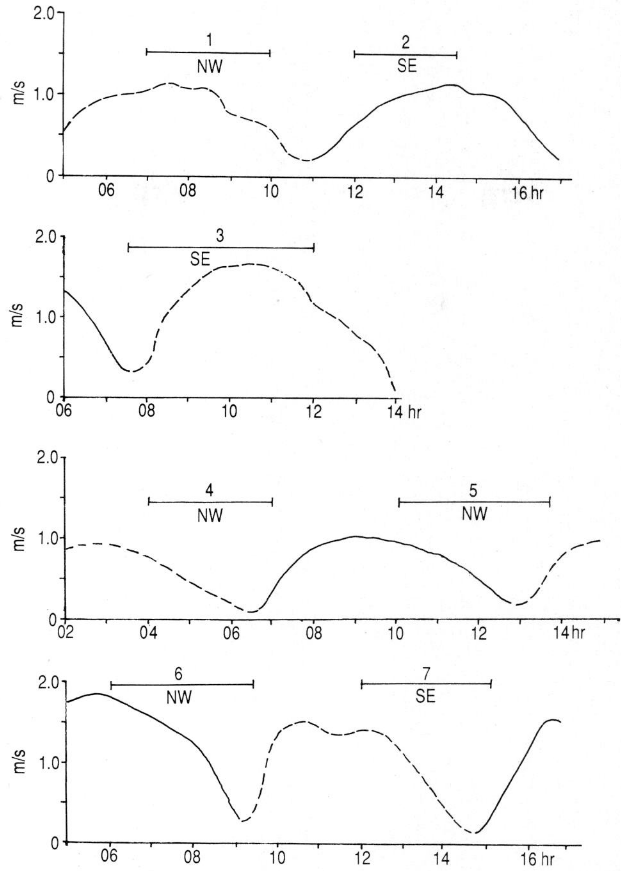

FIGURE 4. Time-spans of Transect-run Nos. 1-7 vs current speed records from the mooring station (solid lines denote ebbing and dashed lines flooding). NW or SE indicates direction of traversing of each transect-run.

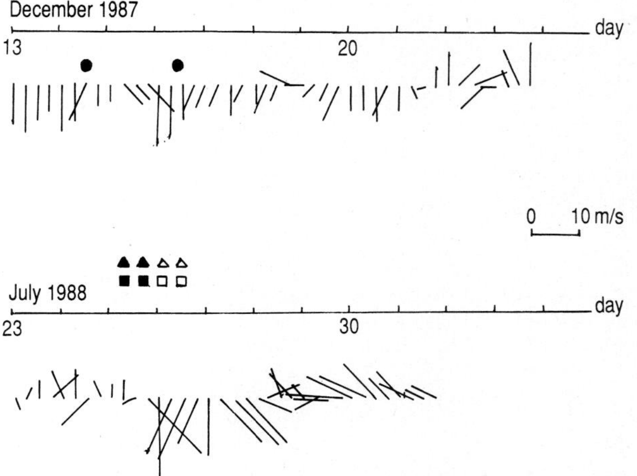

FIGURE 5. Periods of float-trackings vs 6-hourly Dajishan wind vector stick diagrams. Circles, triangles, and squares denote the start and end of each tracking (see figure 11).

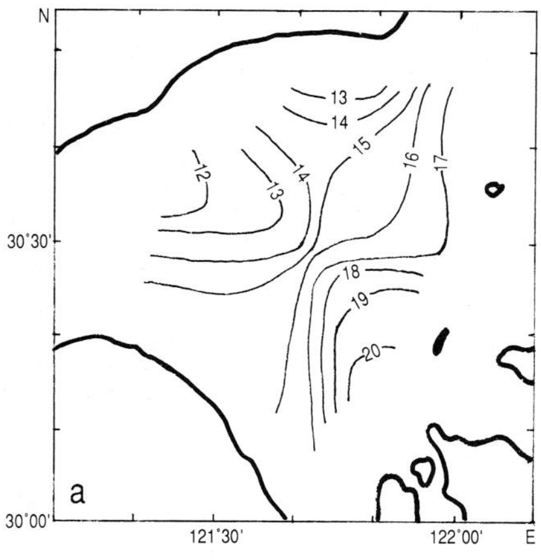

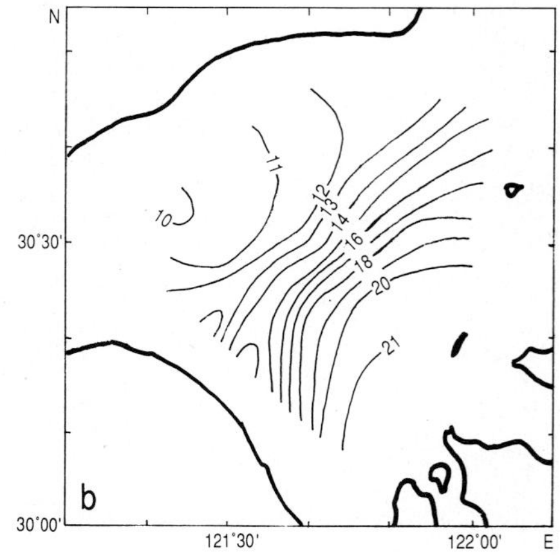

FIGURE 6. Bottom salinity distributions inside Hangzhou Bay in winter during (a) neap tide (Transect-run No. 2) and (b) spring tide (Transect-run No. 3).

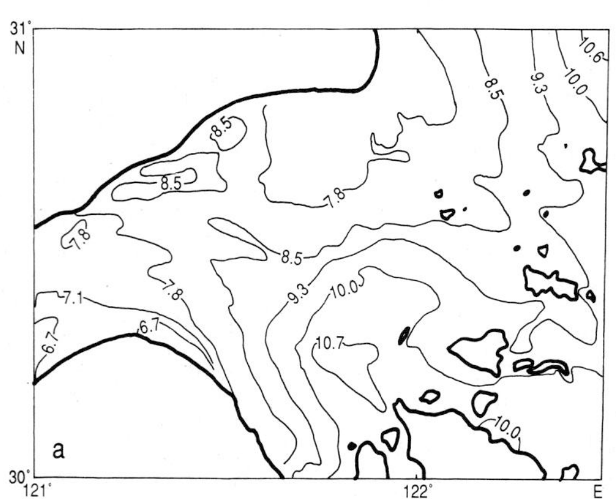

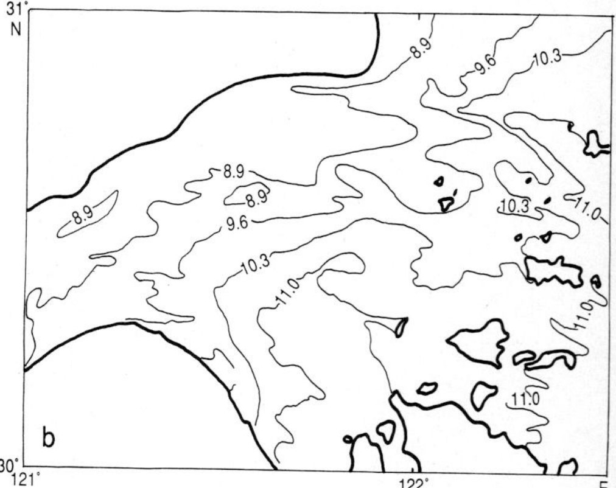

FIGURE 7. Sea surface temperature (°C) in Hangzhou Bay based on infrared satellite imageries of NOAA-9 at (a) 14:17 on 18 December 1987 and (b) 15:09 on 23 December 1987.

massive tidal flats there, over which the water tends, in winter, to be colder. Figure 7b was taken around the end of flood tide at the mooring station two days after Transect-run No. 3 was completed. We see that the isolines in figures 6b and 7b are compatible.

As figure 8a shows, in summer the water near the north shore of Hangzhou Bay is less saline than the winter case, because runoffs from both Changjiang and Qiantang rivers are larger in summer. The large runoffs also resulted in the merging of the secondary Changjiang river plume and the Qiantang river plume even during neap tides (figure 8a). Since, as disscussed earlier, the main Changjiang plume is turned northeastward in summer due to both southerly winds and a stronger Taiwan Warm Current, influence of the saline Taiwan Warm Current inside Hangzhou Bay is thus stronger in summer than in winter. Consequently the maximum salinity value of the bottom water inside the bay is higher than the winter values. This strong saltier offshore water also pushes the summer salinity front closer to the north shore of the bay than the winter front. Figure 8b shows the sea surface temperature distribution in Hangzhou Bay from an infrared satellite imagery taken 15 days before the salinity map in figure 8a was observed. The tidal phase of this imagery probably corresponds to the earlier stage of an ebbing neap tide at the mooring station. Again, the patterns in figures 8a and 8b are compatible.

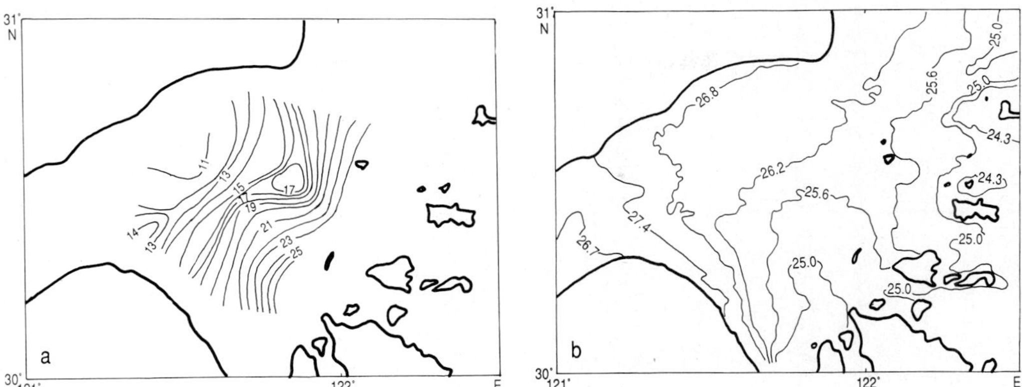

FIGURE 8. (a) Bottom salinity distribution inside Hangzhou Bay in summer during neap tide (Transect-run No.5); (b) sea surface temperature (°C) in Hangzhou Bay based on infrared satellite imagery of NOAA-9 at 08:04 on 8 July 1988.

Frontal structures can readily be identified in salinity profiles along all transects of all runs except for Transect H during the two neap-tide runs in winter. In figures 9 and 10 we give the salinity profiles of Transect D and F from four transect-runs, one during neap and one during spring in both seasons. In winter (figure 9) the salinity is mostly vertically homogeneous except for stations at the southeast ends of the transects at neap tides. In summer (figure 10) the water is strongly stratified at neap tides which becomes well-mixed at spring tides. To show the fronts more clearly transect-runs conducted during the ebb tides are used in figure 10 because the front is closer to the north shore of the bay in summer than in winter (figures 6-8). We note that in summer lenses of fresher water are found at the surface near the northern end of the bay mouth during neap tides, due to large Changjiang runoff and weak wind stirring.

IV.2 Residual currents

Paths of the drifting floats are depicted in figure 11. In winter (figure 11a) the floats were first deployed in the frontal area at the northern end of the bay mouth.The area was located by its sharp salinity gradient ($\Delta S = 4$ in 4.5 km). Surface salinity near the floats was around 12.5 when tracking began and rose rapidly to around 15 in 10 hours, indicating that the northerly wind then was driving the floats across the sharp salinity gradient zone. The surface salinity along the track then fluctuated between 15 and 16 for the next 29 hours until a strong northerly wind event drove the floats rapidly to an area with salinity around 17.5.

In summer the floats were dropped near the winter tracks. At neap tides the surface salinity near the floats in the NE area fluctuated between 14.5 and 16 along the first daytime track and varied only slightly around 14.3 along the second daytime track, while in the SW area it fluctuated over 13-13.8 and 14.2-15.3, respectively. We did only one spring tracking which was done in summer over just one daytime period. This time the surface salinity along the track stayed almost constant at 13.7.

All the floats deployed during neap tides show a net south-west drift (figure 11). In the first two tidal periods (25 hours) of the winter neap-tide tracking the floats drifted 9.6 km, i.e. 0.11 m/s average speed, in the SSW direction. Average

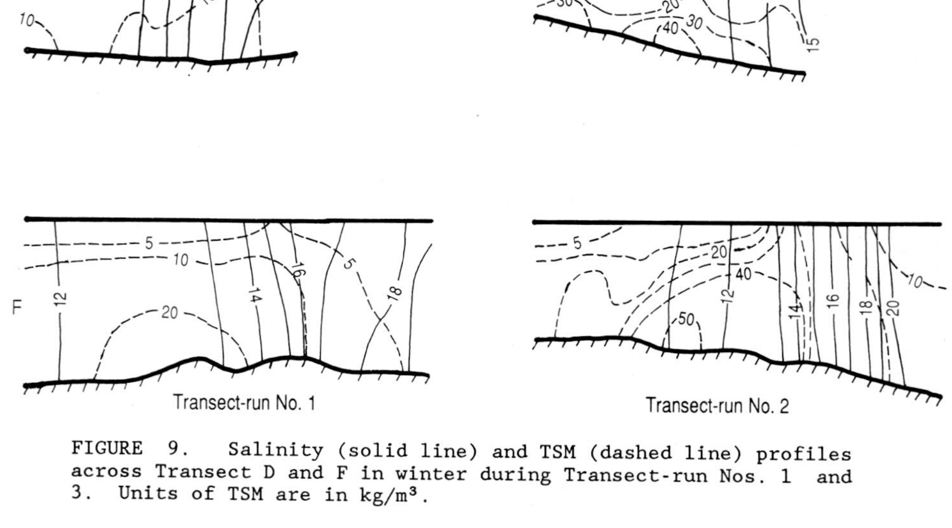

FIGURE 9. Salinity (solid line) and TSM (dashed line) profiles across Transect D and F in winter during Transect-run Nos. 1 and 3. Units of TSM are in kg/m³.

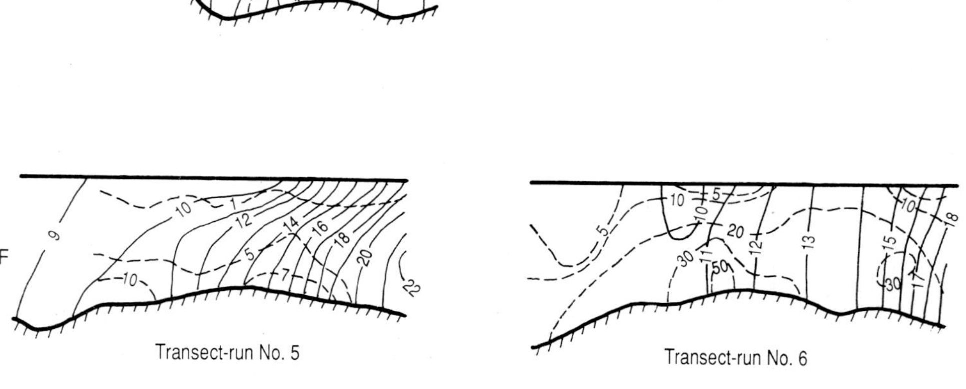

FIGURE 10. Salinity (solid line) and TSM (dashed line) profiles across Transect D and F in summer during Transect-run Nos. 5 and 6. Units of TSM are in kg/m³.

wind speed over the same period was 5 m/s from the north which yielded a wind-induced surface drift of about 0.05 m/s (Wu, 1975). Wave-induced surface drift was around 0.02 m/s, again using an equation provided by Wu (1975). We shall take both these drift velocity to be southward. A two-dimensional numerical M2-tidal model with neap-tide amplitudes predicts a tidal residual velocity of 0.03 m/s directed at 16° from true North in this area. This numerical results agrees with the general trend of the low-passed current residuals of the mooring data of both the winter

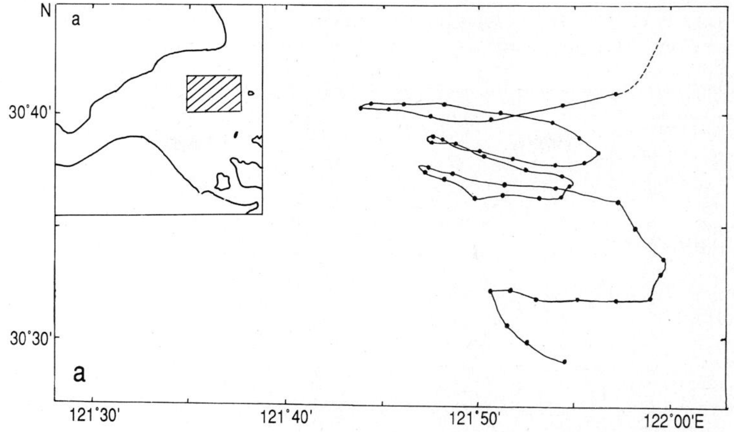

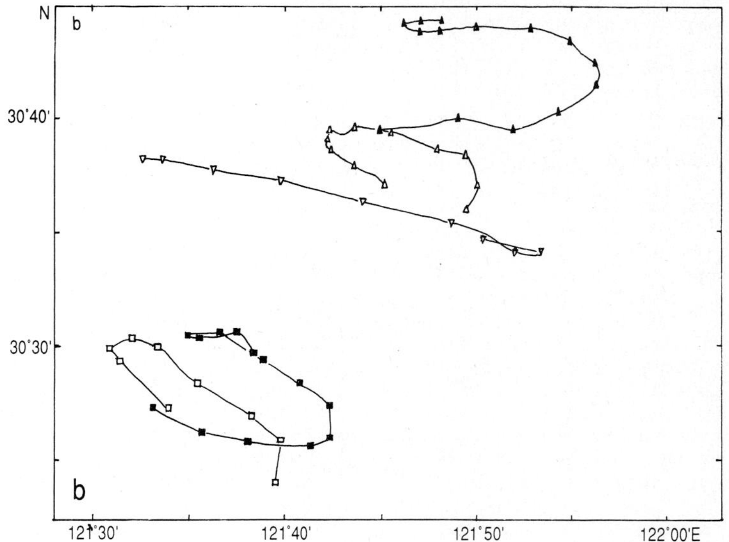

FIGURE 11. Tracks of drifting floats: (a) winter 1987; (b) summer 1988. Inverted triangles denote floats tracked from 09:00 to 17:00 on 2 August 1988. For other symbols see figure 5 for deployment time interval. Time span between two consecutive symbols is one hour.

and summer observations. The predicted Stokes drift is about 0.03 m/s at 296° from true North. The above analysis gives an estimated net drift of 0.08 m/s SSW, i.e., slightly to the south of the along-front direction.

Float-tracking durations were short in summer. The first neap-tide tracking duration lasted 13 hours, i.e., one tidal period. The floats in the NE area drifted 10.2 km, i.e., 0.22 m/s average speed, in the SSW direction, while those in the SW area drifted at an average speed of 0.21 m/s in the WSW direction. Average wind during both periods were from the south. There seems no significant net drift of the floats during the spring tide (figure 11b).

Tidal average surface velocities at the winter anchor stations are given in Table 1. At neap tides, the residuals were generally towards the south. At C3 and C6 the residuals were close to one knot and directed SSW. At spring tides the

Table 1. Tidal residual surface current speed (10^{-2} m/s) and direction (degree from true north) at anchor stations

Station	C1	C2	C3	C4	C5	C6
Winter 1987-neap tides*						
speed	18.5	16.0	46.7		5.8	53.9
direction	208	285	210		140	202
Winter 1987-spring tides**						
speed	21.1	29.3	49.5	26.6	35.6	39.2
direction	145	181	44	17	198	219

* Observation at C4 was aborted half way

** (C1, C3, C5) observed on 21-22 December and (C2, C4, C6) on 22-23 December

residuals were mostly southward except those at C3 and C4. Whether these were due to the prevailing southerly wind then (figure 5), as well as large northerly tidal residual currents at spring tides, is not certain. The speeds of the residuals at the two lower levels of the anchor stations are mostly smaller than or around 0.1 m/s. We judge them to be within the error of tidal residual computation with a 25-hour data set of large tidal currents.

V. Suspended Sediment in Hangzhou Bay

V.1 Fronts and TSM

Transect profiles of TSM are also plotted along with the salinity profiles in figures 9 and 10. High bottom TSM is found near the salinity frontal face across each transect. Similar correlation is also found in all other transect profiles of salinity and TSM except for Transect H during Transect-run Nos. 1 and 2 when the salinity front cannot be indentified there. Since stratification is weak in winter the high bottom TSM near the frontal face seems to be reflected in the surface TSM as well. Figure 12 shows the surface TSM based on visible light satellite imagery taken at the same time as the imagery in figure 7a. It is seen that the northern part of the temperature front more or less coincides with areas where the surface TSM are higher than neighboring waters. The same is also true, though less obvious, between, respectively, figures 7b and 8b and their companion TSM imageries (figures omitted).

Figure 9 shows that, in winter, the TSM level near the salinity frontal face on Transect F is higher than that on Transect D. In fact, in winter the TSM level near the salinity frontal face increases as we move from the bay mouth to the south shore of the bay. In contrast to the winter observations, no such clear trend can be found among the transect profiles of TSM in summer. TSM inside the bay in general is smaller in summer than in winter, probably due to stronger winds in winter. The only exception is during Transect-run No.6 when the TSM values were unusually high along Transect D (figure 10). No satisfactory explanation can be found.

V.2 Anchored station results

The bay-mouth section (C1-C3) and the middle-bay section (C4-C6) are chosen so that they cut across the front in Hangzhou Bay. They are also more or less aligned with the major axis of the elongated tidal ellipses at every station. Tidal average salinity and TSM at these stations are listed in Table 2.

A few features about the salinity front and TSM distribution can be seen from Table 2. At neap tides stratification is significant only at the southeast end of either section in winter but across the whole section in summer. At spring tides the bay water is basically well-mixed in either season. In winter, the mean position of the salinity frontal face lies between the middle and the southeast-end stations of either section at neap tides, while during spring tides it shifts northward near the northwest end of both sections. In summer the mean position of the salinity frontal face seems to be located beyond the northwest end of either section, i.e., closer to the north shore of the bay.

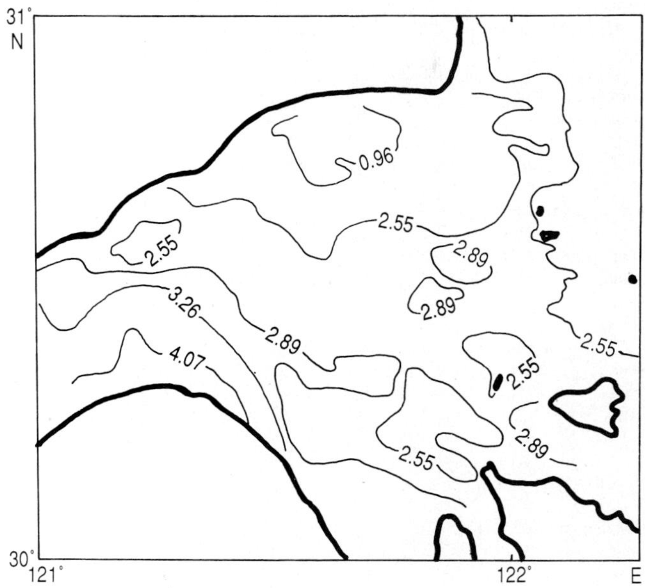

FIGURE 12. Surface TSM (kg/m^3) based on visible light satellite imagery of NOAA-9 at 14:17 on 18 December 1987.

At neap tides in winter, the mean bottom TSM level across the bay-mouth section increases towards the southeast end and that across the middle-bay section has the opposite trend. The same trends are also found for the mean TSM throughout the water column across respective sections during spring tides in winter. In addition the maximum mean TSM level across the middle-bay section is much higher than that across the bay-mouth section.

The summer observation program was not as complete as the winter one. We find that at neap tides the mean TSM is in general much reduced than the corresponding winter neap-tide value. In contrast to the winter case the mean TSM throughout the water column increases towards the northwest end across the bay-mouth section. At spring tides the mean TSM distribution across the middle-bay section shows the same characteristics as in winter, except that the summer surface values are noticeably lower than the winter values.

VI. A Model of Suspended Sediment Transport in Hangzhou Bay

James (1984) computed the three-dimensional current field of three types of two-dimensional fronts, using both a Munk-Anderson model and a turbulence closure model for the eddy coefficients. One of his model type (figure 13) is similar to the frontal structure in Hangzhou Bay except during neap tides in summer. Compared with James' model the front in Hangzhou Bay is stronger in cross-front density difference but broader in width. The water depth used in James' computation is 40 m while average depth here is 10 m. We shall use the qualitative features of James' model result to discuss the role of the front in suspended sediment transport in Hangzhou Bay.

Table 2. Tidal average salinity and TSM (kg/m^3) at anchored stations

Station	C1	C2	C3	C4	C5	C6
Winter 1987-neap tides*						
Tidal average salinity						
surface	13.8	14.4	15.6		12.9	14.0
0.6H	14.2	15.1	17.7		13.0	17.0
H	14.4	15.6	18.5		13.4	18.5
Tidal average TSM						
surface	0.23	0.22	0.18		0.34	0.17
0.6 H	0.54	0.42	0.55		0.38	0.40
H	0.85	0.85	1.06		1.23	0.75
Winter 1987-spring tides**						
Tidal average salinity						
surface	13.3	15.7	16.4	11.9	14.5	19.3
0.6 H	13.4	15.9	16.6	11.7	14.7	20.1
H	13.4	16.1	16.7	11.6	14.8	20.2
Tidal average TSM						
surface	0.59	0.68	0.81	1.13	1.43	0.71
0.6 H	1.13	1.36	1.49	3.69	3.28	2.00
H	1.52	2.34	2.28	4.79	3.91	2.43
Summer 1988***						
Tidal average salinity						
surface	13.3	14.3	19.3	11.3	13.9	19.1
0.6 H	14.9	19.0	22.8	11.2	14.0	20.3
H	15.2	19.6	23.4	11.1	14.1	20.7
Tidal average TSM						
surface	0.07	0.05	0.02	0.71	0.76	0.46
0.6 H	0.22	0.13	0.05	2.72	2.96	1.88
H	0.73	0.39	0.14	4.15	4.11	2.85

* Observation at C4 was aborted half way
** (C1, C3, C5) observed on 21-22 December and (C2, C4, C6) on 22-23 December
*** C1-C3 observed at neap tides and C4-C6 at spring tides

In figure 13 the current field of the front similar to our cases is sketched after James' turbulence closure result. Strong cyclonic along-front currents centered near the surface area of the low-density interface are induced, while close to the bottom there is a weak counter-current. In the lower layer the cross-front circulation concentrates material on both side of the front. There is also a tendency for suspended matter to disperse slowly through the front from the high-density side of the front to the low-density side.

As mentioned in the Introduction, sediment is transported from offshore to Hangzhou Bay and Changjiang is believed to be the primary source. There are two possible ways that suspended sediment from Changjiang may enter Hangzhou Bay. One is by the sediment laden flow near the south shore of Changjiang (Su and Wang, 1986), which entered directly into the bay at its northeast end. Suspended sediment entering the bay through the rest of its mouth is borne by the primary Changjiang plume water whose front lies 50 km or more off Hangzhou Bay, farther in summer. In summer southward excursion of the primary Changjiang plume is limited and most of its suspended sediment is believed to deposit off the Changjiang river mouth and resuspended by winter storms (Milliman et al., 1985). In winter this plume carries the resuspended material all the way south. Consequently, more suspended sediment is carried into the bay by the primary Changjiang plume water in winter than in summer.

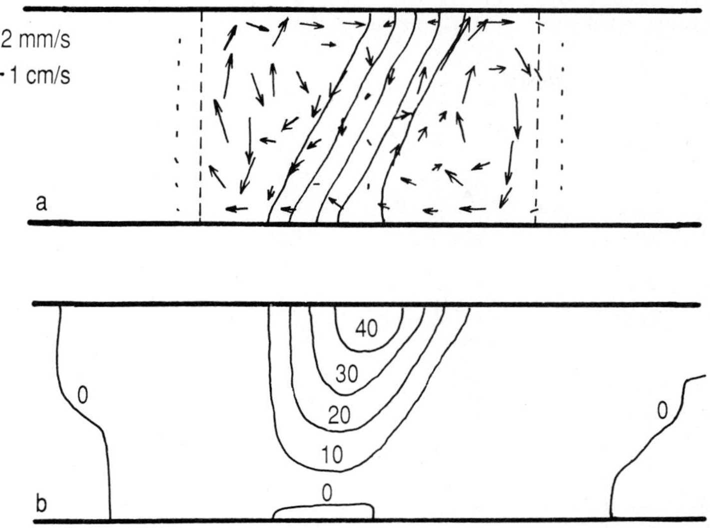

FIGURE 13. Schematic diagram of current field of a front: (a) cross-front circulation (solid lines denote isopycnals and dashed lines limits of stratification), (b) along-front velocity (10^{-2} m/s contour). (after James, 1984)

There are two plumes in Hangzhou Bay. One is the plume of the less saline water discharged near the south shore of Changjiang, i.e., the secondary Changjiang plume. The other is the Qiantang river plume. At neap tides during low runoff seasons these two plumes are somewhat distinct, otherwise they merge together to form a single NE-SW oriented front (figure 14).

As pointed out earlier, most of the sediment from Qiantang is believed to be trapped upstream. However, the secondary Changjiang plume carries significant amount of suspended matter into the Hangzhou Bay all year around (Su and Wang, 1989). Thus there is always high concentration of TSM in the lower layer on the low-density side of the front in Hangzhou Bay. Significant suspended matter is carried into the bay by the primary Changjiang plume water only in winter. As a result, high TSM in the lower layer on the high-density side of the front is found only in winter. Stratification is usually weaker on the low-density side of the front, where penetration of tidally resuspended material into upper layers is more frequent. When tidal resuspension does bring the suspended matter into the upper water column, the along-front current will move it in the southwest direction into the bay. Therefore, TSM increase as we move into the bay along the front in spite of the fact that the bed material there is coarser than that near the bay mouth (see figure 13 of Su and Wang, 1989).

One major shortcoming in the above discussion is that no distinguishment has been made between the contributions of frontal dynamics and tidal dynamics to the TSM distributions. Because of the strong tidal currents in Hangzhou Bay, at any instant tidal resuspension probably contributes to most part of the TSM at a station (figure 15). As shown in figure 4 the seven transect-runs begin at various tidal stages of both neap and spring tides in both winter and summer. The transects also have been traversed in opposite directions. Nevertheless, all the transect profiles across Transect D to G show that high TSM levels are found near the frontal face. In addition, the distributions of the mean salinity and TSM at the two anchored-station sections also support our supposition of the dynamic role that the front plays in the suspended sediment transport in Hangzhou Bay. Finally, we have also analyzed 5, 11, 12, and 5 visible light satellite imageries of Hangzhou Bay in spring, summer autumn, and winter, respectively. These imageries were take at various tidal phases between 1976 and 1987. For each imagery we grade the bay area into two classes, namely, the turbid area (assigned a value of unity) and a less turbid area (assigned a value of zero). If there are more than one imagery at any particular tidal phase an average is taken. These assigned numbers are then added

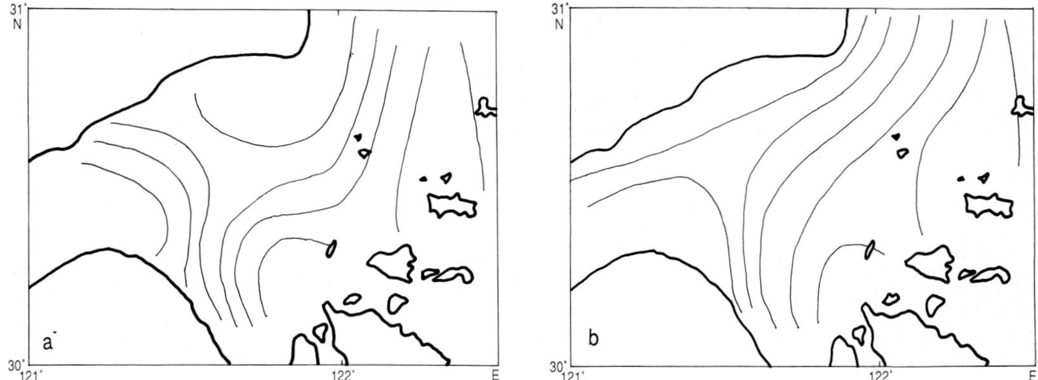

FIGURE 14. Schematic diagrams of the isopycnals of the plume front in Hangzhou Bay: (a) at neap tides during low runoff seasons; (b) at normal and high runoff seasons or at medium and spring tides.

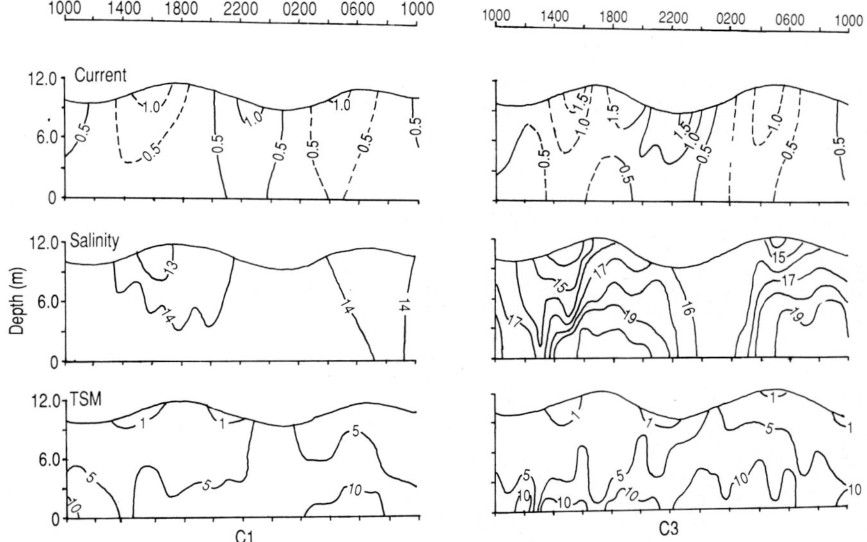

FIGURE 15. Time-series profiles of current speed (ebb-solid line; flood-dashed line), salinity, and TSM at stations C1 and C3 during neap tides in winter. (Speed in m/s and TSM in kg/m³).

together over various tidal phases for each season. The results show that in both spring and summer, i.e., the high run-off seasons, the turbid zone (areas with high number) is located closer to the north shore than that in autumn and winter. The turbid zone is always aligned in a generally NE-SW direction in all seasons. These results are in general agreement with the conclusions drawn from our observation data.

VII. Summary

Observations under different seasonal and tidal conditions all showed the existence of a NE-SW oriented front in Hangzhou Bay. Its part next to the bay mouth is actually the secondary Changjiang plume while the rest is the Qiantang plume. This two plumes may be distinct at neap tides in low runoff seasons. High concentration of suspended matter are found to be related to the front. There are also evidences of a net drift along the front with the low density water on the right side.

Our findings suggest that suspended sediment is transported into the bay through the northern end by the secondary Changjiang plume and through the rest of the baymouth by the primary Changjiang plume water, the latter being significant only during winter. The front acts to concentrate suspended sediment and transport it southwestward into the bay.

Acknowledgement

This work is supported by both Grant-4860244 from the National Natural Science Foundation of China and Grant-188011 from the Zhejiang Provincial Natural Science Foundation.

VIII. References

Hu, F. X., G. C. Gu, D. A. Pan, L. R. Wang, J. T. Xu and Z. C. Zhang, 1986: A numerical solution for 2-D salinity diffusion in Hangzhou Bay. Oceanologia et limnologia Sinica, **17**, 329-338 (in Chinese).

James, I. D., 1984: A three-dimensional numerical shelf-sea front model with variable eddy viscosity and diffusivity. Continental Shelf Research, **3**, 69-98.

Milliman, J. D., H. T. Shen, Z. S. Yang and R. H. Meade, 1985: Transport and deposition of river sediment in the Changjiang estuary and adjacent continental shelf. Continental Shelf Research, **4**, 37-46.

Su, J. L. and Y. Q. Pan, 1987: On the shelf circulation north of Taiwan. Acta Oceanologica Sinica, **6** (supp. I), 1-20.

Su, J. L. and K. S. Wang, 1986: The suspended sediment balance in Changjiang Estuary. Estuarine, Coastal and Shelf Science, **23**, 81-98.

Su, J. L. and K. S. Wang, 1989: Changjiang river plume and suspended sediment transport in Hangzhou Bay. Continental Shelf Research, **9**, 93-111.

Su, J. L. and W. Y. Xu, 1984: Modelling of the depositional pattern in Hangzhou Bay. In: Proceedings of the Nineteenth International Coastal Engineering Conference, American Society of Civil Engineers, New York, pp. 2181-2191.

Sun, Y. and W. S. Huang, 1984: The siltation process and silt source of the Zhejiang coast. Donghai Marine Science, **2**(4), 34-42 (in Chinese).

Wu, J., 1975: Sea-surface drift currents. Offshore Technology Conference, Houston, Texas, Paper# OTC 2294.

Yan, S. Z. and F. X. Hu, 1986: Characteristics of minerals in Qiantangjiang Estuary and its relationship with the hydrodynamic environment. Journal of the East China Normal University, No. 4, pp. 83-90 (in Chinese).

Yu, H. H., D. C. Zheng and J. Z. Jiang, 1983: Basic hydrographic characteristics of the study area. In: Proceedings of the International Symposium on Sedimentation on the Coastal Shelf, China Ocean Press, Beijing.

Currents and Sediment Transport in the Seto Inland Sea, Japan

Tetsuo Yanagi

Department of Ocean Engineering
Ehime University
Matsuyama 790, Japan

ABSTRACT

The characteristics of currents and sediment transport in the Seto Inland Sea, Japan and their interaction are discussed in this paper. The M_2 tidal current is dominant in the Seto Inland Sea and its amplitude exceeds 70 cm/sec at the straits. Caldrons are developed around the straits due to the erosion by strong tidal current. Pebble and sand mainly provided from caldrons are sorted by tidal current and are deposited near the straits. On the other hand, silt and clay are mainly provided from rivers and they are transported by the residual flow in the long term and accumulate in the central part of the bay or Nada (basin) where the tidal current is weak.

I. Introduction

The Seto Inland Sea is situated in the western part of Japan (Fig. 1) and is well known for the beautiful landscape of its 600 islands. It is about 500 km long and 5-50 km wide, with an average depth about 40 m (Table 1). The total area of the Seto Inland Sea is 21,676 km^3, which is nearly the same as that of Lake Ontario in North America. The sea level was about 80 m lower 20,000 years ago, increasing monotonically from then, and it was about 3 m higher 6,000 years ago in the Seto Inland Sea. Therefore many materials have been deposited to the bottom of the Seto Inland Sea for the last 10,000 years.

Many rivers flow into the Seto Inland Sea (e.g. Table 2); the total river discharge was 62,000 x $10^6 m^3$/year (about 2,000 m^3/s) in 1985. A large quantity of sediment is deposited into the Seto Inland Sea from rivers, the open ocean, air and the sea bottom itself, and it accumulates on the sea bottom after dispersion by the current system in the Seto Inland Sea. Therefore, the characteristics of sediment reflect the long term transport system by the current in the Seto Inland Sea.

However, the detailed relationship between the current system and the characteristics of sediment in the Seto Inland Sea has not been clarified yet because the current has been studied separately by physical oceanographers and the sediment by geologists. In this paper I will synthesize information from both fields and show the quantitative relationship between the current system and the sediment transport in the Seto Inland Sea.

II. Currents

The variability in the Seto Inland Sea is schematically shown in Fig. 2 (Yanagi, 1982a). The current system in the Seto Inland Sea is accompanied by such variability, and it has wide ranges of frequency and wave number from wind waves to year-to-year variation. The most dominant current is the M_2 tidal current; the distribution of its amplitude is shown in Fig. 3 (Yanagi, et al. 1982). The tidal current in the Seto Inland Sea is governed by the incoming tidal waves from the

Coastal and Estuarine Studies, Vol. 38
R. T. Cheng (Ed.)
Residual Currents and Long-term Transport

Pacific Ocean through the Bungo Channel and the Kii Channel. The tidal waves show the character of progressive waves in the Bungo Channel and the Kii Channel but the standing tidal wave is generated in the central part of the Seto Inland Sea between the Hayasui Strait and the Tomogashima Strait by the superposition of two progressive tidal waves. The amplitude of the M_2 tidal current exceeds 70 cm/s around the narrow straits and that is under 30 cm/s at the back of the bay or in the central part of Nada (basin in Japanese) as shown in Fig. 3.

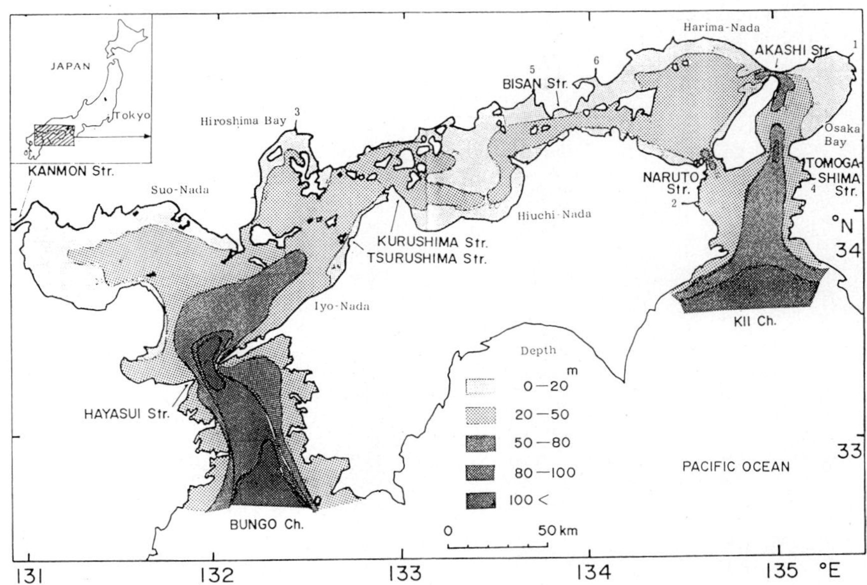

FIGURE 1. Bottom topography of the Seto Inland Sea. Numbers show the main rivers which are shown in Table 2.

There are one or two residual circulations at the bay or Nada in the Seto Inland Sea and the main components of such residual circulation are considered to be the tide-induced residual current, the buoyancy driven current and the wind-driven current (Yanagi, 1982a).

The stokes drift due to wind waves is dominant near the coast, especially in winter when the strong monsoon westerly blows.

TABLE 1. The Seto Inland Sea

Total volume:	893.62 km^3
surface area:	21.68 km^2
average cross-sectional area:	1.35 km^2
average width:	32.8 km
average depth:	41.1 m

III. Sediment

The Seto Inland Sea consists of some bays and Nadas and most parts are shallower than 50 m except the neighboring straits and in the central part of Iyo-Nada, the Bungo Channel and the Kii Channel. Many caldrons develop around the straits because of the erosion by the strong tidal current. There are two types of caldrons in the Seto Inland Sea: the long caldron along the axis of the strait, and the twin caldrons which exist at both sides of the strait. The former is seen in the Kanmon, the Tsurushima, the Kurushima, the Bisan, the Akashi, and the Tomogashima Straits and the latter in the Hayasui and the Naruto Straits.

TABLE 2. Main Rivers Feeding the Seto Inland Sea

	drainage area (km^2)	average discharge (m^3/s) in 1985
1. Yodo River	7,869	264
2. Yoshino River	3,700	116
3. Oota River	1,870	115
4. Kino River	1,978	47
5. Takahasi River	3,110	84
6. Yoshii River	2,360	68
Total	53,500 km^2	62 km^3/year

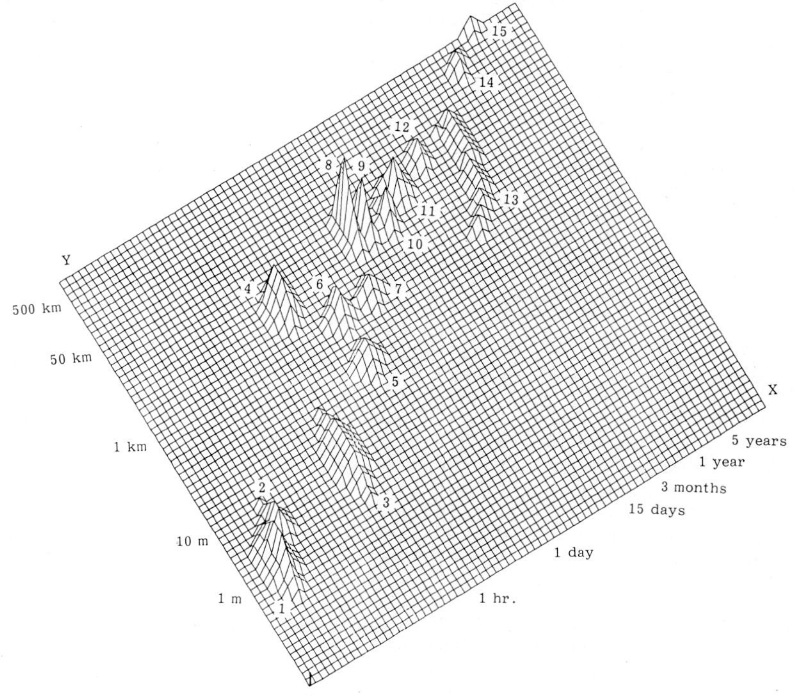

FIGURE 2. Schematic representation of the current variability in the Seto Inland Sea. 1: wind waves, 2: swell, 3: seich in harbor, 4: tsunami, 5: seich in bay, 6: local eddy, 7: quarter-diurnal tide, 8: semi-diurnal tide, 9: diurnal tide, 10: surge, 11: wind-driven current, 12: tide-induced residual current, 13: buoyancy driven current, 14: seasonal variation, 15: year-to-year variation (Yanagi, 1982a).

The horizontal distribution of mean grain diameter of surface bottom sediment, which is expressed by Md_ϕ ($\phi = -\log_2 d$, d is the mean grain diameter in mm), is shown in Fig. 4 (Inouchi, 1982). The distance between sampling stations is 5 or 10 km and there is no station within 5 km of the coast. Therefore the horizontal distribution of sand along the coast, which we can usually see on the beach, is not resolved in Fig. 4. No samples were taken in the white areas near the Kurushima, the Bisan and the Naruto Straits.

The rocks and pebbles whose Md_ϕ is smaller than 0 (grain diameter > 1 mm) are distributed near the straits, and their regions coincide well with the regions where the M_2 tidal current amplitude is larger than 70 cm/s shown in Fig. 3.

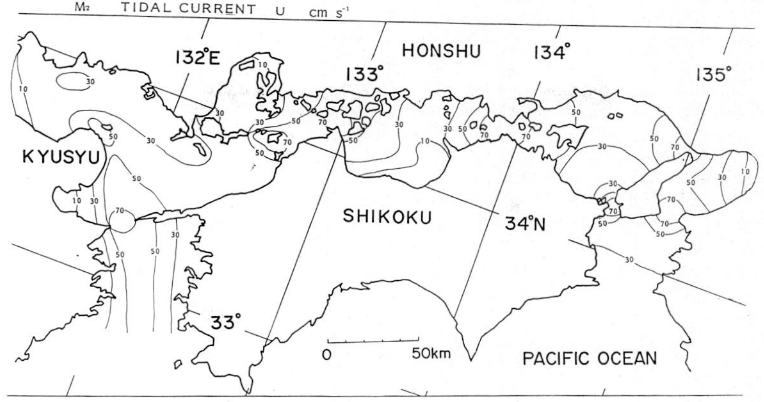

FIGURE 3. Horizontal distribution in the amplitude of M_2 tidal current (Yanagi et al. 1982).

The development of sand waves are observed at some places in the Seto Inland Sea, as shown in Fig. 5. Sand waves are developed at the region where Md_ϕ is between 0 and 2, and their regions coincide well with the area where the M_2 tidal current amplitude is between 50 cm/s and 70 cm/s, shown in Fig. 3. The generation mechanism of sand waves by the tidal current is obscure, but it is an important problem to be studied in the future.

The mud whose Md_ϕ is larger than six (6) is distributed near the head of the bay or in the central part of Nada and their regions well coincide with the areas where the M_2 tidal current amplitude is smaller than 30 cm/s.

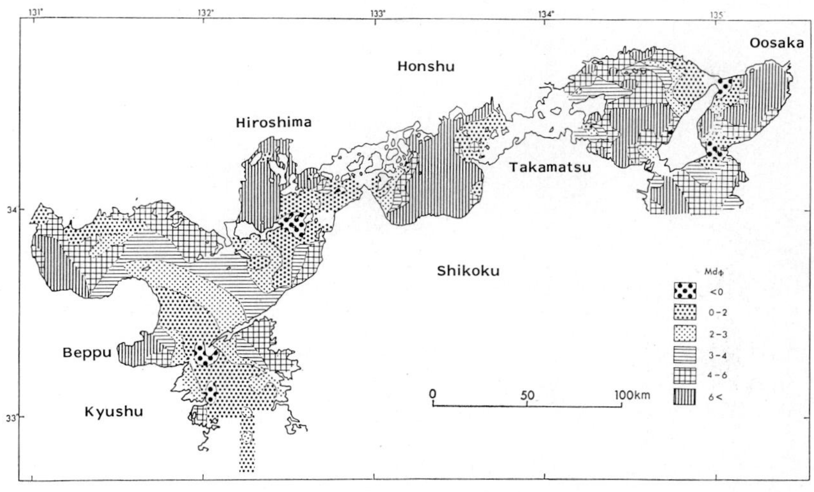

FIGURE 4. Horizontal distribution in Md_ϕ of surface sediment in the Seto Inland Sea (Inouchi 1982).

The horizontal distribution of oxidation-reduction potential Eh in the surface sediment is shown in Fig. 6 (Shiozawa <u>et al</u>., 1979). Eh near the straits is high and it shows that the oxidized state in the surface sediment exists near the straits. Much oxygen is provided from the surface to the bottom by the large vertical mixing due to the strong tidal current and the little accumulated organic matter may keep the bottom surface oxidized. On the other hand, Eh near the head of the bay or in the central part of Nada is less than 0 mv, that is, the anoxic state exists there. Little transport of oxygen from the surface to the bottom by the weak vertical mixing due to the weak tidal current and the large consumption of oxygen due to the decomposition of accumulated organic matter may make the bottom surface anoxic.

The horizontal distributions of sorting coefficient (Folk and Ward, 1957) are shown in Fig. 7 (Inouchi, 1982). The well-sorted sediment whose S_0 is smaller than 1.0 exists near the straits, but the poor-sorted sediment whose S_0 is larger than 2.0 exists in the central part of the bay or Nada.

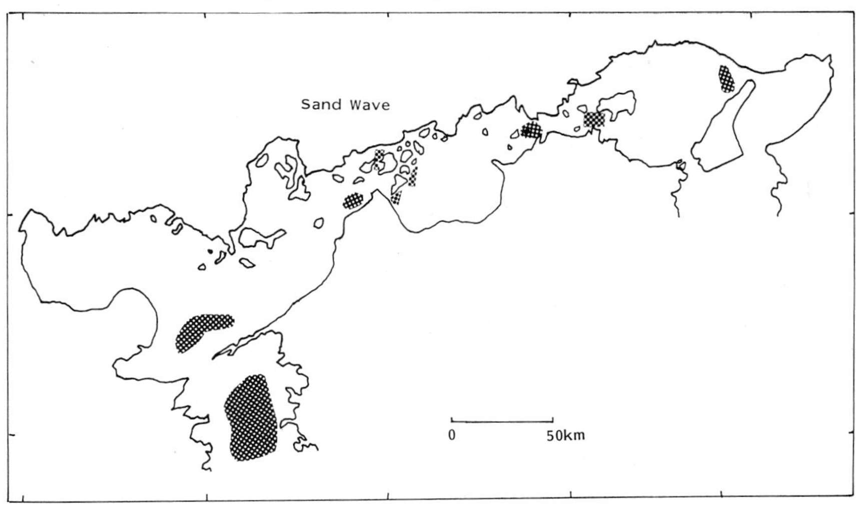

FIGURE 5. Horizontal distribution in the region where the sand wave is developed.

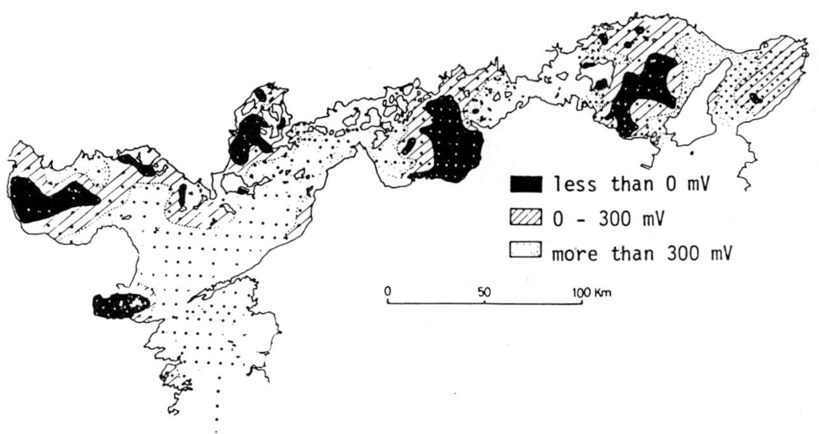

FIGURE 6. Horizontal distribution in the oxidation-reduction potential of surface sediment in the Seto Inland Sea (Shiozawa et al., 1979).

A consistent pattern is seen in the characteristics of surface sediment shown in Figs. 4 and 7; that is, Md_ϕ and S_0 increase as the distance from the straits increases, and this distribution pattern is not influenced by the positions of rivers and coastal features.

IV. Discussion

The sediments are mainly provided from rivers and caldrons, since the supply from the air or open ocean is so small. A uniform distribution of Md_ϕ is found within about 30 km of the Hayasui Strait and within about 10 km of other straits in the Seto Inland Sea. On the other hand, the continuous pattern of Md_ϕ from the coast remains within only several km. Pebbles and sand from rivers cannot move near the straits by jumping over the mud region. Such facts may suggest that the main source of coarse sediment such as pebbles and sand is the straits, and not the rivers in the Seto Inland Sea, which was already pointed out by Inouchi (1982). The largest caldron, whose depth attains 300 m, exists in the northern part of the Hayasui Strait. Coarse sediments such as pebbles and sand are produced in the caldron due to the erosion by the strong tidal current, and they are sorted by the tidal current to deposit near the straits. The sand region whose Md_ϕ is between 0 and 3 in Iyo-Nada is about 600 km^2, and its thickness is about 10 m from the 3.5 kHz acoustic stratigraphy (Inouchi, 1982). This indicates that 6 km^3 of sand accumulates in Iyo-Nada. On the other hand, the volume of the caldron in the northern part of the Hayasui Straits is about 8 km^3 (the radius is 5 km and the depth is 300 m). Both volumes coincide well, and this fact suggests that sand in Iyo-Nada is mainly provided from the caldron in the northern part of the Hayasui Straits.

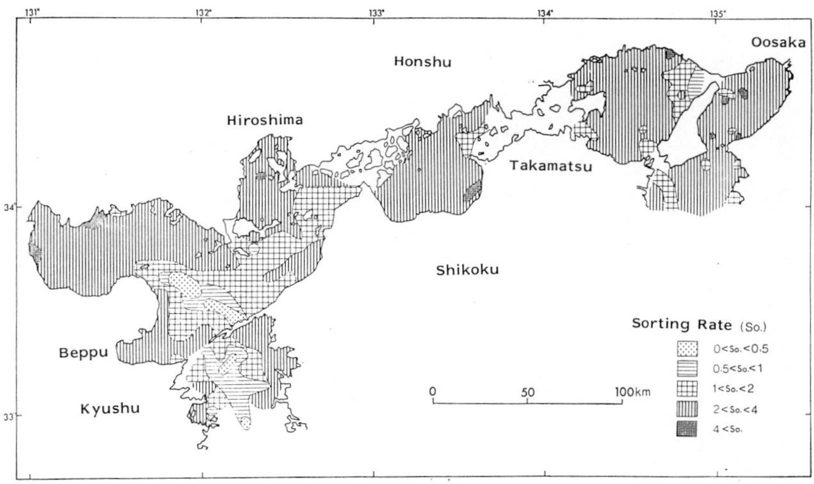

FIGURE 7. Horizontal distribution in the sorting rate of surface sediment in the Seto Inland Sea (Inouchi, 1982).

Fine sediment such as silt and clay from rivers cannot accumulate near the coast due to strong currents associated with wind waves. They deposit to the bottom and are resuspended many times by tidal currents near the coast. They are transported by the residual current to the central part of the bay or Nada in the long term and accumulate in the region where the tidal current amplitude is small.

The transport of such fine sediment is traced with use of heavy metals as the tracer as shown in Fig. 8 (Yanagi, 1982b). In the eastern part of Hiuchi-Nada, there are large pulp plants at the southern coast which are shown by the shadowed area in the upper (d) of Fig. 8. The M_2 tidal current flows nearly along the coast and the residual flow is fairly weak and makes a counterclockwise circulation. The horizontal distribution of heavy metal (Cu) concentration in fine particle fraction in the surface sediment whose diameter is smaller than 76 μm is shown in the upper (d) of Fig. 8. Heavy metal (Cu) supplied from the pulp plants mainly spreads to the north-east direction after the residual flow. The main component of residual flow in Hiuchi-Nada is the tide-induced residual current because there is no prevailing wind or a large river there.

Another example of fine sediment transport is found at Harima-Nada (the lower panel of Fig. 8). There are oil, iron and chemical plants in the northern coast of Harima-Nada which are shown by the shadowed area in the lower (d) of Fig. 8. The M_2 tidal current flows in an almost east-west direction. The residual flow has a clockwise circulation in the northern part and a counterclockwise flow in the southern part of Harima-Nada. The main components of residual flow in Harima-Nada are considered to be the tide-induced residual current and the buoyancy driven current due to the river discharge from the northern coast. Heavy metal (Cr) supplied from land is seen to spread after the flow direction of residual flow.

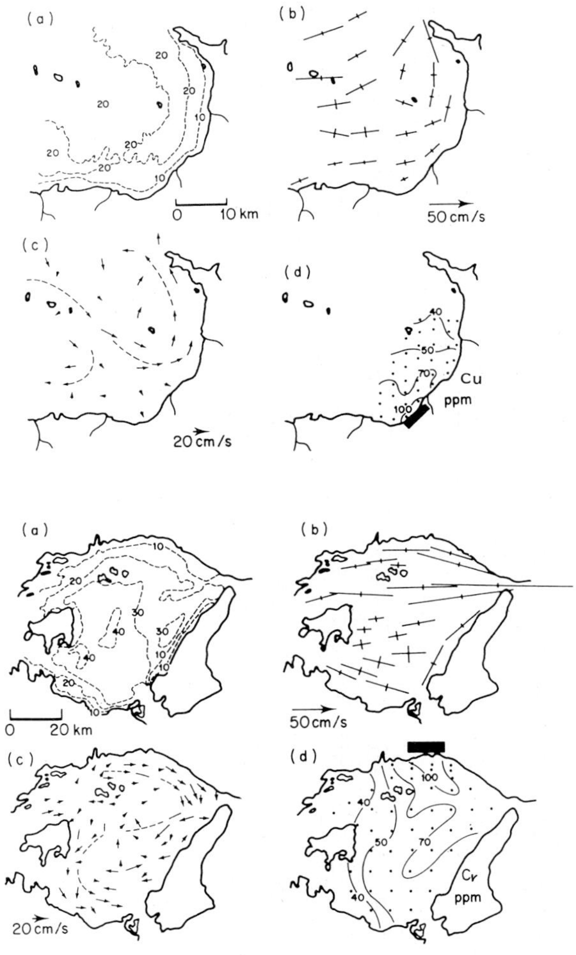

FIGURE 8. Bottom topography (a), major and minor axes of M_2 tidal current ellipse (b), observed residual flow (c) and the concentration of heavy metal in the surface sediment (d) in the eastern part of Hiuchi-Nada (upper) and in Harima-Nada (lower) (Yanagi, 1982b).

The accumulation rate of fine sediment such as silt and clay was measured to be about 0.5 cm/year in the central part of Osaka Bay (Matsumoto and Yokota, 1978).

The transport system of sediment in the Seto Inland Sea is schematically shown in Fig. 9. The supply of sediment from rivers has significantly decreased recently because many dams for sand have been constructed in the upper river. Sediments with large diameter such as sand are mostly stopped by these dams and only a small size sediment such as silt or clay is deposited from rivers. Thus the area of sand beach along the coast has rapidly decreased in recent years.

The transport system of sand in the field is not clarified yet because we have no appropriate tracer. The tracer experiment of sand movement in deep area will be necessary in the future.

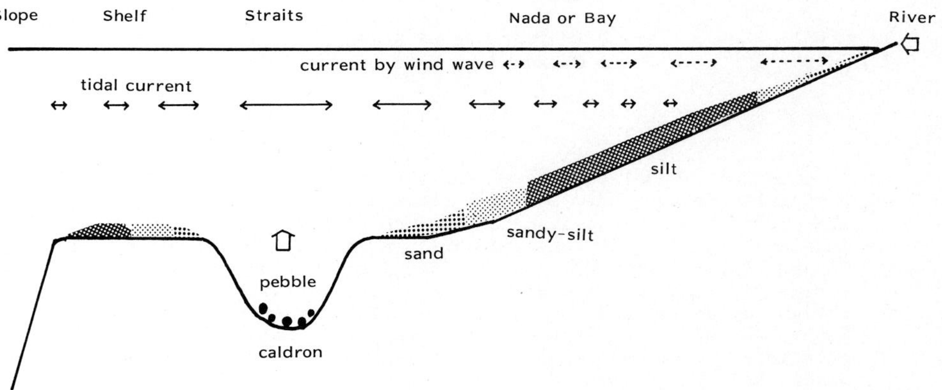

FIGURE 9. Schematic representation of sediment transport system in the Seto Inland Sea.

The author expresses his sincere thanks to Dr. H. Takeoka of Ehime University for his fruitful discussion.

V. References

Folk, R. L. and W. C. Ward, 1957: Braxos River bar: A study in the significance of grain size parameters. J. Sed. Petrol., **27**, 3-26.

Inouchi, Y., 1982: Distribution of bottom sediments in the Seto Inland Sea. J. Geological Soc. Japan, **8**, 665-681.

Matsumoto, E. and S. Yokota, 1978: Accumulation rate and heavy metal pollution in Osaka Bay sediments. J. Oceanogr. Soc. Japan, **34**, 108-115.

Shiozawa, T., K. Kawana, A. Hoshika, T. Tanimoto and O. Takimura, 1979: The bottom sediment in the Seto Inland Sea, Reports of the Government Industrial Research Institute, Chugoku, **4**, 1-24.

Yanagi, T., 1982a: Variability of oceanic condition in the Seto Inland Sea. Bulletin on Coastal Oceanography, **20**, 191-197.

Yanagi, T., 1982b: Sediment transport by tidal residual flow in the bay. in "Sea Bed Mechanics" ed. by B. Denness, Graham & Trotman, London, 207- 214.

Yanagi, T., H. Takeoka and H. Tsukamoto, 1982: Tidal energy balance in the Seto Inland Sea. J. Oceanogr. Soc. Japan, **38**, 293-299.

Simulation of Suspended Sediment Dispersion in the North Sea

Walter Puls[1] and Jürgen Sündermann[2]

[1]GKSS Research Center
2054 Geesthacht, West Germany

[2]University of Hamburg
2000 Hamburg, West Germany

ABSTRACT

A numerical model calculates the transport, the erosion and the deposition of suspended solids in the North Sea by a Monte Carlo tracer method. Suspended solids enter the North Sea from adjacent seas (English Channel, North Atlantic), from land (river input, cliff erosion) and by the erosion of old deposited sediment. Transport is determined mainly by wind and density driven currents and by residual tidal currents. Erosion and deposition of solids depend on the bed shear velocity which is calculated from near-bed current velocities and from surface water waves. Suspended matter concentrations computed for December 1979 are compared with measured data. The model computations result in long term deposition in the Norwegian Trough, the Outer Silver Pit and the Fladen Ground.

I. Introduction

This paper deals with the simulation of the transport of suspended terrigeneous matter only. The production and the decay of organic matter (phytoplankton) is not included. It should be stated that even the terrigeneous suspended matter transport is still under development. This report describes an intermediate stage of the model.

A definition of 'suspended solid matter' in terms of grain sizes or settling velocities is not possible. Sand grains, for instance, are usually transported as bed load, but during a severe storm they belong to the suspended load. Median grain sizes of suspended matter in the Elbe estuary (one of the main suppliers of river-borne suspended solids) are about 10 μm. Eisma and Kalf (1979) measured grain sizes in the Southern Bight, where the bulk of the suspended particles had diameters between 5 and 20 μm. It can thus be argued that suspended sediment in the North Sea has (fine) grain sizes which are smaller than about 20 μm. Only during storms, when bottom sediment is eroded, suspended matter may consist of considerable amounts of larger grain sizes. It must be stressed that the model is only concerned with fine suspended matter and not with sand.

II. Model Area

Backhaus (1985) uses a three-dimensional baroclinic model for routine calculations of the current velocity in the North Sea and the shelf area around the British Isles. This 'shelf model' yields daily patterns of the residual currents, which are based on tidal (M_2) currents, wind-induced currents and density-driven currents. The horizontal grid spacing is about 20 km. The summer version has 12 vertical layers, and the winter version has 7. The summer-version has more layers for a better approximation of the thermocline. The model decribed in this report is

Coastal and Estuarine Studies, Vol. 38
R. T. Cheng (Ed.)
Residual Currents and Long-term Transport

a sub-model of the Backhaus shelf model. The area and the bathymetry of the model are shown in Figure 1.

By means of the horizontal grid and the vertical layers, the model (summer version) is partitioned into about 7000 grid boxes. In the deepest part (600 m) of the Skagerrak the water column consists of 10 grid boxes, the uppermost has a height of 10 m, the lowest a height of 250 m.

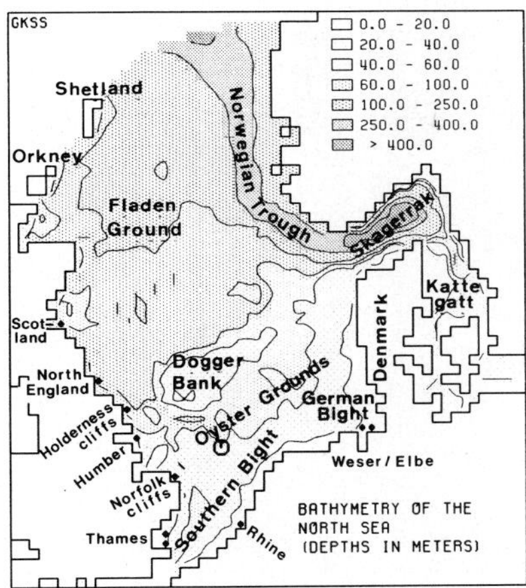

FIGURE 1. Model area with the bathymetry of the North Sea; markers indicate the input points of land-borne suspended matter.

Figure 2 shows the areas of mud, sand or gravel in the North Sea; the map is based on DHI (1915/1952), DHI (1981) and U.S. Naval Oceanographic Office (1965). Figure 2 and data of Kersten and Klatt (1987) were combined to yield Figure 3, which shows the content $c_{20\mu m}$ of sediment with grain sizes smaller than 20 μm in the upper 2 cm of the bed; $c_{20\mu m}$ is in % of dry mass.

III. Transport of Suspended Matter

The numerical technique used in the model for transport computations is a Monte-Carlo tracer method. The suspended matter in the North Sea is represented by 50,000 to 100,000 tracer particles. Each particle is 'loaded' with a certain mass of suspended matter. New particles enter the model area from land (by rivers or cliff erosion), from the adjacent seas and by bottom erosion. Particles can loose their 'load' partly or totally when deposition of suspended matter occurs. Finally particles can leave the North Sea by being transported over one of the seaward boundaries. The time step for the transport computations is 40 minutes.

The concentration of suspended matter in a grid box is simply determined by summing up the mass of all particles within the box and dividing by the box volume. Within the model area, particles migrate through the three-dimensional grid by the following movements:

(a) by the residual currents obtained from the Backhaus shelf model. For every model grid box the daily residual currents in the two horizontal and in the vertical direction are provided. An improved model version uses the time-resolved (time step: 40 minutes) currents of the M_2- and the S_2-tide in addition to the daily wind- and density-driven currents. The results presented in this paper, however, are calculated by using daily residual currents only.

(b) by settling. The settling velocity w_s for river-borne and sea-borne particles (see below) is $w_s = 4x10^{-5}$ m/s (McCave, 1970) in winter and twice this value in summer (the factor 2 takes into account the increased settling due to planktonic flocculation and due to fecal pellets produced by pelagic filter-feeders). The settling velocity of eroded <u>bottom</u> material is $w_s = 3x10^{-4}$ m/s; this value was deduced from measurements by Konig (1987). The settling velocity of eroded bottom sediment is the same both in summer and in winter.

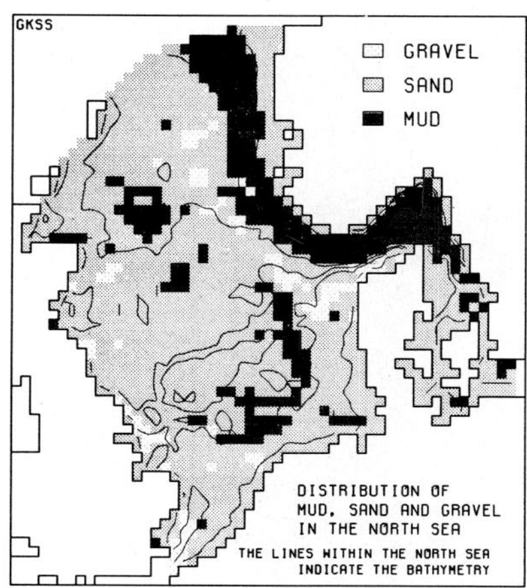

FIGURE 2. Distribution of mud, sand and gravel in the North Sea.

The influence of the water temperature on w_s is not taken into account. In the Central North Sea, the water temperature below 40 m is rather constant and varies between 6 °C and 10 °C (Tomczak und Godecke 1962). In coastal areas, temperature variations are larger. In this rough model, however, these are not considered.

The settling velocity w_s influences (i) the vertical profile of the suspended matter concentration and (ii) the deposition of suspended matter. The model results are sensitive to the settling velocities which are in the range of $10^{-6} < w_s < 10^{-3}$ m/s. The model's w_s-values are within this sensitivity range. Because of the poor information about settling velocities in the North Sea, the model's w_s-values can quite easily differ from the true value by one order of magnitude.

The potential effects of a w_s-error on the model results are:

(i) the vertical concentration profile is a contributing factor to the transport characteristics of suspended matter. If for instance suspended matter is accumulated near the bottom instead of being uniformly distributed over the water column, it is advected by the residual current of the bed-nearest layer only, which in the 'worst case' may be in the opposite direction of the main residual current. Generally, the vertical concentration profile is more sensitive to an over-estimation of w_s than to an under-estimation.

(ii) overall deposition can be over- or under-estimated, depending on whether the model's w_s is larger or smaller than the true value. Another effect is that of changing the pattern and contours of deposition zones: An over-estimated settling velocity makes deposition zones smaller and more sharply defined than in nature.

Apart from the uncertainty about the absolute w_s-values, the use of a single settling velocity instead of a range of settling velocities is another source of error. The potential errors resulting from the use of single settling velocities mainly concerns deposition: a suspended matter fraction with $w_s < 10^{-6}$ m/s behaves almost like dissolved matter and is hardly deposited. This non-deposition of a certain fraction is not adequately represented by model computations using only a single settling velocity.

(c) by horizontal diffusion. The horizontal diffusion coefficient is $A_H = 0.1$ m^2/s; A_H is constant in space and time. The time dependent advective currents generate an effective horizontal diffusion which corresponds to a diffusion process with a coefficient of about 10 m²/s. Taking this into consideration, horizontal diffusion with a coefficient A_H = 0.1 m/s is actually superfluous; it is only included in the model for the sake of 'completeness'. Diffusion processes are simulated in the model by a random walk of particles (e.g. Maier-Reimer and Sundermann 1982).

(d) by vertical diffusion. The vertical diffusion coefficient A_z is 50×10^{-4} m²/s in 'winter' (November to April) and 10×10^{-4} m²/s in 'summer' (May to October). Vertical diffusion in summer is smaller because of stratification and because of the absence of strong winds. A_z is constant in space.

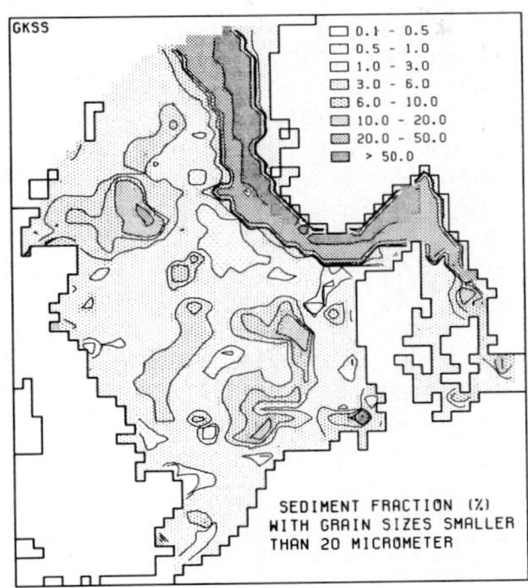

FIGURE 3. Distribution of fine sediment content (grains smaller than 20 μm) in the upper 2 cm of the bed.

The main role of vertical diffusion is to provide a correct vertical distribution of suspended matter. This distribution depends on the relation w_s/A_z. As long as the information about w_s is poor and the model does not include stratification, it is of only limited benefit to describe A_z in a sophisticated manner. In view of this fact and in view of the many shortcomings concerning the knowledge about A_z, it was decided not to give the illusion of a sophisticated (and wrong) approach, but rather to choose a crude but clearly defined and well assessable approximation.

The main error resulting from the crude A_z-formulation concerns the Southern North Sea in summer. For $A_z = 10 \times 10^{-4}$ m²/s, $w_s = 8 \times 10^{-5}$ m/s, and a water depth of 40 m, the surface concentration of suspended matter is only 4 % of the bottom concentration. Measurements, however, show that the vertical distribution of suspended matter concentration is approximately uniform. Considering that tidal currents in the Southern North Sea have a velocity of more than 0.3 m/s, $A_z = 10 \times 10^{-4}$ m/s is certainly too small. The main error resulting from the wrong vertical profile is a horizontal transport of suspended matter decreased by about 20 %, because residual currents are usually smaller in the lower than in the upper half of the water column.

IV. Sediment-water Interaction

Many investigators (e.g. Drake and Cacchione, 1989) postulate a direct relationship between the concentration of suspended matter on the one hand and the bed characteristics and the present bed shear velocity v_* on the other hand. In particular, if v_* is smaller than the threshold shear velocity $v_{*cr,e}$ for erosion, the concentration of suspended matter is assumed to be zero. In the model presented in this paper, the behaviour of suspended matter is time-dependent. This means for instance that there might be suspended matter at a certain site even though $v_* < v_{*cr,e}$. This can happen when an eroding storm is displaced by a calm sea and when the suspended matter is deposited with a time lag.

IV.1 Bed shear velocity

Erosion and deposition of suspended matter depend on the bed shear velocity v_*. The bed shear velocity is calculated from

-- wind and density driven currents plus M_2 and S_2 tidal currents. All currents were computed by the Backhaus shelf model. The shear velocity is calculated from the near-bed currents depending on the water viscosity (which is $1.5x10^{-6}$ m^2/s) and the bed roughness, which is 2 mm for gravel, 200 μm for sand and 100 μm for pure mud.

-- sea and swell waves; the wave data was calculated by the Amt für Wehrgeophysik of the German Armed Forces. The wave spectrum is available every two hours at 60 points in the North Sea. No data is available for the Kattegatt, Figure 1, and the Western Baltic Sea; the wave data in these areas was calculated separately by formulas which give the significant wave height and wave frequency as a function of the wind velocity, the fetch length and the water depth (Stephan, 1978). Because of their longer wave period, swell waves usually contribute more to v_* than sea waves.

Figure 4 shows the time history (August 1979) of the bed shear velocity for a grid point between the Oyster Grounds and the Southern Bight. The position of the grid point is marked in Figure 1 (circle). Because of calm weather, the semidiurnal tidal cycle and the fortnightly spring-neap-cycle are predominant.

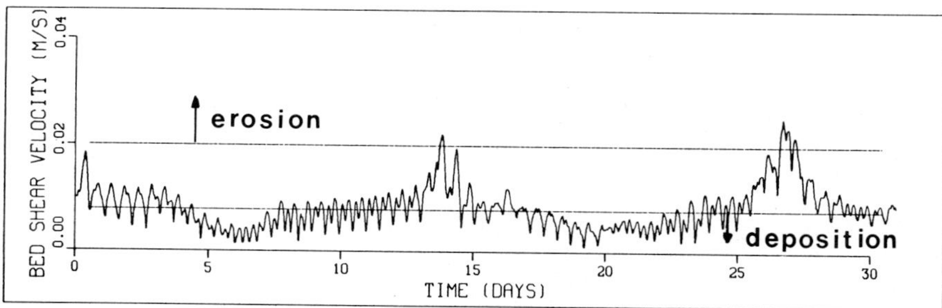

FIGURE 4. Time series (August 1979) of the bed shear velocity at a position in the Souther Bight/Oyster Grounds (circle in Figure 1).

Figure 5 shows the v_*-time history at the same location, but during December 1979; the influence of tides is dominated by the wave influence. The horizontal lines in Figures 4 and 5 are the critical shear velocities for erosion and deposition of suspended matter, respectively (see below).

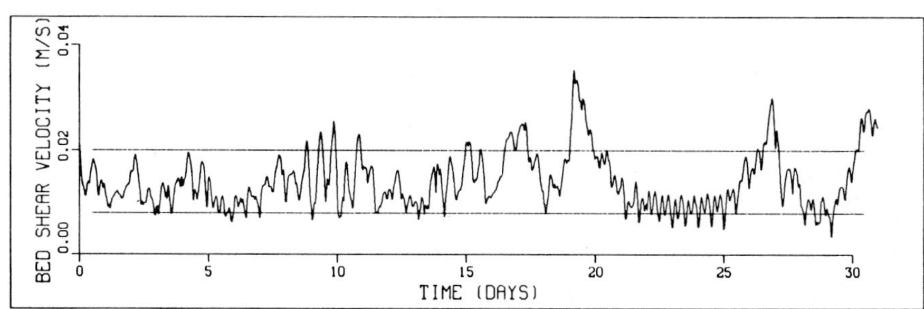

FIGURE 5. Time series (December 1979) of the bed shear velocity at a position in the Southern Bight/Oyster Grounds (see circle in Figure 1). The shear velocity at the bed is calculated from the combined action of waves and currents (Grant and Madsen, 1979).

Figures 6 and 7 show maximum shear velocities v_*^{max} in the North Sea for December 1979. Figure 6 shows v_*^{max} due to currents only, while Figure (7) shows

v_*^{max} due to the combined action of waves and currents. It is emphasized that the v_*^{max}-distribution in Figures 6 and 7 are not synoptic.

IV.2 Bed characteristics and erosion

The bed in the model is divided into 12 layers with thicknesses between $5x10^{-4}$ (bed surface) and $2.5x10^{-2}$ m; the total (initial) thickness of the bed is 0.06 m. The bed in the North Sea is partitioned into two classes: mud and non-mud areas. The limit between these two bed classes is at a fine sediment fraction of $c_{20\mu m}$ = 25 %.

The <u>mud bed</u> has a uniform dry density of 0.5 tons/m³. The content of fine sediment is 100%, irrespective of the actual $c_{20\mu m}$-value. The value 0.5 tons/m³ is based on own density measurements of undisturbed Skagerrak-mud.

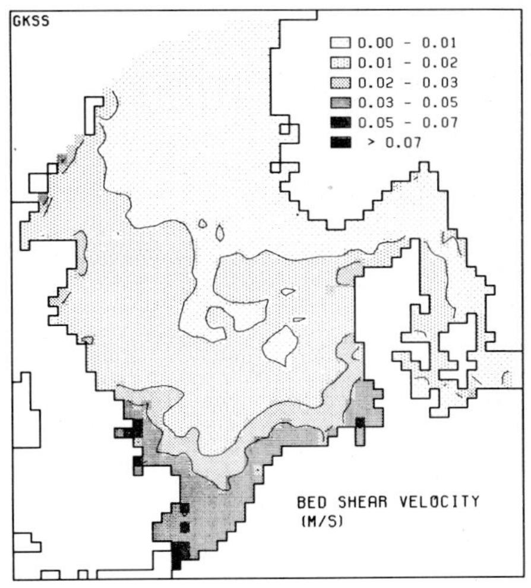

FIGURE 6. Maximum shear velocities in December 1979, generated by currents only.

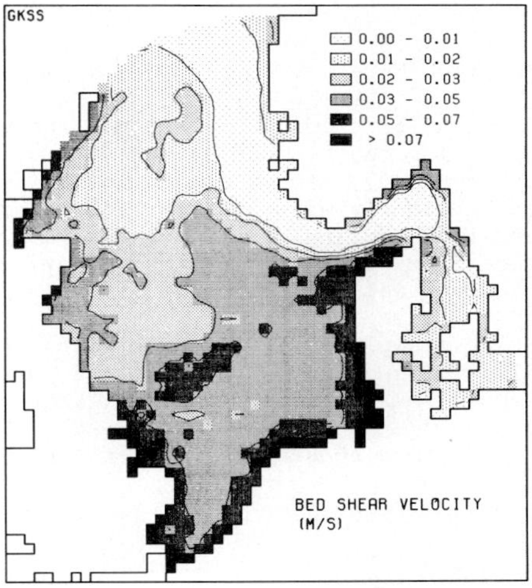

FIGURE 7. Maximum shear velocites in December 1979, genrated by the combination of waves and currents.

Erosion of a mud bed is a time dependent process. Mud erosion is calculated by the formula

$$\epsilon = Mx[v_*^2 - (v_*^{cr,e})^2], \qquad v_* > v_*^{cr,e} \qquad (1)$$

where ϵ is the erosion rate (tons/m²s). The formula is based on flume experiments carried out by Hydraulics Research (UK) with estuarine muds (Puls, 1984; Rodger and Odd, 1985). The critical shear velocity for erosion of a mud bed in the North Sea is $v_*^{cr,e}$ = 0.02 m/s. M is the erosion constant and is equal to 10^{-4} tons•s/m⁴. Both the critical erosion shear velocity and the erosion constant were obtained from own erosion tests with Skagerrak mud.

A <u>non-mud bed</u> has a uniform dry density of 2 tons/m³ (McCave 1987). The content of fine sediment is given by the $c_{20\mu m}$-value (Figure 3). For example, the dry mass of fine sediment in one cubic metre of non-mud with $c_{20\mu m}$ = 2% is 0.04 tons.

The model conception for the erosion of a non-mud bed is: sand is stirred up when v_* exceeds $v_*^{cr,e}$. Only the present content of fine sediment leaves the bed and is transported away by the currents. The fate of sand is not considered because sand is not a 'pollution carrier'. Only the fine sediment is of interest for the model computations.

The threshold bed shear velocity for erosion is the same as that for a mud bed: $v_*^{cr,e}$ = 0.02 m/s. This value was obtained from own erosion tests with a sediment core from the Oyster Grounds and from the time history of suspended matter concentration measured on the German North Sea platform (Konig, 1987).

The erosion formula for non-mud is:

$$D_e = 0.21x[v_*^2 - (v_*^{cr,e})^2]^{0.8}, \quad v_* > v_*^{cr,e} \tag{2}$$

where D_e is the erosion depth in metres and v_* is in m/s. Erosion of non-mud is an instantaneous rather than a time dependent process. The erosion depth formula was deduced from suspended matter concentrations measured by Konig (1987) during a storm at the German North Sea platform.

Apart from the distribution of mud, sand and gravel, there is no information about grain sizes in the North Sea. Therefore the grain size is not used for determining the erosion characteristics of a non-mud bed. Moreover, there is bed armouring by current and wave action and by biological influences, which makes the use of the Shields-curve questionable. An example of bed armouring is: the current velocities needed for the erosion of a natural sand bed (German Wadden Sea) was about three to five times higher than for eroding 'sterile' sand of the same grain size (Fuhrboter, 1983).

Both for the mud bed and for the non-mud bed there is no increase in $v_*^{cr,e}$ with increasing depth (though this is implicitly included in the erosion depth formula for non-mud), which is a crude assumption. Another crude assumption is that no distinction is made between old and newly deposited sediment. Sediment just deposited during tidal slack water should be more easily re-suspended by the succeeding tidal current than older (consolidated) sediment. Concerning this point, the model will be improved.

An example for erosion is given in Figure 8, which shows a time series (August 1979) of suspended matter at the position marked in Figure 1 with a circle. Figure 4 shows the corresponding time series of the bed shear velocity. At August 13 and August 27 the shear velocity exceeds the threshold shear velocity for erosion which results in a sudden increase of the concentration of suspended matter in Figure 8.

IV.3 Deposition

Suspended matter reaches the bed by settling. If c_b is the suspended matter concentration near the bed, the mass $c_b \cdot w_s$ reaches the bed per second and square metre. Whether deposition occurs or not depends on the bed shear velocity v_* and the critical bed shear velocity for deposition $v_*^{cr,d}$: if v_* is greater than $v_*^{cr,d}$, no deposition occurs; if v_* is smaller than $v_*^{cr,d}$, the portion $1-v_*/(v_*^{cr,d})$ of $c_b \cdot w_s$ is deposited (Krone, 1962). Deposition tests of Creutzberg and Postma (1979) with Oyster Ground mud yielded $v_*^{cr,d} = 0.008$ m/s. This is in agreement with flume tests of Hydraulics Research (1979) which yielded $v_*^{cr,d} = 0.007$ m/s. Rodger and Odd (1985) used $v_*^{cr,d} = 0.01$ m/s for modelling suspended matter transport in Liverpool Bay. The value $v_*^{cr,d} = 0.008$ m/s is used in the North Sea model.

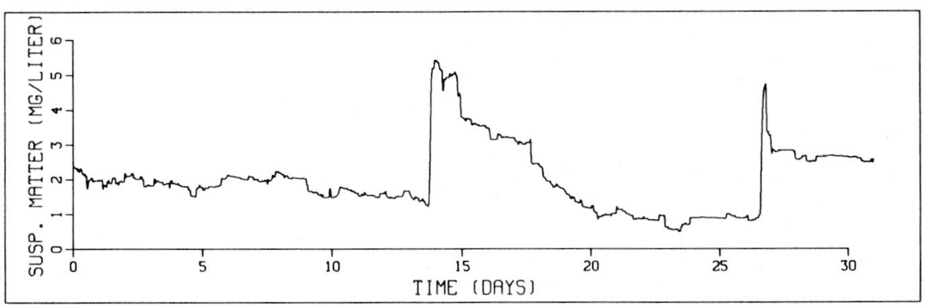

FIGURE 8. Time series (August 1979) of suspended matter at a position in the Southern Bight/Oyster Grounds [circle in Fig 1]

In addition to deposition by settling, there is biodeposition due to benthic filter-feeders, which produce feces and pseudofeces. The water filtration rate of benthic filter-feeders is assumed to be 10 litre/m^{2}day (Gast, 1987). This value is constant in space and time.

V. Bioturbation of the Bed

The bed in the North Sea is continuously reworked by bottom fauna. This leads to vertical mixing of the bed. The mixing coefficient in the upper few centimetres of Long Island Sound (USA) is in the order of 10^{-10} m^2/s (Lee II and Swartz, 1980). This value results in deposited suspended matter being 'immediately' mixed into deeper bed layers. Only a fractional part of that material is then available for the next erosion event. The uncertainties concerning bioturbation meant that mixing was switched off for the model computations presented here.

VI. Input of Suspended Matter

Suspended matter enters the North Sea

(a) from rivers; data was obtained from Rat von Sachverständigen (1980) and from McCave (1987). The following input masses were used in the model (see Figure 1 for the location of the input points; the values are given in tons per year (tons/yr)):

Scotland:	$55x10^3$ tons/yr
North England:	$135x10^3$ tons/yr
Humber:	$50x10^3$ tons/yr
Thames:	$217x10^3$ tons/yr (two input points)
Rhine:	$1260x10^3$ tons/yr
Weser:	$268x10^3$ tons/yr
Elbe:	$500x10^3$ tons/yr

(b) by cliff erosion (McCave 1987) (input points: Figure 1):

Holderness:	$700x10^3$ tons/yr
Norfolk:	$560x10^3$ tons/yr

(c) From the sea; input through the seaward boundaries is achieved by prescribing the concentration of suspended matter at the boundaries. Whether suspended matter enters the North Sea or not depends on the direction of the residual currents and on the efficiency of diffusion. The prescribed concentrations of suspended matter were taken from Eisma and Kalf (1987):

Dover Strait: 1.9 mg/litre
Fair Isle (between Orkney Islands and Shetland Islands) and Pentland Firth (between Orkney Islands and Scotland): 0.4 mg/litre

No suspended matter concentrations are prescribed at the seaward boundary between Shetland and Norway or at the seaward boundary in the Baltic Sea.

VII. Initial Conditions for the Model Computations

The model simulates the fate of suspended matter in the year 1979. For a 'hot start' on the Jan 1, 1979, preliminary model computations started in January 1978 with a North Sea containing no suspended matter. The model simulated the years 1978 and 1979 and subsequently once again for the year 1978. The achieved concentration of suspended solid matter obtained on the Dec 31, 1978 was the 'hot start' condition for the 'true' simulation period January-December 1979. The 'true' calculations started with a virgin bottom, i.e. previous erosion and deposition was not considered.

VIII. Model Results

VIII.1 Concentration of suspended matter

In January 1980 Eisma and Kalf (1987) measured the concentration of suspended matter in the North Sea. The measured surface and bottom data is shown in Figure 9.

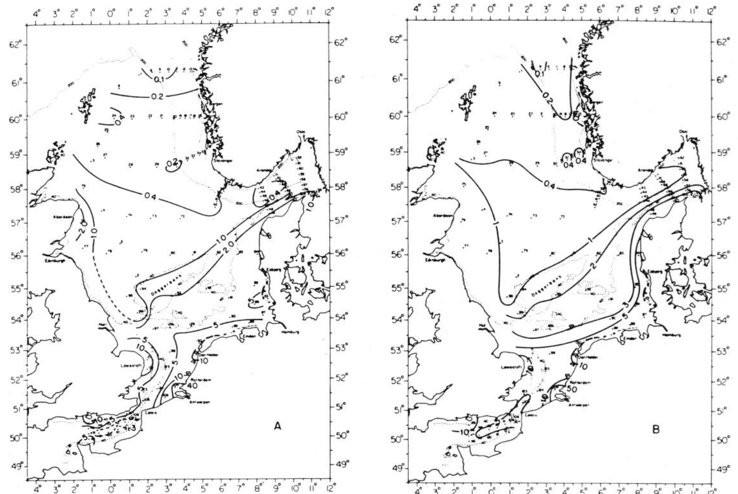

FIGURE 9. Concentration of suspended matter in the surface water (A) and in the bottom water (B) of the North Sea, measured in January 1980 (in mg/l). Reproduced by permission of the Geological Society from 'Dispersal, concentration and deposition matter in the North Sea', D. Eisma and J. Kalf, in Journal of the Geological Society, London, Vol. 144, 1987.

The measured data is compared with computed data. Figure 10 shows the depth mean concentration of terrigeneous suspended matter computed for the Dec 31, 1979. Figure 10 contains both land-borne and sea-borne suspended matter and eroded bottom sediment. Concerning the gross features, the measured and the computed suspended matter concentrations are rather similar; in both Figures 9 and 10 there is a general increase of suspended matter from the north-west towards the south-east of the North Sea. Maximum measured concentrations at East Anglia, Aberdeen and along the coasts of the Netherlands and of Germany are reproduced by the model.

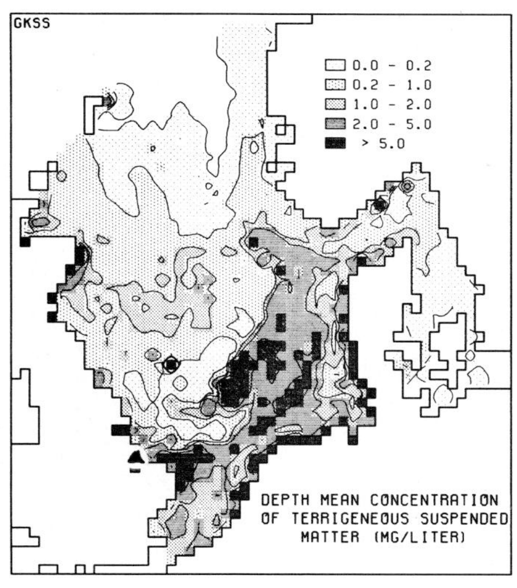

FIGURE 10. Distribution of suspended matter concentration calculated for December 31, 1979.

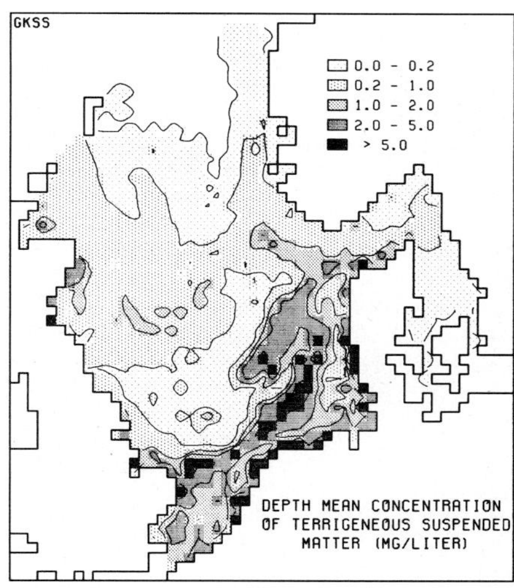

FIGIUR 11. Distribution of suspended matter concentration calculated for December 31, 1979; without bottom sediment

South-east of Dogger Bank, however, the model calculates concentrations which are greater than 5 mg/l. Such large measured concentrations were not found in that area. Further, the model yields concentrations of more than 2 mg/l south of the Norwegian Trough, which were also not found in this region.

The high concentrations south-east of Dogger Bank and south of the Norwegian Trough result partly from bottom erosion, due to storms which ocurred in November and in December 1979. Bottom erosion is also responsible for some of the isolated spots with high concentration (e.g. north-west of Dogger Bank).

Figure 11 shows the suspended matter concentration in December 1979 without bottom sediments. Compared with Figure 10, the concentrations south-east of Dogger Bank are lower, but they are still too high.

The partial non-agreement of measured and computed data leads to the conclusion that the model simulation of erosion, deposition or bioturbation (or all of them) needs to be improved. On the whole, however, there is an encouraging agreement of measured and computed data.

Concerning Figure 11 it is interesting to know how the various suspended matter input points (or input from the sea) contribute to the concentration pattern. This information is given in Figures 12 to 17. The sources are marked by arrowheads.

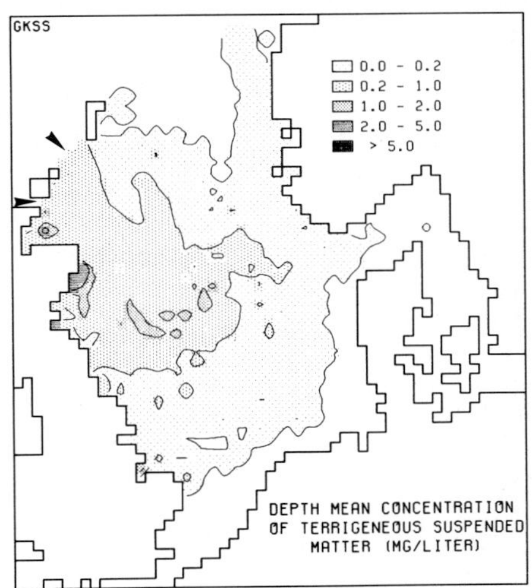

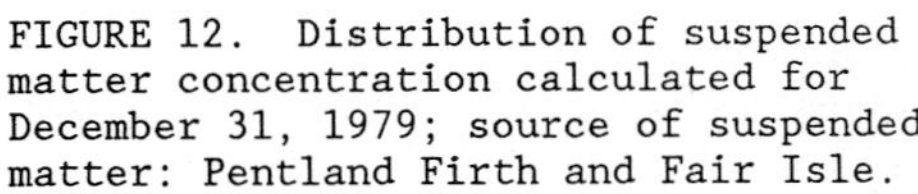

FIGURE 12. Distribution of suspended matter concentration calculated for December 31, 1979; source of suspended matter: Pentland Firth and Fair Isle.

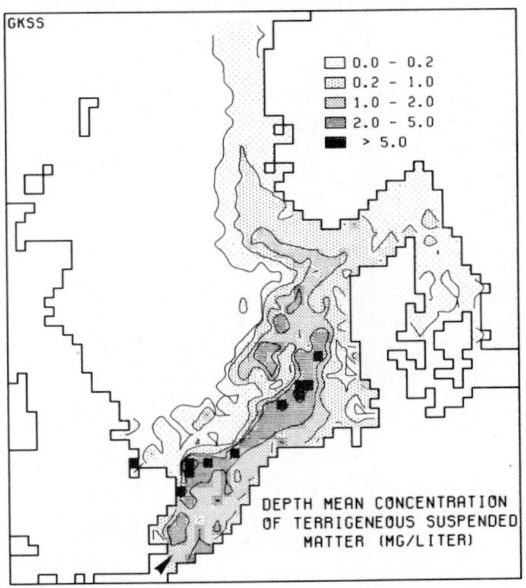

FIGURE 13. Distribution of suspended matter concentration calculated for December 31, 1979; source of suspended matter : English Channel.

VIII.2 Distribution of deposited matter

The distribution of deposited suspended matter (no bottom sediments) at the end of December 1979 is shown in Figure 18. Taking the bottom sediments into account, which were eroded and which were deposited after a phase of transport, results in Figure 19. North of Dogger Bank there is a great area of deposited sediment in Figure 18, but only small spots of deposited sediment in Figure 19. The explanation is that bottom sediment was eroded in this area which over-compensated for the deposited sediment mass of Figure 18. From both figures it is obvious that according to the model computations, the Norwegian Trough, the Fladen Ground and the Outer Silver Pit (southwest of Dogger Bank) are the long term deposition areas of the North Sea.

A typical value of dry mass deposited during 1979 in the Norwegian Trough is 400 g/m^2 (100 g/m^2 without bottom sediments). With a dry bed density of 0.5 tons/m^3

this implies an increase of the bed thickness of 8×10^{-4} m/yr, or an increase of 8 cm in 100 years. This value does not contradict to measured data (Eisma, 1981).

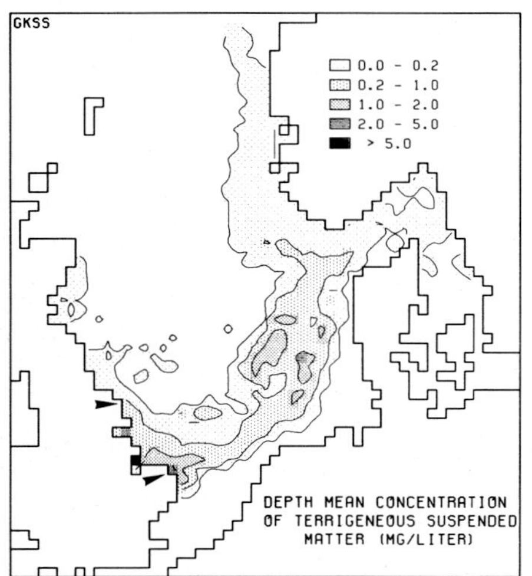

FIGURE 14. Distribution of suspended matter concentration calculated for December 31, 1979; source of suspended matter: Holderness and Norfolk cliffs.

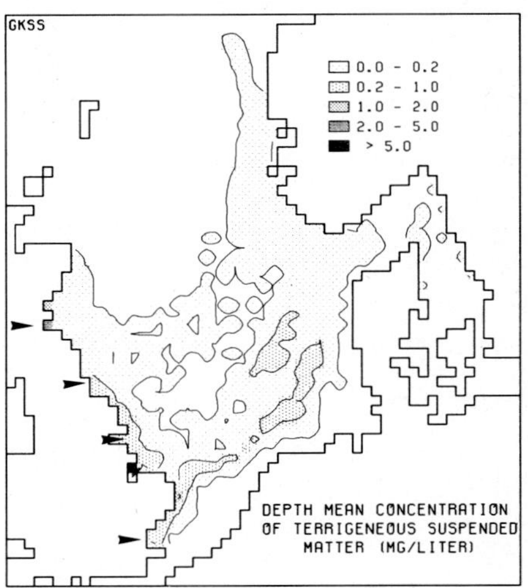

FIGURE 15. Distribution of suspended matter concentration calculated for December 31, 1979; source of suspended matter: UK rivers.

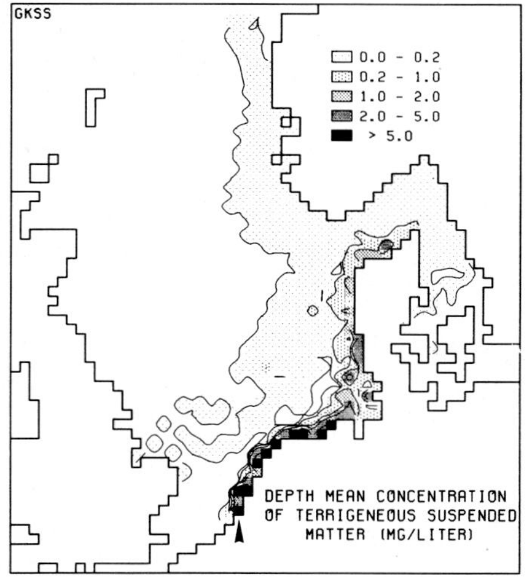

FIGURE 16. Distribution of suspended matter concentration calculated for December 31, 1979; source of suspended matter: Rhine river.

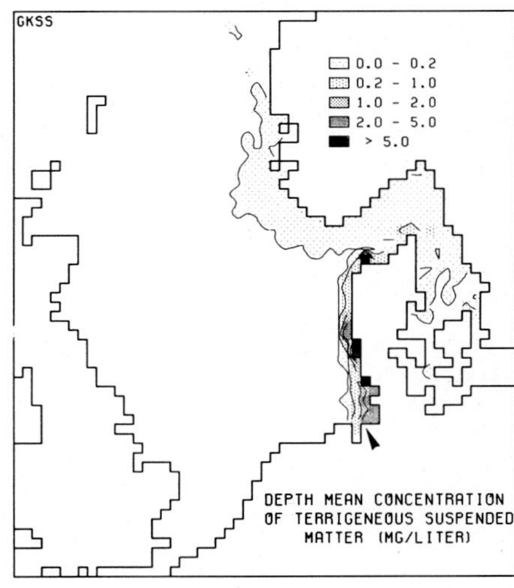

FIGURE 17. Distribution of suspended matter concentration calculated for December 31, 1979; source of suspended matter: Elbe river and Weser river.

It must be considered that Figure 18 or 19 represents storm conditions. During calm weather suspended matter may well be deposited in the southern parts of the North Sea. Figure 20 shows the distribution of dry mass deposited during the period from January to July 1979; bottom sediment is not included (for bottom sediment in July 1979 see Figures 29 and 30). During calm conditions the whole North Sea is a deposition area apart from the Dogger Bank, the Southern Bight and the German Bight.

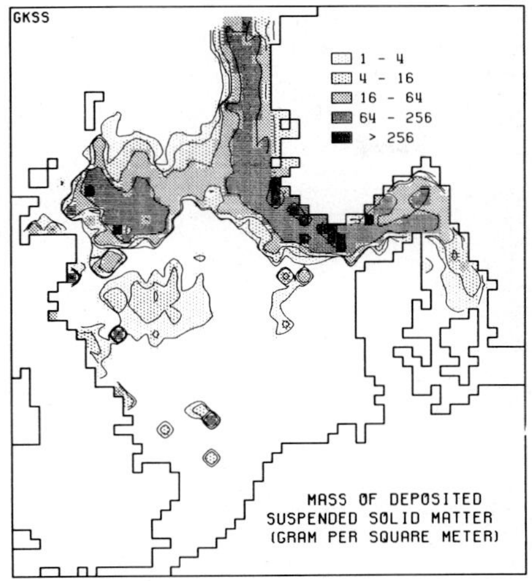

FIGURE 18. Distribution of the mass of deposited suspended matter, calculated for December 31, 1979; without bottom sediment.

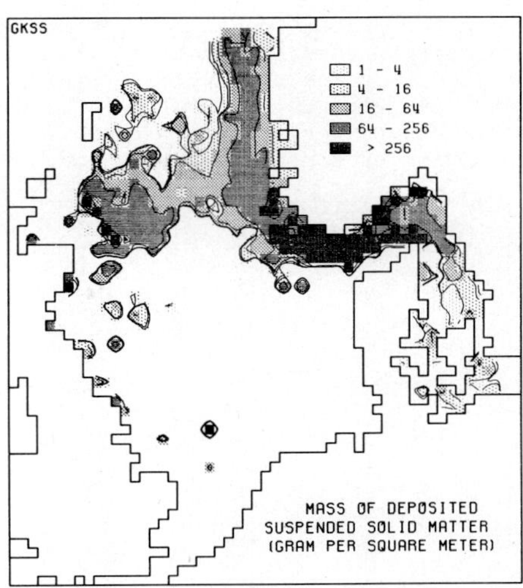

FIGURE 19. Distribution of the mass of deposited suspended matter, calculated for December 31, 1979; with bottom sediment.

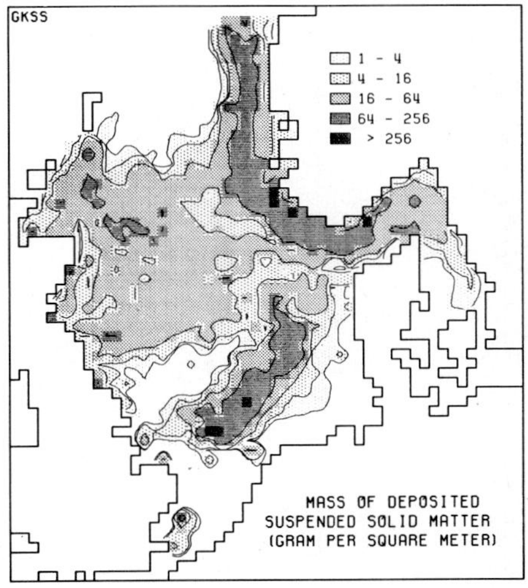

FIGURE 20. Distribution of the mass of deposited suspended matter, calculated for July 1979, without bottom sediments.

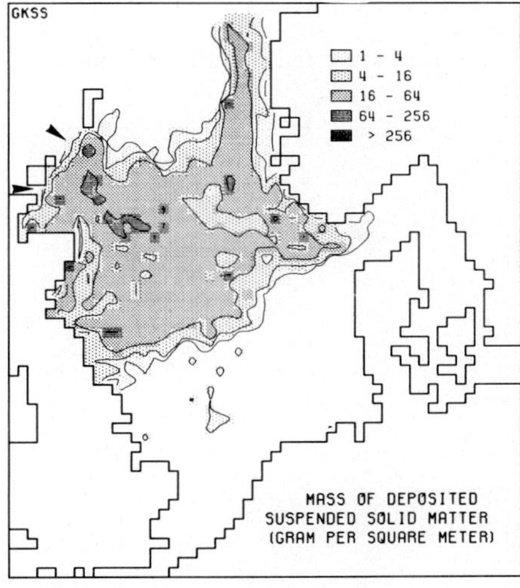

FIGURE 21. Distribution of the mass of deposited suspended matter, calculated for July 31, 1979; source of suspended matter: Pentland Firth and Fair Isle.

The deposited suspended matter is a feeding source for benthic animals. This means that contaminants enter the food web which are attached to suspended matter. It is thus interesting to know the origin of the deposited sediment shown in Figure 20. This is shown in Figures 21 to 26; arrowheads mark the sources of suspended matter.

In the present model computations the deposited suspended matter is not mixed into deeper bed layers (bioturbation is switched off). If bioturbation were to be included (e.g. in the Oyster Grounds), a certain amount of the deposited sediment shown in Figure 20 would not be resuspended during the autumn of 1979, but would contribute to the deposited suspended matter in Figure 18. There would be long term deposition of contaminated suspended matter in most parts of the North Sea.

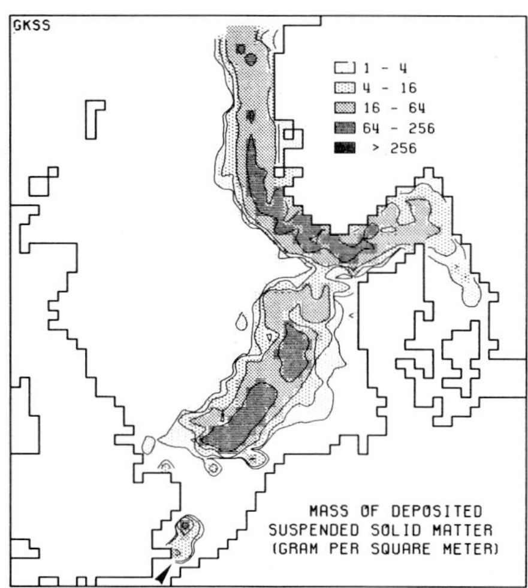

FIGURE 22. Distribution of the mass of deposited suspended matter, calculated for July 31, 1979; source of suspended matter: English Channel.

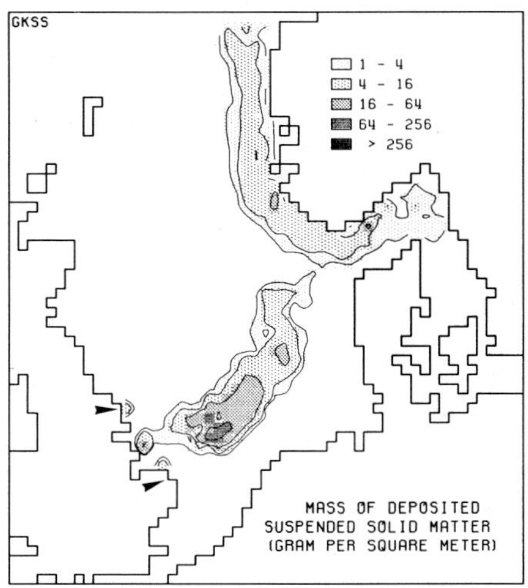

FIGURE 23. Distribution of the mass of deposited suspended matter, calculated for July 31, 1979; source of suspended matter: Holderness and Norfolk cliffs.

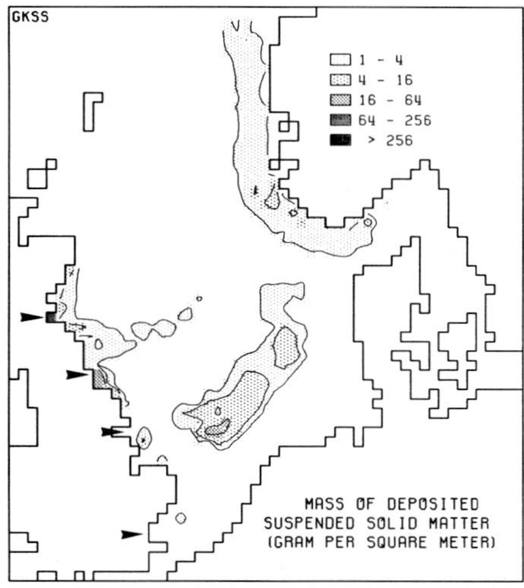

FIGURE 24. Distribution of the mass of deposited suspended matter, calculated for July 31, 1979; source of suspended matter: UK rivers.

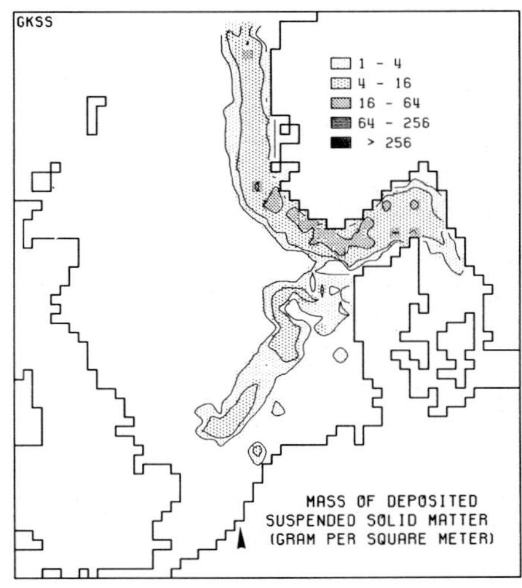

FIGURE 25. Distribution of the mass deposited suspended matter, calculated for July 31, 1979; source of suspended matter: Rhine river.

Eisma and Kalf (1987) report on deposition areas in the North Sea. According to Eisma and Kalf, there is deposition in the Skagerrak and in the Outer Silver Pit, but also deposition in estuaries, in the wadden sea and in the German Bight (southeast of Helgoland). Concerning the estuaries and the wadden sea, the model's scale is too large for simulating those areas correctly. As regards the German Bight, the model prohibits the deposition of suspended matter. From the point of the model, the mud area south-east of Helgoland is a puzzling phenomenon. Deposition in the Fladen Ground is not reported by Eisma and Kalf (1987). Cadee (1986), however, deduces from measurements a deposition rate of 73 gramme per square metre and year (4 mm in 100 years). This is consistent with the model results.

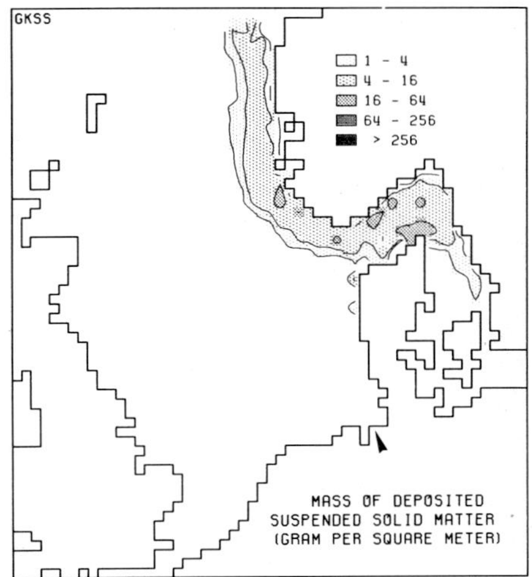

FIGURE 26. Distribution of the mass of deposited suspended matter, calculated for July 31, 1979; source of suspended matter; Elbe river and Weser river.

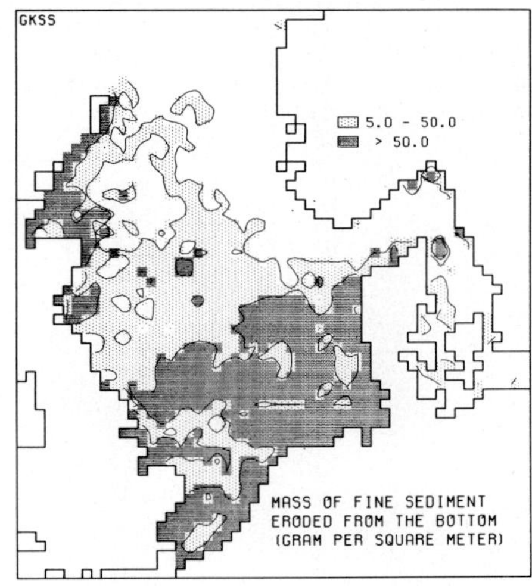

FIGURE 27. Distribution of the mass of eroded bottom sediments, calculated for December 31, 1979.

VIII.3 Erosion and deposition of bottom sediments

Figure 27 gives the distribution of eroded fine bottom sediment in December 1979, Figure 28 gives the distribution of fine bottom sediments deposited after

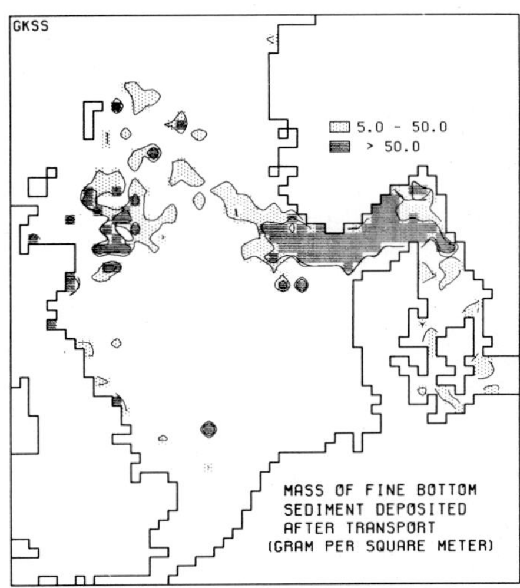

FIGURE 28. Distribution of the mass of bottom sediments, deposited after transport; calculated for December 31, 1979.

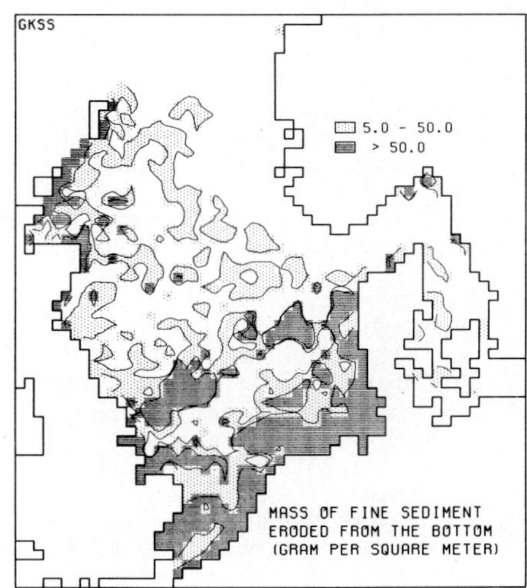

FIGURE 29. Distribution of the mass of eroded bottom sediments, calculated for July 31, 1979.

transport for the same month. To avoid a plot which is too patchy, two shadings only are used in the figures. For 'storm' conditions as in December 1979 the North Sea bottom can be divided into 3 zones: (1) an erosion zone south of the line Holderness cliffs - Skagerrak, (2) a deposition zone in the Norwegian Trough and in the Fladen Ground and (3) an intermediate zone with slight erosion.

This picture, however, is different when there are 'calm' conditions such as in July 1979 (Figures 29 and 30). The erosion zone south of Holderness - Skagerrak is still there, but within this zone there is a big deposition area south of Dogger Bank. The intermediate zone north of Dogger Bank now shows an irregular pattern of erosion and deposition.

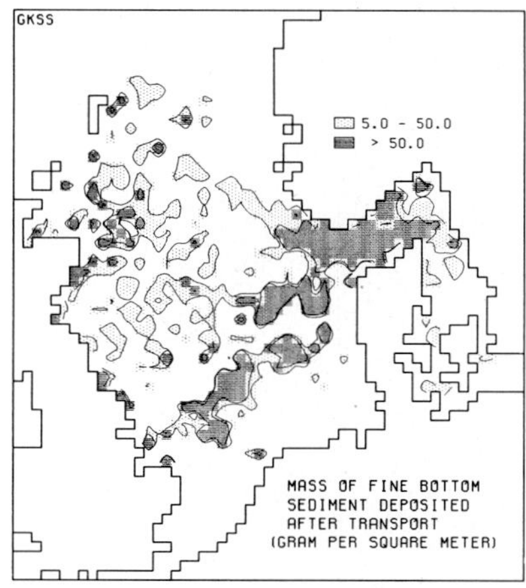

FIGURE 30. Distribution of the mass of bottom sediments, deposited after transport; calculated for July 31, 1979.

VIII.4 Assessment of model results

The actual aim of the model is to compute the fate of contaminants in the North Sea. Of special interest is the residence time of contaminants which are adsorbed to fine sediment particles. With regard to this residence time, the model results are mostly sensitive (i) to a correct evaluation of the bed shear velocity v_*, generated by currents and waves, and (ii) to deposition, bioturbation and erosion. The model results are to a lesser extent sensitive to the transport processes within the water column. The (intentional) absence of bioturbation means that the model results presented in this paper are nothing more than provisional results.

The following is a selection of shortcomings and missing processes which is furthermore indicative of the provisional character of the model results.

(1) The input of suspended matter by rivers is assumed to be constant throughout the year, which is far from realistic. At mean and low fluvial discharge the turbidity zone of an estuary is a trap both for marine and fluviatile suspended matter. At high fluvial discharge the accumulated suspended matter is 'pushed' into the North Sea.

(2) The input of suspended matter through the seaward boundaries is only roughly known. Even for the 'well documented' Strait of Dover information about suspended matter transport differs: Eisma and Kalf (1987) give an input of about 10 million tons per year, while van Alphen (1990) gives a value of 17 million tons per year.

(3) The bed roughness is a decisive parameter for computing the shear velocity generated by surface water waves. This bed roughness, however, can only roughly be estimated.

(4) All the model parameters that characterize suspended matter and sediment properties are only approximately known. Beyond the 'normal' processes of deposition, bioturbation and erosion, two special events are mentioned here which are important for suspended matter dynamics:

-- During and after a bloom the settling phytoplankton scavenges terrigeneous suspended matter from the water column. This was observed by Kempe and Nies (1987) by sediment trap sampling in the Skagerrak
-- Rauck (1988) states that every square metre of the Southern North Sea is 'ploughed up' by beam trawling several times per year. This affects the bed properties and induces additional erosion.

IX. Conclusion

The input data for the model computations consists of the 'Backhaus-currents' and of the wave data obtained from the German Armed Forces. These two powerful 'tools' allow it to tackle suspended matter transport in the North Sea. On the other hand there are (and will remain) many uncertainties concerning the behaviour of suspended matter and bottom sediments. An example is bioturbation of the bed.

On the whole, however, the model presented is a real step foreward towards a quantitatively correct description of the fate of suspended matter in the North Sea. The partial agreement of calculated and measured suspended matter concentrations and deposition reported in this paper is encouraging.

Acknowledgements

We thank the Bundesamt für Wehrtechnik und Beschaffung (Koblenz) and the Amt für Wehrgeophysik (Traben-Trarbach) for their support concerning the wave data.

X. References

Backhaus, J. O., 1985: A three-dimensional model for the simulation of shelf sea dynamics. Deutsche Hydrographische Zeitschrift, **38**, 165-187.

Cadee, G. C., 1986: Organic carbon in the water column and its sedimentation, Fladen Ground (North Sea), May 1983. Netherlands Journal of Sea Research, **20(4)**, 347-358.

Creutzberg, F. and Postma, H., 1979: An experimental approach to the distribution of mud in the Southern North Sea. Netherlands J. of Sea Research, **13(1)**, 99-116.

DHI, 1915/1952: Fischereikarte der Nordsee. Deutsches Hydrographisches Institut, Hamburg.

DHI, 1981: Karte der Sedimentverteilung in der Deutschen Bucht, Deutsches Hydrographisches Institut, Hamburg.

Drake D. E. and Cacchione, D. A., 1989: Estimates of the suspended sediment reference concentration (C_a) and resuspension coefficient (γ_0) from near-bottom observations on the California shelf. Continental Shelf Research, **9(1)**, 51-64.

Eisma, D., 1981: Supply and deposition of suspended matter in the North Sea. Spec. Publs. Int. Ass. Sediment., **5**, 415-428.

Eisma, D. and Kalf, J., 1979: Distribution and particle size of suspended matter in the Southern Bight of the North Sea and the Eastern Channel. Netherlands Journal of Sea Research, **13**2, 298-324.

Eisma, D. and Kalf, J., 1987: Dispersal, concentration and deposition of suspended matter in the North Sea. Journal of the Geological Society, London, vol 144, 161-178.

Führböter, A., 1983: Über mikrobiologische Einflüsse auf den Erosionsbeginn bei Sandwatten. Wasser und Boden, Heft 3, 106-116

Gast, G., 1987: private communication

Grant, W. D. and Madsen, O. S., 1979: Combined wave and current interaction with a rough bottom. J. of Geophysical Research, No. C4, 1797-1808.

Hydraulics Research, 1979: Properties of Belawan mud. Report No. EX 880.

Kempe, S. and Nies, H., 1987: Chernobyl nuclide record from North Sea sediment trap. Nature, **329**, 828-831.

Kersten, M. and Klatt, V., 1988: Trace metal inventory and geochemistry of the North Sea shelf sediments. In: S. Kempe, G. Liebezeit, V. Dethlefsen, U. Harms (editors), Biogeochemisty and distribution of suspended matter in the North Sea and implications to fisheries biology. Mitteilungen aus dem Geol.-Palaontol. Institut der Univ. Hamburg, Heft 65, 289-311

König, P., 1987: private communication

Krone, R. B., 1962: Flume studies of the transport of sediment in estuarial shoaling processes. Final Report, Hydraulic Eng. Lab., Univ. of Berkeley, USA

Lee II, H. and Swartz, R. C., 1980: Biological processes affecting the distribution of pollutants in marine sediments, Part II Biodeposition and bioturbation. In: R. A. Baker (editor), Contaminants and Sediments. Ann Arbor Science Publishers, 555-606

Maier-Reimer, E. and Sündermann, J., 1982: On tracer methods in computational hydrodynamics. In: M. B. Abbott and J. A. Cunge (editors), Engineering applications of computational hydraulics, vol. 1., Pitman Advanced Publishing Program, Boston, London, Melbourne.

McCave, I. N., 1970: Deposition of fine-grained suspended sediments from tidal currents. Journal of Geophysical Research, Vol 75, No. 21, 4151-4159.

McCave, I.N., 1987: Fine sediment sources and sinks around the East Anglian Coast (UK). J. of the Geological Society, London, Vol. 144, 149-152.

Puls, W., 1984: Erosion characteristics of estuarine muds. Hydraulics Research Report No. IT 265.

Rat von Sachverständigen für Umweltfragen, 1980: Umweltprobleme der Nordsee. Sondergutachten, Verlag W. Kohlhammer, Stuttgart und Mainz.

Rauck, G., 1988: Welchen Einfluss haben Grundschleppnetze auf den Meeresboden und Bodentiere?. Infn. Fischw., **35(3)**, 104-106.

Rodger, J. G. and Odd, N. V. M., 1985: Sludge disposal in Liverpool Bay. Hydraulics Research Report No. SR 70.

Stephan, H. -J., 1978: Verfahren zur Voraussage der kennzeichnenden Grössen des Seegangs. Mitteilungen des Leichtweiss-Instituts fur Wasserbau der TU Braunschweig, Heft 63.

Tomczak, G. and Gödecke, E., 1962: Monatskarten der Temperatur der Nordsee, dargestellt für verschiedene Tiefenhorizonte. Deutsche Hydrographische Zeitschrift, Erganzungsheft Reihe B, No. 7

U.S. Naval Oceanographic Office, 1965: Oceanographic Atlas of the North Atlantic Ocean, Section V, Marine Geology, Pub. No. 700.

van Alphen J., 1990: A mud balance for the Belgian Dutch coastal waters between 1969 and 1986. Netherlands Journal of Sea Research, 25(1).

Part V

Field Observations and Other Applications

Variations in Horizontal Density Gradient Forcing at the Mouth of an Estuary

Glenn A. Cannon

Pacific Marine Environmental Laboratory
National Oceanic and Atmospheric Administration
7600 Sand Point Way NE
Seattle, WA 98115-0070
(206-526-6779)

ABSTRACT

Puget Sound is a fjord-like estuary in which bottom-water intrusions are a dominant circulation feature that play a major role in the replacement of water below sill depth within the estuary. New observations on the inside and outside of the entrance sill show that while intrusions occur during neap tides as previously thought, the onset of the intrusions occurs before the minimum in the neap tides. Calculations using a model with a balance between the longitudinal pressure gradient and vertical mixing suggest the onset is influenced by fluctuations in the horizontal density gradient caused by salinity variations across the sill. They also show that the resultant bottom inflow can be further modified by strong surface winds. Deep salinity changes seaward of the sill in the Strait of Juan de Fuca estuary appear to be the result of storms on the Pacific coast causing reversals of surface flow and variations in deep flow more than 135 km from the coast. Although Puget Sound is a glacially carved estuary, flow over the entrance sill in Admiralty Inlet has characteristics which resemble coastal plain estuaries; thus, these studies may have more general applications to time variations in lower layer flow at the mouth of other estuaries.

I. Introduction

Bottom water renewal in fjords is a dominant physical process of particular importance in controlling the physical and chemical environments below sill depth. The time scale of renewal varies greatly from one fjord to another, reflecting differences in external forcing parameters as well as variations in circulation dynamics between fjords. Puget Sound is the southernmost glacially carved estuarine system in western North America, and it has a relatively long and topographically complex entrance sill which plays a major role in regulating the replacement of its deep water. Bottom-water replacement here is sufficiently frequent to keep the deep water below sill depth from becoming anoxic. Early studies, however, suggested replacement occurred yearly following the coastal upwelling cycle when the most dense water entered the Strait of Juan de Fuca (Barnes and Collias, 1958). Later work implied that inflow occurred throughout the year on very large spring tides when the tidal excursion could transit the sill on a single flood cycle (Farmer and Rattray, 1963; Cannon and Ebbesmeyer, 1978; Cannon and Laird, 1978). More recent studies have documented that major bottom-water inflow events occur during neap tides when mixing over the sill is least and bottom gravitational circulation can develop (Geyer and Cannon, 1982). Because the earlier work concerning spring-tide intrusions was based only on observations at a single location within the basin, there was some uncertainty in its interpretation. An experiment was designed to resolve this spring-neap discrepancy and to determine if the onset of the intrusions could be predicted. This paper first reviews some aspects of the intrusion process using data from various experiments, and it then

Coastal and Estuarine Studies, Vol. 38
R. T. Cheng (Ed.)
Residual Currents and Long-term Transport

presents some new ideas on variations in the onset of bottom-water intrusions from the recent experiment.

II. Oceanographic Setting

The entrance to Puget Sound is 135 km from the Pacific Ocean at Port Townsend (Fig.1). The entrance sill (Admiralty Inlet) is about 30 km long and consists of two sills (65 and 105 m) bracketing a deeper area (Fig.2). The main basin of the Sound with an average depth of about 200 m extends for 75 km between this seaward sill and a landward sill (The Narrows; 45 m) which connects to a large southern basin. There are two major branches of Puget Sound: one going south from the entrance sill (Hood Canal) and one heading northward from just inside the sill (Saratoga Passage). The main basin has been the focus of extensive study because it provides the pathway for transport of contaminants from the major urban areas of Seattle and Tacoma (Cannon, 1983).

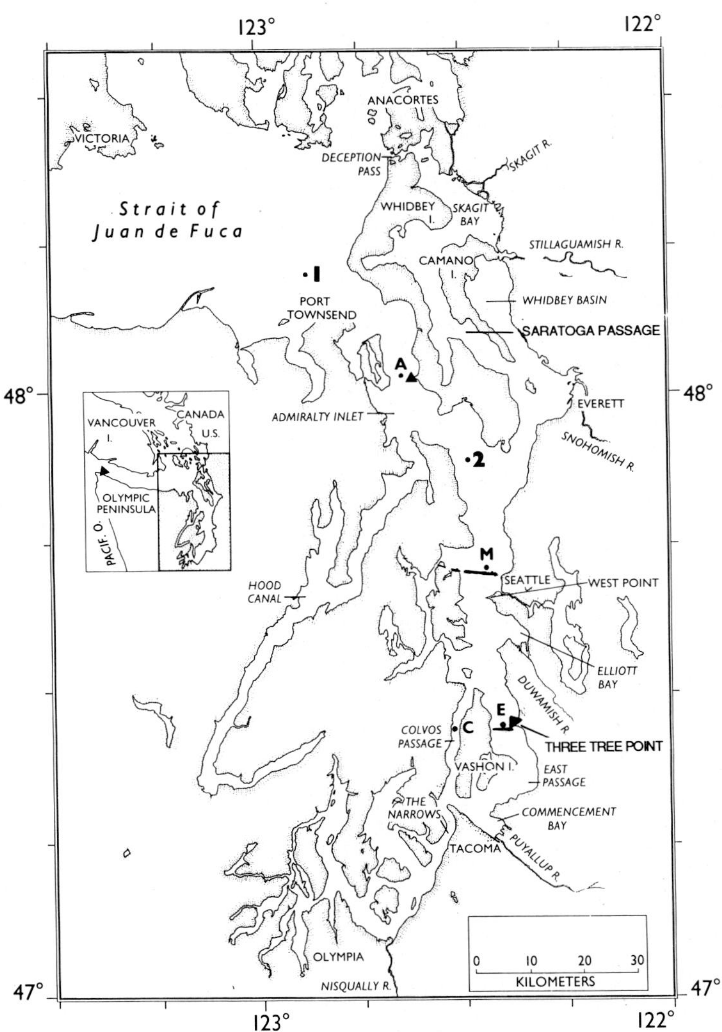

FIGURE 1. Chart of Puget Sound showing mooring locations and two cross sections of moorings. Triangles show the locations of wind observations in Admiralty Inlet and East Passage and of tide measurements at the mouth of the Strait of Juan de Fuca.

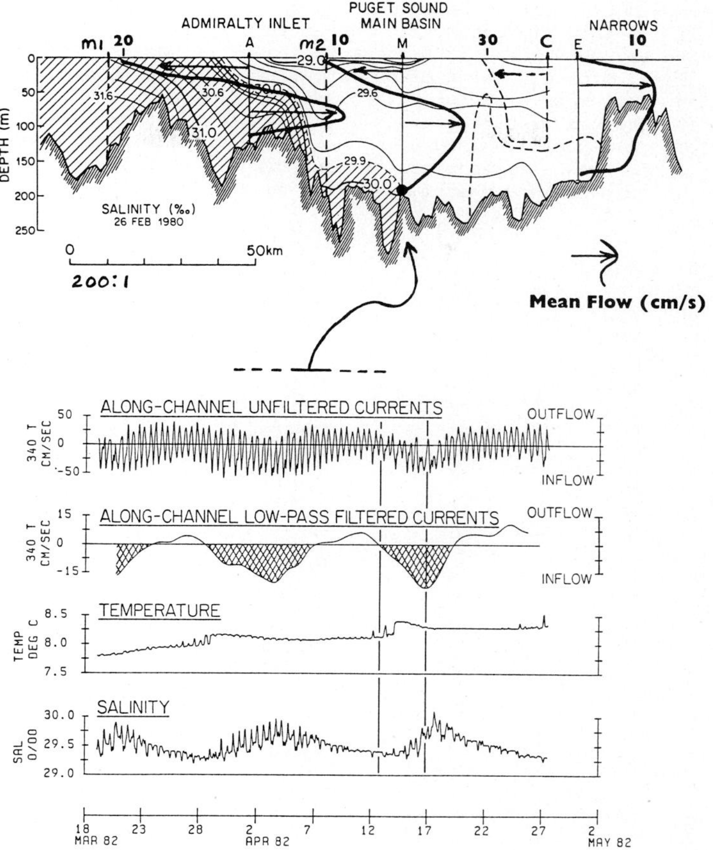

FIGURE 2. Salinity section along Puget Sound with superimposed sample average current profiles at moorings A, M, C, and E (upper) and time series of currents, temperature, and salinity near the bottom of mooring M in the main basin (lower). Positive flow is seaward. Undirectional flows at moorings C and E are along sides of intervening Vashon Island. Vertical exaggeration of the section is 200:1. Mooring locations are shown along top of section and in Fig.(1).

Flow in the estuary is mixed tidal with a superimposed time-varying gravitational circulation (Fig.2). Vigorous vertical mixing of water occurs in the sill zones where amplitudes of the tidal currents are about twice as large as those shown for the deep basin. There also is a discernible tidal component in salinity which decreases with distance from the sill. Larger net flows and higher salinities generally occur during neap tides, and these are associated with bottom water intrusions into the deep basin.

The salinity section shows strong horizontal stratification through Admiralty Inlet, while that in the deeper basin shows a relatively weaker horizontal stratification. The effects of freshwater input from the largest river, the Skagit (maximum flux about 1000 m^3/s), may be seen as it enters the main basin at the confluence with the entrance sill. Other smaller rivers enter at various places, and there is no relatively uniform freshwater lens as is characteristic of many fjords. The deep water is uniformly dense and shows an ongoing bottom-water intrusion.

When the tides are filtered from the observations, the net flow in most of the estuary is seaward near the surface and landward near the bottom (Fig.2). In the northern half of the main basin, the change from inflow to outflow occurs at about 50 m in a depth of 200 m, or 25% of the total depth. This is intermediate between the deeper level for partially mixed estuaries of 40-50% (shown here in Admiralty Inlet) and the shallower level for more classical fjords of 10-15% (not shown here, but occurs in Saratoga Passage; Cannon, 1983). The net flow in most of the estuary is two-layered. An exception occurs in the southern part of the main basin where Vashon Island causes a predominantly southward mean flow on its wider east side (East Passage) and northward mean flow on its west side (Colvos Passage).

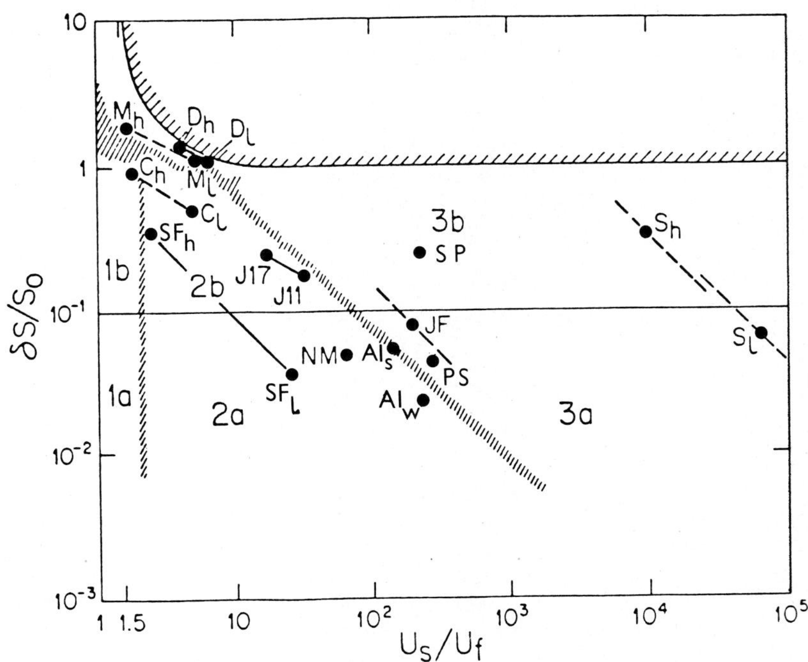

FIGURE 3. Stratification-circulation classification diagram showing characteristics of different parts of Puget Sound relative to other estuaries. Adapted from Hansen and Rattray (1966), Barnes and Ebbesmeyer (1978), Partch (1981), and Conomos (1979). Places within the Puget Sound system include: main basin off Seattle (PS), Admiralty Inlet (AI), Saratoga Passage (SP), Duwamish River (D), and Strait of Juan de Fuca (JF). Other estuaries include: Silver Bay, Alaska (S), Narrows of the Mersey (NM), James River (J), San Francisco Bay (SF), Columbia River (C), and Mississippi River (M). Subscripts h and l denote high and low runoff.

The differing oceanographic characteristics of parts of Puget Sound can be seen by plotting the above data on the stratification-circulation diagram of Hansen and Rattray (1966) (Fig.3). These data show that Admiralty Inlet (AI) falls within the range of partially mixed estuaries while the main basin of Puget Sound (PS) and the Strait of Juan de Fuca (Puget Sound's connection to the ocean, JF) fall within the range of fjords. It is interesting to note that other parts of Puget Sound show more widely varying characteristics spanning the whole range of estuarine types. Saratoga Passage (SP) is highly stratified like a classical fjord due to the runoff from the Skagit River. The relatively small Duwamish River (maximum flux about 100 m^3/s) (D) entering the Sound at Seattle has a pronounced salt wedge at both high and low runoff.

Mean along-channel current velocities through two sections show considerable lateral variability (Fig.4). The northern section has a well-defined two layer current regime which varies in depth from 55 to 75 m across channel. A relatively strong horizontal core of seaward flow is evident between 10 and 30 m, and a region of strong lateral current shear is evident on the east side of the channel where the nearshore surface flow is reversed from the normal estuarine flow. A diagonal band of high landward flow is seen in the lower layer extending from 100 m to the bottom. The southern section only has a very shallow seaward flowing layer overlying a deeper landward flowing layer extending to the bottom. A core of maximum velocity is evident at mid-depth above 100 m which becomes weaker toward the western side.

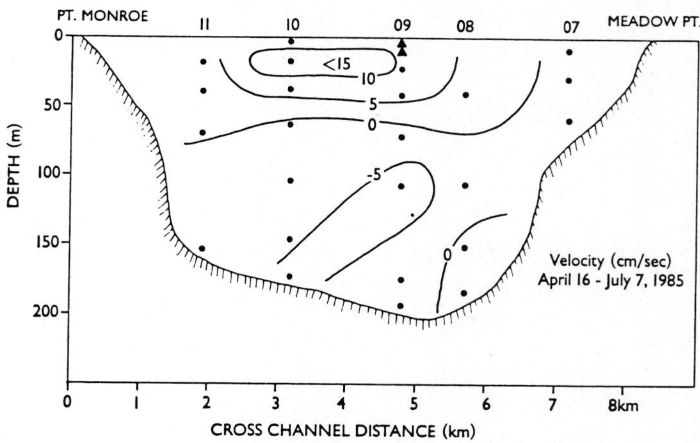

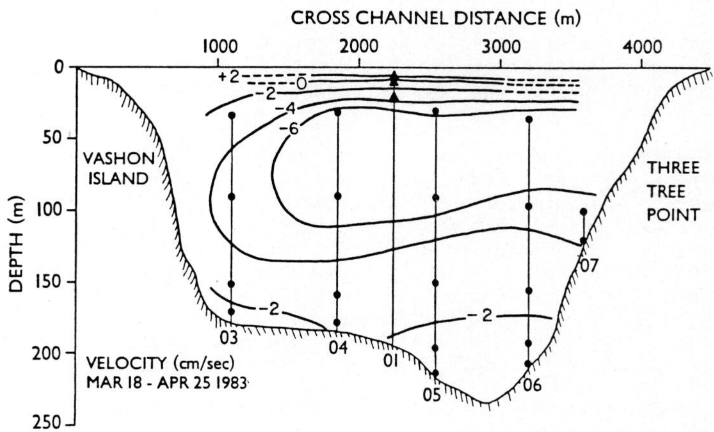

FIGURE 4. Mean along-channel currents in two sections across Puget Sound showing two-layer flow in the north and approximately one-layer flow in the south. Positive flow is seaward. The landward flux through the sections is approximately the same at about 2×10^4 m^3/s. Lower section from Bretschneider et al., (1985). Vertical exaggerations are different; top is 20:1; bottom is 10:1.

To further investigate the spatial structure of subtidal current fluctuations, empirical orthogonal functions (EOF's) have been used. EOF's provide three kinds of information: (1) spatial patterns, (2) percent of variance accounted for by these patterns, and (3) time variations. The EOF analysis of the southern section was separated into three significant modes which accounted for 79% of the variance of current data from the 23 meters in the section (Bretschneider et al., 1986). The second mode, which accounted for 21% of the variance, was concentrated near the bottom bottom (Fig.5). The EOF time series indicates weak and variable seaward flow

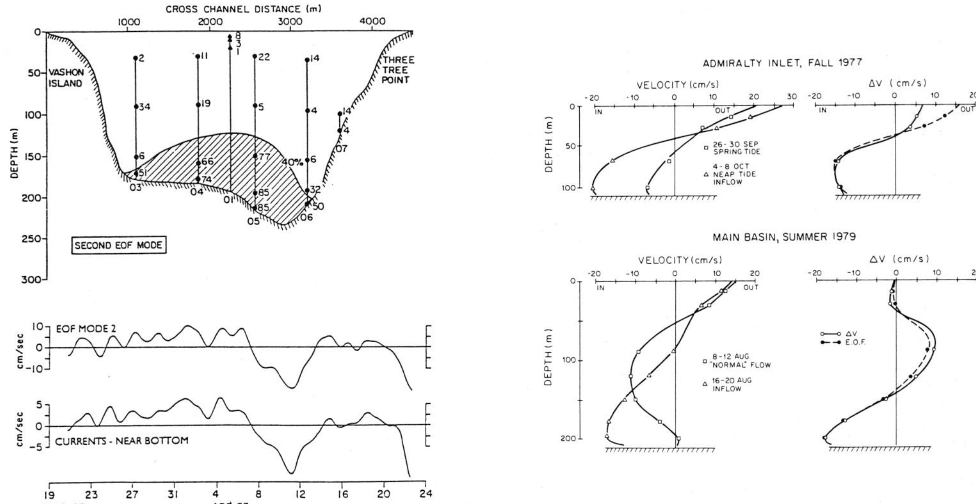

FIGURE 5. EOF mode two showing percentage of variance explained at each meter and, EOF time series for the entire section compared to independently measured near-bottom currents (left), and velocity profiles compared during normal flow and during intrusions with EOF analysis over the sill and in the main basin (right). Total variance explained for the section is 21%, and the correlation coefficient for the two time series is 0.8. Left from Bretschneider et al., (1985); right from Geyer and Cannon (1982).

followed by two strong landward events. This series is highly correlated (r = 0.82) with the low-pass-filtered current data measured at an independent bottom boundary layer experiment approximately 0.1 km south of the section. The velocity record there clearly showed the effects of two density current intrusions that propagated into East Passage. The section also had a relatively high percentage of variance explained at middepths, which is likely to represent compensating flow adjustment. In fact, EOF's calculated using a covariance matrix showed the two regions with opposite signs. EOF analysis of the northern section showed two significant modes with the second mode being the same as here. The first mode in both cases was wind-dominated, and accounted for about 50% of the variance (not shown here).

Earlier modal analysis of single moorings showed compensating flow adjustments for the bottom water intrusions were centered at 90-100 m with little change in flow in the upper 50 m (Geyer and Cannon, 1982) (Fig.5, right). The adjustment appeared as reduced inflow at middepth levels. Over the entrance sill, however, the compensating flow appeared as increased outflow in the surface layer, more like a coastal plain estuary.

III. Across-sill Observations

An experiment was designed with moorings bracketing the Admiralty Inlet entrance sill, which had not been done previously, to better describe the onset of the intrusions and to determine if they could be predicted (Fig.1). Current-meter moorings were deployed from December 1983 to April 1984 on either side of the sill as well as along the main basin including the mid-basin site of numerous earlier studies. Aanderaa current meters were utilized with temperature and conductivity sensors. These sensors were field calibrated to CTD measurements, and their usefulness has been demonstrated in other estuarine studies (See e.g., Cannon and Bretschneider, 1986). Velocity vectors were resolved into an along-channel (positive seaward) and cross-channel right-handed coordinate system based on considerations of the mean velocity vector, axis of maximum variance, and local bathymetry. Tidal and higher-frequency fluctuations were removed using a 35-hr cosine-Lanczos filter. Winds were measured on two points of land in Admiralty Inlet and the main basin.

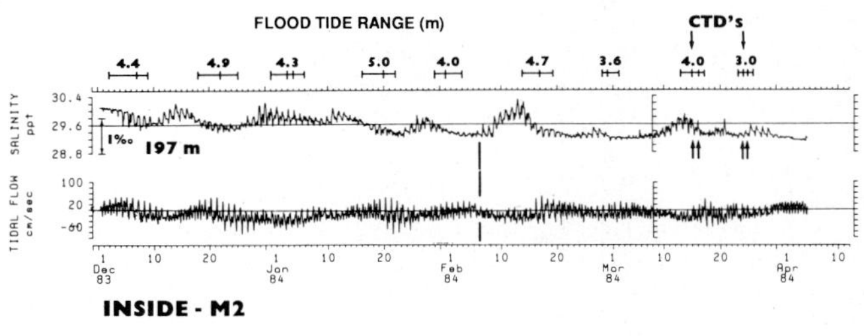

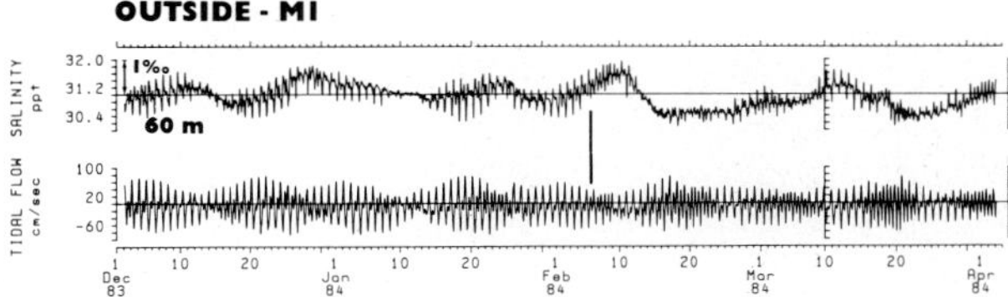

FIGURE 6. Time series of along-channel flow and salinity at mooring 2 inside and below the entrance sill to Puget Sound (upper) and at mooring 1 outside the sill at sill depth in the Strait of Juan de Fuca (lower). Flood-tide ranges (HHW-LLW) for Puget Sound greater than 3.5 m with maximum noted are shown at the top. Positive flow is seaward. Vertical line at 6 February shows salinity increases occur first outside and are not during neap tides. Locations are shown in Fig.(1).

III.1 Current-meter time series

Figure 6 shows times series of along-channel currents and salinity inside the estuary below the sill and outside at sill depth. Tidal currents are mixed semidiurnal with maximum and minimum currents occurring at about fortnightly intervals. Although maximum currents inside the sill typically are 100 cm/s, currents exceeding 200 cm/s have been measured over the sill. The salinity records also have a mixed tidal signal with maximum oscillations up to about 0.6 o/oo. There is a pronounced increase in the overall salinity of up to 0.6 o/oo ($\Delta\sigma_t = 0.6$ g/l) at about fortnightly intervals which is the dominant characteristic of inflow of new bottom water described in the earlier referenced studies. Each salinity increase is followed by a decrease that must be due to mixing within the basin. It is uncertain whether the decrease in salinity over the entire record is real or is a result of fouling from long-term deployment. The annual cycle in bottom salinity usually peaks in late summer and is minimum in midwinter (Cannon, 1983).

In all cases the onset of each salinity increase (an intrusion) occurred before minimum neap tidal currents, while the maximum salinity occurred at approximately the same time as the minimum. Inflow during the strongest intrusions was sufficiently large that the tidal currents did not reverse the flow at the bottom. Large spring fortnightly flood-tide ranges (greater than 3.5 m) occurred following the maximum salinity and generally occurred during decreasing salinity in the basin. However, increase in bottom water salinity inside occurred on specific large flood tides, and then it decreased probably due to mixing in the basin. Bottom tidal currents reversed direction during springs which would increase mixing. Thus, it appears that major inflows occurred only during neap tides, but the onset occurred before minimum neaps.

Outside the sill, the tidal currents also were mixed and had a fortnightly component. Maximum along-channel currents were about the same as inside at about 100 cm/s. Salinity variations were larger than inside the estuary with maximum tidal oscillations of about 1 o/oo at and below sill depth (~60 m). There also was a large low-frequency variation in the salinity of the deeper water of 0.8-1.3 o/oo ($\Delta\sigma_t$ = 0.6-1.0 g/l) which did not occur nearer the surface (not shown). This large variation in outside bottom salinity (density) is, as far as we know, a new observation. In addition, it was not correlated with the fortnightly variation in flow. The salinity minima occurred outside before inside and the outside increases during spring tidal conditions (see, e.g., the changes occurring near 6 Feb. on Fig. 6).

III.2 Horizontal density gradient

Because of the relatively large variation in outside salinity and because increasing inside salinity occurred before neap tides, we felt variations in the longitudinal pressure gradient might play a role in the onset of the intrusions. The horizontal salinity gradients through Admiralty Inlet are much larger than most fjords (see Figs. 2 and 3), and variations in the magnitude of this gradient have not been previously assessed. Within Puget Sound the horizontal salinity gradient is much smaller, which is more characteristic of fjords (Rattray, 1977).

Figure 7 compares conditions outside the estuary at sill depth (60 m) and inside near the bottom (200 m). These records have been low-pass filtered to remove the tidal signal. The flow at the bottom was almost always southerly or into the

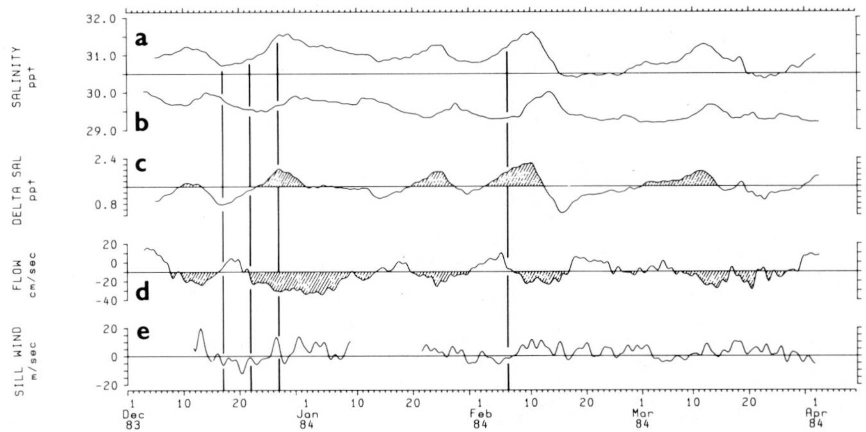

FIGURE 7. Low-pass filtered salinity at sill depth outside the sill (a); salinity at the bottom in Puget Sound (b); the salinity differences between the pairs (c); bottom flow in the Sound (d); and along-channel winds (e). Inflow greater than 10 cm/s and salinity differences greater than 1.4 o/oo are shaded. Positive is seaward. Vertical lines in December are times of calculations in Table 2. Vertical line at 6 February shows the sequence of intrusion onset discussed in text.

basin. However, speeds greater than 10 cm/s inward have been shaded to show the increased inflow associated with the increasing salinity. Maximum inflows exceeded 30 cm/s. The flow outside at sill depth (not shown) was relatively constant toward Admiralty Inlet at about 10 cm/s, but below sill depth outside, there were occasional reversals to seaward flow (Fig.8). An example of the sequence of events is highlighted by the vertical line at 6 February: salinity outside at sill depth has already started increasing (a); salinity inside at the bottom starts increasing (b); the salinity difference across the sill has increased to greater than 1.4 o/oo (c); and flow inside at the bottom increases inward greater than 10 cm/s (d).

The salinity difference between inside and outside the estuary showed large variations up to 2 o/oo, and all maxima were coincident with intrusions of new bottom water into Puget Sound. Differences greater than 1.4 o/oo have been arbitrarily shaded. There is a remarkable visual correlation between density (salinity) differences across the sill exceeding this magnitude and inflow greater than 10 cm/s, and regression analysis indicated maximum correlation (r = -0.56, above 95% confidence) with a lag of 2.8 days.

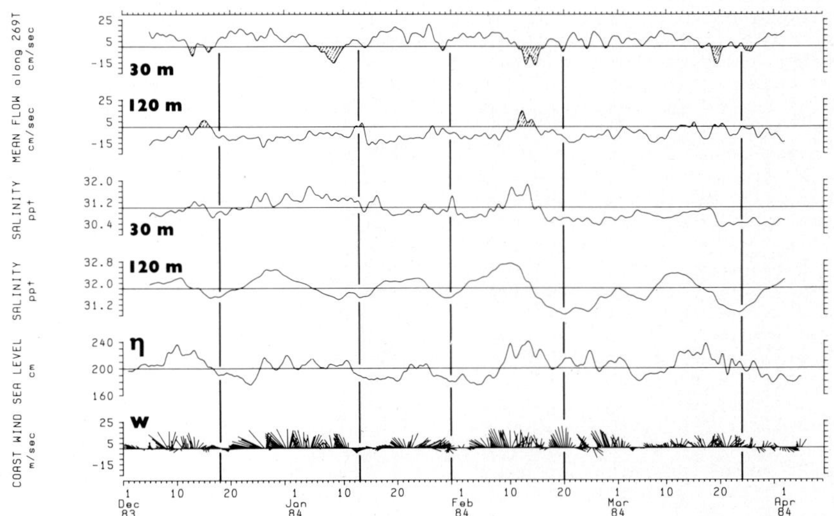

FIGURE 8. Low-pass filtered currents and salinity in the surface (30 m) and bottom (120 m) layers at mooring 1 outside the sill in the Strait of Juan de Fuca; Neah Bay filtered heights (h) at the seaward end of the Strait; and computed coastal winds (w) at 48°N, 125°W. Reversals of surface and bottom currents are shaded. Positive is seaward. Vertical lines show the beginning of bottom layer salinity increases during decreasing sea level.

In order to make an estimate of the effect of the longitudinal density gradient a simple model of gravitational circulation is used with a steady-state balance between the longitudinal pressure gradient and vertical mixing with a constant eddy coefficient. These were the dominant terms determined by Pritchard (1967) for a coastal plain estuary. The characteristics of Admiralty Inlet determined from limited observations (Geyer and Cannon, 1982) using the Hansen and Rattray (1966) stratification-circulation classification of estuaries fall within the range of partially mixed estuaries (Barnes and Ebbesmeyer, 1978).

The problem is formulated following Officer (1976, Eq.4-88) as:

$$\frac{1}{\rho}\frac{\partial p}{\partial x} = -g\frac{\partial \eta}{\partial x} + \frac{gz}{\rho}\frac{\partial \rho}{\partial x} = N_z \frac{d^2u}{dz^2} \qquad (1)$$

where ρ is density, p is pressure, and N_z is the vertical eddy viscosity. The vertical pressure gradient is assumed hydrostatic, the boundary conditions are u = 0 at the bottom and the stress equals that of the wind at the surface, and the horizontal pressure gradient is assumed constant with depth. Combining all these for a uniform channel, Officer (Eq.4-106) gets:

$$u\ (\xi = \frac{z}{H}) = \frac{1}{4} \frac{H}{\rho N_z} [T\ (1 - 4\xi + 3\xi^2) + \frac{gH^2}{12} \lambda\ (1 - 9\xi^2 + 8\xi^3)] \tag{2}$$

where a river runoff term has been neglected as being small, H is the channel depth, T is the wind stress, and λ is the horizontal density gradient $\partial\rho/\partial x$.

An independent set of data is used to estimate N_z when the wind stress was negligible, and the resulting equation is used to estimate currents for various salinity (density) gradients. The data used to determine N_z were a 3-month average current profile midway along Admiralty Inlet (shown on Fig.2) and a CTD section along the sill during 1977 (Cannon, 1983; Geyer and Cannon, 1982). The average density difference across the sill was about 1.2 x 10^{-3} gm/cm³ over about 50 km. Using these data and H = 100 m results in an average N_z of 284 cm²/s. Large N_z is quite possible given the strong currents and rough bottom as inferred from M_2 tidal currents by Mofjeld and Lavelle (1984). This is not exactly correct because N_z is not constant and varies with depth from 174-468 cm²/s using the observed data. The smallest value is at 0.9 depth. However, using a constant value does allow a first estimate of the density gradient effect.

Calculations of near-bottom currents for three times in the 1983-84 data using constant N_z from the 1977 data are shown in Table 1. The three cases correspond to conditions of no inflow, the onset of inflow, and maximum inflow, as shown by salinities which are minimum outside, minimum inside, and maximum outside, respectively (Fig.7). The calculated currents between the two moorings and the observed currents are in good agreement for 12/17 and 12/22 using the average N_z = 284 cm²/s. However, the observations are inside the sill where depths are only about half as much as on the sill. The observed flow inside on 12/27 exceeds the calculated values. Winds during the preceding day were extremely strong to the north (Fig.7), and the wind stress contribution to the near-bottom flow, also shown in Table 1, is comparable to the density flow. Summation of the density and wind effects is closer to what was observed. Winds at the other two times were weaker and had negligible effect.

The Δs value for intrusion onset of 1.4 o/oo is about the average shown in Figure 7. A few historical salinity sections, which are not truly synoptic, showed differences across the sill of about 0.4-1.5 o/oo. However, the time series here showed larger differences, up to 2 o/oo, and record average profiles from the two moorings here showed a salinity difference of about 1.4 o/oo between the bottom inside and most of the middle water column outside. The mean current profiles showed changes from inflow to outflow occurred at about 40 m outside the sill and at 90 m inside the sill. The latter is somewhat deeper than 50-70 m seen further south at the northern cross section in Puget Sound and may result from hydraulic effects as the flow crosses the sill and enters deeper water. The compensating flow adjustments for increased bottom inflow at this mooring near the sill occurred above 100 m, while those of Geyer and Cannon (1982) near the two cross sections showed it centered near 100 m, indicating a possible shrinking height of the intrusion as it propagates along the basin.

III.3 Outside salinity variations

Because the horizontal density gradient variations across the sill contribute to the onset of bottom-water intrusions into Puget Sound, the cause of the salinity variation seaward of the sill needs to be examined. Part of the variation may be caused by tidal mixing over various shoals (~50-60 m) in the Strait, some of which form an effective sill (115 m passage) extending south from Victoria. However, although the currents have a fortnightly-like variation, the deep salinity changes are not correlated with this aspect of the flow, and all the large outside salinity increases commence during spring tides (Fig.6; e.g., on 1 Feb.).

Another possibility is that these salinity changes may be partly caused by offshore winds affecting the flow in the Strait of Juan de Fuca estuary (Cannon and Holbrook, 1981; Holbrook and Halpern, 1982; Cannon and Bretschneider, 1986). Coastal storm winds, particularly in winter, drive surface coastal water onshore at

TABLE 1. Estimates of near-bottom horizontal velocity for different horizontal density gradients (λ) compared to observed currents. Corrections for local wind effects (T) at these depths are given at the bottom. Plus is seaward.

Date	12/17	12/22	12/27	
σ_t status	minimum outside	minimum inside	maximum outside	
Δs	.9	1.4	1.8	o/oo
$\Delta\sigma_t$	.75	1.15	1.52	g/ℓ
u calculated (λ) N_z = 284 cm²/s	-6	-9	-12	cm/s
u observed	-5	-10	-30	cm/s
wind speed	-1	-2	13	m/s
stress	~0	~0	5	dyn/cm²
u calculated (T) N_z = 284 cm²/s	0	0	-9	cm/s

the mouth of the Strait. This results in reversal of the surface pressure gradient in the Strait causing inflow at the surface. During some of these events, the effects are sufficiently large to suppress the deeper inflow of saltier water and/or to reverse the bottom flow. Decreases in bottom salinity of at least 0.75 o/oo have been observed 90 km from the coast with similar increases following the events (Cannon and Bretschneider, 1986). Figure 8 shows low-pass filtered currents and salinity for the surface (30 m) and bottom (120 m) layers in the Strait just outside Admiralty Inlet (mooring 1). Five occurrences of reversed surface flow are shown. Four of these are accompanied by reversed bottom flow, and the fifth, by decreased bottom flow. Although we did not have wind records from the coast, we calculated winds at 48°N, 125°W about 25 km off the coast (Bakun, 1975) and obtained a sea level record from Neah Bay at the mouth of the Strait, which have been shown to be highly correlated (Holbrook and Halpern, 1982) (Fig.7). Here they also were significantly correlated ($r = 0.63$) with sea level lagging wind by 18 hours. All five events of surface flow reversal followed storms and were accompanied by decreases in bottom salinity. There also is a significant correlation ($r = -0.78$) between sea level at the coast and bottom salinity at the other end of the Strait with a lag of 6.5 days. Following the coastal storms (decreasing sea level), bottom salinity increased along with the return of surface outflow and bottom inflow at mooring 1 in the Strait. It is also interesting to note that the northward winds produce countering effects on the flow in Admiralty Inlet. Local north winds augment the flow at the bottom, and non-local forcing, described here, retards it.

The increasing salinity gradient across the Admiralty Inlet sill and the onset of intrusions were coincident with increasing salinity in the Strait following the storms. Conversely, the cessation of an intrusion may be augmented by the start of a storm causing decreasing salinity in the Strait as shown at the end of the February event. In addition, the lack of (or weak) intrusions in mid January and late February could be attributed to the small salinity gradients following storms. That is, the higher salinities had not yet propagated back up the Strait. This is the first time these wind effects have been seen at the easternmost end of the Strait both in the surface and bottom layers, and it appears they randomly contribute to the variability of inflow of bottom water from the Strait of Juan de Fuca into Puget Sound during winter (fall through spring).

During summer this effect probably is small because storms only rarely are of sufficient magnitude and duration. However, at that time the direct effect of strong southward winds in Admiralty Inlet could retard inflows. Taking the ratio of the two effects in Eq. (2) shows winds of 12-15 m/s would be required to stop an inflow driven by $\Delta\sigma_t$= 1 g/l, which is not likely to occur.

IV. Discussion

Bottom-water intrusions are one of the major circulation features of Puget Sound, and they play a major role in flushing some contaminants. Observations presented here show the onset of the intrusions are associated with neap tides and with the horizontal density gradient across the sill. This spring-neap cycle is the result of mixing effects over the sill proposed by Geyer and Cannon (1982). However, wind events in winter on the Pacific coast are capable of reversing circulation far inside the Strait of Juan de Fuca, and consequently can reduce the density at sill depth outside Admiralty Inlet. We do not yet have sufficient summer measurements to assess the magnitude and variations in the density gradient at that time.

Laboratory studies by Linden and Simpson (1986) demonstrate that turbulent mixing can control the inflow of dense water through a channel. They derived a characteristic length scale over which a density current can be arrested. It would be about 15 km for Admiralty Inlet. Thus, it is approximately the right length for tidal mixing, particularly during spring tides, to retard or stop density inflows. They also calculated a characteristic time scale over which retardation would occur, which for this case is about 12 hr. These length and time scales indicate another important factor for Puget Sound. If the Admiralty Inlet sill was deeper, inflows would be more common, but if it was shallower, they would be less frequent, drastically changing the exchange processes with the ocean. They further point out that when mixing is sufficiently strong to retard density currents, a gravitational circulation dependent on the horizontal density gradient results. Its magnitude is the same as that for the density gradient part of Eq.(2) at the surface.

Other studies of deep-water renewal in fjords have shown various causes for the increase in outside density resulting in new water entering the fjord (see, e.g., Gade and Edwards, 1980). None, as far as we know, have shown a time-dependency in the range shown here (storms to seasonal), although this range is known to be important for surface water exchanges. The variations here may be further complicated because the entrance sill is not at the coast, and these processes have not been studied extensively in the intervening Strait of Juan de Fuca estuary. These results indicate the necessity of understanding both estuaries as a system. Observations presented at a conference on Chesapeake Bay (March, 1988) showed possible similar variations at the entrance to the Patuxent River estuary, a tributary of the Chesapeake (Kincaid et al., 1988).

What happens to the intrusions once they enter the basin? In this set of observations only three events in January, February and March propagated the entire length of the basin with decreasing density changes with distance along axis. Using the onset of the density increases as an indicator of the arrival of the intrusion at a mooring, propagation speeds were about 10 cm/s. Current observations showed maximum speeds during these intrusions of 10-20 cm/s. Geyer and Cannon (1982) suggested intrusions take the form of gravity currents propagating at about 20 cm/s along the deep basin. Holbrook et al. (1983) indicated renewal characteristics diminished in intensity with distance from the sill, and the density difference of the front varied inversely with distance due to entrainment. A more thorough discussion of the propagation characteristics will be in a separate paper (Holbrook and Cannon, in progress).

Contribution No. 1103 from NOAA/Pacific Marine Environmental Laboratory.

V. References

Bakun, A., 1975: Daily and weekly upwelling indices, west coast of North America, 1969-73. NOAA Tech. Rep. NMFS SSRF-693, U.S. Dept. Commerce, 114 pp.

Barnes, C. A. and E. E. Collias, 1958: Some considerations of oxygen utilization rates in Puget Sound. J. Mar. Res., **17**, 68-80.

Barnes, C. A. and C. C. Ebbesmeyer, 1978: Some aspects of Puget Sound's circulation and water properties. In: Estuarine Transport Processes, B. Kjerfve (ed.), Univ. South Carolina Press, Columbia, 209-228.

Bretschneider, D. E., G. A. Cannon, J. R. Holbrook and D. J. Pashinski, 1985: Variability of subtidal current structure in a fjord estuary: Puget Sound, Washington. J. Geophys. Res., **90**, 11949-11958.

Cannon, G. A., 1983: An overview of circulation in the Puget Sound estuarine system. NOAA Tech. Memo. ERL PMEL-48, 30 pp.

Cannon, G. A. and D. E. Bretschneider, 1986: Interchanges between coastal and fjord circulation. In: Contaminant Fluxes Through the Coastal Zone, G. Kullenberg (ed.), Rapp. P.-v. Reun., Cons. int. Explor. Mer, Copenhagen, **186**, 33-48.

Cannon, G. A. and C. C. Ebbesmeyer, 1978: Winter replacement of bottom water in Puget Sound. In: Estuarine Transport Processes, B. Kjerfve (ed.), Univ. South Carolinia Press, Columbia, S.C., 229-238.

Cannon, G. A. and J. R. Holbrook, 1981: Wind induced seasonal interactions between coastal and fjord circulation. In: The Norwegian Coastal Current, R. Saetre and M. Mork (eds.), Univ. of Bergen, Norway, 131-151.

Cannon, G. A. and N. P. Laird, 1978: Variability of currents and water properties from year-long observation in a fjord estuary. In: Hydrodynamics of Estuaries and Fjords, J. D. J. Nichoul (ed.), Elsevier, Amsterdam, 515-535.

Conomos, T. J., 1979: Properties and circulation of San Francisco Bay Waters. In: San Francisco Bay: The Urbanized Estuary, T. J. Conomos (ed.), Pac. Div., Am. Assoc. Adv. Sci., San Francisco, 47-84.

Farmer, H. G. and M. Rattray, Jr., 1963: A model of the steady-state salinity distribution in Puget Sound. University of Washington, Department of Oceanography, Technical Report 85, 33 pp.

Gade, H. G. and A. Edwards, 1980: Deep-water renewal in fjords. In: Fjord Oceanography, H. Freeland, D. Farmer, and C. Levings (eds.), Plenum Press, New York, 453-489.

Geyer, W. R. and G. A. Cannon, 1982: Sill processes related to deep-water renewal in a fjord. J. Geophys. Res., **97**, 7985-7996.

Hansen, D. V. and M. Rattray, Jr., 1966: New dimensions in estuary classification. Limnol. Oceanog., **11**, 319-326.

Holbrook, J. R., G. A. Cannon and E. T. Baker, 1983: Propagation of renewed deep water in the Puget Sound estuary. EOS, Trans. Am. Geophys. Un., 64, 1033 (Abstract; manuscript in progress).

Holbrook, J. R. and D. Halpern, 1982: Wintertime near-surface currents in the Strait of Juan de Fuca. Atmosphere Ocean, **20**, 327-339.

Kincaid, C., P. Olson and H. Wang, 1988: A numerical investigation of circulation and salt distribution in the Patuxent River estuary. In: Understanding the Estuary: Advances in Chesapeake Bay Research, M. P. Lynch and E. C. Krome (eds.), Chesapeake Research Consortium, Publ. No. 129, Solomons, Maryland, 323-251.

Linden, P. F. and J. E. Simpson, 1986: Gravity-driven flows in a turbulent fluid. J. Fluid Mech., **172**, 481-497.

Mofjeld, H. O. and J. W. Lavelle, 1984: Setting the length scale in a second-order closure model of the unstratified bottom boundary layer. J. Phys. Oceanogr., **14**, 833-839.

Officer, C. B, 1976: Physical Oceanography of Estuaries and Associated Coastal Waters, Wiley and Sons, New York, NY., 465 pp.

Partch, E. N., 1981: Time dependent mixing in a salt-wedge estuary. Ph.D. Thesis, University of Washington, Seattle.

Pritchard, D. W., 1967: Observations of circulation in coastal plain estuaries. In: Estuaries, G. H. Lauff (ed.), Am. Assoc. Adv. Sci., Washington, D.C., Publ. 83, 37-44.

Rattray, M., Jr., 1977: Fjord and salt-wedge circulation. In: Estuaries, Geophysics, and the Environment, National Academy of Sciences, Wash., D.C., 36-45.

Persistence of Residual Currents in the James River Estuary and its Implication to Mass Transport

Albert Y. Kuo, John M. Hamrick, and Gamble M. Sisson

Virginia Institute of Marine Science
College of William and Mary
Gloucester Point, VA 23062

ABSTRACT

The distribution and persistence of Eulerian and Lagrangian residual velocity in a cross sectional transect of the James River estuary, Virginia are analyzed. The Eulerian residual velocity has the characteristic two-layered estuarine circulation in the northern half of the transect, however, the net flow is directed downriver at all depths in the shallower southern half of the transect. In the deep channel, the two-layered Eulerian residual circulation is highly persistent over the six month study duration, with disruptions occurring less than 10% of the time when meteorological forcings are intense. No spring-neap tidal cycle variation is apparent. The magnitude of the long-term advective mass transport (calculated as the lowest order approximation to the Lagrangian residual velocity) is approximately twice that of the Eulerian residual velocity and in the same direction in the deep channel. The Stokes drift velocity contribution to the Lagrangian residual velocity enhances the upriver Eulerian residual velocity transport on the north side of the transect and only slightly increases the downriver transport on the south side.

I. Introduction

The James River is the southernmost major tributary of Chesapeake Bay. Approximately one-third of its length, the 160 km segment seaward of Richmond, is tidal and about one-third of this is estuarine. The central half of the estuarine portion contains some of the most productive seed oyster beds in the Chesapeake Bay system. Oysters spawn in the James River in summer and early fall. Their larvae stay suspended in the water column and drift with currents for about 10 to 20 days before settling to the bottom. Successful recruitment requires, among other things, a large number of larvae over a favorable hard substrate (e.g., oyster beds) at the time in their development when they must attach for permanent residence. The paths of water mass movement from the spawning stock to ultimate settling is thus crucial. In fact, a pioneering work on estuarine circulation was conducted in the James River (Pritchard, 1952) for the purpose of understanding the oyster recruitment processes. That investigation led to the classical two-layered estuarine circulation model in which the Eulerian residual velocity direction is downriver in a less saline upper layer and upriver in a saltier bottom layer. As a result of that study, Pritchard (1953) suggested that the upriver residual current in the bottom layer served as a mechanism for transport of oyster larvae from the beds of market oysters, near the mouth of the river, up into the seed bed regions. It was also pointed out by Pritchard (1952) that the non-tidal circulation exhibited a lateral variation across the river. There is a tendency for the downriver flow in the upper layer to crowd towards the southwest side of the river, and for the upriver flow in the lower layer to crowd towards the northeast side of the river. This results in the boundary between the two layers being deeper towards the southwest side, and shallower towards the northeast side of the river.

Since the 1960's, there has been a trend of decreasing oyster production both within the Commonwealth of Virginia and within the Chesapeake Bay system. This

Coastal and Estuarine Studies, Vol. 38
R.T. Cheng (Ed.)
Residual Currents and Long-term Transport

study represents one component of a multi-disciplinary study by the Commonwealth to improve the economics of the oyster industry. The objective of this study was to delineate the transport pathways of oyster larvae from their spawning to settling. In order to attain this objective, a series of field measurements were conducted from 1984 to 1987. This paper describes the residual transport processes through a transect located near the downriver limit of the seed oyster beds.

II. Field Measurements and Data Analyses

Almost all of the active oyster beds in the James River are located in shallow waters upriver of the James River Bridge (Figure 1). After spawning, the oyster

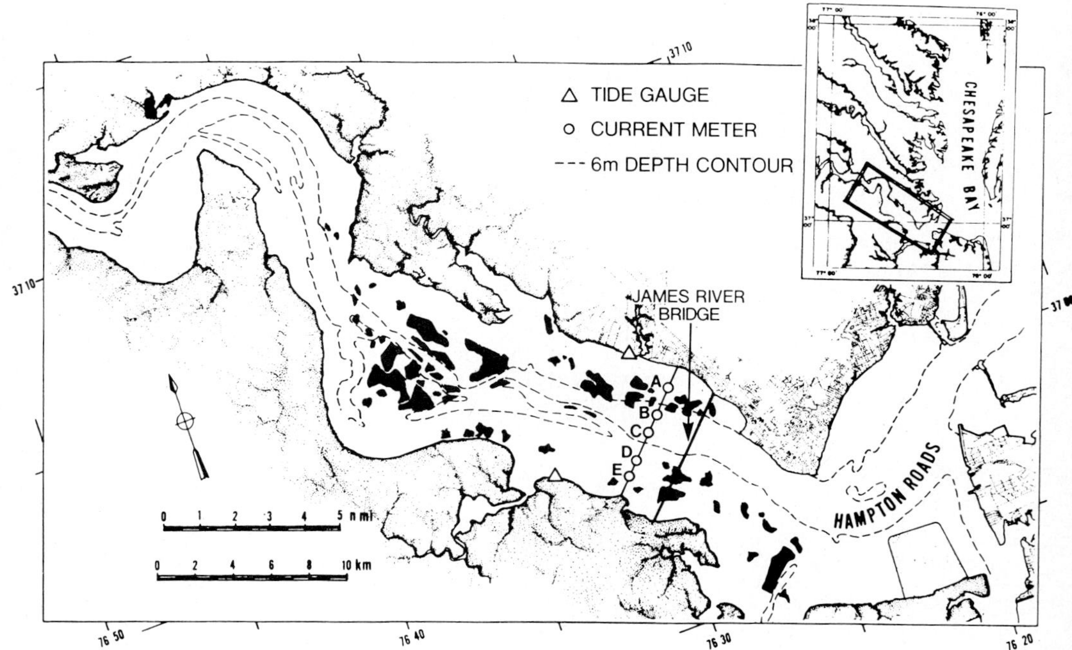

FIGURE 1. The James River Estuary, with tide gauge and current meter stations. Seed oyster beds are shown as irregularly shaped darkened areas.

are suspended in the surface layer of waters that have a net or residual velocity in the downriver direction. To have successful recruitment, there must be net upriver transport pathways at the downriver limit of the oyster beds, otherwise the oyster larvae would be flushed out of the river and be lost from the system. To study the net transport patterns on the downriver side of oyster beds, a total of 14 current meters were moored at 5 stations spanning the river near the James River Bridge. The station locations and the depths of meters at each station are indicated, respectively, in Figs. 1 and 2. The current meter station in the channel was maintained from June to November 1985, and the others for the month of July only. Eight of the current meters were equipped with conductivity and temperature sensors. Average velocity magnitude and direction, and conductivity and temperature if available, were recorded in the meters at half hour intervals. One of the meters at station A failed to provide any data. The meter at the southwest side of the transect provided only 10 days of data.

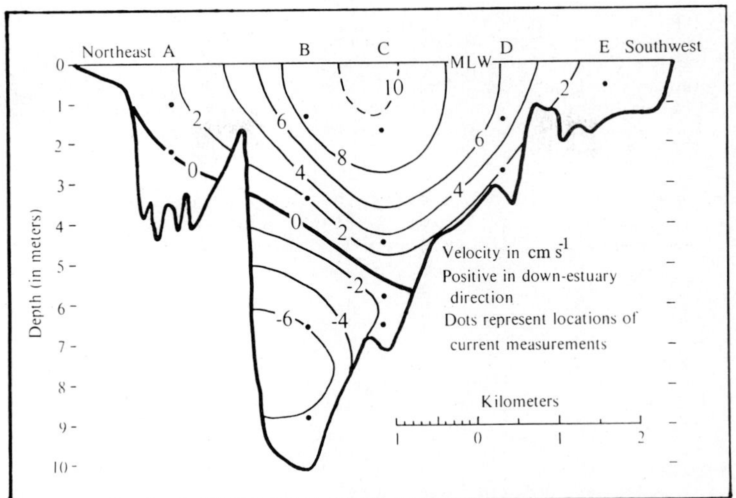

FIGURE 2. Distribution of mean Eulerian residual velocity in a transect across the estuary.

Concurrent with the current meter deployment, a tide gauge was installed on each side of the river. Water surface elevation was measured every 6 minutes and recorded on paper tape. Hourly tidal heights were obtained by averaging 5

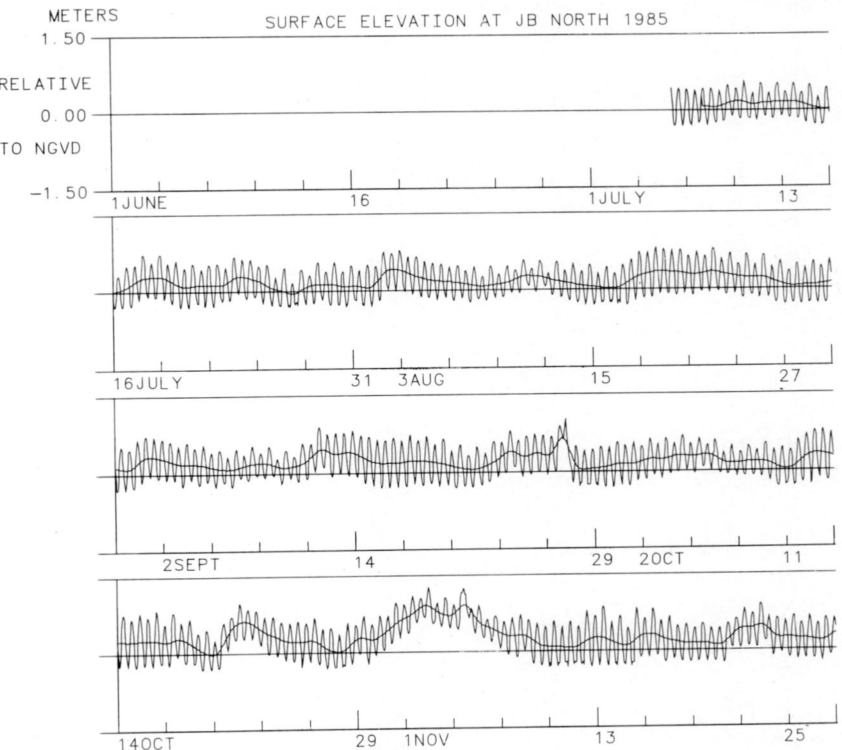

FIGURE 3. Water surface elevation at north side station (both hourly data and low-pass filtered data are shown).

consecutive data points centered on the hour. The hourly data were then adjusted to water surface elevation relative to NGVD (National Geodetic Vertical Datum). The data were then subjected to a low-pass filter to remove the tidal components and other high frequency fluctuations. The filter has a cut-off frequency of $(36\ \text{hours})^{-1}$ and filter window of 96 hours. The hourly heights and filtered water surface variations are plotted as functions of time in Figure 3 for the tide gauge station on the north side of the river.

The current data were analyzed to determine the principal axis, or the dominant flow direction, along which the sum of absolute values of velocity components from all data points was maximum. This axis can be determined for each meter location as follows:

$$P = 0.5 \tan^{-1} \frac{2\overline{uv}}{\overline{v^2} - \overline{u^2}} \tag{1}$$

where P is the angle of principal axis relative to true north, u and v are, respectively, the east and north components of the velocity, and overbars indicate averaging over all data points.

To allow for the possibility that the directions of flood and ebb flows might not be diametrically opposed, the data points were split into two groups by a line perpendicular to the direction with angle P. An angle, E or F, was determined separately for each group of data. The resulting two directions represent the ebb and flood directions, respectively, at the current meter location. The procedure to determine E and F may be repeated by regrouping data points. Those data lying within 90° on either side of direction with angle E (or F) were considered as ebb (or flood) current. The new angles E and F were then determined for each of the new groups of data. It was found that the procedure converged very rapidly, and the second adjustment usually resulted in insignificant refinement.

The most apparent feature of the currents is their strong directionality along principal axes of flood and ebb flows, as illustrated by a polar plot in Figure 4. Table 1 lists the angles P, E and F for each current meter, showing that except

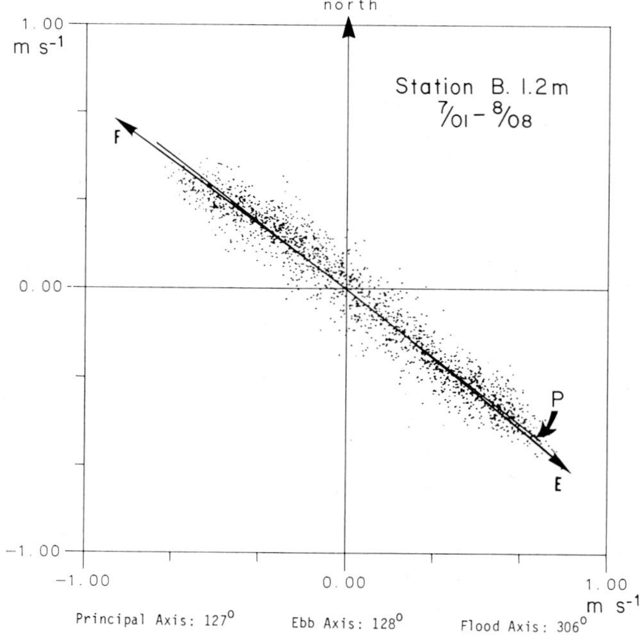

FIGURE 4. Sample of polar plot of current meter data.

station C, the flood and ebb flows were approximately opposite to each other. The principal axes of the surface currents were generally parallel to the shoreline, while data from station B, the station in the channel, show that the principal axes rotated progressively to a direction parallel to the bathymetric contours near the bottom. For the purpose of calculating mass transport, a longitudinal axis was chosen in a direction perpendicular to the transect in which the current meters were deployed. The data were resolved about this direction to produce the longitudinal and transverse components of the velocity series before further analyses were performed.

Table 1. Mean velocity components and principal directions for the July 2 to 31, 1985 data (positive u direction is ebb along the principal axis; the direction angles P, E and F are relative to true north).

Station	Depth m	P	E degree	F	u cm/s	v cm/s
A	0.7	128	130	305	1.1	-1.7
	1.9	126	128	304	-0.04	-0.99
B	1.2	127	128	306	8.9	-1.9
	3.1	120	120	300	1.8	0.28
	6.3	105	106	284	-6.6	-2.5
	8.5	101	108	279	-6.6	-3.8
C	1.5	112	105	307	11.0	8.4
	4.3	115	119	289	3.0	-2.8
	5.6	109	99	301	-0.74	4.8
	6.6	112	105	299	-0.88	2.2
D	1.2	110	113	285	5.4	-1.4
	2.4	110	114	286	1.7	-1.5
E	0.2	125	130	302	0.15	-1.4

III. Eulerian Residual Velocity

Mean Eulerian residual velocities were obtained by averaging the longitudinal and transverse components, respectively, for each current meter over a 29-day period. The cross-sectional distribution of the longitudinal component is presented as isotachs in Figure 2. The distribution shows the characteristic partially mixed estuary circulation pattern in the northeastern half of the transect with the mean current directed downriver in the upper layer and upriver in the lower layer. The level of no net motion tilts downward to the southwest, such that the net flow was directed downriver at all depths on the southwestern half of the transect. The upriver flow existed only in the deeper portion of the transect and near the bottom of the shallow waters along the north shore. This pattern of flow distribution was reported by Pritchard (1952), and may be explained in terms of topographic and geostrophic effects on both the tide induced and density driven portions of the residual circulation (Hamrick, 1986, 1988).

The string of current meters at the station in the channel was maintained for six months, allowing mean longitudinal velocity components to be calculated for several periods representing different hydrological conditions. The results are presented in Figure 5 to demonstrate the vertical distribution of Eulerian residual velocity for each period. It is seen that all velocity distributions possess a distinct feature of two-layered circulation. These data represent freshwater discharges ranging from as low as 35% of annual mean discharge in July to as high as 1200% in early November. A curve representing the velocity profile under low river flow conditions is also presented in Figure 5. This curve was drawn based on the data from three periods, June 1 to July 1, July 2 to August 19, and September 12 to November 3, each having an average river discharge of less than half of the mean annual discharge. The other sets of data may be considered as transient deviations from this typical velocity profile. The high river flow from August 19 to 26, representing 325% of the annual mean discharge, affected only the velocity in the upper layer. A flood flow condition (November 3 to 10) was necessary to noticeably suppress the upriver flow in the lower layer. The post flood period (November 10 to 27) showed a significant increase in upriver flow, which helped to increase salinity depressed by the preceeding flood discharge.

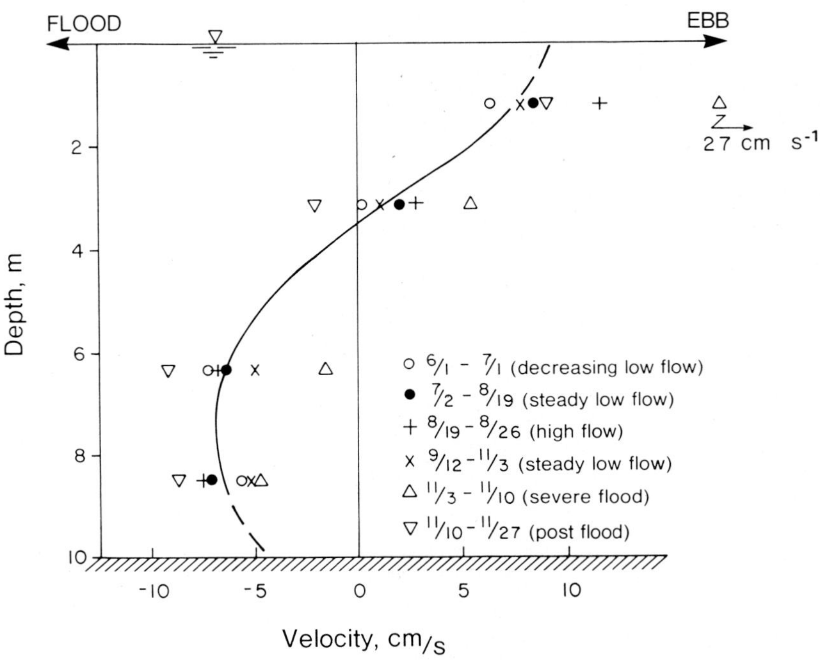

FIGURE 5. Vertical distribution of mean Eulerian residual velocity at the station in the main channel (the curve represents velocity profile in low river flow conditions).

To investigate the temporal variability of the Eulerian residual velocities, the longitudinal components of the half-hourly velocity data for the channel station were subjected to a low-pass filter with a cut-off frequency of (36 hours)$^{-1}$. The filtering process essentially eliminates the diurnal and semi-diurnal tidal constituents, and velocity fluctuations of higher frequencies. The resulting filtered series were considered to be the Eulerian residual velocity, and Figure 6 shows the residual velocities at Station B. The figure shows that the residual currents were directed downriver near the surface and upriver near the bottom, except for several brief periods each lasting less than 3 days. The deviation from the characteristic two-layered circulation usually took the form of upriver flow at all depths for about 2 days followed by strong downriver flow in the upper layer or at all depths. The period from 20 to 23 October typified these periods of deviation from normal condition. Excluding the time when surface or bottom current data were not available, the characteristic two-layered circulation prevailed for 90% of the time. Of the 10% of the time during which other circulation patterns prevailed, half of them were the result of Hurricane Juan, which passed by the area in late October and its induced flooding in early November. This is in contrast with observations reported for the Potomac River (Elliot, 1978) and Chesapeake Bay (Pritchard and Vieira, 1984), which indicated that the characteristic two-layered circulation of the partially mixed estuary was frequently disrupted by meteorological forcings. Elliot (1978) calculated that the characteristic circulation occurred about 43% of the time at a station in the Potomac River.

Though the pattern of the two-layered circulation in the James River estuary was highly persistent, the strength of circulation was quite variable, mainly influenced by the meteorological forcings. No apparent response to spring-neap fortnightly variation was observed. This is consistent with the numerical model results (Cerco and Kuo, 1983) which indicate that depth average salinity did not vary with the spring-neap cycling; and since the longitudinal salinity gradient is the primary driving force of the two-layered circulation, little variability would be expected.

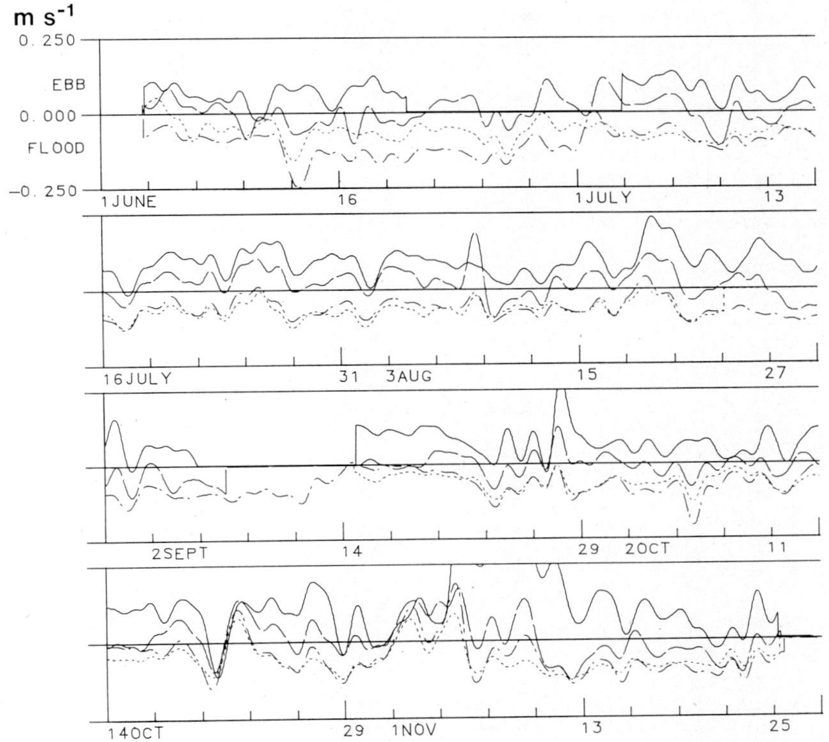

FIGURE 6. Temporal variation of Eulerian residual velocities at station B (— 1.2 m depth, — — 3.1 m depth, —•— 6.3 m depth, ------ 8.5 m depth).

The mean Eulerian velocity distribution presented in Figure 2 was used to calculate the apparent residual transport through the transect. The downriver and upriver transports were calculated separately by integrating the longitudinal Eulerian residual velocity over the cross-sectional area above and below the level of no net motion respectively. To calculate the downriver transport, the velocity was integrated up to mean water level assuming no variation above mean low water. The calculation resulted in apparent residual transports of 1150 and 306 m^3/s, respectively, for the downriver and upriver flows. The fresh water discharge into the river was fairly steady during the period of current meter deployment over the whole transect, having an average value of 80 m^3/s, which is much lower than the difference between the net downriver and upriver flows. Therefore, to maintain mass balance, the cross sectional integral of the longitudinal Stokes drift velocity must contribute about 764 m^3/s to the upriver flow, which was more than twice of that contributed by Eulerian residual velocity. These results suggested that additional analyses accounting for the relationship between the Eulerian residual velocity, the Stokes drift velocity and the Lagrangian residual velocity or residual mass transport velocity be performed.

IV. Residual Mass Transport Velocity

It is now generally accepted that the Lagrangian residual velocity field is the appropriate representation of long-term advective mass transport in long wave dominated estuarine flows. For a vertically uniform distribution of the horizontal velocity components, Longuet-Higgins (1969) showed the two-dimensional lowest order approximation to the Lagrangian residual velocity, termed the residual mass transport velocity, to be the sum of the Eulerian residual velocity and the Stokes drift. Feng et al. (1986a, b) extended this work to second order by adding an additional term called the Lagrangian drift. Hamrick (1987) derived long-term mass transport equations for weak and strong vertical stratification in a vertically stretched and horizontal curvilinear coordinate system. For the weak stratification case, the long-term advective mass transport velocity was shown to be the curvilinear coordinate system generalization of the vertically uniform, two-dimensional horizontal residual mass transport velocity derived by Longuet-Higgins (1969). For the strong stratification case, the long-term advective mass transport velocity was shown to be a three-dimensional generalization of the two-dimensional residual mass transport velocity.

In a vertically stretched, $0 \le z \le 1$, horizontal Cartesian coordinate system, the three-dimensional residual mass transport velocity, or first order approximation to the Lagrangian residual velocity, (u_L, v_L, w_L), is

$$u_L = <u_1> + \frac{<h_1 u_0>}{h_0} + \frac{1}{h_0}\frac{\partial}{\partial y}(h_0 B_z) - \frac{1}{h_0}\frac{\partial}{\partial z}B_y \tag{2}$$

$$v_L = <v_1> + \frac{<h_1 v_0>}{h_0} + \frac{1}{h_0}\frac{\partial}{\partial z}B_x - \frac{1}{h_0}\frac{\partial}{\partial x}(h_0 B_z) \tag{3}$$

$$w_L = <w_1> + 0 + \frac{\partial}{\partial x}B_y - \frac{\partial}{\partial y}B_x \tag{4}$$

$$B_x = <v_0 \int w_0 \, dt_0> \tag{5}$$

$$B_y = <w_0 \int u_0 \, dt_0> \tag{6}$$

$$B_z = <u_0 \int v_0 \, dt_0> \tag{7}$$

with the continuity equations

$$\frac{\partial}{\partial x}(h_0 u_L) + \frac{\partial}{\partial y}(h_0 v_L) + \frac{\partial}{\partial z}w_L = 0 \tag{8}$$

$$\frac{\partial}{\partial x}(h_0<u_1> + <h_1 u_0>) + \frac{\partial}{\partial y}(h_0<v_1> + <h_1 v_0>) + \frac{\partial}{\partial z}<w_1> = 0 \tag{9}$$

also presented for completeness. The operator, $< >$, denotes a low pass filter, while the numerical subscripts indicate ordering in the barotropic long wave Froude number used as the perturbation parameter in deriving Eqs.(2-4). The Eulerian residual velocity, $<u_1, v_1, w_1>$, is formally and operationally equivalent to $<u,v,w>$, the low pass filter of the instantaneous or discretely sampled Eulerian velocity, while h_0 is the equivalent of $<h>$, the low pass filter of the total water column depth h. The long wave or tidal components of the velocity field are (u_0, v_0, w_0) and are equivalent to $(u,v,w) - <u,v,w>$, the high pass filter of the Eulerian velocity, while h_1, tidal surface elevation fluctuation is equivalent to $h - <h>$, the high pass filter of the total water depth.

The last three terms on the right in Eqs.(2-4), respectively, represent the Stokes drift velocity, and the first of these terms, of the form $<h_1 u_0>/h_0$, may be referred to as the wave transport velocity portion of the Stokes drift. The remaining terms involving the variables (B_x, B_y, B_z) may be referred to as the vector potential transport velocity portion of the Stokes drift velocity, since Eqs.(2-4) are the scalar components of

$$\mathbf{q}_L = \mathbf{q}_{ET} + \nabla \times \mathbf{B} \tag{10}$$

in a stretched vertical coordinate system. The sum of the Eulerian residual velocity and the wave transport velocity, q_{ET}, is often termed the Eulerian residual transport velocity which satisfies the Eulerian continuity equation, Eq.(9). It also may be shown that Eq.(10) is equivalent to

$$\mathbf{q}_L = \langle \mathbf{q}_1 \rangle + \langle \int \mathbf{q}_0 dt \cdot \nabla \, \mathbf{q}_0 \rangle \tag{11}$$

the original form presented by Longuet-Higgins (1969). Referral is made to Hamrick (1987, 1989) for derivations and further discussion of the results represented by Eqs.(2-10).

The expressions for the residual mass transport velocity, Eqs.(2-4), derived in a stretched vertical coordinate system, can be applied to the analysis and interpretation of velocity measurements at fixed vertical locations with relative errors not exceeding the ratio of tidal surface displacement amplitude to mean or low pass total depth. For the transect of the James River under consideration, this ratio, which is also an estimate of the barotropic long wave Froude number, has an average value of 0.08. For measurements of horizontal velocity components, (u,v), over the cross sectional transect, now defined as the plane x = 0, it is possible to estimate the first three of the four terms on the right of Eq.(2), contributing to the longitudinal residual mass transport velocity, u_L. Only the first two terms on the right of Eq.(3), contributing to the transverse residual mass transport velocity, v_L, can be estimated, while of course none of the terms in Eq.(4) can be estimated since the vertical velocity w cannot be determined from horizontal velocity measurements in a plane.

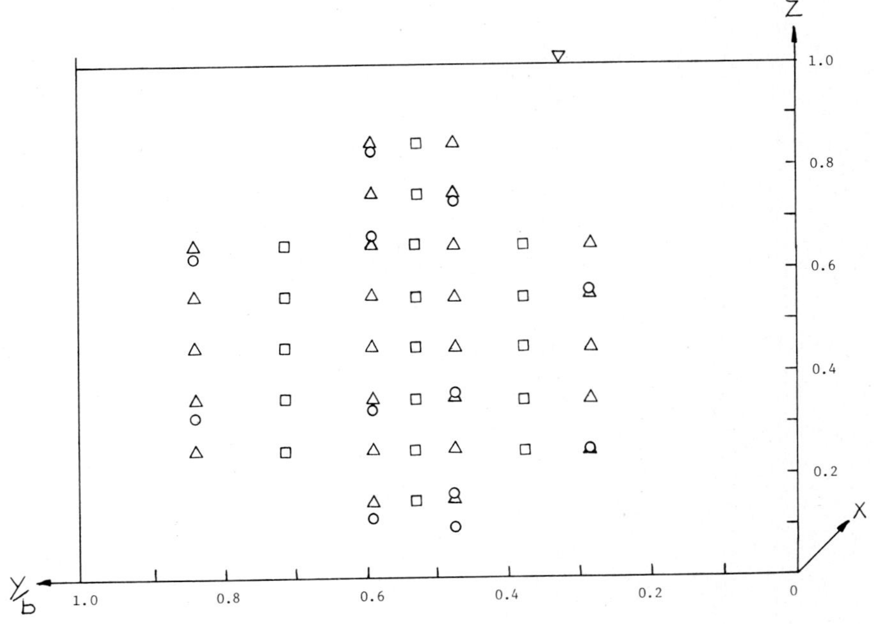

FIGURE 7. Location of points in stretched transect cross section used to compute the residual mass transport velocity. 12 current meter locations, o, 26 interpolation points, **Δ**, 18 residual mass transport velocity points, □ .

The procedure used in estimating the first three terms on the right of Eq.(2) and the first two terms on the right of Eq.(3), first involved extraction of synchronous 512 hour time series of horizontal velocity components for twelve of the current meters and of the water surface elevation for the two tide gauges. Using known depths at the current meter locations and interpolating across the transect between the two tide gauges, time series of total depth at the current meter locations were generated. The horizontal velocity time series, with u being longitudinal, normal to the transect and v being transverse, tangent to the transect, and the total depth time series were filtered to produce the high pass time series, u_0, v_0 and h_1 and the low pass time series $\langle u_1\rangle$, $\langle v_1\rangle$ and h_0.

The filtered velocity time series at the twelve current meter locations were used to generate filtered velocity time series at 26 new, uniform in the vertical, locations using linear interpolation as shown in Figure 7. Filtered time series of the horizontal components of the wave transport velocity, $\langle h_1 u_0\rangle/h_0$ and $\langle h_1 v_0\rangle/h_0$, and the vertical component of the vector potential, B_z, defined by Eq.(7), were calculated at these locations. Time series of the longitudinal components of the Eulerian residual velocity, the wave transport velocity and the transverse, y, gradients of the vector potential component B_z, corresponding to the first three terms on the right of Eq.(2) were then calculated at 18 new locations, as shown in Figure 7, using linear interpolation and finite difference differentiation. To eliminate filter contaminant at the ends of these series, as well as to determine mean values for presentation and interpretation, the mean values of the interior one half series length were determined.

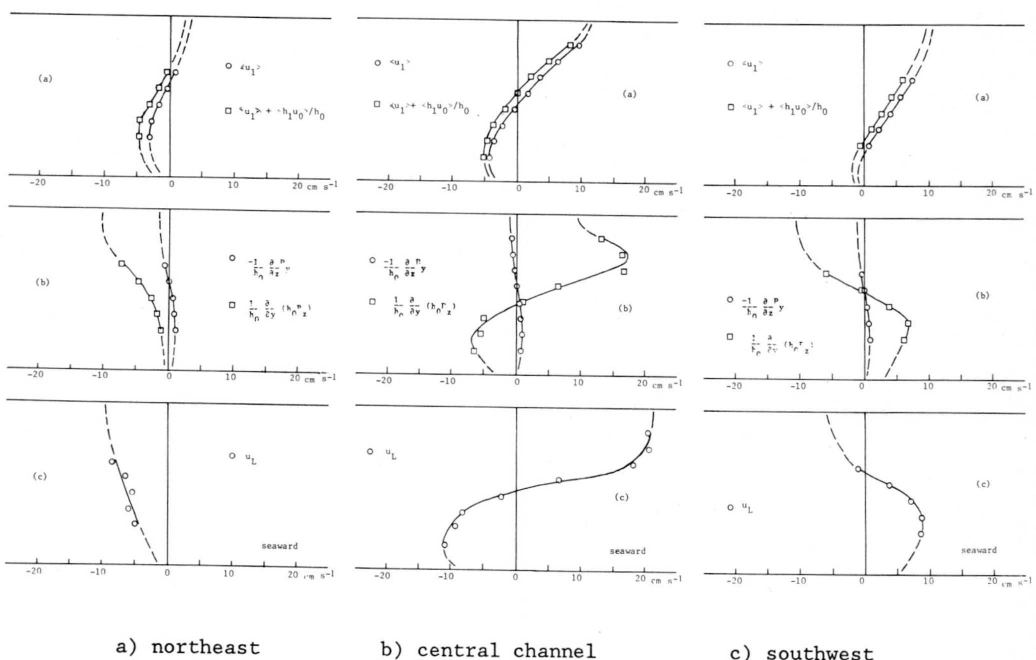

FIGURE 8. Vertical distribution of the Eulerian residual, Stokes drift and residual mass transport velocity, toward the northeast shore (a), the central channel (b), and toward the southwest shore (c). [Eqs.(7) and (15) were used to calculated B_z and B_y respectively].

The final results of the analysis procedure are mean values of the longitudinal components of the Eulerian residual velocity, $\langle u_1 \rangle$, the wave transport velocity, $\langle h_1 u_0 \rangle / h_0$, and the transverse gradient portion, $\partial(h_0 B_z)/h_0 \partial y$, of the vector potential transport velocity at 18 locations in the cross sectional transect. The results are presented in Figure 8a, b, c, which respectively represent vertical profiles based on points 1-5, toward the northeastern shore, (8a), points 6-13, in the central region near the channel, (8b), and points 14-18 toward the southwestern shore, (8c).

The top panels of Figure 8 show the vertical distribution of the Eulerian residual velocity, $\langle u_1 \rangle$, and the Eulerian residual transport velocity, $\langle u_1 \rangle + \langle h_1 u_0 / h_0 \rangle$, composed of the sum of the Eulerian residual velocity and the wave transport velocity. In the central portion of the transect, both the Eulerian residual velocity and the Eulerian residual transport velocity exhibit the classical circulation pattern of seaward flow in the surface layer and landward flow in the bottom layer, with the level of zero transport velocity being near mid-depth. The difference between the two profiles, the wave transport velocity, is nearly uniform over the depth, with a value of 1.2 cm/s, and directed landward consistent with the direction of wave propagation. In the northeastern portion of the transect, both the Eulerian residual and the Eulerian residual transport velocity exhibit the classical circulation pattern. However, the landward transport in the bottom layer is dominant with the level of zero transport velocity being approximately two-thirds of the depth up from the bottom. The wave transport velocity is again nearly uniform over the depth, and directed landward with magnitude of 1.7 cm/s, which is approximately 40% greater than that in the central region. In the southwestern portion of the transect, both the Eulerian residual and the Eulerian residual transport velocity again exhibit the classical circulation pattern. However, the seaward transport in the surface layer is dominant with the level of zero transport velocity being approximately three-fourths of the depth down from the surface. The wave transport velocity is nearly uniform over the depth and directed landward with a magnitude of 1.2 cm/s.

The middle panels of Figure 8 show the vertical distribution of the transverse gradient portion, $\partial(h_0 B_z)/h_0 \partial y$, of the vector potential transport velocity. In the northeastern region of the transect this portion of the vector potential transport velocity is entirely landward. In the central region, there is a two layer distribution with a predominance of seaward transport in the surface layer.. The southwest region shows a two layer distribution, however, the transport in the surface layer is landward.

As previously noted, the vertical gradient portion, $-\partial B_y / h_0 \partial z$, of the vector potential transport velocity cannot be directly determined since B_y, defined by Eq.(6), involves the vertical velocity that cannot be determined from measurements of horizontal velocity in a planar transect. However, the uniformity of the wave transport velocity over the depth suggests that there is little phase variation in the longitudinal high pass velocity, u_0, thus allowing the vertical distribution of u_0 to be approximated by

$$u_0 = \bar{u}_0 [1 + f(z)] \tag{12}$$

where $\bar{u}_0$, is the average of u_0 over the vertically stretched coordinate, $0 \leq z \leq 1$. A similar expression for the transverse high pass velocity may be assumed. The order zero continuity equation relating high pass depth and velocity is

$$\frac{\partial}{\partial t_0} h_1 + \frac{\partial}{\partial x}(h_0 u_0) + \frac{\partial}{\partial y}(h_0 v_0) + \frac{\partial}{\partial z} w_0 = 0 \tag{13}$$

(Hamrick, 1987, 1988). Inserting Eq.(12) and a similar expression for v_0 into Eq.(13) and making use of the depth average of Eq.(13) gives

$$\frac{\partial}{\partial z} w_0 = f(z) \frac{\partial h_1}{\partial t_0} , \tag{14}$$

allowing Eq.(6) to be evaluated as

$$B_y = - \langle h_1 \bar{u}_0 \rangle (1 + f(z)) \int_0^z f(z) dz. \tag{15}$$

Since $\langle h_1\bar{u}_0\rangle$ is the depth integral of the wave transport velocity and can be readily calculated, approximate estimation of B_y and the vertical gradient portion, $-\partial(B_y)/h_0\partial z$, of the vector potential transport velocity is readily accomplished upon choosing a suitable distribution function, f(z). The middle panels in Figure 8 show the vertical distributions obtained using f(z) corresponding to a parabolic velocity profile. The distribution is two-layered with landward transport in the surface layer and the magnitude is substantially less than that of the transverse gradient portion of the vector potential transport velocity.

The vertical distributions of the longitudinal residual mass transport velocity, the lowest order approximation of the longitudinal Lagrangian residual velocity, are shown in the bottom panels of Figure 8. In the northeastern region of the transect, the residual mass transport velocity is landward over the entire depth resulting from the addition of an entirely landward Stokes drift velocity to the classical two-layered Eulerian residual velocity. In the central region of the transect, the residual mass transport velocity has a classical two-layered distribution, with a dominance of seaward transport in the surface layer, resulting from the two-layered distribution of the Stokes drift velocity reinforcing the classical two-layered distribution of the Eulerian residual velocity. In the southwestern region of the transect, the residual mass transport velocity has a reverse two-layered distribution with a dominance of seaward transport in the bottom layer.

V. Discussion and Conclusion

The results of the preceeding section have shown that the distribution of the residual mass transport velocity, or lowest order approximation to the Lagrangian residual velocity, differs considerably from the distribution of the Eulerian residual velocity and the Eulerian residual transport velocity over a transect normal to the longitudinal axis of the James River. For this particular transect, the persistent longitudinal Eulerian residual velocity distribution is significantly influenced by topographic and geostrophic effects, with landward flow favoring the region toward the northeastern shore. The addition of the Stokes drift, and in particular the transverse gradient portion of the vector potential transport velocity, to the Eulerian residual velocity served primarily to intensify the large scale horizontal circulation feature of landward transport along the northeastern shore region and seaward transport along the southwestern shore region, while also greatly intensifying the two-layered transport structure in the central channel region. These general features of the residual mass transport velocity distribution are consistent with hypothesized transport pathways of oyster larvae in the lower James River.

The results of this study are significant in pointing to the necessity of computing the residual mass transport velocity, as well as the Eulerian residual velocity, as part of current meter data analyses. The inability of completely determining the three dimensional structure of the horizontal components of the residual mass transport velocity from horizontal current measurements in a planar transect suggests the need for alternate current meter moorings. One possibility would be to deploy horizontal velocity measuring current meters over two planar transects, on the order of a tidal excursion apart, normal to the longitudinal axis of an estuary channel. Tide gauges would be located near both shores intermediate between the transects. This arrangement would allow complete determination of both horizontal residual mass transport velocity components over the intermediate transect.

In conclusion, this study has shown that field current measurements and their subsequent Eulerian and Lagrangian residual analyses can be used to better understand the transport of materials in an estuary system. In particular, the Eulerian residual current field was shown to be persistent during a time interval critical for oyster spawning and the residual mass transport velocity field, although considerably different from the Eulerian residual velocity field, exhibited an overall pattern favorable to oyster larvae transport. Since the hydrographic conditions in the James River are fairly steady during summer months, it is reasonable to expect that the characteristics of the residual mass transport velocity field calculated from one month of data should be representative.

This paper is contribution number 1553 from the Virginia Institute of Marine Science, College of William and Mary, Gloucester Point, Virginia.

VI. References

Cerco. C. F. and A. Y. Kuo, 1983: Two-dimensional transient behavior of estuarine salinity. Proceedings, Conference on Frontiers in Hydraulic Engineering, ASCE/MIT, Cambridge, Mass. 375-379.

Elliott, A. J., 1978: Observations of the meteorological induced circulation in the Potomac Estuary. Estuarine and Coastal Marine Science, **6**, 285-299.

Feng, S., R. T. Cheng, and P. Xi, 1986a: On tide-induced residual current and residual transport, Part 1, Lagrangian Residual Current. Water Resources Research, **22**(12), 1623-1634.

Feng, S., R. T. Cheng, and P. Xi, 1986b: On tide-induced residual current and residual transport, Part 2, Residual Transport with Application in South San Francisco Bay, California. Water Resources Research, **22**(12), 1635-1646.

Hamrick, J. M., 1986: Subtidal circulation and transport in estuaries. In Advancements in Aerodynamics, Fluid Mechanics and Hydraulics. ASCE.

Hamrick, J. M., 1987: Time averaged estuarine mass transport equations. Proceedings, National Conference on Hydraulic Engineering, ASCE, Williamsburg, VA 624-629.

Hamrick, J. M., 1989: The dynamics of long-term advective transport in estuaries. This volume.

Longuet-Higgins, M.S., 1969: On the transport of mass by time-varying ocean currents. Deep Sea Research, **16**, 431-447.

Pritchard, D. W., 1952: Salinity distribution and circulation in the Chesapeake Bay estuarine system. J. of Marine Research, **1**(2), 106-123.

Pritchard, D. W., 1953: Distribution of oyster larvae in relation to hydrographic condition. Proceedings, Gulf and Caribbean Fisheries Institute, 5th annual session. 123-132.

Pritchard, D. W. and M. E. C. Vieira, 1984: Vertical variations in the residual current response to meteorological forcing in the mid-Chesapeake Bay. In The Estuary as a Filter, V. S. Kennedy (ed.), Academic Press, Inc., Orlando, Florida.

Measurements of Residual Currents in the Coastal Zone with the OSCR HF Radar: A Review of Results From the May 1985 Experiment

John P. Matthews[1], David Prandle[2], and John H. Simpson[2]

[1]School of Ocean Sciences
Menai Bridge
Gwynedd LL59 5EY
United Kingdom

[2]Proudman Oceanographic Laboratory
Bidston Observatory
Birkenhead
Merseyside L43 7RA
United Kingdom

ABSTRACT

Recent results from the OSCR HF radar system are reviewed with the intention of identifying underlying themes in data interpretation and, in the fashion of a technical note, of providing practical advice relating to the HF experimental approach. Outside the constraints of its current measuring capability (~4 cm/s resolution in the along-beam direction as normally operated) the OSCR system provides smoothly varying maps of tidal surface currents which extend and complement the view obtained at greater depths by current meters. Residual currents derived by tidal analysis of the radial current time series acquired during May 1985 show influences from both wind forcing and density gradient terms. For the former, analysis shows a 'slab-like' response to wind forcing associated with a switching back and forth between roughly eastward and westward flows. The density gradient related flows are best interpreted in terms of geostrophy, with a northward flow at the surface, as measured by the OSCR, and a southward flow near the bed, as determined by current meters.

I. Introduction

An understanding of the behaviour of residual currents in coastal zones is of fundamental importance to oceanographic research. Likewise, for coastal engineering, residual currents are significant because they dictate the eventual fate of anthropogenic discharges into the sea. Thus, with a growing scientific and public awareness of environmental issues, particularly in connection with marine pollution, there is an acute need for accurate measurements of tidal and long term residual currents within coastal regions. Such data can be obtained by various modelling and direct measurement techniques. Here we describe recent experience in using HF radar for direct measurements of residuals in UK waters.

For the shelf seas around the British Isles, water motion is known to be governed primarily by strong oscillatory tides and to a lesser extent by residual drifts caused by wind forcing and density gradients. Away from the coast, the main semi-diurnal tidal terms can be accurately predicted by numerical models, though in the shallower regions closer to land, topographic and frictional influences make prediction more difficult. Again, in the shallower coastal regions, the effect of wind-forcing is strongly influenced by the presence of both the coastline and the restricted depth, and it is within such regions that estuarine discharges create

Coastal and Estuarine Studies, Vol. 38
R. T. Cheng (Ed.)
Residual Currents and Long-term Transport

strong density gradients associated with long term drift. Density gradients are also present near complex large-scale frontal systems, some of which lie in close proximity to the coast. In summary, the situation within these coastal regions can be very complex indeed. Because the density fields are continuously changing measurement of the residual currents requires synoptic views with high temporal and spatial resolution.

The HF radar remote sensing technique has the advantage of providing surface current data regularly in time over wide areas. As one radar can only measure radial surface currents along the line of sight of its radar beams, two radar systems must be used if vector currents are required. At HF (3-30 MHz) typical 'bin' or 'cell' sizes over which current data can be obtained is normally of order 1km. This means that present systems cannot be used to investigate currents in small-scale estuarine environments such as minor bays or coves, since there are then too few vector determinations available to build up an overall picture. At the other extreme, for large-scale surveys with a dimension of several hundred kilometers, one may be faced with an awkward trade off between range and resolution or be unable to work at long ranges in the groundwave mode given the limitations of transmitter power. There are nevertheless many applications which fall in between these two extremes and for which bin sizes of ~1 km, spacings between beam crossing points of several km and working ranges of 50 km are quite adequate.

Our practical experience has been with the OSCR (Ocean Surface Current Radar) system. This HF radar was first developed at the Rutherford-Appleton Laboratory (King *et al.*, 1984) and has since been deployed on many occasions for both practical current survey work and for purely 'scientific' applications such as comparisons with current meters, studies of wind forcing and so forth. The important characteristics of the OSCR system are listed in Table 1. The transmitter, working at a frequency of 27 MHz, irradiates a large area of ocean uniformly. In contrast to the CODAR system (Barrick *et al.*, 1977), a beam-forming antenna array is used to receive the sea echoes. This is composed of 16 separate elements equi-spaced over a line 90 m long oriented orthogonally to the centre line of the radar beam coverage. The beam-forming array can differentiate 16 separate beams each 6° wide. Along each beam, signals can be separated into bin lengths of either 1.2 or 2.4 km with a range along each beam of up to 40 km.

Table 1. OSCR PARAMETERS (1985)

Frequency	27 MHz	
Transmitter Power	2 kw (peak)	30 w (mean)
Beam width	6°, 16(12) Beams per site	
Bin length	1.2km (2.4km)	
Range 40km		
Pulses	8(16)μs long repeated at 1.95 KHz	
Measurement of all bins along 1 beam in 2 min		

The received echo spectra, when viewed in the amplitude versus frequency domain, normally display two well-defined peaks corresponding to Bragg resonance with advancing and receding (ocean) surface waves at wavelengths (λ_w) equal to half that of the radar signals. When an underlying current is present these Bragg peaks are offset with respect to the transmitted frequency by an amount f where (Crombie, 1955)

$$f = \frac{U}{\lambda_w} \tag{1}$$

so that measurement of f enables U, the component of the current along the radar line of sight, to be determined. While some uncertainty remains in defining the depth of the surface layer to which radar current measurements pertain, a value of order 1 m (for 27 MHz radar signals) is normally assumed (Stewart & Joy, 1974).

The observational period (or dwell time) used along each beam of the OSCR is in multiples of 67 seconds (i.e. echo spectra are gathered along one beam in multiples of this period). Thus the resolution dU of U is given by (Prandle, 1989) as

$$dU = \lambda_w df = 500/67n \qquad (2)$$

i.e. for n = 2, dU = 3.7 cm/s , for n = 4, dU = 1.9 cm/s etc.

In order to calculate residual currents from the derived radial current time series, a harmonic analysis programme is used to extract the tidal component. It is at this stage that residual current vectors are formed by combination with radial data from the other radar site. Some filtering is normally required to remove high frequency noise and produce the smoothly varying residual current maps used in our discussions below. For the complete removal of components at the tidal frequency a Doodson Xo filter can be used (Doodson and Warburg, 1941).

Ideally it seems the OSCR units should be placed in close proximity to the shore. However, successful deployments have been made from a wide range of locations in the U.K. including rock shelves 100m above sea level (Red Wharf Bay and Whitby experiments) and even from a location 10km inland (Cumbria experiment). The optimum sea-state for OSCR operation is near flat-calm. As wave heights increase, the level of the background fluctuations rises thereby reducing the signal to noise ratio of the Bragg peaks. Since the magnitude of this background level relative to the Bragg peaks is larger for more distant bins, increasing sea-state eventually leads to an effective reduction in the range of OSCR. Precise quantification of the relationship between range and sea-state has not been attempted with this instrument, though work has been reported using other systems (Forget _et al_., 1982). In broad terms, slight data loss may be expected for wind speeds of 10 m/s, while successful observations have continued in winds exceeding 20 m/s. During normal operation one can expect a data retrieval rate of at least 90%, with most of the loss caused by minor logistical difficulties. The only instance in which the sea-state was apparently insufficient to enable the Bragg peaks to be identified occurred in measurements within a pool of sea water at the head of an estuary cut off from the open sea on the ebbing tide.

The sampling rate for conventional moored current meters is generally at 10 minute intervals, whereas hourly intervals may be regarded as sufficient to resolve tidal motions. The sampling rate used in practice for the OSCR is dictated by a number of factors. The present system adopts an interleaving pattern of transmissions that avoids simultaneous operation of both transmitters and thereby reduces the potential for bi-static interference. Within this format, transmissions are stepped sequentially through all beams at each radar site and as analysis of echo data is performed at one site, further transmissions are made at the other. The resulting time interval between data points is decided primarily by the number of beams used and by the sampling interval required along each beam. Hourly surface current values can be obtained if, say, 12 beams from each site are used and in this case a sampling interval of 2x67 s can be adopted. Even in those deployments where only 6 beams are monitored it is usual to retain this same sampling duration to provide half-hourly intervals between data points.

Though the main focus of the article is on residual currents (produced, as described above, from tidal analysis of the radial current time series) we first briefly consider the experience gained in measurements of the main tidal component of the flow field. Further sections will be devoted to a review of wind driven and density gradient related effects observed near Red Wharf Bay (Wales) in 1985.

II. Measurement of Tidal Flows

Early 'proof of concept' experiments with the OSCR concentrated mainly on the ability of the system to measure the main tidal components of the current field. Indeed, the disadvantage of having only a single operational OSCR unit in the early days was overcome by making deployments at intervals spaced one month apart in areas where tidal currents predominate. Current ellipse parameters for the major M2 tidal constituents were derived by Prandle and Ryder (1985) in this way within Liverpool Bay. The spatial distributions of these parameters were found to agree well with values obtained from a numerical model and, when corrected for depth differences, from moored current meters. In general terms, amplitudes agreed to within 10% and phase and direction to within 5°. Later, the dual-radar deployment at Red Wharf Bay illustrated in Figure 1, consolidated the view that the OSCR system was producing a credible picture of surface current motion (Prandle, 1987). There was a high degree of internal consistency within the data (to better than 0.5 cm/s for the M_2 major ellipses) so that M_2 current amplitudes varied smoothly within the experimental region. The major M_4 higher harmonic constituent, by way of an interesting contrast, showed a marked spatial variability.

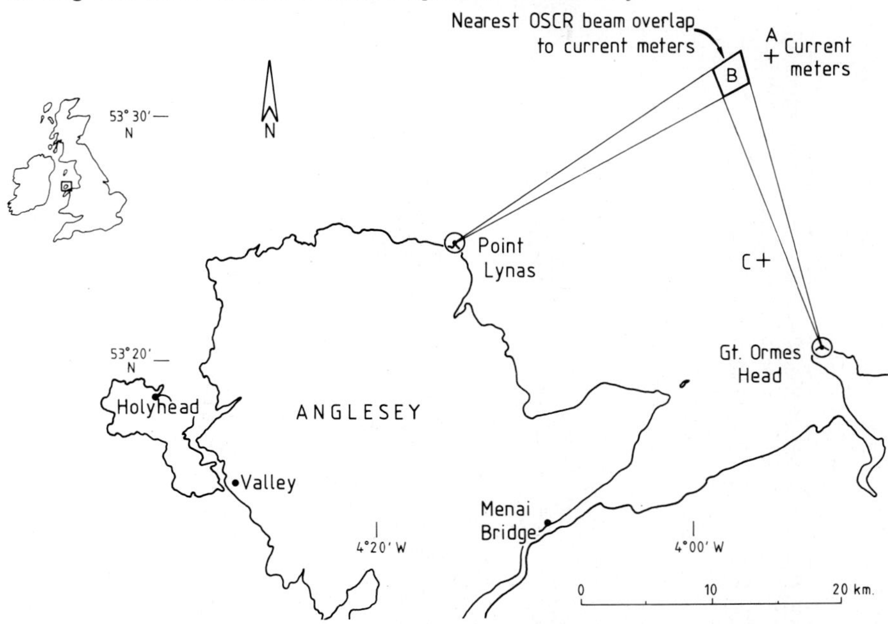

FIGURE 1. Site locations for the May 1985 OSCR experiment. A = Current Meters, C = Acoustic Doppler Current Profiler. The triangular region defined by Point Lynas, Great Orme's Head and beam intersection point B represents approximately, the survey zone in which vector currents were derived by the HF radars. Adapted from Matthews, (1988).

The ability of the OSCR system to detect tidal signals has been compared with that of current meters (Hammond et al., 1987; Matthews et al., 1988), a bed-mounted acoustic doppler current profiler (Griffiths, 1986) and drogued drifters (Collar et al., 1985). A direct comparison with current meter data is of limited value since wave action is known to influence conventional current meters in the vicinity of the surface to where the OSCR currents relate. A more stringent comparison can be made by combining current measurements made at depth with a suitable analytic model of vertical current structure (Prandle, 1982) to provide a prediction for the surface current. This approach was used by Matthews et al.(1988) and some results taken from the Red Wharf Bay campaign of May 1985 are illustrated in Figure (2). Here the comparison involves data from a current meter array at point A, just beyond the edge of the radar survey region. Note that as the model deals with a three dimensional situation, the tidal current ellipse must be resolved into clockwise and anti-clockwise components. The OSCR surface current amplitudes and clockwise phase in Figure (2) follow the trends in the current meter data and are reasonably consistent with the theory. Anticlockwise phase, however, shows a roughly 20° difference with the model and current meter measurements. Both Griffiths (1986) and Matthews et al. (1988) found that, in terms of northward and eastward components, this discrepancy showed up mainly as a larger northward magnitude measured by the OSCR. As the tidal flow in this area is virtually rectilinear with a small northward magnitude of 6 cm/s registered by the near surface current meter and the resolution of northward

OSCR flow determinations from Eq.(2) is 8 cm/s, it is not surprising that the some discrepancy is evident.

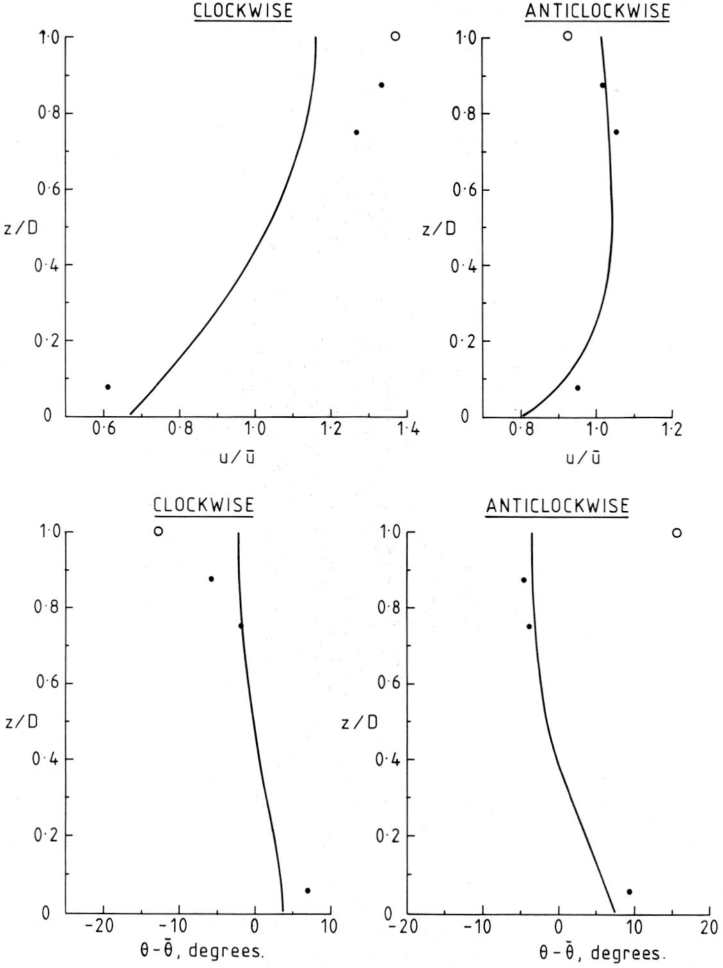

FIGURE 2. Clockwise and anticlockwise amplitudes and phases derived from the radar at beam intersection point B and the three current meters located at 3 , 30 and 35m from the bed in 35m of water (tidal range = 9m). Open circles = OSCR data, solid circles = current meter data, line = prediction based on theory of Prandle (1982). After Matthews et al., (1988).

III. Measurement of Wind Driven Residuals

The measurement of tidal surface currents obtained from OSCR is sufficiently accurate for the requirements of many scientific studies and also for coastal engineers engaged in the design of sea outfalls etc. Of equal importance are the long-term, non-tidal contributions to the flow which can be derived by using standard methods of harmonic analysis to extract the tidal component from the radial current time series. In order to generate the residual currents however, a time series of generally 30 days duration is required for accurate resolution and independent verification. Even then, error estimation is much more difficult for residuals than for the tidal components. Nevertheless, owing to its synoptic long-term mapping capability, the HF radar technique offers a new approach to the study of surface residual currents in coastal regions. For the study of wind driven flows, this is a particularly exciting prospect and one that promises a detailed investigation of so far untested model predictions.

The early analytic work of Ekman (1905) and Jeffries (1923) placed emphasis on the way that the open ocean response to wind forcing was modified by the presence of a coastline. In particular, the presence of land must, in an average sense, prevent any resultant drift across sections parallel to the coast. Consistent with this, early predictions for most wind vectors (onshore hurricanes and so forth apart) required the wind driven flow to run parallel to the coast out to a considerable distance from the shore. Computer based numerical models can provide detailed maps of the circulation in specific regions under given wind conditions, though, until recently, no counterpart in instrumental techniques was available for measuring currents synoptically. Pre-radar methods relied heavily on the use of clusters of drogued drifters which normally provide data in a spatially haphazard fashion. Some success was reported by Murray (1975) in the use of a theodolite tracking technique, though this method is restricted in application to regions in close proximity to the coast. Apart from drogues, the other main method of studying residual flows is by current meters. These have two major disadvantages in that they only give point measurements and are unable to reliably detect flows at the surface, due to the influence of wave action on their mode of operation.

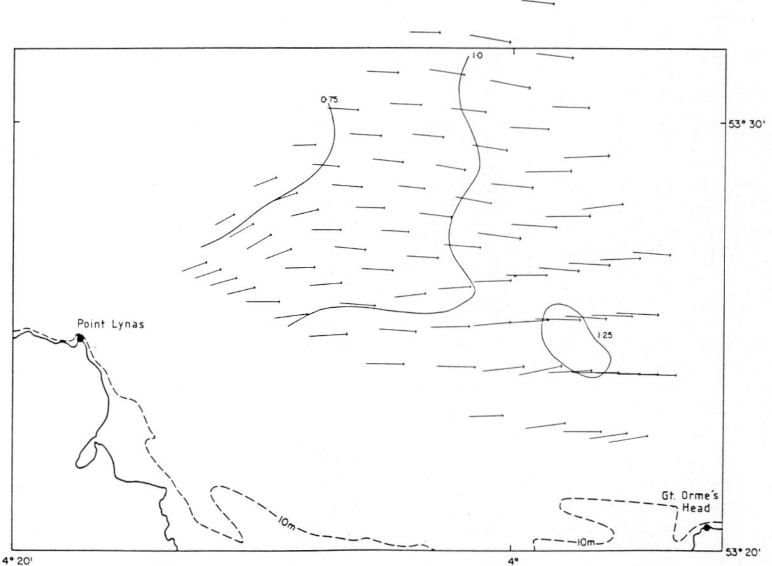

FIGURE 3. Velocity eigenvectors for the dominant EOF mode. The contours show current amplitudes as a fraction of the overall mean. This mode is responsible for, on average, 66% of the variance associated with wind forcing. After Prandle, (1987).

The first thorough investigation of wind driven currents measured by the OSCR system was made by Prandle (1987), using data gathered during the Red Wharf Bay experiment of 1985. A rotary empirical orthogonal function (EOF) analysis on the Doodson Xo filtered residual data showed that a single mode was responsible on average for 66% of the residual current variance. The velocity vectors corresponding to the dominant EOF mode are shown in Figure 3. They are in close alignment increasing in magnitude from east to west and they indicate a 'slab like' response to wind forcing. The time series for the complex amplitude of this mode shows a significant correlation with the wind-stress time series in the east-west direction (0.73) and only a limited correlation (0.32) in the north-south direction.

Further insight into the role of wind forcing as recorded during the Red Wharf Bay campaign was provided in the case study by Matthews (1988). Here attention was concentrated on the total residual field for a period between 12 and 15 May 1985 during which a well defined change in the direction of wind forcing took place (Figure (4)).

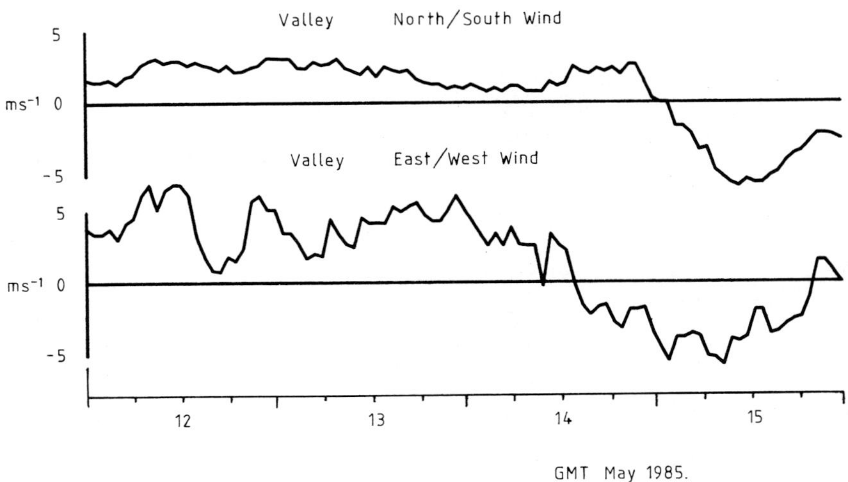

FIGURE 4. North-south and east-west components of wind observed at Valley observatory (Anglesey) during the period 12 - 15 May, 1985. After Matthews (1988).

Examples of the residual vector maps measured during this period are provided in Figure (5a), where the residual current data were recorded following a period of sustained forcing by north-easterly winds at levels of 7 m/s. Water motion tends to follow the depth contours which run, on average, from east north-east to west north-west. This is particularly well defined close to the coast. The data shown in Figure (5b) were recorded after a period of several hours of forcing by north north-westerly winds at a mean speed of 5 m/s. Although the flow field may not yet have reached a final steady state, it is clear that this change in wind direction has caused a complete reversal in the direction of the residuals. Such reversals in association with wind direction changes are quite typically observed within this radar data set. Our final view, Figure (5c), shows the response following a period of forcing by stronger south south-westerly winds at a mean speed of 9 m/s. The alongshore tendency in residual directions remains as water is presumably driven eastwards around the headland of Anglesey. The overall picture emerging from these studies shows that most of wind associated flow variability is related to a switching back and forth of approximately alongshore responses. Given the geography of the region, this must inevitably lead to a close correlation between the wind forcing and east-west residual flow components, as determined in the statistical work of Prandle (1987).

IV. Density Gradient Related Residuals

In this section we show that current components associated with density gradients can be detected by the OSCR radar and that its ability to determine these flows snyoptically is of sufficient accuracy to complement the view obtained at single locations by current meters. For the study of near-shore density gradient related flows, the Red Wharf Bay region, (Figure (1)) is in many ways ideal. Here density gradients are formed as mixed, colder waters of the central Irish Sea meet the low salinity and frequently stratified waters created in the estuarine environments of Liverpool Bay. The area was extensively studied in CTD surveys between 1978 and 1979 (Czitrom, 1982) and it was confirmed that stratification and frontal development occur preferentially in calm periods around neaps when wind and tidal stirring are weakest (Simpson *et al.*, 1988). During such periods, regions of large horizontal density gradient aligned nominally north-south are found near and eastwards of the 4°W meridian. This is convenient from our point of view since the 4°W line passes through the radar survey region employed in the Red Wharf Bay experiment.

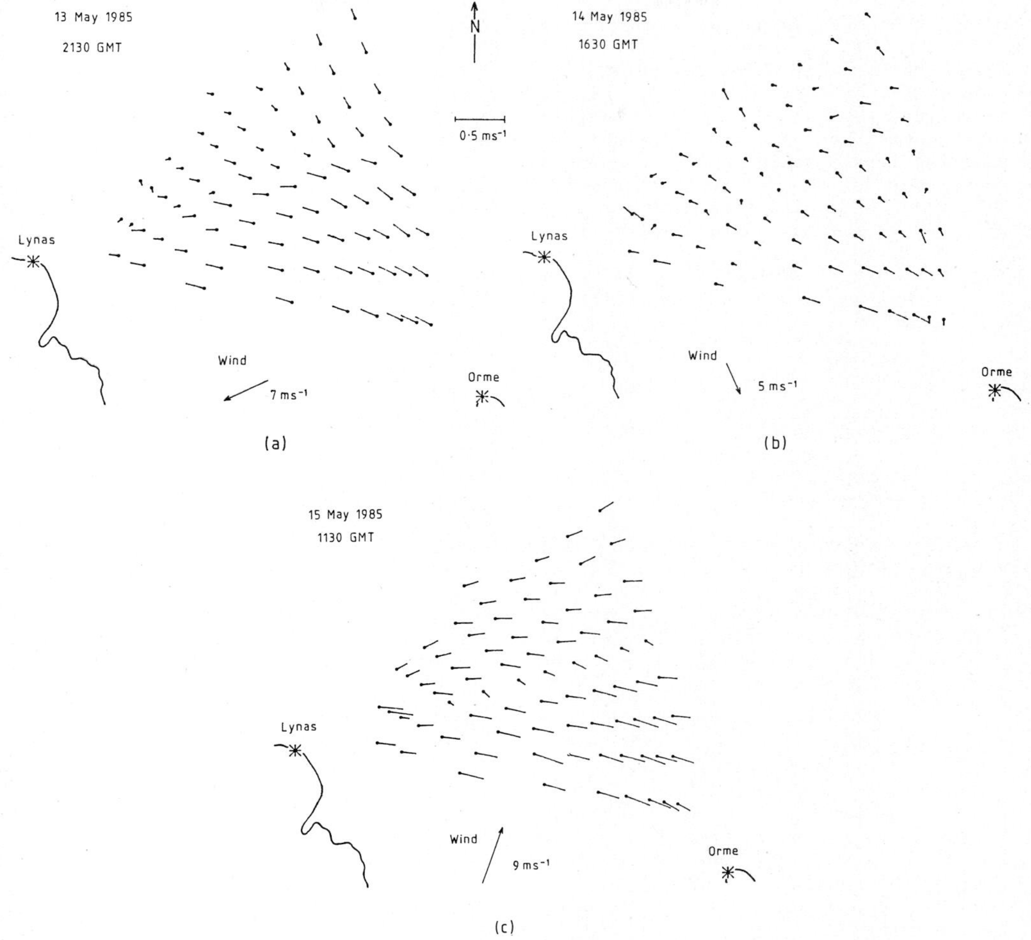

FIGURE 5. Surface residual currents following periods of forcing by (a) northeasterly winds at mean speeds of 7 m/s, (b) north north-westerly winds at mean speeds of 5 m/s (c) south south-westerly winds at mean speeds of 9 m/s. Dots equal zero flow. After Matthews, (1988).

Mean residual current data recorded during this experiment are presented in Table 2 (Matthews et al., 1988). there is a substantial difference between the long term east-west surface residual measured by the radar and that registered by the top current meter. This discrepancy may be attributed to the dominance of wind forcing at the surface in the east-west direction, as sensed by the radar. The current meter (moored at 5 m below LLW and 14 m below HHW) will clearly differ in its response to this forcing. Indeed these two net residuals are in opposite directions. On the other hand, there is close agreement between radar and top current meter for the northward mean residuals, and note that the flow reverses to the south near the bed. Throughout the water column the north-south residual behaviour can be interpreted in terms of a "thermal wind" balance, requiring the isopycnals to be primarily northward in orientation. Previous survey work (Czitrom,

1982) and CTD measurements taken at the start of the Red Wharf Bay campaign indicate that, in an average sense, this is the case. A thermal wind balance requires

$$\frac{\partial \rho}{\partial x} = \frac{f\rho}{g}\frac{\partial \bar{v}}{\partial z} \tag{3}$$

with f the Coriolis parameter and $\frac{\partial \bar{v}}{\partial z}$ the vertical shear in long term residual flow. Using the data in Table 2 we obtain $\frac{\partial \rho}{\partial x} = 2.6 \times 10^{-5}$ kg/m⁴ as a horizontal density gradient which is consistent with the measured gradients. This indicates that a convenient separation in the directions of the main contributions from wind forcing (east-west) and density gradient related flow (north-south) exists. Further, assuming little variation in the density gradient related term near the top of the water column and a largely east-west wind driven contribution, these data indicate that the OSCR senses this mean residual flow to within ~0.3 cm/s of the value recorded by the top current meter.

Table 2. Mean Residual Flows measured at OSCR beam intersection point B and current meters at point A. After Matthews et al., (1988).

	OSCR	Current Meter TOP	Current Meter Middle	Current Meter Bottom
East residual	0.024	-0.007	-0.002	0.018
North residual	0.052	0.049	-0.009	-0.036

The above discussion has been concerned with the mean residual flow recorded at one location. One of the most attractive aspects of the HF radar technique is its potential for mapping mean residuals across the whole of the radar field of view. The synoptic view of residual currents measured during the Red Wharf Bay experiment is provided in Figure (6) (Prandle, 1987). Maximum northward drifts are 17 cm/s in

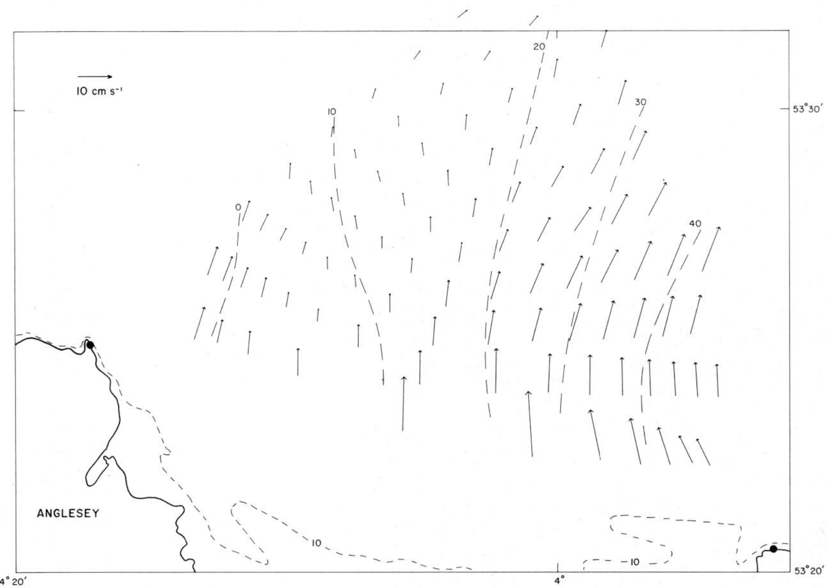

FIGURE 6. Mean surface residual flow over the whole of the Red Wharf Bay experimental period. The contour lines indicate integrated flow in units of 10^3 m³/s, assuming that the surface drift persists through depth. After Prandle, (1987).

the deeper northern areas. The net northward transports calculated on the basis of uniformity of the surface flow throughout the water column are high, lying at a level of $4x10^4 m^3/s$, compared to a long term flow through the Irish Sea of $6x10^4$ m^3/s. Freshwater inflow into this region is at a much lower rate and, as indicated by the data of Table 2, a southward counterflow is therefore required near the bed.

The data in Figure 6 are of interest in that they show an abrupt increase (by a factor of about 2) in the size of the mean northward drift to the east of the radar survey region. This develops close to the 4°W meridian. One speculative explanation of the increase is that it is related to the regions of high horizontal density gradient discussed by Czitrom (1982), which are formed to the east of the 4°W meridian. The pattern in long term currents may then highlight a boundary between a regime in which strong stratification or even frontogenesis can be expected and one in which denser mixed water is present.

V. Conclusion and Discussion

It is clear from this short review that the OSCR HF system proved to be an invaluable tool in the study of large scale tidal currents and the measurement of long term residual flows during the Red Wharf Bay experiment of 1985. A wealth of information is present in the hourly synoptic current maps derived from this technique - we have concentrated here on averaged behaviour and a limited number of case studies to discuss the major features of wind forcing and density gradient related currents. Briefly, the results show a uniform 'slab-like' surface response to wind forcing in which vectors switch back and forth in a roughly alongshore direction (i.e. east-west). In the north-south direction, density gradient related flows are present which can be interpreted in terms of geostrophy. The near bottom current meter shows a reversal in mean residual with respect to that detected at the surface by the radar.

The main problem faced in the interpretation of the May 1985 data set was the lack of simultaneous ship survey data (particularly CTD sections) and the paucity of current meter data. In a most recent experimental campaign (1988), a concerted effort was made to collect sufficient sea truth data to enable accurate interpretation of the radar data to proceed. Indeed, a requirement for remote sensing experiments in general is that sufficient ground/sea truth data be collected.

For workers with a primarily oceanographic background it is useful to bear in mind that rapid progress in instrumentation, experimental techniques and data processing/interpretation is continuing apace in the field of HF radar oceanography (see IEEE Journal of Oceanic Engineering, vol. OE-11, 1986 for a special issue on this topic). Certainly, new developments are required to meet the needs of estuarine research on small spatial scales (~10km) on the one hand, and for larger range systems (working out to distances of several hundred km). Longer range HF radar systems are particularly important, since it is at these scales that comparison and integration with the next generation of remote sensing satellites will be most effective. This approach, of using HF radar together with remotely sensed data gathered by satellites, hold great promise, particularly since there is no means of measuring current by satellite borne sensors to the same temporal or spatial resolution as with HF radar. Direct use of HF radar from space platforms faces major problems (such as ionospheric attenuation and large antenna size) though future developments using a dual frequency microwave system with an HF beat frequency have been suggested.

VI. References

Barrick, D. E., M. W. Evans, and R. L. Weber, 1977: Ocean Surface Currents Mapped by Radar, Science, **198**, 138-144.

Collar, P. G., D. Eccles, M. J. Howarth and N. W. Millard, 1985: An intercomparison of HF radar observations of surface currents with moored current meter data and displacement rates of acoustically-tracked droqued floats, in Advances in Underwater Technology and Offshore Engineering, vol. 4. Evaluation, Comparison and Calibration of Oceanographic Instruments, London, U.K., Graham & Trotman, **9**, 163-182.

Crombie, D. D., 1955, Doppler spectrum of the sea echo at 13.56 Mc.s^{-1}, Nature, **175**, 681-682.

Czitrom, S. P. R., 1982, Density stratification and an associated front in Liverpool Bay. Ph.D. thesis, School of Ocean Sciences, University College of North Wales, Menai Bridge, Gwynedd, U.K.

Doodson, A. T., and H. Warburg, 1941, Admiralty Manual of Tides, H.M.S.O., London, 270 pp.

Ekman, E., 1905, On the influence of the earth's rotation on ocean currents. Ark. Mat. Astron. Fys., **2**, 1-53.

Forget, G., P. Broche, and J. C. de Maistre, 1982, Attenuation with distance and wind speed of HF surface waves over the ocean. Radio Sci., **17**, **3**, 599-610, 1982.

Griffiths, G., 1986, Intercomparison of an acoustic Doppler current profiler with conventional instruments and a tidal flow model. Proc. IEEE Third Working Conference on Current Measurement. Airlie, VA, U.S.A., Jan. 22-24.

Hammond, T. M., C. B. Pattiaratchi, D. Eccles, M. J. Osobrne, L. A. Nash and M. B. Collins, 1987, Ocean Surface Current Radar (OSCR) vector measurements on the inner continental shelf. Cont. Shelf. Res., **7**, **4**, 411-431.

Jeffries, H., 1923, The effect of a steady wind on the sea-level near a straight shore. Philosophical Magazine, Vol. XLVI.

King, J. W., F. D. G. Bennet, R. Blake, D. Eccles, A. J. Gibson, G. M. Howes and K. Slater, 1984, OSCR (Ocean Surface Current Radar) observations off the coast of N. Ireland, England, Wales and Scotland. Paper presented at the conference on current measurements offshore, Society for Underwater Technology, London.

Matthews, J. P., 1988, Advances in Underwater Technology, Ocean Science and Offshore Engineering, **16**, 99-105, 1988, copyright Graham and Trotman.

Matthews, J. P., J. H. Simpson and J. Brown, 1988, Remote sensing of shelf sea currents using a High Frequency Ocean Surface Current Radar System. J. Geophy. Res., **93**, **C3**, 2203-2310.

Murray, S. P., 1975, Trajectories and speeds of wind driven currents near the coast. J. Phys. Oceanogr., **5**, 347-360.

Prandle, D., 1982, The vertical structure of tidal flows and other oscillatory flows. Cont. Shelf Res., **1**, 191-207.

Prandle, D., 1987, The fine structure of near-shore tidal and residual cirulations revealed by HF radar surface current measurements, Journal of Physical Oceanography, **17**, 231-245.

Simpson, J. H., J. Brown, J. Matthews and G. Allen, 1988, Tidal straining, density currents and stirring in the control of estuarine stratification. In press, Estuaries.

Stewart, R. H., and J. W. Joy, 1974, HF radio measurements of surface currents. Deep Sea Res., **21**, 1039.

Analyses of Flux Measurements in the River Mersey

D. Prandle, A. Murray, and R. Johnson

Proudman Oceanographic Laboratory
Bidston Observatory,
Birkenhead, Merseyside, L43 7RA
United Kingdom

ABSTRACT

The present analyses forms part of a major program to improve water quality in the River Mersey and the adjacent Liverpool Bay. Since similar clean-up programmes are planned for many tidal estuaries, the measurements undertaken in the Mersey are examined: (i) to determine their efficacy in calculating contaminant fluxes within the river and (ii) more generally, to consider possible alternative future strategies.

Measurements considered here are limited to currents, suspended sediment concentrations, salinity, temperature and oxygen at a particular section in the Mersey Narrows. Continuous long term measurements were made, both at the surface and at mid-depth, from a moored-buoy. These data were supplemented by 4 cross-sectional surveys, with measurements over depth taken from 5 stations at a section close to the buoy. Each of these surveys extended over a full semidiurnal tidal cycle.

Tidal analyses of the buoy data established the relative accuracy and reliability of the data sets. The degree of consistency between the major tidal constituents derived from successive monthly observations indicated the predominance for each parameter of the tidal component. A simplified tidal analysis was applied to each of the cross-sectional surveys, these results indicated the cross-sectional variation in various tidal parameters and, thereby, the required link with the tidal results from the buoy measurements.

Analyses of the vertical variation in residual currents revealed a pronounced 'classical' estuarine structure with denser (saltier) sea water flowing landwards at the bed, with a compensating seaward flow of fresher river water at the surface. Analysis of the suspended sediment concentrations showed the nature of both the vertical and lateral variability. Likewise, the variation in cross-sectional mean concentrations with varying tidal amplitudes indicated that suspended sediment concentrations were proportional to tidal velocity cubed. By using this latter result, simple analytical derivations indicate the sensitivity of the ned sediment flux to: (i) spring/neap tidal conditions, (ii) net cross-sectional residual velocity, (iii) lateral and vertical variations in residual velocity and (iv) vertical variations in sediment concentration.

I. Introduction

The Mersey is a strongly tidal estuary with elevation ranges at the mouth varying from 4 to 10m over the extremes of the Neap-Spring cycle. The 'Narrows' (see figure 1) is approximately 1km wide with a mean depth of 15 m, tidal currents through this section can exceed 2 m/s. Freshwater flow, Qf, into the Mersey varies from 25 to 200 m^3/s indicating a mean flow ratio (Qf x 12.42 h/volume between high

Coastal and Estuarine Studies, Vol. 38
R. T. Cheng (Ed.)
Residual Currents and Long-term Transport

and low water) of approximately 0.01. Flow ratios of less than 0.1 normally indicate a well-mixed estuary. However in certain sections, during part of the tidal cycle, the Mersey is only partially-mixed. This may be due to the unusual bathymetry with a narrow entrance channel broadening into a wide and shallow upper basin.

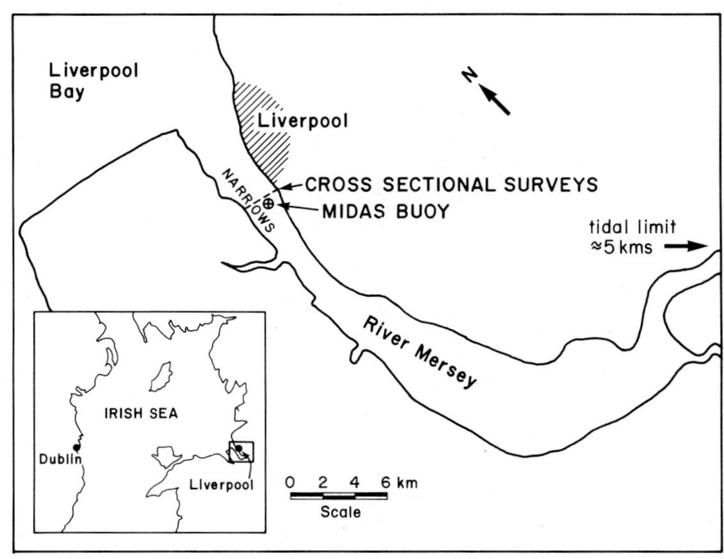

FIGURE 1. River Mersey showing the positions of the cross sectional surveys and the Midas Buoy.

Despite significant dredging over the last century (averaging 5 x 10^6 m^3/a), the volume of the estuary (650 x 10^6 m^3 is reducing at a rate of approximately 10^6 m^3/a (O'Connor, 1987) with a large percentage of the dredged fines returning directly into the River.

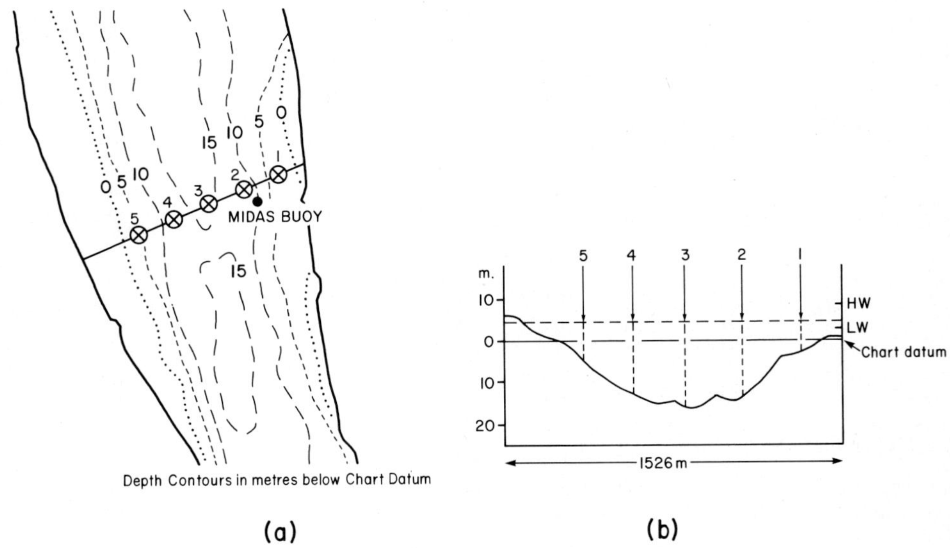

FIGURE 2. Bathymetry in the survey region.

The river banks of the Mersey have long provided attractive locations for various chemical and petro-chemical industries. Historically, the Mersey has been seriously polluted by discharges from such industries and by domestic sewage discharged both directly and at adjacent sea disposal sites. A comprehensive programme is presently being undertaken to improve water quality in the river. As part of this programme, detailed measurements have been made to determine contaminant fluxes and thereby to quantify the direct benefits which would accrue from various remedial actions. The present paper analyses the 'physical components' in these measurements with the aim of assessing the accuracy and difficulty of determining these fluxes.

Measurements were made both continuously and in discrete surveys. A single moored buoy (referred to as MIDAS) was anchored in the Narrows (figure 2), from which continuous measurements were made of current speed and direction, salinity, temperature, suspended sediment and dissolved oxygen. In section II, a three month long set of such observations is examined. In addition to this single-point continuous data set, on four separate occasions complete cross-sectional surveys were made using 5 boats moored at the section shown in figure 2 close to the MIDAS buoy position. Each of these surveys extended over a semi-diurnal tidal cycle and included observations of current speed, salinity and suspended sediment, these are examined in section III.

In section IV simple analytical derivations for sediment fluxes are examined to identify their sensitivity to various parameters and thereby to indicate the required accuracy and resolution required in field measurements.

II. Analysis of Tidal Constitutents from the Midas Moored Buoy Data

Measurement details of the buoy design are shown in figure 3, the buoy was moored at the position shown in figure 2 with a mean depth of 15 m. Pumped sampling systems were fitted, extracting water from (i) 2 m below the surface and (ii) 8 m below the surface. These samples were analysed to determine suspended sediment load (from a turbidity meter), salinity, temperature and dissolved oxygen (only surface samples were taken of the latter two). At these same two depths, current speed and direction were measured using electro-magnetic flow meters. The measured current directions were relative to the axis of the buoy, conversion to orthogonal N-S and

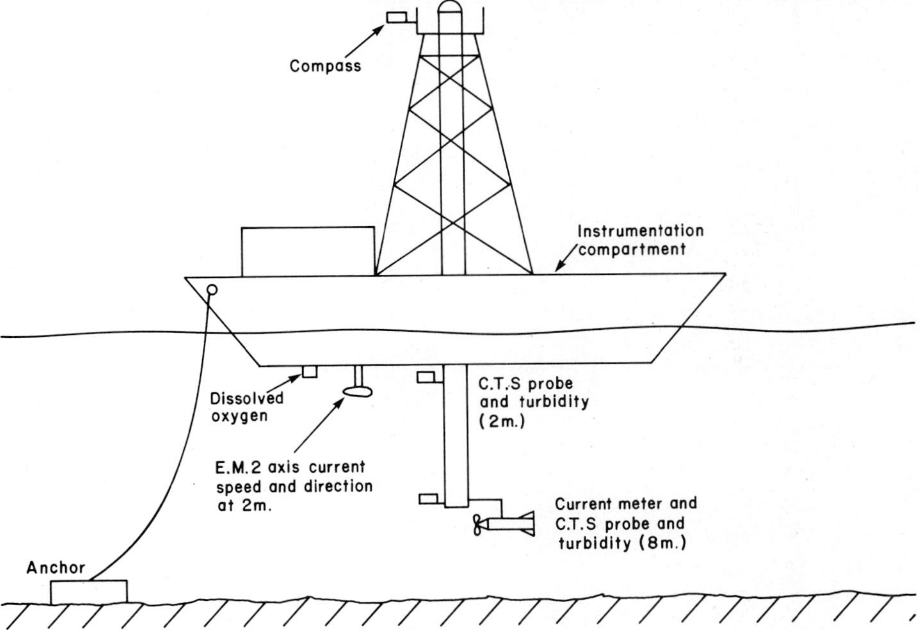

FIGURE 3. Instrumentation of the Midas Buoy.

E-W axes was made from recordings of the compass bearing of the buoy. The sampling frequency used for all observations was at 15 minute intervals, the sampling duration was 2 minutes with all values time-averaged over this period.

Tidal Analyses

Table 1. Results from tidal analyses of 29 days Midas Buoy data Starting on 1: 3/04/86; 2:30/04/86; 3:4/06/86; SD - Standard deviation; Elevation results A refer to a 21 year data set.

	SURFACE SALINITY						MID DEPTH SALINITY					
	AMP PPT			PHASE			AMP PPT			PHASE		
	1	2	3	1	2	3	1	2	3	1	2	3
Zo	24.1	25.9	27.3	-	-	-	25.3	27.1	28.2	-	-	-
Mm	1.7	1.2	1.7	145	302	90	0.8	1.4	0.4	170	318	216
Mfs	0.2	0.5	1.0	104	26	320	0.7	0.6	0.3	222	101	274
MV2	0.5	0.1	0.1	53	148	24	0.3	0.1	0.1	33	120	85
N2	0.7	0.6	0.4	327	315	307	0.4	0.3	0.6	316	313	295
M2	3.3	2.9	1.8	334	330	324	1.8	1.8	1.7	344	337	339
S2	0.9	1.0	0.3	2	18	12	0.5	0.6	0.3	24	35	35
M4	0.4	0.3	0.0	204	180	104	0.2	0.1	0.1	316	299	184
MS4	0.1	0.1	0.1	119	195	318	0.1	0.1	0.0	20	323	102
M6	0.1	0.1	0.1	91	60	200	0.0	0.0	0.0	96	66	187

	TEMPERATURE						O_2					
	AMP °C			PHASE			AMP %			PHASE		
	1	2	3	1	2	3	1	2	3	1	2	3
Zo	6.1	9.8	14.6	-	-	-	65	73	38	-	-	-
Mm	2.5	1.1	1.3	356	338	71	6	7	29	298	305	85
Mfs	1.1	0.8	1.5	104	51	220	3	6	9	316	261	346
MV2	0.4	0.1	0.1	67	109	59	1	1	1	152	153	163
N2	0.6	0.2	0.2	137	129	101	3	2	4	313	289	239
M2	0.9	0.8	1.0	142	152	157	13	13	12	331	330	340
S2	0.5	0.4	0.4	214	180	205	4	4	4	29	19	43
M4	0.3	0.3	0.3	276	304	340	1	2	2	206	164	275
MS4	0.2	0.2	0.2	336	278	356	0	1	2	69	170	264
M6	0.1	0.1	0.1	75	89	164	1	1	1	70	29	112

SD RESIDUAL:SD OBSERVATION (approximates noise:signal ratio)	1	2	3
NORTH-SOUTH SURFACE CURRENT	0.22	0.23	0.19
NORTH-SOUTH MID DEPTH CURRENT	0.23	0.29	0.26
SURFACE SALINITY	0.36	0.31	0.30
MID DEPTH SALINITY	0.51	0.58	0.45
TEMPERATURE	0.35	0.61	0.59
DISSOLVED O_2	0.34	0.38	0.49
ELEVATION	0.05	0.06	0.05

Table 1. Continued --
Results from tidal analyses of 29 days Midas Buoy data
Starting on 1: 3/04/86; 2:30/04/86; 3:4/06/86; SD - Standard deviation; Elevation results A refer to a 21 year data set.

	ELEVATION							
	AMP M				PHASE			
	A	1	2	3	A	1	2	3
Zo	-	5.01	5.20	5.12	-	-	-	-
Mm	0.02	0.23	0.09	0.09	216	307	263	192
Mfs	0.02	0.11	0.07	0.02	232	160	64	16
MV2	0.06	0.08	0.03	0.04	40	79	79	345
N2	0.59	0.60	0.59	0.58	301	299	300	301
M2	3.12	3.07	3.10	3.15	324	322	321	321
S2	1.01	1.00	0.98	1.03	8	6	6	6
M4	0.22	0.21	0.20	0.23	214	215	216	216
MS4	0.13	0.16	0.11	0.07	257	252	246	262
M6	0.55	0.06	0.07	0.07	321	321	305	303

	CURRENT SURFACE											
	AMP CM PER S			DIR			PHASE			ECC		
	1	2	3	1	2	3	1	2	3	1	2	3
Zo	23	30	23	6	25	42	-	-	-	-	-	-
Mm	16	15	22	33	84	88	-13	-26	-129	0.02	-0.24	0.08
Mfs	7	15	11	30	56	118	109	69	- 37	0.25	-0.29	0.50
MV2	7	3	6	201	183	36	23	-18	84	-0.04	-0.68	-0.01
N2	29	26	28	346	350	340	53	38	34	-0.25	-0.05	0.16
M2	157	149	157	348	349	344	62	61	61	-0.03	0.02	0.01
S2	48	48	48	161	162	171	110	109	110	-0.03	-0.05	0.08
M4	30	45	43	347	344	343	- 3	- 8	- 3	-0.08	-0.01	-0.24
MS4	24	24	18	353	341	351	43	30	48	0.01	0.09	-0.08
M6	10	8	13	348	24	348	6	4	10	0.17	0.05	0.32

	CURRENT MID DEPTH											
	AMP CM PER S			DIR			PHASE			ECC		
	1	2	3	1	2	3	1	2	3	1	2	3
Zo	7	15	16	18	42	65	-	-	-	-	-	-
Mm	12	14	19	25	82	87	1	-16	-127	0.07	-0.15	0.06
Mfs	9	14	10	11	101	117	91	89	- 38	0.01	-0.32	0.58
MV2	7	7	6	12	137	18	82	74	66	-0.39	-0.28	-0.17
N2	28	22	21	347	354	353	44	29	29	-0.03	0.00	0.22
M2	130	123	127	348	351	346	58	59	58	-0.01	0.02	-0.01
S2	43	43	45	162	171	169	107	108	111	-0.08	-0.03	0.08
M4	27	40	38	351	348	347	- 1	2	1	-0.01	-0.12	-0.24
MS4	25	25	17	353	345	350	44	30	41	0.03	-0.03	-0.08
M6	7	3	7	336	38	347	0	-38	6	0.56	0.41	0.32

Following a period of technical development of the buoy's instruments, reasonably continuous data were obtained over the period 3rd April to 3rd July 1986. For tidal analyses three periods, each of 29 days duration, were selected starting on: (A) 3rd April, (B) 30th April and (C) 4th June. Occasional short periods of missing data were approximated using an average of values observed 12.5 hours before and after.

Tidal analyses were carried out using the TIRA programme developed at the Proudman Oceanographic Laboratory. A total of 35 constituents were resolved including the mean, Zo, and eight related constituents (i.e. close-frequency constituents not independently resolved in a 29 day analysis but derived by assuming the same amplitude and phase relationships exist as determined from longer-period analysis of elevation data from Liverpool). Table 1 shows these results for 10 major constituents. For comparative purposes, analyses from three corresponding periods of tidal elevation data recorded at Liverpool are included.

II.1 Currents

The data are transformed into ellipse parameters namely: amplitude, phase and direction (all of the semi-major axis) and eccentricity (the ratio of semi-minor to semi-major axes with a +ve sign indicating anticlockwise rotation).

The predominance of the semi-diurnal M_2 constituent is clearly evident in Table 1. Next largest are the other major semi-diurnal constituents S_2 and N_2 with amplitudes approximately one-third and one-fifth that of M_2. These same ratios are shown for the elevation data indicating the primarily semidiurnal nature of tides in the Mersey with an almost constant ratio of current to elevation amplitudes for these major constituents. The S_2:M_2 ratio indicates a significant fortnightly Neap-Spring variability with amplitudes varying by a factor of 2, the N_2:M_2 ratio indicates an additional monthly cycle of amplitude variations by a factor of 3:2. Over the three successive months, the M_2 current amplitude varies by ± 3% of the mean compared with ± 1% for elevations indicating a slightly higher noise:signal ratio for the current measurements but suggesting a satisfactory level of instrument performance. Likewise monthly variations in direction, phase and eccentricity for M_2 are all minimal. However, larger variations shown for all four ellipse parameters for S_2 and N_2 (relative to those for M_2) reflect the poorer resolution of these smaller constituents. The surface amplitudes for these three major constituents are up to 20% larger than the mid-depth values, this accords with theoretical descriptions of vertical current structure (Prandle, 1982). The M_2 phase difference between elevation and current is 262° indicating a near 'standing wave' propagation.

The quarter-diurnal constituents M_4 and MS_4 are generally produced by non-linear aspects of tidal propagation (the shallow water, advective and frictional terms in the shallow water wave equations). The amplitudes of these constituents are often larger in currents than in elevation, thus in table 1, for elevation M_4 is 7% of M_2 and MS_4 4%, while for current M_4 ranges from 20 to 30% of M_2 and MS_4 from 13 to 20%. The determination of these constituents is generally less accurate than for the semi-diurnal constituents described previously though table 1 indicates reasonable consistency in the direction and phase values.

The large monthly variations in Mm and MSf indicate that these constituents reflect mainly non-tidal energy, a much longer observational data set is required to determine these tidal constituents. Large variations are also shown for Zo, however there is a persistently strong seaward flow at the surface of about 25cm/s, reducing by 50% at mid-depth and veering about 20° clockwise.

II.2 Salinity, Temperature and Oxygen

Examination of Table 1 shows that all three of these parameters include a significant tidal component. The predominance of the three major semidiurnal constituents is maintained. However, the ratios between semi-diurnal amplitudes are not equal to those constant values found for both elevation and currents. This might be attributed to the much higher noise: signal ratios as shown in Table 1 or to non-advective processes.

The large values shown for the longer-period constituents Mm and MSf indicate the effect of extraneous non-tidal influences. The respective values of Zo show a steady increase in both temperature and salinity as the summer approaches while the oxygen levels peak during the middle period. Consideration of the cross-sectionally averaged equation for mass conservation provides a useful check on the validity of these tidal constituents and also an indication of the time-averaged axial distributions along the river.

For a conservative substance, cross-sectionally averaged mass conservation requires:

$$\frac{\partial C}{\partial t} + U\frac{\partial C}{\partial x} = Dx \tag{1}$$

where C is concentration, U velocity, t time, x distance downstream and Dx an axial dispersion term. Assuming the time variations can be represented by an expansion involving only the tidal constituents Z_0, N_2, M_2 and S_2, Eq.(1) simplifies to

$$i(\omega\hat{C}_{N_2} + \omega\hat{C}_{M_2} + \omega\hat{C}_{S_2}) + (U_{Z_0} + \hat{U}_{N_2} + \hat{U}_{M_2} + \hat{U}_{S_2}) \left(\frac{\partial C}{\partial x}_{Z_0} + \frac{\partial C}{\partial x}_{N_2} + \frac{\partial C}{\partial x}_{M_2} + \frac{\partial C}{\partial x}_{S_2}\right) = Dx \tag{2}$$

where ω is tidal frequency, ^ denotes tidal amplitude, and the complex number i indicates a phase advance of 90 . Equating terms in Eq.(2) involving time variations at the periods of N_2, M_2 and S_2 we obtain

$$\text{for } N_2 \quad i\omega\hat{C}_{N_2} = -\hat{U}_{N_2}\frac{\partial C}{\partial x}_{Z_0} + Dx_{N_2} \tag{3}$$

$$\text{for } M_2 \quad i\omega\hat{C}_{M_2} = -\hat{U}_{M_2}\frac{\partial C}{\partial x}_{Z_0} + Dx_{M_2} \tag{4}$$

$$\text{for } S_2 \quad i\omega\hat{C}_{S_2} = -\hat{U}_{S_2}\frac{\partial C}{\partial x}_{Z_0} + Dx_{S_2} \tag{5}$$

Assuming that terms such as $U_{Zo}\frac{\partial C_{M2}}{\partial x} << U_{M2}\frac{\partial C_{Zo}}{\partial x}$ as suggested from Table 1.

The dispersion term in Eq.(1) is generally found to be much smaller than the other two terms. By neglecting this term in Eqs.(3) to (5) the time-averaged longitudinal concentration gradients can be calculated. Two checks can be made on the validity of these results (and thereby indirectly on the accuracy of the original constituent values). The first check is the consistency of the three estimates for $\frac{\partial C_{Zo}}{\partial x}$ obtained independently from Eqs.(3) to (5) and the second is the expectation of a 90° phase advance of $\hat{U}$ relative to $\hat{C}$ for each constituent for all three tracers.

Table 2 shows that these phase differences are all reasonably close to 90° (or 270°). Likewise the longitudinal gradients calculated from the data pertaining to the three separate constituents show a reasonable level of agreement. These gradients indicate a net increase seawards over the 10km long 'Narrows' section of approximately 2o/oo in salinity, 14% in oxygen and -1.3° in temperature (the -ve sign in the latter being obtained by correcting the phases from 270° to 90°).

As an additional verification of the above approach, Eqs.(3), (4) and (5) were solved simultaneously for $\frac{\partial C_{Zo}}{\partial x}$ by a least squares method involving the minimisation of $\sum_{i=i}^{3} Dx^2$. The values of the time-averaged concentration gradients found by this method were all in close agreement with the values for the M_2 constituent shown in Table 2. The magnitude of the dispersion term as a percentage of (either of) the other terms was computed as: (i) for salinity, 11%, (ii) for oxygen 13% and (iii) for temperature 16%. These results indicate that variability of salinity, temperature and oxygen in the semi-diurnal frequency band is predominantly due to tidal advection.

II.3 Residual time series

An analysis of residual (i.e. tidal component removed) time series obtained from the Midas Buoy data was made. This will be reported elsewhere. However, the major conclusion of interest here is the additional evidence of the small influence of wind and river flow in comparison with the predominant tidal forcing.

Table 2. Time-averaged concentration gradients for salinity, oxygen and temperature computed (separately) from N_2, M_2 and S_2 tidal constituents for current $\hat{U}$ and concentration $\hat{C}$. (θ is the phase difference between $\hat{U}$ and $\hat{C}$.

	N_2			M_2			S_2		
$\hat{U}$ m s^{-1}	(0.26,38°)			(1.40,60°)			(0.46, 109°)		
	$\hat{C}$	θ	$\partial C_{zo}/\partial x$	$\hat{C}$	θ	$\partial C_{zo}/\partial x$	$\hat{C}$	θ	$\partial C_{zo}/\partial x$
SALINITY °/oo	(0.5,314°)	84°	0.26	(2.2,335°)	85°	0.22	(0.6,21°)	88°	0.19
OXYGEN %	(3,280°)	118°	1.6	(13,334°)	86°	1.3	(4,30°)	79°	1.3
TEMP °C	(0.3,125°)	273°	0.16	(0.9,150°)	270°	0.09	(0.4,200°)	269°	0.12
			km^{-1}			km^{-1}			km^{-1}

III. Cross Sectional Surveys

Cross-section surveys were repeated at the same location (figure 2) on (A) 16th June 1982 (B) 16th September 1982, (C) 21st September 1982 and (D) 9th August 1984. The tidal ranges on these days were, 5.35, 8.14, 7.22 and 6.17m respectively while the river flows, measured just beyond the tidal limits, were 40, 10, 87.5 and 8.3 m^3/s. (In general, river flow data averaged over a preceding period of the order of the estuarine flushing time might be more representative - on examination, these daily values were found to be close to the mean values over the preceding week.) Each survey extended over a complete semi-diurnal tidal period. Operating from five boats (four boats in survey C), vertical profiles were taken measuring: current speed and direction (by DRCM using 2 min averages), salinity and suspended sediment concentration (both from bottle samples). These profiles were repeated at half hourly intervals with typically six values of current and three values of salinity and sediment concentration. In each case, the depth of the observation below the surface was recorded.

III.1 Current data

In this analysis, the directions recorded were used only to indicate seaward or landward flow, the speeds were then assumed to be either positive or negative respectively. An alternative scalar time series could be obtained by resolving into upstream and downstream current components, however the short sampling duration of the current measurements (2 minutes) was considered likely to make the observed directions inaccurate.

III.2 Tidal analyses

A simplifed tidal analysis of the currents at each location was carried out using the technique described by Prandle and Ryder (1985). This involves predicting the tidal elevations (using the five largest constituents namely M2, S2, N2, K2 and ν_2 derived from long term tide gauge recordings from the Port of Liverpool) for the day of each survey. The phase of these elevation data, $\zeta(t)$, are then successively shifted to maximise the correlation with the current observations, U(t). Thus λ and Δt in Eq.(6) are determined, (by least squares fitting and successive substitution respectively) to minimise ϵ where

$$\epsilon = \sum_{i=1}^{N} [U(t) - \bar{U} - \lambda(\zeta(t+\Delta t) - \bar{\zeta})] \qquad (6)$$

N is the number of observations.

This time-shift Δt is then assumed to indicate the phase difference between current and elevation for the predominant M_2 tidal constituent. (While corresponding amplitude and phase values could be calculated for the currents associated with the other semi-diurnal constituents (S_2,N_2,K_2 and ν_2), this is effectively only an arithmetic exercise since only one phase lag and one amplitude ratio (λ)are obtained from the correlation procedure.)

III.2.1 Depth averaging

Since the above analysis considers only a single relationship between elevation and current, the observed depth-varying currents must be reduced to a depth-averaged form. A depth profile varying from zero velocity at the bed to a maximum at the surface according to some power, n, of the fractional depth z, was assumed (Veen, 1938)

$$U(z) = \bar{U}^D z^n (1+n) \qquad (7)$$

where $\bar{U}^D$ is the depth-averaged value.

Table 3. Cross-Sectional Surveys, analyses of currents

	M_2 amplitude					X-sectional mean	M_2 phase					X-sectional mean (area weighted)	Tidal Range
VESSEL / SURVEY	1	2	3	4	5		1	2	3	4	5		(m)
A	0.96	1.03	1.40	1.03	1.03	1.12	53	49	45	47	45	48	5.3
B	1.01	1.24	1.38	1.32	1.11	1.25	55	55	57	59	59	57	8.1
C	0.68	1.18	1.42		0.87	1.18	43	45	45		43	44	7.2
D	0.55	1.16	1.33	1.22	0.76	1.06	47	40	42	40	34	41	6.2

(a) M_2 Tidal Currents (m s^{-1}) depth-averaged

VESSEL / SURVEY	1	2	3	4	5	X-sectional mean (area weighted)	Net Residual Flow	Qf	Tidal Range
A	7.2	3.7	8.0	6.4	2.2	5.8	1030	40	5.3
B	-10.8	-5.0	3.0	-1.6	1.3	-1.8	-330	10	8.1
C	-11.3	7.7	5.3		0.0	3.2	560	87.5	7.2
D	-15.0	2.5	6.2	5.5	8.5	2.5	440	8.3	6.2
							$m^3 s^{-1}$	$m^3 s^{-1}$	m

(b) Depth-averaged Residual Currents (cm s^{-1}) (-ve landward)

In each separate analysis of observations from a particular vessel a range of values of n were investigated. The value which minimised the root mean-square time-varying residual component between the observed currents and predicted tidal currents was adopted (see section III.3 for details). Values of n ranged from 0.15 to 0.5 wth a mean of 0.25, U_R ranged from 0.17 to 0.40 m/s with a mean of 0.25.

III.2.2 M_2 currents

Table 3(a) shows the results from these analyses, the location of the vessels varied in each survey and hence direct comparison is complicated. Overall the M_2 current amplitude is largest in the deepest section in mid-channel (up to 1.4m/s) decreasing significantly in the shallower sections on the east and west shores. The cross-sectional mean values were calculated by applying appropriate weightings to each vessel's results according to the proportion of adjacent cross-section. The cross-sectional mean M_2 current amplitudes vary from 1.06 m/s to 1.2 m/s or 0.10% of the mean of 1.15 m/s. The M_2 current phase values show an average value of 48 with a standard deviation of 7°.

III.2.3 Comparison of cross sectional tidal currents with Midas Buoy measurements.

In Section 2, the M_2 constituent values calculated from the Midas data were (i) at 2m below the surface (1.54, 61°), (ii) at mid-depth (1.27, 58°). These values may be compared with the survey data from location 2 (closest to the MIDAS buoy) for which the mean value is (1.15, 48°). The earlier phase values recorded at position 2 may be partially explained by vertical phase differences - in estuaries near-bed phases may be in advance of surface values by up to 20°.

The Midas M_2 amplitudes can be converted to depth-averaged values using equation (7) and the mean value of n = 0.25 found at location 2. The amplitudes 1.54 and 1.27 then correspond to depth-mean values 1.26 and 1.23 respectively - in reasonable agreement with the survey mean of 1.15. Thus the M_2 phase and amplitude agreement between survey data at location 2 and the Midas data suggests that the simple tidal analysis used for analysing the survey data provides a reasonable first order value for the major tidal constituent.

Since the depth-mean M_2 tidal amplitude at position 2 is close to the cross-sectional mean value, a reasonable approximation to net tidal fluxes of any contaminant (evenly distributed across the section) can be derived by simple multiplication of the cross sectional area with the depth-mean values derived from measurements at the Midas Buoy. (Note, from table 1, that phases of elevation and currents are close to quadrature for the major constituents.) Taking the cross-sectional area as 17.750 m^2, the M_2 flow amplitude is then 20,400 m^3/s with a net ebb and flood volume of 290 x 10^6 m^3. O'Connor (1987) estimated the volume of the Mersey Estuary as approximately 650 x 10^6 m^3. The strong tidal flows across the mouth of the Mersey ensure effective mixing of estuarine water in Liverpool Bay. Hence the large ratio of tidal flux: net volume within the estuary, reflects the extent of tidal flushing in the Mersey.

III.3 Residual Currents

To determine the residual component, U_R, from the observed currents, U_B, the tidal current component, U_T, was first calculated using the tidal elevation predictions (with appropriate phase and amplitude adjustments) and the vertical current structure, Eq.(7). These tidal currents were then subtracted from the observed currents to compute 'residual' values. Thus

$$U_R(z,t) = U_B(z,t) - [z^n(1+n)]\, U_T(t) \tag{8}$$

where $U_T(t) = \bar{U} + \lambda[\zeta(t+\Delta t) - \bar{\zeta}]$.

These residuals contain a significant quarter-diurnal component since the simple tidal prediction technique used here fails to remove this component. However, by averaging these residuals over a semi-diurnal period these shorter-period constituents can be effectively eliminated.

The method of least squares was used to fit a time invariant vertical profile

$$U_o(Z) = \alpha z^3 + \beta z^2 + \gamma z + \delta \tag{9}$$

to the complete set of $U_R(z,t)$ from each vessel. This equation follows from Prandle (1985) where the analytical form, in narrow estuaries, of residual vertical profiles associated with freshwater flow, wind and well-mixed longitudinal density gradients are derived.

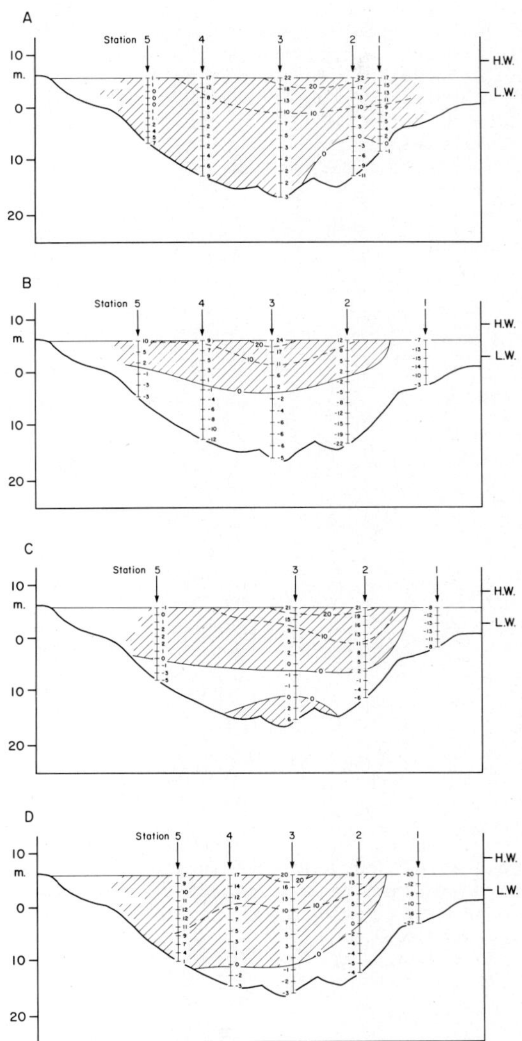

FIGURE 4. Residual velocity distributions cm/s; surveys A, B, C, and D.

In fitting Eq.(9) by a least squares approach to the calculated residuals, the paucity of near-bed data often produced profiles with unrealistically large near-bed residuals. To overcome this problem Eq.(9) was truncated, removing the term z3. The resulting residual profiles are indicated in figure 4. Table 3(b) shows the depth-mean residual flows together with estimates of the net cross-sectional residual transports. These calculated residual transports are between one and two orders of magnitude greater than the river flows and their relative magnitudes are uncorrelated.

The more significant correlation is with the spring-neap tidal cycle. Thus for Survey A with a neap tide of 5.3 m, a large net seaward flow (average 5.8 cm/s) occurs everywhere except for a small near-bed region close to section 2 (figure 4). For Surveys C and D with intermediate tidal ranges of amplitudes 7.2 and 6.2 m respectively the seaward flow is partially balanced by a landward inflow both near the bed and on the eastern side. For Survey B with a spring tide of 8.1m the extent of this landward inflow increases producing a net inflow of 1.8 cm/s. The vertical velocity distributions are consistent with the theoretical distributions for a well-mixed estuary (Prandle, 1985). The variation in net residual transport might be attributed to a spring-neap-spring (MSf) tidal current cycle of amplitude 3.5 cm/s. However, accurate determination of this constituent requires long periods of observational data. This explanation would also leave an unresolved net seaward current of 2.3cm/s i.e. transport of 400 m^3/s.

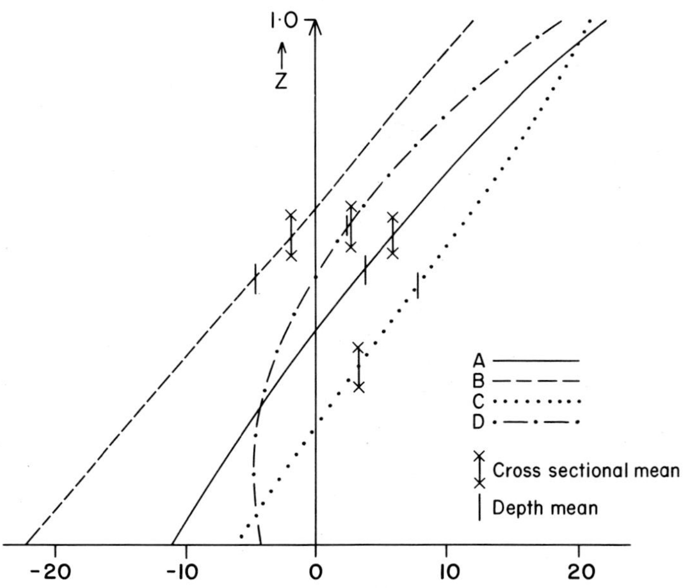

FIGURE 5. Residual vertical current structure at position 2 cm/s, surveys A, B, C and D.

Figure 5 shows the vertical current structure at section 2 for the four surveys, also indicated are the depth-mean residuals and cross-sectional mean residuals. For Surveys A, B and D the observed residuals at location 2 for $z = 0.6$ reasonably approximate the cross-sectional-mean values. These results suggest that residuals derived from current measurements at the Midas-Buoy can provide an approximate indication of net cross-sectional currents.

III.4 Salinity data

Tidal analyses of the salinity data were performed in similar manner to that described for current. Vertical variations in salinity were generally small and hence the results shown ignore verical structure. Table 4 shows (i) the time-averaged and (ii) the M_2 tidal variations at each location together with cross-sectional mean values (area weighted).

Although both the mean and M_2 values vary significantly from one survey to another, cross-sectional variations for specific surveys are much smaller. The latter indicates that any conservative dissolved contaminant originating either in the Irish Sea or in the upper reaches of the Mersey should be well-mixed cross-sectionally at this location and hence a single point measurement would be representative. Moreover, the mean salinity of approximately 28.5 o/oo compared with 30 o/oo in Liverpool Bay indicates that river water is diluted by a factor of 20 at this location. The largest tidal variation in salinity occurs on Survey C and coincides with the minimum mean salinity, while the opposite occurs for Survey D.

Table 4. Salinity, time-averaged S and M_2 tidal constituent (A,g) (o/oo)

VESSEL	1			2			3			4			5			Cross-sectional mean (area weighted)		
SURVEY	$\overline{S}$	A	g	$\overline{S}$	A	g	$\overline{S}$	A	g	$\overline{S}$	A	g	$\overline{S}$	A	g	$\overline{S}$	A	g
A	29.03	1.97	320°	28.98	1.91	322°	29.02	1.77	322°	29.18	1.73	318°	29.53	1.78	318°	29.13	1.83	320°
B	29.31	1.66	297°	29.12	1.72	309°	29.40	1.69	313°	29.58	1.74	313°	29.44	1.71	309°	29.37	1.71	308°
C	27.31	3.35	301°	27.40	2.87	315°	27.92	3.08	320°				27.38	3.55	313°	27.63	3.15	312°
D	29.46	1.67	293°	29.54	1.60	313°	29.74	1.53	317°	29.74	1.53	317°	29.83	1.58	315°	29.67	1.57	312°

This lowest mean salinity also coincides with the highest freshwater flow and likewise the maximum salinity coincides with the minimum river flow. These relationships between the tidal amplitude of salinity and the mean salinity reflect the increase in the longitudinal salt gradient for larger river flows at this location.

IV. Suspended Sediments

IV.1 Observations from Midas Buoy

Figure 6 shows suspended sediment concentrations recorded, at the surface and at mid-depth, over a 30 day period in April 1986. Features immediately evident are:

(a) Concentrations at mid-depth are significantly larger than at the surface.
(b) The neap-spring variation is much more pronounced than for currents.
(c) There is a pronounced flood/ebb asymmetry with higher concentrations on the ebb tide.

These features suggest:

(a) The sediment includes coarser fractions with significant inter-tidal settling

(b) The erosion/transporting capacity is proportional to some power (> 1) of the current
(c) Sediments transported past the buoy are more readily available upstream.

IV.2 Observations from cross sectional surveys

For each set of synoptic profiles a vertical distribution

$$C(z) = C_o e^{-bz} \quad . \qquad (10)$$

was assumed. An optimum value of b (for the complete set of profiles) was determined by successive substitution such that the r.m.s. error between observed concentrations and Eq.(10) (expressed as a fraction of the depth-mean concentration) was minimised. Table 5 shows these values of b together with the time-averaged value of the depth-mean concentrations i.e. the time average of $-C_o(e^{-b}-1)/b$.

Table 5 shows b ranges from 0.5 to 4.0 with a mean of about 1.5, the latter corresponding to a 4.5:1 ratio between bed and surface concentrations. Highest depth and time averaged concentrations are always found on the Liverpool side of the River, with an average ratio of 3.5:1 between the east and west sides. Cross sectional mean concentrations (calculated by applying 'area weightings') are also shown in table 5. For all 4 Surveys, concentrations at location 2 are reasonably close to the cross sectional mean value.

Table 5. Time and depth-averaged suspended sediment concentrations $\bar{C}$ (mg/l) vertical variation parambet b, $[C(z) = C_{bed}e^{-bz}]$.

Vessel	1		2		3		4		5		Cross-sectional mean (area weighted)
Survey	$\bar{c}$	b	$\bar{c}$	b	$\bar{c}$	b	$\bar{c}$	b	$\bar{c}$	b	$\bar{c}$
A	55	1.5	42	2.0	36	2.5	15	1.0	11	0.5	32
B	299	0.5	217	0.5	300	2.0	142	1.5	86	0.5	213
C	299	1.0	248	1.5	199	1.5			95	0.5	200
D	161	2.0	89	1.0	118	3.0	84	4.0	42	1.5	100

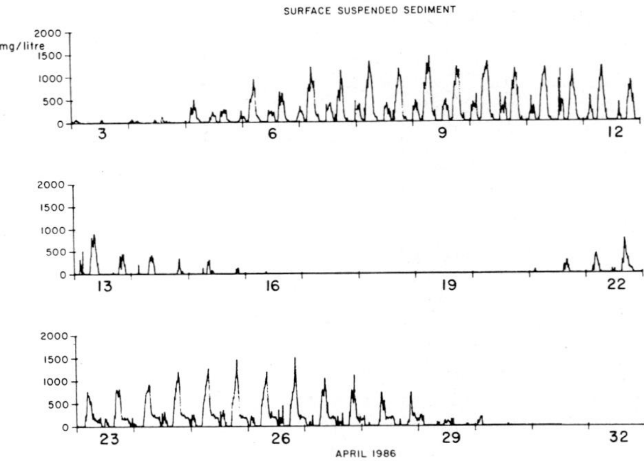

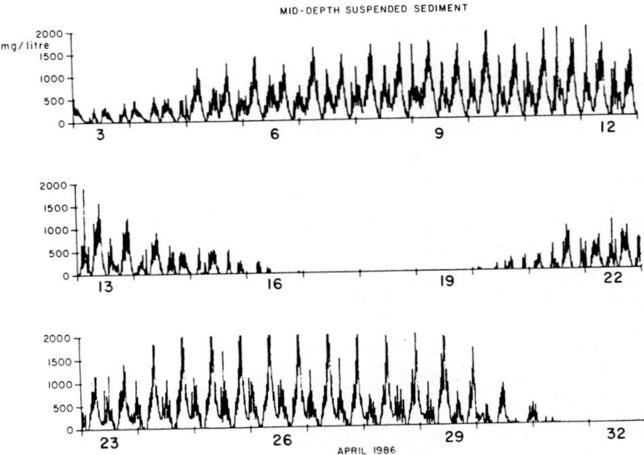

FIGURE 6. Observed suspended sediment concentrations. (a) near-surface, (b) mid-depth.

IV.3 Relating Sediment Concentration to Tidal Velocity

Correlating the cross-sectional and tidally averaged concentrations $\overline{C}_A^{\,T}$ with tidal elevation amplitude , we derive the relationship

$$\overline{C}_A^{\,T} = 3.4\hat{Z}^3 \qquad (11)$$

where C_A is in mg/l, $\hat{Z}$ in m.

Replacing $\hat{Z}$ with $\hat{U}$ in Eq. (11) by assuming their M_2 ratio of 1.15/3.1 and furthermore replacing tidally-averaged values with instantaneous values, we obtain

$$C_A(t) = 158\ U(t)^3 \qquad (12)$$

in which $\int_0^{\pi/w} \sin^3 wt dt = 4/3\pi$.
Likewise assuming the exponential depth variation e^{-bz} with b=1.5 (section IV.2) we obtain

$$C(z,t) = 305\ U(t)^3 e^{-1.5z} \qquad (13)$$

To compare Eq.(13), derived from the cross sectional survey data, with the Midas buoy data shown in figure 6 we substitute a spring-tide current of 1.8 m/s and a neap-tide current of 0.9 m/s. These currents yield concentrations of: surface 397, mid-depth 840 on springs and: surface 50, mid-depth 105 (mg/l) on neaps. These calculated values are in reasonable agreement with the observations, although there is observational evidence of greater vertical homogeneity on spring tides - the latter deduction is also supported by the values for b in table 5.

IV.4 Tidal Flux

The net cross-sectional instantaneous flux, FA is

$$F_A = C_A\ UA \qquad (14)$$

or from Eq.(12) $F_A = 158\ U^4 A$, with A = 17,750 m², $\hat{U}_{M_2} = 1.15$ m/s, the amplitude of the M_2 cross-sectional sediment flux is

$$F_{M_2} = 158 \times 17.750 \times 1.15^4 = 4.9 \text{ tonnes/s} \qquad (15)$$

The net ebb or flood mass V_{M_2} is ($\int_0^{\pi/w} \sin^4 wdt = 3/8$)

$$V_{M_2} = F_{M_2} \times 3/8 \times 44712/2 = 41{,}000 \text{ tonnes.} \qquad (16)$$

On an extreme spring tide the velocity may be $1.5 \times \overline{U}_{M_2}$ and hence the extreme mass is $(1.5)^4 = 5 \times V_{M_2}$, on extreme neap tides the currents may be $0.5 \times U_{M_2}$ reducing the mass flux to $(0.5)^4 = V_{M_2}/16$. Thus the ratio of total mass flux between extreme spring and neap tides is approximately 80, thus much of the sediment mobilised on these spring tides (>200,000 tonnes) must be temporarily deposited on neap tides.

IV.5 Residual Fluxes

Calculations of residual fluxes from various sources are described in an appendix. The salient results considered here are as follows

(i) F_o, flux due to a residual axial velocity U_o,

$$F_o = 4\ F_{M^2}\ U_o/\hat{U} \qquad (17)$$

(ii) F_v, flux due to a residual vertical current structure Ud(2z-1) combined with a vertical concentration gradient $C_o e^{-bz}$

$$F_v = -0.6b\ F_{M_2}\ Ud/\hat{U} \tag{18}$$

(iii) F_w, flux due to lateral changes in : (i) concentration of ratio M across the estuary and (ii) residual velocity from U_L to $-U_L$

$$F_w = 4\ F_{M_2} \frac{(M-1)}{(M+1)}\ U_L/\hat{U} \tag{19}$$

These simplified results indicate that while lateral variations are unlikely to make a major contribution to net fluxes of sediment, vertical variations can form a major component. Thus vertical variations in residual velocity and sediment concentration must be as accurately determined as the net residual velocity U_o.

V. Conclusions

For a strongly tidal estuary, this study examines the difficulties of combining continuous single point recordings with short period cross sectional surveys to provide estimates of both tidal and residual fluxes.

By comparing constituents from a tidal analysis of currents, oxygen, temperature and salinity, it is shown that tidal advection predominates at the semi-diurnal frequencies. This enables time-averaged axial gradients of these tracers to be calculated directly.

A simplified tidal analysis technique used for short period (13h) cross sectional data enables the vertical and horizontal variations in both tidal and residual parameters to be determined. Removal of the tidal component using results from this simplified analysis reduces the sensitivity of the estimate of residual components to the assumed duration of the semi-diurnal cycle.

The resulting cross sectional profiles of residual velocity were consistent with the classical near bed saline intrusion, the extent of which depends more on tidal amplitude than river discharge. Analyses of sediment distributions were combined with the current analyses to calculate sediment fluxes associated with (i) Spring, Neap and mean tides together with (ii) a net seaward current, (iii) vertical structure and (iv) lateral variability (of both residual currents and sediment concentrations).

This study emphasizes the difficulties in determining net sediment fluxes by direct measurement. Since conditions in the Mersey appear particularly favourable for this objective (i.e. overwhelming tidal predominance, a straight, narrow and shallow channel, well-mixed vertically, tidal elevation and current close to quadrature), these difficulties are likely to be even greater in other estuaries.

By contrast, reasonably consistent estimates of the ebb and flood tidal fluxes were obtained. Moreover, the enormous variability over the Spring-Neap cycle of these tidal sediment fluxes (80:1 or 3^4:1 indicates this cycle may be more significant than net fluxes for water quality studies. Whereas net sediment fluxes determine net accretion/scour processes, in water quality studies it may be more important to determine: (i) the exchange of sediments seawards of the mouth (i.e. the replacment of sediments of river-origin with sediments of other origin between both the ebb and flood tides and over the Spring-Neap cycle) and (ii) the exchange or degree of remobilisation of sediments within the estuary over the same intervals.

There is a clear requirement to increase the spatial and temporal resolution of the measurements. Exciting prospects are offered by the Development of ADCP (Acoustic Doppler Current Profiler) for vertical profiles of both currents and suspended sediment and also by H.F. Radar for synoptic mapping of surface currents. However, it seems likely that accurate estimation of residual fluxes will only be obtained by using such advanced measurement techniques indirectly, namely to develop and validate appropriate models.

Acknowledgements

The design and implementation of the observational program were carried out by Drs P. Head and P.D. Jones of the North West Water Authority, together with staff from the Water Research Centre, Medmenham.

VI. References

Amin, M., 1982: On analysis and prediction of tides in the west coast of Britain. Geophys. J. R. Astr. Soc., **68**, 57-78.

O'Connor, B., 1987: Short and long term changes in estuary capacity. J. Geol. Soc. **144(1)**, 187-195.

Prandle, D., 1982: The vertical structure of tidal currents and other oscillatory flows. Continental Shelf Research, **1**, 191-207.

Prandle, D., and Ryder, D. K., 1985: Measurement of surface currents in Liverpool Bay by high-frequency radar. Nature, **315**, **6015**, 128-131.

Prandle, D., 1985: On salinity regimes and the vertical structure of residual flows in narrow tidal estuaries. Estuarine, Coastal and Shelf Science, **20**, 615-635.

Veen, J-van, 1938: Water movements in the Straits of Dover. J. Cons. Perm. Int. Explor. Mer., **13**, 7.

APPENDIX. SEDIMENT FLUX CALCULATIONS

Residual Flux Associated with a Residual Axial Velocity U_o

In the Mersey Narrows, the phase difference between tidal elevation and current is close to 90°. This allows, to first order accuracy, net fluxes to be calculated without explicit consideration of changes of depth over the tidal cycle.

Assuming a cross-sectionally averaged net residual current U_o, superimposed on the tidal component U, Eq.(14) gives a net flux

$$F_A = A.158 \mid (U + U_o)^3 \mid (U + U_o) \qquad \text{(A1)}$$

Assuming Uo << $\hat{U}$, the tidal component remains unchanged while the residual flux, Fo, is approximately

$$F_o = A.158.4\, U_o \quad \bar{U}^3 \quad = 4\, F_{M2} U_o / \hat{U} \qquad \text{(A2)}$$

Residual Flux Associated with Vertical Structure

The classical residual estuarine circulation shown in figure 4 with upstream flow of saline water at the bed compensated by seaward flow of fresher water at the surface can produce a net flux of sediments. Assuming a simple structure

$$U_o(z) = U_o + U_d(2z - 1) \qquad \text{(A3)}$$

where U_d is the upstream current at the bed (z = 0) and Ud the seaward flow at the surface (z = 1).

In calculating the residual flux, we must now consider the depth variation in concentration. Assuming, the depth-averaged concentration is given as in Eq.(14), with $(U + U_o)$ the velocity, we calculate the depth-varying flux F(z) as

$$F(z) = b \frac{158}{1-e^{-b}} A \mid (\hat{U}+U_o)^3 \mid e^{-bz} [\hat{U}_A + U_o + U_d(2z-1)] \qquad \text{(A4)}$$

Integration of (A4) yields a net residual flux of

$$F_v = F_o + 158\frac{Ab}{1-e^{-b}}\hat{U}^3 U_d \int_o^1 (2z-1)e^{-bz}dz \qquad (A5)$$

$$= F_o\{1 + [\frac{2}{b} - \frac{(1+e^{-b})}{(1-e^{-b})}]\frac{U_d}{U_o}\}$$

or for $b \leq 3$

$$= F_o[1-0.15b\frac{U_d}{U_o}]$$

Thus the values pertaining to U_o, in Eq.(A2), may be multiplied by the factor shown in (A5) to calculate the fluxes associated with Ud. Figures 5(a) to (d) indicate a typical value for Ud of about 10 cm/s. Assuming b = 1.5, Eq.(A5) indicates that net fluxes due to (i) residual vertical current structure and (ii) residual axial velocity will be in the ratio $-0.225\ U_d/U_o$.

Residual flux associated with lateral variations in both velocity and concentration

A combination of lateral variations in concentrations and residual currents will produce a net movement of sediment. Cross sectional surveys B, C and D indicate (table 3(b)), on average, a greater inflow on the Liverpool side of 9 cm/s while table 5 indicates mean concentrations a factor of 3.5 higher on this side. Assuming this variation is linear across the estuary,

$$U_o(y) = U_o + U_L[1 - 2y/w] \qquad (A6)$$

where y is the transverse axis and y = 0 on Liverpool side, y = W on the Wirral side.

Assuming a similar linear variation, by a factor M, in the concentrations across the estuary,

$$C(y) = 2\frac{[1-(1-M)y/w]}{1+M} \qquad (A7)$$

Substituting for C_A from Eq.(14) the net flux is given by

$$F_W = \frac{A}{W}\int_0^W 158.2\frac{[1-(1-M)y/w]}{1+M}[\hat{U} + U_o + U_L(1-\frac{2y}{w})]^4 \qquad (A8)$$

For residual fluxes this yields

$$F_w = F_o[1 + \frac{(M-1)}{3(1+M)})\frac{U_L}{U_o}] \qquad (A9)$$

For the case of M = 3.5, Eq.(A9) indicates that the transverse residual circulation produces a net flux 5/27 U_L/U_o of that due to residual axial velocity.

Dynamics of Peel-Harvey Estuary, Southwest Australia

C.J. Hearn[1] and R.J. Lukatelich[2]

[1]Department of Geography and Oceanography
University College, University of New South Wales
Australian Defence Force Academy
Canberra, ACT 2600
Australia

[2]Center for Water Research and Department of Botany
The University of Western Australia
Nedlands, Western Australia 6009

ABSTRACT

Peel-Harvey Estuary is an ultra-shallow multi-basin estuary in which salinity varies from almost fresh in winter to hypersaline in autumn. The physical dynamics of the estuary are controlled by riverflow, low-frequency variations of ocean water level, wind forcing, evaporation, and diurnal tidal currents in the channels connecting the basins. A study of these processes is presented with examples of observational data and numerical models. A discussion is also given of the importance of physical processes to the occurrence of annual blooms of *Nodularia spumigena* and the cycling of phosphorus, (which is input to the estuary by riverflow), between plankton and sediments.

I. Introduction

Peel-Harvey Estuary (Hodgkin *et al*., 1980) is typical of the estuaries of southwest Australia. They consist of inter-connected shallow basins with one central basin connected by a long narrow channel to the ocean. Prior to engineering modifications during the present century (and in most cases during the last few decades) many of the estuaries were seasonally closed to the ocean by sand bars.

Estuaries of this type, with topographic variations in the way the basins are interconnected and joined to the ocean, are found world-wide and may be dominantly fresh, marine or hypersaline depending on a whole range of physical factors, of which riverflow is the most important.

The present study of the Peel-Harvey Estuary, and a related investigation of Songkhla Lake in Thailand (Meetam and Hearn, 1988), have been motivated by environmental problems. The usual trend is for such estuaries to become more marine under man's activities. This results firstly from dredging of the channel connecting the estuary to the ocean and secondly from the diversion or damming of riverflow. Dredging is usually undertaken to improve navigation through the channels and also to reduce flooding of the estuary; river diversion may also be required for flood prevention. The result is an estuary of increased salinity and tidal range; in many cases a sudden change in regime may disrupt the ecology of the system and also damage fisheries and farming activities. For example, in Songkhla Lake, dredging of the main channel to the ocean has increased salinities in portions of the lake where the water was used extensively for rice cultivation. Altered

Coastal and Estuarine Studies, Vol. 38
R. T. Cheng (Ed.)
Residual Currents and Long-term Transport

tidal ranges and reduction in riverflow can also cause major changes in flow regimes in the channels with associated large scale movements of sediment.

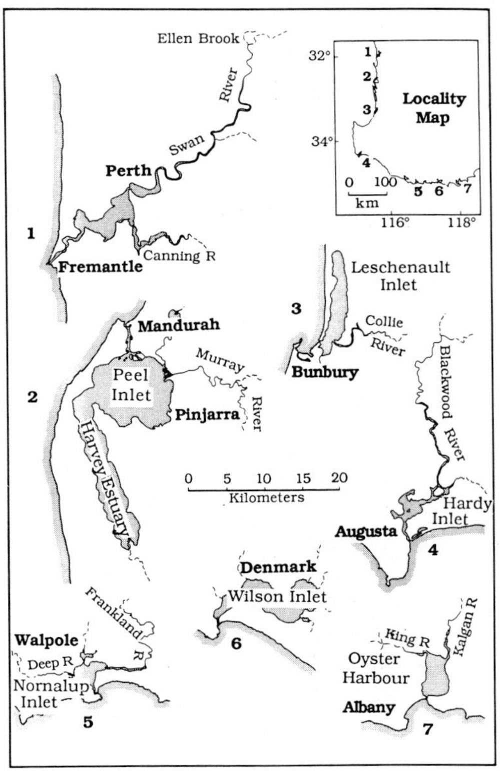

FIGURE 1. The estuaries of southwest Australia: 1 Swan; 2 Peel-Harvey; 3 Leschenault; 4 Hardy; 5 Nornalup; 6 Wilson; 7 Oyster Harbour.

The natural estuaries are usually microtidal because of tidal attenuation by the channels. Consequently the entry of marine water is a slow process and during periods of low riverflow ocean flushing times may be of order several months. The estuaries therefore tend to accumulate most types of man-made pollutants, and are liable to become eutrophic. For example, Peel-Harvey Estuary has become hyper-eutrophic in the last two decades due to phosphorus input from riverflow which arises from increased agricultural use of superphosphates in the coastal plain portion of the river catchments. The algal problems take the form of large accumulations of macroalgal weed in Peel Inlet and massive blooms of the blue-green microalga, Nodularia in Harvey Estuary.

Any attempt to manage the estuaries is dependent on an understanding of their physical dynamics. The present paper provides the results of some detailed studies of water flow in Peel-Harvey Estuary which may serve as a guide to some of the general features of this type of shallow multi-basin estuary.

II. Structure

Peel-Harvey Estuary (Fig. 2) consists of an approximately circular basin called Peel Inlet and a long narrow basin known as Harvey Estuary. The two basins are joined by a shallow sill through which passes a very narrow L-shaped natural channel. Peel Inlet is joined to the ocean by a long narrow passage called Mandurah channel. Maximum water depth in Peel Inlet is 2.5 m at mean water level. However, about 60% of the area of the basin consists of marginal regions with mean water depths below 1 m. As Fig. 2 shows, Mandurah channel has been extended across the marginal shelf of the basin to beyond the 1m contour. This portion of the channel is very narrow and convoluted; it has recently been redredged to form a straighter and wider channel. Harvey Estuary has a central axis with depths just greater than 2 m and parallel isobaths. Its eastern shore also contains extensive margins below 1 m in depth.

Peel Inlet receives riverflow from the Murray and Serpentine Rivers, and Harvey Estuary is fed by Harvey River. The latter two rivers are dammed, so that their effective catchment areas have been greatly reduced and limited to the coastal plain; Harvey River also has a major diversion drain to the ocean some 30 km south of the estuary. Both basins have numerous agricultural drains.

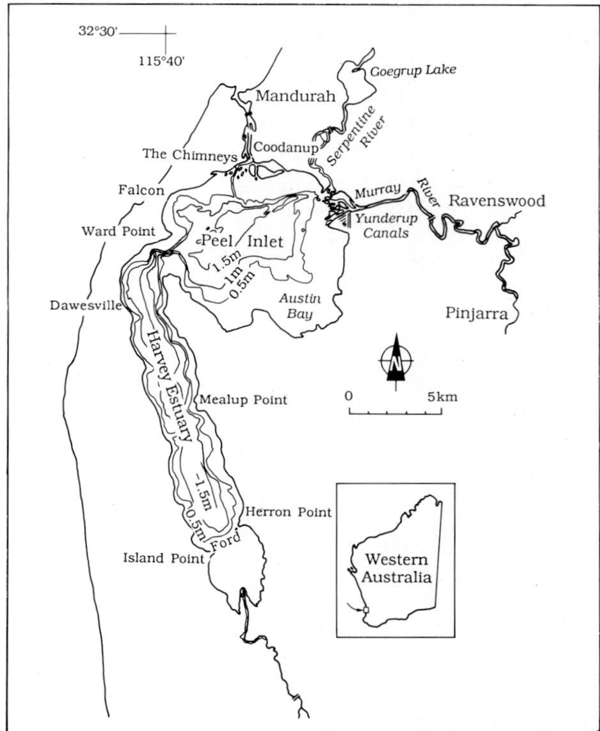

FIGURE 2. Peel-Harvey Estuary. Mean water level is 0.5 m above datum.

III. Tides

Figure 3 shows typical tidal records from Peel-Harvey Estuary in summer. Table 1 shows a harmonic analysis of tides in the ocean and at the center of Peel Inlet (before and after the recent redredging of Mandurah Channel). Before dredging, the predominantly diurnal ocean tides were attenuated to 9.8% in Peel Inlet, whilst the smaller, and higher frequency, semi-diurnal tides were reduced to 5.7% of their ocean value. Dredging has increased these values to 15.0% and 6.8% respectively. However, the estuary remains microtidal with a mean range of only 0.07 m. Tides in Harvey Estuary do not respond well to harmonic analysis presumably because of various non-linear effects (mainly within the channel from Peel Inlet) and effects of wind stress (to be discussed later).

The basins of the estuary are small compared to the wavelength of a standing diurnal wave and so there is little tidal attenuation along a basin of fairly uniform cross-sectional area, such as Harvey Estuary, so that the surface moves approximately as a horizontal plane. However, near to the channel entrance in a circular basin such as Peel Inlet a surface slope is required to overcome both inertial and frictional forces. Neglecting these effects the tidal elevation ξ in a basin connected only to the ocean with elevation ξ_o is given by

$$A \frac{d\xi}{dt} = -\gamma(\xi - \xi_o) \tag{1}$$

where A is the surface area of the estuary, and γ is a constant which relates the flux of water in the channel to the elevation difference across its ends. The right-hand side of Eq.(1) is a linear approximation to the non-linear flow within the channel. It can be replaced by other forms which may be more realistic; the only real physical difference is some interesting singular behavior at the zero flow condition in the channel (e.g. Hearn _et al_., 1985). If

$$\begin{aligned} \xi_o &= \sin\omega t, \\ \xi &= \nu \sin(\omega t - \phi), \\ \nu &= [1 + (\omega t)^2]^{-1/2}, \\ \tan\phi &= \omega\tau, \\ \tau &= A/\gamma, \end{aligned} \tag{2}$$

where τ is the characteristic exchange time for the basin and ocean levels to reach equilibrium by flow through the channel; such flow may be a consequence of riverflow or changes in ocean level.

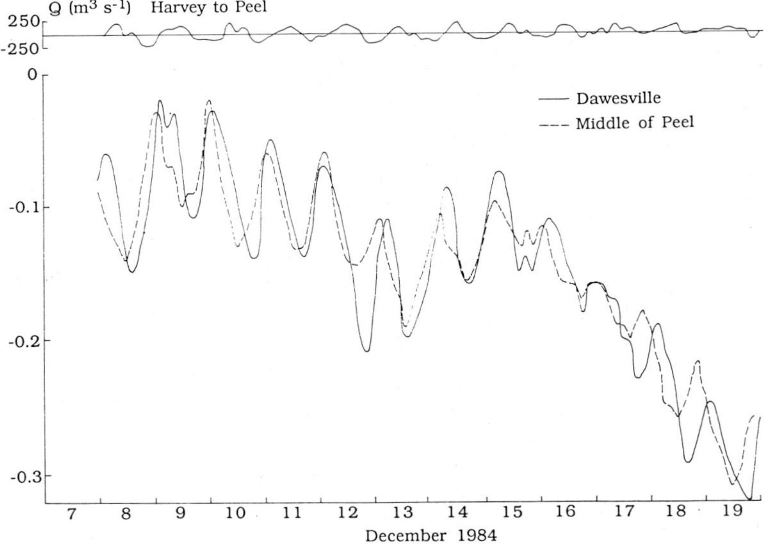

FIGURE 3. Comparison of water elevation in Harvey Estuary (Dawesville) and the middle of Peel Inlet. Also shown is the flow Q along the channel from Harvey Estuary to Peel Inlet based on the rate of change of mean water level at two stations at either end of Harvey Estuary.

For an estuary with two basins of area A_1 and A_2 joined by a channel with flux constant γ_2 in which the first basin is joined to the ocean by a channel with flux constant γ_1

$$\nu_1^2 = \frac{(1 + \omega^2\tau_2^2)}{(1-\omega^2\tau_1\tau_2)^2 + \omega^2\{\tau_2 + (1+\beta)\tau_1\}^2}$$

$$\beta = A_2/A_1$$

$$\tau_i = A_i/\gamma_i \quad ; \quad i = 1,2 \tag{3}$$

$$\tan \theta_1 = \omega\tau_1 \frac{1 + \beta + \omega^2\tau_2{}^2}{1 + \omega^2\tau_2^2(1 + \beta\tau_1/\tau_2)}$$

$$\nu_2/\nu_1 = [1 + (\omega\tau_2)^2]^{-1/2}$$

$$\tan(\theta_2 - \theta_1) = \omega\tau_2$$

For Peel-Harvey Estuary $\tau_2 << \tau_1$ because the interconnecting channel is relatively short, and so to a first approximation, it behaves like a single basin with characteristic time $\tau = (1 + \beta)\tau_1$. Values of ν for the estuary can be deduced from the tidal data which are presented in Table 1. The approximation $\nu \sim 1/(\omega t)$ which follows from Eq.(2) for $\nu << 1$, is in accord with the observed pre-dredging values of ν deduced from Table 1, of 0.06 (semi-diurnal) and 0.1 (diurnal) if $\tau = 1.6$ days.

Table 1. Amplitude and phase of the main astronomical tidal components in the ocean and middle of Peel Inlet (before and after the 1987 dredging of Mandurah Channel). The maximum range is defined as twice the sum of amplitudes of all components and represents an optimum spring tide. The mean range is a yearly average of the daily astronomical tidal range.

	Amplitude (mm) (Phase in degrees) Components						Range (mm)	
	M_2	S_2	N_2	O_1	K_1	P_1	max	mean
Ocean	53 (289)	48 (290)	15 (338)	118 (289)	169 (303)	51 (295)	908	493
Peel Inlet 1986 (before dredging)	3.8 (002)	1.5 (003)	1.3 (042)	12.8 (011)	16.0 (022)	4.4 (352)	80	45
Peel Inlet 1988 (after dredging)	4.8 (339)	2.2 (042)	0.9 (005)	18.4 (357)	25.5 (021)	6.8 (354)	118	69

Comparisons of the two-dimensional model and data for flow through the interconnecting channel suggests that the tidal amplitude in Harvey Estuary is about 60% of that in Peel Inlet and this requires $\tau_2 \sim 0.2$ days. For $\tau_2 << \tau$, Eq.(3) gives

$$\nu_1 \sim \frac{1}{\omega\tau}\left[1 - \frac{\beta\tau_2}{(1 + \beta)\tau}\right] \tag{4}$$

For a single basin with $\omega\tau >> 1$, i.e. $\nu_1 << 1$, Eq.(2) shows that $\theta \rightarrow \pi/2$ and this result applies to Peel Inlet (since $\tau_2 << \tau_1$). The phases shown in table 1 confirm that the (diurnal) estuarine tide lags the ocean tide by about 90° (6 hours). For the above estimate of the value of τ_2, Eq.(3) gives $\theta_2 - \theta_1 \sim 52°$(3.5 hours) which is consistent with Fig. 3.

If $\nu << 1$, in a single basin, tidal flow through the channel from the ocean is essentially independent of the estuarine tide. The tidal prism ν AR $\sim R\gamma/\omega$ (where R is the range of the ocean tide) is then just the volume of water which can flow through the channel in half the tidal cycle with the estuarine level fixed, it is independent of the area, and geometry of the basin. If the cross-sectional area of the channel is denoted by α, the channel constant γ can be expressed as,

$$\gamma = \alpha/T \tag{5}$$

where T is some constant with the dimension of time. If a is the ocean tidal amplitude, the amplitude of the tidal velocity in the channel is a/T. The value of γ which leads to τ = 1.6 day is equivalent to T ~ 1 s for Mandurah Channel, and so the amplitudes in table 1, give velocities of 0.2 to 0.5 m/s, which is consistent with observations (Hodgkin et al., 1980). Thus the channels in the estuary have strong tidal currents in spite of the basins being microtidal; this is due to the large ratio of basin surface area to channel cross-section. T is proportional to the length of the channel.

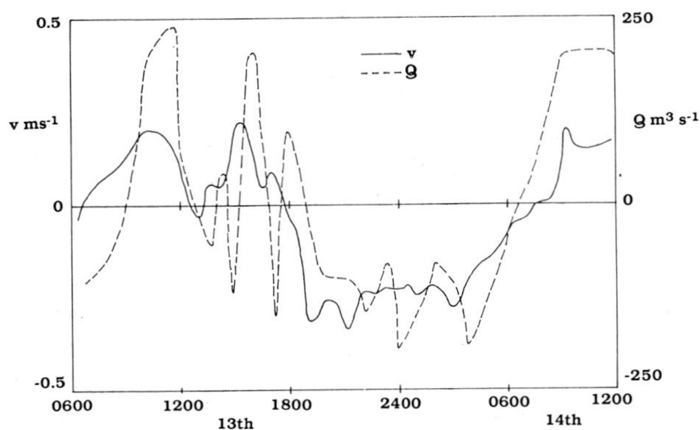

FIGURE 4. Full line shows the velocity measured at the center of the Harvey to Peel channel (using a Marsh-McBirney electromagnetic current meter mounted on a stand with mean currents determined every 30 minutes using a ten minute average). The broken line is the volume flow Q calculated from the rate of change of mean water level in Harvey Estuary deduced from tide gauges at Dawesville and Ford.

Figure 4 shows 30 hours of measured current in the inter-connecting channel between the basins together with the volume flux based on tidal gauges at two stations at either end of Harvey Estuary; it establishes that about 50% of the flux is carried by the deep portion of the channel since this has a cross-sectional area of about 300 m². For linear bottom friction, and ignoring acceleration and inertial terms the depth-average velocity is simply proportion to depth so that flux varies as the square of depth. The sill at the sides of the channel has maximum depth of 1 m over a width of 2 km whilst the channel is 3 m deep over about 100 m; this produces a maximum effective cross-section (ratio of total flux to current in the center of the deep channel) of 660 m² which is in accord with Fig. 4.

The ocean tides (Table 1) are dominated by the tidal components O_1 and K_1; semi-diurnal components are much smaller. If K_1 had a period of exactly one cycle per day the time of the (diurnal) high waters at spring, and neap tides, would be unchanged through the year. However, K_1 has 366 cycles per year and so the times of high water become two hours earlier each month. Fig. 5 shows the annual variation of the time of high water in the ocean outside Peel-Harvey Estuary. Because K1 is phase-locked to the annual cycle i.e. its phase is always about the same value at midnight on 1st January, the annual variation shown in Fig. 5 is virtually the same each year. Since the tides in Peel Inlet are about six hours later than the ocean, Fig. 5 shows that high water occurs at night during the first half of the year, i.e. an ebbing tide during darkness in summer and autumn; this is important to the migration of crab larvae and prawns. Therefore, major dredging can change the phase of the estuarine tide by up to 6 hours (for a basin connected directly to the ocean) and so displace the period of night-time ebb by up to 3 months with potentially serious ecological effects.

IV. Barotropic Models

Hydrodynamic models have been implemented for Peel-Harvey Estuary. The finite-difference models, which are detailed by Hearn and Hunter (1987), consist of a fully non-linear two-dimensional model (with drying algorithms) and a linear three-dimensional (sigma co-ordinate) model. Both models use an empirical bottom friction parameter which is designed to allow for the effect of ambient surface waves. This is of major importance in an ultra-shallow estuary and is akin to the original work of Grant and Madsen (1979). The sigma co-ordinate model uses a linear bottom slip condition and a constant vertical eddy viscosity. Both models incorporate riverflow, precipitation and evaporation into the continuity equation.

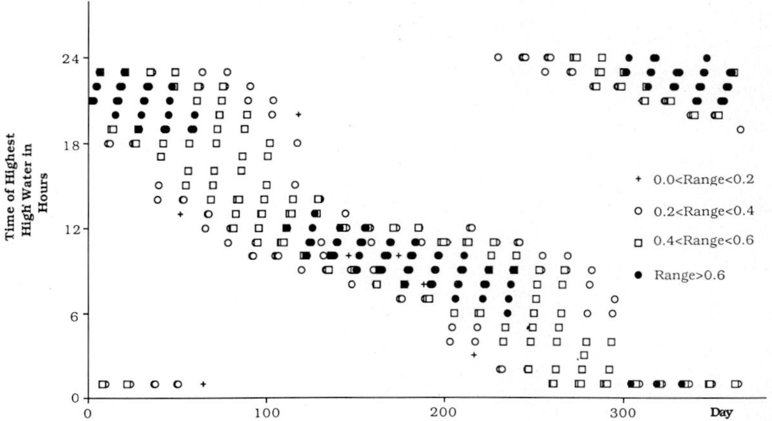

FIGURE 5. Variation of time of daily highest water level in ocean with day of the year for various tidal ranges (in meters).

Because of the difference in spatial scales and flow regimes between the open basins and channels it is not practical at present to attempt to model both simultaneously (although nested models could be used). For this reason, the Peel to Harvey channel was approximated by a simple linear connection between individual spatial cells in Peel Inlet and Harvey Estuary. This involves a single parameter γ_2 as in the analytic model of the last section. Similarly the ocean is connected linearly to a cell within Peel Inlet using a parameter γ_1. These two parameter

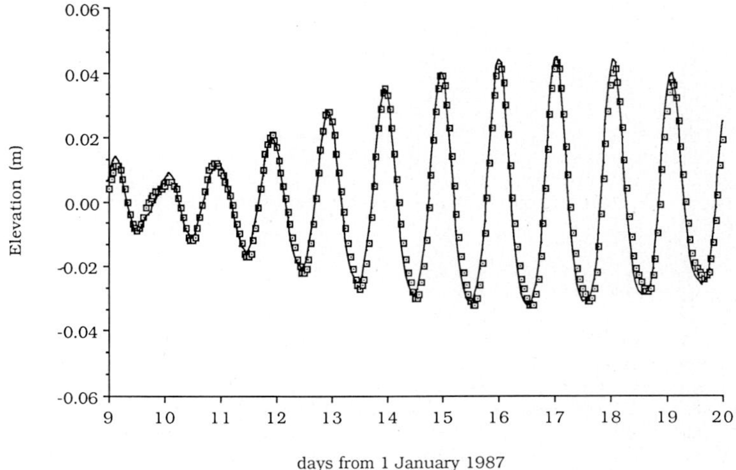

FIGURE 6. Comparison of two-dimensional model elevation (solid line) and harmonic tidal predictions (points) for the center of Peel Inlet (days from 1 January 1987).

s allow the model to be calibrated against tidal data. From the simple analysis of astronomical tides in the last section, initial estimates are $\gamma_1 \sim 950\ m^2/s$, $\gamma_2 \sim 3200\ m^2/s$ using areas of 75 and 56 km^2 for Peel Inlet and Harvey Estuary, respectively.

Figure 6 shows the surface elevation at the center of Peel Inlet as derived from the two-dimensional tidal model when driven by the astronomical ocean tide . Also shown is the predicted astronomical tide in Peel Inlet; these astronomical tides are based on the harmonic constants given in Table 1 (prior to dredging). The value of the channel flux parameter has been adjusted to give a best fit at $\gamma_1 = 700\ m^2/s$. This value is close to that deduced from the previous section which suggests that the simple model gives a reasonable representation of the tide; the small differences are due to phase and elevation changes across the basins. The revised value of the characteristic time is $\gamma_1 = 1.2$ days i.e. $\gamma = (1 + \beta)\tau_1 = 2.2$ days. Comparison between the modeled and observed tides shows that the value of γ_1 has been increased by dredging to $1050\ m^2/s$ i.e. $\gamma = 1.4$ days.

Model tidal currents corresponding to ocean high water i.e. maximum flow into the estuary are shown in Fig 7. The maximum tidal excursion in Peel Inlet is 1.6 km radially from Mandurah Channel, and in Harvey Estuary 700 m at the northern end.

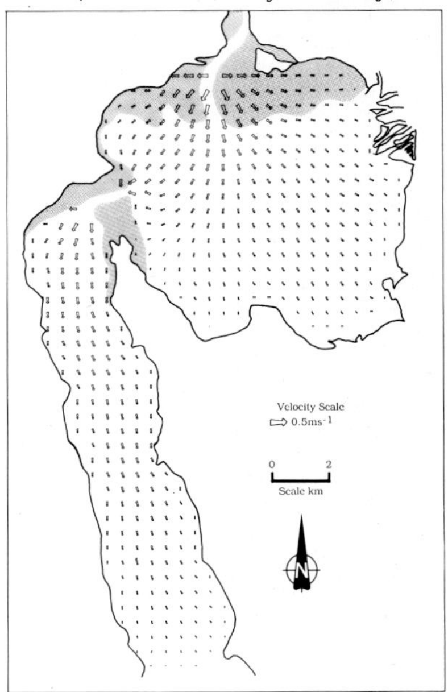

FIGURE 7. Currents from two-dimensional tidal model corresponding to ocean high water at spring tide on 17 January 1987. To optimise the display of velocities of greatly varying magnitude the arrow lengths are proportional to the cube root of speed.

V. Barometric Tides

Evident in Fig. 3 is a typical low frequency variation in estuarine water level resulting from 'barometric' ocean tides. These arise from meteorological changes such as pressure variations and wind forcing. They are spread over the 'weather-band' with periods from one to 10 days or more, and have a range of up to 0.3 m. They are important in estuaries like Peel-Harvey because they are a significant fraction of the estuarine depth and are better able to penetrate into the estuary than astronomical tides because of their lower frequency.

Using the value of $\gamma = 2.2$ days derived in section IV for the characteristic exchange time between the estuary and ocean (prior to dredging), Eq.(2) gives the relative basin amplitude ν for a barometric tide with a period of 5 days as 0.34. Dredging has significantly increased this value to 0.49. The simple analytic model represented by Eq.(1) was applied to ocean tidal residues over a period of one year. For $\gamma = 2.2$ days, 8.3 estuarine volumes were exchanged whilst for $\gamma = 1.4$ days this was significantly increased to 20.2 volumes. This is to be compared with the astronomical annual tidal exchange of 11.8, and 18.1 volumes, respectively.

VI. Wind Driven Motion

The barotropic models (section IV) can be used to determine the wind-driven flows in Peel-Harvey Estuary. Because of the ultra-shallow nature of the basins, bottom friction has a major effect. The model bottom frictional stress used in these models has the form,

$$F = \rho C_d u_f u \tag{6}$$

where ρ is density, C_d bottom drag coefficient ($\sim 2.5 \times 10^{-3}$), u_f is the background velocity due to ambient surface waves and u is the depth-averaged water speed (Hearn and Hunter, 1987). The time taken to reach a steady-state (in which a constant wind-stress is balanced by bottom friction and surface set-up) is of the order of the frictional damping time,

$$\tau_f = \frac{h}{C_d u_f} \tag{7}$$

where h is a mean depth.

Figure 8 shows the topographic gyres produced by the two-dimensional model due to a typical daytime seabreeze from the southwest. Water moves windward in the shallower margins and returns along the deeper central regions. There are three distinct gyres in Peel Inlet and a further pair in each of the topographic sub-

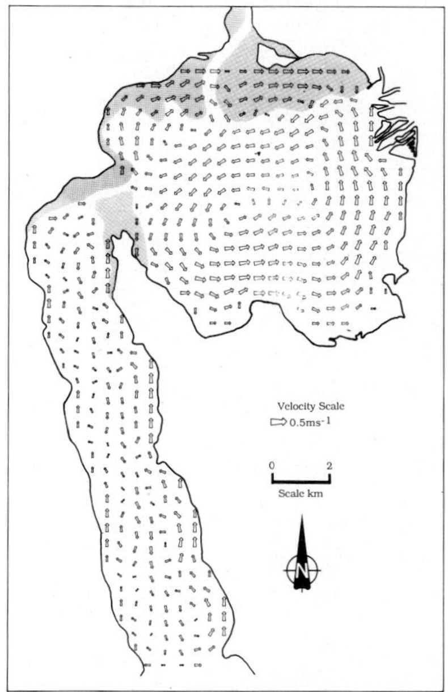

FIGURE 8. Currents from two-dimensional model due to a southwesterly wind at mid-day The speed and direction of the wind had followed a typical sea-breeze cycle for several days being easterly at 3 m/s from 23.00 hours until 09.00 hours and then swinging to the south-west and increasing to a maximum speed of 8 m/s at 15.00 hours. A cubic velocity scale is used as in Fig. 7.

basins within Harvey Estuary. These model results are derived with a value u_f= 0.1 m/s (τ_f = 1.7 hours). This value was chosen from model runs with various values of u_f as being the value of uf which gives a mean speed over Harvey Estuary which is the best fit to a linear regression of drogue and wind speeds (Hearn et al., 1986). These drogue data were collected over 10 days in late spring under moderate sea-breeze conditions; in reality u_f is not constant but should be adjusted for prevailing wave conditions (Grant and Madsen, 1979).

In very shallow basins (in which bottom friction is a dominant term) the conventional two-dimensional bottom friction law Eq.(6) may introduce serious errors into model determinations of currents and wind setup (e.g. Davies, 1988). For the high Ekman numbers encountered in such estuaries (~1 in Peel-Harvey Estuary), the set-up determined from conventional friction may be in error by up to 50% (Simons, 1980). This can be overcome by using a three-dimensional model such as a sigma co-ordinate model (Hearn and Hunter, 1987) the results of which are shown in Fig. 9; u in Eq.(6) is then the speed in the bottom layer. An alternative to using a three-dimensional model is to adopt the single relaxation approximation for bottom friction in a two-dimensional model (Hearn and Hunter, 1988; Hunter and Hearn, 1989). This exactly reproduces the steady-state currents and surface elevations for any profile of vertical eddy viscosity (from which vertical current shear can then be computed). It is ideally suited to ultra-shallow basins where the steady-state is a fair approximation to the actual wind-induced motion.

The vertical shear associated with surface water moving downwind and returning in the bottom layers can be observed in Fig. 9. Topographic gyres continue to exist in the depth-average motion but are much less obvious in the currents at any particular level (which is the circulation mapped by drogues). In a closed basin which has a perfectly flat bottom there is no steady-state vertically-averaged wind motion; only a vertical overturning flow exists in which surface water moves downwind and returns in the lower water column. Hunter and Hearn (1987) have analysed the comparative contribution of vertical overturning and depth-averaged motions to the wind-induced circulation. A speed, V_T, is defined representing total wind-driven transport per unit cross-sectional area of the basin, with V_L denoting the depth-averaged contribution. The ratio (V_L/V_T) is modeled for an infinitely

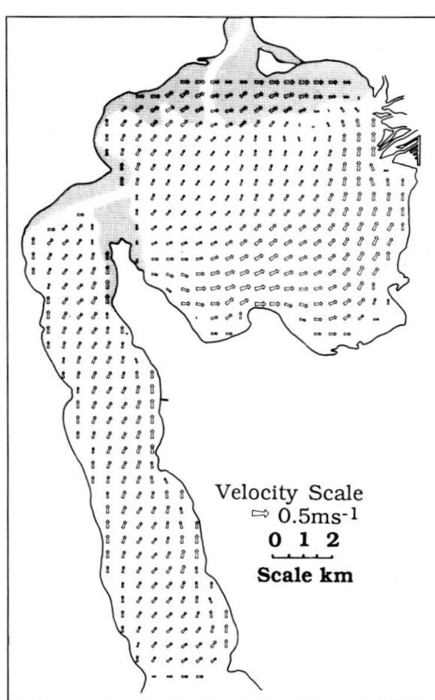

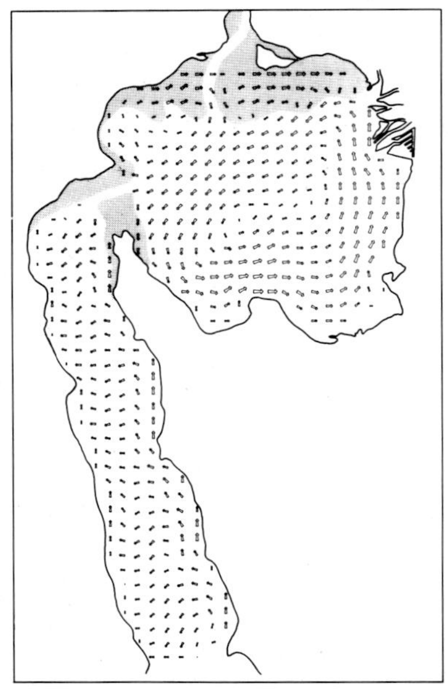

FIGURE 9. Currents in surface (left) and bottom (right) layers of three-dimensional model corresponding to the same southwesterly wind conditions as Fig. 8; a cubic velocity scale is used as in Fig.7.

long basin assuming a cross-sectional bathymetry

$$h = h_{max} \left[1 - \left(\frac{x}{B}\right)^n\right] \tag{8}$$

in which B is the half-width of the basin, h_{max} is the maximum depth, and coordinate x is measured perpendicular to its central axis. For Harvey Estuary, Hunter and Hearn (1987) show that $n \sim 2.8$, so that $(V_L/V_T) \sim 0.8$, whilst for Peel Inlet, $n \sim 1.1$, $(V_L/V_T) \sim 0.9$. So, within the confines of that model, most of the transport is due to the two-dimensional motion. However, models of wind-driven circulation in ultra-shallow estuaries predict that the marginal areas of the basins generate most of the down-wind transport. If that is invalid, due for example to the effects of vegetation, algae etc., the basin is effectively much flatter, the value of n is greatly reduced and so (V_L/V_T) is decreased.

VII. Wind Induced Tides

Because of the ultra-shallow nature of the basins, surface set-up by wind stress can be large (and a significant fraction of water depth). The order of magnitude of this effect can be assessed from a balance between wind stress and the pressure gradient created by the gradient of surface elevation ξ,

$$\rho_A C_W W^2 = \rho\, hg\Delta\xi/L \tag{9}$$

where ρ_A is the density of air, C_W the surface drag coefficient, W the wind speed, ρ the density of water, h the estuarine depth and $\Delta\xi$ the change in elevation along the length L of the basin (measured parallel to the wind direction). This balance does not apply precisely because of the influence of bottom friction, bathymetric variations and the Coriolis force; it is also affected by the acceleration of the water column but in ultra-shallow estuaries a steady-state tends to be reached in a time τ_f which is sufficiently short to make the effects of acceleration negligibly small. Then, since the system is linear,

$$\Delta\xi = \beta C_W \frac{sL}{gh} W^2 \;, \quad s \equiv \rho_A/\rho \tag{11}$$

where β is a coefficient, and according to Eq.(9), $\beta \sim 1$. Thus for the Harvey Estuary with L = 20 km, h = 1.5 m, W = 7 m/s (annual mean), $\Delta\xi$ = 0.13 m; in a storm with W = 20 m/s, Eq.(9) predicts $\Delta\xi$ = 1 m so that non linear effects would become important. These large values of $\Delta\xi$ are just a consequence of the inverse proportionality of $\Delta\xi$ to h.

Figure 10 shows differences in elevation at two tide gauge stations in the Harvey Estuary as a function of WU where U is the component of wind speed along the estuary; the strongest winds are northwesterly (during storms). The tide gauge stations are located at the extreme ends of the estuary (Ford and Dawesville as shown in Fig.2); wind data are from a Woufle anemometer mounted on a navigation pole at Dawesville. Fig. 10 shows a linear regression line which has a correlation coefficient of 0.83; if the mean depth h is taken as 1.3 m (Hodgkin et al., 1980). the slope of the line is equivalent to $\beta C_W = 0.93 \times 10^{-4}$ in Eq.(10). Harvey Estuary is an especially easy basin in which to detect the linear trend Eq.(9) in water elevation data because its seiching period is short compared to the astronomical tidal period. At high Ekman numbers, β is mainly dependent on bathymetry; using the long-basin model of Hunter and Hearn (1987) for Harvey Estuary gives $\beta = 0.81$ so that $C_W = 1.2 \times 10^{-3}$ which is in agreement with the value from Garratt (1977) for the mean annual wind speed of 7 m/s.

A wind stress from the south increases the water elevation at the northern end of Harvey Estuary, relative to the elevation at its southern end, and this is denoted by $\Delta\xi$. In Peel Inlet there will also be an elevation gradient due to wind stress. Suppose that under a southerly wind the elevation at Roberts Bay (on the Peel Inlet end of the Peel-to-Harvey channel) drops by $\delta\xi$ relative to the elevation at the station called 'The Chimneys' in Fig. 2; this is located near the Mandurah Channel which runs from Peel Inlet to the ocean. If the wind is suddenly turned-on the volumes of the two basins will initially be conserved and so the level at the northern end of Harvey Estuary will rise in relation to that at Roberts Bay; similarly Chimneys will have a higher level than the ocean. As a result, water will flow from Harvey Estuary into Peel Inlet and there will also be a flow (of different magnitude) to the ocean. In principle these flows will continue until a steady state is reached in which the levels (excluding astronomical tides) at Dawesville and Roberts Bay are equal, and similarly levels are equalized along the Mandurah Channel.

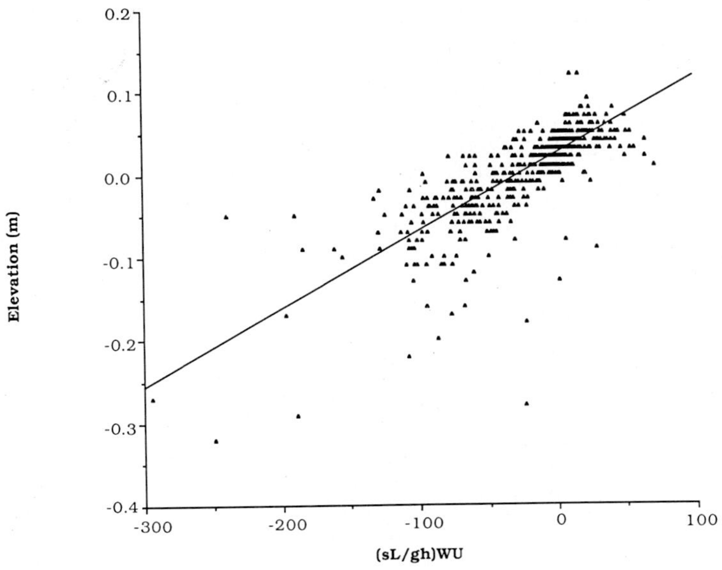

FIGURE 10. Elevation difference in Harvey Estuary versus wind stress using notation of section VII; W is wind speed and U the component of wind velocity along the basin from south to north.

The mean water level in Harvey Estuary therefore drops under a southerly wind (provided that $\Delta\xi > 0$) so that the basin as a whole experiences a wind-induced tide, in addition to the surface elevation gradient. Based on the values of $\Delta\xi$ given above for mean wind strengths the variation in water level would be estimated to be of order 0.1 m which is comparable with the maximum astronomical tidal range. Wind-induced tides are associated with the quasi-diurnal sea breeze cycle and there are strong components that have periods which span the weather-band and also produce variations in water level on seasonal time-scales.

VIII. Salinity Variations

Figure 11 shows the changes in salinity of Peel-Harvey Estuary for the year May 1981 to April 1982; it represents a fairly typical annual variation. The three stations shown in Fig. 11 are at the center of Peel Inlet and the northern and southern ends of Harvey Estuary. The lower part of Fig. 11 provides the cumulative flow from the Harvey, Serpentine and Murray Rivers expressed as a multiple of the total estuarine volume (at mean water level).

Southwest Australia normally experiences drought conditions throughout summer (January to March) with rainfall mainly concentrated in winter (June to September). Riverflow as shown in Fig. 11 follows a similar pattern; a very atypical summer storm occurred in January 1982. During the summer drought evaporation is high and so the estuary is usually hypersaline in autumn. Winter riverflow reduces the salinity to a minimum of a few parts per thousand in late winter after a total seasonal riverflow of several estuarine volumes. This shows that the mixing of riverwater must be almost totally complete; freshwater also enters the estuary through drains. During spring, riverflow gradually ceases and so the estuary returns to marine salinity by the middle of summer; this did not occur in the summer of 1982 due to the atypical summer storm and as a result the estuary did not became hypersaline in the autumn of that year.

Figure 11 shows that there is a salinity difference between Peel Inlet and Harvey Estuary and a salinity gradient along the axis of Harvey Estuary. The simplest possible model of the estuary (Black and Hearn, 1987) assumes complete vertical mixing and an essentially one-dimensional path from the southern end of Harvey Estuary, through Peel Inlet to the ocean. Salt flow through the open basins is approximated by a constant dispersion coefficient. The model includes the riverflows, mean monthly evaporation, and precipitation. Fig. 11 shows the predictions of such a model for the salinity at the southern end of Harvey Estuary. The dispersion coefficient in the two basins is 50 m^2/s and the conductance of each channel has been derived from mean tidal exchange with a finite return coefficient. The model reproduces the gross features of the annual salinity variation.

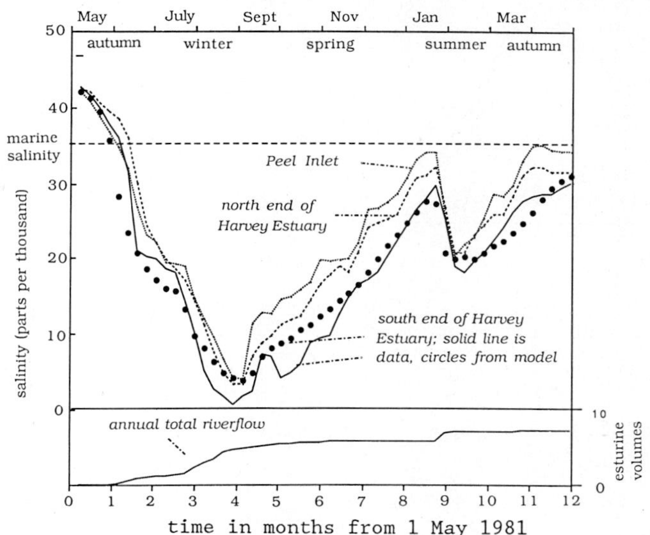

FIGURE 11. Annual variation of salinity for 1981/82 at three stations in Peel-Harvey Estuary. The lower portion of the figure shows the cumulative riverflow

Most winter rainfall occurs during storms as frontal systems cross the coast from the Indian Ocean. There is usually a period of a week or so between major fronts and this consists of calm high pressure conditions. The rivers flowing into Peel-Harvey Estuary rise and fall very rapidly after these storm events so that riverflow occurs in short pulses often under calm conditions. Most of the annual rainfall is associated with a few very major storms each of which may bring a flow of order one or more estuarine volumes. There is considerable variance between the relative storm flows of the rivers because of the spatial separation of their catchment areas.

Figure 12 shows the 1981 to 1982 annual cycle of stratification (difference between top and bottom salinities) at stations in Peel Inlet, and the northern, and southern ends of Harvey Estuary. The southern station shows peak stratification after each storm; the peaks are greatest in autumn when Harvey River water flows over the existing hypersaline estuarine water. The station at the northern end of Harvey Estuary also shows stratification peaks after the storms and it is noted that after the salinity minimum has been passed (in late winter) it is normally the most stratified station in the Estuary until the following autumn. This is often the result of the more saline water from Peel Inlet passing through the channel to Harvey Estuary and then forming a thin bottom salt wedge. Figs 13 to 15 show salinity contours in Harvey Estuary immediately after a storm and during one of the calms between storms; the salt wedge from Peel Inlet is observed passing down Harvey Estuary and becoming pinched-off. This bottom layer is very thin so that the wedge carries only small quantities of salt (relative to the salt content of the basin at marine salinity). Fig. 14 shows a sloping halocline due to the usual decreasing barometric tide following a storm.

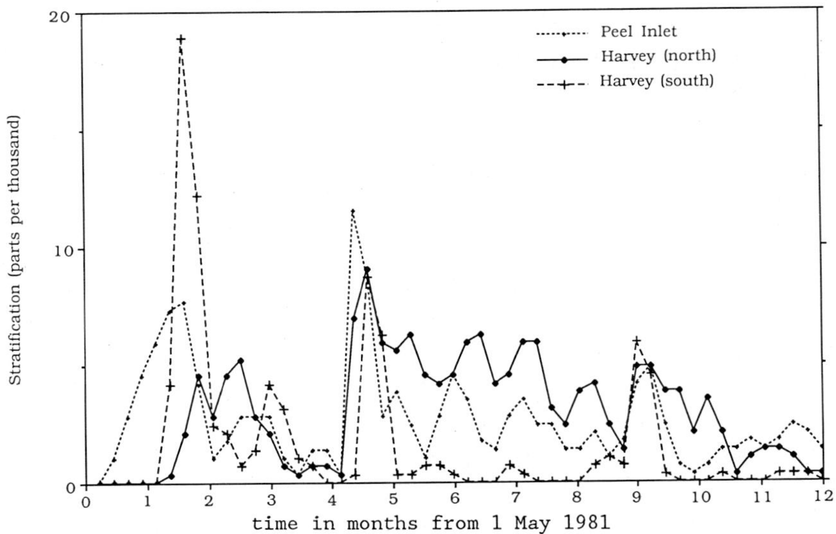

FIGURE 12. Difference between bottom and surface salinities at three stations in Peel-Harvey Estuary for 1981/82 based on weekly sampling in early morning.

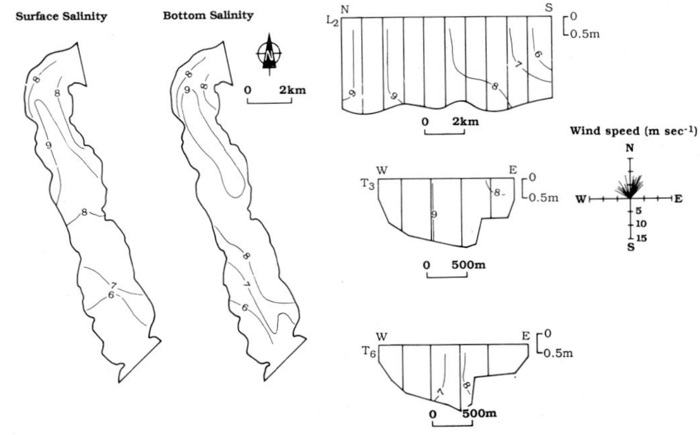

FIGURE 13. Salinity contours over the surface and bottom layers, and an axial and two transverse vertical sections of Harvey Estuary on 3 August 1983. Hourly wind source vectors are shown for the previous 24 hours. L2 is an axial section. T3 and T6 are transverse sections at the northern and southern ends of Harvey Estuary, respectively.

Harvey Estuary also stratifies due to the axial salinity gradient in the absence of wind mixing. Apart from the winter cycle of fronts and intervening anticyclonic systems, the wind pattern on this coast is dominated by the daily sea breeze cycle. During night-time, winds are weak and stratification occurs; this coincides in spring with the astronomical tidal flood from Peel Inlet (compare the discussion of section III and fig. 5). The salt which is transported by the night-time gravitationally driven motion in the basin is vertically mixed next day by the seabreeze. This mechanism makes a major contribution to the longitudinal dispersion coefficient of Harvey Estuary which is used in the model discussed above. The dispersion transport has therefore been adapted to include a dependence of the coefficient on salinity gradient (Chatwin, 1976; Smith, 1976; Thatcher and Harleman, 1981; Kraneuberg, 1986) with some minor improvement in model predictions.

IX. The Mandurah Front

There is a sharp density front between ocean and estuarine water which can propagate back and forth through Mandurah Channel with the diurnal astronomical tide. Mixing across this front is the major process by which salt enters Peel-Harvey Estuary. The front is unable to penetrate into Peel Inlet during strong riverflow and presumably this occurs when the daily riverflow is of the same order as the tidal prism. Using data from Table 1, this gives a combined critical flow of order 100 m^3/s; such flows occur after storms. Dredging has increased this value by 50% so effectively enhancing the input of salt. Movement of the front into the estuary is also strongly affected by the barometric tides. The position of the front can, to a good first approximation, be determined by driving the barotropic models of section IV with measured riverflows, winds, and ocean water levels.

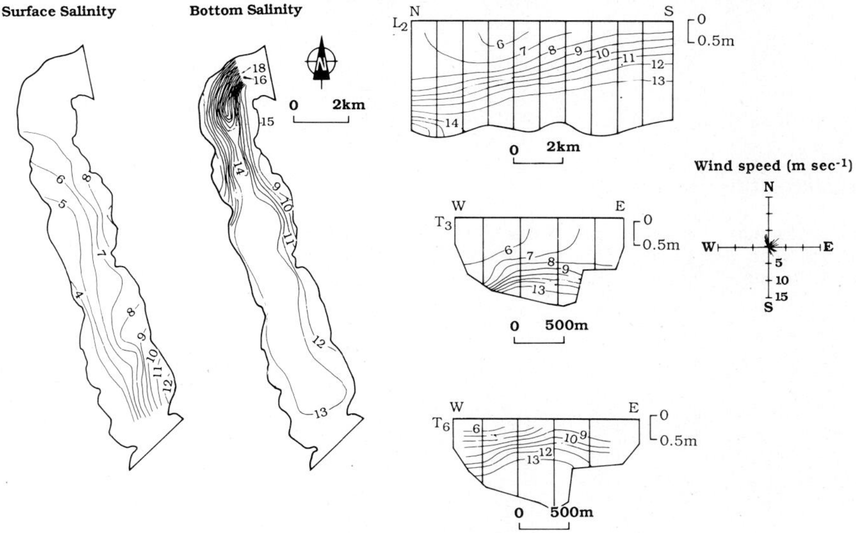

FIGURE 14. As Fig. 13 on 28 July 1983.

When the front enters Peel Inlet it remains clearly visible over much of the flood period of a tidal cycle. The maximum astronomical spring tidal prism (from Table 1) is 10.5 x 10^6 m^3 (now increased to 15.5 x 10^6 m^3) which is equivalent to a semi-circular front of radius 2.1 km measured from the location marked 'The Chimneys' in Fig. 2. After dredging this radius is increased to 2.7 km. Fig. 16 shows that a thin bottom layer of marine water spreads forward of the front across the center of Peel Inlet and towards the channel to Harvey Estuary. This layer seems to fill the deeper portions of the basin. The volume of water necessary to fill the bottom of Peel Inlet to the 2 m contour is 0.9 x 10^6 m^3 which is small compared to the maximum tidal prism. To fill to the 1.5 m contour requires 8.7 x 10^6 m^3 which would produce a bottom layer of maximum thickness 0.8 m. The limited thickness of the bottom marine layer observed in Peel Inlet is evidently due to the restricted volume of marine water which enters the estuary behind the front.

The movement of salt into Peel-Harvey Estuary which is modeled macroscopically in Fig. 11 really involves a very delicate balance of riverflow, astronomical, and barometric tides; coupled to these factors is the dynamics of the bottom layer which is affected by wind mixing. The annual cycle can receive major perturbation from single meteorological events such as the summer storm shown by Fig. 11 or a large barometric tide.

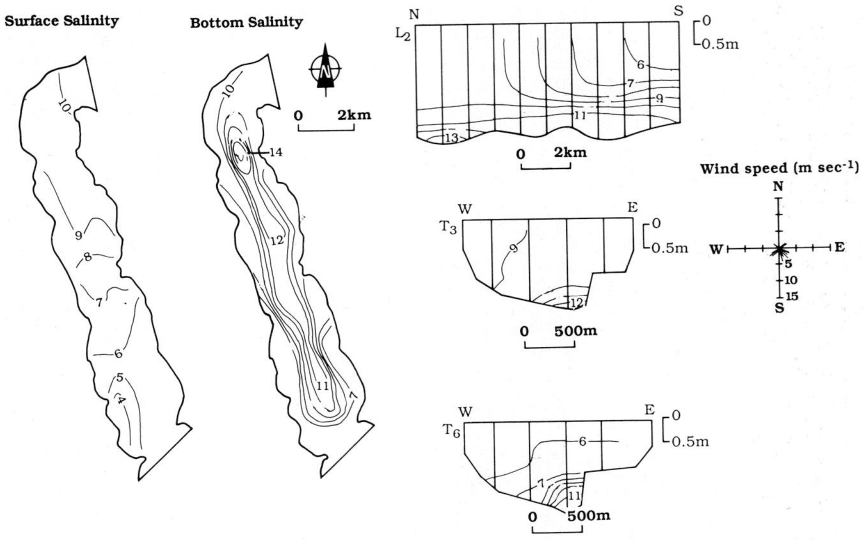

FIGURE 15. As Fig. 13 on 31 July 1983.

X. Eutrophication in Harvey Estuary

Harvey Estuary has become highly eutrophic since 1970 with large blooms of the blue-green alga Nodularia spumigena occurring in the spring and early summer of most years. An extensive study of the estuary (Hodgkin et al., 1980) has shown that

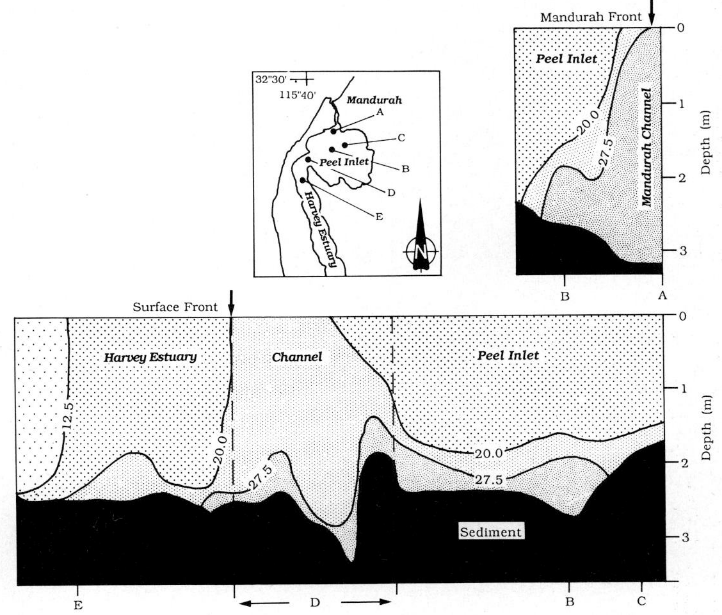

FIGURE 16. Vertical section of salinity in Peel-Harvey Estuary on 21 October 1988.

these blooms result from excessive phosphorus input to the estuary in riverwater; this is due to the application of fertilizers to the sandy soils of the coastal-plain portions of the catchments. Because of the long flushing times the nutrients accumulate in the water and sediments of the estuary (McComb _et al._, 1981). Lukatelich (1987) has shown that phosphorus is taken up by diatom blooms and absorbed by the sediments. As water temperatures increase during spring _Nodularia_ are able to grow by utilising phosphorus within the water body much of which has been released from the sediments under anoxic conditions. The resultant _Nodularia_ bloom only collapses when the estuarine salinity approaches marine values in early summer; at the height of the bloom the _Nodularia_ biomass contains a significant fraction of the phosphorus input to the estuary during the previous winter.

The cycling of phosphorus between the water column, sediments, and plankton is controlled by physical conditions within the estuary and a numerical model has been developed which helps quantify the processes which are involved. The estuary is treated as three horizontal layers representing the surface, and bottom waters, and sediments respectively. Each of the three rows of boxes in Fig. 17 belongs to one of these layers. The cross-hatched box at the bottom of Fig. 17 represents the sediments; there are analogous boxes in the other layers to allow for sediment which is suspended in the water column by wind mixing (Gabrielson and Lukatelich, 1985). Associated with the surface and bottom _Nodularia_ are boxes representing intracellular phosphorus which is an important mechanism by which _Nodularia_ can store phosphorus for later use (Lukatelich, 1987; Huber, 1984). The model determines the stratification of the water column by calculating the total salt load of the surface and bottom waters and there is a box in each of these two layers for salt.

Many of the processes indicated by arrows in Fig. 17 are controlled by the current value of physical variables such as wind strength (W), stratification (S), solar radiation (L), longitudinal density gradient along the axis of the estuary (G) and the concentration of suspended sediment (P). Such dependence is indicated in the figure by the appropriate letter adjoining an arrow e.g. one of the factors which control photosynthesis is solar radiation (L) whilst vertical mixing of oxygen depends on wind strength (W) and stratification (S). The variables W, L and G are input to the model externally (as functions of time) whilst S is calculated from the two salt boxes; P (as a function of time) is determined from a differential equation involving W. Temperature also appears as an externally input variable which controls many of the processes. The arrow from surface to bottom salt bearing a G represents gravitationally-driven circulation which tends to increase vertical stratification.

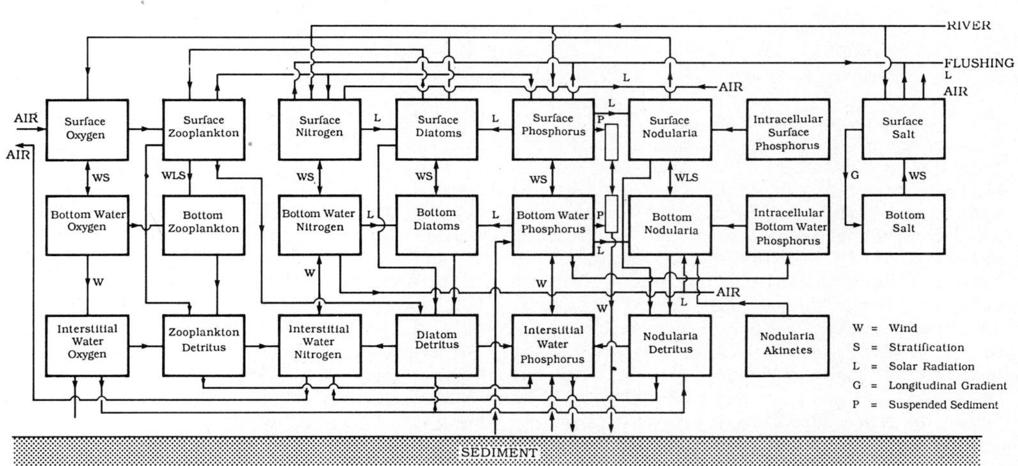

FIGURE 17. Major compartments and processes in the combined physical-biological model of Harvey Estuary.

The model integrates the rate equations through the light-dark cycle. Light intensity throughout the water column is calculated from a diurnal solar input at the surface with attenuation by all plankton, their detritus and suspended sediments uniformly distributed over a wind-mixed surface region (which may extend to the bottom). Photosynthesis can occur in both the surface and bottom layers; the layers are only used to simplify the physical mixing processes.

The model subjects Nodularia to both ordinary vertical mixing due to wind-induced turbulence and a buoyancy/sinking rate controlled by light intensity. Thus Nodularia can sink at night and rise during daylight to optimise both the uptake of phosphorus released by sediments and their rate of photosynthesis.

Figure 18 shows a sample output from the model corresponding to a single storm event which gives high riverflow for days 1 to 2, with all nutrient and plankton loads initially zero; all the loads in Fig. 18 are expressed in terms of phosphorus. The pulse of riverflow inputs 5 tonnes of phosphorus into the estuary. The diatom bloom peaks after about 20 days at which time its further growth is nutrient limited; it produces a zooplankton bloom which peaks at about 30 days. This is followed by Nodularia growth which becomes significant after about 60 days. The model was run with a constant salinity of 10 o/oo at a temperature of 25°C.

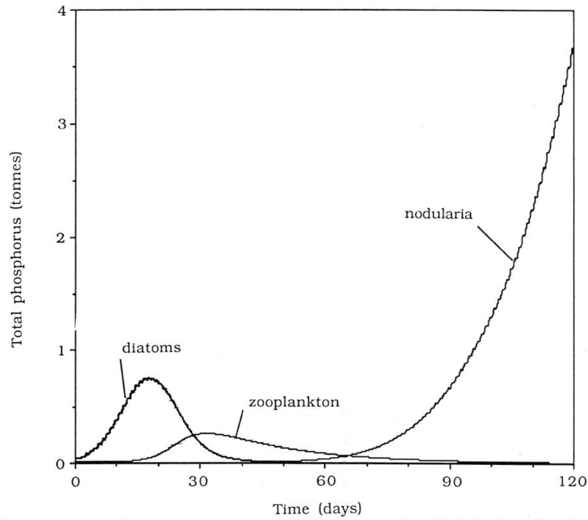

FIGURE 18. Sample output from the physical-biological model of Harvey Estuary.

The wind strength in the sample model run was maintained at zero for days 1 and 2 corresponding to the normal calm during riverflow. The estuary was assumed to have been initially well-mixed by the storm winds and the model showed that it became increasingly stratified by the riverflow. The normal wind conditions on this coast consist of a diurnal seabreeze cycle (throughout the year), and this was adopted by the model after day 2. A longitudinal salinity gradient of 10^{-6}/m was used, i.e. 20 ^{o}/oo over a distance of 20 km, and the model showed a vertical mixing time reaching a minimum in late afternoon and a maximum at night (as the estuary restratifies); note the diurnal modulation of Fig. 18 due to the light-dark and seabreeze cycles. The model results show that the phosphorus released from sediments due to anoxic conditions at night is mostly taken up by Nodularia (which sink in the absence of light and wind mixing) and then converted to organic phosphorus. Thus at the end of the model run Nodularia have taken up most of the original phosphorus input by the riverflow.

XI. Discussion

The dynamics of Peel-Harvey Estuary are probably representative of many estuaries of its type i.e. ultra-shallow basins interconnected by narrow channels. The tidal range in such estuaries is often indicative of the extent to which the channels have been dredged; in their natural state the estuaries are usually microtidal. However, tidal currents in the channels are always strong so that nearly all water exchange between the basins is due to barotropic tidal motions. Because the basins are ultra-shallow, a very significant exchange is induced by barometric tides i.e. changes in ocean level which result from variations in atmospheric pressure, wind stress, and shelf waves. For Peel-Harvey Estuary these tides are coupled to the major storm events which bring most of the riverflow.

Wind mixing can change the basins from a vertically mixed to stratified state in a few hours. Structures related to baroclinic effects are always ephemeral but basin-scale salinity gradients are stable over seasonal time-scales. Wind stress produces steady-state closed circulation cells with considerable vertical shear; the motion is not simply two-dimensional. It also creates surface elevations which are a significant fraction of water depth and induce flow between the basins; quasi-diurnal variations are produced in Harvey Estuary by the sea-breeze cycle.

Fronts between basins, or a basin and the ocean, form downstream of the strong tidal advection through the channels (into microtidal basins). From these fronts are formed thin bottom layer intrusions.

The biological dynamics of the estuary is directly related to the physical dynamics. A particularly important aspect is the fast stratification/mixing cycle which in the case of Harvey Estuary allows rapid cycling of phosphorus. Flushing times of the basins are very slow and material which does not leave the estuary during strong riverflow is only removed by slow dispersion through the basins and mixing across the inter-basin and basin-ocean fronts.

Acknowledgements

Financial support was provided by a Marine Science and Technology Grant, the Australian Research Grant Scheme and Waterways Commission of Western Australia. The authors are grateful to Professor A. J. McComb for his constant help and encouragement, and to Dr. J. R. Hunter for his extensive input to the project particularly in regard to the hydrodynamic models. Tidal data were supplied by the WA Department of Marine and Harbours.

XII. References

Black, R. E., and C. J. Hearn, 1987: Management of a eutrophic estuary: modelling the effects of a new outlet to the sea. 8th Australian Conference on Coastal and Ocean Engineering (The Institution of Engineers, Australia), 284-287.

Black, R .E., R. J. Lukatelich, A. J. McComb, and J. E. Rosher, 1981: Exchange of water, salt, nutrients and phytoplankton between Peel Inlet, Western Australia and the Indian Ocean. Aust. J. Mar. Freshw. Res., **32**, 709-720.

Chatwin, P. C., 1976: Some remarks on the maintenance of the salinity distribution in estuaries. Est. Coast. Mar. Sci., **4**, 555-566.

Davies, A., M., 1988: On formulating two dimensional vertically integrated numerical models with an enhanced representation of bed stress. J. Geophys., **93**, 1241-1263.

Gabrielson, J., O., and R., J. Lukatelich, 1985: Wind-related resuspension of sediments in the Peel-Harvey Estuarine System. Est. Coast. Shelf Sci., **20**, 135-145.

Garratt, J. R., 1977: Review of drag coefficients over oceans and continents. Mon. Wea. Rev., **105**, 915-929.

Grant, W. D., and O. S. Madsen, 1979: Combined wave and current interaction with a rough bottom. J. Geophys. Res., **84**, 1797-1808.

Hearn, C. J. and J. R. Hunter, 1987: Modelling wind-driven flow in shallow systems on the southwest Australian coast. In: Numerical Modelling: Applications to Marine Systems. (Ed J Noye), pp 47-58 Elsevier/North Holland, Amsterdam.

Hearn, C. J. and J. R. Hunter, 1988: A new method of describing bottom stress in two-dimensional hydrodynamical models of shallow homogeneous seas and estuaries. App. Math. Modelling, **12**, 573-580.

Hearn, C. J., J. R. Hunter, J. Imberger, and D. van Senden, 1985: Tidally induced jet in Koombana Bay, Western Australia. Aust. J. Mar. Freshw. Res., **36**, 453-479

Hearn, C. J., R. J. Lukatelich, and J. R. Hunter, 1986: Dynamical models of the Peel-Harvey estuarine system. Environmental Dynamics Report ED-86-152. The University of Western Australia, Nedlands 6009, Western Australia

Hodgkin, E. P., P. B. Birch, R. E. Black, and R. B. Humphries, 1980: The Peel-Harvey Estuarine System Study (1976-1980). Report No 9, Department of Conservation and Environment, Perth, Western Australia.

Huber, A. L., 1984: Physiology and ecology of cyanobacteria in the Peel-Harvey Estuarine System, Western Australia with particular reference to Nodularia spumigena. Ph.D. thesis, The University of Western Australia. Nedlands 6009, Western Australia.

Hunter, J. R. and C. J. Hearn, 1987: Lateral and vertical variations in the wind-driven circulation in long, shallow lakes. J. Geophys. Res., **92**, 13106-13114.

Hunter, J. R. and C. J. Hearn, 1989: The single relaxation approximation for bottom stress in two-dimensional hydrodynamic models of shallow seas. Continental Shelf Res., **9**, 465-478.

Kranenburg, C. 1986: A time scale for long-term salt intrusion in well-mixed estuaries. J. Phys. Oceanogr., **16**, 1329-1331.

Lukatelich, R. J., 1987: Nutrients and phytoplankton in the Peel-Harvey Estuarine System, Western Australia. Ph.D. thesis, The University of Western Australia, Nedlands 6009, Western Australia.

McComb, A. J., R. P. Atkins, P.B. Birch, D. M. Gordon, and R. J. Lukatelich, 1981: Eutrophication in the Peel-Harvey Estuarine System, Western Australia. In: Nutrient Enrichment in Estuaries. (Eds B J Neilsen and L E Cronin), pp 332-342. Humana Press, New Jersey

Meetam, P., and C. J. Hearn, 1988: A tidal model of Lake Songkhla, Thailand. Department of Mathematics Report, The University of Western Australia, Nedlands 6009, Western Australia.

Simons, T. J., 1980: Circulation models of lakes and inland seas. Can. Bull. Fish. Aquat. Sci., **203**, 146pp.

Smith, R., 1976: Longitudinal dispersion of a buoyant contaminant in a shallow channel. J. Fluid Mech., **78**, 677-688.

Thatcher, M. L. and D. R. F. Harleman, 1981: Long-term salinity calculation in Delaware estuary. J. Environ. Eng. Div. (Am. Soc. Civ. Eng.), **107**, 11-27.

On the Summer Circulation in New York Bight and Contiguous Estuarine Waters

Alan F. Blumberg[1] and Boris Galperin[1,2]

[1]HydroQual, Inc.,
1 Lethbridge Plaza,
Mahwah, NJ 07430

[2]Department of Marine Science
University of South Florida
St. Petersburg, FL 33701

ABSTRACT

The summer average circulation in the New York Bight, Long Island Sound and New York Harbor complex is investigated using a three-dimensional hydrodynamic circulation model. Unlike previous studies that have only concentrated on decoupled parts of the system, the present study emphasizes the importance of the coupling between them. Here the mean summer circulation is determined by mean temperature and salinity fields prescribed from climatological data. In addition, the model is forced by free surface elevation gradients along the northeastern and southwestern cross-shelf open boundaries; the elevation along the shelf-break is set to zero. The mean climatological summer wind stress is also imposed. A sensitivity study was conducted for the circulation and the hydrodynamic structure of the bight as functions of the mean wind and the cross-shelf elevation gradients. All the conclusions are based on simulations during which the whole system has spun up and reached a steady-state (after two days), wherein the total flows through the northeast and southwest cross-shelf open boundaries were approximately equal. Without winds and density structure, the circulation is a consequence of conservation of potential vorticity and follows f/H isopleths. When the density field is imposed, the circulation changes markedly. Clearly evident in the surface currents is the Hudson plume emanating out of New York Harbor. The plume makes an anticyclonic turn as it flows out of the harbor onto the continental shelf and forms a small eddy of size of the baroclinic Rossby radius of deformation. The near-bottom flow is onshore along the Hudson Canyon. This circulation is extremely sensitive to the wind. The southwesterly winds reverse the flow close to the New Jersey shore northeastward and lead to a significant intensification of the fresh water eddy structure. The water column in the Long Island Sound is strongly stratified and the classical, two-layer estuarine circulation pattern prevails. However, this pattern is also sensitive to the winds. Even moderate winds can change the magnitude and direction of the total transport through the East River. An analysis of the model results using mean summer forcing functions shows a total transport of approximately 160 m^3/s from Long Island Sound to New York Harbor.

Coastal and Estuarine Studies, Vol. 38
R. T. Cheng (Ed.)
Residual Currents and Long-term Transport

I. Introduction

The deterioration of the water quality in the different areas of the New York Bight complex has attracted the attention of scientists and federal, state and local authorities. From the analysis of data it has become clear (HydroQual, 1989) that local water quality is only partially determined by local conditions; the hydrodynamic communication between interconnected areas of the New York Bight complex is equally important. It has also been progressively realized that hydrodynamic circulation is the carrier for many of the processes determining water quality and a firm understanding of the circulation needs to be established prior to conducting water quality studies.

The strategy of a combined and interacting hydrodynamic-water quality modeling study has been conceptualized for the New York Bight system (HydroQual, 1989). Specifically, hydrodynamic analyses are necessary for defining the transport into and through New York Bight due to various hydrographical (runoff, estuarine circulation), meteorological (surface wind, surface heat and salt fluxes), open ocean (large scale ocean circulation) and internal (density gradients) forcing mechanisms. The general purpose of these analyses are to provide estimates of circulation and mixing in New York Bight which are directly usable in subsequent studies of contaminant fate. Estimates of the circulation are also necessary to determine pollutant inputs to the bight from flows along the continental shelf and New York - New Jersey Harbor.

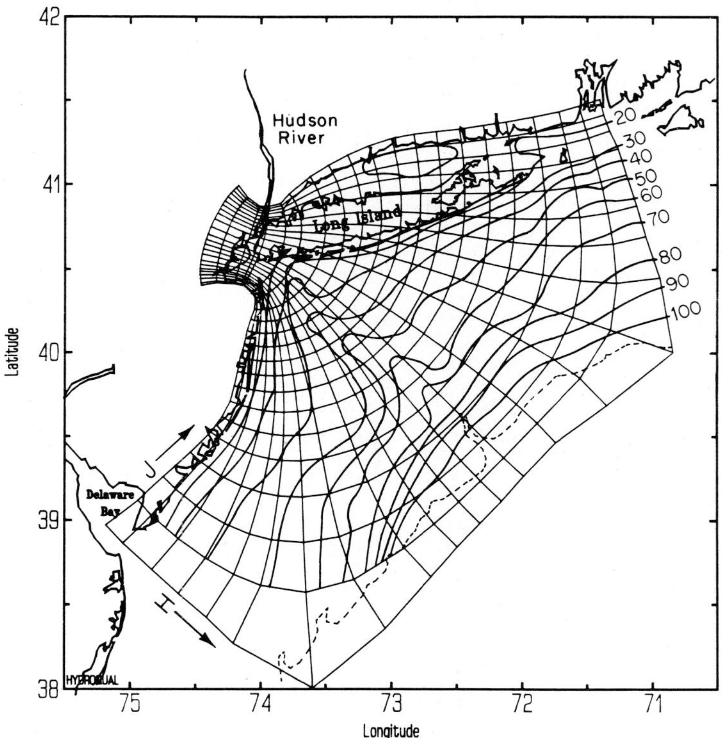

FIGURE 1. The study area of the New York Bight, New York Harbor, Long Island Sound and Hudson River along with the bathymetry, computational domain and curvilinear orthogonal grid.

This work presents the results of the hydrodynamic analysis, in which the circulation in New York Bight, Long Island Sound and the New York - New Jersey Harbor complex (Figure 1) is determined by application of a three-dimensional hydrodynamic circulation model. Despite the importance of this natural water system, it has not been studied as a whole in the past; previous investigations have concentrated on different parts of the system in a decoupled fashion. The present study highlights the mean summer circulation, with mean temperature and salinity fields prescribed from the historical data base. A series of numerical sensitivity experiments will be used to understand the nature of the system's response to the various forcing mechanisms.

II. Previous Observational and Modeling Studies

A comprehensive discussion of the circulation in the New York Bight and surrounding coastal waters can be found in Beardsley and Boicourt (1981). The numerous measurements which have been made in the region make it among the most extensively investigated regions in the world. Recent experiments such as the Nantucket Shoals Flux Experiment (Beardsley et al., 1985; Brown et al., 1985), Shelf Edge Exchange Processes (SEEP), Warm Core Rings Program (Joyce, 1985) and Middle Atlantic Shelf and Rise Study (MASARS) have added a wealth of information on the circulation. Based upon the analysis of Beardsley and Boicourt (1981), Beardsley, Boicourt and Hansen (1976), Beardsley and Winant (1979), Han, Hansen and Galt (1980) and Mayer, Han and Hansen (1982) as well as the drifter-bottle studies of Bumpus (1973), it is apparent that the characteristic feature of the circulation is a southwestward mean flow along the shelf of roughly 2 to 10 cm/s with a mean transport of the order of 2.5x105 m^3/s. The average currents generally increase in magnitude offshore and decrease with depth. This flow is thought to be driven by an alongshore sea level slope of approximately 2cm/100 km at the shelf break presumably due to the large scale circulation offshore. The observational data and analysis described in Chapman et al. (1986) do not support this hypothesis for the genesis of the alongshore pressure gradient. They suggest that the flow is forced by upstream inflow from the Scotian Shelf and/or downstream outflow at Cape Hatteras where the shelf water may be entrained by the Gulf Stream. Superimposed on these mean currents are fluctuations due to the wind stresses. Interaction with the deep ocean at the shelf-break helps to keep the shelf water on the shelf. Flow off the shelf break exhibits large variability associated with the meanders and warm-core eddies of the Gulf Stream.

Simulations of the long-term mean circulation in the Middle Atlantic Bight have been conducted by Blumberg and Mellor (1980), and discussed subsequently by Blumberg and Oey (1985). The major conclusion from their three-dimensional simulations was that the presence of the Gulf Stream is an important component of the circulation along the continental shelf in the New York Bight. Moreover, a series of observations reported by Garvine et al. (1988), using buoy tracking, intensive hydrography, satellite thermal imagery and moored current meters have shown that eddy-like features are almost always present along the shelf break. The eddies typically appear either as plumes of less saline shelf water that protrudes into slope water, curling "backwards" opposite the direction of the mean shelf flow, or as eddies with warmer, saltier slope water in their cores, partly or wholly encircled by the plumes. While these plumes can potentially produce vigorous cross-shelf transfer of heat, salt and nutrients, their influence is for the most part, restricted to the outer shelf regions.

Diagnostic studies of the mean barotropic and baroclinic circulation in New York Bight have been conducted by Han, Hansen and Galt (1980) and Hopkins and Dieterle (1983, 1987) using a steady, vertically integrated vorticity equation approach. Despite the simplicity of the approach, essentially solving for the free surface elevation only, the investigators were able to capture the major features of the circulation and its response to wind forcing. A more detailed discussion of their results will be presented in the sections which follow.

Oey et al. (1985b,c,d) have undertaken a detailed study of the New York Harbor using the fully three-dimensional, time dependent primitive equation model developed by Blumberg and Mellor (1980, 1987). A fine spatial resolution (500m x 500m) enabled Oey et al. to avoid the use of any horizontal viscosity.

They demonstrated good agreement with data on tidal and subtidal time scales. Those authors emphasize that the inclusion in their model of a portion of the adjacent continental shelf will further improve their predictions. Highlighting this aspect, one can note that in the existing body of literature dealing with the New York Bight and adjacent estuarine complex, the entire system has not yet been treated as a coupled entity; the present work is the first step in that direction.

III. Computational Domain

The computational grid superimposed on the bathymetry map is shown on Figure 1. The curvilinear, shoreline fitted, orthogonal coordinate system used in this study contains 23 points in the I and 20 points in the J directions, with coarser resolution towards the shelf-break and finer resolution in the New York Harbor area. Although the resolution in Long Island Sound is relatively coarse, it will be shown later that the model provides insight into the circulation and transport processes in Long Island Sound and the East River. The Hudson River is parameterized as two-dimensional, laterally-integrated body of water, similar to the methodology used in the studies by Oey et al. (1985a).

IV. Description of the Model

The model used for this study is the time-dependent, three-dimensional estuarine and coastal circulation model developed by Blumberg and Mellor (1980, 1987). This model has been used in numerous studies on various important natural water systems; among them one should mention works by Blumberg and Mellor (1983) on the South Atlantic Bight, Oey et al. (1985b,c,d) on the Hudson Raritan estuary, Galperin and Mellor (1989a,b,c) on Delaware Bay, River and adjacent continental shelf and Blumberg and Goodrich (1990) on Chesapeake Bay. In all these studies the model performances were assessed via extensive comparisons with data and a confidence has been established that the predominant physics are realistically reproduced by the model.

Briefly, the model consists of the ensemble mean, primitive equations describing conservation of mass, momentum, temperature and salinity in the hydrostatic and Boussinesque approximations. An important enhancement of the present model over that used earlier is the use of a horizontal, orthogonal, curvilinear coordinate system. The full set of the equations is given in Blumberg and Herring (1987); for the present, diagnostic, study only the continuity and Reynolds equations were solved. These equations, re-written in a more conservative form than in Blumberg and Herring (1987), are:

The Continuity Equation:

$$\frac{\partial}{\partial \xi_1}(h_2 U_1) + \frac{\partial}{\partial \xi_2}(h_1 U_2) + h_1 h_2 \frac{\partial W}{\partial z} = 0 \,, \tag{1}$$

The Reynolds Equations:

$$\frac{\partial}{\partial t}(h_2 U_1) + \frac{1}{h_1}\frac{\partial}{\partial \xi_1}[(h_2 U_1) U_1] + \frac{1}{h_2}\frac{\partial}{\partial \xi_2}[(h_2 U_1) U_2] + \frac{\partial}{\partial z}[(h_2 U_1) W] + U_2\left(\frac{2U_1}{h_1}\frac{\partial h_1}{\partial \xi_2} - \frac{U_1}{h_2}\frac{\partial h_2}{\partial \xi_2} - \frac{U_2}{h_1}\frac{\partial h_2}{\partial \xi_1} - f h_2\right) = -\frac{1}{\rho_o}\frac{h_2}{h_1}\frac{\partial P}{\partial \xi_1} + \frac{\partial}{\partial z}[-\overline{(h_2 u_1) w}] + F_1 h_2 \,, \tag{2}$$

$$\frac{\partial}{\partial t}(h_1 U_2) + \frac{1}{h_1}\frac{\partial}{\partial \xi}[(h_1 U_2) U_1] + \frac{1}{h_2}\frac{\partial}{\partial \xi_2}[(h_1 U_2) U_2] + \frac{\partial}{\partial z}[(h_1 U_2) W] + U_1\left(\frac{2U_2}{h_2}\frac{\partial h_2}{\partial \xi_1} - \frac{U_2}{h_1}\frac{\partial h_1}{\partial \xi_1} - \frac{U_1}{h_2}\frac{\partial h_1}{\partial \xi_2} + f h_1\right) = -\frac{1}{\rho_o}\frac{h_1}{h_2}\frac{\partial P}{\partial \xi_2} + \frac{\partial}{\partial z}[-\overline{(h_1 u_2) w}] + F_2 h_1 \,, \tag{3}$$

where P is found from the vertical integral of the hydrostatic equation,

$$P(\xi_1, \xi_2, z, t) = P_{atm} + \rho_o g \eta + g \int_z^0 \rho(\xi_1, \xi_2, s, t) ds \,, \tag{4}$$

and ρ_o is a reference density, ρ is the density, g is the gravity acceleration, η is the free surface elevation, ξ_1 and ξ_2 are arbitrary horizontal curvilinear orthogonal coordinates, z is the vertical coordinate, h_1 and h_2 are the metric coefficients, P_{atm} is the atmospheric pressure, and f is the Coriolis parameter. The terms, F_1 and F_2, are related to small-scale mixing processes which are not resolved explicitly in the model; they are parameterized as horizontal diffusion terms. The horizontal viscosity in the present model is given according to the Smagorinsky (1963) formulation and the coefficient varied between 1 to 50 m^2/s in the simulations described below. The Reynolds stresses, $\overline{u_1w}$ and $\overline{u_2w}$, are evaluated using the level 2-1/2 turbulence closure model of Mellor and Yamada (1982) wherein:

$$-[\overline{(h_2u_1)w}, \overline{(h_1u_2)w}] = K_M \frac{\partial}{\partial z}[(h_2U_1), (h_1U_2)] , \tag{5}$$

where the eddy viscosity, K_M, is given through the use of the closure model. The level 2-1/2 closure model supplies two more prognostic equations to this model. They describe the evolution of turbulence energy and turbulence macroscale. The entire set of governing equations is then transformed using the bottom and free-surface following, σ-coordinate system:

$$\sigma = \frac{z - \eta}{H + \eta} , \tag{6}$$

where H is the depth relative to the geoid. Obviously, σ=0 at the free surface (where $z=\eta$) and σ=-1 at the bottom (where z = -H).

The resulting equations are vertically integrated to extract barotropic variables and a mode splitting technique is introduced such that fast, barotropic, two-dimensional "external mode" and slow, baroclinic, three-dimensional "internal mode" variables are calculated by prognostic equations with different time steps. The ratio of the internal and external mode time steps for this particular application was set to 4. The set of equations are approximated by a finite difference scheme using a spatially staggered "C" grid (Arakawa and Lamb, 1977). The stability criterion of the numerical scheme, the CFL condition, is similar to that described by Blumberg and Mellor (1987) as modified by Blumberg and Herring (1987) for curvilinear coordinates. The model also employs a technique described by Shuman et al. (1972) affording an increase to the time step limited by the CFL condition. The external mode time step was set to 108 s and the corresponding internal mode time step was 432 s.

The same vertical spacing in the σ-coordinate was used in this model as in the previous study of Oey et al. (1985b,c,d) wherein σ=0, -0.03125, -0.0625, -0.125, -0.25, -0.5, -0.75, -0.875, -0.9375, -0.96875, -1.0.

The model through its boundary conditions responds to surface wind stress, heat flux, salinity flux and to the specification of tidal forcing and freshwater discharges. Bottom friction is incorporated into the modeling framework by matching U_1 and U_2 to the classical near bottom, logarithmic velocity distribution.

IV.1 The Model forcing functions

The model forcing information consists of:

- cross-shelf elevation gradients;
- shelf-break elevation;
- three-dimensional fields of temperature and salinity;
- surface wind stress;
- fresh water inflow.

Since the mean summer circulation is sought here, the forcing information is also summer-averaged.

The cross-shelf elevation gradients are obtained from the study of Hopkins and Dieterle (1987). They showed that for August 1978, a typical summertime period, a gradient of 13 cm across a Narragansett Bay shelf-break section and a 11 cm gradient across a Cape May shelf-break section can produce the observed summer along-shelf flux of water. The sensitivity of the circulation to these two prescriptions will be commented upon subsequently.

The effect of an along-shelf elevation gradient imposed at the shelf-break on the barotropic circulation in New York Bight has been studied by Hopkins and Dieterle (1983). That analysis used a model that was driven externally by cross-shelf elevation gradients and by wind stresses. That study emphasized that if the adjacent deep ocean were quiescent, the zero-elevation shelf-break boundary condition would be adequate. However, if the southwestward shelf flow is assumed to be established by the along-shelf (parabathic) pressure gradient, set up by the deep ocean circulation (Csanady, 1978; Hsueh and Peng, 1978), it should be reflected in the elevation boundary condition. Observations (Beardsley <u>et al</u>., 1985; Chapman <u>et al</u>., 1986) indicate little net transport across the shelf-break, although circulation on the outer shelf can still be affected by active water mass exchange due to eddies. These eddies can either originate directly from the Gulf Stream (the warm-core rings; Bisagni, 1983; Houghton, Olson and Celone, 1986), or indirectly by breakdown of the shelf/slope front perturbed by rings (Ramp, Beardsley and Legeckis, 1983) and winds (Oey, 1986). The sensitivity of the proposed model to parabathic pressure gradient will be discussed later.

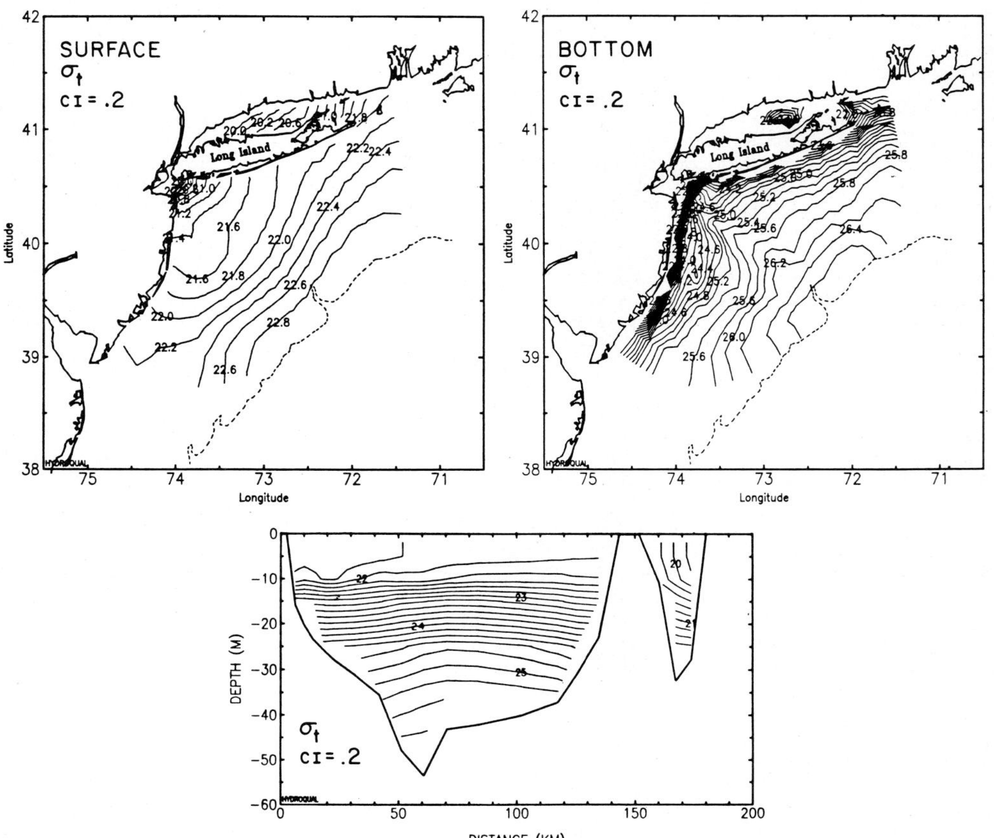

FIGURE 2. Maps of the climatological summer density distribution in σ_t units. Vertical section is along J=9.

Hopkins and Dieterle (1983) have studied the sensitivity of their barotropic model to different elevation boundary conditions, bottom friction and wind forcing. They found that the parabathic elevation gradient at the shelf-break affects the total transport through the cross-shelf boundaries. However, this effect decreased toward shore. An important implication of this result is a presumably weak sensitivity of the near-coastal circulation and total transport through the East River to the elevation boundary condition at the shelf-break.

To produce a simulation of the summer mean circulation including the effects of baroclinicity, it is necessary to prepare a data bank containing the historical thermodynamic information collected in New York Bight. The source of the data used in this study is temperature and salinity data on file at the National Oceanographic Data Center (NODC) collected through 1987. Some of the data extend back to 1897. The relatively intensive hydrographic sampling programs sponsored by the National Oceanographic and Atmospheric Administration's (NOAA) Marine EcoSystems Analysis (MESA) New York Bight project have produced reliable high quality data sets that have been incorporated into the present data base. From this data, fields averaged for July, August and September have been produced on a 6′ x 6′ latitude-longitude grid at standard analysis levels between the surface and 500m depth. Then the salinity and temperature fields are analyzed as described in Kantha et al. (1981) such that data is interpolated to every cell in the model grid. The resulting surface and bottom density fields (σ_t) are shown on Figure 2. A vertical section along J=9 is also shown on Figure 2. This figure illustrates the distinctive water masses of the New York Bight. There is a shallow (~ 5 to 10 m), strong (3 σ_t units) pycnocline typical of summer conditions. The Hudson River discharge generates significant lateral structure and its plume is clearly depicted in the density fields. In addition to the general offshore increase in density, there are several mesoscale features present. In the deeper waters of the New York Bight, the isopycnals follow contours of bathymetry as pointed out by Han, Hansen and Galt (1980). It perhaps should be remarked here that the data cannot be considered true summer averages since in any 6′ square there may be data from certain years while another nearby square may contain only data from other years. If there are long-term trends or significant interannual variations present, the data cannot be expected to represent the long-term mean.

The surface wind stress was adopted from the study of Saunders (1977); a northeastward stress with a magnitude of 0.14 dyne/cm^2 is applied at the water surface. A Hudson River summer fresh-water flow of 200 m^3/s is also imposed.

V. Results

To assess the effects of different factors on circulation and transport in the New York Bight system, the model was exercised under a variety of forcing conditions. For meaningful comparisons, one needs, at first, to establish a "benchmark case" which then can be modified by controlled changes in forcing and boundary conditions. The benchmark case for the present study includes cross-shelf elevation gradients, fresh water run-off from the Hudson River, steady density structure and wind stress. The shelf-break elevation is set to zero thus eliminating the parabathic pressure gradient. Cross-shelf elevations are distributed linearly reaching 11 cm at Cape May and 13 cm at Narragansett Bay. The combination of all these forcing components produces a southwestward mean flow with a total transport in agreement with data.

Figure 3 shows surface and bottom currents and vertical sections (along J=9) of primarily along-shore velocity and turbulent mixing (K_M) for the benchmark case. Clearly evident in the surface currents is the Hudson plume emanating from New York - New Jersey Harbor. The currents associated with this plume of fresh, lighter water make a clockwise turn as they flow out of the harbor onto the continental shelf and form a small eddy with a radius of approximately 10 km, the size of the Rossby baroclinic radius. The near-bottom flow appears to be onshore along the Hudson Canyon.

The water column in Long Island Sound is strongly stratified and a classical, two-layer estuarine circulation pattern prevails. The surface current flows oceanward along the southern bank of the sound and flows southward or southwestward closer to its northern bank. The weaker, bottom currents flow oceanward in the vicinity of the southern bank while in the middle and northern parts the bottom currents are northward and northwestward. An analysis of these flows shows a total transport through the East River of approximately 160 m^3/s from Long Island Sound to New York Harbor.

It is useful to compare this transport with the estimate given by Jay and Bowman (1975) and Bowman (1975). Based upon measurements and an analysis of the non-tidal hydraulic head across the East River, the seasonal wind and baroclincity and the differences in ebb and flood cross-sectional areas, the mean non-tidal flux of water was estimated to be 350 $m^3/s \pm 75\ m^3/s$ into New York Harbor. A major contribution to this flux was due to the tidally induced residual. Tides have been neglected in this study; however, it is anticipated that their inclusion in this modeling framework, perhaps with better grid resolution, could eliminate much of the differences between the fluxes deduced here and those of Jay and Bowman (1975) and Bowman (1975). In any case, what determines the total flux in reality is a complicated interplay of all of the different forcing mechanisms. Therefore, a realistic prediction of the total transport through the East River and of the related water quality on various time scales should be performed in the framework of a fully prognostic model which can naturally accommodate the entire set of forcing parameters.

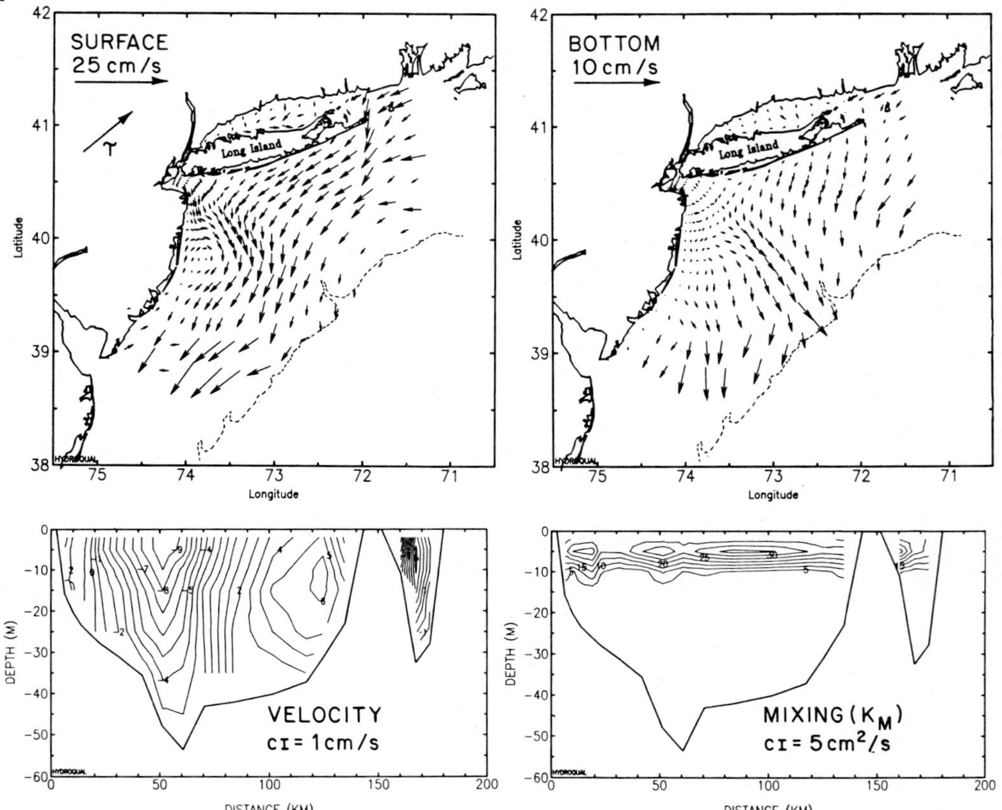

FIGURE 3. Circulation and mixing from the benchmark case (Experiment 1) using cross-shelf elevation gradients, density structure, wind stress and Hudson River inflow as forcing functions. Vertical section is along J=9.

The circulation in New York Bight is extremely sensitive to winds. The strongest wind effects are exerted in the near-shore regions and in Long Island Sound. As will be shown below, strong southwesterly winds can produce northerly flows along the New Jersey coast and reduce the eastward flow along the bight side of Long Island. Interesting enough, increasing winds can reverse the net flow of water in the East River; doubling the benchmark value of the wind stress produces a flow of some 473 m^3/s to Long Island Sound from New York Harbor. A wind of approximately 3.5 ms appears to be the threshold value controlling the direction of the net flow as determined by a theoretical analysis and confirmed through a sequence of numerical experiments. It should be remarked that the model grid resolution is somewhat coarse in these estuarine regions. Doubling the wind stress would also reverse the flow between New York Harbor and the offshore shelf region. While in the benchmark case, the flow of 375 m^3/s was onto the shelf, the flow produced by double winds was into New York Harbor with a total flux of 270 m^3/s. This wind driven response of the flux of water at the New York Harbor entrance has been well documented by Oey <u>et al</u>. (1985b,c,d).

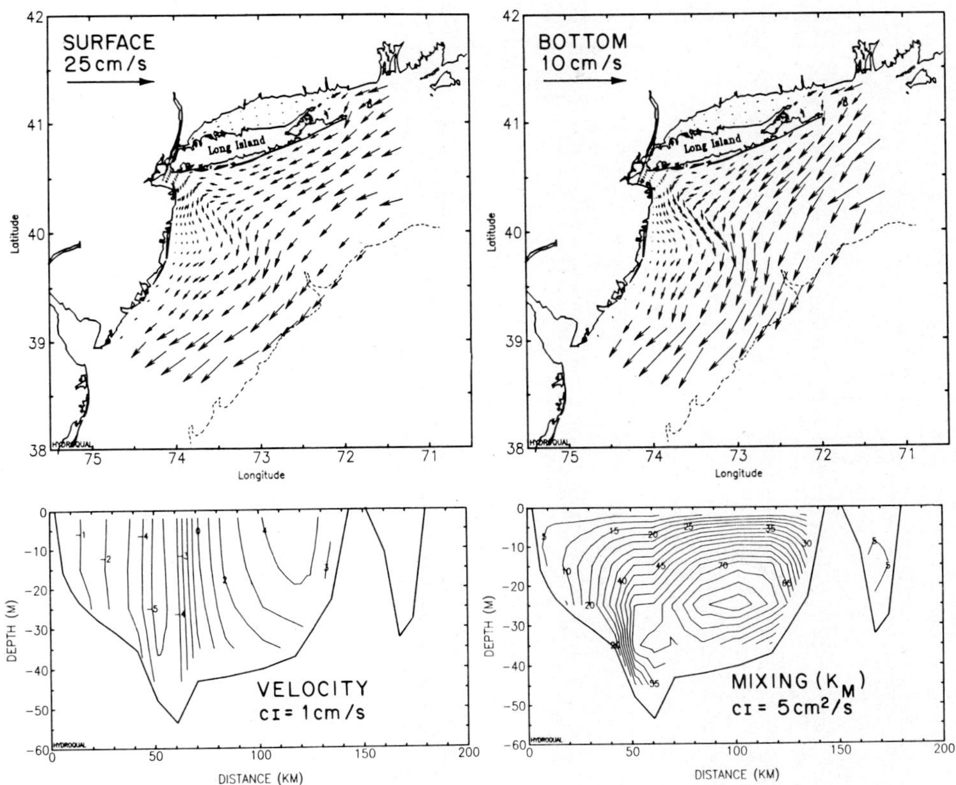

FIGURE 4. Circulation and mixing from Experiment 2. Forcing is only from cross-shelf elevation gradients. Vertical section is along J=9.

The numerical experiments designed to isolate the effects of the different forcing mechanisms on the circulation in the New York Bight complex are listed in Table 1. The benchmark case is denoted as Experiment 1. It should be noted that the winds used in Experiment 1 represent nearshore climatology. If one is interested in the circulation of the entire bight, it is probably better to use a larger value of the wind. Saunders (1977) has shown that the mean and seasonal wind stress in the Middle Atlantic Bight exhibit a significant increase in magnitude with increasing distance from the coast. Therefore, for numerical experimentation purposes, the wind stress will be taken as double that of Experiment 1.

The simplest case, Experiment 2, is that with zero wind forcing and with no density structure. The elevation along the shelf-break is set to zero; the elevation is distributed linearly in the cross-shelf directions reaching 11 cm at Cape May and 13 cm at Narragansett Bay. As shown on Figure 4, the shelf flow with these forcings is close to being in geostrophic balance; the elevation and currents follow the isopleths of f/H, as expected from conservation of potential vorticity. The total flow is southwestward, as dictated by the imposed cross-shelf elevation gradients.

The surface and bottom currents in Long Island Sound are weak. The vertical shear of the mean currents leads to the intense production of turbulence energy and vertical mixing throughout the water column as shown in the vertical sections of horizontal velocity and vertical eddy viscosity, K_M, of Figure 4.

Table 1. Catalogue of the various numerical experiments

Cross-Shelf Elevation Difference (cm)			Wind Stress	Observed Density Structure	Parabathic Shelf-Break Elevation Gradient
Experiment	Southwestern	Northeastern	dyne/cm^2		
1	11	13	0.14	Yes	0.0
2	11	13	0.0	No	0.0
3	11	13	0.0	Yes	0.0
4	11	13	0.28	No	0.0
5	11	13	0.28	Yes	0.0
6	12	12	0.28	Yes	0.0
7	13	11	0.28	Yes	0.0
8	13	15	0.28	Yes	0.0
9	11	15	0.28	Yes	0.0
10	11	13	0.28	Yes	$2x10^{-7}$
11	16	13	0.28	Yes	$2x10^{-7}$

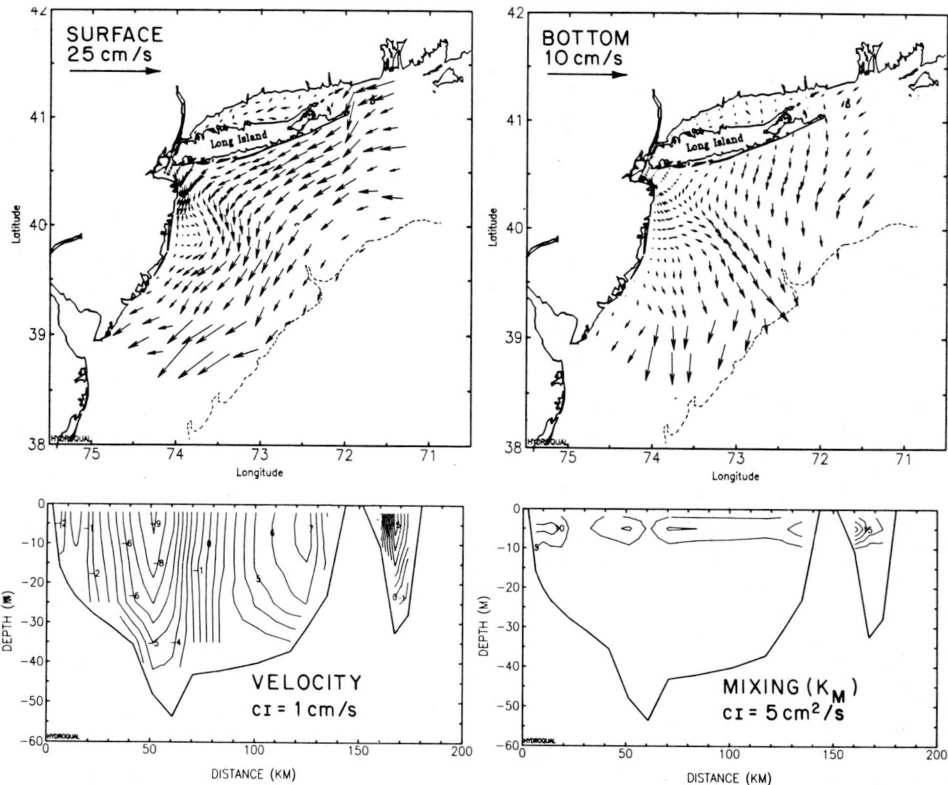

FIGURE 5. Circulation and mixing from Experiment 3. Forcing is from cross-shelf elevation gradients and climatological density fields. Vertical section is along J=9.

Figure 5 shows the results in the same format for Experiment 3, in which the density structure was added. As a result, south of New York Harbor the surface currents reveal a well defined eddy formed by fresher water emanating from the harbor. The circulation in Long Island Sound becomes quite complicated, with strong eastward surface currents in the central and northern parts and westward currents near the southern bank. Weaker bottom currents generally flow to the north, northeast. The normal velocity cross-section confirms the presence of the eddy in the vicinity of the New Jersey coast, and a stagnation region (V=0) has developed. A two-layer, classical estuarine circulation is well defined in the Long Island Sound portion of this cross-section. The vertical mixing is strongly suppressed below a depth of 10m by the imposed stable stratification.

Experiment 4 is performed with no density structure, but with the summer mean wind forcing doubled compared to the benchmark case as explained above (τ_s=0.28 dyne/cm^2, τ_s being the wind stress). Figure 6 shows the strong influence of the wind on the currents. The wind causes a flow reversal near the New Jersey shore, a feature also noted by Hopkins and Dieterle (1983) using a barotropic model forced by southwesterly winds. The circulation in Long Island Sound is affected in a most profound way: the surface currents flow westward driven by the wind while bottom currents flow eastward driven by the externally imposed southwestward shelf flow. In a way, the wind forcing alone, with no density structure, establishes a two-layer circulation pattern (with respect to the velocity profile, not density) in Long Island Sound. The net result of the wind in Long Island Sound is to change the direction of the total transport through the East River. The vertical distribution of turbulent mixing indicates that the surface wind stress generates considerable vertical mixing.

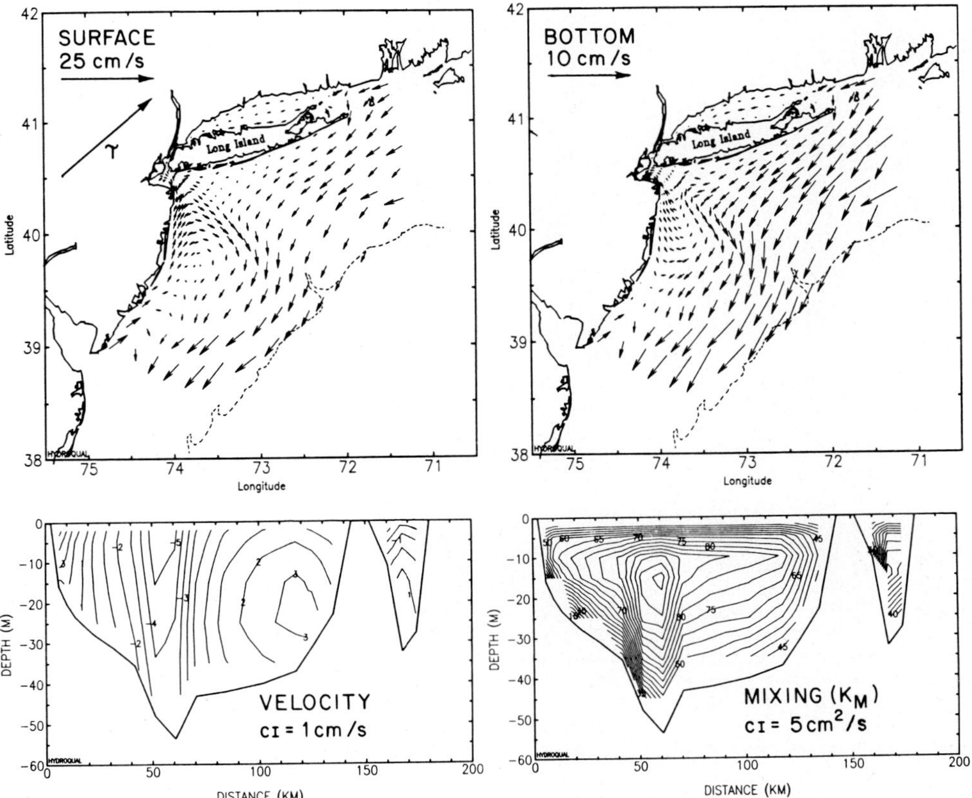

FIGURE 6. Circulation and mixing from Experiment 4. Forcing is from cross-shelf elevation gradients and wind forcing (doubled). Vertical section is along J=9.

In Experiment 5, the forcing is a combination of elevation gradients, wind stress, density structure and fresh water inflow. The only difference between this case and that of the benchmark case (Experiment 1) is the doubling of the surface wind stress. Surface and bottom currents shown on Figure 7 clearly demonstrate the Hudson plume. Compared to the benchmark case, increased wind stress causes significant intensification of the plume circulation. An additional factor enhancing the plume currents is the wind generated flow reversal near the New Jersey shore. As mentioned earlier, increased winds also reverse the total transport through East River.

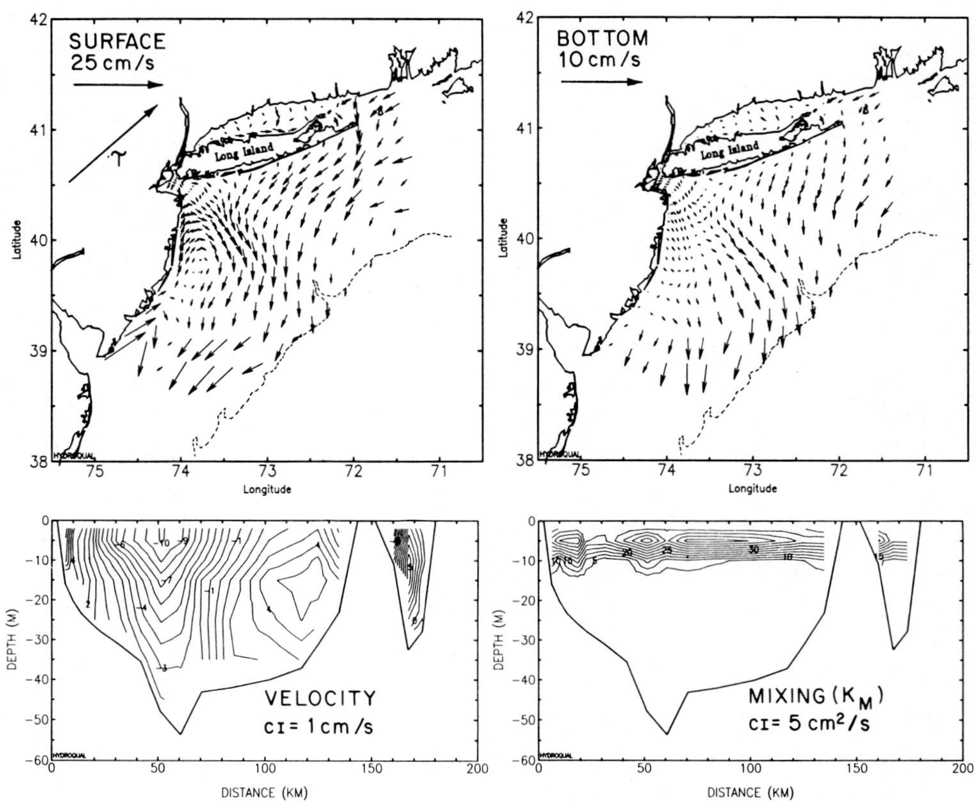

FIGURE 7. Circulation and mixing from Experiment 5. Forcing is from cross-shelf elevation gradients, density structure and winds (doubled). Vertical section is along J=9.

Figure 8 compares free surface elevations for Experiments 2 through 5. One can see that for Experiment 2, elevation isopleths follow isobaths, as dictated by the conservation of potential vorticity. For Experiment 3, the density structure lifts elevations and enhances elevation gradients in the vicinity of the New Jersey shore. When only winds are imposed, as was the case in Experiment 4, elevations are reduced and elevation gradients are significantly smoother close to the shore and almost unaffected in the deeper regions. Finally, in the benchmark case, free surface elevations reflect superposition of winds and density structure in that elevation isopleths have more complicated structure near the shore than in other runs.

Figure 9 compares Experiments 6, 7, 8 and 9 in which the free surface elevations at Cape May and at Narragansett Bay varied while the shelf-break elevation was kept at zero. Again, the wind is double that of the benchmark case. These runs illustrate the sensitivity of the circulation to cross-shelf elevation gradients. Besides modifications of the surface elevation in the vicinity of New York Harbor, the results are not much different when compared to the benchmark case. The total transport through the East River does not change significantly. Generally, the cross-shelf pressure gradient has little affect on the distribution of free surface elevation and current strength in the interior of the modeled region.

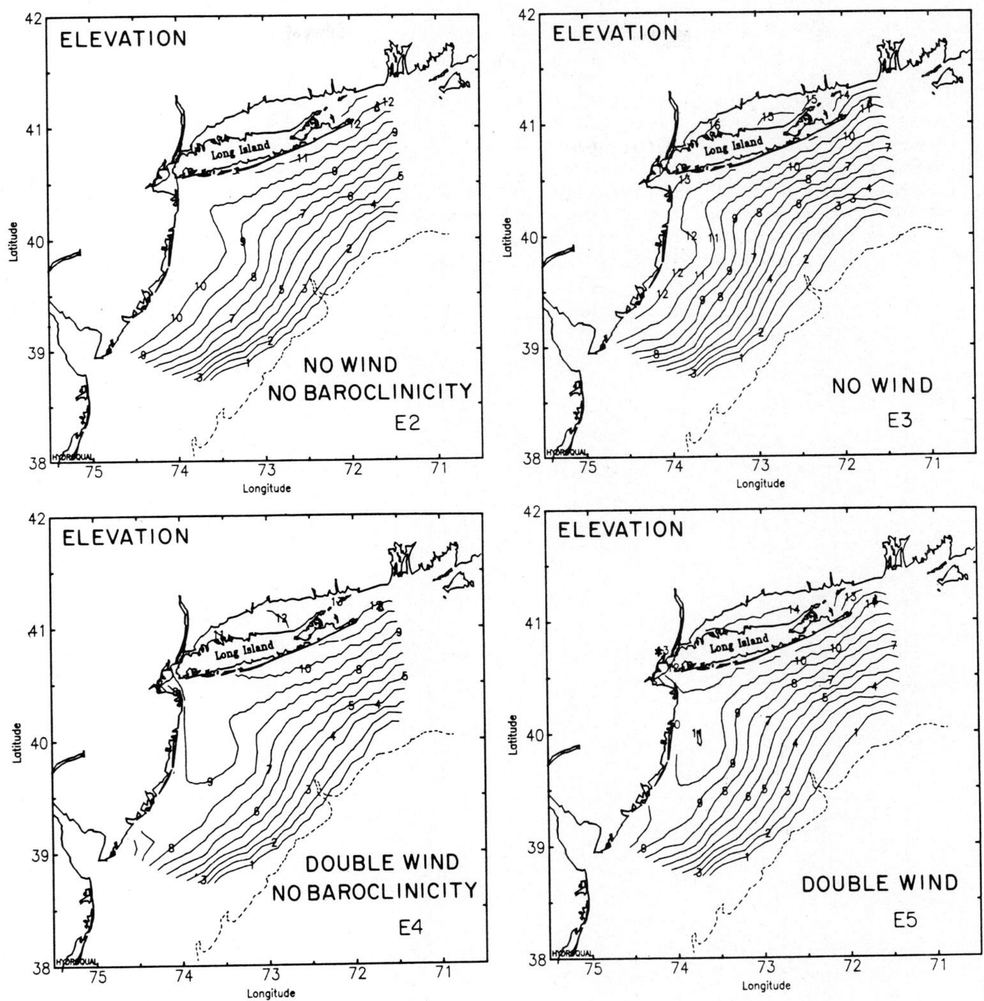

FIGURE 8. Comparison of elevation maps from experiments 2-5 investigating senitivity of circulation to various forcings (see Table 1 for description fo each case). CI = 1 cm.

The last two cases to be described here, Experiments 10 and 11, are designed to study the sensitivity of the circulation to a parabathic elevation gradient at the shelf-break. The along-shelf break gradient is set to approximately 2 x 10-7 (2 cm in 100 km). Figure 10 indicates that although the circulation is affected near the shelf-break, the effect of the parabathic elevation gradient quickly diminishes toward the coastline. There is an increased cross-shelf elevation gradient along the southwestern open boundary which increases the southwestward velocities and even overwhelms the wind-driven flow reversal near the shore. However, neither the circulation in Long Island Sound nor the total transport through the East River are affected significantly.

VI. Conclusions

This study shows that major features of the summer circulation in New York Bight and contiguous estuarine waters can be reproduced by a three-dimensional model forced by observed climatological temperature and salinity, winds and sea level elevations. Circulation and mixing in the study area is complex because of the interplay among tides, wind-induced flows and density-driven flows due to the Hudson plume, fronts and eddies. Of special significance is the important role that New York - New Jersey Harbor and Long Island Sound play in determining the circulation in New York Bight. The diagnostic steady-state analysis demonstrates a general southwesterly current flow in New York Bight which is in agreement with observations. The major features of the Hudson plume that transports pollutants southward out of New York - New Jersey Harbor can be understood in terms of interaction between eddy circulation and local discharges off the New Jersey coast. There is an indication that circulation processes produced by plumes and fronts, as well as their interaction with winds, can readily transport bottom waters onshore. This factor has important implications in water quality considerations.

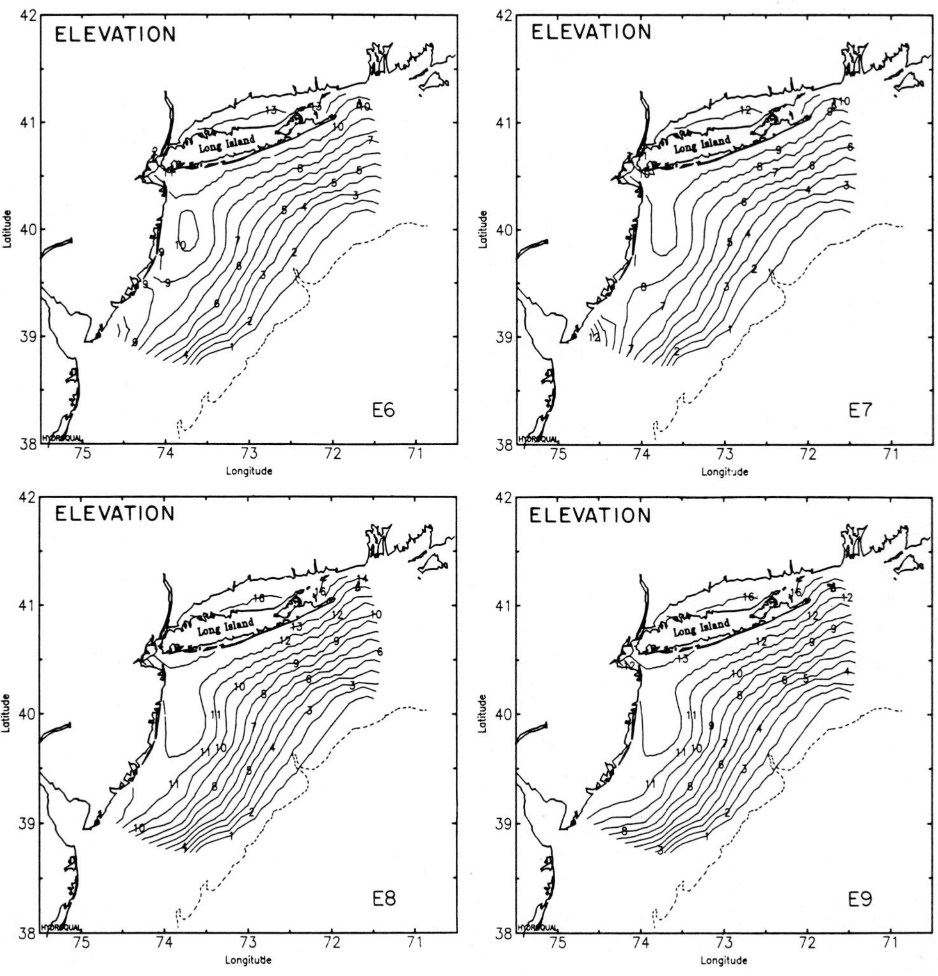

FIGURE 9. Comparison of elevation maps from Experiments 6 through 9 investigating sensitivity of circulation to various cross-shelf elevation gradient forcings (see Table 1 for description of each case). CI = 1 cm.

The summer average conditions produce non-tidal flow from Long Island Sound through New York - New Jersey Harbor to New York Bight. The calculations indicate that this net transport is very sensitive to wind. Increasing the wind by 40 percent reverses the flow direction from the bight through the harbor into the sound. The direction and magnitude of this net flow has implications with regard to pollutant mass transport and fate. In addition, the eddy circulation off the New Jersey coastline which may trap or entrain pollutants in this region is intensified by summer winds.

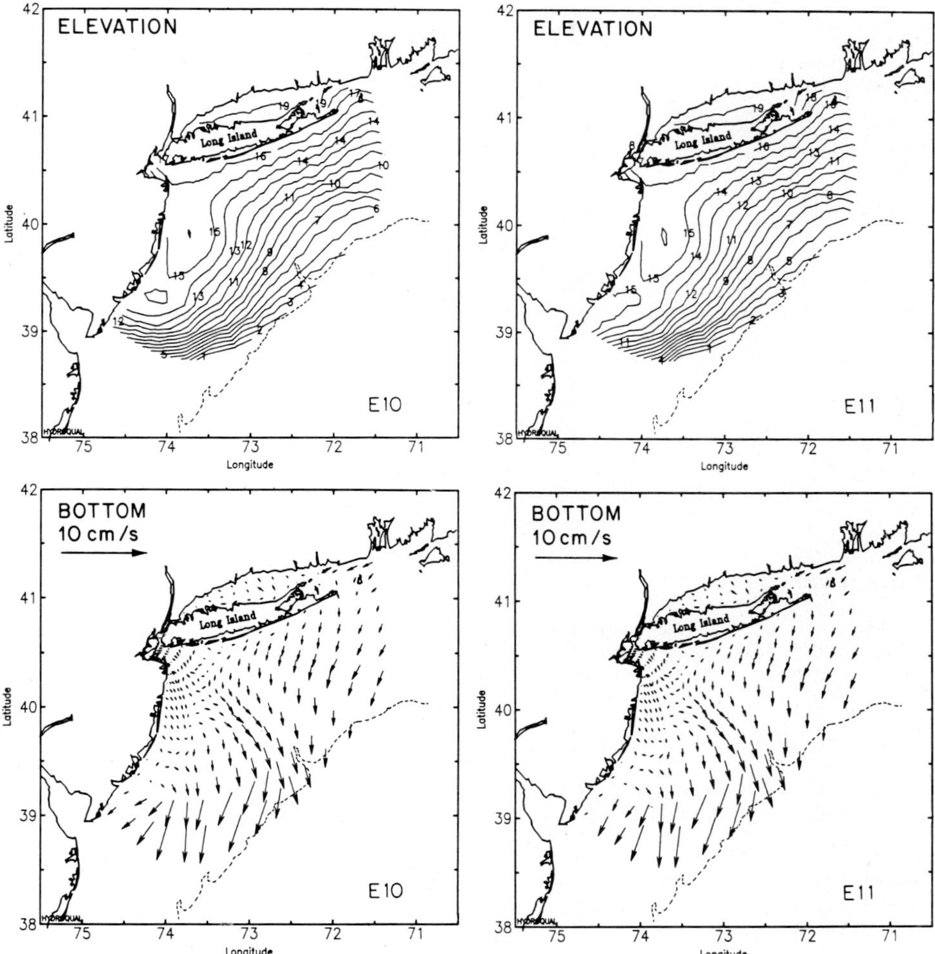

FIGURE 10. Compariosn of elevation maps and bottom currents from Experiments 10 and 11 investigating sensitivity of circulation to parabathic elevation gradients imposed at the shelf-break (see Table 1 for description of each case). CI = 1 cm.

The results of this effort indicate the importance of hydrodynamic interactions among New York Bight, New York - New Jersey Harbor and Long Island Sound. For the future studies, the model grid segmentations should have a finer resolution in the key areas, including the East River and the Hudson River. Tidal variation should be included as an additional forcing mechanism. To investigate the evolution of bight circulation patterns under the action of time-dependent forcing, the hydrodynamic model should be used in its prognostic mode. Identification of the effects of variable winds on plumes, fronts and circulation eddies in the bight may prove to be very valuable for water quality studies. Additional simulations are necessary to determine the circulation on seasonal time scales. The responses of the system to prolonged periods of wind or to the absence of wind are important factors for effective water quality analyses.

Acknowledgements

The authors gratefully acknowledge Donald J. O'Connor and John P. St. John for their role in the development of this project and for their constructive criticisms which considerably strengthened this effort. George L. Mellor suggested the particular conservative form of the governing equations used here. This effort has been funded by the U.S. Environmental Protection Agency under Work Assignment No. L01-5 through Contract No. 68-03-3533 with Dynamac Corporation. The results of the effort do not reflect the Agency's views, and no official endorsement should be inferred from them.

VII. References

Arakawa, A. and V. R. Lamb, 1977: Computational design of the basic dynamical process of the UCLA general circulation model. Methods in Computational Physics, **17**, 173-265.

Beardsley, R. C., W. C. Boicourt and D. V. Hansen, 1976: Physical oceanography of the middle Atlantic Bight in middle Atlantic Continental Shelf and the New York Bight. M.G. Gross ed., Amer. Soc. Limnol. Oceanogr., Special Symposia, 2, 20-34.

Beardsley, R. C. and C. D. Winant, 1979: On the mean circulation in the middle Atlantic Bight. J. Phys. Oceanogr., **9**, 612-619.

Beardsley, R. C. and W. C. Boicourt, 1981: On estuarine and continental-shelf circulation in the middle Atlantic Bight. In: Evolution of Physical Oceanography, 198-233.

Beardsley, R. C., D. C. Chapman, K. H. Brink, S. R. Ramp and R. Schlitz, 1985: The Nantucket Shoals flux experiment (NSFE79). Part I: A basic description of the current and temperature variability. J. Phys. Oceanogr., **15**, 713-748.

Bisagni, J. J., 1983: Lagrangian current measurements within the eastern margin of a warm-core gulf stream ring, J. Phys. Oceanogr., **13**, 709-715.

Blumberg, A. F. and D. M. Goodrich, 1990: Modeling of wind-induced destratification in Chesapeake Bay. Estuaries, 13.

Blumberg, A. F. and H. J. Herring, 1987: Circulation modeling using orthogonal curvilinear coordinates. In: Three-dimensional models of marine and estuarine dynamics, J. C. J. Nihoul and B. M. Jamart, eds., Elsevier Scientific Publishing Co., Amsterdam, 55-88.

Blumberg, A. F. and G. L. Mellor, 1980: A coastal ocean numerical model. In: Mathematical modelling of estuarine physics, Proceedings of an international symposium, Hamburg, August 24-26, 1978. J. Sundermann and K. P. Holz, eds., Springer-Verlag, Berlin.

Blumberg, A. F. and G. L. Mellor, 1983: Diagnostic and prognostic numerical circulation studies of the south Atlantic Bight. J. Geophys. Res., **88**, 4579-4592.

Blumberg, A. F., and G. L. Mellor, 1987: A description of a three-dimensional coastal ocean circulation model. In: Three-dimensional coastal ocean models, coastal and estuarine sciences, Vol. 4, N.S. Heaps, Editor, AGU, Washington, DC, 1-16.

Blumberg, A. F. and L.-Y. Oey, 1985: Modeling circulation and mixing in estuaries and coastal oceans. Adv. in Geophysics, **28A**, 525-547.

Bowman, M. J., 1975: Scales of variability of nontidal characteristics of the East River, New York. EOS Trans AGU, 56, 382.

Brown, W. S., N. R. Pettigrew and J. D. Irish, 1985: The Nantucket Shoals flux experiment (NSFE79). Part II: The structure and variability of across shelf pressure gradients. J. Phys. Oceanogr., **15**, 749-771.

Bumpus, D. F., 1973: A description of the circulation on the continental shelf of the east coast of the United States. Prog. Oceanogr., **6**, 111-157.

Chapman, J., A. Barth, R. C. Beardsley and R. G. Fairbanks, 1986: On the continuity of mean flow between the Scotian Shelf and the middle Atlantic Bight. J. Phys. Oceanogr., **16**, 758-722.

Csanady, G. T., 1978: The arrested topographic wave. J. Phys. Oceanogr., **8**, 47-62.

Galperin, B. and G. L. Mellor, 1989a: Time-dependent, three-dimensional model of the Delaware Bay and River System. Part 1: Description of the model and tidal analysis. Submitted.

Galperin, B. and G. L. Mellor, 1989b: Time-dependent, three-dimensional model of the Delaware Bay and River System, Part 2: Three-dimensional flow fields and residual circulation. Submitted.

Galperin, B. and G. L. Mellor, 1989c: Salinity intrusion and residual circulation in Delaware Bay during the drought of 1984. This volume.

Garvine, R. W., K.-C. Wong, G. G. Gawarkiewicz, R. K. McCarthy, R. W. Houghton and F. Aikman III, 1988: The morphology of shelf-break eddies. J. Geo. Res., **93**, 15, 593-15, 607, December.

Han, G., D. V. Hansen and J. A. Galt, 1980: Steady-state diagnostic model of the New York Bight. J. Phys. Oceanogr., **10**, 1998-2020.

Hopkins, T. S. and D. A. Dieterle, 1983: An externally forced barotropic circulation model for the New York Bight. Cont. Shelf Res., **2**, 49-73.

Hopkins, T. S. and D. A. Dieterle, 1987: Analysis of the baroclinic circulation of the New York Bight with a 3-D diagnostic model. Cont. Shelf Res., **7**, 237-265.

Houghton, R. W., D. B. Olson and P. J. Celone, 1986: Observation of an anticyclonic eddy near the Continental Shelf-break south of New England. J. Phys. Oceanogr., **16**, 60-71.

Hsueh, Y. and C. Y. Peng, 1978: A diagnostic model of Continental Shelf circulation. J. Geophys. Res., **83**, 3033-3041.

HydroQual, Inc., 1989: Assessment of pollutant fate in New York Bight. HydroQual Report DYNM0010, Mahwah, New Jersey.

Jay, D. A. and M. J. Bowman, 1975: The hydrography and circulation of New York Harbor and western Long Island Sound: Implications for water quality. Tech. Rep. 24, Mar. Sci. Res. Center, State University of New York, Stony Brook.

Joyce, T. M. (Editor), 1985: Warm-core rings collection. J. Geophys. Res. (Special Section), **90**, 8801-8951.

Kantha, L. H., A. F. Blumberg, H. J. Herring and G. L. Mellor, 1985: The physical oceanography and sea surface flux climatology of the south Atlantic Bight. Dynalysis of Princeton, Report 70, Princeton, New Jersey.

Mayer, D. A., G. C. Han and D. V. Hansen, 1982: The structure of circulation: MESA physical oceanographic studies in New York Bight, 2. J. Geophys. Res. **87**, 9579-9588.

Mellor, G. L. and T. Yamada, 1982: Development of a turbulence closure model for geophysical fluid problems, Rev. Geophys. Space Phys., **20**, 851-875.

Oey, L.-Y., 1986: The formation and maintenance of density fronts on U.S. southeastern Continental Shelf during winter. J. Phys. Oceanogr., **16**, 1121-1135.

Oey, L.-Y., G. L. Mellor and R. I. Hires, 1985a: Tidal modeling of the Hudson-Raritan Estuary. Estuarine Coastal Shelf Sci., **20**, 511-527.

Oey, L.-Y., G. L. Mellor and R. I. Hires, 1985b: A three dimensional simulation of the Hudson-Raritan Estuary. Part I: Description of the model and model simulations. J. Phys. Oceanogr., **15**, 1676-1692.

Oey, L.-Y., G. L. Mellor and R. I. Hires, 1985c: A three dimensional simulation of the Hudson-Raritan Estuary. Part II: Comparison with observations. J. Phys. Oceanogr., **15**, 1693-1709.

Oey, L.-Y., G. L. Mellor and R. I. Hires, 1985d: A three-dimensional simulation of the Hudson-Raritan Estuary, Part III: Salt flux analyses. J. Phys. Oceanogr., **15**, 1711-1720.

Ramp, S. R., R. C. Beardsley and R. Legeckis, 1983: An observation of frontal wave development on a shelf-slope/warm core ring front near the shelf-break south of New England. J. Phys. Oceanogr., **13**, 907-912.

Saunders, P. M., 1977: Wind stress on the ocean over the eastern Continental Shelf of North America. J. Phys. Oceanogr., **7**, 555-566.

Shuman, F. G., J. A. Brown, Jr., and K. A. Campana, 1972: A new explicit differencing system for primitive equations. Trans. Amer. Geophys. Union, **53**, 306.

Smagorinsky, J., 1963: General circulation experiments with the primitive equations, I. The basic experiment. Mon. Weather Rev., **91**, 99-164.

Salinity Intrusion and Residual Circulation in Delaware Bay During the Drought of 1984

Boris Galperin[1] and George L. Mellor[2]

[1]Program in Applied and Computational Mathematics
Princeton University
Princeton, NJ 08544

[2]Program in Atmospheric and Oceanic Sciences
Princeton University
Princeton, NJ 08544

ABSTRACT

A three-dimensional, time-dependent numerical model is used to simulate the dynamics and thermodynamics of Delaware Bay and River and the adjacent continental shelf. The model forcing includes observed winds, observed surface heat flux, observed free surface elevation at the mouth of the Bay, observed fresh water run-off from 16 rivers and tributaries, climatological temperatures and salinities at the shelf open boundaries, and predicted tides at the shelf break. The model was started at 00:00 am January 1, 1984 and ran through the entire year.

Since the fall of 1984 was dry, the model predictions for the late autumn months are an indication of model performance in simulating droughts. The present study compares daily salinity intrusions on April 1, June 1, August 1 and October 1 of 1984 and shows substantial salinity penetration into the Bay during this period, (about 30-40 km). The strongest stratification is found in April (as much as 5-8 ppt), and the weakest is in October; this is consistent with observations.

Salinity distributions in the Bay and on the shelf are affected by such factors as the freshwater run-off, open boundary conditions and the dominant winds. During April through July, the salinity approached zero values southward of Artificial Island, while in October, the salinity penetrated all the way up to Chester, Pennsylvania which is generally consistent with observations. The study also concentrates on some aspects of the monthly mean circulation.

I. Introduction

Increasing availability of computational resources has made three-dimensional, time dependent modeling of natural water systems more effective during the last decade. Numerical models, designed for modern supercomputers, are able to combine sophisticated physics with high spatial and temporal resolution. Particular success has been achieved in modeling relatively small bodies of water, such as estuaries and lakes, which can be simulated on a dense mesh with decreased horizontal viscosity. A model developed by Blumberg and Mellor (1980,1987) has been used and verified in numerous applications. Most recently, the model was employed to study Delaware Bay, River and the adjacent continental shelf (Galperin and Mellor, 1989a,b). Model calculations were extensively compared with data, particularly on tidal time scales, for different periods of 1984. The dynamics and thermodynamics

Coastal and Estuarine Studies, Vol. 38
R. T. Cheng (Ed.)
Residual Currents and Long-term Transport

of this coupled estuary-shelf system were also investigated on longer time scales. For example, the frictional sub-tidal response to strong winds could enhance or destroy the classical, two-layer estuarine circulation pattern. The winds can generate strong salinity fronts inside the Bay and they can affect the behavior of the fresh water plume over the shelf.

The present paper extends the analysis to another important long term process, salinity intrusion into Delaware Bay. Usually, one-dimensional dispersion models are used for routine predictions of salinity intrusion (see Chatwin and Allen, 1985, for a review). These models are calibrated against observations in a particular estuary and then are believed to have satisfactory predictive capabilities for that estuary. However, one-dimensional models are limited in scope since they exclude important 3-D physics and therefore probably do not possess universality for a wide range of parameters, such as fresh water run-off, winds, shelf conditions etc. This situation has motivated the present study of salinity intrusion into Delaware Bay using a fully three-dimensional, time-dependent numerical model. The study also concentrates on the wind-driven monthly circulation of the Delaware Bay-shelf complex.

II. Description of The Numerical Model and Simulations

The model has been formulated elsewhere (Blumberg and Mellor, 1980,1987; Oey et al., 1985). Particularly, the application of the model to Delaware Bay, River and the adjacent continental shelf was described by Galperin and Mellor (1989a,b) (referred to as GM1 and GM2 hereafter). Very briefly, the model solves prognostic equations for free surface elevation (continuity equation), for both horizontal velocity components, and for temperature and salinity. The model includes a turbulence closure sub-model (the level 2 1/2 model of Mellor and Yamada, 1982) to describe turbulent transport processes, particularly bottom and surface boundary layers. The closure model adds two more transport equations for: (1) turbulence kinetic energy, and (2)turbulence macroscale. The vertical velocity is then found diagnostically. The hydrodynamics model recognizes fast, surface (external) waves and slow, internal waves and uses a time splitting technique in which prognostic equations for external (barotropic) and internal (baroclinic) variables are solved

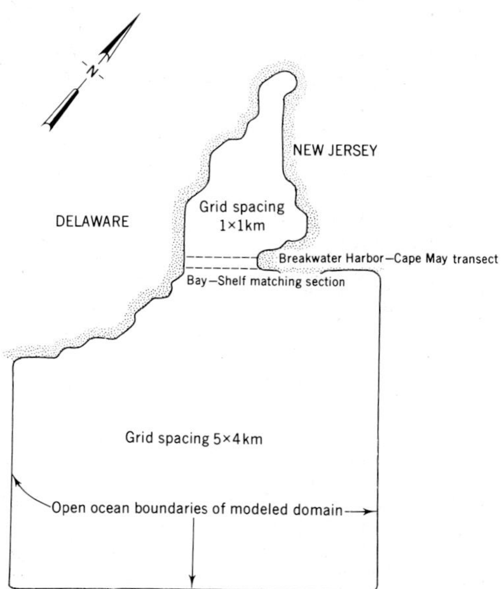

FIGURE 1. Schematic description of the modeled area. Observed elevations were inserted at the extremities of the Breakwater Harbor-Cape May transect and then linearly interpolated along the transect. Bay and shelf were coupled along the Bay-shelf matching section.

with different time steps. For the present application, the external time step was 24s and the internal time step was 240s. The model also has a bottom and free-surface-following coordinate system. Smagorinsky's horizontal viscosity was incorporated in the model. The full set of equations was approximated by a finite difference scheme and solved by a "leap frog" method in time, explicit in the horizontal and implicit in the vertical. One month of a real time simulation on CDC Cyber-205 supercomputer required about 2 hours of CPU time.

Schematically, the computational domain is shown in Figure 1 (note that the actual grid extends all the way up to the head of the tide at Trenton). Different spatial resolutions over the shelf (5 x 4 km) and over the Bay (1 x 1 km) were employed. The Bay and the shelf are matched across the mouth using the continuity equation in such a way that the mass flux across the Breakwater Harbor-Cape May transect is conserved. Where the width of Delaware River becomes narrower than 1 km the river is described by a two-dimensional, laterally-integrated model with spatial resolution of 1 km along the river.

The actual computational domain is shown in Figure 2 which also includes tidal gage and current meter stations used by the National Oceanic Service.The portion of the continental shelf used by the model is placed in the top left corner, initially occupied by land, for computational efficiency. Similar methodology governs the model representation of the C&D Canal and the upper portion of the River. The computational domain extends 58 points in x, 100 points in y and 11 levels in σ (for more details, see GM1).

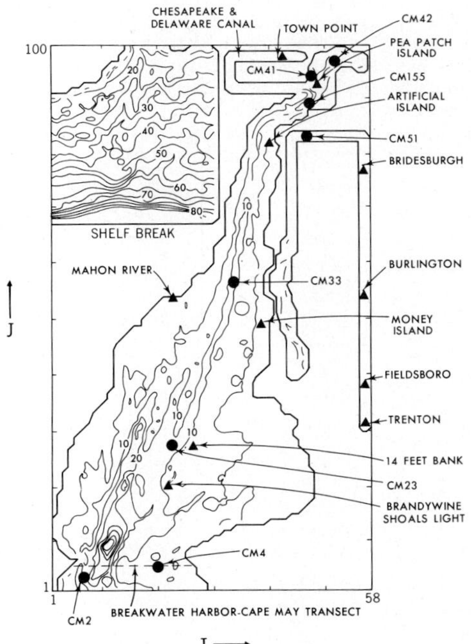

FIGURE 2. A plan view of the computational domain with digitized images of the Delaware Bay, River and adjacent con-tinental shelf. Depths are in meters. The grid has 58 points in the x-direction (index I), 100 points in the y-direction (index J) and 11-levels in the vertical. The included portion of the con-tinental shelf resolved on a coarser grid is placed in the top left corner initially occupied by land for computational efficiency. Δ mark the locations of tidal gages and • provide the locations of current meter moorings in the grid. The narrow portion of the river and C & D Canal are shown with constant width, however, the calculations make use of the local river width.

A significant effort was required to provide an external forcing environment for the model as close to reality as possible. Thus, the model was forced by: a) tidal elevations (containing 37 astronomical constituents) at the shelf break that were calculated in such a way that they produced observed tidal elevations at the mouth of the Bay (Breakwater Harbor - Cape May transection) in a two-dimensional, vertically integrated model; b) measured tidal elevations at Breakwater Harbor and Cape May that were inserted in the model to override predicted elevations. The resulting cross-mouth mass flux was conserved through the continuity equation; c)surface wind stress and surface heat flux including the diurnal cycle calculated from the meteorological information at the airport of Atlantic City; d) climatological boundary conditions for temperature and salinity at all three open shelf boundaries, e) seasonal temperature (and null salinity) at Town Point and Trenton (the Chesapeake Bay water near Town Point is usually fresh); f) freshwater run-off from 16 major rivers and tributaries. The year 1984 was cooler than the climatology so that the model predictions for temperature were off-set by several degrees C, especially on the shelf. However, realistic representation of the surface heat flux led to a gradual improvement of the results along the bay, as discussed in GM2. Furthermore, the results for temperature are not expected to have much of the effect on salinity distributions, the major concern of the present study.

The model run was started at 00:00 January 1, 1984 and covered the entire year 1984. The first two months were required for spin up; the subsequent results were used to assess the model performance via comparisons with data and to study different aspects of circulation in the Delaware Bay complex. The previous two papers (GM1 and GM2) have focused on the spring-summer period of 1984. The present study concentrates on autumn of 1984 when drought conditions prevailed. Along with daily salinity intrusion curves for April 1, June 1, August 1 and October 1 and daily salinities at Reedy Island Jetty, circulation patterns in April, July and October are also shown and discussed.

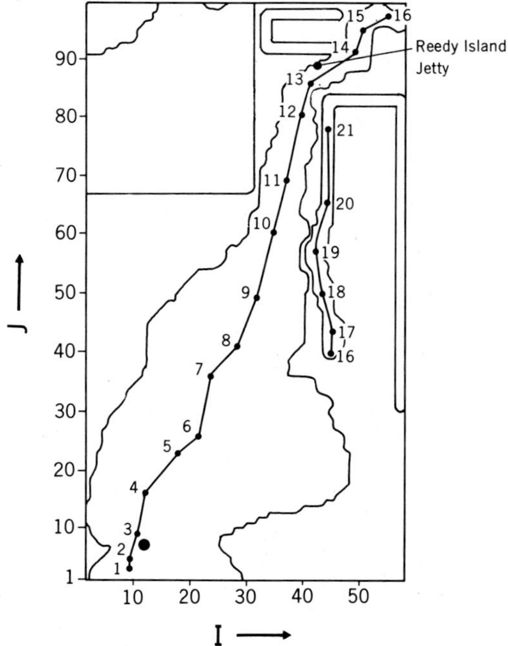

FIGURE 3. Locations of salinity monitoring stations along the deep channel in the numerical model of Delaware Bay. A large black point at the entrance indicates the position of the Delaware Bay mouth according to the Delaware River Basin Commission definition.

III. Study of the Salinity Intrustion

To trace the extent of salinity penetration into Delaware Bay, 21 locations in the numerical grid were designated for monitoring salinity as a function of time. These salinity stations are shown in Figure 3. Table 1 provides distances between the salinity stations and the standard zero-location at the mouth of the Bay. Figure 4 shows surface and near-bottom daily-averaged salinity intrusions for April 1, June 1, August 1 and October 1. One sees that the intrusion grows with time; it is minimal in April and maximal in October; the difference in the intrusion between those months is about 50 km. Maximum stratification is observed in April and it can reach as much as 8 ppt over the water column; in October, stratification is at its minimum of less than 5 ppt.

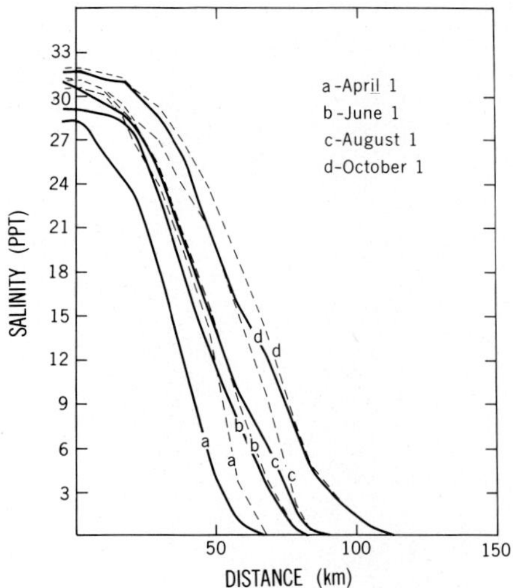

FIGURE 4. Daily averaged, surface (solid) and near-bottom (dashed) salinity intrusions for April 1, June 1, August 1, and October 1, 1984.

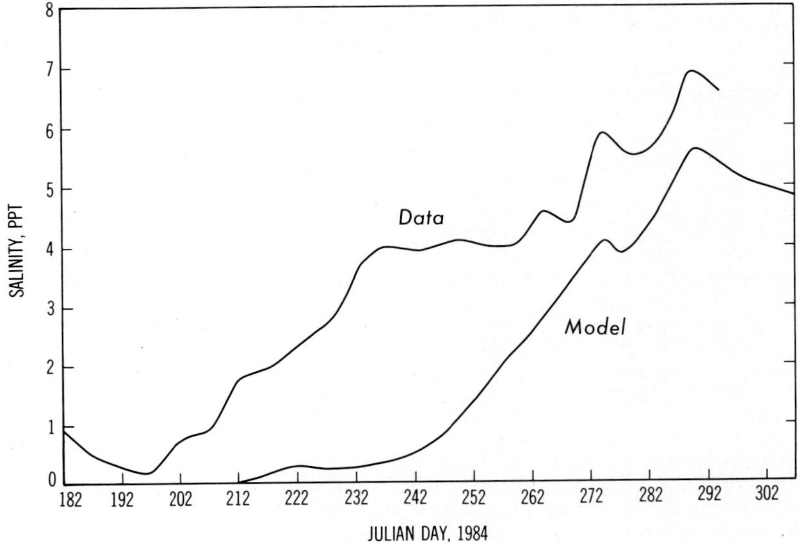

FIGURE 5. Daily averaged, modeled and observed salinities at Reedy Island Jetty.

Figure 5 compares daily averaged salinity at Reedy Island Jetty, Delaware as a function of time (one can see from Figure 4 that, at this distance, surface and bottom salinities are practically equal). This station is located in the upper Bay (see Figure 3) and provides a good indication of salinity changes during dry periods. Unfortunately, other salinity data were unavailable.

TABLE 1. Distances between the model salinity stations and the Delaware Bay mouth position along the deep channel.*

Station	Distance relative to the mouth of the Bay, KM	Latitude	Longitude
1	-5.21	38°47.84′	75°02.25′
2	-3.76	38°48.50′	75°03.33′
3	1.97	38°50.70′	75°04.65′
4	9.68	38°53.79′	75°08.55′
5	16.89	38°58.38′	75°09.26′
6	20.86	39°00.67′	75°09.61′
7	29.89	39°04.53′	75°13.63′
8	37.95	39°09.11′	75°14.34′
9	47.49	39°13.40′	75°17.94′
10	57.80	39°17.59′	75°22.50′
11	67.95	39°21.79′	75°27.06′
12	78.08	39°25.98′	75°31.62′
13	83.32	39°28.08′	75°33.90′
14	94.00	39°33.62′	75°32.80′
15	98.46	39°35.81′	75°34.12′
16	104.96	39°38.97′	75°33.63′
17	110.29	39°41.12′	75°31.51′
18	114.77	39°43.51′	75°30.90′
19	121.67	39°46.85′	75°28.48′
20	132.18	39°49.63′	75°24.02′
21+	145.00	--------	--------

* The position of the Delaware Bay mouth (38 50′32"N, 75 03′18"W) is given according to the Delaware River Basin Commission definition. The distances are calculated by a computer program created at the University of Delaware and kindly provided by Dr. C. Culberson.

+ The coordinates for station 21 are not shown; this station is located in the part of the Delaware River which is represented in the model by laterally-integrated, two-dimensional body of water.

One can see in Figure 5 that although the model reproduces low-frequency trends in salinity at Reedy Island Jetty, it systematically underpredicts the absolute values of salinity compared to the data. As was emphasized in GM1, the numerical model requires high values of horizontal viscosity in the narrow parts of the estuary and in the River to prevent numerical instability during high fresh water run-off events (the viscosity values reached as much as 50 m/s and more in Delaware River while in the Bay interior it was about 10 m/s). In addition, fine topographical features, such as the deep navigational channels were not resolved in the model. All of this caused some underprediction of the model tidal currents. As a consequence, residual currents and related advection processes were also underpredicted so that the model underpredicts salinity intrusion on tidal and subtidal time scales. Another source of the discrepancy between simulated and measured salinites, in the upper Bay, is numerical truncation error resulting from the rapid decrease of salinity in this region. As was emphasized in GM2, one of the ways to improve the prediction of currents and salinity in the upper Bay and in the River would be to use a finer resolution grid and possibly a curvilinear coordinate system which would allow one to better resolve this area and to decrease horizontal viscosity. Oey et al,. (1985) in their simulations of the Hudson-Raritan estuary with a spatial resolution of 500m x 500m succeeded in reducing horizontal viscosity to null.

IV. Monthly Mean Circulation and Salinity Distributions

Figures (6-9) compare monthly averaged currents and salinity distributions in Delaware Bay and in the Bay-shelf complex in April, July and October of 1984. These months represent conditions typical for spring, summer and fall and therefore the results provide an insight into various processes taking place in the estuary during those seasons.

Figure 6 shows surface and near-bottom currents inside the Bay along with the wind stress. For all three months considered, the classical two-layer circulation in the estuary is well defined. The surface, onshelf currents are most intensive along the deep channels reaching 20 cm/s while near-bottom, landward currents are weaker by a factor of 5. The near-bottom currents are also strongest along the deep channels. Generally, the currents are stronger in July as a result of strong winds and are weakest in October when the winds were practically null. One can see that in April and July, near-bottom, landward currents terminate in the vicinity of Artificial Island, which is also a bound for the salinity intrusion. In October, the near-bottom currents penetrate deeper into the estuary, and so does the zero-salinity front (see Figure 7). Along with strong seaward and landward currents, Figure 6 also reveals several little gyres caused by the topographical features.

Figure 7 shows surface and near-bottom salinity distributions in the Bay. A distinctive feature of the near-bottom salinities is an intense intrusion along the deep channels which is intensified by winds in April and July. This intensification is partially due to the upwelling shelf circulation caused by the southwesterly winds and partially due to the local enhancement of the two-layer estuarine circulation caused by the wind friction over the Bay (see GM2). As a result of the prevailing wind pattern the fresh water in the Bay is moved to the New Jersey shore of the estuary so that the halocline becomes tilted to the north-east and higher isohalines dominate the Delaware part of the Bay. Consistent with Figure 6, the terminus of the salinity intrusion is near Artificial Island in April and July and much farther north in October.

Figure 8 shows the surface and near-bottom currents in the Bay-shelf complex. One can see that in April and July the shelf surface currents are set up in the direction of the prevailing winds. In October, due to weak winds, shelf circulation is weak, too; one can distinguish several gyres, probably associated with the density structure and bathymetry. In GM2 it was emphasized that the shelf portion of the model is wind driven; it excludes alongshelf pressure gradient setting up the southwestward shelf flow as well as the interaction with the deep ocean and Gulf Stream rings at the shelf break. The net model response to the local northeastward winds is the generation of northeastward shelf currents opposing the observed mean shelf flow. It should be noted that such currents have been indeed observed even on monthly time scales during late spring - summer when winds are favorable, and Allen et al., (1983) call them "rather the rule than the exception." These local currents are sometimes referred to as "the inner shelf flow reversals", as opposed to the mean shelf current. These reversals were also mentioned by Pape and Garvine (1982) and by Beardsley and Boicourt (1981). Due to the Coriolis effect, the flow reversals cause the surface, lighter currents to deflect offshore and compensating, saline bottom currents to flow onshore, thus creating an upwelling circulation. One can see that this pattern is captured by the model.

Figure 9 shows monthly salinity distributions in the Bay-shelf system. Due to the higher salinity gradients in the Bay than on the shelf (which would blur salinity contours inside the Bay), the last isohaline shown there is 26 ppt. Generally, for all three months shown the near-bottom salinities are, as expected, higher than the corresponding surface values; the top-to-bottom salinity difference is about 2 ppt. In April, the plume of fresh water leaves the Bay and turns southwestward. The plume is distinct, consistent with the high fresh water run-off in spring. The plume disappears in July and in October. The disappearance of the plume in July is probably caused by strong southwesterly winds driving the surface waters to the northeast (an indication of this is a little eddy in the vicinity of the northeast open boundary which probably shed from the Bay plume due to the winds). In October, when the winds are weak, the shrinkage of the coastal plume is probably due to the reduction in the freshwater run-off (for comparison, the run-off at Trenton was 988 m/s in April, 451 m/s in July and only 106 m/s in October of

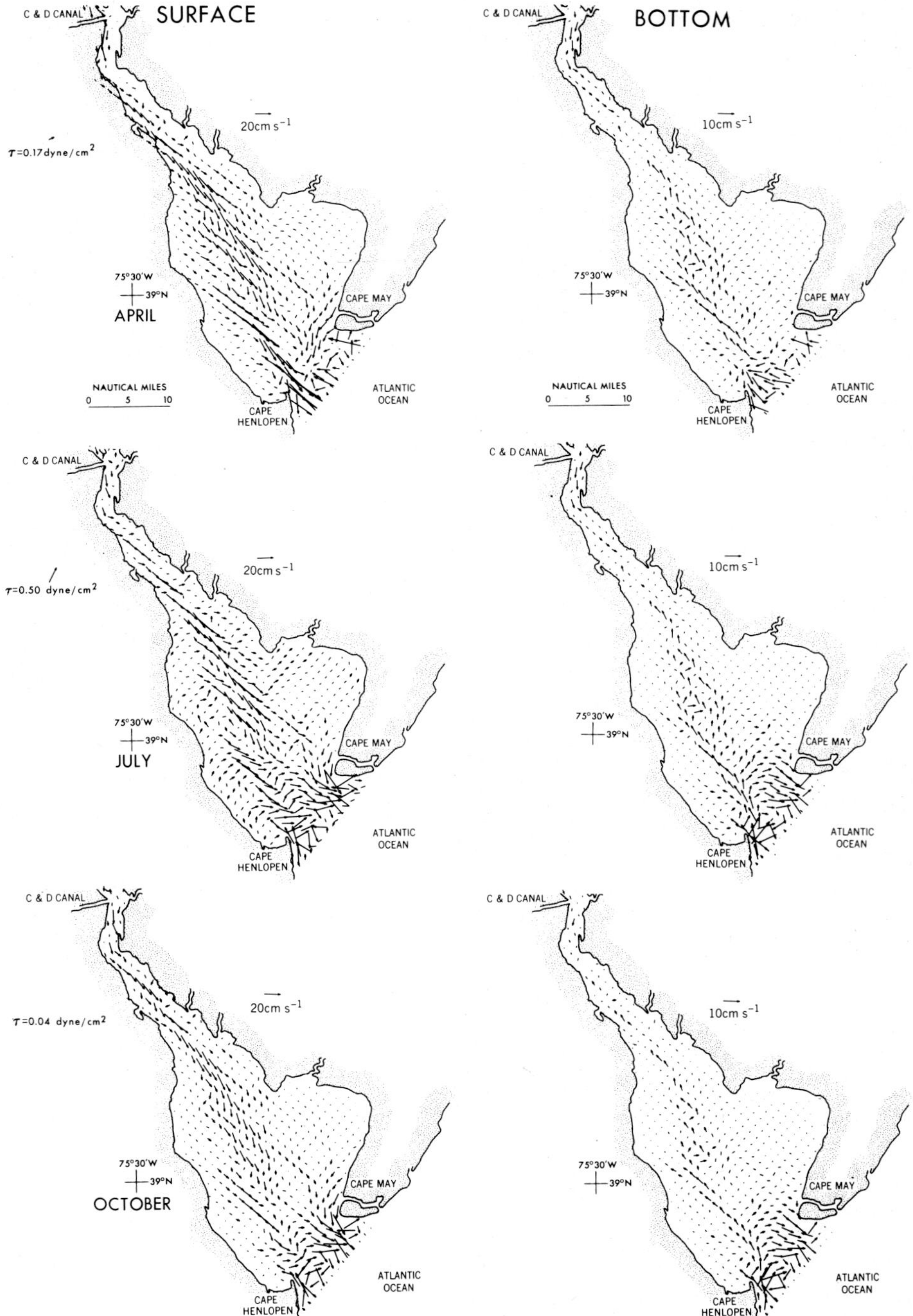

FIGURE 6. Monthly averaged surface (left column) and near-bottom (right column) currents in Delaware Bay for the months of April (top), July (middle) and October (bottom). Corresponding monthly averaged wind stresses are also shown.

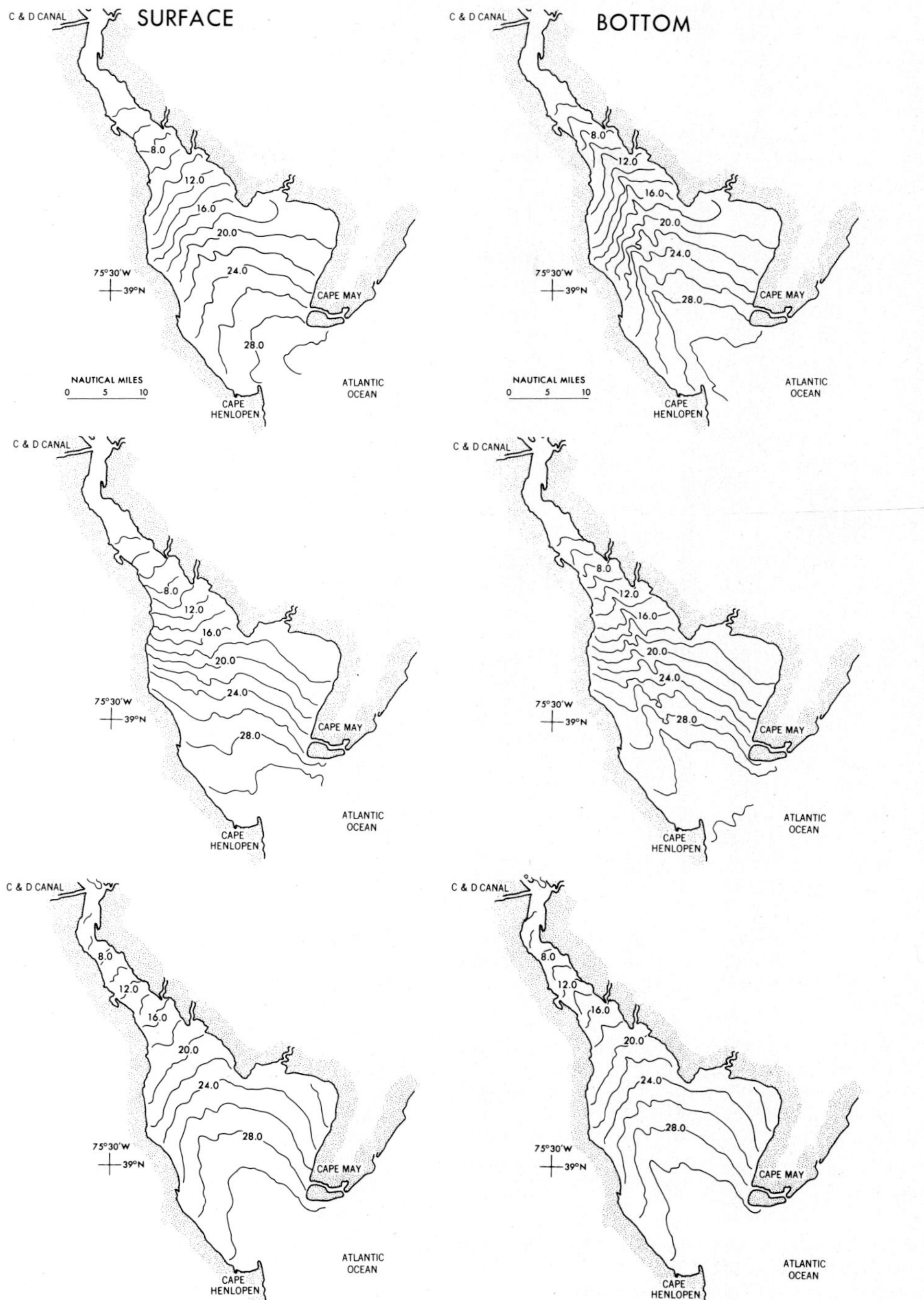

FIGURE 7. Same as in Figure 6 but for salinity distributions. Contour interval is 2.0 ppt.

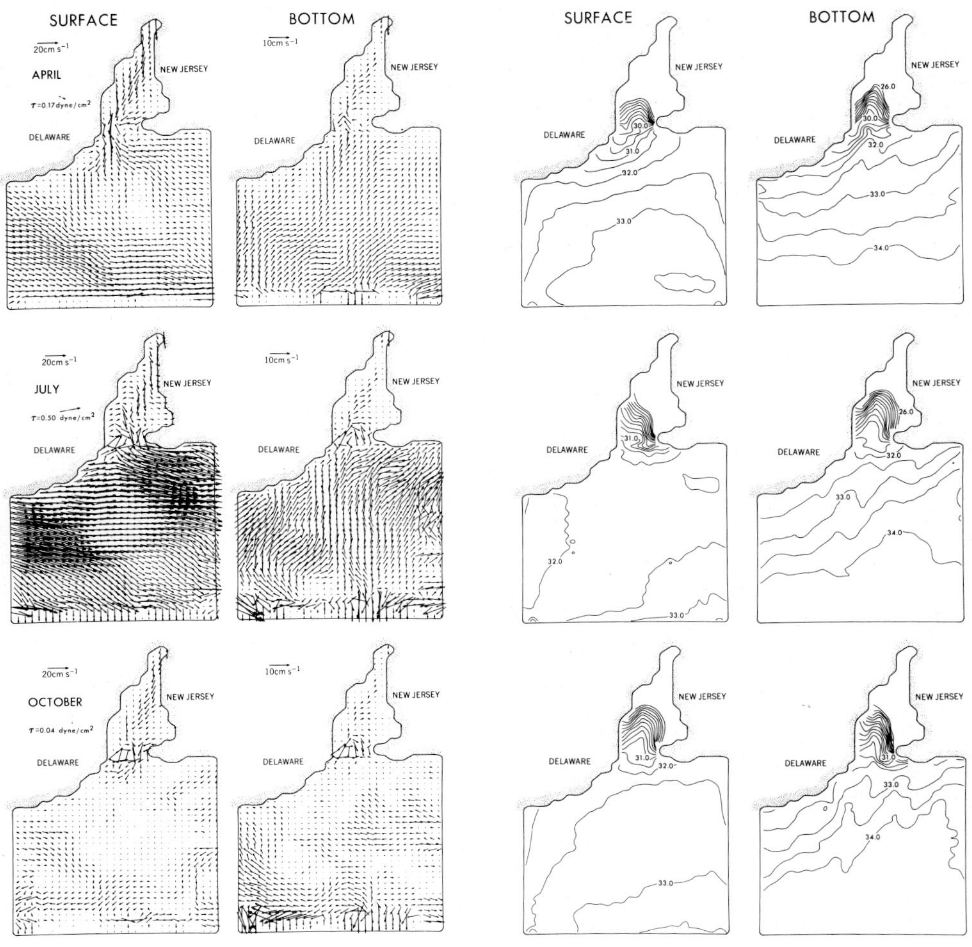

FIGURE 8. Same as in Figure 6 but for the Delaware Bay-shelf coupled system.

FIGURE 9. Same as in Figure 6 but for the Delaware Bay-shelf coupled system. Contour interval is 0.5 ppt.

1984). Comparison of surface isohalines also indicates that in October the 26 ppt isohaline has penetrated much further up the Bay than in April or in July. This means that during the drought of 1984 salinity has not only penetrated deeper into the Bay than in other, not drought seasons, but also the surface salinity was higher than usual throughout the estuary.

V. Conclusions

The study has demonstrated that the three-dimensional, time-dependent numerical model of Delaware Bay, River and adjacent continental shelf is capable of simulating salinity intrusions, mean circulation and dispersion processes. The model indicates that the smallest intrusion and strongest stratification occur in spring when freshwater run-off is large. On the other hand, the deepest salinity intrusion and weakest stratification are characteristic of autumn, when freshwater run-off is the smallest. In drought years, salinity may penetrate as far as Chester and even to Philadelphia, thus removing a source of fresh water in these regions.

Monthly mean circulation in the Bay is in accordance with the classical, two-layer estuarine pattern for all seasons. Lighter, faster surface currents flow out of the Bay onshore while heavier, slower near-bottom currents flow in the opposite direction. This circulation is sensitive to the wind direction and magnitude that can substantially affect the circulation pattern and salinity distribution. The model also provides important information on the coastal, fresh water plume and its interaction with winds and shelf currents. The shelf circulation in the model is predominantly wind-driven as it excludes the alongshelf pressure gradient responsible for the mean southwestward flow and the interaction with the deep ocean off the shelf break and Gulf Stream variability. The model does reproduce the inner shelf flow reversals caused by the southwesterly winds and related upwelling circulation. Such a circulation pattern leads to the bottom salinity increase close to the shore and the mouth of the bay. Fortunately, the reversals mostly happen in spring-summer periods; in fall the wind direction usually changes to southwestward. If the reversals prevailed in the autumn, particularly in drought years, they could significantly enhance salinity intrusions and produce major problems to the users of the Delaware River.

A limitation of the model is inadequate resolution in the upper Bay and in Delaware River. This makes it difficult to simulate the processes taking place inside the deep, navigational channels because a relatively high horizontal viscosity is required in that region. This leads to underprediction of tidal and residual currents, related advective processes and, as a result, salinity intrusion. An enhancement in the model performance can be expected with further increase in resolution and with the use of a curvilinear coordinate system (Blumberg and Herring, 1987) to locally improve resolution where necessary.

Acknowledgments

The National Ocean Service supported the project and made observational data available. Further support was provided by GFDL under the NOAA Grant NA84EAD00057, by Princeton University and by the New Jersey Commission on Science and Technology. Excellent drafting and figure processing were performed by the GFDL Graphics Department and by Mr. J. Conner. Ms. Johann Callan typed the manuscript.

VI. References

Allen. J. S., R. C. Beardsley, J. O. Blanton, W. C. Boicourt, B. Butman, L. K. Coachman, A. Huyer, T. H. Kinder, T. C. Royer, J. D. Schumacher, R. L. Samith, W. Sturges, and C. D. Winant, 1983: Physical oceanography of continental shelves. Rev. Geophys. Space Phys., **21**, 1149-1181.

Beardsley, R. C., and W. C. Boicourt, 1981: On estuarine and continental-shelf circulationa in the Middle Atlantic Bight, in Evolution of Physical Oceanography: Scientific Surveys in Honor of Henry Stommel, edited by B. A. Warren and C. Wunsch, pp. 198-233, The MIT Press, Cambridge, MA.

Blumberg, A. F., and H. J. Herring, 1987: Circulation modeling using orthogonal curvilinear coordinates, In: Three Dimensional Models of Marine and Estuarine Dynamics, J. C. J. Nihoul and B. M. Jamart, Ed., Elsevier Oceanography Series, 45, 55-88.

Blumberg, A. F., and G. L. Mellor, 1980: A coastal ocean numerical model. In: Mathematical Modeling of Estuarine Physics, Proc. Int. Symp., Hamburg, August 24-26, Edited by J. Sundermann and K. -P. Holz, 203-214, Springer Verlag, Berlin.

Blumberg, A. F., and G. L. Mellor, 1987: A description of a three-dimensional coastal ocean circulation model. In: Three-dimensional Coastal Ocean Models, Coastal and Estuarine Sciences, Vol. 4, N.S. Heaps, Editor, AGU, Washington, DC, 1-16.

Chatwin, P. C., and C. M. Allen, 1985: Mathematical models of dispersion in rivers and estuaries. Ann. Rev. Fluid Mech., **17**, 119-149.

Galperin, B., and G. L. Mellor, 1989a: A time-dependent, three-dimensional model of the Delaware Bay and River system. Part 1: Description of the model and tidal analysis. Submitted to Estuarine Coastal Shelf Sci.

Galperin, B., and G. L. Mellor, 1989: A time-dependent, three-dimensional model of the Delaware Bay and River system. Part 2: Three-dimensional flow fields and residual circulation. Submitted to Estuarine Coastal Shelf Sci.

Mellor, G. L. and T. Yamada, 1982: Development of a turbulence closure model for geophysical fluid problems. Rev. Geophys. Space Phys., **20**, 851-875.

Oey, L. -Y., G. L. Mellor, and R. I. Hires, 1985: A three-dimensional simulation of the Hudson-Raritan estuary. Part 1: Description of the model and model simulations. J. Phys. Oceanog., **15**, 1676-1692.

Pape, E. H. III, and R. W. Garvine, 1982: The subtidal circulation in Delaware Bay and adjacent shelf waters. J. Geophys. Res., **87**, 1955-1970.

Wind and River Induced Fluctuations in a Small, Shallow, Tributary Estuary

William W. Schroeder[1], William J. Wiseman, Jr.[2], and Scott P. Dinnel[3]

[1]Marine Science Program
The University of Alabama
Dauphin Island, AL 36528

[2]Coastal Studies Institute and Department of Geology and Geophysics
Louisiana State University
Baton Rouge, LA 70803

[3]Coastal Studies Institute and Department of Marine Sciences
Louisiana State University
Baton Rouge, LA 70803

ABSTRACT

Two years of wind and water level data and 18 months of runoff data from Weeks Bay, Alabama, a small tributary estuary of Mobile Bay have been analyzed. These data were supplemented with occasional current meter records of a few months' duration. The subtidal barotropic exchange of water between Weeks Bay and Mobile Bay is predominantly a response to the shelf-estuarine exchange of waters between the Gulf of Mexico and Mobile Bay. The local north-south wind stress modifies this response at very low frequencies. Runoff to Weeks Bay is episodic and of short duration. It drives short-lived exhanges between the two bays. These exchanges are modified by the local winds and are probably strongly baroclinic.

I. Introduction

Weeks Bay, Alabama, one of the National Estuarine Research Reserve Systems, is a small tributary estuary on the eastern shore of Mobile Bay (Fig. 1) in the northern Gulf of Mexico. It is nearly diamond shaped, with a 3.4-km longitudinal axis running north-south from the head of the bay, where the Fish River discharges, to its mouth, where it is connected to Mobile Bay (Fig. 1). Weeks Bay has a lateral axis oriented east-west; the widest section, 3.1 km, occurs in the central region, where Magnolia River discharge enters along the eastern shore. A 5- to 7-m-deep depression is located in the narrow mouth of the bay, and a 3- to 4-m-deep depression occurs at the head of the bay. Depths in the 2- to 3-m range are common in the lower bay, whereas depths in the upper bay are often 1 m or less. The bay has a mean depth of 1.4 m, a surface area of 6.95×10^6 m^2, and a volume of 9.73×10^6 m^3 (Crance, 1971).

The tides are principally diurnal and have a mean range of approximately 0.4 m. Direct fresh-water discharge to the bay comes from the Fish and Magnolia Rivers. Mean combined discharge is estimated at 9 m^3/s, with mean freshets up to 4 times larger. The Fish River is the principal source of fresh water, accounting for approximately 73% of the inflow. The extreme flooding event on record for the Fish River is estimated to be 243 m^3/s.

** Current address: Center for Marine Science, University of Southern Mississippi, Stennis Space Center, MS 39529

Coastal and Estuarine Studies, Vol. 38
R. T. Cheng (Ed.)
Residual Currents and Long-term Transport

The small volume of Weeks Bay limits the volume of water that can be exchanged with Mobile Bay in a barotropic mode. Hydrographic data and current meter records from a 70-day deployment suggest that baroclinic flow patterns are intermittently important. However, because most of the current meter deployments involved only a single meter, we will focus our discussion on the barotropic exchanges.

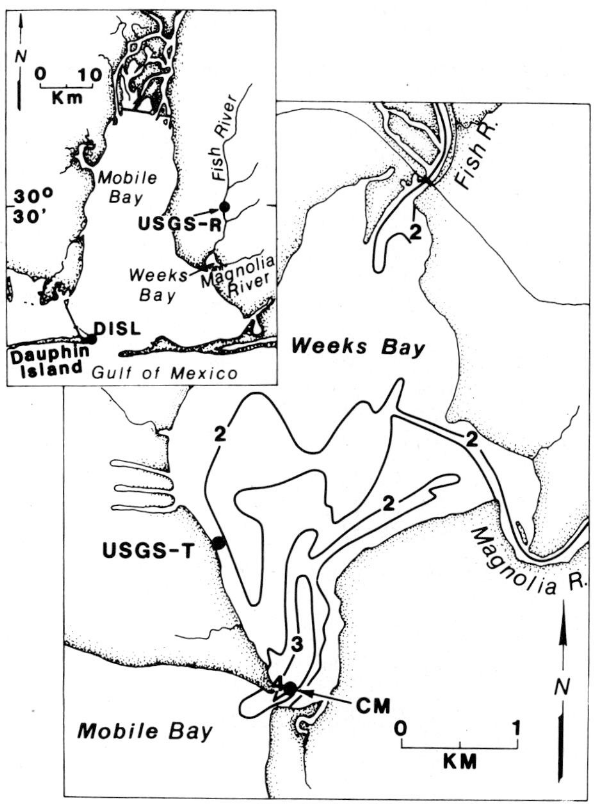

FIGURE 1. Location map of Weeks Bay showing location of current meter, CM, and Weeks Bay water level gauge, USGS-T. Insert shows location of wind station and NOAA/NOS Mobile Bay water level at DISL and Fish River discharge gauge at Silverhill, USGS-R.

Most previous studies of subtidal estuarine exchange processes have been unable to fully resolve river-induced effects on subtidal exchange because of the long time scales involved (e.g., Wong and Garvine, 1984; Schroeder and Wiseman, 1986; Wiseman et al., 1988). The small drainage basin of the Fish River (394 km^2) should respond rapidly to local rainfall. Freshets occur on time scales of a few days. These time scales are the same as those associated with strong wind forcing during the fall, winter, and spring. During summer, synoptic weather time scales are somewhat longer, of the order of 10-12 days. In most estuaries, river-induced and wind-induced exchanges are broadly separated in time scale. This is not the case in Weeks Bay; the two responses are confounded in this system. Simple estimates of the replacement time-scales for water within Weeks Bay are such that a normal freshet (30-50 m^3/s) can completely refill the bay within two or three days, normal tidal exchange can do the same on time scales of three days, and wind-driven subtidal exchanges can replace the water within the bay, assuming a barotropic flow, in two or three days. An understanding of the linear dynamics governing the estuary requires separation of the responses to the different forcings using multiple input time-series analysis techniques and detailed investigation of individual events. These are the approaches taken in the present study.

II. Observations and Analysis Techniques

A field monitoring program was in place over the period April 1986 to March 1988. A single current meter mooring was intermittently maintained (deployments ranged from 52 to 109 days) in a water depth of 4.2 m at the mouth of the bay (Fig. 1). The mooring had either one meter at a depth of 2.1 m or two meters at depths of 1.0 m and 3.6 m. Endeco-105 current meters were used throughout the field program. Hourly values for wind speed and direction were obtained from the Dauphin Island Sea Lab meteorological station on the east end of Dauphin Island (Fig. 1). Hourly sea levels were obtained from a USGS gauge (#87503201) in Weeks Bay and from a NOAA/NOS gauge (#873-5180) operated in Mobile Bay at Dauphin Island (Fig. 1). Daily average river discharge measurements for the Fish River were taken from the USGS gauge (#02378565) at Silverhill (Fig. 1).

Wind velocity time-series from Dauphin Island were converted to north-south and east-west wind stress before further analysis. In previous work (Schroeder and Wiseman, 1986), this coordinate system was found to be excellent for explaining low-frequency water level fluctuations in Mobile Bay. Visual comparison of Mobile Bay and Weeks Bay water levels showed striking similarity, suggesting co-oscillation. Therefore, velocity time-series were rotated to a natural coordinate system for analysis. Our primary concern with the currents was low-frequency flushing events. For this reason, we utilized a natural coordinate system different from that of the winds. Since tidal currents were the most energetic current signals in the records, the currents were rotated into a representation defined by the principal axes of the currents within the tidal band (0.6-2.4 cpd). Along-channel was defined to be the direction of the major principal axis (50°M) and cross-channel was defined to be 90° to the left of this direction.

All time series were initially low-pass filtered (eliminating signals with periods shorter than 40 hours). This was accomplished by fast Fourier transforming the time-series, setting the coefficients at all frequencies to be eliminated to zero, and inverse transforming. Periodograms of the resultant series were constructed, averaged over non-overlapping bands to increase stability, and divided by the band width to obtain estimates of the power spectra. Confidence intervals were determined from chi-square tables. Phase and coherence estimates were calculated from the fast Fourier transforms for non-overlapping adjacent band-widths. Significance levels for the coherence-squared were also estimated (Jenkins and Watts, 1968).

Multiple coherence-squared between three variables was estimated following Jenkins and Watts (1968) and significance levels again determined as before. Multiple coherence analysis attempts to sort out the individual impacts of multiple inputs on the output of a linear system. Suppose a system has one output, say bay water level, and two inputs, perhaps wind stress components. Assume the output is coherent with both inputs and these are coherent with one another. If the true dynamics relate water level to only one wind component, we cannot determine this relation from simple coherence analysis. The other stress component is also coherent with water level, not because of the dynamics but because it resembles the dynamically important component. Multiple-input spectrum analysis, or multiple coherence analysis, attempts to circumvent this problem. It finds the optimum linear prediction of the output using the first input and of the second input using the first input. It then computes the coherence between the residuals of the output and second input after subtraction of the linear predictions. If this coherence is significant, it cannot be attributable to similarity between inputs one and two. A dynamical cause may reasonably be sought. For the shorter time-series, in order to reduce the variability between neighboring estimates, a simple triangular smoothing filter, a Hanning filter, was passed over the coherence estimates.

Complex empirical orthogonal function (EOF) analysis (Legler, 1983; Wiseman et al., 1988) was applied to concurrent groups of time series using the correlation matrix rather than the covariance matrix. When these groups included scalar time-series, the series were synthetically treated as complex by assuming the imaginary portion of the signal was always zero. Although the third complex EOF was generally significantly different from random noise (Overland and Preisendorfer, 1982), the amount of variance explained was so slight that we present only the first two components.

III. Results

III.1 Spectra

Two long time-series (9 May 1986 until 31 March 1988) of water level in Weeks Bay and at Dauphin Island in Mobile Bay were available as well as a concurrent time-series of wind stress at Dauphin Island. The spectra estimated from the two water level records were nearly identical (Figs. 2 and 3). Both water levels were forced by the same external processes. Both spectra increased rapidly toward low frequencies. There was an indication of elevated energy levels between 0.1 and 0.4 cpd (cycles per day). This is the frequency band that contains the strongest signals of winter cold-front passages (Fernandez-Partagas and Mooers, 1975; DiMego et al., 1976). The wind stress spectra also increased toward low frequencies (Fig. 4), but the spectra were not as energetic at low frequencies as those from the water level time-series. The spectra exhibited a broad plateau between frequencies of 0.02 and 0.3 cpd, a frequency range that contains the synoptic weather band. Within this frequency band the north-south wind stress exceeded the east-west wind stress by a factor of about 3.

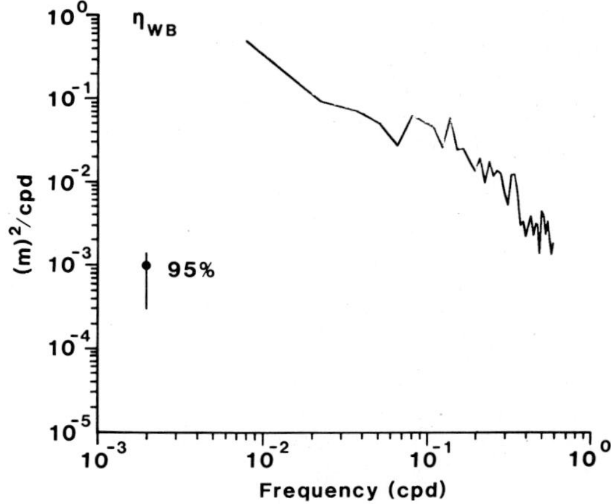

FIGURE 2. Spectrum of Weeks Bay water level.

The spectrum of Fish River runoff to Weeks Bay (not shown) is moderately flat at the lowest frequencies resolved. As frequency increases, so does the rate of decay of the variance in the signal. At the highest subtidal frequencies, the spectrum decays as F^{-3}

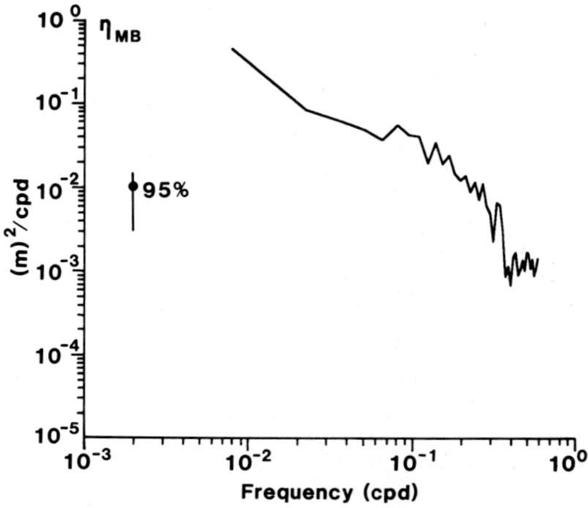

FIGURE 3. Spectrum of Mobile Bay water level.

The current meter records from the mouth of Weeks Bay were too short to allow us to obtain the same spectrum resolution and stability as shown in the previous spectra. Stability was sacrificed for increased resolution, i.e., the number of degrees of freedom in each spectrum estimate was reduced. The along-channel components were always more energetic than the cross-channel components, often by more than an order of magnitude. There was a slight tendency for the currents to increase in variance at lower frequencies. The trend, though, was weak. The low number of degrees of freedom of the spectrum estimates precluded accurate estimation of the spectrum slope or identification of significant spectrum peaks. The spectrum of the longest record, in summer 1987 (not shown), decayed at a rate of approximately f^{-1} for the along-channel component, and a significant peak occurred near 0.075 cpd in the cross-channel spectrum. We will generally confine our attention to the along-channel component, which controls the exchange of water between Weeks Bay and Mobile Bay.

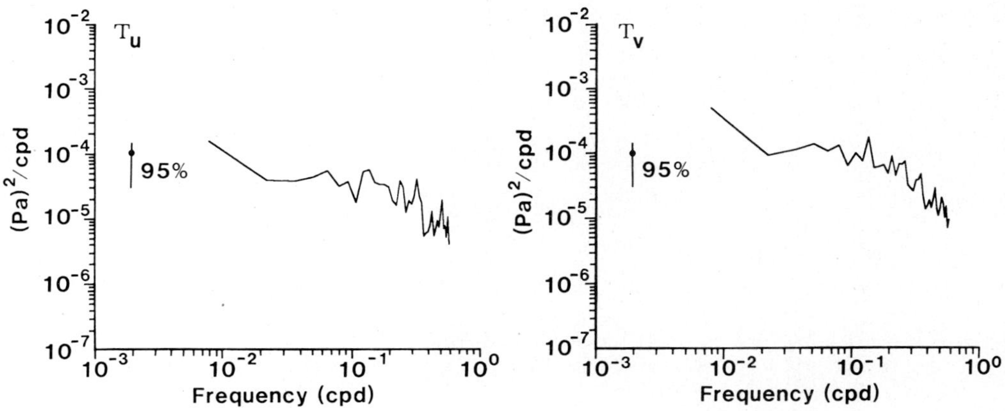

FIGURE 4. Spectra of east-west (left) and north-south (right) wind stress.

III.2 Coherence Analysis

Cross-spectrum analysis of the longest available time-series is enlightening. The water level series from Mobile Bay and Weeks Bay are significantly coherent at the 95% level for all frequencies analyzed. Unit coherence is contained within the 95% confidence intervals at most frequencies. The coherence decays with increasing frequency and the rate of decay appears to increase somewhat near frequencies of 0.1 - 0.2 cpd. The lag between the two signals is near zero. The dominant control of water level in Weeks Bay and, therefore, barotropic subtidal exchange with Mobile Bay is a co-oscillation with Mobile Bay.

Because of the flashy nature of the runoff time-series, we had initially expected runoff to be coherent with indicators of storminess and the latter to be responsible for increases of water level within Weeks Bay. Somewhat surprisingly, we find that, aside from a few isolated spikes, the coherence spectrum between runoff and north-south wind stress, east-west wind stress, and Weeks Bay water level (not shown) are all insignificant at the 95% level.

The two wind-stress components exhibit weak, but significant, coherence in two narrow bands, near 0.12 and 0.35 cpd (not shown). In what follows we consider the two wind stress components to be essentially incoherent at subtidal frequencies.

Weeks Bay water level is significantly coherent with east-west wind stress at frequencies between 0.1 and 0.5 cpd, with a peak near 0.35 cpd (Fig. 5, left). The coherence between Mobile Bay water level and the same wind stress time series is nearly identical, again reflecting the forcing of Weeks Bay water level by changes in Mobile Bay water level. These changes are the result of estuarine-shelf exchange induced by wind-driven Ekman convergence/divergence at the coastline (Schroeder and Wiseman, 1986). Weeks Bay water level is significantly coherent with the north-south wind stress at frequencies lower than about 0.5 cpd (Fig. 5, right). The coherence spectrum is similar to that between Mobile Bay water level and north-south wind stress, but the coherence levels are higher for Weeks Bay. Estuarine-shelf exchange in response to north-south wind forcing is generally restricted to the highest subtidal frequencies in Mobile Bay (Schroeder and Wiseman, 1986). The exchange of waters between Weeks Bay and Mobile Bay is related to this shelf-estuarine exchange, but also to local forcing by the north-south wind stress.

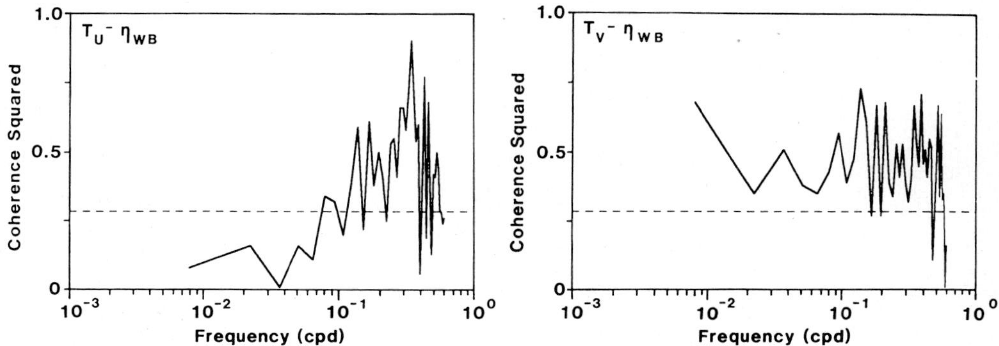

FIGURE 5. Coherence-squared between east-west wind stress and Weeks bay water level (left). Coherence-squared between north-south wind stress and Weeks Bay water level (right). Dashed line represents 95% significance level.

The coherence estimates from the short records appropriate to individual current meter deployment periods were less illuminating. The short records required low stability in the estimates in order to retain reasonable frequency resolution. Significance levels were extremely high. Somewhere in the band from 0.2 to 0.5 cpd, we generally found a region of significant coherence between east-west wind stress and along-channel current and between along-channel current and Weeks Bay water level. These coherences were not very much above the significance level, nor were they at a consistent frequency from deployment to deployment. Occasionally, a relationship was found between runoff, along-channel flow, and Weeks Bay water level. At the lowest frequencies, though, it is not immediately apparent that high coherence is indicative of an important relationship if one of the time series consists of isolated events, as in our runoff record. A single large event of only one or two days' duration might dominate the low-frequency coherence analysis of a two-month record. Even when modest and potentially meaningful cross-spectrum relationships were observed, these did not occur consistently between years or seasons. It is not at all clear that spectrum and cross-spectrum analyses are the statistical techniques of choice when dealing with such signals. Furthermore, the various forcing processes were clearly not independent. We attempted to resolve certain of these difficulties by utilizing multiple-input spectrum analysis techniques.

III.3 Multiple Coherence Analysis

Once again, the most informative results were derived from the longest concurrent records. The coherence between Weeks Bay water level and east-west wind stress, having removed the effects of north-south wind stress, was generally significant below frequencies of about 0.5 cpd (Fig. 6, left). This is very similar to what was found when Mobile Bay water level was substituted for Weeks Bay water level (Fig. 7, left), again suggesting a co-oscillation of the two bays with the east-west wind stress forcing Mobile Bay water levels. The coherence between Weeks Bay water level and north-south wind stress, having removed the effects of east-west wind stress (Fig. 6, right), is significant below 0.5 cpd. When Mobile Bay water level (Fig. 7, right) is substituted for Weeks Bay water level, this coherence is only marginally significant at the highest subtidal frequencies.

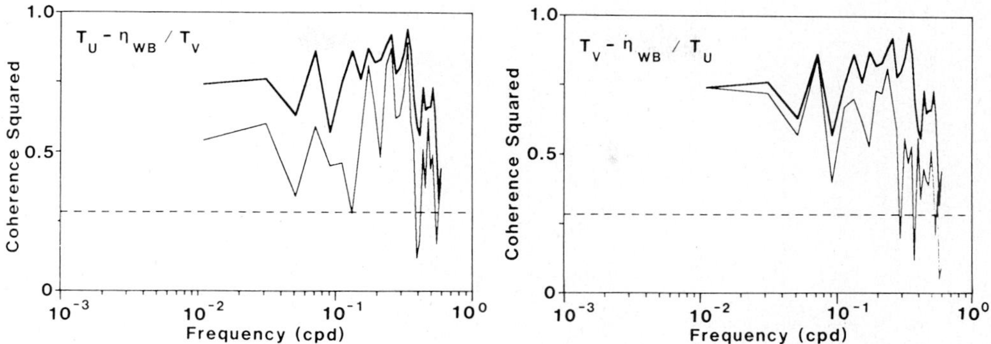

FIGURE 6. Partial coherence-squared (light line) between east-west wind stress and Weeks Bay water level after the effect of north-south wind stress is removed (left). Partial coherence-squared (light line) between north-south wind stress and Weeks Bay water level after the effects of east-west wind stress are removed (right). Dark line is total coherence-squared. Dashed line represents the 95% significance level.

Co-oscillation of the two bays is further supported by the coherence between the two water level records after removing the effects caused by either wind stress component (Figs. 8, right and 9, right). The resultant coherences are significant at all subtidal frequencies investigated. Coherence between Weeks Bay water level and north-south wind stress after removal of Mobile Bay water level effects is significant in a narrow band near 0.4 cpd and at frequencies below 0.35 cpd (Fig. 8, left). A similar analysis using east-west wind stress produces significant coherences at essentially all subtidal frequencies analyzed (Fig. 9, left). These analyses suggest that the local north-south wind stress is important in driving exchange between the two bays at periods longer than 3 days and the east-west wind stress at periods longer than 1.7 days.

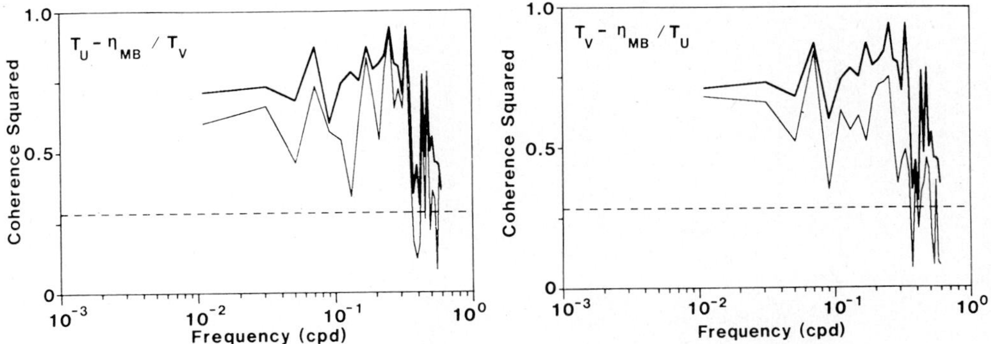

FIGURE 7. Partial coherence-squared (light line) between east-west wind stress and Mobile Bay water level after the effect of north-south wind stress is removed (left). Partial coherence-squared (light line) between north-south wind stress and Mobile Bay water level after the effects of east-west wind stress are removed (right). Dark line is total coherence-squared. Dashed line represents the 95% significance level.

Other than a few isolated spikes of coherence, Weeks Bay water level is insignificantly coherent with runoff after removal of the effects of Mobile Bay water level or of either wind stress component. If river flow is important to the functioning of the bay, the effects are confined principally to driving a baroclinic flow. As we shall see, the existing records are probably too short to allow extraction of evidence for such flow from the data.

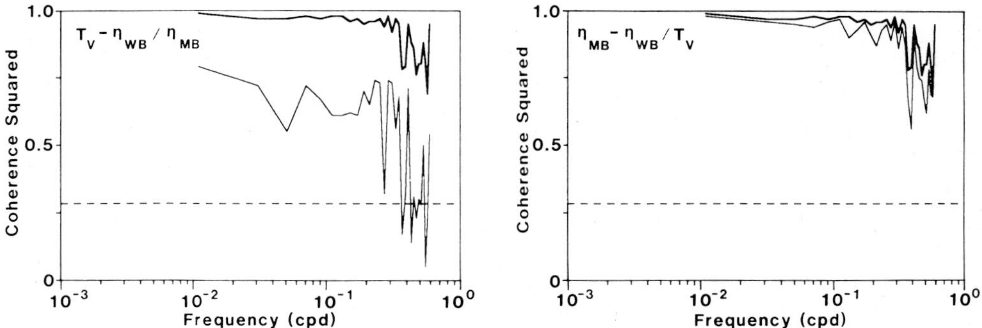

FIGURE 8. Partial coherence-squared (light line) between north-south wind stress and Weeks Bay water level after the effects of Mobile Bay water level are removed (left). Partial coherence-squared (light line) between Mobile Bay and Weeks Bay water levels after the effect of north-south wind stressis removed (right). Dark line is total coherence-squared. Dashed line represents the 95% significance level.

Current meter records were only a few months in duration. Coherence and multiple coherence analyses were unreliable because of the nonstationary nature of the signals. For example, a long summer and fall record from 1987 exhibited strong coherence between along-channel flow and runoff after removal of the effect of other inputs. This broad-band response in the frequency domain, though, is due to a single short-lived event (Fig. 10, left), to be discussed below. With so many spectrum analysis packages available, a certain amount of caution must be exercised in drawing conclusions concerning the parent distributions from analysis of short records.

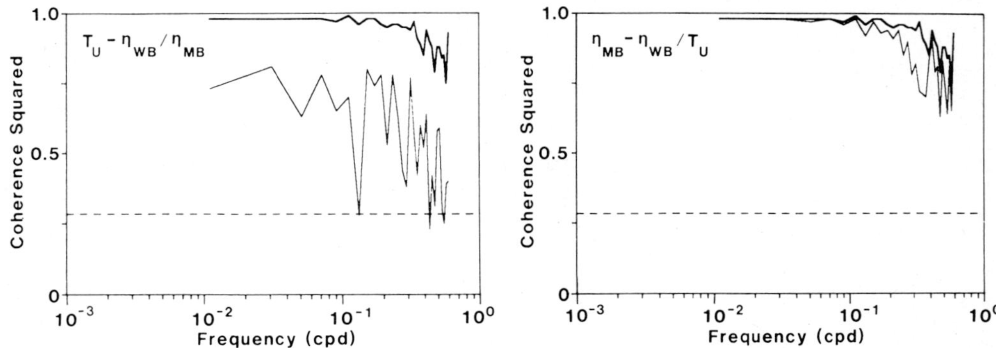

FIGURE 9. Partial coherence-squared (light line) between east-west wind stress and Weeks Bay water level after the effects of Mobile Bay water level are removed (left). Partial coherence-squared (light line) between Mobile Bay and Weeks Bay water levels after the effect of east-west wind stress is removed (right). Dark line is total coherence-squared. Dashed line represents the 95% significance level.

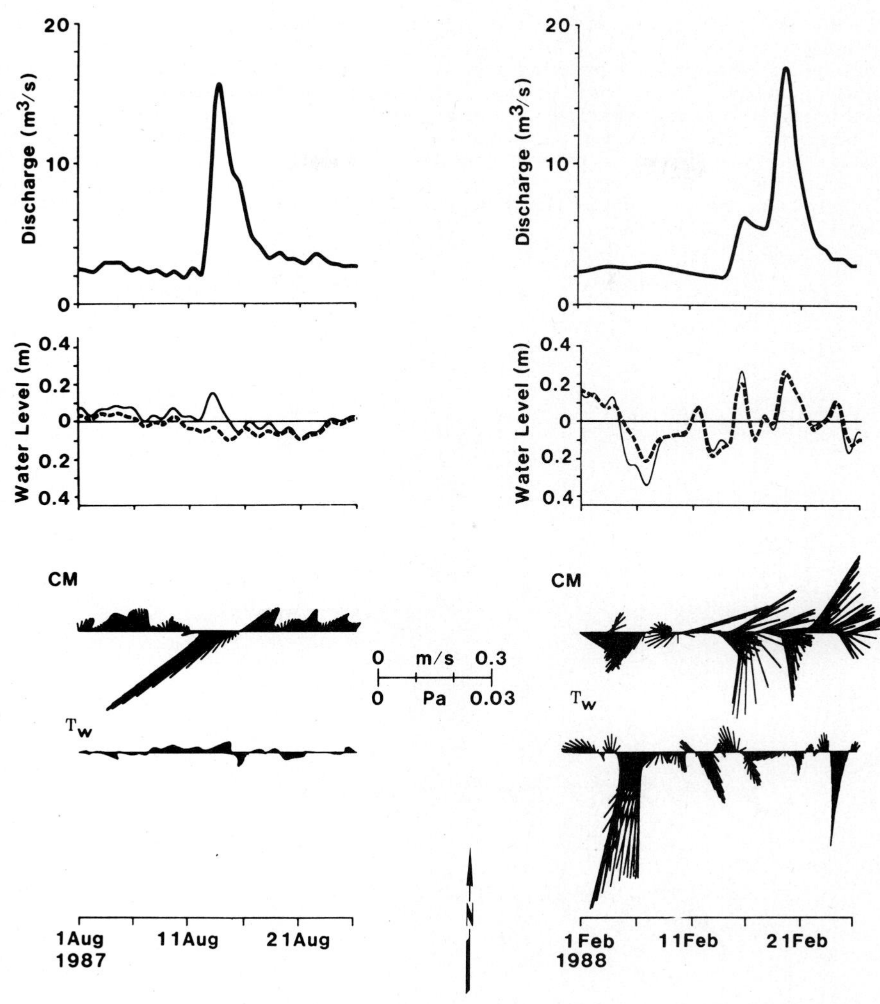

FIGURE 10. Fish River discharge, Mobile Bay (dash line) and Weeks Bay (solid line) water level, current records, and wind stress, for 1 August to 26 August 1987 (left) and 1 to 21 February 1988 (right).

III.4 Discharge Events

It has been suggested above that spectrum analysis is not the appropriate statistical methodology to apply to determine the role of an event-dominated process in driving some other process. Most of the cross-spectra and multiple cross-spectra estimated were erratic in character, exhibited low coherence, or were difficult to interpret because the time-scales of significant coherence did not correspond to a dominant and identifiable period in the data time-series. Nevertheless, visual inspection of the individual time series suggests important physical relationships.

During the winter of 1987, winds were strong and variable in the first half of the current meter deployment, i.e., from mid-January until about 10 February (Fig. 11). Throughout this period, the wind stress controlled the water level and the currents, even though a weak increase in discharge occurred from 23 to 29 January. The second half of the deployment was a period of weak winds, and a strong flooding event occurred from 17 February until 3 March. During this event, observed along-channel currents were clearly out of the estuary. Runoff appears to be an important control of the current field when the wind stress is not dominant.

From 13 to 15 August 1987, during a period of weak winds, there was a short-lived discharge event on the Fish River (Fig. 10, left). Daily discharge rates attained values of 17 m^3/s. This event was accompanied, once again, by a strong along-channel outflow from the estuary. It was also accompanied by a significant rise in water level within Weeks Bay at a time when water level in Mobile Bay was falling. Whereas the February 1987 discharge event was not accompanied by large apparent water level differences between the two bays (Fig. 11), the August event was (Fig. 10, left). It would appear that the February 1987 event drove a baroclinic flow response, while the August event was, at least partially, barotropic.

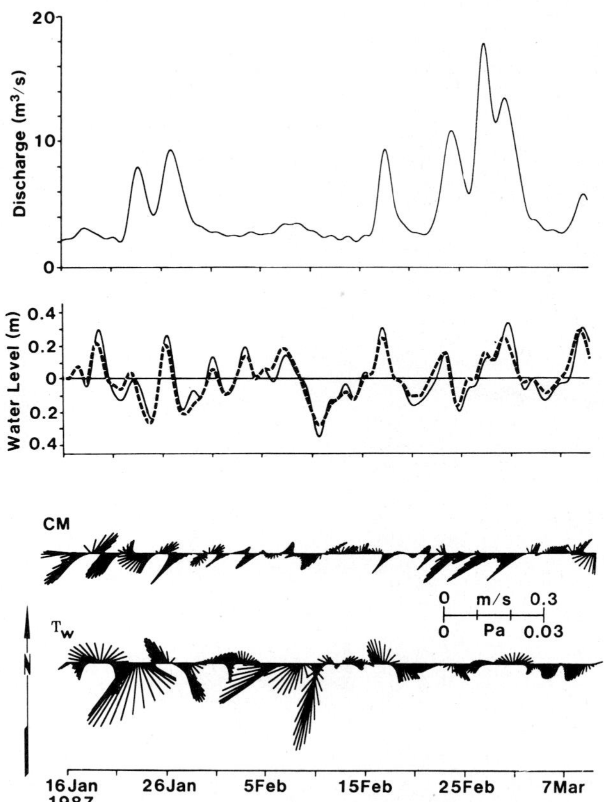

FIGURE 11. Fish River discharge, Mobile Bay (dash line) and Weeks Bay (solid line) water level, current meter records, and wind stress, for 16 January to 12 March 1, 1987.

Two significant events occurred in early and mid-February, 1988 (Fig. 10, right). A strong north wind pulse affected the region from 3 to 6 February, resulting in outflow from Weeks Bay and a significantly larger water level drop in Weeks Bay than was oberved in Mobile Bay. This was due to the local effect of the north-south wind stress described above. The Fish River daily discharge peaked at 17 m^3/s on 19 February. No change in Weeks Bay water level could be directly attributed to this event. The currents, though, were out of the estuary during periods of rising water level. This was a direct response to local forcing by the river discharge. Again, the currents responded to local river discharge, but the water level signal was predominantly affected by co-oscillation with Mobile Bay.

Visual inspection of the various deployment records is suggestive of runoff driving both barotropic and baroclinic exchanges between Weeks Bay and Mobile Bay. The records, though, are insufficient to determine the dynamical balances involved.

III.5 Complex Orthogonal Function Analysis

Because of the short nature of the available records and their event-dominated character, we analyzed the data from each deployment period using empirical orthogonal functions (Table 1). The first empirical mode was dominated by the variation of the water level in the two bays. It usually included a significant amount of wind stress variation as well. During the winter of 1987, though, the wind stress was not strongly involved in this mode. Rather, current variations played the role normally played by wind stress.

TABLE 1. Statistics for Mode 1 and Mode 2 complex orthogonal function analysis results for four time periods associated with four current meter deployments: percent total variance explained, $\%S_t$, and amplitudes, phase angles, and percent variance explained, $\%S_i$, for individual inputs. Inputs are low-passed Weeks Bay water level, WB, Mobile Bay water level, MB, current velocity rotated along-channel, CM, wind stress, T_w, and Fish River discharge, Q.

Input	Mode 1 Amp.	Mode 1 Phase+	Mode 1 $\%S_i$	Mode 2 Amp.	Mode 2 Phase+	Mode 2 $\%S_i$
Summer 1986						
	$\%S_t = 55.41$			$\%S_t = 31.55$		
WB	0.65	2	93	0.18	-22	2
MB	0.61	-2	81	0.34	16	15
CM	0.21	-143	10	0.75	85	71
T_w	0.41	111	38	0.54	-97	37
Q	-	-	-	-	-	-
Winter 1987						
	$\%S_t = 46.59$			$\%S_t = 27.67$		
WB	0.62	1	88	0.20	1	5
MB	0.55	-1	71	0.39	-3	21
CM	0.40	-166	37	0.53	39	39
T_w	0.23	62	13	0.60	159	50
Q	0.33	5	25	0.41	-155	23
Summer 1987						
	$\%S_t = 42.65$			$\%S_t = 32.56$		
WB	0.67	-1	95	0.11	58	2
MB	0.61	-4	79	0.22	132	8
CM	0.05	-135	1	0.68	167	75
T_w	0.42	83	37	0.19	3	6
Q	0.08	46	2	0.67	-17	72
Winter 1988						
	$\%S_t = 47.23$			$\%S_t = 24.04$		
WB	0.60	-1	86	0.31	-11	12
MB	0.59	-3	81	0.29	-4	10
CM	0.25	22	15	0.70	169	59
T_w	0.43	95	44	0.25	-56	8
Q	0.21	7	11	0.52	-169	32

+ Positive phase measured counterclockwise from positive east.

During summer 1986 and winter 1987, the second mode was dominated by currents and wind stress. In summer 1987 and winter 1988, though, currents and runoff were the two variables most explained by this mode.

The third mode (not shown), while generally different from random noise, explained less than 18% of the correlation in each data set. During the winters, the mode principally explained the runoff signal, whereas it was dominated by wind stress variation during the summers.

IV. Conclusions

The two years of available data from Weeks Bay allow us to begin to understand the mechanisms of subtidal barotropic exchange with Mobile Bay. The available current meter records, though, are not sufficiently long to permit us to discern the relationships between baroclinic flow between the two water bodies and the external driving forces.

It is clear from the long water level and wind stress time series that subtidal barotropic exchange is dominated by a co-oscillation of the two bays. The east-west wind stress causes filling/emptying of Mobile Bay (Schroeder and Wiseman, 1986) and concurrent filling/emptying of Weeks Bay. The north-south wind stress also drives an exchange of water between Mobile Bay and the adjacent shelf. This exchange is also reflected in the water level records at Weeks Bay, but there is an additional signal in Weeks Bay that appears to be driven locally by the north-south wind stress. This is probably due to the winds blowing over the shallow waters of southeastern Mobile Bay, immediately adjacent to Weeks Bay, similar to the effects seen near Atchafalaya Bay, on the west Louisiana shelf, by Chuang and Wiseman (1983).

The time-series available and spectrum analyses performed did not permit determination of a statistically significant relationship between river discharge and Weeks Bay water level. Nevertheless, consideration of individual flooding events does suggest that large runoff events will both alter the water level within Weeks Bay and drive a strong exchange with the adjacent waters of Mobile Bay. These relationships may be reflected in the seasonal cross-spectrum analyses and also in the complex EOF analyses, but the relationships are weak. This reflects the paucity of events during current meter deployment periods, the complexity of the relationships between the various parmameters considered in this study, and the inappropriateness of spectrum analysis for the investigation of event-dominated processes.

Acknowledgements

This work is the result of research sponsored by the U.S. Department of Commerce, NOAA, NOS, Office of Ocean and Coastal Resource Management, Marine and Estuarine Management Division under Grant No. NA86AA-D-CZ017, The University of Alabama Marine Science Program, the Marine Environmental Sciences Consortium of Alabama, the Mississippi-Alabama Sea Grant Program (Grant No. NA85AA-D-SG005; Projects R/ER-16 and E/O-16) and the Louisiana State University Sea Grant College Program (Grant No. NN85AA-D-SG141, Project R/EMP-3). T. R. Matthews served as the Graduate Research Assistant for the project and was responsible for field surveys and laboratory analyses. J. L. Pearman and H. H. Jeffcoat coordinated our co-operative program with the USGS.

Contribution No. 127 of the Aquatic Biology Program, The University of Alabama; Contribution No. 161 of the Marine Environmental Sciences Consortium, Dauphin Island, Alabama; and Contribution No. 4 from the School of Geoscience, Louisiana State University.

V. References

Chuang, W. S. and Wm. J. Wiseman, Jr., 1983: Coastal sea level response to frontal passages on the Louisiana-Texas shelf. J. Geophys. Res., **88**, 2615-2620.

Crance, J. H., 1971: Description of Alabama estuarine areas - cooperative Gulf of Mexico estuarine inventory. Alabama Mar. Res. Bull., **6**, 85 pp.

DiMego, G. W., L. F. Bosart and G. W. Endersen, 1976: An examination of the frequency and mean conditions surrounding frontal incursions into the Gulf of Mexico and Caribbean Sea. Mon. Weather Rev., 104, 709-718.

Fernandez-Partagas, J. and C. N. K. Mooers, 1975: Some front characteristics over the eastern Gulf of Mexico and surrounding land areas. Final report to BLM under contract 08550-CT4-L6.

Jenkins, G. M. and D. G. Watts, 1968: Spectral analysis and its applications. Holden-Day, 525 pp.

Legler, D. M., 1983: Empirical orthogonal function analysis of wind vectors over the tropical Pacific region. Bull. Amer. Meteor. Soc., **64**, 234-241.

Overland, J. E. and R. W. Preisendorfer, 1982: A significance test for the principal components applied to a cyclone climatology. Mon. Weather Rev., **110**, 1-4.

Schroeder, W. W. and Wm. J. Wiseman, Jr., 1986: Low-frequency shelf-estuarine exchange processes in Mobile Bay and other estuarine systems on the northern Gulf of Mexico. Estuarine Variability, D. Wolfe, Ed., Academic Press, 355-367.

Wiseman, Wm. J., Jr., W. W. Schroeder and S. P. Dinnel, 1988: Shelf-estuarine water exchanges between the Gulf of Mexico and Mobile Bay, Alabama. Amer. Fish. Soc. Symp., **3**, 1-8.

Wong, K. C. and R. W. Garvine, 1984: Observation of wind-induced subtidal variability in the Delaware Estuary. J. Geophys. Res., **89**, 10589-10597.

Circulation Dynamics in Mt. Hope Bay and the Lower Taunton River

Malcolm L. Spaulding and Frank M. White

Department of Ocean Engineering
University of Rhode Island
Kingston, RI 02881

ABSTRACT

A field study was conducted in Mt. Hope Bay and the lower Taunton River to describe the circulation in response to tide, wind and density induced forcing. Five current meter deployments were made from December 1985 to January 1987 with a typical duration of 60 days. Measurements were made at surface and bottom locations at three stations: Mt. Hope Bridge, Brightman St. Bridge and at the entrance to the Sakonnet River. Time series measurements of currents, conductivity and temperature were made at each of the stations. Tidal elevations were recorded at Mt. Hope Bridge, Steep Brook and Touisset. River flow data were obtained for the Taunton River from USGS and wind speed and direction from NOAA National Weather Station at Green Airport, Warwick, Rhode Island.

Analysis of the field data indicate that the sea surface elevation and currents are dominated by the semi-diurnal tide (80-90% current variance) in an approximate standing wave pattern. Tidal currents are a strong function of location and in narrow passages reach peak values of 25 cm/s, but have more typical maxima of 10 cm/s in the bay proper. Low frequency currents (30 hr filtered) show little wind induced forcing and are dominated by steady flows. For the lower portion of the Taunton River (Brightman St. Bridge) a classical stratified estuarine circulation pattern is observed. For the Sakonnet River and Mt. Hope Bridge sites the flow is consistently out of Mt. Hope Bay. The hydrography shows that the water column is well mixed throughout Mt. Hope Bay. As one approaches the lower end of the Taunton River the water column becomes stratified because of the river discharge of freshwater.

I. Introduction

Pollution in Mt. Hope Bay and the Taunton River located adjacent to the east passage of Narragansett Bay (Figure 1) has long been recognized as a serious interstate water quality problem concerning both Massachusetts and Rhode Island (EPA, 1971). Seven municipal waste water treatment plants discharge into the Mt. Hope Bay watershed. In addition there are six thermal electric generating stations which discharge water into the basin.

Unfortunately little is known about how this material is transported through the bay or its ultimate flux into Narragansett Bay. Previous investigations of circulation in the region are extremely meager (see Bay Bib, Dunn _et al._, 1979); restricted to short duration measurements (Haight, 1936) or coarse grid two-dimensional, vertically-averaged hydrodynamic modeling (Swanson and Spaulding, 1974; Gordon, 1982 and Hess and White, 1974). The objective of the present paper is to describe the circulation dynamics in the Mt. Hope Bay-Taunton River system using field measurements and data analysis.

Coastal and Estuarine Studies, Vol. 38
R. T. Cheng (Ed.)
Residual Currents and Long-term Transport

II. Description of Study Area

The bay is relatively shallow with 70% of the bay having a mean low water (MLW) depth of 6 m or less. Mt. Hope Bay is connected to the east passage of Narragansett Bay at Bristol Ferry (Figure 1). A restricted connection is made to Rhode Island Sound through the Sakonnet River at the southeast corner of the bay. The primary tidal entrance to the bay is at Bristol Ferry. Starting at Mt. Hope Bridge, a natural channel extends to a point between Mt. Hope and Common Fence Points. From this point a man-made channel 120 m wide with a MLW depth of 10.5 m extends to the northeast and up the Taunton River. A second channel with similar dimensions originates at Common Fence Point, extends easterly and then northerly toward Fall River. Finally a small channel, approximately 3 km south of the Taunton River mouth extends northwesterly from the central channel to the Brayton Point power plant.

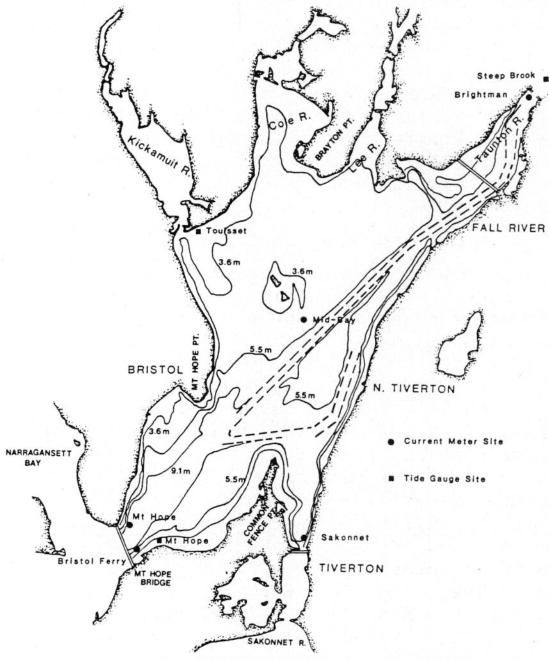

FIGURE 1. Study area and location of current and sea elevation observation stations in Mt. Hope Bay and the lower Taunton River. The location of the dredged shipping channels, minimum 10 m depths, are shown by the dashed lines.

Tidal fluctuations for Mt. Hope Bay, measured at Fall River, indicate a mean tidal range of 1.34 m, a maximum of 1.68 m for spring conditions and a minimum of 1.0 m at neap. The water column is generally well mixed throughout the year, although significant stratification may occur in salinity in the Taunton River during spring runoff (Hicks, 1959). Maximum tidal currents are typically 10-25 cm/s in the bay but can reach 2 m/s in the narrows connecting Mt. Hope Bay to the Sakonnet River.

Five rivers discharge into Mt. Hope Bay - the Taunton, Cole, Lee, Kickamuit and Quequechan. Together they represent 35% of the watershed for Narragansett Bay (Pilson, 1984). The Taunton has the most significant input with an annual average discharge of 13.14 m^3/s over a 46 year period (1929-1976) measured at the State Farm gauge in Bridgewater, Massachusetts. Peak flows occur in the spring with values on the order of 20 m^3/s and minimum flows in the fall of 3 m^3/s. Combining this with the 4.8 m^3/s annually discharged by the Three Mile River gives a discharge of 17.94 m^3/s into Mt. Hope Bay. The Mill, Forge, and Assonet Rivers, not shown in Figure 1, contribute significantly smaller amounts of fresh water to the tidal portion of the Taunton.

II.1 Measurement program

The primary purpose of the measurement program was to obtain data that would allow insight into how the movement of water and salt in the study area are influenced by tidal, meteorological, and density forcing. The seasonal and spatial variation of these processes in the system is also of interest. The second purpose was to obtain data sets for the calibration and verification of hydrodynamic models of the area.

Six current meters were used in the study: five Endeco type 174 meters and one Marsh-McBirney model ARC-585 meter. The Endeco meters were equipped with sensors to measure temperature and conductivity. Current speeds were averaged over 5 minute (Endeco 174) or 2 minute (Marsh McBirney) intervals and then recorded at 5 minute (Endeco 174) or 10 minute (Marsh McBirney) intervals. Recent calibration data was available for all meters and expected instrument errors were small for current speed (<0.4 %), temperature (<0.1 °C) and conductivity (<1.0 μmho/cm).

Wind and atmospheric data were obtained at hourly intervals from the National Weather Service Station at T.F. Green Airport, Warwick, Rhode Island, approximately 8 km west of Mt. Hope Bay. Weisberg (1974) found from an analysis of wind data at Green Airport and Quonset Point, 22 km distant, that the wind field was nearly uniform at periods greater than 12 hours. Winds were therefore assumed to be uniform in the Mt. Hope Bay-Green Airport area.

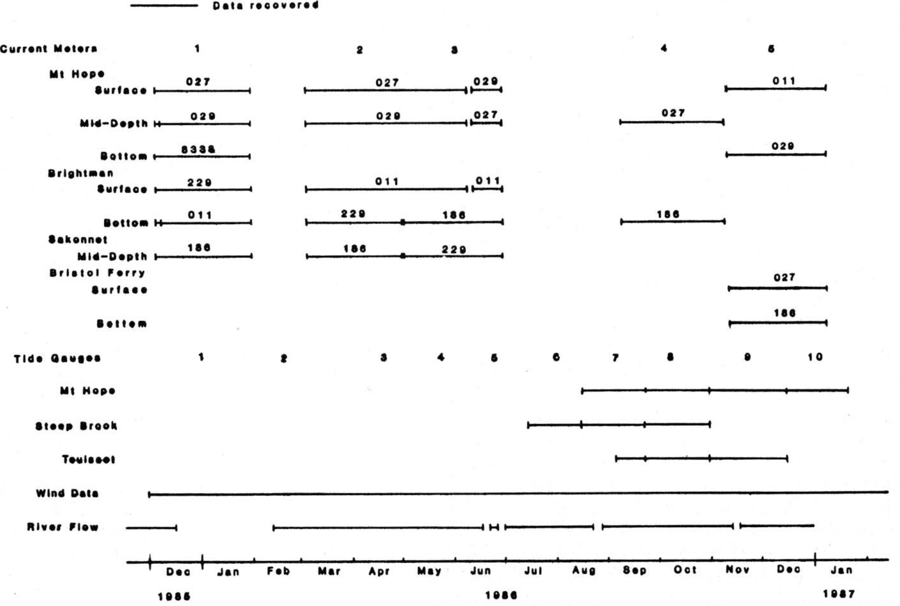

FIGURE 2. Time schedule for field measurement program.

Daily averaged river flow data were obtained from the U.S. Geological Survey for the Taunton River at the State Farm gauge at Bridgewater, Massachusetts. The data were spotty throughout the study period. Measurements of tidal height were made using atmospherically compensated submerged pressure transducers. Pressure was recorded on solid state memory at five minute intervals. Current measurements were made at the surface (15.9 m, referenced to the bottom), mid-depth (9.47 m), and bottom (3 m) at the Mt. Hope Bridge transect. At the Brightman Street transect measurements were made at the surface (6.86 m) and bottom (3 m). Only mid-depth measurements (7.7 m) were made at the Sakonnet River at the Hummocks. Figure 1 shows the location of the observation stations.

Four deployments were performed roughly corresponding to each of the four seasons. Figure 2 shows the times and lengths of the deployments for each station. The deployment number is given at the top of the figure. Noted above the line is the serial number for the current meter (011, 027, 029, 186, 229 - Endeco and 8338 - Marsh-McBirney) used at that location. All deployments were the same except the fourth where the Sakonnet River meter was moved to a location in the middle of the bay at mid-depth (2.9 m). This change was made to gather data from the central part of Mt. Hope Bay, since the Sakonnet measurement station showed extremely small tidal currents. This meter was stolen and no data was recovered.

A fifth deployment was made to gather additional data on the lateral variability of flows through the Mt. Hope Bridge transect. The meters were located in two arrays, one on either side of the shipping channel. The surface meters were at 2.5 m below the surface and the bottom meters at 17 m (north, Mt. Hope) and 10 m (south, Bristol Ferry), from the surface, respectively. Water depths at the north and south stations were 22 m and 17 m, respectively.

Analysis of the current meter data was focused on identifying the response to tidal, meteorological, and density induced forcing. Free surface and current field response to tidal forcing were determined using a modified version (Carrier et al., 1982) of the harmonic analysis method of Dennis and Long (1971). The M_2, S_2, N_2, O_1, K_1, M_4, M_6, M_8, S_4 and S_6 constituents were of primary interest. Tidal constituent phases were determined in modified epoch form.

A multiple input, single output, linear analysis technique, developed by Weisberg (1974), was used to relate low frequency current fluctuations with the local wind field. In this method, the one-dimensional output (current) series is expressed as the sum of the frequency domain filtering of the two input series (the North-South and East-West components of wind stress) by two transfer functions which are assumed to be time-invariant. The difference between this method and the more common one-input/one-output technique is that the two inputs are allowed to be partially coherent.

Low-pass filtering of the data was accomplished using a fourth order Butterworth filter (Roberts and Roberts, 1978). The phase shift inherent in the Butterworth filter is eliminated by passing the data forward and then backward through the filter.

III. Analysis of Field Measurements

III.1 Tidal circulation

The harmonic method was used to extract tidal constituent amplitudes and phases from the sea surface elevation and current meter data. Analysis of sea surface elevation tide height records reveals that tidal forcing accounts for about 80 to 95 % of the free surface fluctuations in Mt. Hope Bay. Tidal fluctuations are predictably dominated by the M_2 constituent with an amplitude of 57.6 cm at Mt. Hope Bay entrance and 56.7 cm at Steep Brook. The Touisset record shows an amplitude of only 49.5 cm but the record length is too short to provide an accurate measure of the M_2 tide. The M_2 constituent, which effectively represents the mean semi-diurnal tidal wave is modulated by the S_2 (17.6 cm) and N_2 (8.1 cm) constituents at periods of 14.8 and 27.6 days, respectively. The interaction of the three constituents causes the semi-diurnal tidal amplitude to vary between 37 cm at extreme neap conditions to 78 cm at extreme spring conditions.

The amplitudes of the two significant diurnal constituents, O_1 (4.1 cm) and K_1 (9.6 cm), are about 7-16 % that of the M_2. Their only significant contribution is to create a diurnal inequality. The M_4 overtide is the fourth largest constituent in the Taunton River, with an amplitude of approximately 10.1 cm. In spite of its relatively small amplitude, the M_4 constituent has a significant impact on the tidal nature of the area. Tidal forcing also dominates the current variance. Tidal forcing accounted for 87% of the variance for the Mt. Hope Bridge records, and 93 to 58% of the Brightman Street and Sakonnet record variances, respectively (Table 1).

Tidal currents in the bay are again dominated by the M_2 constituent, which have a vertically averaged mean amplitude of 20.3 cm/s, 22.6 cm/s, 30.1 cm/s and 4.6 cm/s at Mt. Hope Bridge, Brightman St. Bridge, Bristol Ferry, and Sakonnet, respectively. The near surface currents at Mt. Hope Bridge and Brightman St. Bridge are twice as large as the near bottom currents but only 50% higher at Bristol Ferry. The N_2 and S_2 constituents have approximately one fifth the magnitude of the M_2 and interact with it to create the neap-spring variation in the semi-diurnal band. The magnitudes of the diurnal constituents (O_1 and K_1) are quite small, on the order of 1-2 cm/s, while the M_4 and M_6 overtides are quite significant, 7-10 and 3-5 cm/s, respectively for the Mt. Hope and Brightman St. Bridge stations. The ratio of M_2/M_4 increases from 0.28 at Mt. Hope Bridge to 0.37 at Brightman St. Bridge and indicates that tidal energy at the M_2 frequency is transferred to its harmonic (M_4) as the tide passes through the bay. Numerical hydrodynamic model simulations (Spaulding et al, 1988) give similar results and indicate that the primary cause of the energy transfer within the bay is the quadratic bottom friction in the lower Taunton River.

Table 1. Partitioning of Current Meter Energies.

Location and Deployment	---- Percent of Total Energy ----		
	Low Frequency	Tidal Frequency	High Frequency
Near Brightman:			
1.011 bottom	3%	92%	5%
1.229 surface	2%	96%	2%
2.011 surface	2%	96%	2%
2.229 bottom	5%	88%	7%
3.011 surface	2%	96%	2%
3.186 bottom	5%	87%	8%
4.185 bottom	2%	93%	5%
Near Mt. Hope Bridge:			
1.027 surface	9%	86%	5%
1.029 mid-depth	6%	89%	5%
2.027 surface	7%	89%	4%
2.029 mid-depth	8%	87%	5%
3.027 mid-depth	3%	92%	5%
3.029 surface	8%	87%	5%
4.027 mid-depth	7%	83%	10%
5.011 surface	6%	89%	5%
5.029 bottom	7%	77%	16%
5.027 surface	5%	92%	3%
5.186 bottom	5%	90%	5%
Near Sakonnet:			
1.186 mid-depth	19%	60%	21%
2.186 mid-depth	27%	54%	19%
3.229 mid-depth	24%	59%	17%

The M_4 and M_6 constituents combine with the semi-diurnal constituents to create a double peak in current velocity during the flood portion of the tidal cycle. This effect is shown graphically in Haight (1936) and Turner (1984). The double flood effect is most apparent in the neap tide condition, since the sum of the M_2, N_2, and S_2 constituents minimizes the magnitude of the semi-diurnal signal at that time.

Between the Mt. Hope Bridge and Steep Brook stations, the tidal constituents based on surface elevation show very little change in phase and amplitude. The lunar semi-diurnal tide changes by 4 degrees in phase but by less than 1% in amplitude. The M_4 constituent phase changes by 1 degree and its amplitude increases by 33%. The diurnal constituents are constant but relatively noisy.

The uncertainty in the tidal constituent amplitudes and phases which result form non-tidal fluctuations in the data (Turner, 1984) are probably equal to or greater than their variations in the study area. The tidal motion of the free surface may therefore be approximated as simultaneous throughout the bay. At Mt. Hope Bridge the averaged tidal surface current leads the tidal free surface elevation by about 89 degrees. Both measures indicate that the tide in the bay is approximately a standing wave (Officer, 1976). This is consistent with the length scale of the area relative to the tidal wave length (Dronkers, 1964; Prandle and Rahman, 1980).

III.2 Wind-driven circulation

Narragansett Bay is a windy environment throughout the year. Wind speeds average 5 m/s during the winter and spring (from the northwest) and 4 m/s (from the southwest and west) during summer and fall.

Power spectra of the Green Airport winds show a peak wind energy at periods ranging from 3 to 10 days, corresponding to the passage of weather systems through the area. There is also a smaller peak in summer at a frequency of 0.042 cycles/hr (cph) corresponding to the diurnal seabreeze. Integration of these wind spectra shows that approximately 75% of the wind energy is contained at frequencies less than 0.033 cycles/hr, or periods greater than 30 hours. One therefore expects these low-frequency wind events to have the greatest influence on Mt. Hope Bay currents.

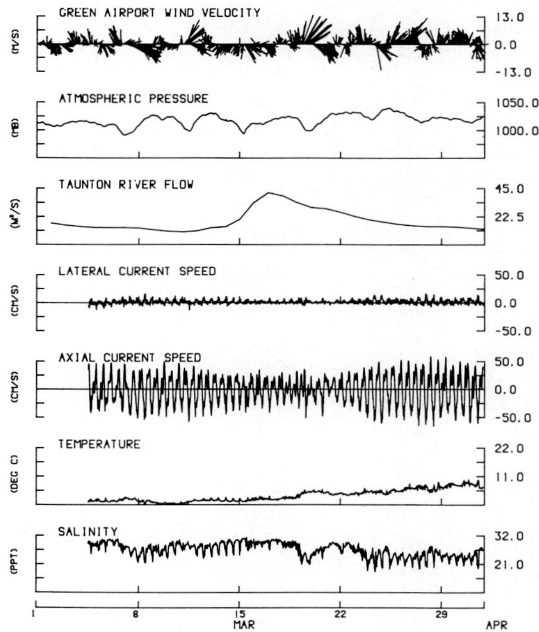

FIGURE 3. Wind, barometric pressure, and river-flow inputs compared with raw meter data at the surface at Brightman station for deployment 2.011 during March 1986.

Let us examine a particular station in detail. Figure 3 shows winds, barometric pressure, and Taunton River freshwater flow combined in a stackplot with the raw current meter, salinity, and temperature data during March 1986: surface, Brightman St. station, deployment 2.011. The current speeds and salinity variations are dominated by tidal oscillations. Temperature is only moderately affected by the tides. The raw data does not illuminate any obvious wind effects, although there is probably a connection between the high river flow on March 17th and the sharp drop in river salinity two days later.

Figure 4 shows the low-pass filtered (30 hr) version of Figure 3. Residual currents from 2 to 5 cm/s persist in the filtered record and vary with periods from

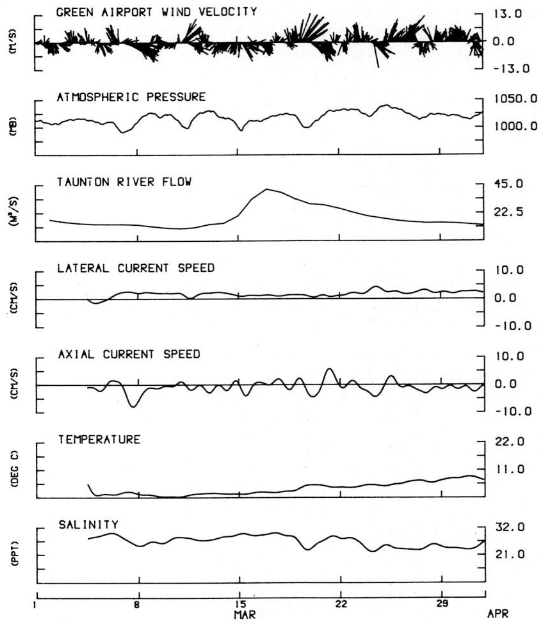

FIGURE 4. Low-passed (30 hour) filtered data for deployment 2.011 at Brightman station (Figure 3).

Table 2. Peak low-frequency multiple coherence between the wind and local current components.

Location and Deployment		--- Peak Coherence --- Axial	Lateral
Brightman:			
1.011 bottom		0.91	0.61
1.229 surface		0.77	0.72
2.011 surface		0.43	0.38
2.229 bottom		0.83	0.64
3.011 surface		0.66	0.51
3.186 bottom		0.65	0.48
4.186 bottom		0.33	0.45
	Average:	0.65	0.54
Mt. Hope:			
1.027 surface		0.67	0.71
1.029 mid-depth		0.68	0.72
2.027 surface		0.60	0.66
2.029 mid-depth		0.68	0.47
3.027 mid-depth		0.76	0.80
3.029 surface		0.89	0.89
4.027 mid-depth		0.71	0.35
5.011 surface		0.60	0.75
5.029 bottom		0.36	0.45
5.027 surface		0.85	0.59
5.186 bottom		0.68	0.42
	Average:	0.68	0.62

1 to 3 days. Axial (along channel) residual currents show only low correlation with the wind (Table 2) and little, based on a visual comparison, with river flow. These low frequency motions are probably caused by non local wind forcing in Narragansett Bay and adjacent R.I. Sound (Wang, 1979; Swenson and Chuang, 1983; Elliot, 1976, Garvine, 1985; Schroeder and Wiseman, 1986).

An indication of the overall effect of winds on the measured currents in Mt. Hope Bay can be found by examining the power spectra of the current data. Figure 5a shows the frequency spectrum of N/S and E/W currents from the Brightman deployment 2.011. A moderate amount of low-frequency energy, sharp peaks at the dominant tidal frequencies (O_1, K_1, N_2, M_2, M_4, M_6), and slight energy at high frequencies are noted. Figure 5b shows the cumulative energy computed by integrating Figure 5a over frequency to a normalized total of 1.00 or 100%. Low (<0.0333 cycles/hr) and high (>0.322 cycles/hr) frequencies in this figure are seen to contribute only a few % of the total energy. The dominant contribution comes from the tidal components, with the M_2 tide being responsible for about 70%. If we assume that all the low-frequency energy comes from the wind, this contribution is only two % of the total.

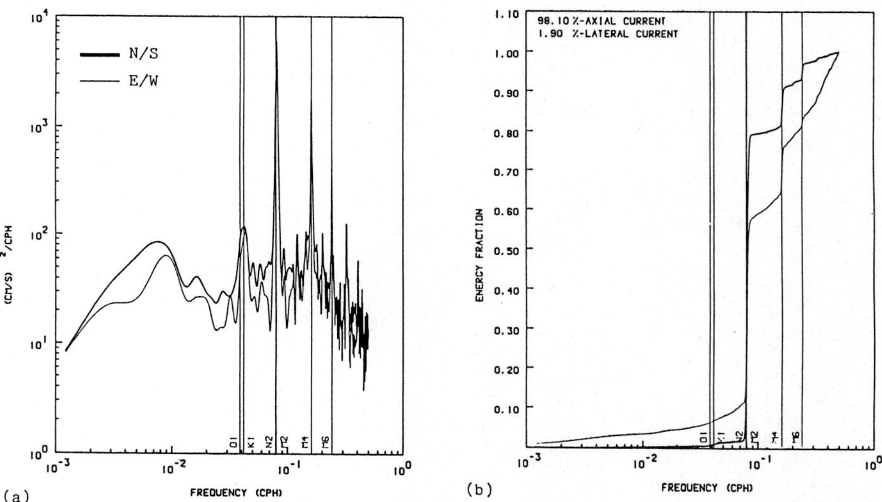

FIGURE 5. Power spectra of currents measured at the Brightman surface station during deployment 2.011: (a) energy at each frequency and (b) cumulative energy.

The partitioning of current energy among low, tidal, and high frequencies are listed in Table 1 for all of the 21 meter deployments of the study. The average of these values may be summarized as follows:

Frequencies: Location	Low	Tidal	High
Brightman:	3%	93%	4%
Mt. Hope:	7%	87%	6%
Sakonnet:	23%	58%	19%

Clearly the tides are dominant at Brightman and Mt. Hope but less so at Sakonnet. The lack of tidal activity at Sakonnet makes the low and high frequency currents appear more significant when they are really quite small in magnitude.

Following Weisberg (1974) and Turner (1984), the multiple coherence between the wind and current data is determined. The time series of wind stresses were computed using the standard quadratic procedure based on the Green Airport wind record. The two orthogonal directions were selected to be axial and lateral to the main channel where the current data are collected. Figure 6 shows the multiple coherences at the Brightman surface-meter station, deployment 2.011. Coherences are low everywhere, with a moderate peak at the diurnal frequency (0.042 cycles/hr) possibly attributable to the sea-breeze. The coherence scales range from zero to unity and represent the fraction of current energy which may be attributed to direct wind input at that frequency. The dashed line marked "95%" represents the level below which a 95% probability exists that there is zero coherence between the wind and the current.

Restricting ourselves in Figure 6 to the frequency range 0.002 to 0.03 cycles/hr where wind energy is strongest (Turner 1984), we see that this record shows very low coherence between wind and current. Recalling from Figure 5 that the current energy in this range is quite small, it is concluded that the direct linear contribution of wind to the currents for this deployment is truly small.

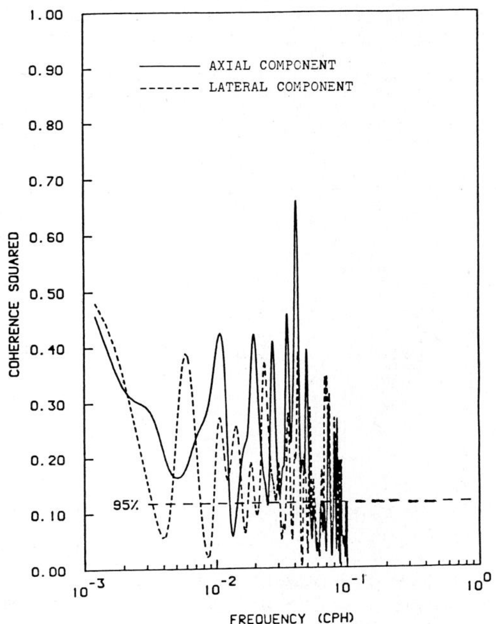

FIGURE 6. The computed multiple coherence squared between wind stress and current velocity components at the Brightman surface station during deployment 2.011.

Other current-meter records taken during this study show more coherence between wind and water velocity. Table 2 summarizes the peak coherences found in the wind-frequency ranges for the seven Brightman and eleven Mt. Hope meter records. There is considerable variability so that it is difficult even to conclude whether the surface or the bottom is more responsive to the wind.

The peaks measured and tabulated in Table 2 average to about 60% coherence. This is low compared to the data of Turner (1984) in the Providence River, where peak coherences at low frequency were about 80%. Even higher coherences exceeding 90% were reported by Weisberg (1974) for upper Narragansett Bay. The low values of coherence and low-frequency current energy lead to the conclusion that the overall effect of wind excitation on Mt. Hope Bay currents is very small. Therefore no wind/current statistical transfer function was attempted for the present Brightman and Mt. Hope data records.

III.3 Density induced circulation

The current meter records can be filtered (30 hr cutoff) to yield the residual current at a given station. As an example let us examine some current records from the Brightman station in the Taunton River, taken on the side of the ship channel where the water depth is 14 m. Figure 7 shows the filtered currents 3 meters from the surface for the period March 4, 1986 to June 9, 1986. One sees considerable variability even at periods greater than 30 hours. The average east-west current is approximately zero, but there is clearly a non-zero southward average current of about 2 cm/s.

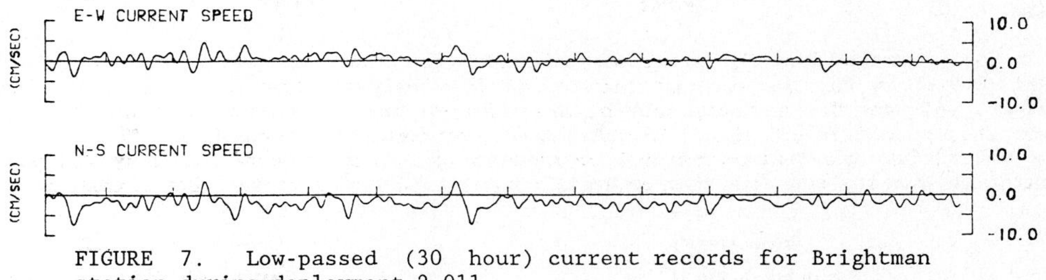

FIGURE 7. Low-passed (30 hour) current records for Brightman station during deployment 2.011.

The filtered record for the same period, at 3 meters from the bottom, is shown in Figure 8. Here a near-zero north/south current but a definite eastward average speed of 2 cm/s are observed. Since the river is slightly stratified, with surface salinity of 26 o/oo and bottom salinity of 28 o/oo both these mean currents are consistent with density-driven flows, down-estuary at the surface and up-estuary at the bottom.

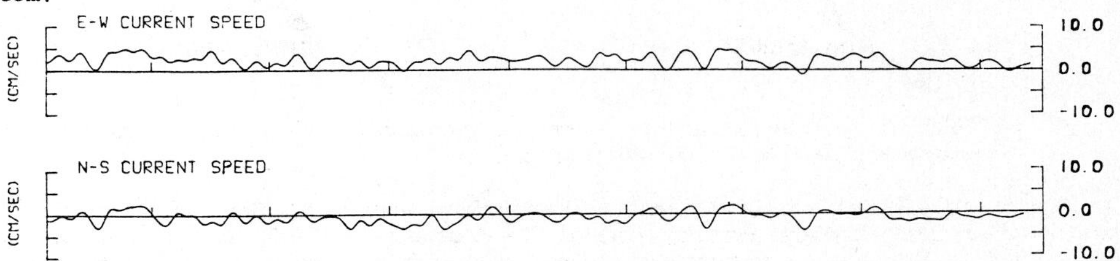

FIGURE 8. Low-passed (30 hour) current records for Brightman station during deployment 3.186.

Table 3. Mean current speed and direction for each deployment.

Deployment			$\lvert\vec{V}\rvert$ (cm/s)	θ (° clockwise from true north)
BRIGHT	1.011	bottom	2.48	74.6
BRIGHT	1.229	surface	3.81	184.2
SAKON	1.186	mid	3.85	178.8
MTHOPE	1.027	surface	11.09	214.0
MTHOPE	1.028	mid	10.63	205.9
BRIGHT	2.011	surface	1.84	168.4
BRIGHT	2.229	bottom	2.07	81.7
SAKON	2.186	mid	3.33	168.9
MTHOPE	2.027	surface	9.56	205.1
MTHOPE	2.029	mid	8.59	208.1
BRIGHT	3.011	surface	1.86	170.9
BRIGHT	3.186	bottom	2.32	107.3
SAKON	3.299	mid	3.10	181.8
MTHOPE	3.027	mid	10.40	205.7
MTHOPE	3.027	surface	10.89	205.3
MTHOPE	4.027	mid	11.64	205.4
MTHOPE	5.011	surface	8.76	196.4
MTHOPE	5.029	bottom	13.71	199.1
BFERRY	5.027	surface	4.38	209.1
BFERRY	5.186	bottom	6.62	161.1

The average N/S and E/W current speed for each meter and each deployment were computed and resolved into net mean velocity vectors. The results are listed in Table 3. The mean current speeds vary from 1.84 to 13.71 cm/s. The residual velocities are thus significant and are comparable to estimates of density-driven currents (see later section). The variability about the mean is quite large typically equivalent in magnitude to the mean. This is clearly illustrated in Figures 7 and 8.

The vectors from Table 3 can be plotted for visual inspection of each station. Figure 9 shows the mean current vectors at the Brightman station in the Taunton River, (a) near the surface, and (b) near the bottom. The axial components of these vectors are consistent with a classic two-layer density-driven flow. The directions of the surface and bottom residual components appear to be controlled by the shallow water bathymetry and the deep center channel topography of the river, respectively.

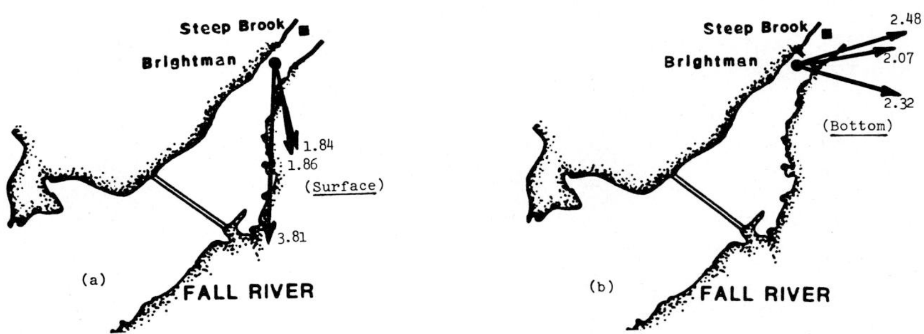

FIGURE 9. Mean residual current vectors (cm/s) for three deployments at the Brightman station.

Figure 10 shows residual current vectors at the Mt. Hope and Bristol Ferry stations near the Mt. Hope Bridge, (a) at the surface, and (b) at mid-depth and near the bottom. It is seen that all measured mean flows except possibly the Bristol Ferry bottom, have a strong component out under the bridge into Narragansett Bay. The component velocities of 4 to 14 cm/s are substantial. No residual inflow was measured. Taking the average residual flow to be 10 cm/s, with a flow cross-section under the bridge of approximately 10^4 m^2, these data imply a mean outflow of 1000 m^3/s or sixty times more than the Taunton River fresh water flow. Stokes induced transport (Dyke, 1980; Lewis and Lewis, 1983; Uncles _et al_., 1985; 1986; and Uncles and Jordan, 1979) is small since the tidal range is small compared with the water depth. It follows that there must be residual inflow at other (unmeasured) stations near the Mt. Hope Bridge. The present data imply a topographical effect, possibly associated with eddy currents near the bridge structure or eddies caused by the sudden contraction of the land boundaries. More detailed observations however need to be collected to determine the lateral variability in currents (Kjerfve _et al_., 1981, 1982).

FIGURE 10. Mean residual current vectors (cm/s) for four deployments near the Mt. Hope Bridge.

Figure 11 shows residual current vectors at the Hummocks near the Sakonnet River Bridge. All three meter deployments are at mid-depth and show an outward flow of about 3 cm/s. Since the cross-sectional area near this deployment point is about 5000 m^2, these measurements imply an outward mean flow of 150 m^3/s from Mt. Hope Bay into the Sakonnet River, or eight times greater than the Taunton River flow. Like the Mt. Hope Bridge result above, these estimated flow rates are highly implausible since they lead to long term loss of water from the bay. There must be points of inflow elsewhere along these boundary cross-sections.

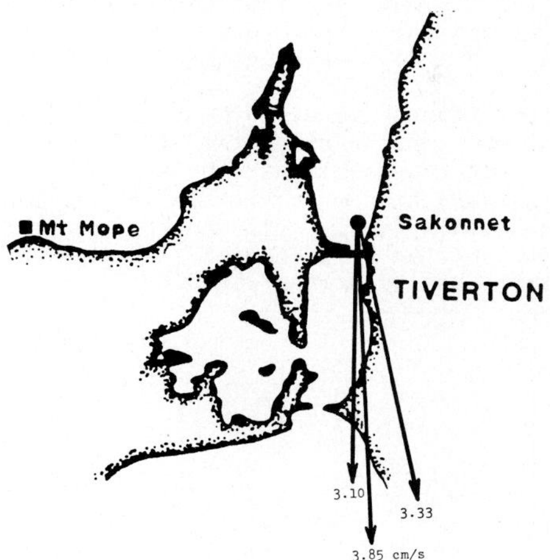

FIGURE 11. Mean residual current vectors (cm/s) for three deployments at mid-depth in the Sakonnet River near the Hummocks.

To visualize the residual flow variability progressive vector plots are used. Figure 12 shows progressive vector plots for four meter deployments near the surface at the Mt. Hope station, ranging in length from 18 to 97 days. The starting points are separated by 100 kilometers for clarity. All the raw data at 5-minute intervals was used to construct these plots, yet the displacement lines show almost no variability and proceed in nearly a straight line. Tidal excursions are essentially eliminated by these computations when plotted on these 100-kilometer scales. The residual flow at this point is nearly constant, at rates from 9 to 11 cm/s and a direction of about 205° clockwise from true north.

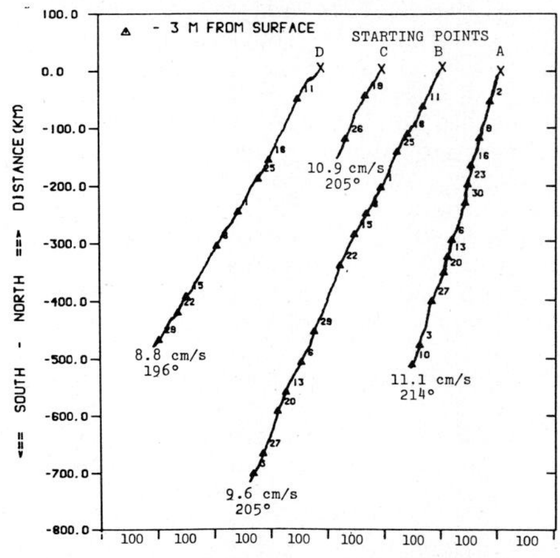

FIGURE 12. Progressive vector plots for four deployments near the surface of Mt. Hope station (a) 1.027; (b) 2.027; (c) 3.029; (d) 5.011.

Progressive vector plots, not shown, at mid-depth for the Mt. Hope station are remarkably similar to the surface plots in Figure 12. Again the displacements are relentlessly uniform, with little tidal excursion or direction change. The average residual flow is 10 cm/s at an angle of 205°. Progressive vector plots at the surface and bottom during the 5th deployment at Bristol Ferry show that the residual flows are less on the southeast side of the ship channel near the Mt. Hope Bridge.

Progressive vector plots for the Sakonnet station shows some variability but generally the residual flow is south and into the Sakonnet River. Flow rates are less, averaging 3.4 cm/s, and the average direction is 177°. General flow activity is less at Sakonnet than at Mt. Hope or Brightman.

Figure 13 shows vectors for three deployments near the surface at the Brightman station in the Taunton River. In the river the tidal velocities are higher than at the other stations and not necessarily in the same direction as the low frequency currents. Tidal excursions are therefore visible in the displacement plots. But the residual motion is still fairly uniform and toward the south, that is, down-estuary and into the center of the ship channel. The mean residual surface speed of 2.5 cm/s is consistent with density-driven circulation.

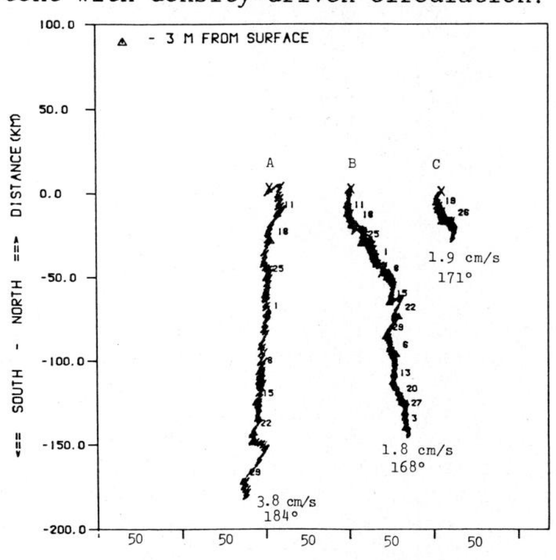

FIGURE 13. Progressive vector plots for three deployments at three meters from the surface at the Brightman Street Bridge station: (a) 1.229; (b) 2.011 and (c) 3.011. For clarity, starting points are separated by 100 kilometers.

Progressive vector plots of the bottom flow at the Brightman station are given in Figure 14. Like the surface flows, the displacements are fairly uniform but show the tidal excursions. The average residual speed of 2.3 cm/s is consistent with a density-driven flow up-estuary and toward the center of the ship channel. The almost orthogonal relationship between the mean surface and bottom velocities are caused by bathymetric effects. The bottom direction is controlled by the local orientation of the dredged channel wall and the surface direction by the shallower bathymetry. The tidal current directions are relatively independent of depth with direction controlled by the local orientation of the river topography.

III.4 Comparison with density current theory

The residual flow measurements at the Brightman station in the Taunton River support the concept of a two-layer density current driven by salinity variations in the river.

From a short-term field study on May 20, 1986 at the Brightman St. station, the surface to bottom temperature difference is 5.1 °C, the salinity change is 3.3 o/oo, and the density difference is 3.7 mg/cm^3. On that particular day the residual currents were about 3 cm/s down-estuary at the surface and 4 cm/s up-estuary at the bottom. We may compare these estimates with the classic theory of two-layer density driven flow.

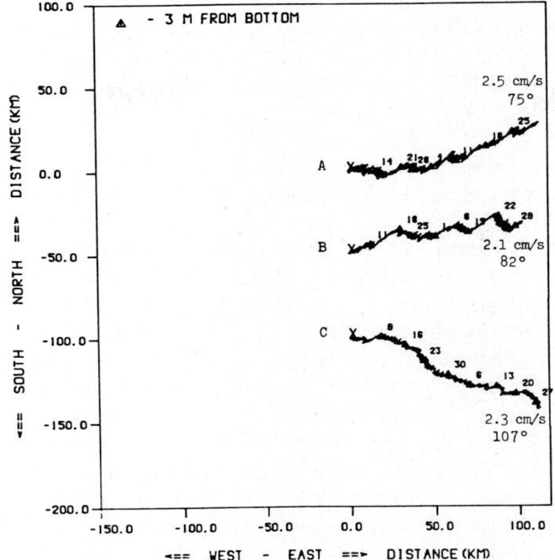

FIGURE 14. Progressive vector plots for three deployments three meters from the bottom at the Brightman Street Bridge station (a) 1.011; (b) 2.229 and (c) 3.186. Data points (triangles) are shown at weekly intervals.

For a simple two-layer density current with no lateral effects, the dominating long term mean vertical force balance on an element is between the vertical turbulent shear gradient and the horizontal density gradient (Officer 1976; Dyer, 1973). Using a mean (tidal invariant) density gradient along the ship channel of 2×10^{-7} determined from an intensive series of field measurements, a channel depth near Brightman St. station of 14 m and a vertical eddy viscosity of 22.4 cm^2/s based on Hess's (1974, 1976) study gives a density induced surface current of 5 cm/s and a bottom current of 3.4 cm/s. These are comparable to, or larger than, the measured Brighton residual flows. Since the field measurements did not measure mean density gradients or vertical eddy viscosities, it is difficult to determine if the discrepancy is due to non-local, wind, or topographic effects.

The simple theory assumes a very wide estuary with negligible horizontal diffusion. Hansen and Rattray (1965) modified the theory to include a finite width and eddy diffusion of salt (density). Their refined theory gives velocity profiles very similar to the simple approach above, but the maximum up- and down-estuary velocities are related, not to mean density gradient, but the mean freshwater velocity. Use of Hansen and Rattray's (1965) approach however does not improve our ability to predict the observed residual flows. The primary culprit appears to be topographic effects. It was impossible to test the density induced flow argument further since no data were availale to estimate the longitudinal pressure gradient for other time periods.

IV. Conclusions

Based on the analysis of the field data and modeling investigations the following conclusions evolve.

1. The field data and hydrodynamic modeling (Spaulding et al., 1987) show that the sea surface elevation and currents in the area are dominated by the M_2 tide in approximately a standing wave pattern, with tidal currents leading surface elevation by approximately 90° or 3 hours. The M_2 currents typically account for 80-90% of the current variance and 90% of the surface elevation variance. The M_4 tidal components while relatively small are important and generate a double peaked flood, single peaked ebb similar to that observed in Narragansett

Bay proper. The M_4 tide is also observed to increase in magnitude relative to the M_2 as one moves from Mt. Hope Bridge to the lower Taunton River.

2. Tidal currents (M_2) are generally weak in the Sakonnet River (4.6 cm/s) (on the Mt. Hope Bay side of the bridges) and stronger under Mt. Hope Bridge (20.3 cm/s) and near the Brightman Street Bridge (22.6 cm/s). Typical maximum mid-bay model predicted values are 10 cm/s.

3. The tidal range (1.14 m) shows no amplification from the Mt. Hope Bridge to Steep Brook (lower Taunton River) stations.

4. Low frequency currents (30 hour filtered) show little wind induced forcing and are dominated by relatively steady flows. For the Brightman Street Bridge station a classical stratified, but strongly topographically controlled, estuarine pattern is displayed with flow up river at the bottom and down river at the surface. For the Sakonnet River and Mt. Hope Bridge measurement sites the flow is consistently out of Mt. Hope Bay. More detailed measurements at the Mt. Hope Bridge transect in deployment 5 show consistent outflow of 4.4 to 13.7 cm/s at all four measurement stations. Using the transect observed mean velocities/direction and the associated cross sectional areas, for the three boundary stations (Brightman Street Bridge, Mt. Hope Bridge, and Sakonnet River Bridge) a strong net outflow from Mt. Hope Bay; is predicted about two orders of magnitude larger than the river flow. Strong lateral variations in the Mt. Hope Bridge currents are a potential cause of this imbalance in net flows through the system.

5. Two layer net estuarine flow is observed in the lower Taunton River. Use of simple analytic models for density induced flows give estimates that are qualitatively consistent with observations but predicted values are a factor of two or more larger than measurements. The differences appear to be attributable to the variable topography and bathymetry of the area that is not represented in the simple uniform channel analytic models.

Acknowledgements

This two-year study was jointly funded by the University of Rhode Island Sea Grant program and the EPA-sponsored Narragansett Bay Project. Martha Marcy Simoneau, Paul Heinmiller, Jei Kok Choi, and Shin Jye Liang performed the data gathering and analysis tasks. Mr. Larry Simoneau, Ocean Engineering, provided oceanographic instrumentation support. Computational resources were provided by the University of Rhode Island (URI) Academic Computer Center (ACC). Dr. David Shonting of the Naval Underwater Systems Center (NUSC), Newport, Rhode Island kindly allowed us to use his Endeco current meter reader on numerous occasions. Meteorological data were provided by the NOAA, U.S. Weather Service office at Green Airport, Warwick, Rhode Island and river flow data by the United States Geological Survey (USGS), Providence office.

V. References

Carrier, R., W. Brown and J. Irish, 1982: TIHARC User's Guide, unpublished manuscript, University of New Hampshire, Durham, New Hampshire.

Dennis, R. E., and E. E. Long, 1971: A User's Guide to a Computer Program for Harmonic Analysis of Data at Tidal Frequencies, NOAA Technical Report, NOS 41, U.S. Dept. of Commerce, NOAA, National Ocean Survey, Rockville, Maryland.

Dronkers, J. J., 1964: Tidal Computations in Rivers and Coastal Waters, North-Holland Publishing Company, Amsterdam, 516 pp.

Dyer, K. R., 1973: Estuaries: A Physical Introduction, John Wiley & Sons, New York.

Dyke, P. P. G., 1980: On the Stokes drift induced by tidal motions in a wide estuary, Estuarine and Coastal Marine Science, II, 17-25.

Dunn, C. Q., L. Z. Hale and A. Bucci, 1979: The Bay Bib: Rhode Island Marine Bibliography, Revised Edition, Vols. I and II. University of Rhode Island, Mar. Tech. Reports Nos. 70 and 71.

Elliot, A. J., 1978: Observations of the meteorologically induced circulation in the Potomac Estuary, Estuarine and Coastal Marine Science, **6**, 285-299.

EPA, 1971. Report on Pollution of the Interstate Waters of Mt. Hope Bay and It's Tributary Basins. Environmental Protection Agency, Washington, D.C. 70 pp (plus Append.).

Garvine, R. W., 1985: A simple model of estuarine subtidal fluctuations forced by local and remote wind stress. Journal of Geophysical Research, **90**, 11945-11948.

Gordon, R. B., 1982: Wind Driven Circulation in Narragansett Bay, Ph.D. Dissertation, Department of Ocean Engineering, University of Rhode Island, Kingston, Rhode Island.

Haight, F. J., 1936: Currents in Narragansett Bay, Buzzards Bay and Nantucket and Vineyard Sounds, Special Pub. No. 208, Washington: U.S. Dept. of Commerce, Coast and Geodetic Survey.

Hansen, D. V. and M. Rattray, 1965: Gravitational Circulation in Straits and Estuaries, J. Marine Research, **23**, 104-122.

Hess, K. W., 1976: A three-dimensional numerical model of the estuary circulation and salinity in Narragansett Bay, Estuarine and Coastal Marine Science, **4**, 325-338.

Hess, K. W. and F. M. White, 1974: A Numerical Tidal Model of Narragansett Bay, Ocean Engineering, University of Rhode Island, Marine Technical Report No. 20 141 p.

Hicks, S. D., 1959: The Physical Oceanography of Narragansett Bay, Limnology Oceanography, **4**, 316-328.

Kjerve, B., J. A. Prochl, F. B. Schuring, H. E. Seisma, and M. Marozas, 1982: Temporal and spatial considerations in measuring estuarine water fluxes. Estuarine Comparisons (V.S. Kennedy, ed). Proceedings of the Sixth Biennial International Estuarine Research Conference. Oregon, November 1-6, 1981, Academic Press, pp. 37-51.

Kjerve, F., L. H. Stevenson, J. A. Prochl, T. H. Chrzanowski and W. M. Kitchens, 1981. Estimation of material fluxes in an estuarine cross section: A critical analysis of spatial measurement density and errors, Limnology and Oceanography, **26**, 325-335.

Lewis, R. E. and J. O. Lewis, 1983: The principal factors contributing to the flux of salt in a narrow, partially stratified estuary, Estuarine, Coastal and Marine Science, **16**, 699-626.

McDowell, D. M. and B. A. O'Connor, 1977: Hydraulic Behavior of Estuaries, John Wiley & Sons, New York.

Officer, C. B., 1976: Physical Oceanography of Estuaries (and Associated Coastal Waters), New York: Wiley.

Pilson, M. E. Q., 1985: On the residence times of water in Narragansett Bay, Estuaries, Vol 8, No 1, 2-14.

Prandle, D. and M. Rahman, 1980: Tidal response in estuaries, Journal of Physical Oceanography, **10**, 1552-1573.

Roberts, J. and T. D. Roberts, 1978: Use of Butterworth low pass filter for oceanographic data, Journal of Geophysical Research x3, 5510-5514.

Schroeder, W. W. and W. J. Wiseman, Jr., 1986: Low frequency shelf estuarine exchange processes in Mobile Bay and other estuarine systems on the northern Gulf of Mexico, in Estuarine Variability (editor) D.A. Wolfe. Academic Press New York, 355-368.

Spaulding, M. L., F. M. White, P. Heinmiller, M. Simoneau, S. Liang and J. K. Choi, 1988: Circulation Dynamics in Mt. Hope Bay and the Lower Taunton River, EPA Narragansett Project.

Swanson, J. C. and M. L. Spaulding, 1974: Tides and Tidal Currents of Narragansett Bay, Ocean Engineering, Sea Grant, University of Rhode Island, Marine Technical Report 35.

Swenson, E. M. and W. Chuang, 1983: Tidal and subtidal water volume exchange in an estuarine system, Estuarine, Coastal and Shelf Science, **16**, 229-240.

Turner, A. C., 1984: Tidal and Subtidal Circulation in the Providence River. M.S. Thesis, Department of Ocean Engineering, University of Rhode Island, Kingston, Rhode Island.

Uncles, R. A., R. C. A. Elliott and S. A. Weston, 1986: Observed and computed lateral circulation patterns in a partly mixed estuary, Estuarine, Coastal and Shelf Science, **22**, 439-457.

Uncles, R. J., R. C. A. Elliott and S. A. Eston, 1985: Observed fluxes of water, salt and suspended sediment in a partly mixed estuary, Estuarine, Coastal and Shelf Science, **20**, 147-168.

Uncles, R. J. and M. B. Jordan, 1979: Residual fluxes of water and salt at two stations in the Severn Estuary, Estuarine and Coastal Marine Science, **9**, 287-302.

Wang, D. P., 1979: Wind driven circulation in the Chesapeake Bay and Potomac River; evidence for non local forcing, Journal of Physical Oceanography, **8**, 225-232.

Weisberg, R. H., 1974: The Non-Tidal Flow in the Providence River of Narragansett Bay: A Stochastic Approach to Estuarine Circulation, Journal Physical Oceanography, **6**, 721-734.

Part VI

Interdisciplinary Studies

Recruitment in Estuarine Benthic Communities: The Role of Physical Processes

Timothy R. Jacobsen[1], James D. Milutinovic[2], and James R. Miller[2]

[1]Department of Oyster Culture
Cook College, Rutgers University
New Brunswick, NJ 08903

[2]Department of Meteorology and Physical Oceanography
Cook College, Rutgers University.
New Brunswick, NJ 08903

ABSTRACT

The mechanisms for retention of invertebrate larvae are governed by the interaction of several physical processes within Delaware Bay. Work during the last two years has combined numerical models of estuarine circulation with biological sampling and with field studies of small scale fronts and tidal circulation. The combination of tidal currents, wind-induced currents and the interactions of frontal systems are important factors governing the recruitment of oyster larvae into the benthic community. Two-dimensional and three-dimensionals models are used to calculate particle trajectories for one tidal cycle and for twelve-day periods. The results indicate that under certain conditions, primarily controlled by the wind, the estuarine circulation may favor the retention of oyster larvae within the oyster beds during most of their two-week larval phase. However, the particle trajectories are critically dependent on the vertical distribution of larvae within the water column.

I. Introduction

Estuarine ecosystems are among the most biologically productive regions on earth (Odum 1971). The movement and/or export of planktonic material in an estuary is not thoroughly understood. The lack of knowledge about biological processes and interactions within the water column of an estuary is due largely to the inability to define and understand the spatial and temporal variability of physical processes and how they affect biological processes.

Estuarine planktonic and benthic communities are dramatically affected by physical processes. Benthic organisms are a dominant component of living biomass in an estuarine ecosystem. Biologists have long been perplexed by the erratic fluctuations in recruitment (addition of new members to the community) of bivalve species. Many of these species undergo a complex life cycle (two or more stages) that has significant effects on their recruitment success. For instance, bivalves undergo a planktonic larval stage, and its movement within the water column affects larval retention and export from, or import into the estuary.

Marine biologists have long grappled with understanding and explaining the processes that govern bivalve recruitment in estuarine ecosystems (Galstoff 1964). For many years the question of how bivalve larvae are retained within estuaries has been intractable for both biologists and physical oceanographers alike (Boicourt 1982). Biologists have long known that the movement of planktonic larvae within estuaries is controlled ultimately by the circulation within the estuary (Galstoff 1964). Initially, many biologists believed that larval behavior was an important factor in the retention of bivalve larvae within estuaries (Nelson, 1921; Prytherch, 1928; Loosanoff, 1932). Butman (1987), in a good review of the subject, has

Coastal and Estuarine Studies, Vol. 38
R. T. Cheng (Ed.)
Residual Currents and Long-term Transport

discussed the importance of active habitat selection and hydrodynamical processes for invertebrate recruitment and suggests that larval behavior is of secondary importance to physical processes.

Pritchard (1952; 1953) first proposed a general hydrographic scheme to account for larval retention in estuaries. He proposed a two-layered circulation system in which there is net (tidally averaged) upstream movement of higher salinity bottom water and net downstream movement of lower salinity surface waters. This two layered circulation scheme cannot account for total retention of larvae within an estuary during their approximately two-week development from fertilized egg to settled spat. Early stage larvae were shown to have a homogeneous distribution in the seawater column (Carriker 1951; Nelson 1955; Kunkle 1958; Haskin 1964) that would result in some larval loss due to flushing of surface waters from the estuary. However, it was thought that late stage larvae increased their chances of being retained in an estuary, by rising off the bottom on flood tide (on an increase in salinity) and sinking to or near the bottom on ebb tide (on a decrease in salinity) (Nelson 1921; Carriker 1951; Manning and Whaley, 1954; Nelson 1955; Kunkle 1958; Haskin 1964; Wood and Hargis, 1971).

Studies of fine-scale circulation in Chesapeake Bay have shown that specific hydrographic features are important to successful recruitment of oysters (Boicourt, 1982; Seliger et al., 1982). They hypothesized a tidally-pumped two-layer flow to explain how surface waters of Chesapeake Bay containing early stage larvae are carried below lower density surface waters of Choptank River, MD. Furthermore, they found that differences in the number of oysters that settled on the bottom between two adjacent sub-estuaries of the Choptank River could be due to differences in the vertical stability or stratification of the water column. The system with the greater degree of stratification has a greater potential for retaining planktonic larvae.

Roughgarden et al. (1988) demonstrated the influence of a wide spectrum of physical processes on governing recruitment success in barnacles occupying west coast rocky intertidal communities. They provided convincing evidence that the community ecology at a particular site is governed by the regional patterns of circulation. Physical processes within estuarine ecosystems are also important but operate on different temporal and spatial scales than those observed by Roughgarden et al. (1988). In order to understand the spatial variability and productivity of these estuarine communities, we need to understand the biological and physical interactions that occur within the estuarine ecosystem.

Rothlisberg et al. (1983) used two-dimensional and three-dimensional numerical modeling to study the advection of penaeid shrimp. They found that the horizontal advection of shrimp larvae was enhanced by diurnal vertical migrations in the Gulf of Carpentaria (Northern Australia) and that larvae could be advected for distances up to 165 km during the two to three week planktonic larval stage. They also found that there was little recruitment during the highest reproductive period because of offshore advection, whereas there was greater recruitment during lower reproductive periods when there was onshsore advection to the nursery grounds.

Johnson et al. (1984) used numerical modeling and determined that the success of blue crab larval recruitment in Chesapeake Bay was dependent upon physical circulation patterns during the development of the larvae. Crab larvae are released by adults near maximum ebb tide. The larvae then ascend to surface waters where they are carried out of the bay and spend about 30 to 40 days developing offshore. They stay near the mouth of the bay in the upper 1 m of water in spite of the southward surface current because a wind stress toward the north is usually present during this critical period. Because the larvae are retained near the mouth, the post larval forms can then move to the bottom and move northward into the bay on the bottom currents.

Bartsch (1988) used a three-dimensional shelf circulation model to study the movement of herring larvae in the North Sea. He found that variability of drift routes for herring larvae was caused by changes in circulation due to meteorological variability. He found that the vertical migration of herring larvae, which is triggered by light availability, was important in determining drifter routes.

These studies demonstrate that modeling of larval advection is a tractable means of addressing recruitment questions. However, these studies do not resolve questions of larval recruitment within estuaries. Herring larvae and other fish larvae have well-developed sensory and locomotion systems to allow them to migrate within a water column. The other studies dealt with crustaceans that travel out of estuaries into nearshore environments during their planktonic stages. In the case of blue crabs, later larval stages are returned to the estuary (Boicourt, 1982; Epifanio and Dittel, 1982). The adults of the larvae used in these two studies are mobile after settling to the bottom. This scheme for crustacean larval recruitment is not typical of many benthic larval forms. Molluscan larvae have shorter development times (10-14 days) than crab or shrimp larvae but are unable to migrate like juvenile crabs to suitable substrates.

The purpose of this paper is to examine the role of physical processes on the advection, retention and recruitment of oyster larvae in Delaware Bay. Our analysis of records of benthic recruitment during the last 28 years shows that oyster larvae recruitment has been both spatially and temporally variable. Successful recruitment has not been well correlated with the numbers of early stage larvae present within

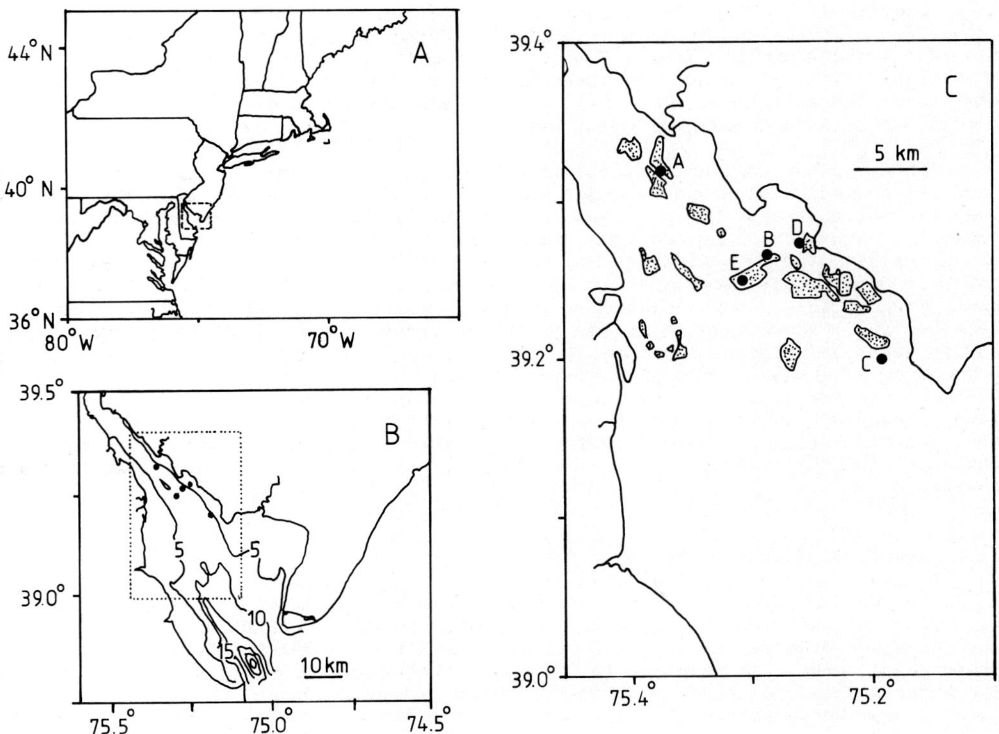

FIGURE 1. A) Overview of the study area. B) Map of Delaware Bay showing bottom topography. The boxed region indicates the area where oyster beds can be found. C) Location of the natural oyster seed beds. Letters show the locations of the field sampling stations.

the water column. Numerical models and field programs have been integrated to study the role of physical processes in structuring benthic communities in Delaware Bay, with particular emphasis on oyster recruitment. Figure 1 shows Delaware Bay and the location of the oyster beds.

II. Circulation of Delaware Bay

The circulation within Delaware Bay has been described by Hires et al. (1983). The description of the circulation in this section is based on their paper. The estuary is partially mixed with a net seaward flow in the upper layer and a net inflow of higher salinity bottom water from the ocean. The dominant constituent of the astronomical tide is the M2 principal lunar component which has a period of 12.42 hours and an amplitude of about 0.6 m at the mouth of the estuary. Peak ebb and flood tidal currents can exceed 100 cm/s, whereas the residual tidal currents are in the 0 to 10 cm/s range. The interactions of the tidal currents with variations in bottom topography, shoreline geometry or Coriolis force affect the tidally-induced residual circulation. The high-water phase of the tide propagates from the ocean entrance to Trenton in about seven hours.

Subtidal circulation, which is the residual circulation remaining after removal of the tidal currents, is driven by meteorological conditions amd stream flow. Precipitation within the watershed of Delaware Bay affects the gravitational estuarine circulation driven by density differences between the freshwater stream flow and inflow from the ocean. The total volume flux of water entering the bay on the flood tide is about 300 times larger than the average freshwater inflow. In most cases the effect of the wind on the subtidal currects dominate that of precipitation-induced changes in run-off.

Within the bay the surface wind stress transfers momentum to the water. The wind-driven water transports can lead to changes in water elevation in the longitudinal or transverse directions, which in turn produce pressure gradients and currents. However, the regional winds over the continental shelf can also affect estuarine circulation. Winds toward the north on the shelf move surface waters offshore and lower sea level at the Delaware Bay entrance, while winds toward the south move surface waters onshore and increase sea level at the entrance. Wong and Garvine (1984) found that these changes in sea level at the bay mouth caused by wind stress over the adjacent continental shelf could affect currents within the estuary and have a greater effect on the subtidal circulation than the local effects of winds within the bay. Pape and Garvine (1982) used surface and bottom drifters to study residual currents in the estuary and found similar results. These studies show that the effects of wind variability within the bay and offshore can affect the currents in the estuary, and that the subtidal currents within Delaware Bay are complex.

III. Numerical Models of Delaware Bay

Both two-dimensional and three-dimensional numerical models are used in this study. The two-dimensional model is a vertically averaged version of the more complex three-dimensional model of Galperin and Mellor (1989a, 1989b). The three-dimensional model is similar to Oey et al. (1985) and Blumberg and Mellor (1987). The three-dimensional model contains 11 levels (sigma surfaces) with an inner grid located inside the bay having a resolution of 1 km x 1 km and an outer grid for the adjacent shelf water with a resolution of 4 km x 5 km.

The model is driven by specifying the time dependent observed winds, tidal elevations and river discharge. Climatological temperature and salinity are also specified at the grid boundaries throughout the simulation. Initial values of temperature and salinity are also specified throughout the grid.

The governing equations for the prediction of velocity and surface height are the continuity and momentum equations. The turbulent viscosity and diffusivity are calculated as described in Blumberg and Mellor (1987). The conservation of heat and salt are used to obtain equations for temperature and salinity.

Current fields from the two-dimensional tidal model of Delaware Bay was obtained from the National Oceanic Service (NOS) of NOAA. This model does not include wind or river forcing. The three-dimensional model of Galperin and Mellor (1989a, 1989b) incorporates tides, wind forcing and river run-off. Currents from this model were obtained from Galperin and Mellor. A two-dimensional version of the Galperin and Mellor model that includes wind forcing is also available. Current fields from each of the models were obtained at hourly intervals for the simulations.

Figures 2a and 2b show the instantaneous surface currents for a transect across the bay at the maximum ebb tide for the two-dimensional NOS model and three-dimensional Galperin and Mellor model. Since the current is uniform with depth in the two-dimensional model, the surface current and mean current in the water column are identical for the two-dimensional model. The maximum currents and the

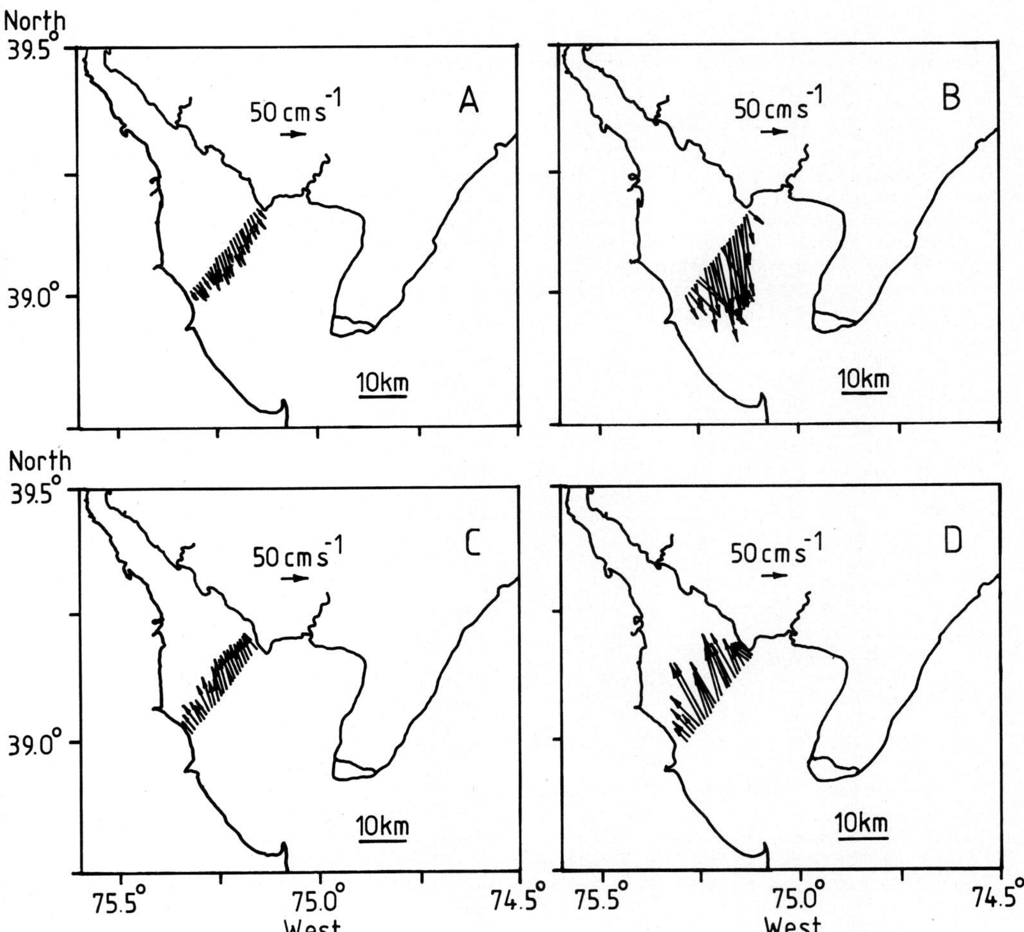

FIGURE 2. Current vectors on a cross bay transect near the lower oyster beds from the NOS two-dimensional model (15 June 1977) and the Galperin and Mellor three-dimensional model (30 June 1984). (A) maximum ebb tide, 2-D model, (B) maximum ebb tide, 3-D model, (C) maximum flood tide, 2-D model, (D) maximum flood tide, 3-D model.

variability of the currents in the three-dimensional model are larger than for the two-dimensional model. In the shallower areas near the sides of the estuary, the two models are in better agreement, primarily because there is less vertical structure in the three-dimensional model in the shallower water. Since the oyster beds are in the shallower areas (less than 5m), the two-dimensional model might be useful for doing Lagrangian tracer studies in cases where movement is confined within the shallow region. Figures 2c and 2d show the corresponding surface currents at the maximum flood tide.

IV. Larval Field Studies

A field sampling study was conducted to determine the spatial and temporal variability of oyster larvae over the oyster beds. Figure 1 shows the locations of the natural oyster beds and the five sampling stations (letters) occupied during the study. Stations were occupied during one tidal cycle on 28 July 1986. Plankton samples were collected 1 m off the bottom and 1 m from the surface at 1/2 hour intervals. Measurements of temperature and salinity were also taken. Plankton samples were analyzed for the abundance and age structure of oyster larvae present within a sample.

Larval abundances within the water column varied significantly between stations during the tidal cycle. The larval population was divided into different developmental stages as shown in Table 1. Figure 3 shows the variability of the larval population in these stages at the three stations in a longitudinal transect during one tidal cycle. All stages of larvae are normalized to the maximum abundance found for that stage (100% being the numerically highest value found for that larval stage in sampling). Typically, early stage larvae (straight hinge, early umbo and late umbo) were in tenfold greater abundance.

Table 1. Characteristics of the developmental stages of oyster larvae after Galstoff (1964). The term early stage larvae is the sum of straight hinged, early and late umbo stages, while late stage larvae refers to the sum of mature and eyed larvae. The highest abundance of each stage of larvae found in the tidal sampling study is shown in the last column.

Stage	Size	Age	Abundance # m-3
Straight Hinged	65-75 μm	0-1 days	188
Early Umbo	80-130 μm	2-6 days	2969
Late Umbo	150-200 μm	7-9 days	904
Mature	230-280 μm	10-12 days	193
Eyed	300 μm	13-14 days	235

The presence and abundance of late stage larvae in the surface and bottom sample were more variable than the early stages. We questioned if larval behavior (settling of late stage larvae to the bottom of the ebb tide) could explain the variation in late stage larval abundance. A chi-square analysis showed that there was no preferential pattern of clustering of late stage larvae (eyed and mature) near the bottom during the ebb tide (hour 1 to 5). When late stage larvae were present, they tended to be near the bottom throughout the tidal cycle. This analysis suggests that the late stage larval patchiness observed in the study was the result of physical processes and not governed by larval behavior.

Density fronts may affect larval distribution in Delaware Bay on scales smaller than the large scale circulation. Klemas and Polis (1977) used satellite imagery to identify frontal systems throughout Delaware Bay. They found a relationship between when the fronts occur during the tidal cycle and the geographic location of the front within Delaware Bay. They found that flood associated fronts are more prominent on the shallow New Jersey side of the bay, while fronts in the deeper channel occur mostly on the ebb tide on the Delaware side. Jacobsen et al. (1989) observed frontal features during two field studies in 1988. The fronts observed in these studies trapped surface drifters.

The role of fronts in oyster larvae dispersal within Delaware Bay is not known. During our field studies, axial fronts, spaced about 1 km apart, were observed to form on the flood tide nearly parallel to the New Jersey shore and the main channel. These fronts tended to trap surface drifters at the frontal interface, and they may be important in concentrating oyster larvae, too.

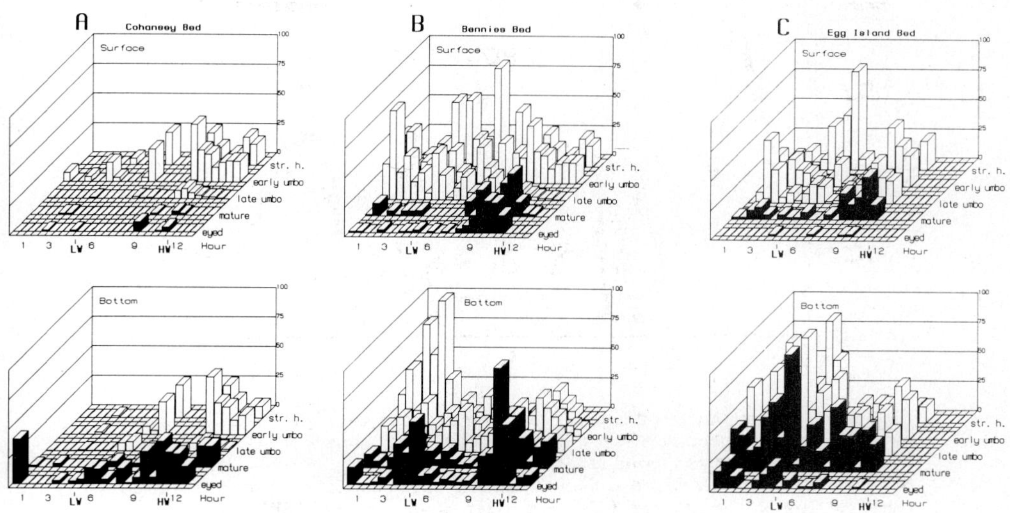

FIGURE 3. Relative abundance of oyster larval stages over a tidal cycle at three stations on a down bay transect (See figure 1 for locations). Sampling was conducted on 28 July 1986. Samples were taken 1 m from the surface and 1 m from the bottom at 30 minute intervals. Larval abundance is normalized to the highest concentration found for each particular stage (See table 1). Early stage larvae are straight hinged, early and late umbo (open bars). Late stage larvae are mature and eyed larvae (solid bars). The time of low slack water (LW) and high slack water (HW) are also indicated.

Some evidence of the importance of fronts in concentrating oyster larvae was found during the field sampling study. Larval abundance at the center station (B) was influenced by the passage of a front. Figure 4 shows that larval concentration varied more than ten-fold when the slick of a front passed the sampling station. The increase of late stage larvae within the water shown in Figure 4 occurred during the passage of the front. The front passing this station, although oriented in the along shore direction, did not reach any of the other sampling stations.

V. Tidal Residual Circulation and Particle Trajectories

To examine the role of physical processes on oyster recruitment, it is important to examine the currents and circulation within the oyster beds. As a first step, Eulerian tidal residuals are calculated from the NOS two-dimensional model at each grid point within the estuary. The Eulerian tidal residual is obtained by averaging the currents during one tidal cycle. Although these residuals are less important in regions where the total distance traveled during a tidal cycle is large, the overall pattern of the residuals gives some insight into the transport of larvae within different regions of the bay.

Tidal residuals are calculated from the two-dimensional NOS tidal model at each grid point inside the bay for a two-week period from June 15-30, 1977. Th e maximum residual currents are about 5 cm/s. Within the oyster seed beds, there are some residual currents that are directed on shore or upbay. In such cases there is a suggestion that oyster larvae could be retained in the area. Surface tidal residuals calculuated from the three-dimensional model showed similar residual patterns near the oyster beds.

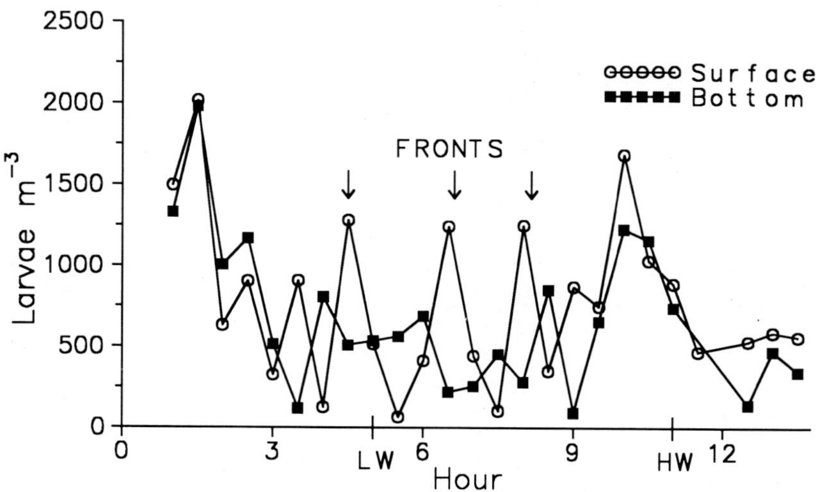

FIGURE 4. Abundance of oyster larvae in the surface and bottom waters at station B (See Figure 1 for positions) over one tidal cycle. Sampling was conducted on 28 July 1986. Arrows indicate times when a slick from a front passed the sampling station. The time of low slack water (LW) and high slack water (HW) are also indicated.

The discovery of these residual features led to a reconsideration of the mechanisms governing the distribution of oyster larvae within the bay and on the oyster beds. One hypothesis is that oysters spawn and release their larvae to be carried down Delaware Bay by the tides. A week to ten days later the larvae sink to the middle and lower levels of the water column and are carried back up bay by bottom currents. However, the model residuals indicate that oyster larvae might not undergo such a route, but rather remain near the region where they were spawned. The Eulerian residuals cannot be used directly to determine the movement of oyster larvae. Lagrangian particle trajectories are needed to determine the fate of oyster larvae in Delaware Bay.

The velocities from both the two-dimensional and three-dimensional models of Galperin and Mellor (1989a, 1989b) are used to calculate Lagrangian trajectories in Delaware Bay. Although the vertical distribution of oyster larvae within the water column and the variation of this distribution with time is not known, the Lagrangian trajectories give some insight about the movement of passive drifters within the estuary.

The trajectories are calculated with a second order scheme which includes the Stoke's drift term as discussed by Longuet-Higgins (1969). The change in position of a particle during a time step, dt, is given by

$$dx = udt + \frac{1}{2}\left(u\frac{\partial u}{\partial x} + v\frac{\partial u}{\partial y}\right)(dt)^2 ,$$

$$dy = vdt + \frac{1}{2}\left(u\frac{\partial v}{\partial x} + v\frac{\partial v}{\partial y}\right)(dt)^2 ,$$

where u and v are the velocities at any point in the estuary interpolated from the four surrounding grid points. The hourly velocities are also interpolated to produce new velocity fields at shorter time intervals for dt, which is set equal to six minutes in the following calculations.

Figure 5a shows surface trajectories during one tidal cycle beginning at high slack water. The solid squares indicate the starting points which correspond to the oyster seed beds shown in figure 1. Wind speeds (m/s) and directions from observations at Atlantic City International Airport are shown on the bottom of the figure at three hour intervals. The Lagrangian tidal residual is not shown but would be represented by a vector from the starting point to the end point for each of the three locations. Jacobsen et al. (1989) showed that the net surface drift during a tidal cycle is strongly dependent on the wind. Figure 5b shows the corresponding trajectories for the two-dimensional version of the model. The trajectories are similar to those in figure 5a.

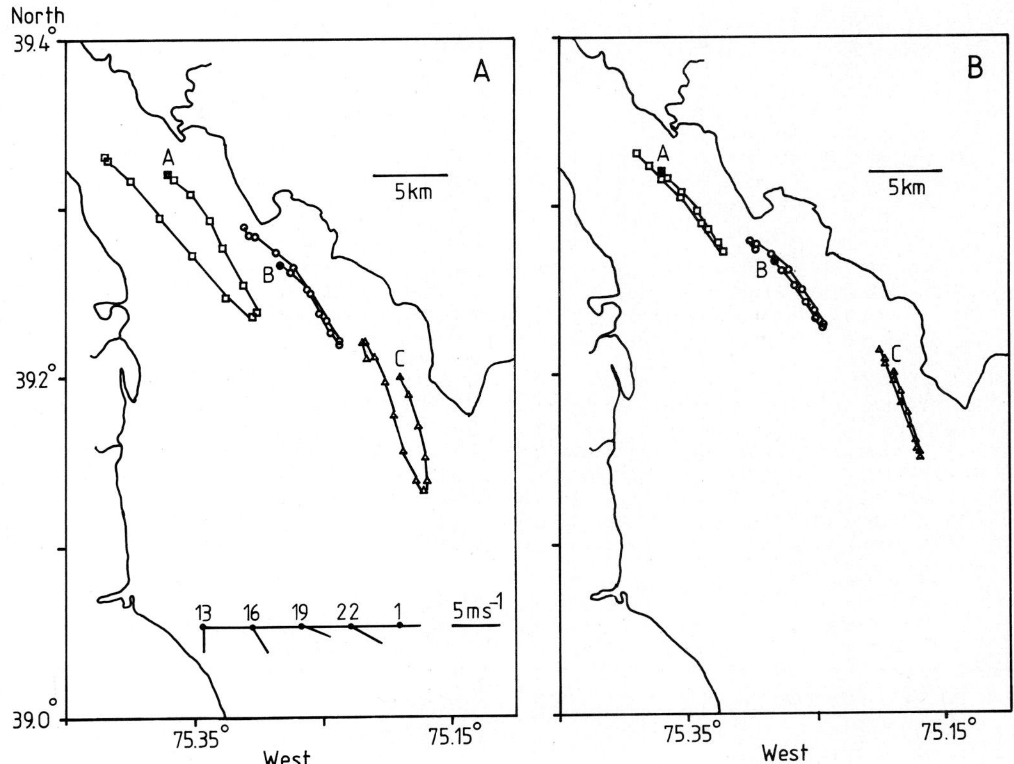

FIGURE 5. Particle trajectories for one tidal cycle. A) surface trajectories from three-dimensional model for June 30, 1984. (B) trajectories from the two-dimensional version of the Galperin and Mellor model. Solid squares represent initial positions and letters correspond to seed beds in figure 1. Wind speeds at three-hour intervals are shown on bottom.

Since oyster larvae remain in the water column 10 to 15 days before settling to the bottom, it is useful to examine trajectories of passive drifters for a similar period. Figures 6, 7 and 8 show twelve-day particle trajectories from the Galperin and Mellor two-dimensional and three-dimensional models for three different starting times and three different starting positions. Subsequent particle positions are shown at the end of each tidal cycle. Figure 6a shows a trajectory from the two-dimensional model in which the particle remains within the seed beds near its initial postion. The surface trajectory shown in Figure 6b is quite different. After several days, the particle reaches the main channel of the estuary and is swept to the mouth of the estuary at the end of the period.

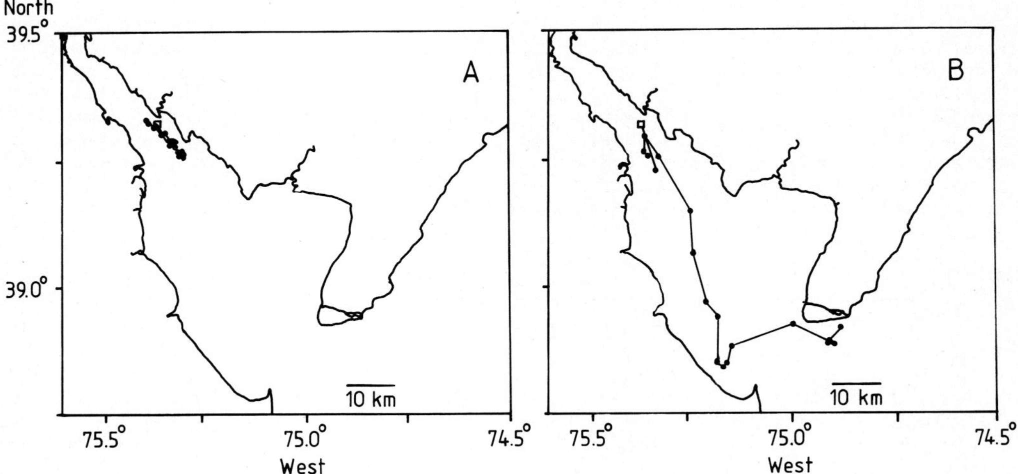

FIGURE 6. Twelve-day trajectories for passive particles originating within the oyster seed beds starting June 16, 1984. (A) two-dimensional model (B) three-dimensional model. Open squares represent initial positions and circles indicate the end of a tidal cycle.

In figures 7 and 8 the final position in the two-dimensional case is much closer to the initial position than for the three-dimensional model. In figure 7 the surface trajectory shows that particles move down bay but stay near shore where there are other oyster beds. The two-dimensional and three-dimensional trajectories are quite different. In figure 8, the surface trajectory shows that particles initially move toward shore and remain trapped there for many tidal cycles. The two-dimensional case shows the particle moving steadily down estuary.

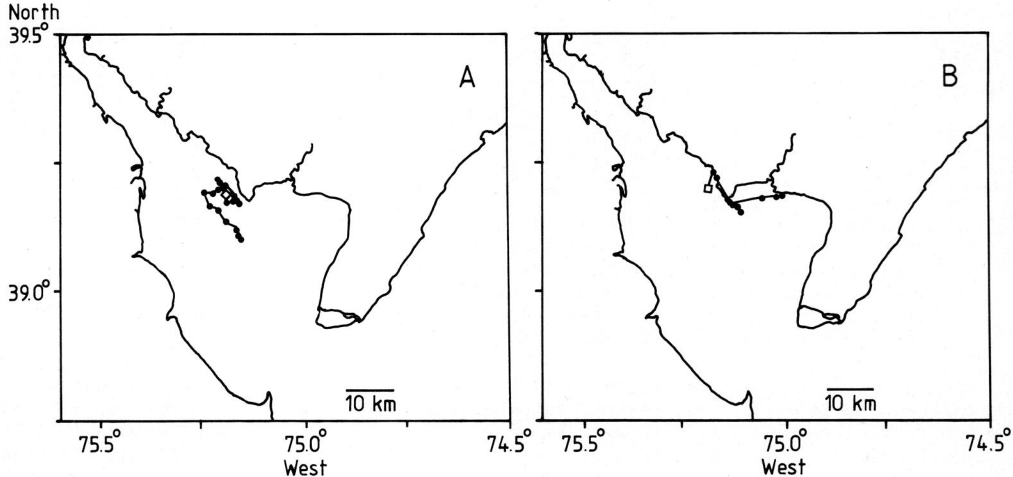

FIGURE 7. Same as Figure 6 but for June 23, 1984.

VI. Conclusions and Discussion

Our results indicate that tidal residual circulation in Delaware Bay may be important in the long-term recruitment within communities. There is some evidence from two-dimensional tidal models that oyster larvae are retained within the oyster beds along the New Jersey shore. However, during the period when larvae are in the water column, the winds are critical in determining the route of passive drifters. Strong winds from the northeast, even for a short period, may drive nearshore larvae toward the channel where currents will transport them downstream. Light or southwesterly winds tend to keep larvae nearshore in the seed beds.

This study has helped formulate more questions rather than provide answers. How sensitive are model trajectories to different interpolation and advection-diffusion schemes, to random perturbations in the flow field, and to the grid size? The significant differences between the two-dimensional and three-dimensional model trajectories indicate the need for a better model of the vertical distribution of the larvae within the water column. It may be necessary to incorporate a frontal dynamics submodel within the numerical circulation model to account for the effects of axial fronts that occur during the tidal cycle. Such frontal models have been discussed by Nunes and Simpson (1985) and Huzzey and Brubaker (1988). Continued progress in our understanding of benthic recruitment in Delaware Bay requires an integrated modeling and field sampling program that will identify the physical and biological processes that may govern larval distribution in Delaware Bay. It would also be useful to do model simulations for other years since 1984 was not a good recruitment year for oysters, and that is the only year for which the three-dimensional model has been run.

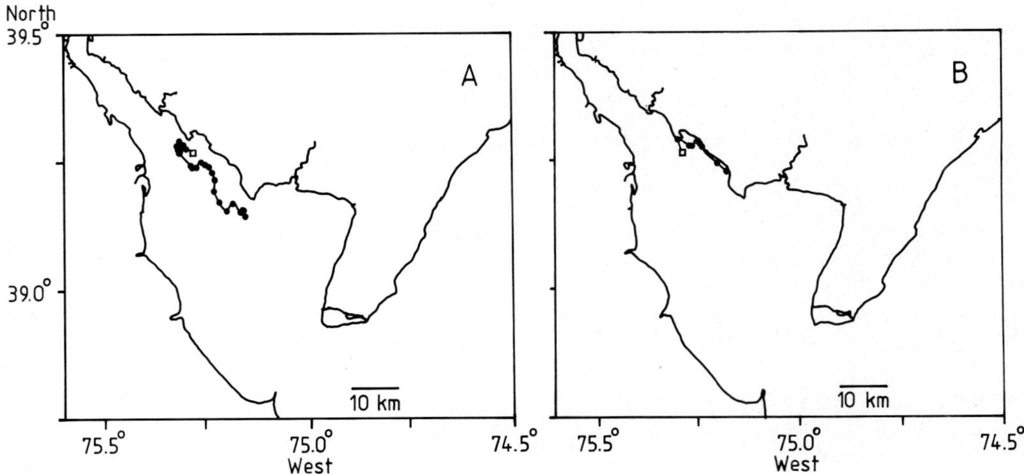

FIGURE 8. Same as Figure 6 but for June 29, 1984.

Acknowledgments

New Jersey Agricultural Experiment Station, Publication F-32001-1-88, Supported by State Funds (Project 13452), NJDEP Bureau of Shellfisheries, NJMSC minigrant program, and NOAA/Sea Grant. The two-dimensional simulation was obtained from NOAA/NOS. We would like to thank James Johnson and Richard Patchen (NOAA/NOS) for their help with the two-dimensional model, George Mellor and Boris Galperin for providing the data from the three-dimensional simulation and for helpful discussions about the model, and David Aubrey and Ralph Cheng for providing useful comments on the manuscript.

VII. References

Bartsch, J., 1988: Numerical simulation of the advection of vertically migrating herring larvae in the North Sea. Meeresforsch, **32**, 30-45.

Blumberg, A. F. and G. L. Mellor, 1987: A description of a three-dimensional coastal ocean circulation model. Three dimensional Coastal Ocean Models, N.S. Heap ed. Am. Geophysical Union, Washington, D.C., pp. 1-16.

Boicourt, W. C., 1982: Estuarine larval retention mechanisms on two scales. Estuarine Comparisons, V.S. Kennedy ed. Academic Press, New York, pp. 445-458.

Butman, C. A, 1987: Larval settlement of soft-sediment invertebrates: The spatial scales of pattern explained by active habitat selection and the emerging role of hydrodynamical processes. Oceanog. Mar. Biol. Ann. Rev., **25**, 113-165.

Carriker, M. R., 1951: Ecological observations on the distribution of oyster larvae in N. J. estuaries. Ecol. Monog., **21**, 19-38.

Epifanio, C. E. and A. I. Dittel, 1982: Comparison of dispersal in crab larvae in Delaware Bay, USA, and the Gulf of Nicoya, Central America. Estuarine Comparisons, V. S. Kennedy ed. Academic Press, New York, pp. 477-488.

Galperin, B. B. and G. L. Mellor, 1989a: A time dependent, three-dimensional model of the Delaware Bay and river system. Part 1: Description of the model and tidal analysis. Submitted to Est. Coast. Shelf Sci.

Galperin, B. B. and G. L. Mellor, 1989b: A time dependent, three-dimensional model of the Delaware Bay and river system. Part 2: Three dimensional flow fields and residual circulation. Submitted to Est. Coast. Shelf Sci.

Galstoff, P. S., 1964: The American oyster, Crassostrea virginica Gmelin. Fishery Bulletin, **64**, 1-480. Englewood Cliffs, NJ.

Haskin, H. H., 1964: The distribution of oyster larvae. pp. 76-80. Proceedings Symposium Experimental Marine Ecology, Rhode Island. Occasional Publ. No. 2, Narragansett Mar. Lab., Kingston: University of Rhode Island.

Hires, R. I., G. L. Mellor, L. Y. Oey, and R. W. Garvine, 1983: Chapter 3. Circulation of the estuary, pp. 27-48. In: The Delaware Estuary: Research as background for estuarine management and development. A report to the Delaware River and Bay Authority. J. S. Sharp [ed.] , Univ. Delaware and N. J. Mar. Sci. Cons., Lewes, DE.

Huzzey, L. M. and J. M. Brubaker, 1988: The formation of longitudinal fronts in a coastal plain estuary. J. Geophys. Res., **93**, 1329-1334.

Jacobsen, T. R., J. D. Milutinovic and J. R. Miller, 1989: Lagrangian tracer studies in Delaware Bay: Observations and models. submitted to J. Geophys. Res.

Johnson, D. R., B. S. Hester and J. R. McConaugha, 1984: Studies of a wind mechanism influencing the recruitment of blue crabs in the Middle Atlantic Bight. Cont. Shelf Res., **3**, 425-437.

Klemas, V. and D. F. Polis, 1977: A study of density fronts and their effects on coastal pollutants. Remote Sens. Environ., **6**, 95-126.

Longuet-Higgins, M. S., 1969: On the transport of mass by time varying ocean currents. Deep-Sea Res., 431-437.

Loosanoff, V. L., 1932: Observations on propagation of oysters in James and Corrotoman Rivers and the seaside of Virginia. Virginia Fisheries Commision, Newport News, VA. 45 pp.

Manning, J. H. and H. H. Whaley, 1954: Distribution of oyster larvae and spat in relation to some environmental factors in a tidal estuary. Proc. Nat. Shellfish. Assoc., **45**, 56-65.

Nelson, T. C., 1921: Aids to successful oyster culture. I. Procuring the seed, N. J. Agr. Exper. Sta., Rutgers Univ., New Brunswick, NJ. Bull. 351. 59 pp.

Nelson, T. C., 1955: Observations on the behavior and distribution of oyster larvae. Proc. Nat. Shellfish Assoc., **45**, 23-28.

Nunes, R. A. and J. H. Simpson, 1985: Axial convergence in a well-mixed estuary, Est. Coast. Shelf Sci., **20**, 637-649.

Odum, E. P., 1971: Fundamentals of Ecology, Sanders, Philadelphia. pp. 352-362.

Oey L., G. L. Mellor and R. I. Hires, 1985: A three-dimensional simulation of the Hudson-Raritan estuary. Part I: Description of the model and model simulations. J. Phys. Oceanogr., **15**, 1676-1692.

Pape, E. H. and R. W. Garvine, 1982: The subtidal circulation in Delaware Bay and adjacent shelf waters. J. Geophys. Res., **87**, 7955-7970.

Pritchard, D. W., 1952: Salinity distribution and circulation in the Chesapeake Bay estuarine system. J. Mar. Res., **11**, 106-123.

Pritchard, D. W., 1953: Distribution of oyster larvae in relation to hydrographic conditions. Proc. Gulf Carib. Fish. Inst., **5**, 123-132.

Prytherch, H. F., 1928: Investigation of the physical conditions controlling spawning of oysters and the occurrence, distribution and setting of oyster larvae in Milford Harbor, Conn. Bull. Bur. Fish., **44**, 429-503.

Rothlisberg, P. C., J. A. Church and A. M. G. Forbes, 1983: Modelling the advection of vertically migrating shrimp larvae, J. Mar. Res., **41**, 511-538.

Roughgarden, J., S. Gaines and H. Possingham, 1988: Recruitment dynamics in complex life cycles, Science, **241**, 1460-1466.

Seliger, H. H., J. A. Boggs, W. H. Biggley, and K. R. H. Aspden, 1982: The transport of oyster larvae in an estuary, Mar. Biol., **71**, 57-72.

Wong, K-C. and R. W. Garvine, 1984: Observations of wind-induced, subtidal variability in the Delaware Estuary, J. Geophys. Res., **89**, 10589-10597.

Wood, L. and W. J. Hargis, Jr., 1971: Transport of bivalve larvae in a tidal estuary. Fourth Eur. Mar. Biol. Symp., D. J. Crisp. ed., Cambridge Univ. Press, Cambridge, pp. 21-44.

A 3-D Eutrophication-diffusion Model of the Venice Lagoon: Some Applications

Camillo Dejak, Davide Franco,
Roberto Pastres and Giovanni Pecenik

University of Venice, Department of Physical Chemistry
Section of Environmental Physical Chemistry
Calle Larga S. Marta, 2137, 30123 Venice, Italy

ABSTRACT

A discussion is presented of the problems encountered in the formulation, calibration and validation of a truly interdisciplinary 3-D eutrophication-diffusion macromodel of a complex ecosystem. In water quality modelling too much emphasis may be placed on transport phenomena in comparison with other processes, such as meteoclimatic, chemical, and ecological processes, having inherent larger stochastic fluctuations.

Interdisciplinary contributions are outlined, and approximations are made for reproducing transport process, accounting for the effects of residual currents. The latest macromodel's improvements are described as well as its major applications: analysis of eutrophication phenomenon, optimization of a water-quality monitoring network, assessment of thermal impact, and prediction of the effect of the temporary closure of the inlets on lagoon water quality.

I. Introduction

Building truly interdisciplinary models of complex ecosystems is still one of the greatest problems in environmental studies. A monodisciplinary model, such as a pure transport model, which has limited contributions from other disciplines, is of doubtful value. Equally inadequate is the mere combination of several distinct monodisciplinary models, built with arbitrary coordinated interaction amongst the various modelers. The following types of model may be devised for a waterbody:

i) purely hydrodynamic models;

ii) hydrodynamic models which include some selected environmental variables;

iii) a trophic-dynamic reactor combined with a hydrodynamic model;

iv) an environmental model intrinsically linked with transport phenomena.

Real interdisciplinarity may be assured only by type (iv) models, as the former types may represent just intermediate stages in developing global models. In building a truly interdisciplinary model, attention must be paid constantly to the objectives for which the model is being conceived. In effect, based upon these objectives and to build such models, one needs to:

1) identify those disciplines which are fundamental, those which are less significant, and those which may be considered negligible.

2) undertake appropriate analyses of information provided by each contribution in order to avoid placing undue relevance upon information provided by one discipline while contributions obtained by others do not present a comparable level of accuracy.

Coastal and Estuarine Studies, Vol. 38
R. T. Cheng (Ed.)
Residual Currents and Long-term Transport

As far as models of waterbodies open to the sea are concerned, these models should be designed in order to achieve some of the following objectives:

a) assisting in a rational management of engineering projects, such as the temporary closure of the inlets when high floods are foreseen;

b) predicting the consequences of modifications of the hydraulics of the system, sometimes necessary to protect the waterbodies either from sediment erosion or for increasing water mixing;

c) describing the dispersion of persistent pollutants released from continuous sources, their transport and possible accumulation; and

d) examining the interactions among chemical pollutants, and/or nutrients, as well as their effects on other biological variables (phyto- and zoo-plankton, macrophytes, bacteria) of the ecosystem under investigation.

For each of these objectives, there will be different criteria for selecting possible contributions and for estimating the incidence of statistical fluctuations affecting the processes investigated by each discipline. For example, in a transport model of a conservative pollutant, hydrodynamic variables would play a major role, while in a eutrophication model, less emphasis might be given to these variables. The difficulties involved in performing valuable "in situ" measurements would reveal the correct formulation of chemical and biological processes and statistically estimate all related parameters. Additional peculiarities may complicate the modeling efforts, particularly when the devised objectives are:

a) to assess the impact on the waterbody of a specific pollutant such as that released from a chemical plant;

b) to predict ecosystem responses to thermal discharges from a power plant;

c) to protect fisheries resources in a polluted or eutrophic waterbody; and

d) to safeguard sanitary quality and recreational uses of lagoon waters.

In some instances, the mere investigation of first order effects may still be insufficient, as higher order effects must also be taken into account. For example, it is indispensable not only to study the spatial distribution of temperature, but also its effects on chemical and biochemical reactions (second order), together with all synergisms arising from the combined effects of temperature and nutrients, in the presence of other perturbing pollutants, such as toxic substances, (third order). Modeling activity becomes even more difficult whenever multiple objectives need to be achieved, since higher order effects must be considered for each of the objectives and for all their reciprocal interactions.

Modeling work inevitably demands experimental data; if these data are unavailable, it is compulsory to plan rather expensive campaigns for collecting them. In general, three types of experimental data are required, which should permit:

1) establishing the basis of the model, indentifying all primary and secondary processes, choosing the best formulation for each process;

2) calibrating each process in the model; (the acquisition of these type of data requires expensive and time consuming "functional" and "structural" field researches, at all apparently significant sites within the modeled area);

3) validating the model using historical series of relevant water quality parameters, thus overcoming the episodicity of each year to enable revealing through statistical treatment, the multiannual trend, the seasonal behavior of a typical average year, and the stochastic fluctuations. The distributions of the latter should be tested for normality before proceeding to any transformation of the data sets.

It is unlikely that an efficent model, having sufficent predictive potentiality, may be developed without these experimental data, each of which plays a specific role, and should be weighted separately.

II. 3-D Eutrophication-diffusion model of the Venice Lagoon

For the past 15 years our research group has been working towards the development of a macromodel of the central part of the Venice Lagoon. The modeled area includes the highest Italian industrial concentration and the historic town of Venice (Fig.1). The main objectives of this model are to describe the eutrophication phenomenon periodically occurring in the lagoon and the effect of the emission of nutrients (BOD, nitrogen and orthophosphates), and also to assess the impact of heat discharged by a thousand MWs thermal power plant, as well as of the prospected closure of sea entrances to the Venice Lagoon.

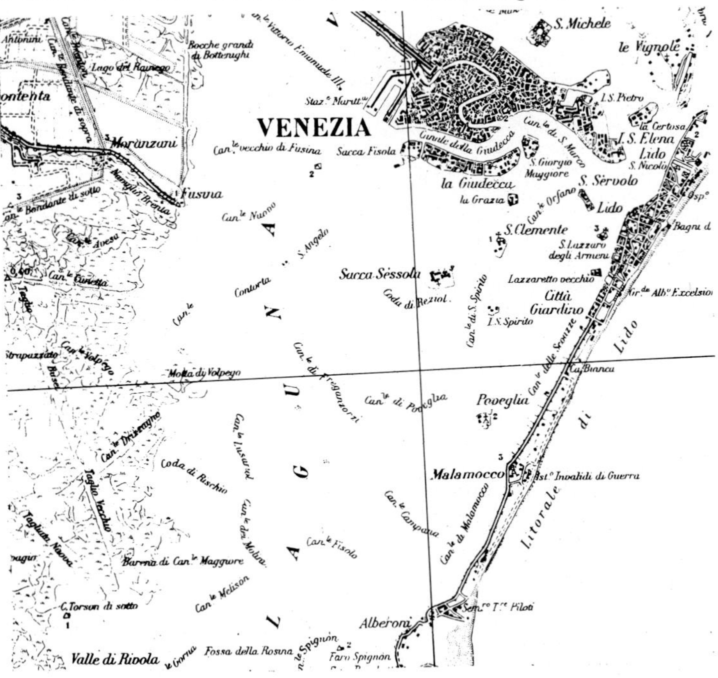

FIGURE 1. Map of the modeled lagoon area, (Scale 1:100000).

When the modeling activity began, many hydrodynamic data were available, which were collected after the "high tide" of 1966 and later used for the calibration of one and 2-D, finite-difference, hydrodynamic models (D'Alpaos and Di Silvio, 1979). On the contrary, there was little published information on the chemical and biological conditions, only few indirect, (mostly qualitative) structural measurements of pollutant concentrations in sediments and in benthic fauna were available. In particular, functional properties of the lagoon ecosystem were completely unknown, and since these specifically deal with the rate of flows of energy and matter through the ecosystem, (Matthews, et al., 1982), their knowledge is essential for a correct modeling approach.

From a preliminary analysis, it became clear that pollutant transport in the lagoon was mainly governed by tidal currents, which if averaged over a suitable time period, (at least one tidal cycle), present a zero average velocity. The presence in the lagoon of residual currents has never been observed experimentally, but only revealed by comparing results yielded by running, under two different conditions, a somewhat complex 2-D mathematical model, whose conception was not free from subjective influences. The appearance of such residual currents seemed to be attributed more to thermohaline effects rather than to the mechanics of the lagoon water circulation.

In fact, while the more saline and often colder sea water enters the lagoon as a saline wedge, along the bottom of the deeper navigation channels (which are their preferential paths), ebbing water, being less salted, also expands over the shallower lagoon areas, which are characterized by depths sometime lower than the tidal amplitude (the mean lagoon depth is about one meter). This behavior is confirmed by the fact that, at flood, the flow in the main channels is much higher than that observed during ebbing. For this reason, residual currents might also present typical seasonal variations. The reproduction of such a complex behavior could not be attempted merely through deductive actions or through 2-D hydrodynamic models, but through the construction of a 3-D, baroclinic model, accounting for salinity and temperature variations. Clearly, the task would have been disproportional relative to the aim of modeling the lagoon eutrophication phenomenon, both because of the difficulty in finding financial support and of the lack of specific expertise. Consequently, the attempt has been to model the transport phenomenon while adopting constant eddy-diffusivities, and leave the possibility of introducing some minor modifications in order to consider the effects of residual currents even if these are marginal.

Thus, the modeling work started by developing a purely 2-D finite-difference diffusion model for a conservative pollutant with a horizontal step of 100 m. and 140 x 128 grid points. With this model, a steady state was never achieved, since a continuous accumulation at the source was observed, although appropriate conditions at the open boundaries were established (Dejak et al., 1987a). In fact, for extrapolating the boundary values a method was devised, based upon a finite difference analysis of the logarithmic concentrations at interior points, which led to a typical Gaussian pattern, integral function of the general 1-D, 2-D, and 3-D diffusion equations including turbulent diffusion. The method allows the extrapolation of boundary values, at all times and points, based on three interior values as required for computing the second differences.

With an advection-diffusion model, the same conclusions were reached when, for the first time, tidal currents were computed from experimental velocity fields, (Ghetti et al., 1979) with a simple log-log regression of velocities on the distances of grid points from sea entrances (Dejak et al., 1987b). A steady state was instead achieved, (Dejak et al., 1987c) by introducing, in a second time, tidal currents provided by a model which, although being 2-D, also included effects derived from the sharply varying bathymetry and calibrated with tidal elevation data taken from the many tidal gauges placed in the lagoon after the 1966 flood, (D'Alpaos and Di Silvio, 1979).

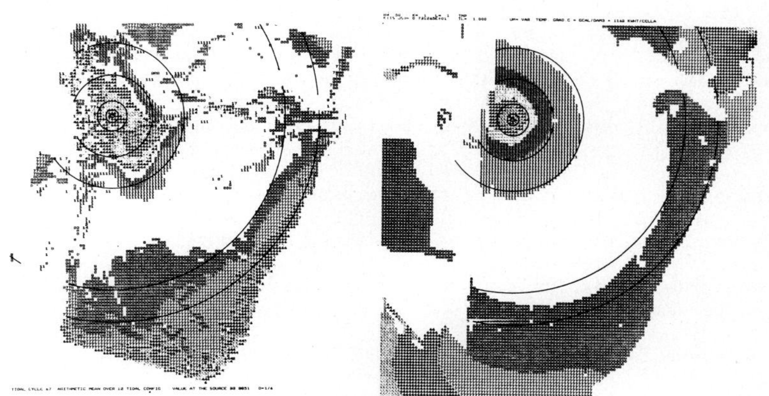

FIGURE 2. Steady-state surface distributions of pollutant resulting from the 2-D advection-diffusion model (left), and from the 3-D diffusion model, embodying tidal mixing (right). Circles evidence the radial symmetry of the two diffusional patterns.

The steady state concentration patterns exhibited the characteristic centrosymmetric distribution of a conservative pollutant when it is continuously released from a point source of polluting power $\dot{Q}$, (with zero initial values at all points) and was comparable to that yielded by a purely diffusion model. However, to reach a steady state, it is inevitable to resort to a 3-D eddy-diffusion model which is, as mathematically demonstrated in the Appendix, the only one converging asymptotically, in time, to a stationary distribution. By comparing the surface spatial steady state patterns of the two models, (Fig. 2) it therefore becomes possible to estimate an eddy-diffusivity, of such a value as to assure a reasonable coincidence of the two distributions. The effect of the residual currents could only be revealed by the displacement of the actual source to a virtual point 1.5 km North-Eastward. Thus, by considering this virtual point as the location of the point source, it has been possible to take into account the effects of residual current and hence their rather small influence in this peculiar waterbody (Dejak et al., 1987c). A 3-D model has therefore been designed, with the same horizontal steps and a vertical step of 1 m., thus leading to about 360,000 mesh points. Vertical diffusivity was tentatively taken one order of magnitude lower than in the horizontal, in adherence to what was found under comparable conditions by Kullenberg (1971) and Kullenberg et al., (1973).

Furthermore, a method had to be found for the numerical integration of the governing differential equation, which could adequately address the abruptly changing lagoon bathymetry that could lead to severe numerical instabilities if the usual Crank-Nicolson (1947) method were adopted. A Fourier analysis of the integral functions of the diffusion equation showed the quasi-instability, at lower wavelengths, to be produced mainly by the very indented boundaries, whose irregularities led to the failure of the more commonly used integration schemes. This approach helped in the selection of the method more appropriate for the present application. Through comparative analyses of the most important explicit and implicit two or three time steps integration schemes, (Richtmyer and Morton, 1967) it has been found that the two step Laasonen method (1949) showed the greater stability in the presence of continuous sources and at higher diffusion numbers (product of diffusivity and time step divided by the square of the space step). Moreover, this method required less computational time and hence reduced costs (Dejak et al., 1987d) and assured stability, convergence and consistency. It also guarantees the conservativity and transportivity properties, (Roache, 1972).

The model, in which the integration is performed by the fractional step method, owing to turbulent diffusion, may be applied with the same eddy diffusivity for intensive state variables, such as concentrations, temperature, and plankton densities (Okubo, 1980), whose transport is followed in time. Moreover, the model allowed for the strict linking with a trophic-dynamic reactor in which all state variables interact by means of physical, chemical and biological processes. In Table 1, the system of differential equations is reported, describing interrelationships among these variables. The complete macromodel is run, assigning as initial values, at all cells, the average lagoon phyto- and zooplankton densities, $N\text{-}NO_3^-$, $N\text{-}NH_4^+$, $P\text{-}PO_4^{3-}$ concentrations. Temperature initial values are computed with respect to depth by a purposely prepared 1-D submodel. The oxygen saturation value is considered given at all cells.

III. Additional implementations of the macromodel

Improvements successively introduced in the macromodel mainly concerned with:

1) program reorganization and optimization for use of vectorial and parallel supercomputers;

2) elaboration of submodels, adopted to verify and test all improvements;

3) seasonal properties of the macromodel;

4) reproduction of thermal stratification and of its effect on vertical diffusion mixing;

5) reformulation of some of the processes included in the model;

6) introduction of additional variables, such as macrophytes and bacteria; and

7) validation of the macromodel by using statistically treated historical series of experimental data.

TABLE 1. Differential equations interconnecting the state variables included in the biological compartment and functional expressions.

Connected ordinary differential equations:

$$d[F]/dt = \{\beta - (K_{mf}+K_{sedf}+K_{rf}+K_{escf})\ [F]\} - K_{grz}\ f(F)\ [Z] \qquad (1)$$

$$d[Z]/dt = (K_{grz}\ f(F) - K_{mz} - K_{escz})[Z] \qquad (2)$$

$$d[NO_x^-]/dt = K_{nit}\ [NH_4^+] - RCN\ \beta\ ([NO_x^-]/[N_{tot}]) \qquad (3)$$

$$d[NH_4^+]/dt = -K_{nit}[NH_4^+] - RCN\{(\beta[NH_4^+]/[N_{tot}]+K_{rf}+K_{escf})[F]+K_{escz}\ [Z]\}+K_{rsn} \qquad (4)$$

$$d[PO_4^{3-}]/dt = -RCP\{(\beta+(K_{rf}+K_{escf})[F] + K_{escz}\ [Z])\} + K_{rsp} \qquad (5)$$

$$d[O_2]/dt = K_{rear}\ DO_2+ROC(\beta-K_{rf})[F]-K_{dec}\ [BOD]-RON\ (K_{nit}\ [NH_4^+])-K_{osed} \qquad (6)$$

$$d[BOD]/dt = -\ K_{dec}\ [BOD] + ROC\ (K_{mf} + K_{sedf})\ [F] + ROC\ K_{mz}\ [Z] \qquad (7)$$

Variables:

$[F]$, $[Z]$ = phyto- and zooplankton densities	(mgC/l)
$[PO_4^{3-}]$ = orthophospate	(mgP/l)
$[NO_x^-]$, $[NH_4^+]$, $[N_{tot}] = [NO_x^-]+[NH_4^+]$	(mgN/l)
$[BOD]$ = Biochemical Oxygen Demand	(mgO_2/l)
$[O_2]_t$ = dissolved oxygen at time t	(mgO_2/l)
T = temperature	$(°C)$

Constants:

K_n, K_p, K_f = halfsaturation for $[N]$, $[P]$, and $[F]$;	(mg/l)
K_{rf}, K_{sedf} = respiration and sedimentation for $[F]$;	(h^{-1})
K_{mf}, K_{mz} = phyto- and zooplankton mortality;	"
K_{escf}, K_{escz} = " " " excretion;	"
K_{nit}, K_{rear} = nitrification and reaeration;	"
K_{dec} = B.O.D decay;	"
K_{rsn}, K_{rsp} = sediment release of $[NH_4^+]$ and $[PO_4^{3-}]$;	(mg/hl)
K_{grz} = grazing;	(h^{-1})
K_{osed} = sediment oxygen demand;	(mgO_2/hl)
ROC, RON, RCP, RCN = conversion factors O_2/C, O_2/N, P/C, N/C;	(dimensionless)

Functional expressions:

$$\beta = \mu(T)\ f(N_{tot})\ f(PO_4^{3-})\ f(I)$$

$$\mu(T) = \mu(T_o)\ [(T_a-T)/(T_a-T_o)]^{b(T_a-T_o)}\ \exp[b(T-T_o)]$$

$$f(N_{tot}) = [N_{tot}]/(K_n+[N_{tot}]) \text{ and } f(PO_4^{3-}) = [PO_4^{3-}]/(K_p+[PO_4^{3-}]);$$

$$f(I) = e/\alpha_1\ \{\exp[-\alpha_2\ \exp(-\alpha_1)] - \exp(-\alpha_2)\}$$

$$\alpha_1 = K_h\ z, \alpha_2 = I/I_o \text{ and } K_h = .4[\tfrac{1}{2}\ F(1)+F(2)+\ldots+F(h-1)+\tfrac{1}{2}\ F(h)$$

$$\text{where}: F = [F] \text{ for } h = z$$

$$f(F) = [F]/(K_f+[F])$$

$$DO_2 = [O_{2sat}] - [O_2]_t = \text{oxygen deficit}; \qquad (mg/l)$$

$$[O_{2sat}] = 14.6244-0.367134T+0.0044972T^2-0.0966S+0.00205TS+0.0002739S^2$$

and related definitions:

z = depth;	(m)
I, I_o = incident and optimal growth light intensities;	(lux)
K_h = extintion coefficient;	(m^{-1})
$\mu(T)$, $\mu(T_o)$ = phytoplankton specific growth rate at T and T_o	(h^{-1})
T_o, T_a = phytoplankton optimal growth and lethal temperatures	$(°C)$
b = Phytoplankton growth thermal damping coefficient	$(°C^{-1})$
S = average lagoon salinity	$30°/_{\circ\circ})$

The installation of supercomputers in Italy in 1984 and 1987 made it possible, through the syntactical and structural reorganization of the program, to drastically reduce both computation time and expenses. Actually, the CPU time required for simulation of one year decreased from 20 days (as required by the originally used CDC 7600) to about 1.5 days with the Cray X-MP/12 and finally to nearly 3 hours elapsed time, by exploiting multitasking with the Cray X-MP/48. Using the Cray X-MP/48, 90% of vectorization and 80% of parallellism were achieved, (Dejak et al., 1988).

A series of submodels, meanwhile, have been developed using either 0-D or vertical 1-D and 2-D, as well as two 3-D models of smaller size (one of 4x4x14 cells and the other of 43x40x20 cells based on the actual lagoon bathymetry). These submodels constituted valuable tools for testing, at reduced costs, all improvements later introduced in the macromodel. Accounting for all factors affecting heat-exchanges between air-water interfaces, (Mariotti and Dejak, 1982), the space-time variations of temperature are also computed, based on a quantification of the global heat balance in the lagoon. Also, based upon hundreds of experimentally determined temperature vertical profiles, depth varying eddy diffusivities were estimated, which exhibit a minimum at the thermocline, and whose onset and seasonal evolution have also been reproduced (Dejak and Pecenik, 1987).

Using "in situ" measurements of phytoplankton growth rate, within a large temperature range, it has been possible not only to calibrate the more flexible formulations found in literature, (Joergensen, 1974), but also to propose additional modifications to the model, leading to further applications such as the understanding of the spatial distributions of phytoplankton species' dependence on temperature.

In recent years, abnormal growth of macrophytes has been observed at specific lagoon areas in competition with microalgae which are usually more abundant. It is therefore intended to map such sites by remote sensing after calibration with a set of field observations. This variable will be soon introduced into the macromodel.

Decennial unpublished historical time series of data relative to important water quality parameters have recently been gathered. Although the series exhibit irregularities and sometime lack of data for extended periods, (i.e.; they are unevenly spaced), statistical methods have been elaborated to compile more regular data sets, thus facilitating the decomposition of the series into their essential components, with the aim of validating the macromodel. With appropriate extensions of the methodologies adopted, time series, referring to important locations in the modeled area, may also be analyzed, though these consist of only an initial and a final three year period of observations.

IV. Applications of the 3-D macromodel

Since the macromodel achieved its earlier configuration, it soon was used for urgent applications; for example, these are some of the more important cases:

i) to comprehend and simulate lagoon eutrophication phenomenon;

ii) to optimize the sites for placing water quality parametèrs monitoring stations;

iii) to assess thermal impact upon the lagoon ecosystem;

iv) to predict the effects of the closure of sea entrances on lagoon water quality.

As far as the eutrophication phenomenon is concerned, the earlier results obtained by the combined 3-D eutrophication-diffusion macromodel have already been reported elsewhere, (Dejak et al., 1987e). This involved running the model up to steady state, under both eutrophic and hypereutrophic conditions. An example reported here is the steady state spatial distribution of ammonia concentration under stressed conditions, (Fig.3), i.e., simulating both a high input from an industrial point source and a strong release from the sediments. These results reasonably agree with the available data sets, but a definitive validation will be performed when the statistical treatment of the historical series is completed.

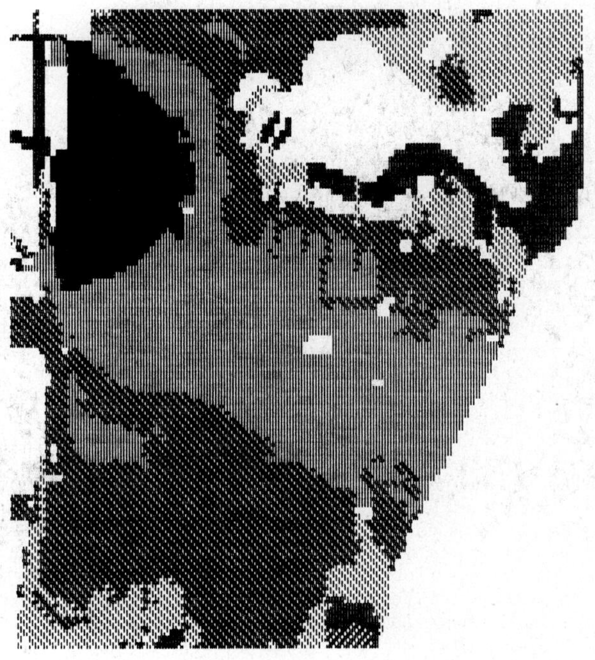

FIGURE 3. Steady-state surface spatial distribution of ammonia.

By running the model under normal and stressed conditions, the differences among values of variables at each grid point may be computed and their distributions may then be utilized for a statistical analysis enabling to devise a grid of the more appropriate sites in a lagoon for monitoring water quality parameter. Differences and their distributions may be reasonably well reproduced with a generalized Gamma function. Also, by dividing the surface of the waterbody in squares, each of 10x10 points (1 km^2), it is easy to detect which of them present the higher variability and hence establish the best position inside the squares themselves. The statistical treatment involves the adoption of appropriate weights for the variability range of each single variable in order to account for its relative influence in the lagoon's ecosystem. This makes it possible to design a score of the capability of each selected site to yield the highest amount of "information." The rational choice of the monitoring stations is very important because of the rather high costs of the sensors.

The application of the macromodel had made the greatest contribution in the study of the thermal impact derived from heat discharges of a thermal power plant about one thousand MWs (Dejak, 1979a, 1979b). The macromodel was found to be an indispensable tool for planning long-term field works, including functional research on relevant processes, (Battaglia _et al_., 1984, Bertonati, _et al_., 1987). This encouraged further development of the model by coordinating theoretical and experimental works, which enabled mapping the surface temperature fields as well as vertical temperature profiles. These vertical temperature profiles are very important in the description of thermo-halyne stratification and are important for examining thermal effects on the rate of chemical and biochemical reactions.

In this particular application, the model really became a management tool, since, based on its results, the "Commissione Regionale di Controllo" of the Fusina

Power Plant could plan two consecutive three yearly field surveys. This work made it clear that no significant damage to the ecosystem could be ascribed to cooling water discharges and the local authorities were persuaded to allow the Ente Nazionale Energia Elettrica (E.N.E.L.) to exploit the plant at full capacity.

Finally, a specific application recently undertaken involves predicting the impact on the lagoon of the proposed closure of the sea entrances during exceptionally "high tides" or floods, which are becoming more frequent. This application could be realized only by firstly drastically reducing the computational time and costs through the optimization of the program as mentioned above and secondly, by collaborating with the Consorzio Venezia Nuova and the Hydraulic Danish Institute. In fact, eddy diffusivities varying in time and in space have been introduced into the macromodel, with the aid of the 2-D finite-difference advection model developed by the just quoted institutions. From the components of tidal velocities and the related kinetic energy, the diagonal diffusivities tensors may be computed, assuming a proportionality relation. To this end, the capability of the 2-D hydraulic model is exploited, to be zoomed from its space step of 300m. to 100m. as needed for this application. Also, in order to achieve a perfect adherence with the macromodel, use has been made of the possibility of rotating the 2-D model along the North-South direction. Fig. 4 shows the velocity fields of the model at maximum flood and ebb tides. After performing a series of operations upon the velocity data in order to homogenize the boundaries and the associated values, these have been used to provide a more accurate description of the dispersion of conservative pollutants and then in the global eutrophication-diffusion macromodel. Modifications and improvements are still being undertaken with the objective of ascertaining whether the proportionality relation of diffusivities is more realistic when performed with respect to the tidal kinetic energies itself, (Price, 1982) or to the squared kinetic energies, (Coachman et al., 1988) or to the product of the latter with the eddy dimensions.

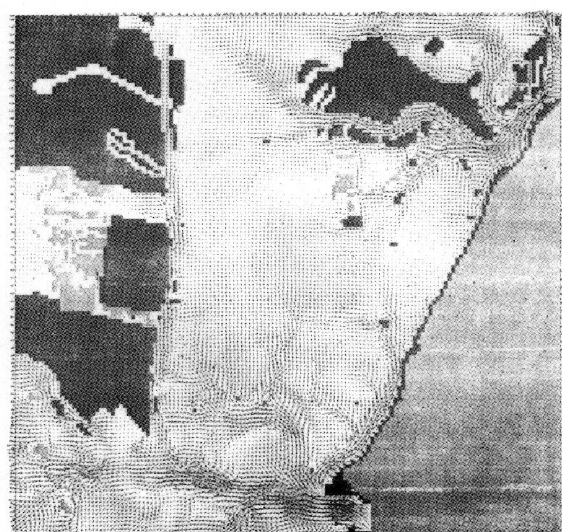

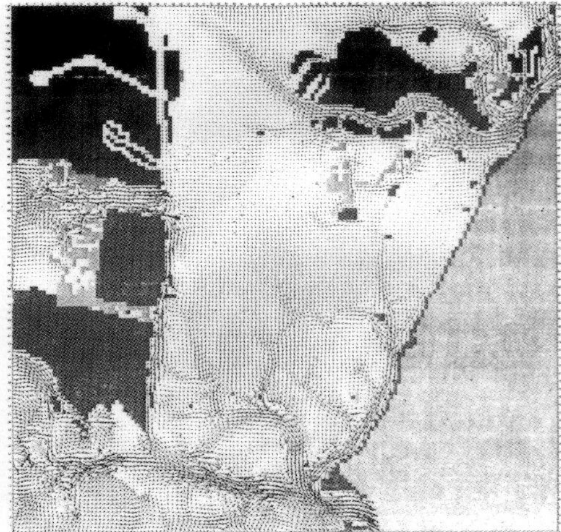

FIGURE 4. Surface velocities field obtained with the 2-D hydraulic model, at maximum flood, (left) and at maximum ebb, (right), 1.2 cm = 1 m/s.

V. Conclusions

An interdisciplinary approach has played a major role in the development of the 3-D macromodel described in this paper. In effect, its design has been made possible by including contributions from different fields of study, including hydraulics, thermodynamics, physico-technical, physico-chemical, biochemical and ecological disciplines. Contributions were not treated separately, but integrated.

Hence, it has been possible to avoid emphasizing more advanced fields, such as hydraulics, while errors higher than one order of magnitude in other less known fields could be included from research dealing with processes of more difficult and less detailed description. With such a macromodel it is thus possible to analyze ecosystem behavior and to utilize it for management of productive activities and/or researches concerning either present or future problems. It has been attempted to give more emphasis to the reproduction of the phenomena occurring along the vertical, both with respect to the thermal-hydrodynamics and biological aspects (photosynthetic activity). Only by considering the third dimension in the model, conditions are guaranteed for the mathematical convergence of the integration of the diffusion equation.

Because of the marginal influence of the residual currents in this water body, the simpler solutions have been sought in the surface plane. However, efforts are being made to enhance the macromodel's potentialities with the introduction of diffusivities which are not only tensorial, but also vary in time and space. By appropriate selection of methods for reproducing the transport process, we are confident that the model will cope with any application posed by the prospective engineering interventions proposed for the Venice Lagoon, without overemphasizing a particular discipline, unless a specific demand is made to the macromodel.

Acknowledgments

The work was funded by the Projects: Progetto Finalizzato Energetica 2, (PFE) of C.N.R., (Italy) and EUROTRAC. The authors thank Saviano Enzo and Bonello Arnaldo for their technical helps.

VI. References

Abramowitz, M., Stegun, I., 1964: Handbook of Mathematical Functions. Dover, New York, pp. 1046.

Battaglia, B., Datei, C., Dejak, C., Gambaretto, G., Guarise, G.B., Perin, G., Vianello, E., Zingales, L., 1984: Indagine per la valutazione dei riflessi ambientali del funzionamento a piena potenza della centrale termoelettrica di Fusina. Commissione Tecnico-Scientifica per la sperimentazione ed i controlli periodici sulla Centrale Termoelettrica dell'ENEL sita in localit Fusina di Porto Marghera, Venezia. Provvedimenti Regione Veneto N 810 and N 811, May, 10, 1979.

Bertonati, M., Dejak, C., Mazzei Lalatta, I., Pecenik, G., 1987: Eutrophication model of the Venice Lagoon: statistical treatments of phytoplankton growth parameters. Ecol. modeling, **37(1/2)**, 103-130.

Coachman, L.K., McRoy, C.P., Nihoul, J.C.J., Walsh, J.J., 1988: "Model of the Northern Bering Sea Ecohydrodynamics" Paper presented at the Joint Oceanographic Assembly, Acapulco Aug. 23-31, 1988. (In press).

Crank, J., Nicolson, P., 1947: A practical method for numerical integration of solution of partial differential equations of heat-conduction type. Proc. Cambridge Philos. Soc., **43(50)**, 50-67.

D'Alpaos, L., Di Silvio, G., 1979: Utilizzazione dei risultati per la taratura dei modelli matematici. In: "Le correnti di marea nella laguna di Venezia." Ibidem: pp. 82-95.

Dejak, C., Mazzei Lalatta, I., Meregalli, L., Pecenik, G., 1987a: Development of a mathematical eutrophication model of the lagoon of Venice. Ecol. modeling, **37(1/2)**, 1-20.

Dejak, C., Mazzei Lalatta, I., Messina, E., Pecenik, G., 1987b: A two-dimensional diffusion model of the Venice Lagoon and relative open boundaries conditions. Ecol. modeling, **37(1/2)**, 21-45.

Dejak, C., Mazzei Lalatta, I., Messina, E., Pecenik, G., 1987c: An advection diffusion pollution model of the lagoon of Venice. Ecol. modeling, **37(1/2)**, 47-57.

Dejak, C., Mazzei Lalatta, I., Messina, E., Pecenik, G., 1987d: Steady state achievement by introduction of true tidal velocities in a pollution model of the Venice Lagoon. Ecol. modeling, **37(1/2)**, 59-79.

Dejak, C., Mazzei Lalatta, I., Molin, M., Pecenik, G., 1987e: Tidal three-dimensional diffusion model of the lagoon of Venice and reliability conditions for its numerical integration. Ecol. modeling, **37(1/2)**, 81-101.

Dejak, C., 1979a: Problemi riguardanti gli scarichi di acque di raffredamento di centrali termoelettriche e di altri impianti industriali nella laguna veneta. Report for Ente Zona Industriale Porto Marghera. **n.A2**, 1-17.

Dejak, C., 1979b: Verifica sperimentale sugli scarichi termici della Centrale di Fusina. Ibidem, **A3**, 1-4.

Dejak, C., Pecenik, G., 1987: A trophic-diffusion 3D model of the Venice Lagoon. In: "Three-dimensional models of marine and estuarine dynamics. Nihoul, J.C.J. & Jamart, B.M., Eds. Elsevier Oceanography Series, **45**, 391-404.

Dejak, C., Pecenik, G., Pastres, R., Zane, E., 1988: Optimizing an eutrophication-diffusion 3-D macromodel for the CRAY X-MP/12. Environmental Software, **3(2)**, 56-64.

Ghetti, A., Bartoletti, U., Lippe, E., D'Alpaos, L., Di Silvio, G., 1979: Le correnti di marea nella laguna di Venezia. Introduzione e Parte Ia. Ministero del Lavori Pubblici. Comitato per lo Studio dei Provvedimenti a Difesa della Citt di Venezia ed a Salvaguardia dei suoi Caratteri Ambientali e Monumentali. Istituto di Idraulica; Universit di Padova, Ed.; pp. 3-66.

Joergensen, S.E., 1979: "Handbook of environmental data and ecological parameters." In: Biswas, A.K., Biswas, M.R., Eds., Environmental Sciences and Applications. Pergamon Press, Oxford, Vol.6. 1162 pp.

Kullenberg, G.E., 1971: Results of diffusion experiments in the upper region of the sea. Report 12, Kobenhavns Universitet, Institut for Fysisk Oceanografi, 66 pp.

Kullenberg, G.E., Murthy, C.R., Westerbuy, H., 1973: An experimental study of diffusion characteristics in the thermocline and hypolimnion regions of Lake Ontario. In: Proc. 16th Conf. Great Lakes Res., pp: 774-790.

Laasonen, P., 1949: Ueber eine Methode zur Loesung der Waermeleitungsgleichung., Acta Math., **81**, 309-314.

Mariotti, M., Dejak, C., 1982: Valutazione dei Flussi energetici all'interfaccia acqua-aria nella laguna di Venezia. In: Atti del 10 Convegno "Ambiente e Risorse", Bressanone, 30 Agosto-4 Settembre, 1982. Sect. I5:15 pp.

Matthews, R.A., Buikema, A.L. Jr., Cairns, J. Jr., Rodgers, J.H. Jr., 1982: Biological monitoring. Part IIA - Receiving system functional methods. Relationships and Indices. Water Res., **16**, 129-139.

Okubo, A., 1980: "Diffusion and Ecological Problems: Mathematical Models." Biomathematics. Springer-Verlag, Vol.10, pp.254.

Price, J.F., 1982: Eddies in Marine Science. Robertson Ed. Springer Verlag, 1982. p.609.

Richtmyer, D.R., Morton, K.W., 1967: Difference methods for initial value problems., 2nd Ed. Interscience Publishers, London, 405 pp.

Roache, P.J., 1972: Computational fluid dynamics. Hermosa Publishers Albuquerque, New Mexico 87108, USA. 446 pp.

Appendix

Considering the well known purely diffusion equation as applied in an centro-symmetric system, in n = 1, 2, and 3 dimensions, and assuming an instantaneous input of an amount $\dot{Q}$ of pollutant, the conservation equation, i.e., the integral of the concentration must equal to $\dot{Q}$ over the entire space, is

$$\frac{\partial c}{\partial t} = k\nabla^2 c = \frac{k(n-1)}{r}\frac{\partial c}{\partial r} + k\frac{\partial^2 c}{\partial r^2} \qquad (1)$$

By integration, it gives

$$c = (4k\pi t)^{-n/2}\exp(-r^2/4kt) \qquad (2)$$

where c is the concentration and r is the distance from the source.

This solution may be generalized, through convolution, to a continuous source, of $\dot{Q}$, (polluting power) which is constant in time and a delta function in space to give

$$c = \frac{\dot{Q}}{(4k\pi t)^{n/2}}\int_o^t \exp\left(\frac{-r^2}{4k(t-\tau)}\right)\frac{d\tau}{(t-\tau)^{n/2}}$$

$$= \frac{\dot{Q}}{4k\pi^{n/2}r^{n-2}}\int_{r^2/4kt}^{\infty} \exp(-x)x^{n/2-2}\,dx \qquad (3)$$

For each single case the following integrals are obtained:

for n = 1

$$c = \dot{Q}k^{-1/2}\{(t/\pi)^{1/2}\exp[-r^2/(4kt)] - (4k)^{-1/2}r\cdot\mathrm{erfc}[r(4kt)^{-1/2}]\}$$

$$= \dot{Q}(k\pi)^{-1/2}[t^{1/2} - r(\pi/4k)^{1/2} + r^2t^{-1/2}/4k - r^4k^{-2}t^{-3/2}/96 + ..]$$

$$\to \infty \quad \text{as } t \to \infty,$$

for n = 2

$$c = \dot{Q}/(4k\pi)\,E_1[r^2/(4kt)]$$

$$= \dot{Q}/(4k\pi)\{\ln(t) - \ln(r^2\Gamma/4k) + r^2/(4kt) - r^4/(8kt)^2 + ..\}$$

$$\to \infty \quad \text{as } t \to \infty,$$

for n = 3

$$c = \dot{Q}/(4k\pi r)\,\mathrm{erfc}[r(4kt)^{-1/2}]$$

$$= \dot{Q}(16k\pi)^{-1/2}[(k\pi)^{-1/2}/r - (k^2\pi^2t)^{-1/2} + r^2/(12k^2\pi t^{3/2}) - ..]$$

$$\to \dot{Q}/(4k\pi r) \quad \text{as } t \to \infty.$$

Expansion of these solutions in power series of the dimensionless product $r^2/(kt)$ shows that, besides trivial soultions which are either constant or vanishing in time, the solutions contain terms in squared and logarithmic of t, respectively for n = 1 and n = 2. These terms diverge as t tends to infinity. Only the three dimensional integral converges, while approaching the steady state, to a finite value which is inversely proportional to r.

By deducing from these integrals, and for the same cases the amount, q_o, which accumulates within the geometrical space of the points at a distance σ, whatever small, one obtains:

for n = 1

$$q_o = \dot{Q}t\{[1 + \sigma^2/(2kt)]\ \mathrm{erf}[\sigma(4kt)^{-1/2}] + \sigma(k\pi t)^{-1/2}\exp[-\sigma^2/(4kt)] - \sigma^2/(2kt)\}$$

$$= \dot{Q}\sigma(4k\pi)^{-1/2}[4t^{1/2} - \sigma(\pi/k)^{1/2} + \sigma^2 t^{-1/2}/(3k) - \sigma^4 t^{-3/2}/(120k^2) - \quad]$$

$$\to \infty \quad \text{as } t \to \infty \quad ,$$

for n = 2,

$$q_o = \dot{Q}t\{\sigma^2/(4kt)\bullet E_1[\sigma^2/(4kt)] + 1 - \exp[-\sigma^2/(4kt)]\}$$

$$= \dot{Q}\sigma^2/(4k)\{\ln(t) + 1 - \ln(\sigma^2\Gamma/4k)] + \sigma^2/(8kt) - \sigma^4(kt)^{-2}/192\} + (\sigma^2$$

$$\to \infty \quad \text{as } t \to \infty \quad ,$$

for n = 3

$$q_o = \dot{Q}t\{-\sigma(k\pi t)^{-1/2}\exp[-\sigma^2/(4kt)] + [1-\sigma^2/(2kt)]\ \mathrm{erf}[\sigma(4kt)^{-1/2}] + \sigma^2/(2kt)\}$$

$$= \dot{Q}\sigma^2/k[1/2 - \sigma(9k\pi t)^{-1/2} + \sigma^3(k^3\pi)^{-1/2}t^{-3/2}/60 - \quad]$$

$$\to \dot{Q}\sigma^2/(2k).$$

It is seen that values diverges either in one or two dimensions, whereas in three dimensions a limit is reached which is time invariant even in the presence of a continuous input. In this situation, the amount of substance introduced equals the amount which diffuses outside the system, i.e., the greater is the distance from the source, the substance is more diluted. In these formulae, k is the diffusivity in m^2/s, while erf(x), erfc(x), E_1, and $\Gamma = \exp(\gamma)$ are the standard notations for error function, complementary error function, exponential integral, and γ is the Euler's constant according to Abramowitz and Stegun, (1964, pp 914, 991, 877).

List of Contributors

List of Contributors

J.S.L.J. van Alphen
Rijkswwaterstaat
North Sea Directorate
P.O. Box 5807
2280 HV Rijswijk
The Netherlands

David G. Aubrey
Woods Hole Oceanographic Institution
Woods Hole, MA 02543
USA

Alan F. Blumberg
Hydroqual, Inc.
1 Lethbridge Plaza
Mahwah, NJ 07430
USA

J.C. Borst
Rijkswaterstaat
Tidal Water Division
P.O. Box 20907
2500 EX The Hague
The Netherlands

Malcolm J. Bowman
Marine Sciences Research Center
State University of New York
at Stony Brook
Stony Brook, NY 11794

Glenn A. Cannon
Pacific Marine Environmental
Laboratory
National Oceanic and
Atmospheric Administration
7600 Sand Point Way NE
Seattle, WA 98115-0070
USA

Ralph T. Cheng
Water Resources Division, MS-496
U.S. Geological Survey
345 Middlefield Road
Menlo Park, CA 94025
USA

William R. Crawford
Institute of Ocean Sciences
Dept. of Fisheries and Oceans
P.O. Box 6000
Sidney, BC V8L 4B2
Canada

Camillo Dejak
University of Venice
Department of Physical Chemistry
Section of Environmental Physical
Chemistry
Calle Larga S. Marta, 2137
30123 Venice
Italy

Scott P. Dinnel
Coastal Studies Institute
Department of Marine Sciences
Louisiana State University
Baton Rouge, LA 70803
USA

J. Dronkers
Rijkswaterstaat
Tidal Waters Division
P.O. Box 20907
2500 EX The Hague
The Netherlands

Shizuo Feng
Institute of Physical Oceanography
Ocean University of Qingdao
Qingdao
China

Davide Franco
University of Venice
Department of Physical Chemistry
Section of Environmental Physical
Chemistry
Calle Larga S. Marta, 2137
30123 Venice
Italy

Carl T. Friedrichs
Woods Hole Oceanographic Institution
Woods Hole, MA 02543
USA

Boris Galperin
Department of Marine Science
University of South Florida
St. Petersburg, FL 33701
USA

W. Rockwell Geyer
Woods Hole Oceanographic Institution
Woods Hole, MA 02543
USA

John M. Hamrick
Virginia Institute of Marine Science
College of William and Mary
Gloucester Point, VA 23062
USA

E.J. Hayter
College of Engineering
Clemson University
Clemson, SC 29634
USA

C.J. Hearn
Department of Geography
and Oceanography
University College,
University of New South Wales
Australian Defence Force Academy
Canberra, ACT 2600
Australia

Timothy R. Jacobsen
Department of Oyster Culture
Cook College, Rutgers University
New Brunswick, NJ 08903
USA

David A. Jay
Geophysics Program
University of Washington
Seattle, WA 98195
USA

Gerhard Jirka
DeFrees Hydraulics Laboratory
Cornell University
Ithaca, NY 14853
USA

R. Johnson
Proudman Oceanographic Laboratory
Bidston Observatory
Birkenhead, Merseyside, L43 7RA
United Kingdom

F.J. Kelly
Department of Environmental
Engineering
Texas A & M University
College Station, TX 77843
USA

J. van de Kreeke
Division of Applied Marine Physics
Rosenstiel School of Marine
and Atmospheric Science
University of Miami
4600 Rickenbacker Causeway
Miami, FL 33149
USA

Albert Y. Kuo
Virginia Institute of Marine Science
College of William and Mary
Gloucester Point, VA 23062
USA

Denis Lefaivre
Institut Maurice Lamontagne
Departmemt of Fisheries and Oceans
P.O. Box 1000
Mont Joli, Quebec
Canada G5H 3ZH

Wim van Leussen
Rijkswaterstaat
Tidal Water Division
P.O. Box 20907
2500 EX The Hague
The Netherlands

Li Yan
Second Institute of Oceanography
P.O. Box 507
Hanzhou
Zheijiang 310012
China

R.J. Lukatelich
Center for Water Research
and Department of Botany
The University of Western Australia
Nedlands, Western Australia 6009

John P. Matthews
School of Ocean Sciences
University College of North Wales
Menai Bridge
Gwynedd LL59 5EY
United Kingdom

W.H. McAnally, Jr.
U.S. Army Corps of Engineers
Waterways Experiment Station
Vicksburg, MS 39180
USA

Ashish J. Mehta
Coastal and Oceanographic Engineering
University of Florida
Gainesville, FL 32611
USA

George L. Mellor
Program in Atmospheric
and Oceanic Sciences
Princeton University
Princeton, NJ 08540
USA

James R. Miller
Department of Meteorology
and Physical Oceanography
Cook College, Rutgers University
New Brunswick, NJ 08903
USA

James D. Milutinovic
Department of Meteorology
and Physical Oceanography
Cook College, Rutgers University
New Brunswick, NJ 08903
USA

A. Murray
Proudman Oceanographic Laboratory
Bidston Observatory
Birkenhead, Merseyside, L43 7RA
United Kingdom

Roberto Pastres
University of Venice
Department of Physical Chemistry
Section of Environmental Physical
Chemistry
Calle Larga S. Marta, 2137
30123 Venice
Italy

Giovanni Pecenik
University of Venice
Department of Physical Chemistry
Section of Environmental Physical
Chemistry
Calle Larga S. Marta, 2137
30123 Venice
Italy

David Prandle
Proudman Oceanographic Laboratory
Bidston Observatory
Birkenhead, Merseyside, L43 7RA
United Kingdom

Walter Puls
GKSS Research Center
2054 Geesthacht
West Germany

Peter Ridd
Australian Institute of Marine Sciences
P.M.B. No. 3, Townsville, M.C.
Queensland 4810
Australia

H. Ridderinkhof
Netherlands Institute for Sea Research
P.O. Box 59
1790 AB Texel
The Netherlands

Panagiotis D. Scarlatos
South Florida Water Management District
P.O. Box 24680
West Palm Beach, FL 33416-4680
USA

William W. Schroeder
Marine Science Program
The University of Alabama
Dauphin Island, AL 36528
USA

Richard P. Signell
Woods Hole Oceanographic Institution
Woods Hole, MA 02543
USA

John H. Simpson
Proudman Oceanographic Laboratory
Bidston Observatory
Birkenhead
Merseyside L43 7RA
United Kingdom

Gamble M. Sisson
Virginia Institute of Marine Science
College of William and Mary
Gloucester Point, VA 23062
USA

Ned P. Smith
Harbor Branch Oceanographic Institute
5600 Old Dixie Highway
Fort Pierce, FL 34946
USA

Malcolm L. Spaulding
Department of Ocean Engineering
University of Rhode Island
Kingston, RI 02881
USA

Paul E. Speer
Center for Naval Analyses
Alexandria, VA
USA

J.A. Stephens
Plymouth Marine Laboratory
Prospect Place
The Hoe
Plymouth,. PL1 3DH
United Kingdom

Su Jilan
Second Institute of Oceanography
State Oceanic Adminstration
Hanzhou, Zheijiang
China

Jürgen Sündermann
University of Hamburg
2000 Hamburg
West Germany

E.M. Swenson
Coastal Ecology Institute
Center for Weland Resources
Louisiana State University
Baton Rouge, LA 70803
USA

Kim-Tai Tee
Physical and Chemical Sciences
Department of Fisheries and Oceans
Bedford Institute of Oceanography
Dartmouth, Nova Scotia
Canada B2Y 4A2

R.J. Uncles
Plymouth Marine Laboratory
Prospect Place
The Hoe
Plymouth, PL1 3DH
United Kingdom

Andre W. Visser
Marine Sciences Research Center
State University of New York
at Stony Brook
Stony Brook, NY 11794
USA

H.J. de Vriend
Delft Hydraulics
P.O. Box 152
8300 Ad Emmeloord
The Netherlands

Wang Kangshan
Second Institute of Oceanography
P.O. Box 507
Hanzhou
Zheijiang 310012
China

Frank M. White
Department of Ocean Engineering
University of Rhode Island
Kingston, RI 02881
USA

Johan C. Winterwerp
Delft Hydraulics
Estuaries and Seas Division
P.O. Box 177
2600 MH Delft
The Netherlands

William J. Wiseman, Jr.
Coastal Studies Institute
Department of Geology & Geophysics
Louisiana State University
Baton Rouge, LA 70803
USA

Eric Wolanski
Australian Institute of Marine Science
P.M.B. No. 3, Townsville M.C.
Queensland 4810
Australia

Tetsuo Yanagi
Department of Ocean Engineering
Ehime University
Matsuyama 790
Japan

J.T.F. Zimmerman
Netherlands Institute for Sea Research
P.O. Box 59
1790 AB Texel
The Netherlands

Coastal and Estuarine Studies

(formerly Lecture Notes on Coastal and Estuarine Studies)

Vol. 1: J. Sündermann, K.-P. Holz (Eds.), Mathematical Modelling of Estuarine Physics. Proceedings, 1978. 265 pages. 1980.

Vol. 2: D.P. Finn, Managing the Ocean Resources of the United States: The Role of the Federal Marine Sanctuaries Program. 193 pages. 1982.

Vol. 3: M. Tomczak Jr., W. Cuff (Eds.), Synthesis and Modelling of Intermittent Estuaries. 302 pages. 1983.

Vol. 4: H.R. Gordon, A.Y. Morel, Remote Assessment of Ocean Color for Interpretation of Satellite Visible Imagery. 114 pages. 1983.

Vol. 5: D.C.L. Lam, C.R. Murthy, R.B. Simpson, Effluent Transport and Diffusion Models for the Coastal Zone. 168 pages. 1984.

Vol. 6: M.J. Kennish, R.A. Lutz (Eds.), Ecology of Barnegat Bay, New Jersey. 396 pages. 1984.

Vol. 7: W.R. Edeson, J.-F. Pulvenis, The Legal Regime of Fisheries in the Caribbean Region. 204 pages. 1983.

Vol. 8: O. Holm-Hansen, L. Bolis, R. Gilles (Eds.), Marine Phytoplankton and Productivity. 175 pages. 1984.

Vol. 9: A. Pequeux, R. Gilles, L. Bolis (Eds.), Osmoregulation in Estuarine and Marine Animals. 221 pages. 1984.

Vol. 10: J.L. McHugh, Fishery Management. 207 pages. 1984.

Vol. 11: J.D. Davis, D. Merriman (Eds.), Observations on the Ecology and Biology of Western Cape Cod Bay, Massachusetts. 289 pages. 1984.

Vol. 12: P.P.G. Dyke, A.O. Moscardini, E.H. Robson (Eds.), Offshore and Coastal Modelling. 399 pages. 1985.

Vol. 13: J. Rumohr, E. Walger, B. Zeitzschel (Eds.), Seawater-Sediment Interactions in Coastal Waters. An Interdisciplinary Approach. 338 pages. 1987.

Vol. 14: A.J. Mehta (Ed.), Estuarine Cohesive Sediment Dynamics. 473 pages. 1986.

Vol. 15: R.W. Eppley (Ed.), Plankton Dynamics of the Southern California Bight. 373 pages. 1986.

Vol. 16: J. van de Kreeke (Ed.), Physics of Shallow Estuaries and Bays. 280 pages. 1986.

Vol. 17: M.J. Bowman, C.M. Yentsch, W.T. Peterson (Eds.), Tidal Mixing and Plankton Dynamics. 502 pages. 1986.

Vol. 18: F. Bo Pedersen, Environmental Hydraulics: Stratified Flows. 278 pages. 1986.

Vol. 19: K.N. Fedorov, The Physical Nature and Structure of Oceanic Fronts. 333 pages. 1986.

Vol. 20: A. Rieser, J. Spiller, D. VanderZwaag (Eds.), Environmental Decisionmaking in a Transboundary Region. 209 pages. 1986.

Vol. 21: Th. Stocker, K. Hutter, Topographic Waves in Channels and Lakes on the f-Plane. 176 pages. 1987.

Vol. 22: B.-O. Jansson (Ed.), Coastal Offshore Ecosystem Interactions. 367 pages. 1988.

Vol. 23: K. Heck, Jr. (Ed.), Ecological Studies in the Middle Reach of Chesapeake Bay. 287 pages. 1987.

Vol. 24: D.G. Shaw, M.J. Hameedi (Eds.), Environmental Studies in Port Valdez, Alaska. 423 pages. 1988.

Vol. 25: C.M. Yentsch, F.C. Mague, P.K. Horan (Eds.), Immunochemical Approaches to Coastal, Estuarine and Oceanographic Questions. 399 pages. 1988.

Vol. 26: E.H. Schumann (Ed.), Coastal Ocean Studies off Natal, South Africa. 271 pages. 1988.

Vol. 27: E. Gold (Ed.), A Law of the Sea for the Caribbean: An Examination of Marine Law and Policy Issues in the Lesser Antilles. 507 pages. 1988.

Vol. 28: W.S. Wooster (Ed.), Fishery Science and Management. 339 pages. 1988.

Vol. 29: D.G. Aubrey, L. Weishar (Eds.), Hydrodynamics and Sediment Dynamics of Tidal Inlets. 456 pages. 1988.

Vol. 30: P.B. Crean, T.S. Murty, J.A. Stronach, Mathematical Modelling of Tides and Estuarine Circulation. 471 pages. 1988.

Vol. 31: G. Lopez, G. Taghon, J. Levinton (Eds.), Ecology of Marine Deposit Feeders. 322 pages. 1989.

Vol. 32: F. Wulff, J.G. Field, K.H. Mann (Eds.), Network Analysis in Marine Ecology. 284 pages. 1989.

Vol. 33: M.L. Khandekar, Operational Analysis and Prediction of Ocean Wind Waves. 214 pages. 1989.

Vol. 34: S.J. Neshyba, Ch.N.K. Mooers, R.L. Smith, R.T. Barber (Eds.), Poleward Flows Along Eastern Ocean Boundaries. 374 pages. 1989.

Vol. 35: E.M. Cosper, V.M. Bricelj, E.J. Carpenter (Eds.), Novel Phytoplankton Blooms. 799 pages. 1989.

Vol. 36: W. Michaelis (Ed.), Estuarine Water Quality Management. 478 pages. 1990.

Vol. 37: C.M. Lalli (Ed.), Enclosed Experimental Marine Ecosystems: A Review and Recommendations. X, 218 pages. 1990.

Vol. 38: R.T. Cheng (Ed.), Residual Currents and Longterm Transport. XI, 544 pages. 1990.